Ohio State University Mathematical Research Institute Publications 1

Editors: Gregory R. Baker, Walter D. Neumann, Karl Rubin

Topology '90

Editors

Boris Apanasov
Walter D. Neumann
Alan W. Reid
Laurent Siebenmann

W DE G

Walter de Gruyter · Berlin · New York 1992

Editors:

BORIS APANASOV
Department of Mathematics
The University of Oklahoma
Norman, OK 73019, USA

WALTER D. NEUMANN
Department of Mathematics
The Ohio State University
Columbus, Ohio 43210-1174, USA

ALAN W. REID
Department of Mathematics
The Ohio State University
Columbus, Ohio 43210-1174, USA

LAURENT SIEBENMANN
Mathématique, Bât. 425
Université de Paris-Sud
F-91405 Orsay, France

Series Editors:

Gregory R. Baker, Walter D. Neumann, Karl Rubin
Department of Mathematics, The Ohio State University, Columbus, Ohio 43210-1174, USA

1991 Mathematics Subject Classification: Primary: 57-06; 11F06; 14H50; 20F32, 20H10, 20H15; 30F40; 32Gxx; 55R40; 57Mxx, 57N10; 81R50, 81T40

∞ Printed on acid-free paper which falls within the guidelines of the ANSI to ensure permanence and durability.

Library of Congress Cataloging in Publication Data

Topology '90 / editors, Boris Apanasov ... [et al.].
p. cm. – (Ohio State University Mathematical Research Institute Publications ; 1)
Contributions from a research semester in low dimensional topology held under the auspices of the International Mathematical Research Institute at Ohio State University from Feb. through June 1990.
Includes bibliographical references.
ISBN 3-11-012598-6 (acid-free paper)
1. Low-dimensional topology – Congresses. I. Apanasov, B. N. (Boris Nikolaevich) II. Ohio State University. International Mathematical Research Institute. III. Series.
QA612.14.T67 1992
514'.2–dc20 92-16406

Die Deutsche Bibliothek – Cataloging in Publication Data

Topology '90 / ed. Boris Apanasov ... – Berlin ; New York : de Gruyter, 1992
(Ohio State University Mathematical Research Institute Publications ; 1)
ISBN 3-11-012598-6
NE: Apanasov, Boris; International Mathematical Research Institute <Columbus, Ohio>: Ohio State Mathematical ...

Printing: Gerike GmbH, Berlin. Binding: Dieter Mikolai, Berlin. Cover design: Thomas Bonnie, Hamburg. Printed in Germany.

Preface

This volume consists of contributions from participants in a Research Semester in Low Dimensional Topology which took place under the auspices of the International Mathematical Research Institute at Ohio State University from February through June 1990. The Research Semester was funded by The Ohio State University through a grant from University Challenge and included an international conference in March 1990.

The main topics of the Research Semester included: the geometry and topology of 3-manifolds, with particular emphasis on hyperbolic 3-manifolds and their interactions with number theory; the "new" invariants of 3-manifolds related to quantum field theory; plane algebraic curves.

A number of long term visitors (2–3 months) were in residence at any given time. These visitors were: L. Siebenmann (Orsay), T. Yoshida (Tokyo), B. Apanasov (Novosibirsk, now Oklahoma), V. Turaev (Leningrad, now Strasbourg) and S. Orevkov (Moscow). In addition each week saw additional short term visitors. A list of these visitors and all talks is given after this Preface.

The Research Institute. The International Mathematical Research Institute at Ohio State University was founded in 1989 to support a program of visiting research scholars in mathematics at Ohio State and to run Workshops and Special Emphasis Programs on topics of particular importance and timeliness. The Research Semester on Low Dimensional Topology was the first major program of the Institute. Since then the Institute has supported workshops on, among others, Nearly Integrable Wave Phenomena in Nonlinear Optics, Quantized Geometry, Arithmetic of Function Fields, and L-Functions Associated to Automorphic Forms, and a workshop on Geometric Group Theory will take place from May to June 1992. The Institute is currently supporting about 20–30 other research visitors (mostly short term) per year. The Institute publishes a preprint series as well as this book series, which is devoted to research monographs, lecture notes, proceedings, and other mathematical works arising from activities of the Research Institute.

Acknowledgements. First and foremost, the editors thank The Ohio State University for its support of this program through the Research Institute. We thank our visitors and our fellow topologists at Ohio State for their contributions to the success of this project. We also thank the non-academic staff of the Mathematics Department for their help in the organization and running of the Research Semester, particularly Marilyn Howard (administration and visas), Marilyn Radcliff (expenses), Gena Dacons (administration) and Terry England (typing).

All contributions to this volume were refereed, and the editors thank the referees for their invaluable service. Many authors helped us by preparing their contributions in TeX. The TeX macros were written by Larry Siebenmann and edited by Walter Neumann, who benefited from the TeXpertise of many unnamed people.

Walter D. Neumann and Alan W. Reid, for the editors, March 1992.

Program of the Research Semester

February to June 1990

The visitors to the program included the speakers listed below and additional short term visitors: A. Broughton, T. Chinburg, M. Feighn, J. Gilman, C. Hodgson, K. Millet, U. Oertel, J. Harer, J. Ratcliffe, M. Scharlemann, R. Skora, A. Thompson, J. Weeks.

In addition to the talks listed below, a seminar was run through weeks 13–19 on the new 3-manifold invariants of Turaev, Reshetikhin, Viro, and Witten, with talks by D. Burghelea, V. Turaev and D. Yetter.

Week 1 (February 4–10)
H. M. Hilden, Universal Groups (3 lectures).

Week 2 (February 11–17)
L. Siebenmann, Knot complements—the Tietze conjecture revisited.
C. Frohman, Knot invariants via intersection homology.

Week 3 (February 18–24)
R. Lee, Topological invariants from conformal field theory (2 lectures).
B. Apanasov, Hyperbolic cobordisms and conformal structures, I.

Week 4 (February 25–March3)
W. Whitten, Imbeddings of 3-manifold groups.
L. Mosher, Dynamical systems and the homology norm of a 3-manifold (2 lectures).
F. Bonahon, The circle at infinity of a surface and applications.
B. Apanasov, Hyperbolic cobordisms and conformal structures, II.

Week 5 (March 4–10)
Y. Xia, Tate-Farrell cohomology of mapping class groups.
B. Apanasov, Hyperbolic cobordisms and conformal structures, III.

Week 6 (March 11–17)
D. McCullough, A conjectural picture of 3-manifold mapping class groups.
Lee Rudolph, Generalized Jones' polynomial, symplectic topology, and complex plane curves.
R. Meyerhoff, Anti-length spectrum of hyperbolic 3-manifolds.

Week 7 (CONFERENCE IN LOW-DIMENSIONAL TOPOLOGY, March 17–20)
C. Adams, Noncompact hyperbolic 3-orbifolds of small volume.
B. Apanasov, Nonstandard conformal 3-manifolds and 4-dimensional topology.
M. Bestvina, The boundary of negatively curved groups.
L. Kauffman, Combinatorial version of the SL(2)q 3-manifold invariant—spin networks and quantum groups.

D. Long, Peripheral separability.

P. Melvin, Evaluations of the 3-manifold invariants of Witten and Reshetikhin-Turaev.

A. Reid, Commensurators of hyperbolic 3-manifolds.

N. Reshetikhin, Invariants of 3-manifolds connected with finite dimensional Hopf algebras.

H. Rubinstein, Polyhedral metrics of non-positive curvature on three and four-manifolds.

P. Shalen, Patterson measures, Margulis numbers, and volumes of hyperbolic 3-manifolds.

O. Viro, Combinatorial construction of quantum invariants of 3-manifolds.

T. Yoshida, Floer homology and splittings of manifolds.

Week 8 (March 25–31)

R. Brooks, Low eigenvalues of arithmetic manifolds.

R. Brooks, Isospectral manifolds.

W. Menasco, Developing a calculus on links in S^3.

Week 9 (April 1–7)

V. Poenaru, The 3-dimensional Poincaré conjecture, I, II.

V. Poenaru, Almost convex groups, combable groups and π_1^∞ of universal covering spaces of 3-manifolds.

C. Maclachlan, Fuchsian subgroups of Bianchi groups.

Week 10 (April 8–14)

V. Poenaru, The 3-dimensional Poincaré conjecture, III, IV.

V. Poenaru, Killing stable 1-handles in π_1^∞ of open 3-manifolds.

J. Przytycki, Skein module of handlebodies.

J. Hass, Flows and intersections of curves on surfaces.

P. Scott, Least area surfaces in 3-manifolds (2 lectures).

H. Rubinstein, Cubulated 3-manifolds (2 lectures).

T. Yoshida, A splitting formula of spectral flow and calculation of Floer homology of some special homology 3-spheres, I.

Week 11 (April 15–21)

C. McA. Gordon, Reducible manifolds and Dehn surgery.

T. Yoshida, A splitting formula of spectral flow and calculation of Floer homology of some special homology 3-spheres, II.

Week 12 (April 22–29)

S. Morita, On the structure of the mapping class group and the Casson Invariant.

M. Shapiro, Automatic structures and 3-manifold groups.

Week 13 (April 30–May 5)

V. G. Turaev, New invariants of links and 3-manifolds.

T. Yoshida, On ideal points of deformation curves of hyperbolic 3-manifolds with 1 cusp.

Week 14 (May 6–12)

M. Baker, Finding homology in covers of 3-manifolds.

M. Baker, Reminisces on the Cuspidal Cohomology Problem and arithmetic links.

Week 15 (May 13–19)

A. Libgober, Topology of affine surfaces and trigonometric sums.

S. Orevkov, Fundamental group of plane curve complements and the Zariski conjecture, I.

S. Kerckhoff, Local rigidity and the representation space of link complements.

Week 16 (May 20–26)

A. Pazhitnov, Morse-Novikov theory for closed 1-forms.

S. Orekov, Fundamental group of plane curve complements and the Zariski conjecture, II.

S. Kaliman, On the classification of polynomials in 2 variables.

D. Yetter, Tangles in cobordisms.

Week 17 (May 27–June 2)

S. Orevkov, Some approaches to the Jacobian conjecture.

J. Corson, Two-complexes of groups.

W. Neumann, Amalgamation and the invariant trace-field of Kleinian groups.

Week 18 (June 3–9)

No talks

Week 19 (June 10–16)

R. Penner, Decorated Teichmüller theory (2 lectures).

Contents

Noncompact Hyperbolic 3-Orbifolds of Small Volume

*Colin C. Adams**

Abstract. We determine the six noncompact orientable and the six noncompact nonorientable hyperbolic 3-orbifolds of least volume. This extends previous results of Meyerhoff where he determined the unique noncompact orientable and nonorientable hyperbolic 3-orbifolds of least volume. Our results are obtained through analysis of horoball diagrams for rigid cusps.

1. Introduction

In **[6]**, the noncompact orientable and nonorientable hyperbolic 3-orbifolds of least volume were determined. We extend those results here to determine the six noncompact orientable and nonorientable hyperbolic 3-orbifolds of least volume. See Theorem 6.1 and Corollary 6.2 for the volumes; the corresponding orbifolds are described in the paragraphs preceding the statements of Theorems 3.3, 4.2, and 5.2.

The idea is to analyze horoball patterns in order to list all the possible volumes of maximal cusps below a given volume. These results, in conjunction with known cusp density results, then yield the noncompact hyperbolic 3-orbifolds of least volume. Similar techniques were used in **[3]** to determine the three smallest limit volumes of hyperbolic 3-orbifolds.

In what follows, v_0 will denote the volume of an ideal tetrahedron in $\mathbb{H}^3$ with all dihedral angles equal to $\pi/3$. In particular, $v_0 = 1.01494146\ldots$. We will let v_1 denote the volume of an ideal tetrahedron in $\mathbb{H}^3$ with dihedral angles $\pi/2$, $\pi/4$ and $\pi/4$. In particular, $v_1 = 0.91596544\ldots$.

We will work in the upper-half space model of hyperbolic 3-space, denoted $\mathbb{H}^3$. We will think of a hyperbolic 3-orbifold as being obtained by taking the quotient of $\mathbb{H}^3$ by a discrete group G of hyperbolic isometries. Since all of the orbifolds which we will be interested in have a single cusp, we will always assume that the single point at ∞ on the boundary of the upper-half-space model of $\mathbb{H}^3$ is a parabolic fixed point for this cusp.

The pre-image in $\mathbb{H}^3$ of the cusp will be a set of disjoint horoballs, including one centered about ∞. (The center of a horoball is the point on the boundary of hyperbolic space where the horoball is tangent.) The horoball centered about ∞ appears as a horizontal plane with positive z coordinate together with all the points above it.

We will maximize the single cusp in the orbifold by expanding it until it touches itself on the boundary. This corresponds in $\mathbb{H}^3$ to expanding the set of horoballs equivariantly until they first touch one another.

* Supported in part by NSF Grants DMS-8711495 and DMS-9000937.

It is convenient to normalize the picture so that the bounding plane of the horoball centered about ∞ has Euclidean height 1 above the $x - y$ plane. Since the cusp has been maximized, there are horoballs tangent to the horoball centered about ∞. These horoballs will have Euclidean diameter 1 and will be the horoballs of largest Euclidean diameter. We call these horoballs ***full-sized horoballs***.

We call the view of the set of all horoballs looking from ∞ down toward the $x - y$ plane a ***horoball diagram*** or a ***cusp diagram***. See **[4]** for examples of horoball diagrams corresponding to knots.

Additional background on hyperbolic 3-orbifolds appears in **[8]**. Additional background on horoball diagrams and definitions of terms can be found in **[1]**, **[2]** and **[3]**. All of the hyperbolic 3-orbifolds under consideration will be assumed orientable until the last section.

We will utilize the following lemma which follows from work of Robert Meyerhoff in **[5]**.

Lemma 1.1. *If a finite volume hyperbolic 3-orbifold Q contains a set of cusps with disjoint interiors such that their total volume is β, then* $\mathrm{vol}(Q) \geq \beta(2v_0/\sqrt{3})$.

We include one geometric lemma from **[3]** which will be helpful in what follows.

Lemma 1.2. *If two horoballs H_x and H_y have Euclidean diameters a and b respectively and their centers x and y are a distance c apart, then there exists a horoball with diameter ab/c^2.* □

Proof. The shortest hyperbolic distance between the two horoballs can be determined by rotating $180°$ about a geodesic which is a semi-circle of radius c with one endpoint at y and with its high point directly above x. Since the hyperbolic distance from H_x to the high point on the geodesic is $\ln(c/a)$, H_x is sent by the rotation to a horoball centered at ∞ with boundary a horizontal plane at height c^2/a. The rotation fixes H_y. The shortest distance between the two horoballs is unaffected by the rotation and is therefore $\ln(c^2/ab)$.

Since all horoballs are identified in the quotient, there is an isometry in the orbifold group which sends H_x to H_∞. It must then send H_y to a horoball which is a hyperbolic distance $\ln(c^2/ab)$ from H_∞. Since the boundary of H_∞ is a plane at height 1, H_y must be sent to a horoball of Euclidean diameter ab/c^2. □

2. Rigid cusps

A cusp in a hyperbolic 3-orbifold is called ***rigid*** if Dehn filling cannot be performed on it and otherwise it is called ***non-rigid***. In **[3]**, it was shown that a cusp is rigid if and only if there are singular curves of order other than two going all the way out the cusp.

Since a cusp is isometric to the quotient of the set of all of the points above a horizontal plane in the upper-half-space model of hyperbolic space by a Euclidean group of transformations, there are exactly three possibilities for the structure of the singularities in a rigid cusp. In all three cases, the cusp contains three singular axes. Denoting the orders of the three singular axes as $\{a, b, c\}$, the possibilities are $\{3, 3, 3\}$, $\{4, 4, 2\}$ or $\{6, 3, 2\}$.

Lemma 2.1. *If the action of the orbifold subgroup fixing ∞ identifies all of the full-sized horoballs in the horoball diagram of the orbifold, then every point of tangency between two horoballs lies on the axis of an order two elliptic isometry in the orbifold group such that the axis is tangent to both the horoballs.*

Proof. Let H_x be a full-sized horoball centered at the point x in the $x-y$ plane and let H_∞ be the horoball centered at ∞. Then there is an isometry j in the fundamental group of the orbifold such that $j(x)=\infty$. Let $y=j(\infty)$. Then $j(H_\infty)$ is a full-sized horoball, denoted H_y. By hypothesis, there is an isometry k which fixes ∞ and which sends y to x.

Then $k\circ j$ sends x to ∞ and ∞ to x, while fixing the point of tangency of H_x and H_∞. Thus $k\circ j$ is the order two elliptic isometry with axis through the point of tangency between H_x and H_∞. Since elements of the fundamental group of the orbifold identify all tangency points between pairs of horoballs to tangency points on H_∞, there must be corresponding order two elliptic axes through all points of tangency. □

Note that once we have normalized so that the boundary of the horoball centered at ∞ has Euclidean height one, the volume of the cusp is one half of the area of a fundamental domain in the $x-y$ plane for the Euclidean subgroup of the orbifold group which fixes ∞.

In the next three sections, we will only be interested in orbifolds with volume less than $v_0/4$. By Lemma 1.1, a maximal cusp in such an orbifold has volume less than $\sqrt{3}/8$. Hence, our goal will be to list those cusps with volumes less than $\sqrt{3}/8$.

The singularities in the rigid cusps will restrict the placement of the full-sized horoballs in the horoball diagrams. Since none of the horoballs overlap in their interiors, the only way a vertical singular axis in the upper-half-space model can intersect the interior of a horoball is if the horoball is centered at the end of the vertical axis in the $x-y$ plane.

3. $\{6,3,2\}$-cusp

We examine first the case where the cusp under consideration is a $\{6,3,2\}$-cusp. If there is more than one equivalence class of full-sized horoballs under the action of the subgroup of isometries fixing ∞, the least volume for the cusp occurs when a pair of full-sized horoballs are centered at the 3-fold and 6-fold singularities, yielding a volume in the cusp of at least $\sqrt{3}/8$. Hence, we will assume from now on that there is only one equivalence class of full-sized horoballs in the cusp diagram.

If there is a full-sized horoball which is not centered at any of the singular points in the cusp diagram, the least volume cusp appears as in Figure 1(a), with a cusp volume of $(3/2+\sqrt{3})/6=.5386\ldots$.

If there is a full-sized ball centered at the 2-fold singularity, then the smallest possible area for the cusp diagram occurs when the full-sized horoballs at each of the 2-fold singularities touch each other as in Figure 1(b). This gives an area for the fundamental domain in the plane of at least $1/\sqrt{3}$ and a cusp volume of at least $\sqrt{3}/6=.2887\ldots$.

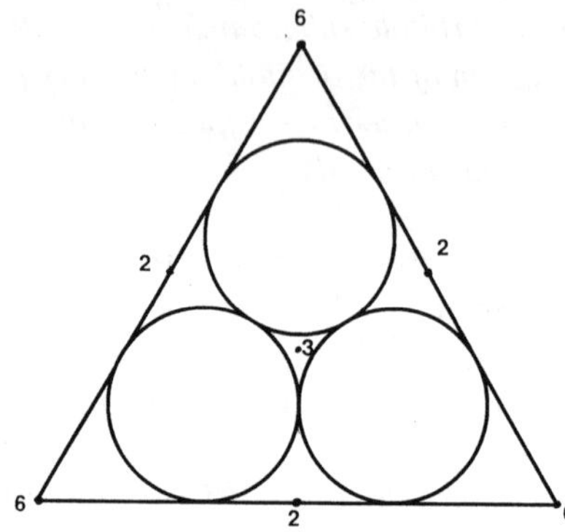

Figure 1(a)

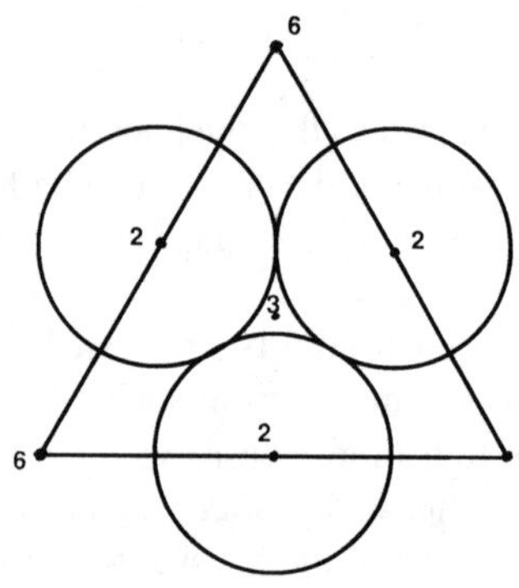

Figure 1(b)

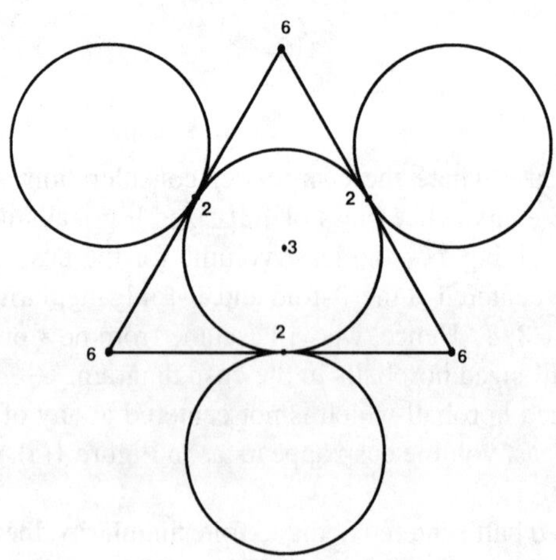

Figure 1(c)

If there is a full-sized ball at the 3-fold singularity, the volume is at least $\sqrt{3}/8$ as in Figure 1(c). Note that this volume is also the least volume for the case that we have full-sized balls at both the 6-fold and 3-fold singularities.

The rest of our time will be spent on the case that there is a full-sized horoball at the 6-fold singularity and nowhere else. The least such volume that can occur is in the case that these full-sized balls touch one another, yielding a cusp volume of $\sqrt{3}/24$. This is the smallest possible volume of an orientable cusp in an orbifold and it occurs in the smallest noncompact orientable orbifold, which was determined in **[4]**. It is not hard to prove that this smallest noncompact orientable orbifold is unique.

Now, we look at the situation where the full-sized balls centered at the 6-fold singularities do not touch. Let the shortest distance between the centers of a pair of full-sized balls be given by d. Throughout the rest of the section, we are assuming $d > 1$. We will find the following lemma useful.

Lemma 3.1. *If $d > 1$, there cannot be a set of three horoballs which are pairwise tangent.*

Proof. If there were such a set of three balls, an isometry exists which would take the center of one of them to ∞ and therefore send the other two balls to two full-sized horoballs which are tangent to each other. This implies that the minimal tangency distance d is in fact equal to 1, contradicting our assumption that $d > 1$. □

Since we are assuming that there is only one full-sized horoball up to the group action, we know from Lemma 2.1 that there is an order two axis perpendicular to the order six axis and through the point of tangency at height 1. Rotation about the order two axis sends the full-sized ball to a ball centered at ∞ and sends the ball centered at ∞ to the full-sized ball. It also sends the six neighboring full-sized balls, each a distance d away from our original full-sized ball, to six smaller balls. We call these smaller balls ***$(1/d)$-balls*** as each of their centers will be a distance $1/d$ from the center of the full-sized ball. See Lemma 4.3 and the paragraphs following Lemma 4.6 of **[3]** for more details. The Euclidean diameter of these $(1/d)$-balls is $1/d^2$ by Lemma 1.2.

Define the distances u, v and w as in Figure 2. We can determine each of them in terms of d and β utilizing the law of cosines.

By symmetry, we will always assume that $0 \leq \beta \leq \pi/6$. We obtain the following equations:

$$u^2 = d^2 + 3/d^2 - \sqrt{3}\sin\beta - 3\cos\beta \tag{1}$$

$$v^2 = 4\left((d/2)^2 + 1/d^2 - \cos\beta\right) \tag{2}$$

$$w^2 = d^2 + 1/d^2 - 2\cos\beta\,. \tag{3}$$

If the center of a $(1/d)$-ball comes within a distance 1 of the center of a full-sized ball without the two balls touching or if the centers of two $(1/d)$-balls come within $(1/d)$ of each other without the two balls touching, then when one of the two balls is put at ∞, the other will form a ball intermediate in size between a full-sized ball and a $(1/d)$-ball, by Lemma 1.2. Since $(1/d)$-balls are the biggest balls tangent to full-sized balls, there must then be a ball tangent to no ball bigger than itself. The upper hemisphere of this ball

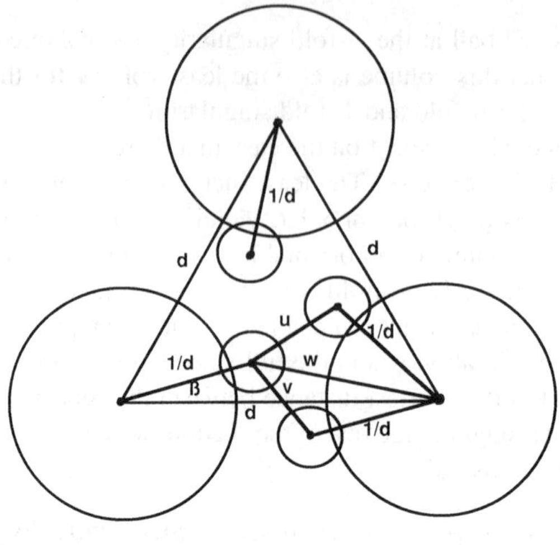

Figure 2

forms a disk of no tangency. (See the proof of Lemma 4.1 in **[2]** or Lemma 4.6 and the proof of Lemma 4.8 in **[3]**). Hence, there is the equivalent of another full-sized ball in the diagram. The least possible volume results when the additional ball is centered at the 3-fold singularity, yielding a volume of $\sqrt{3}/8$.

From now on, we will only investigate the situation where no such disk of no tangency exists. Suppose first that a single $(1/d)$-ball is shared by a pair of neighboring full-sized balls and the $(1/d)$-ball does not touch a third full-sized ball. Since the angles between the centers of the $(1/d)$-balls as measured from the center of a full-sized ball that they touch must be a multiple of $\pi/3$, it is easy to see that the center of the single $(1/d)$-ball is in line with the centers of the two full-sized balls that it touches. Hence $d = 2/d$, yielding $d = \sqrt{2}$ and a cusp volume equal to $\sqrt{3}/12 = .1443\ldots$. In this case, the horoball diagram appears as in Figure $3(a)$.

Suppose now that a single $(1/d)$-ball is shared by three full-sized balls. Then $1/d = d/\sqrt{3}$ yielding $d = \sqrt[4]{3}$. The cusp volume is then $1/8 = 0.125$, with a horoball diagram as in Figure 3(b).

From this point on, we assume that each $(1/d)$-ball touches a unique full-sized ball. In order that no intermediate balls are created, it must be that the center of each $(1/d)$-ball stays a distance at least 1 away from the center of the nearest full-sized ball. This forces $w \geq 1$ and hence

$$\cos\beta \leq \frac{d^2 + 1/d^2 - 1}{2}. \tag{4}$$

Since we are assuming by symmetry that $0 \leq \beta \leq \pi/6$, we have $\sqrt{3}/2 \leq \cos\beta \leq 1$. This can only occur when

$$d \geq \frac{\sqrt{1 + \sqrt{3} + \sqrt{2\sqrt{3}}}}{2} = 1.51546\ldots.$$

We will restrict ourselves to these values of d.

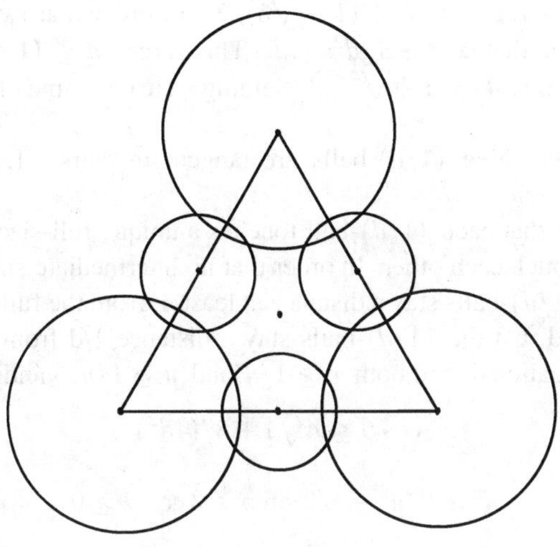

Figure 3(a)

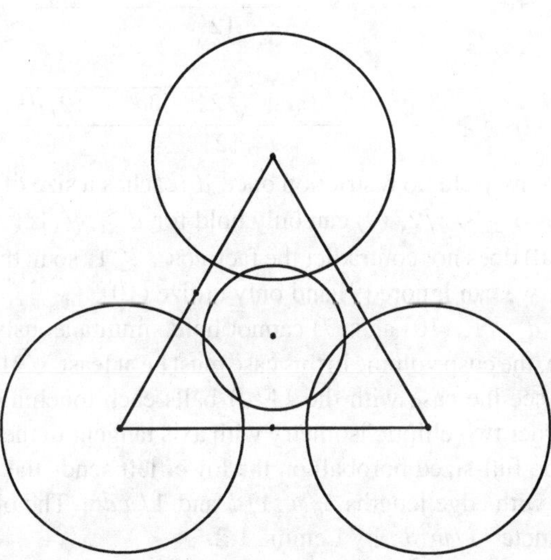

Figure 3(b)

Suppose first that a pair of $(1/d)$-balls touch each other and no other $(1/d)$-ball touches this pair. The point of tangency of the two balls must then be directly over a 2-fold singularity. It follows that

$$\cos\beta = d^2/4 + 1/d^2 - 1/(4d^4)\,. \tag{5}$$

Then (4) and (5) together yield the inequality

$$d^6 - 2d^4 - 2d^2 + 1 \geq 0\,. \tag{6}$$

This has no solutions for $1 < d < (1+\sqrt{5})/2$. In order that two $(1/d)$-balls touch each other, it must be that $2/d + 1/d^2 \geq d$. This forces $d \leq (1+\sqrt{5})/2$. Thus, the only possible solution is $d = (1+\sqrt{5})/2$ yielding a cusp volume of $\sqrt{3}(3+\sqrt{5})/48 = 0.18894\ldots$.

Suppose now that three $(1/d)$-balls are tangent in pairs. Lemma 3.1 yields an immediate contradiction.

We now suppose that each $(1/d)$-ball touches a unique full-sized ball and that none of the $(1/d)$-balls touch each other. In order that no intermediate sized balls are created, it must be that the $(1/d)$-balls stay a distance at least 1 from the full-sized balls that they are not touching and that the $(1/d)$-balls stay a distance 1/d from each other. Hence, $w \geq 1$, yielding Equation 4, and both $v \geq 1/d$ and $u \geq 1/d$, yielding:

$$\cos\beta \leq d^2/4 + 3/(4d^2) \tag{7}$$

$$d^2 + 2/d^2 - \sqrt{3}\sin\beta - 3\cos\beta \geq 0. \tag{8}$$

From (8) we have the following possibilities

$$\cos\beta \leq \frac{3(d^2+2/d^2) - \sqrt{24 - 3d^4 - 12/d^4}}{12} \tag{9}$$

or

$$\cos\beta \geq \frac{3(d^2+2/d^2) + \sqrt{24 - 3d^4 - 12/d^4}}{12}. \tag{10}$$

These two equations yield no restriction once d reaches a size of $1.65289\ldots$. Since we are assuming $\cos\beta \geq \sqrt{3}/2$, (9) can only hold for $d \geq \sqrt{(1+\sqrt{3})} = 1.65289\ldots$. For any $d \geq \sqrt{2}$, (10) does not contradict the fact $\cos\beta \leq 1$, so in the range $1.51546\ldots \leq d \leq 1.65289\ldots$, we can ignore (9) and only utilize (10).

However, when $d > 1$, (10) and (7) cannot hold simultaneously unless $d \geq \sqrt[4]{7} = 1.62658\ldots$. Hence, the cusp volume in this case must be at least $\sqrt{21}/24 = 0.19094\ldots$.

In Figure 4, we see the case with the $(1/d)$-balls each touching a unique full-sized ball. Applying an order two elliptic isometry with axis tangent to the horoball centered at ∞ and tangent to the full-sized horoball on the lower left sends the $(d, w, 1/d)$-triangle to a similar triangle with edge lengths $1/w, 1/d$ and $1/wd^2$. The old $(1/d)$-ball is sent to a new ball of diameter $1/w^2d^2$ by Lemma 1.2.

Call this new ball a $(1/w)$-ball. Suppose first that the three $(1/w)$-balls corresponding to three distinct $(1/d)$-balls are in fact the same ball. Then the single $(1/w)$-ball has its center at the center of the equilateral triangle with edge lengths d. Hence, it must be that $1/w = \sqrt{3}/d$. This fact, together with the law of cosines and the law of sines then forces $d = \sqrt[4]{7}$. The resulting cusp has volume exactly $\sqrt{21}/24$, and is depicted in Figure 5.

We assume now that the $(1/w)$-balls are distinct. If the $(1/w)$-balls touched full-sized balls, this would force two $(1/d)$-balls to touch, contradicting our assumption that they do not. The three $(1/w)$-balls cannot touch each other pairwise by Lemma 3.1.

In order to prevent the creation of any intermediate sized balls and the resulting disks of no tangency, it must be that the $(1/w)$-balls stay a distance at least $(1/w)$ from the full-sized balls and that the $(1/w)$-balls stay a distance $1/dw^2$ from each other by

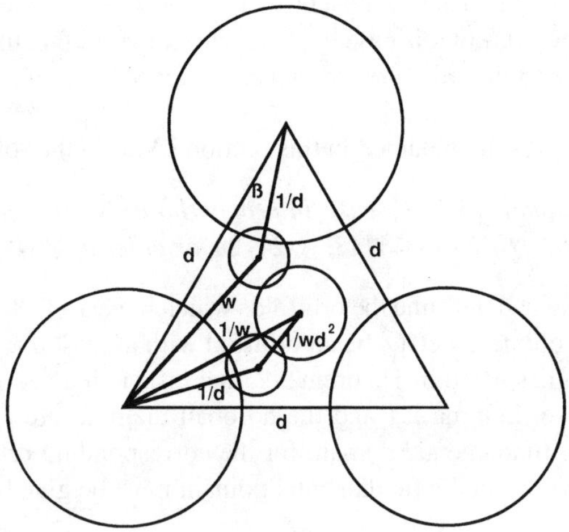

Figure 4

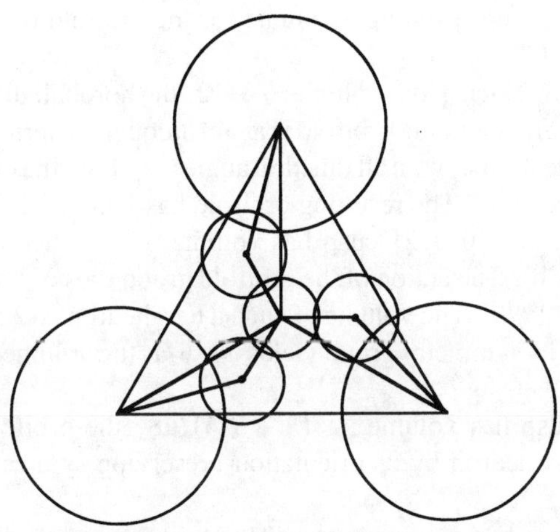

Figure 5

Lemma 1.2. This yields the following two equations:

$$\cos\beta \leq 1/2 + 1/d^2 \tag{11}$$

$$36 - 228cd^2 + 93d^4 + 484c^2d^4 - 376cd^6 - 392c^3d^6 + 70d^8 + 420c^2d^8 + 112c^4d^8 - 138cd^{10} - 152c^3d^{10} + 13d^{12} + 72c^2d^{12} - 14cd^{14} + d^{16} \geq 0, \tag{12}$$

where $c = \cos\beta$

In fact, (12) will not be utilized in the argument. Since we are assuming that $\cos\beta \geq \sqrt{3}/2$, (11) implies that $d \leq \sqrt{1+\sqrt{3}}$. However, (11) together with (10) yields

$$48 + 72d^2 - 24d^4 - 36d^6 + 12d^8 \geq 0. \tag{13}$$

This forces $d \geq \sqrt{1+\sqrt{3}}$. Thus, it must be that $d = \sqrt{1+\sqrt{3}}$, yielding a cusp volume of $(\sqrt{3}+3)/24$. However, in this case, one can check that in fact the three $(1/w)$-balls at the center of the triangle are pairwise tangent. Hence, Lemma 3.1 implies no such orbifold can exist.

We summarize the results obtained in this section so far in the following theorem.

Theorem 3.2. *A maximal* $\{6,3,2\}$*-cusp in a hyperbolic* 3*-orbifold has volume either* $\sqrt{3}/24$, $\sqrt{3}/12$, $1/8$, $\sqrt{3}(3+\sqrt{5})/48$, $\sqrt{21}/24$ *or at least* $\sqrt{3}/8$. □

We would now like to determine the orbifolds which have a $\{6,3,2\}$-cusp with one of these possible cusp volumes. Let Q be an orbifold with at least one $\{6,3,2\}$-cusp. For each of the cusp volumes listed in Theorem 3.2, there is a unique corresponding horoball diagram. In each case, Lemma 2.1 and the horoball diagram are enough to determine the singular set and a fundamental domain for the corresponding orbifold. The singular axes determine how faces on the fundamental domain must be glued in order to yield the corresponding orbifold.

In the case that the $\{6,3,2\}$-cusp has volume $\sqrt{3}/24$, the corresponding horoball diagram forces Q to be the quotient of an ideal regular tetrahedron by its orientation-preserving symmetry group, yielding a volume for the orbifold of $v_0/12$. The orbifold is uniquely determined.

If instead, the $\{6,3,2\}$-cusp has volume $\sqrt{3}/12$, the horoball diagram must appear as in Figure 3(a). The corresponding orbifold is again uniquely determined and corresponds to the quotient of an ideal cube, with all dihedral angles $\pi/3$, by the orientation-preserving symmetry group of the cube. The resulting orbifold has volume $5v_0/24$.

Suppose now that the $\{6,3,2\}$-cusp has volume $1/8$. Then the horoball diagram appears as in Figure 3(b). The action of the orbifold group tiles $\mathbb{H}^3$ with tetrahedra, all of angles $\pi/6, \pi/6$ and $2\pi/3$. The volume of such a tetrahedron is $2v_0/3$. The quotient of such a tetrahedron by its symmetry group yields $v_0/6$ as the volume of the corresponding orbifold.

If the $\{6,3,2\}$-cusp has volume $\sqrt{3}(3+\sqrt{5})/48$, the orbifold is the quotient of an ideal regular dodecahedron by its orientation-preserving symmetry group, yielding a volume of $0.3430\ldots$.

If the $\{6,3,2\}$-cusp has a volume of $\sqrt{21}/24$, the horoball diagram appears as in Figure 4. The action of the orbifold group tiles all of $\mathbb{H}^3$ with two ideal tetrahedra. The first tetrahedron has dihedral angles a, b and c where $\cos(a) = 5/(2\sqrt{7})$ and $\cos(b) = 9/(2\sqrt{21})$. The second tetrahedron has dihedral angles d, e and f where $d = \pi/3$ and $e = a$. A fundamental domain is obtained by taking half of each of these two tetrahedra, yielding a volume of approximately 0.47 .

Theorem 3.3. *A hyperbolic* 3*-orbifold with a* $\{6,3,2\}$*-cusp has volume either* $v_0/12$, $v_0/6$, $5v_0/24$ *or at least* $v_0/4$.

Proof. If such a hyperbolic 3-orbifold does not have one of the first three volumes, it has a cusp volume of either $\sqrt{3}(3+\sqrt{5})/48$, $\sqrt{21}/24$, or at least $\sqrt{3}/8$. In the case that the cusp volume is $\sqrt{3}(3+\sqrt{5})/48$ or $\sqrt{21}/24$, we have seen that the corresponding orbifold has volume $0.3430\ldots$ or $.47\ldots$, both of which are greater than $v_0/4$. Otherwise, the cusp volume must be at least $\sqrt{3}/8$, which by Lemma 1.1 implies that the corresponding orbifold has volume at least $v_0/4$. □

4. {3,3,3}-cusp

A fundamental domain in the x-y plane for a $\{3,3,3\}$-cusp is a rhombus. Assume first of all that there is a full-sized horoball which is not centered at one of the singular points in the plane. Then the shortest distance between two singular points which are identified by the Euclidean group action is at least 2. The resulting cusp has volume at least $\sqrt{3}/3$.

Suppose now that there is a full-sized ball at one of the three 3-fold singularites and nowhere else. The pattern of $(1/d)$-balls will resemble the pattern of $(1/d)$-balls which we obtained for the $\{6,3,2\}$-cusp when the cusp has one full-sized ball located at the 6-fold singularity. Hence we will have exactly twice the cusp volumes we found in that case.

Corollary 4.1. *A maximal* $\{3,3,3\}$*-cusp in a hyperbolic 3-orbifold has volume either* $\sqrt{3}/12$, $\sqrt{3}/6$, $1/4$, $\sqrt{3}(3+\sqrt{5})/24$, $\sqrt{21}/12$ *or at least* $\sqrt{3}/4$. □

For each of these possible volumes of a $\{3,3,3\}$-cusp, there is a unique orbifold, each such orbifold corresponding to the double cover of an orbifold in the $\{6,3,2\}$-cusp case. The following theorem is then immediate.

Theorem 4.2. *A hyperbolic 3-orbifold with a* $\{3,3,3\}$*-cusp has volume either* $v_0/6$, $v_0/3$, $5v_0/12$ *or at least* $v_0/2$. □

5. {4,4,2}-cusp

Suppose first that there is a full-sized ball in the horoball diagram which is not centered at one of the singularities. The volume in the cusp is then at least $1/2$ as in Figure 6(a). If instead, there is one full-sized ball at the 2-fold singularity, the volume in the cusp is at least $1/4$, as in Figure 6(b).

Note that if there is more than one full-sized horoball in the cusp, then the least volume occurs when there are two full-sized balls and they occur at the two 4-fold singularities, giving a volume of at least $1/4$, as in Figure 6(c).

Henceforth, we will assume that there is one full-sized ball which is centered at one of the 4-fold singularities. Suppose first that the full-sized balls in the cusp diagram touch each other. The volume in the cusp is then exactly $1/8$.

From now on, we will assume that the full-sized balls do not touch. Let d again be the shortest distance between full-sized horoballs. Then there is a set of four $(1/d)$-balls touching each full-sized ball.

Suppose first that four full-sized balls share a single $(1/d)$-ball which is centered at the other 4-fold singularity. Then $2/d = \sqrt{2}d$ and $d = \sqrt[4]{2}$. The volume of the corresponding cusp is then $\sqrt{2}/8$.

If instead, a $(1/d)$-ball is shared by exactly two full-sized balls, it must be that $2/d = d$ and $d = \sqrt{2}$, yielding a volume of $1/4$.

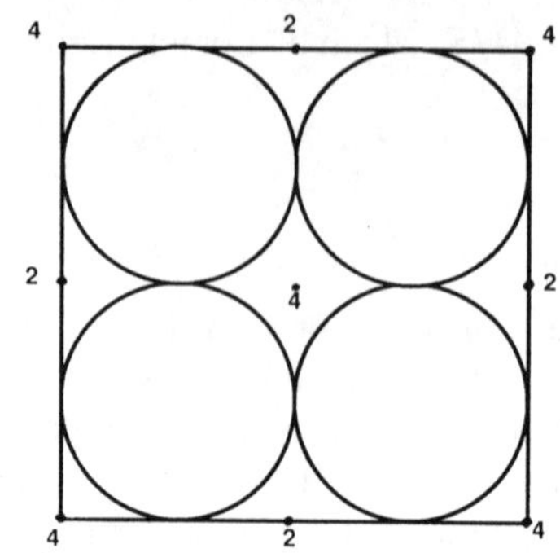

Figure 6(a)

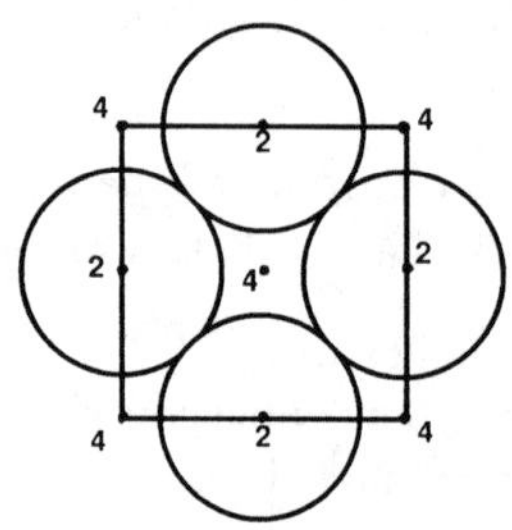

Figure 6(b)

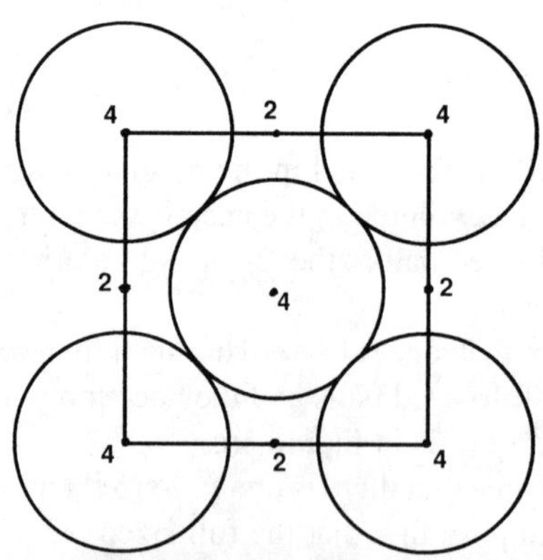

Figure 6(c)

From now on, we will assume that no $(1/d)$-balls are shared by full-sized balls. In order that disks of no tangency are not created, we will assume that all $(1/d)$-balls stay a distance at least 1 from any full-sized balls they are not tangent to. Thus, (4) from Section 3 holds.

Suppose first that a pair of $(1/d)$-balls corresponding to two distinct full-sized balls are tangent, and no other $(1/d)$-ball is tangent to the pair. Then (5) from Section 3 must hold. As in Section 3, the only possible value for d is $(1+\sqrt{5})/2$. The resulting cusp has volume $(3+\sqrt{5})/16 = 0.32725\ldots$.

Suppose now that there are four $(1/d)$-balls such that each is tangent to two others and they are symmetrically placed around one of the 4-fold singularites. Apply an isometry of the group which takes one of the $(1/d)$-balls to the horoball at ∞. The two $(1/d)$-balls which were tangent to the first $(1/d)$-ball are sent to full-sized balls. The fourth $(1/d)$-ball is sent to a horoball tangent to each of these full-sized balls with center in line with their centers. Since the centers of the four $(1/d)$-balls formed a square, this new ball will have center a distance $1/\sqrt{2}$ from each of the full-sized balls. Hence the full-sized balls have centers a distance $\sqrt{2}$ apart. However, this forces $d=\sqrt{2}$ and $1/d = 1/\sqrt{2}$. Thus, this fourth ball must be a $(1/d)$-ball. This contradicts the fact that every $(1/d)$-ball is tangent to two others.

Suppose now that none of the $(1/d)$-balls touch each other. Then, in order that no intermediate sized balls are introduced, it must be that the centers of the $(1/d)$-balls stay a distance at least $1/d$ apart by Lemma 1.2. This yields both (4) from Section 3 and

$$d^4 - 2(\sin\beta + \cos\beta)d^2 + 1 \geq 0. \tag{14}$$

This last equation becomes

$$\cos\beta \leq \frac{d^2 + 1/d^2 - \sqrt{6 - d^4 - 1/d^4}}{4} \tag{15}$$

or

$$\cos\beta \geq \frac{d^2 + 1/d^2 + \sqrt{6 - d^4 - 1/d^4}}{4}. \tag{16}$$

By symmetry, we are only interested in the cases where $0 \leq \beta \leq \pi/4$ so $\sqrt{2}/2 \leq \cos\beta \leq 1$. Applying this restriction to (15), we find that (15) does not apply until $d = 1.5537743$ at which time, (15) and (16) no longer restrict $\cos\beta$. Hence, in the range $1 < d \leq 1.5537743$, (16) must hold.

Comparing (16) with (4) yields a contradiction unless

$$d \geq \frac{\sqrt{1+\sqrt{3}+\sqrt{2}\sqrt{3}}}{2} = 1.51546\ldots.$$

This yields a cusp volume of at least $(1+\sqrt{3}+\sqrt{2}\sqrt{3})/16 = 0.28690\ldots$.

Combining the results obtained so far in this section, we have the following.

Theorem 5.1. *A maximal* $\{4,4,2\}$*-cusp in a hyperbolic* 3*-orbifold has volume either* $1/8$, $\sqrt{2}/8$ *or at least* $1/4$. □

The only way for a $\{4,4,2\}$-cusp to have volume $1/8$ is if the corresponding orbifold is the quotient of an ideal regular octahedron by its orientation preserving symmetry group. A fundamental domain will be one sixth of an ideal tetrahedron with dihedral angles $\pi/4$, $\pi/4$ and $\pi/2$. Hence, the orbifold has volume $v_1/6$.

If the $\{4,4,2\}$-cusp in a hyperbolic 3-orbifold has volume $\sqrt{2}/8$, the corresponding orbifold must come from the quotient of an ideal tetrahedron with dihedral angles $\pi/4$, $\pi/4$ and $\pi/2$ by its orientation-preserving symmetry group, yielding a volume of $v_1/4$.

Otherwise, the $\{4,4,2\}$-cusp has a volume of at least $1/4$. Lemma 1.1 then implies that the orbifold has a volume of at least $v_0/(2\sqrt{3})$, yielding the following theorem.

Theorem 5.2. *A hyperbolic 3-orbifold with a $\{4,4,2\}$-cusp has volume either $v_1/6$, $v_1/4$ or at least $v_0/(2\sqrt{3})$.* □

With an analysis of densities of horoball packings, we expect that the lower bound on volumes of $v_0/(2\sqrt{3})$ given in the above theorem could be improved to $v_1/3$.

6. Conclusions

Utilizing the results from Sections 3, 4 and 5 we have the following theorem.

Theorem 6.1. *The six noncompact orientable hyperbolic 3-orbifolds of volume less than $v_0/4$ have volumes $v_0/12$, $v_1/6$, $v_0/6$, $v_0/6$, $5v_0/24$ and $v_1/4$.*

Proof. A noncompact hyperbolic 3-orbifold must have at least one cusp. If any of the cusps are non-rigid, the results of **[3]** show that the orbifold has a volume of at least $v_1/3$. We can therefore assume all the cusps are rigid. However, a rigid cusp must be either a $\{6,3,2\}$-cusp, a $\{3,3,3\}$-cusp or a $\{4,4,2\}$-cusp. The theorem then follows immediately from Theorems 3.3, 4.2 and 5.2. □

Corollary 6.2. *The six noncompact nonorientable hyperbolic 3-orbifolds of volume less than $v_0/8$ have volumes $v_0/24$, $v_1/12$, $v_0/12$, $v_0/12$, $5v_0/48$ and $v_1/8$.*

Proof. The six orientable orbifolds from Theorem 6.1 all double cover nonorientable orbifolds of half their volumes. Any other noncompact nonorientable orbifold will be double covered by an orientable orbifold of volume at least $v_0/4$, and hence will have volume itself of at least $v_0/8$. □

An investigation into the volumes of hyperbolic 3-orbifolds with multiple cusps will appear in a subsequent paper.

References

[1] C. Adams, The noncompact hyperbolic 3-manifold of minimal volume, Proc. A.M.S. 100 (1987), 601–606.

[2] C. Adams, Volumes of N-cusped hyperbolic 3-manifolds, J. London Math. Soc. (2) 38 (1988), 555–565.

[3] C. Adams, Limit volumes of hyperbolic 3-orbifolds, to appear in Journal of Differential Geometry.

[4] C. Adams, M. Hildebrand and J. Weeks, Hyperbolic invariants of knots and links, to appear in Trans. A.M.S.

[5] R. Meyerhoff, Sphere packing and volume in hyperbolic 3-space, Comment. Math. Helv. 61 (1986), 271–278.

[6] R. Meyerhoff, The cusped hyperbolic 3-orbifold of minimum volume, Bull. Amer. Math. Soc. 13 (1985), 154–156.

[7] W. Thurston, The geometry and topology of 3-manifolds, lecture notes, Princeton University, 1978.

Department of Mathematics, Williams College, Williamstown, MA 01267

Combinatorial Cubings, Cusps, and the Dodecahedral Knots

I. R. Aitchison and J. H. Rubinstein

Abstract. There are finitely many tessellations of 3-dimensional space-forms by regular Platonic solids. Explicit examples of constant curvature finite-volume 3-manifolds arising from these are well-known for all possibilities, except for the tessellation $\{5,\ 3,\ 6\}$. We introduce the dodecahedral knots D_f and D_s in S^3 to fill this gap. Techniques used illustrate the results on cusp structures and π_1-injective surfaces of alternating link complements obtained by Aitchison, Lumsden and Rubinstein **[ALR]**.

The Borromean rings and figure-eight knot arise from the tessellation of hyperbolic 3-space by regular ideal octahedra and tetrahedra respectively. We produce exactly four new links in S^3, corresponding to the tessellations $\{4,\ 3,\ 6\}$ and $\{5,\ 3,\ 6\}$ of $\mathbb{H}^3$, and united by a canonical construction from the Platonic solids.

The dodecahedral knot D_f is the third in an infinite sequence of fibred, alternating knots, the first member of which being the figure-eight. The complements of these new links contain π_1-injective surfaces, which remain π_1-injective after 'most' Dehn surgeries. The closed 3-manifolds obtained by such surgeries are determined by their fundamental groups, but are not known to be virtually Haken.

1. Introduction

Regular tessellations of space-forms by Platonic solids have played a significant rôle in the exploration and exposition of 3-dimensional geometries and topology. Table 1, derived from Coxeter **[Co1]**, **[Co2]**, gives all such tessellations, including those by solids with deleted vertices.

Remark 1.1. The tessellations $\{3,\ 3,\ 3\}$, $\{4,\ 3,\ 4\}$ and $\{5,\ 3,\ 5\}$ are self-dual. The links of vertices are respectively tetrahedra, octahedra and icosahedra, the Platonic solids with triangular faces. The corresponding edge degrees — 3, 4 and 5, the most famous Pythagorean triple — encapsulate the notions of positively-curved, flat and negatively-curved geometry.

The corresponding symmetry groups are well-understood in the spherical and flat spaceforms, as are the subgroups acting without fixed points. For the hyperbolic tessellations, finite-index torsion-free subgroups exist by Selberg's theorem, with corresponding quotient 3-manifolds having finite volume. Infinitely many such subgroups exist.

The dodecahedral tessellation $\{5,\ 3,\ 3\}$ gives rise to Poincaré's homology sphere $\mathcal{P}^3$, a manifold ubiquitous in geometric topology, associated with problems of smoothings and triangulations of manifolds. A beautiful description of $\mathcal{P}^3$ in terms of face identifications of a dodecahedron has been given by Seifert and Weber **[SW]**, where another such

Solid	Tessellations of spaceforms by Platonic solids			
	S^3	$\mathbb{R}^3$	$\mathbb{H}^3$, compact	$\mathbb{H}^3$, ideal
tetrahedra	$\{3,\ 3,\ 3\}$	none	none	$\{3,\ 3,\ 6\}$
	$\{3,\ 3,\ 4\}$			
	$\{3,\ 3,\ 5\}$			
icosahedra	none	none	$\{3,\ 5,\ 3\}$	none
octahedra	$\{3,\ 4,\ 3\}$	none	none	$\{3,\ 4,\ 4\}$
cubes	$\{4,\ 3,\ 3\}$	$\{4,\ 3,\ 4\}$	$\{4,\ 3,\ 5\}$	$\{4,\ 3,\ 6\}$
dodecahedra	$\{5,\ 3,\ 3\}$	none	$\{5,\ 3,\ 4\}$	$\{5,\ 3,\ 6\}$
			$\{5,\ 3,\ 5\}$	

Table 1

compact 3-manifold, the hyperbolic Seifert-Weber space, is also described. The latter manifold arises from the dodecahedral tessellation $\{5,\ 3,\ 5\}$. In both cases, opposite faces are identified in a natural fashion.

The cube is distinguished in that it tessellates all spaceforms. The cubical tessellation of $\mathbb{R}^3$ gives rise to the 3-dimensional torus, with flat geometry. Euclidean space does not admit regular tessellations other than by the cube. Nonetheless, the mysterious connections between the Platonic solids allows for an intriguing manifestation of dodecahedra even in this context. Thurston **[Th]** has shown how a dodecahedron can be flattened into a cube, and then allowed to 'tessellate' $\mathbb{R}^3$. Allowing orbifold structures, Thurston then shows how this tessellation induces a singular metric on S^3, with cone angle π concentrated along the Borromean rings. Using the universality of the Borromean rings in the construction of closed orientable 3-manifolds as branched covering spaces, Hilden, Lozano, Montesinos and Whitten **[H*]** demonstrate the significance of the dodecahedral tessellation $\{5,\ 3,\ 4\}$: its group of symmetries is rich enough to produce all closed 3-manifolds.

Similarly, orbifold structures on links in S^3 arise from the Seifert-Weber manifold in the guise of the Whitehead link, and from the tessellation of $\mathbb{H}^3$ by cubes with icosahedral vertex links via the 5_2-knot (**[Be]**, **[AR1]**). Both of these links are universal.

Other *closed* hyperbolic 3-manifolds arising from tessellations of $\mathbb{H}^3$ have been described in Best **[Be]**, and in Richardson and Rubinstein **[RR]**.

The tessellations of hyperbolic space by ideal Platonic solids are of equal interest. The most famous contemporary example is $\{3,\ 3,\ 6\}$, giving the figure-eight knot complement (again universal) as quotient (**[Th]**). An example of a link complement in S^3 whose complement is the quotient of $\{4,\ 3,\ 6\}$ is described in **[AR1]**. Thurston **[Th]** also shows that two octahedra of $\{3,\ 4,\ 4\}$ form the fundamental domain for a discrete subgroup of symmetries, with quotient again the complement of the Borromean rings in S^3.

The remaining tessellation $\{5,\ 3,\ 6\}$ of $\mathbb{H}^3$ by ideal dodecahedra has not been considered previously — no *explicit* link complement in any 3-manifold is known to have such a structure. We will construct two such examples in S^3, obtaining what we call the

dodecahedral knots D_f and D_s. Whether the tessellation $\{5, 3, 6\}$ leads to as rich a domain as the other dodecahedral tessellations remains to be seen.

Complements of alternating links

In each of the cases above, the resulting link in S^3 is alternating. Investigations of the hyperbolic structures of alternating link complements have been given by Lawson **[La]**, Menasco **[Me]**, Takahashi **[Ta]**, and more recently by Weeks **[We]**, seeking to generalize the beautiful constructions of Thurston **[Th]**. In each case, the aim has been to demonstrate the existence of a complete metric of constant curvature -1 on the complement, and to calculate various invariants from such a (unique) structure. This invariably necessitates determining a combinatorial description of the link complement as the union of two 'ideal' polyhedra, with face identifications, and then decomposing these polyhedra into ideal tetrahedra whose shapes and volumes can be calculated. At this stage of the procedure, there is no canonical way to proceed, and any structure hidden in the combinatorics at the polyhedral level is lost.

That some beautiful deeper combinatorial structure may have existed has been remarked in these papers, but neither revealed nor exploited explicitly.

Retrospectively, our starting point is two remarks of Thurston **[Th]**. The first is that the figure-eight knot can be arranged on the 1-skeleton of a tetrahedron, as a 'heuristic' that the complement admits a tetrahedral decomposition. In fact, there are two simple such arrangements, and we develop the second one. Thurston's second remark is that for the Borromean rings, face identifications have a beautiful naturality: "Faces are glued to their corresponding faces with 120° rotations, alternating in direction like gears" **[Th]**.

We describe how, with our arrangement of the figure-eight knot on the tetrahedron, these remarks are related, and generalize to face identifications of two identical polyhedra, producing all of the examples of alternating links considered. We illustrate with each of the ideal regular tessellations of $\mathbb{H}^3$, producing 4 new links in the process. Our favourites, arising from $\{5, 3, 6\}$, are a new fibred knot D_f, and a knot D_s possessing a high degree of symmetry. The existence and simplicity of this combinatorial structure of alternating link complements is described in detail in **[ALR]**. A more general context is described in **[AR2]**.

2. The general construction for 4-valent graphs

We recall the construction of **[ALR]**. Take an arbitrary finite connected planar graph Γ, all of whose vertices having degree 4. We also require that at any vertex, all regions meeting at the vertex are distinct. Two-colour the regions of the plane checker-board fashion using white and black, with the exterior white by convention. Assign signs '$+$' and '$-$' to the white and black regions respectively. Denote the resulting combinatorial polyhedron by Π_Γ^+.

Now take an identical copy of Π_Γ^+, reverse all colours and signs, and denote the resulting polyhedron by Π_Γ^-. Each face ϕ_i of $\Pi_\Gamma^\pm$ is a combinatorial n_i-gon, with sign allocation σ_i, and we identify ϕ_i with the corresponding face ϕ_i' of $\Pi_\Gamma^\mp$ by a rotation of $\sigma_i.2\pi/n_i$, with a '$+$' sign denoting clockwise.

Denote the resulting topological space by $\overline{M}_\Gamma$, and let M_Γ denote $\overline{M}_\Gamma$ with deleted vertices. Finally, let $\mathfrak{L}_\Gamma$ denote the alternating link in S^3 canonically associated to Γ, as in Figure 1. Observe that, viewed from the center of any region, crossings are of the sign assigned to that region.

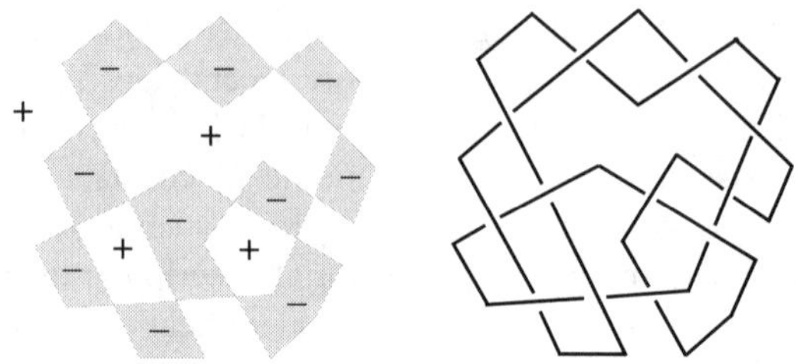

Figure 1

One of the results of **[ALR]** is

Theorem 2.1. *M_Γ is canonically homeomorphic to $S^3 - \mathfrak{L}_\Gamma$. Each edge of M_Γ is of degree 4.*

3. The six examples arising from ideal tessellations

In each case, we describe an alternating link, and face identifications of the corresponding pair of identical polyhedra. That the link complement has a complete metric of constant curvature -1 follows immediately on declaring each polyhedron to be ideal and regular in hyperbolic space.

Example 1. The Borromean rings. Applying this construction to the graph $\Gamma_{\{3,4\}}$ underlying the octahedron, we recover Thurston's description of the complement of the Borromean rings of Figure 2. The universal cover is the tessellation $\{3, 4, 4\}$ of $\mathbb{H}^3$.

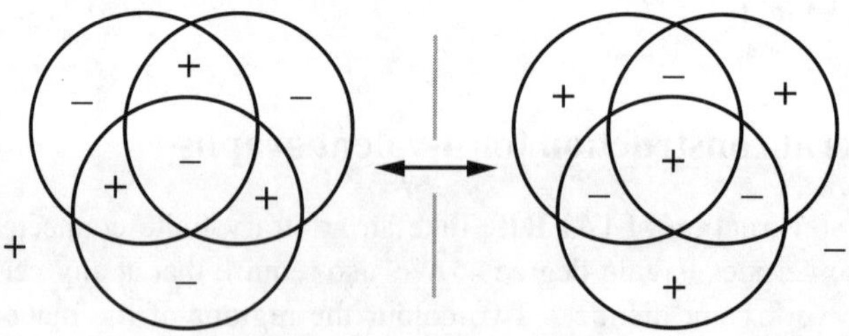

Figure 2

Example 2. The figure-eight knot. Take a tetrahedron, corresponding to the graph $\Gamma_{\{3,3\}}$ and 2-colour its faces black and white in the unique (up to symmetry) way so that no vertex is surrounded by regions all of the same colour. Assign the sign '$+$' to

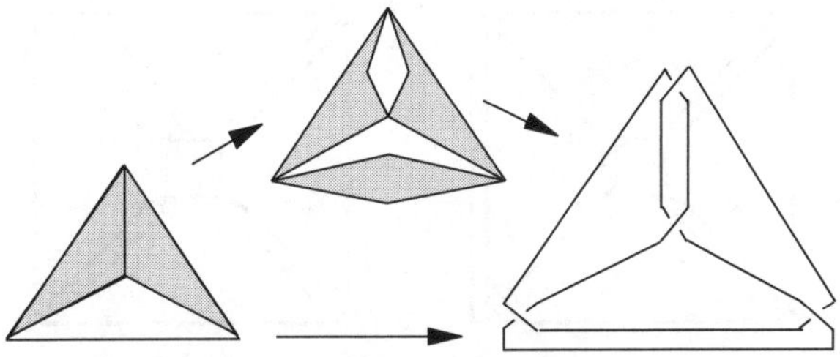

Figure 3

the white regions, '$-$' to the black. Now split each edge separating regions of the same colour to obtain a 4-valent graph 2-coloured as above. (Figure 3.)

Carrying out face identifications yields the figure-eight knot complement. The two resulting 'bigons' can be squeezed back to a single edge to recapture the face identifications of tetrahedra as in Thurston's description. Note that in removing a bigon, two edges are identified in each polyhedron $\Pi_\Gamma^\pm$, from different equivalence classes. Every edge in the quotient is thus of degree 6. The universal cover corresponding to this combinatorial structure is geometrically the tessellation $\{3,3,6\}$ of $\mathbb{H}^3$, giving rise to the complete structure on the knot complement.

Examples 3, 4. Two cubical links. There are *two* ways to 2-colour the regions of the graph $\Gamma_{\{4,3\}}$. These are depicted in Figure 4.

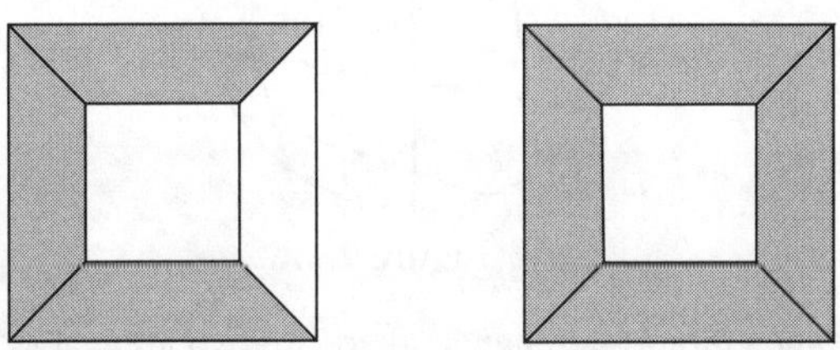

Figure 4

Proceed exactly as in the last example, observing that the introduction and deletion of bigons is unnecessary provided the link associated with such a 2-coloured trivalent graph is interpreted according to Figure 5.

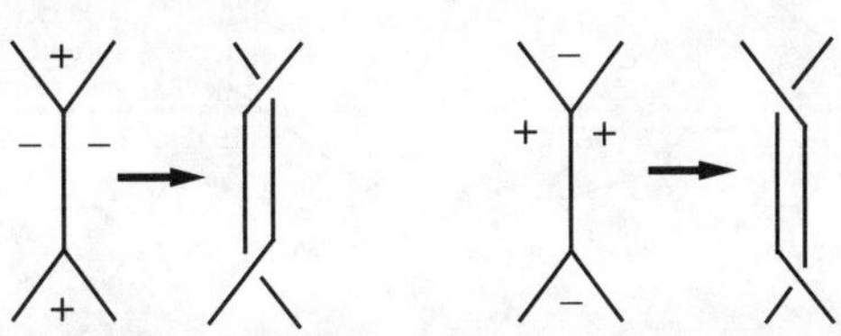

Figure 5

These two links obtained from the cube arise from the tessellation $\{4,3,6\}$ of $\mathbb{H}^3$, and are the links 8_4^3 and 8_1^4 in Rolfsen's book, depicted in Figure 6.

In **[ALR]**, 4-valent graphs admitting a collapse to a 2-coloured 3-valent graph without bigons are called 'balanced': the construction applied here works for all such graphs.

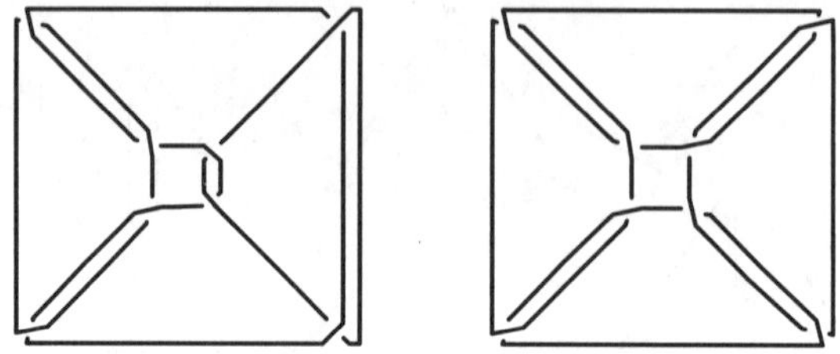

Figure 6

Remark 3.1. These two are the *only* links in Rolfsen's tables which have balanced bigons, in the sense of **[ALR]**, and no triangular regions.

There is another 3-component link also corresponding to the tessellation $\{4,3,6\}$ of $\mathbb{H}^3$, described in **[AR1]**. This does not obviously arise as part of our general construction.

Examples 5, 6. The two dodecahedral knots. The dodecahedron may be depicted combinatorially as in Figure 7.

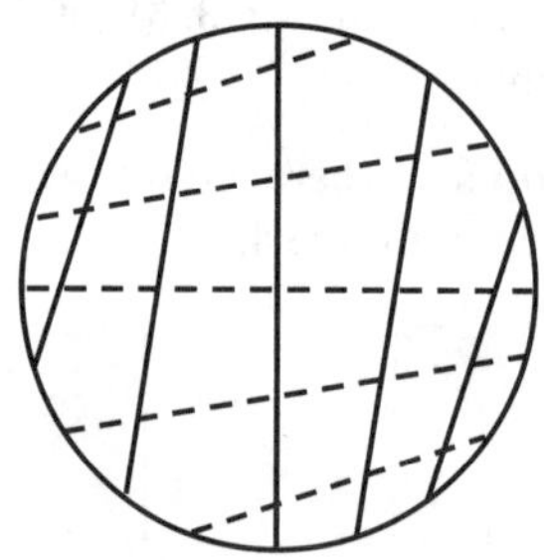

Figure 7

Up to symmetry and colour interchange, there are two allowable 2-colourings. These are depicted in Figure 8, with corresponding knots in Figure 9 denoted D_s and D_f arising from the tessellation $\{5,3,6\}$ of $\mathbb{H}^3$. The knot D_s has considerable symmetry, whereas D_f turns out to be fibred.

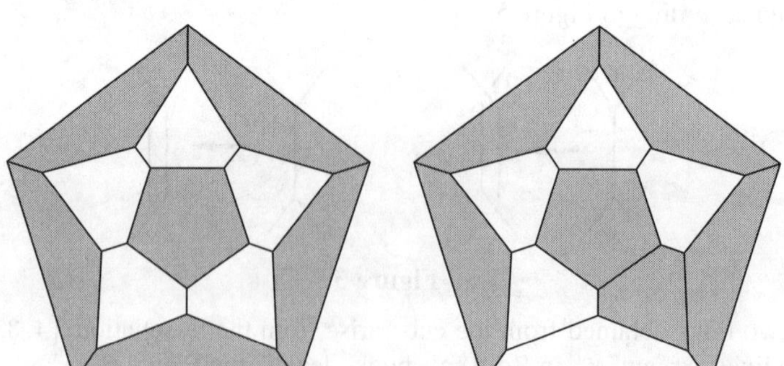

Figure 8

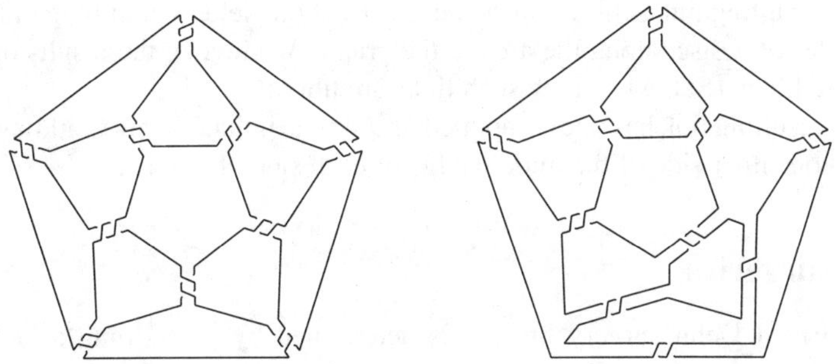

Figure 9

Added in Proof. Alan Reid and Walter Neumann have demonstrated some fascinating properties of these dodecahedral knots, in the context of their beautiful work on arithmetic structures **[NR]**.

Hatcher has used similar ideas in **[Ha]**, and it seems likely that Thurston is aware of the general construction, particularly since we have found the fibred dodecahedral knot in **[Ri]**, referred to by Riley as 'Thurston's knot'. It is clear from the construction above that the complement of D_f admits an orientation reversing involution. The complements of both D_f and D_s contain totally geodesic immersed surfaces with respect to the complete metric of constant curvature.

4. Some fibred alternating knots from balanced links

We begin with a characterization of a class of colourable graphs.

Lemma 4.1. *Suppose Γ is a connected trivalent planar Hamiltonian graph. Then Γ can be 2-coloured with no vertex surrounded by regions of the same colour.*

Such a graph arises by drawing a circle as the equator of the sphere, and adding disjointly embedded arcs with endpoints on the equator. Colour one hemisphere white, the other black.

Remark 4.2. The 2-colourings of the cube and dodecahedron described above show that a graph with Hamiltonian circuit need not have a *unique* 2-colouring, and that the resulting alternating link may have more than one component.

A particularly nice class arises by taking the sequence of graphs Γ_t generalizing Figure 7: instead of 5 arcs in each hemisphere, take $2t-1$ for any natural number t, with t arcs at the back meeting the equator in the left and right regions of the front. The top arc at the back meets the equator between the front t^{th}-and $(t+1)^{\text{st}}$-arcs numbered from the left. Observe that Γ_1 is a tetrahedron, whereas Γ_3 is the dodecahedron.

Proposition 4.3. *Each of the graphs Γ_t gives rise to an alternating fibred knot $K_t = \mathfrak{L}_{\Gamma_t}$. The knot K_1 is the figure-eight knot, and K_3 is the dodecahedral knot D_f.*

Proof. The resulting link is fibred since the construction yields a plumbing of Hopf bands onto two sides of a disc, along the arcs of the graph. We invoke the results of Murasugi **[Mu]** and Stallings **[St]**, who show such links are fibred.

That the resulting link has one component is a simple induction on t, adding additional Hopf bands on either side of the middle edge of each side of the disc. □

5. Dehn surgeries

Every non-trivial Dehn surgery on K_t is determined by prescribing a Dehn surgery coefficient $\rho = (p, q) \neq \infty$. Denote the resulting 3-manifold by $M_{t,\rho}$.

Theorem 5.1. *For each $\rho \neq \infty$ and $t > 2$, $M_{t,\rho}$ is irreducible, has universal cover homeomorphic to $\mathbb{R}^3$, and contains an immersed π_1-injective surface satisfying the 4-plane, 1-line condition. Hence $M_{t,\rho}$ has homotopy type determined by its fundamental group.*

Sketch of Proof. Each of the trivalent polyhedra $\Pi^{\pm}_{\Gamma_t}$ has $2(2t-2)$ pentagonal faces, four $(t+2)$-gons, $(12t-6)$ edges and $(8t-4)$ vertices. Each polyhedron can be decomposed into $(8t-4)$ cubes in the standard manner (see **[AR1]** for example). After face identifications, all edges of $S^3 - K_t$ have degree $(t+2)$, 5, or 6. The former two values occur along introduced edges joining the centers of the polyhedra through points at the center of faces.

Consider a cube in the ideal cubing $\{4,\ 3,\ 6\}$ of $\mathbb{H}^3$, and bisect it symmetrically into 8 isometric subcubes by planes orthogonal at the centre, and orthogonal to the edges. Endow each of the cubes of $\Pi^{\pm}_{\Gamma_t}$ with the geometry of one of these subcubes, with the distinguished vertex at a vertex of $\Pi^{\pm}_{\Gamma_t}$. The resulting singular metric is complete, and has negative curvature at every point. The structure of the cusps is depicted in Figure 10, where there are $(16t-8)$ equilateral triangles in the decomposition of the torus. Such pictures occured originally in **[Th]**. Generators for the homology of the peripheral torus of the knot have been labelled. These are sufficiently long for $t \geq 3$ that any non-trivial Dehn surgery, in the sense of Gromov-Thurston (**[AR]**, **[GT]**) always yields a closed Cartan-Hadamard manifold with negative curvature along the core of the sewn-in solid torus, and with the metric away from the cusp remaining unaltered.

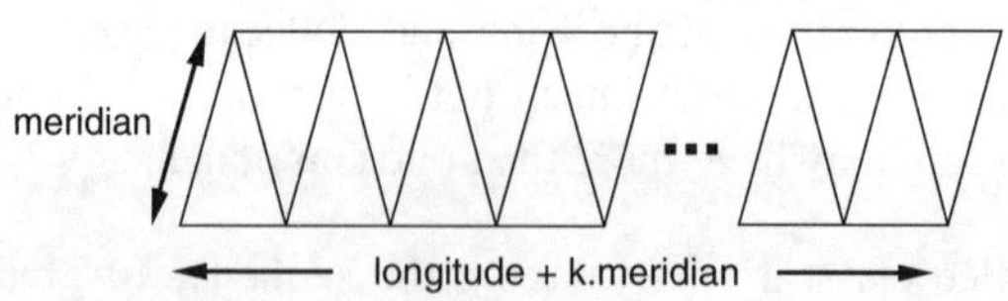

Figure 10

The immersed surface obtained by taking the union of squares bisecting each of the cubes of the decomposition of $S^3 - K_t$ is π_1-injective, being isotopic to a (singular) totally geodesic surface. Since this surface is in the 'thick' part of $S^3 - K_t$, it survives to produce an injective surface after surgery. This surface satisfies the conclusions of the theorem. For further details, see **[ALR]**, **[AR1]** and **[AR2]**. □

Remarks 5.2. The symmetric dodecahedral knot also belongs to an infinite family, obtained from the trivalent graphs $\mathcal{G}_k$, $k \geq 1$. These are obtained by drawing concentric $(k+1)$-gons in the plane, rotated relative to each other, and filling the annular region between them by $2k+2$ pentagons. The results on surgery also apply to this class, when $k > 2$. A similar argument applies to the cubical links described above.

The resulting closed 3-manifolds are not known to be virtually Haken.

Remark 5.3. The 14-sided polyhedron corresponding to $\mathcal{G}_5$ can be realized in hyperbolic space as the fundamental domain of the group action giving rise to Löbell's manifold **[Lö]**, the first closed hyperbolic 3-manifold to appear in the literature.

Acknowledgements

The authors would like to thank Darren Long for an invaluable remark on the Borromean rings, and a question concerning the dodecahedral knots, and Yoav Moriah for correcting an error in an earlier version of this paper.

References

[AR1] I.R. Aitchison and J.H. Rubinstein, An introduction to polyhedral metrics of non-positive curvature on 3-manifolds, Geometry of Low-Dimensional Manifolds, Volume II: Symplectic Manifolds and Jones-Witten Theory, Cambridge University Press, 1990, 127-161.

[AR2] I.R. Aitchison and J.H. Rubinstein, Polyhedral metrics of non-positive curvature on 3-manifolds with cusps, in preparation.

[ALR] I.R. Aitchison, E. Lumsden and J.H. Rubinstein, Cusp structures of alternating links, Research report, University of Melbourne, preprint series #13 (1991).

[Be] L.A. Best, On torsion-free discrete subgroups of $\mathrm{PSL}(2, C)$ with compact orbit space, Can. J. Math. **23** (1971), 451–460.

[Co1] H.S.M. Coxeter, Regular Polytopes, London, 1948.

[Co2] H.S.M. Coxeter, Regular honeycombs in hyperbolic space, Proc. I.C.M. Amsterdam 1954.

[GT] M. Gromov and W.P. Thurston, Pinching constants for hyperbolic manifolds, Invent. Math. **89** (1987), 1–12.

[Ha] A. Hatcher, Hyperbolic structures of arithmetic type on some link complements, J. London Math. Soc. **27** (1983), 345–355.

[H*] H.M. Hilden, M.T. Lozano, J.M. Montesinos and W.C. Whitten, On universal groups and three-manifolds, Invent. math. **87** (1987), 441–456.

[La] T. C. Lawson, Representing link complements by identified polyhedra, preprint.

[Lö] F. Löbell, Beispiele geschlossener drei-dimensionaler Clifford-Kleinscher Räume negativer Krümmung, Ber. Sächs. Akad. Wiss. Leipzig **83** (1931), 167–174.

[Me] W. W. Menasco, Polyhedra representation of link complements, Amer. Math. Soc. Contemporary Math. **20** (1983), 305–325.

[Mu] K. Murasugi, On a certain subgroup of the group of an alternating link, Amer. J. Math. **85** (1963), 544–550.

[NR] W.D. Neumann and A.W. Reid, Arithmetic of hyperbolic manifolds, these Proceedings.

[RR] J.S. Richardson and J.H. Rubinstein, Hyperbolic manifolds from regular polyhedra, preprint 1982.

[Ri] R.F. Riley, Parabolic representations and symmetries of the knot 9_{32}, in Computers in Geometry and Topology, Edited by Martin C. Tangora, Marcel Dekker (1989), 297–313.

[Ro] D. Rolfsen, Knots and Links, Publish or Perish, 1976.

[St] J. Stallings, Construction of fibered knots and links, Proc. Symp. Pure Math. **32** (1976), 55–60.

[Ta] M. Takahashi, On the concrete construction of hyperbolic structures of 3-manifolds, preprint.

[Th] W.P. Thurston, The geometry and topology of 3-manifolds, Princeton University Lecture Notes 1978.

[WS] C. Weber and H. Seifert, Die beiden Dodekaederräume, Math. Z. **37** (1933), 237–253.

[We] J.R. Weeks, Hyperbolic structures on three-manifolds, PhD dissertation, Princeton 1985.

University of Melbourne, Department of Mathematics, Parkville, Victoria 3052, Australia
Email: iain@mundoe.maths.mu.oz.au

University of Melbourne, Department of Mathematics, Parkville, Victoria 3052, Australia
Email: rubin@mundoe.maths.mu.oz.au

Hyperbolic Cobordism and Conformal Structures

*Boris N. Apanasov**

Abstract In this paper, we briefly survey selected recent developments and present some new results in the area of uniformized conformal structures on a complete hyperbolic finite volume n-manifold (even closed) related to $(n+1)$-dimensional homology cobordisms with hyperbolic structures, especially, for the three-dimensional case.

1. Isometric and conformal group actions and maximal balls

Let $\mathbb{H}^n$ be the subspace

$$\{(x_0, \ldots, x_n) \in R^{n+1}: q(x_0, \ldots, x_n) = -x_0^2 + x_1^2 + \cdots + x_n^2 = -1\}$$

and $x_0 > 0$. The quadratic form q restricts to give a positive definite form on each tangent space of $\mathbb{H}^n$ and, consequently, endows $\mathbb{H}^n$ with a Riemannian metric. We call this Riemannian manifold the hyperbolic n-space. It has constant sectional curvature -1 and is homogeneous. Its isometry group is the real linear subgroup $SO(n,1)$ of matrices in $SL_{n+1}(R)$ preserving the form q and $\mathbb{H}^n$. A hyperbolic n-manifold M is a complete Riemannian manifold locally isometric on $\mathbb{H}^n$. In fact M is isometric to the quotient $\mathbb{H}^n/G_M$ where $G_M \simeq \pi_1(M)$ is some discrete torsion free subgroup of $SO(n,1)$ determined by M up to conjugation in $SO(n,1)$.

The hyperbolic metric in $\mathbb{H}^n$ endows the $(n-1)$-sphere at infinity $\partial\mathbb{H}^n$ with a conformal structure where $SO(n,1)$ acts as the group of all conformal automorphisms of the sphere. Taking the Poincaré ball model of the hyperbolic n-space (in the unit ball $B^n(0,1)$), we have the isomorphism (see **[AP1]**):

$$\{\mathbb{H}^n, \partial\mathbb{H}^n, SO(n,1)\} \cong \{B^n, S^{n-1}, \text{Möb}(n-1)\}$$

where $\text{Möb}(n-1)$ is the Möbius group generated by reflections in $(n-2)$-dimensional subspheres of S^{n-1}, the sphere S^{n-1} has the standard conformal structure induced by the Euclidean metric of $\mathbb{R}^n$ and the unit ball B^n has the Poincaré hyperbolic metric with the length differential

$$ds = \frac{2|dx|}{1-|x|^2}, \qquad x \in B^n(0,1).$$

---

* This paper was written while the author was visiting Mittag-Leffler Institut, which he thanks for its hospitality and funding. The author thanks Ravi Kulkarni for his patience and helpfulness in listening to some of the author's ideas.

1.2. Now let us fix a torsion-free, finitely generated group G. An n-dimensional hyperbolic structure on G is determined by a pair $\{M, \psi\}$, where M is a hyperbolic n-manifold and $\psi: G \to \pi_1(M)$ is an isomorphism. We denote the set of all hyperbolic structures on G (defined up to isometries of hyperbolic manifolds and inner automorphisms on G) by $\mathcal{H}^n(G)$. This set is naturally identified with the set of conjugacy classes of faithful representations $\rho: G \to SO(n,1)$ with discrete image and therefore $H^n(G)$ has a natural topology, induce by the algebraic convergence topology on $\mathrm{Hom}(G, SO(n,1))$. Namely, representations are close if they are close on a finite generating set.

As an example, given a closed surface S_g of genus $g > 1$, $\mathcal{H}^2(\pi_1(S_g))$ is the Teichmüller space of $G = \pi_1(S_g)$ and is homeomorphic to $\mathbb{R}^{6g-6}$. On the other hand, given a closed (or finite volume) hyperbolic n-manifold M^n, $n > 2$, $\mathcal{H}^n(\pi_1(M^n))$ is a point due to the Mostow rigidity theorem **[MW]**.

Also there are Morgan-Shalen-Thurston's results which say that $\mathcal{H}^n(G)$ is compact for a wider class of groups G as a weak version of Mostow rigidity for infinite volume hyperbolic manifolds; see **[MO2]**. In fact, one shows

1.3. Theorem. *Let M^3 be a hyperbolic 3-manifold, and suppose that $G = \pi_1(M^3)$ is finitely generated. Then for any $n > 2$, the space $\mathcal{H}^n(G)$ is compact if and only if G does not have a decomposition of one of the following types:*
(i) *$G = A *_C B$ with C cyclic and of infinite index in A and B;*
(ii) *$G = A*_C$ with C cyclic.* □

Notice that it follows that $\mathcal{H}^n(\pi_1(M^3))$ is compact if and only if $\mathcal{H}^3(\pi_1(M^3))$ is compact. In particular, we have the following fact.

1.4. Corollary. *If M is a finite volume hyperbolic 3-manifold, then $\mathcal{H}^n(\pi_i(M)$ is compact for all $n > 2$.* □

1.5. Given a discrete group $G \subset \mathrm{Möb}(n-1)$, we define the Nielsen hull $H_G \subset \mathbb{H}^n \cup \partial\mathbb{H}^n$ as the minimal convex (in $\mathbb{H}^n$) set containing the limit set $L(G) \subset S^{n-1} = \partial\mathbb{H}^n$. Let $\rho: \mathbb{H}^n \cup \partial\mathbb{H}^n \to H_G$ be the G-equivariant retraction where, for $x \in \mathbb{H}^n \backslash H_G$, $\rho(x) \in \partial H_G$ is the point with shortest distance to x and, for $x \in \partial\mathbb{H}^n \backslash L(G)$, $\rho(x)$ is the first point of tangency with H_G of a horosphere in $\mathbb{H}^n$ with the center at x.

For a description of the boundary of the Nielsen hull H_G, we define (following **[AP5]**) the (strictly) maximal balls in the discontinuity set $\Omega(G) \subset S^{n-1}$. Namely, an open ball $B \subset \Omega(G)$ is called a (strictly) maximal ball if the sphere of lowest dimension containing the limit subset $\partial B \cap L(G)$ is the sphere ∂B itself.

For the case $n = 3$, the discontinuity set $\Omega(G) \subset S^2$ of any finitely generated Kleinian group G (G is discrete with non-empty $\Omega(G)$) whose limit set $L(G)$ is not contained in a circle is covered by the family $\mathcal{B}(G)$ of strictly maximal discs, finite modulo G. For the case of geometrically finite quasi-Fuchsian groups $G \subset \mathrm{Möb}(n)$, an almost similar situation holds (see **[AP5]**). There we needed the following fact (**[AP5**, Th. 6.1**]**):

1.6. Theorem. *Let $G \subset \mathrm{Möb}(3)$ be a Kleinian group having at least three maximal balls $B_i \subset \Omega(G)$ with two common limit points $x, y \in \partial B_i$ and let $\mathrm{int}(\cap B_i) \neq \emptyset$. Then*

the boundary of the Nielsen hull is a pleated 3-surface in $\mathbb{H}^4$ *with a conical singularity along the geodesic with the end points* x, y. *Its neighborhood in this surface is an union of dihedral angles with the sum of magnitudes less than* 2π. □

The standard notion of a pleated surface (cf. **[EM]**) arrives from the Krein-Milmann theorem on extreme points of a convex hull in Euclidean space (if we take the projective Klein model of $\mathbb{H}^n$ in the ball $B^n(0,1) \subset \mathbb{R}^n$). Here the pleating locus is a geodesic lamination (partial foliation) whose leaves of co-dimension one or more may be singular (for $n = 4$, as pages of an open book).

2. Uniformized conformal structures on manifolds

2.1. Given an n-manifold M, $n \geq 3$, by a conformal structure (conformally flat structure) on the manifold M we mean a $(S^n, \text{Möb}(n))$-structure on M, i.e., a structure locally modeled on the standard conformal structure of the n-sphere $S^n \cong \mathbb{R}^n \cup \{\infty\}$. In other words, a conformal structure is a maximal atlas on M with all changes of charts in a Möbius group $\text{Möb}(n)$. Extending chart by chart in the universal covering $\widetilde{M}$ of M, we obtain the developing map $d\colon \widetilde{M} \to S^n$ inducing the holonomy homomorphism $d^*\colon \pi_1(M) \to \text{Möb}(n)$.

A conformal structure c on M will be called a uniformized structure (compare **[KP]**) if its development d is not surjective while the holonomy group $G = d^*(\pi_1(M))$ acts discontinuously in the domain $\Omega_0 = d(\widetilde{M})$, i.e., G is a Kleinian group (see **[KM]**); here the manifold Ω_0/G with the natural conformal structure is conformally equivalent to the conformal manifold (M, c). Using the fundamental group $\pi_1(M)$ for the marking of conformal structures on M, we obtain the space $\mathcal{C}(M)$ of uniformized marked conformal structures on the manifold M.

2.2. Especially, for a finite volume hyperbolic manifold M, the space $\mathcal{C}(M)$ is naturally identified with the set of conjugacy classes of faithful representations

$$\rho\colon \pi_1(M) \longrightarrow SO(n+1,1) \tag{2.1}$$

with discrete image which act discontinuously somewhere in the sphere at infinity $S^n = \partial\mathbb{H}^{n+1}$. Namely, if

$$\mathcal{R}(M) \subset \text{Hom}(\pi_1(M), SO(n+1,1)) \tag{2.2}$$

is the subspace of such representations in the representation variety, the group $SO(n+1,1)$ acts on the representation variety by conjugation leaving the subspace $\mathcal{R}(M)$ invariant. The quotient space

$$\mathcal{T}(M) = \mathcal{R}(M)/SO(n+1,1) \tag{2.3}$$

is the desired space of conjugacy classes of representations (2.1) and is naturally identified with the space $\mathcal{C}(M)$ via the holonomy representation d_*; see **[LK]**, **[GM]**. This yields a topology on $\mathcal{C}(M)$ defined by the topology of algebraic convergence in the representation variety $\text{Hom}(\pi_1(M), SO(n+1,1))$. Immediately from this description, the definition of the space $\mathcal{H}^n(G)$ and Corollary 1.4 in the case $n = 4$, we obtain

2.3. Theorem. *Let M be a finite volume hyperbolic 3-manifold. Then the space $\mathcal{C}(M)$ of uniformized marked conformal structures on M has a natural compactification $\overline{\mathcal{C}(M)}$ such that each of its points corresponds to a faithful representation ρ in the corresponding compactification $\overline{\mathcal{T}(M)}$ with discrete image of $\rho(\pi_1(M)) \subset \text{Möb}(3)$.* □

2.4. The space $\mathcal{C}(M)$ contains an open subspace $\mathcal{C}_q(M)$ of quasi-Fuchsian structures on the manifold M which corresponds to an open subspace $\mathcal{T}_q(M) \subset \mathcal{T}(M)$ of quasi-Fuchsian representations, i.e., quasi-conformal conjugations

$$\rho: G \longrightarrow fGf^{-1} \subset \text{Möb}(3) \tag{2.4}$$

where $\pi_1(M) \cong G \subset \text{Isom}\mathbb{H}^3 \subset \text{Isom}\mathbb{H}^4 \cong \text{Möb}(3)$ and $f: S^3 \longrightarrow S^3$ is a quasi-conformal automorphism of the sphere S^3 compatible with the action of G. This fact follows from Sullivan's stability theorem **[SU3]**; see also **[JM]**.

The first results to obtain some boundary points of the space $\mathcal{C}(M)$ as end points of smooth curves in the open subspace $\mathcal{C}_q(M)$, for the case of a closed manifold M, was Theorem B and Corollary 5.2 in **[AP2]**. These boundary points are similar to cusps on the boundary of Teichmüller space $\mathcal{T}(S_g)$ of Riemann surfaces of genus $g > 1$ (they correspond to so-called accidental parabolic elements in the holonomy groups; see **[BR]**) and were obtained as limits of bending deformations of the distinguished conformal (hyperbolic) structure on the manifold M. Here a bending deformation of the manifold $M = \mathbb{H}^n/G$, $G \subset \text{Isom}\mathbb{H}^n$, gives conformal structures $c_{\text{bend}} \in \mathcal{C}_q(M)$ which correspond to singular hyperbolic structures on M obtained by bending of M along a totally geodesic hypersurface through some angles. In fact, such a singular n-structure on M has a pleated n-plane $H_{\text{bend}} \subset \mathbb{H}^{n+1}$ as its universal covering and the structure c_{bend} corresponds to a conformal n-manifold $\Omega_0/d^*_{\text{bend}}(\pi_1(M))$ where the domain $\Omega_0 \subset \mathbb{H}^{n+1} = S^n$ is an invariant component of the holonomy group $d^*_{\text{bend}}(\pi_1(M))$ spanned on the $(n-1)$-sphere at infinity of the n-cell H_{bend}, $\partial H_{\text{bend}} = \partial\Omega_0$; see **[AP4]**, **[JM]** and **[KR]** for details. Here we give only the following well-known property of such bending structures $c_{\text{bend}} \in C_q(M)$ (see **[AP5]**):

The boundary of the Nielsen hull for the holonomy group

$$d^*_{\text{bend}}(\pi_1(M)) \subset \text{Möb}(n) \tag{2.5}$$

can be isometrically developed in the hyperbolic n-plane $\mathbb{H}^n \subset \mathbb{H}^{n+1}$.

Except bending structures in $\mathcal{C}_q(M)$, now for a 3-manifold M, only stamping structures $c_{\text{stamp}} \in C_q(M)$ obtained by stamping deformations of the distinguished structure on M along a geodesic ℓ (either with torsion around ℓ or not) are known; for definitions and details see **[AP6]**. For totally geodesic surfaces $S_i \subset M$ intersecting along the geodesic $\ell \subset M$, such a stamping deformation gives compatible simultaneous bendings along these surfaces S_i and some compression along the geodesic ℓ. The type of difference between bending and stamping structures on M is described by the conic singularity of the boundary of the Nielsen hull for the holonomy group of c_{stamp}, $d^*_{\text{stamp}}(\pi_1(M))$ — like in Theorem 1.6.

Now we formulate the following three related problems about the space $\mathcal{C}(M)$ of uniformized conformal structures (see also **[AP6**, §5] and **[AP8]**):

2.5. Problems. *(A) Is the space* $\mathfrak{C}(M)$ *compact, especially for a closed manifold* M*? If not, what kinds of boundary points distinct from cusps are there?*

(B) Do all points of the subspace $\mathfrak{C}_q(M) \subset \mathfrak{C}(M)$ *of quasi-Fuchsian structures become exhausted by structures obtained by bending, stamping, and stamping-with-torsion deformations of the distinguished structure on* M*, especially for the three-dimensional case?*

(C) Is the space $\mathfrak{C}(M)$ *of uniformized conformal structures a connected space?*

Below, we will give some advances in this direction.

2.6. Theorem. *The space* $\mathfrak{C}(M)$ *of uniformized conformal structures on a closed hyperbolic* n*-manifold* M*,* $n > 2$*, with two distinct totally geodesic hypersurfaces is non-compact.*

Proof. Actually, we shall show that a limit structure on M obtained by a bending deformation of M in opposite directions along distinct totally geodesic hypersurfaces on M does not belong to $\mathfrak{C}(M)$ (this is also true for $n = 2$). Let us consider the closed hyperbolic n-manifold M constructed in **[AP2**, Theorem A], i.e., a manifold with two disjoint totally geodesic hypersurfaces. As in Corollary 5.2 in **[AP2]**, by a similar way, we obtain a smooth curve

$$\beta\colon [0,1) \longrightarrow \mathfrak{C}_q(M) \tag{2.6}$$

such that its end point, $\beta(1)$, is not contained in the space $\mathfrak{C}(M)$ because its holonomy group $G(1)$ has the following properties:

(i) $G(1)$ is a Kleinian group in S^3 with the union of two invariant non-contractible components, Ω_0 and Ω_1, as the discontinuity set $\Omega(G)$.

(ii) The non-contractibility of these components is related to the existence of accidental parabolic elements in the group $G(1)$ which correspond to a loxodromic one in the hyperbolic co-compact group $d^*_{\beta(0)}(\pi_1(M)) \subset \mathrm{Isom}\mathbb{H}^3$, i.e., the holonomy group for the distinguished (hyperbolic) structure on M.

(iii) The quotients $\Omega_0/G(1)$ and $\Omega_1/G(1)$ are non-compact manifolds whose (cusp) ends are conformally equivalent to ends of manifolds $\mathbb{R}^3/\Gamma$, where Γ is some discrete cyclic group of Euclidean isometries.

Therefore, for the closed hyperbolic n-manifold M, the last non-compactness property (iii) contradicts the inclusion $\beta(1) \in \mathfrak{C}(M)$. This completes the proof. □

2.7. Remarks. The point $\beta(1)$ is a boundary cusp point in the sense of L. Bers **[BR]**. So the second part of the question (A) is still open. In particular, the existence of spatial degenerate groups (compare **[BR]**) is unknown.

For question (B), an expected answer will likely be positive, since the above circumscribed bending and stamping structures $c \in \mathfrak{C}_q(M)$ can give any possible local behavior of the limit set (= quasisphere in S^3) for holonomy groups; see **[AP5]**, **[AP6**, §5], **[AP9]**.

The following theorem gives the advance in the problem (C).

2.8. Theorem. *On a closed hyperbolic 3-manifold* M *with a number* $N > 70$ *of disjoint totally geodesic surfaces, there exists an exotic uniformized conformal structure* c^* *that* *can not be approximated by bending, stamping, and stamping-with-torsion structures on* M. □

The proof of this fact is obtained in **[AP8]** where the exotic conformal structure $c^* \in \mathcal{C}(M)$ is obtained as the result of some modification of the author's Block-Building Construction for Kleinian groups in S^3 with wildly (even locally wildly) embedded 2-spheres as the limit sets. This Block-Building method has been developed in **[AT]** and **[AP7]**. Also we remark that an obstruction for approximation of c^* by bending and stamping structures on M is the property of the covering by a family $\mathcal{B}(G)$ of strictly maximal balls of the discontinuity set for the obtained exotic holonomy group $G^* \subset \text{Möb}(3)$ described as the condition of Theorem 1.6. As a result, it gives a conic singularity of the boundary of Nielsen hull H_{G^*} in $\mathbb{H}^4$.

3. Four-dimensional cobordisms with hyperbolic structures

3.1. Let us fix some closed (for simplicity) hyperbolic 3-manifold M and consider the space $\mathcal{W}(M)$ of all 4-dimensional cobordisms $(W; N_0, N_1)$, $\partial W = N_0 \cup N_1$, with the following properties:

(i) W is a geometrically finite hyperbolic cobordism: $\text{int}(W)$ has a complete geometrically finite hyperbolic structure, i.e., there is a decomposition of $\text{int}(W)$ into a cell by means of cutting along a finite set of totally geodesic hypersurfaces; see **[AP3**, Ch. 5**]**.

(ii) W is a homology cobordism: for its boundary components, N_0 and N_1, the relative homology groups are trivial:

$$H_*(W, N_0) = H_*(W, N_1) = 0. \tag{3.1}$$

(iii) The first boundary component of W, N_0, is homeomorphic to the closed hyperbolic manifold M and its inclusion $N_0 \subset W$ induces the homotopy equivalence:

$$\pi_*(W, N_0) = 0. \tag{3.2}$$

We note that due to A. Marden's results **[MD]**, the similar space of 3-dimensional hyperbolic cobordisms consists only of trivial cobordisms that are homeomorphic to the product of a closed surface S_g of genus $g > 1$ and a segment. After usual marking by the fundamental group, its quotient under the homotopy equivalence is isomorphic to $T(S_g) \times T(S_g)$ where $T(S_g)$ is the Teichmüller space for S_g —see Bers decomposition theory **[BR]**.

What kind of properties are there for the space $\mathcal{W}(M)$?

First, each uniformized conformal structure $c \in \mathcal{C}(M)$ with geometrically finite holonomy group $G \subset \text{Möb}(3)$ having an invariant contractible component of the discontinuity set $\Omega(G)$ corresponds to a cobordism $W \in \mathcal{W}(M)$, $\text{int}W \approx \mathbb{H}^4/G$, where $M(G)$ may be non-compact, $M(G) \neq W$; see **[AT**, Theorem 3.2 and Corollary 3.3**]**, **[TE1]**, **[TE2]**.

Second, the following converse statement holds:

3.2. Theorem. *Let* M *be a closed hyperbolic 3-manifold. Then for every cobordism* $(W; N_0, N_1) \in \mathcal{W}(M)$, *there is a Kleinian group* $G \subset \text{Möb}(3)$ *with an invariant* *contractible component* Ω_0 *of the discontinuity set* $\Omega(G)$ *such that either the Kleinian manifold*

$$M(G) = [\mathbb{H}^4 \cup \Omega(G)]/G \tag{3.3}$$

is the manifold $W \in \mathcal{W}(M)$ *itself* (*with* $N_0 = \Omega_0/G$ *and* $N_1 = [\Omega(G)\backslash\Omega_0]/G$) *or the manifold* W *is obtained from a non-compact manifold* $M(G)$ *by the natural compactification of a finite number of its cusp-ends. These cusp-ends are homeomorphic to the product of the strip* $[0,1] \times [0,\infty)$ *and either the cylinder* $S^1 \times [0,1]$, *or the Möbius band.*

Proof. Firstly, the condition (i) for W gives us a discrete action $G \subset \mathrm{Isom}\mathbb{H}^4$ of the fundamental group $\pi_1(W)$. Moreover, this action is discontinuous on the sphere at infinity and its discontinuity set $\Omega(G)$ has a G-invariant contractible component $\Omega_0 \subset \Omega(G)$, due to the condition (iii) for W; see **[AP3**, Ch. 7**]**. Applying this fact and the Tetenov's finiteness theorem (see **[TE1]**, **[TE2**, Theorem 2**]**, or **[AT**, Theorem 3.2 and Corollary 3.3**]**), we complete the proof. □

3.3. Remark. Let $\mathcal{W}_{\mathrm{triv}}(M) \subset \mathcal{W}(M)$ be a set of cobordisms $W \in \mathcal{W}(M)$ for which the following additional condition holds:
(iv) W is homeomorphic to the product $N_0 \times [0,1]$ where $N_0 \subset \partial W$.

For the correspondence between $\mathcal{W}(M)$ and $\mathcal{C}(M)$, i.e., for the holonomy group $G \subset \mathrm{M\ddot{o}b}(3)$ related to a cobordism $W \in \mathcal{W}_{\mathrm{triv}}(M)$, we have that $M(G)$ may be non-compact and non-homeomorphic to W. This is realized for boundary (cusp) points of $\mathcal{C}_q(M)$ related to Kleinan groups with accidental parabolic elements — as in Theorem B in **[AP2]**. However, in the compact case, the following is true:

The holonomy group $G \subset \mathrm{M\ddot{o}b}(3)$ *without cusps related (by Theorem 3.2) to the trivial cobordism* $W \in \mathcal{W}_{\mathrm{triv}}(M)$ *is a quasi-Fuchsian group conjugated by a quasisymmetric embedding*

$$f\colon S^2 = \partial\mathbb{H}^3 \hookrightarrow S^3$$

with the Fuchsian group $\Gamma \subset \mathrm{Isom}\mathbb{H}^3$, $M = \mathbb{H}^3/\Gamma$, *and the limit set* $L(G)$ *is a quasisphere, i.e.,* f *is the restriction to* $S^2 \subset S^3$ *of a quasiconformal automorphism of the sphere* S^3.

This fact is a consequence of a deep result of D. Sullivan **[SU2]*** which is based on the following two conditions:
(1) local unknottedness of the limit set $L(G)$ of a Kleinian group G with two invariant components of the discontinuity set (this follows from the condition (iv) on the cobordism W);
(2) the uniform self-similarity condition for the limit set $L(G)$.

This self-similarity condition says that there exists a uniform constant $K > 0$ such that, for any point $x \in L(G)$ and any small ball $B(x,r)$ with radius $r > 0$ centered at x, there exists a K-quasi-isometry F_r which maps the set

$$h(B(x,r) \cup L(G))\,, \quad \text{for } h\colon \mathbb{R}^3 \longrightarrow \mathbb{R}^3\,, \ \ h(y) = x + (y-x)/r\,,$$

* For an independent approach see **[MT**, Corollary 5.9**]**.

into the limit set $L(G)$.

In particular, this condition holds for Kleinian groups with the limit sets consisting of approximation points **[SU1]**. On the other hand, the exhaustion of the present limit set $L(G)$ by the points of approximation of the group G follows from our conditions on W. Namely, the geometrical finiteness of the group G gives that the limit set $L(G)$ contains the approximation points and parabolic cusp points only (**[BM]** and **[AP3**, Ch. 5**]**); the absence in $L(G)$ of parabolic fixed points is obtained from the compactness condition. This completes the proof. □

3.4. In contrast to the 3-dimensional case, the subset $\mathcal{W}(M)\backslash\mathcal{W}_{\mathrm{triv}}(M)$ of nontrivial 4-dimensional cobordisms $W \in \mathcal{W}(M)$ is non-empty; specifically, for a closed hyperbolic 3-manifold M with a big number of disjoint totally geodesic surfaces. This fact follows from our construction in **[AT**, Theorem 5.1**]**.

Moreover, it is likely true that there exists a 1-1 correspondence between this subset $\mathcal{W}(M)\backslash\mathcal{W}_{\mathrm{triv}}(M)$ of non-trivial cobordisms and the subset $C(M)\backslash\overline{C_q(M)}$—compare Theorem 2.8.

It is interesting to correlate this conjecture with the following three points of view:

First, with the special case of S. P. Novikov's conjecture on triviality of h-cobordisms of the type $K(\pi, 1)$ obtained as the quotients

$$W = M(G) = [\mathbb{H}^4 \cup \Omega(G)]/G$$

where $\Omega(G)$ is the union of two G-invariant contractible components Ω_0 and Ω_1, $\Omega_0/G \approx M$; see **[AT**, p. 408**]**.

Second, with the theorem of F. T. Farrell and L. E. Jones **[FJ]** about K-flatness of the fundamental group $\pi_1(M)$ of a closed hyperbolic 3-manifold M, in particular.

Here a group $G = \pi_1(M)$ is called K-flat if the Whitehead group $\mathrm{Wh}(G \times C^n)$ of any group $G \times C^n$, $n \geq 0$, is trivial (C^n denotes the free abelian group of rank n). Note that $G \times C^0$ is isomorphic to G itself, $\widetilde{K}_0(\mathbb{Z}G)$ is a direct summand of $\mathrm{Wh}(G \times C^1)$ and, for $n > 0$, $K_{-n}(\mathbb{Z}G)$ is a direct summand of $\widetilde{K}_0(\mathbb{Z}(G \times C^n))$. Therefore they are trivial for a K-flat group G; see **[FJ]**.

Third, with the fact that, for the closed hyperbolic manifold $M = \mathbb{H}^3/\Gamma$ from Theorem 2.8, the Chern-Simons and η-invariants for an exotic uniformized conformal structure $c^* \in \mathcal{C}(M)$ on M are the same as for a complete hyperbolic structure on M, namely, they vanish (see **[AP9]**).

Here the Chern-Simons invariant and the η-invariant for (M, c^*) are computed in a special metric on M (we call this metric a "Kobayashi conformal metric") which induces the structure c^*. In fact, this metric corresponds to the conformally invariant metric $k(*, *)$ in the invariant contractible component $\Omega_0 \subset \Omega(G) \subset S^3$ of the holonomy group $G \subset \mathrm{M\ddot{o}b}(3)$ for $c^* \in \mathcal{C}(M)$. For any points $x, y \in \Omega_0$, the Kobayashi conformally invariant metric is defined as follows:

$$k(x, y) = \inf \left\{ \sum_i d_i(x_{i-1}, x_i) \right\} \tag{3.4}$$

where we take inf over all conformal chains, i.e., couples $(x_0 = x, x_1, \ldots, x_n = y)$ of points in Ω_0 and conformal embeddings $f_i\colon B \hookrightarrow B_i \subset \Omega_0$ of the open ball $B =$

$B^3(0,1) \subset \mathbb{R}^3$ such that $x_{i-1}, x_i \in B_i = f_i(B)$ where $d_i(*,*)$ is the Poincaré hyperbolic metric in the i-th ball $B_i \subset \Omega_0$.

Details and the proof of vanishing for the Chern-Simons invariant and the η-invariant are related to the maximal ball cover of the discontinuity set component $\Omega_0 \subset \Omega(G)$ (see **[AP9]**).

References

[AP1] B. N. Apanasov, On certain analytic method in the theory of Kleinian groups in multidimensional space, Soviet Math. Dokl. I6 (1975), no. 3, 553–556.

[AP2] B. N. Apanasov, Nontriviality of Teichmüller space for Kleinian group in space, in "Riemann Surfaces and Related Topics: Proceedings of the 1978 Stony Brook Conference", ed. I. Kra and B. Maskit, Annals of Math. St. 97, Princeton Univ. Press, 1981, 21–31.

[AP3] B. N. Apanasov, Discrete Groups in Space and Uniformization Problems, Kluwer Academic Publishers, Dordrecht, 1991.

[AP4] B. N. Apanasov, Thurston's bends and geometric deformations of conformal structures, in "Complex Analysis and Applications '85: Proc. Intern. Conf. on Complex Analysis, Varna, May 1985", Publ. House of Bulgarian Acad. Sci., Sofia, 1986, 14–28.

[AP5] B. N. Apanasov, The geometry of Nielsen's hull for Kleinian groups in space and quasiconformal mappings, Ann. Global Anal. Geom. 6 (1988), no. 3, 207–230.

[AP6] B. N. Apanasov, Deformations of conformal structures on hyperbolic manifolds, J. of Diff. Geom. 34 (1991), no. 2, to appear (Preprint MSRI-02723-89, Berkeley/Calif., 1989).

[AP7] B. N. Apanasov, Quasisymmetric embeddings of a closed ball inextensible in neighborhoods of any boundary points, Ann. Acad. Sci. Fenn., Ser. AI Matyh. 14 (1989), 243–255.

[AP8] B. N. Apanasov, Non-standard uniformized conformal structures on hyperbolic manifolds, Inventiones Math. 149 (1991), to appear (Mittag-Leffler Institut, Preprint # 1, 1989/1990, Djursholm, 1989).

[AP9] B. N. Apanasov, Kobayashi conformal metric on manifolds, Chern-Simons and η-invariants, Intern. J. of Math. 2 (1991), no. 4, to appear (Univ. Autonoma de Barcelona, Preprint 25/1990).

[AT] B. N. Apanasov, A. V. Tetenov, Nontrivial cobordisms with geometrically finite hyperbolic structures, J. Differential Geom. 28 (1988), 407–422.

[BM] A. F. Beardon, B. Maskit, Limit points of Kleinian groups and finite sided fundamental polyhedra, Acta Math. 132 (1974), 1–12.

[BR] L. Bers, Uniformization, Moduli and Kleinian groups, Bull. London Math. Soc. 4 (1972), 257–300.

[EM] D. B. A. Epstein, A. Marden, Convex hulls in hyperbolic space, a theorem of Sullivan, and measured pleated surfaces. Analytic and geometric aspects of hyperbolic space, London Math. Soc. Lecture Notes Ser. 111, Cambridge Univ. Press, 1987, 113–254.

[FJ] F. T. Farrell, L. E. Jones, K-theory and dynamics. 1, Ann. of Math. 124 (1986), 531-569.

[GM] W. M. Goldman, Geometric structures on manifolds and varieties of representations, in: Geometry of Group Representations, Contemp. Math. 74, Providence, 1988, 169–198.

[JM] D. Johnson, J. J. Millson, Deformation spaces associated to compact hyperbolic manifolds, in "Discrete Groups in Geometry and Analysis: Papers in Honour of G. D. Mostow on His Sixtieth Birthday", ed. R. Howe, Birkhäuser, Boston, 1987, 48–106.

[KM] Y. Kamishima, Conformally flat manifolds whose development maps are not surjective, Trans. Amer. Math. Soc. 294 (1986), 607–623.

[KO] C. Kourouniotis, Deformations of hyperbolic structures, Math. Proc. Cambridge Phil. Soc. 98 (1985), 247–261.

[KP] R. S. Kulkarni, U. Pinkall, Uniformization of geometric structures with applications to conformal geometry, in "Differential Geometry: Proc. Symp. Peniscola, Spain 1985", ed. A. M. Naveira a.o., Lect. Notes in Math. 1209, Springer-Verlag, 1986, 190–210.

[LK] W. L. Lok, Deformations of locally homogeneous spaces and Kleinian groups, Ph.D. Thesis, Columbia Univ., New York, 1984.

[MD] A. Marden, The geometry of finitely generated Kleinian groups, Ann. of Math. 99 (1974), 383–462.

[MT] G. J. Martin, Infinite group actions on spheres. Rev. Mat. Ibereroamericana 4 (1988, no. 314), 407–451.

[MO1] J. Morgan, Group actions on trees and the compactification of the space of classes of $SO(n,1)$-representations, Topology 25 (1986), 1–33.

[MO2] J. Morgan, Trees and hyperbolic geometry, Proc. Intern. Congress Math., Berkeley 1986, Providence, 1987, 590–597.

[MW] G. D. Mostow, Quasiconformal mappings in n-space and the rigidity of hyperbolic space forms, Publ. Math. Inst. Hautes Etudes Sci. 34 (1968), 53–104.

[SU1] D. Sullivan, Seminar on conformal and hyperbolic geometry, IHES Seminar notes, March 1982.

[SU2] D. Sullivan, Oral communications, 1989.

[SU3] D. Sullivan, Quasiconformal homeomorphisms and dynamics, Part 2: structural stability implies hyperbolicity for Kleinian groups, Acta Math. 155 (1985), 243–260.

[TE1] A. V. Tetenov, Kleinian groups in space and their invariant domains, Ph.D. Thesis, Inst. of Math. Acad. Sci. USSR, Novisibirsk, 1982.

[TE2] A. V. Tetenov, The discontinuity set for a Kleinian group and topology of its Kleinian manifold, preprint, 1989.

Department of Mathematics, University of Oklahoma, Norman, OK 73019

Greatly Symmetric Totally Geodesic Surfaces and Closed Hyperbolic 3-Manifolds Which Share a Fundamental Polyhedron

*Boris N. Apanasov** and *Ivan S. Gutsul*

Abstract. Given an integer N, we construct very symmetric compact hyperbolic polyhedron $P_N \subset \mathbb{H}^3$ which allows at least $2N$ different side-pairings. In particular, this gives at least $2N$ (up to homeomorphisms) closed hyperbolic 3-manifolds which share the fundamental polyhedron P_N (i.e. with the same volume) and have a symmetric totally geodesic surface S of genus $2N$ with $|\,\mathrm{ISOM}_+ S| \geq 2N$.

1. Introduction

Well known corollaries of the Jørgensen-Gromov theorem (see **[TH]**) say that:

Firstly, the set of volumes of hyperbolic 3-manifolds is well ordered.

Secondly, the volume is a finite-to-one function on the set of hyperbolic manifolds of finite volume.

Therefore the number of 3-dimensional hyperbolic manifolds with the same volume is finite. But that number is not bounded, as is known from examples of non compact (cusped) hyperbolic 3 manifolds which were constructed by N. Wielenberg **[W]**.

We will discuss some discrete, torsion free subgroups G of $PSL(2,\mathbb{C})$ which are constructed as HNN-extensions of web-groups $G_0 \subset PSL(2,\mathbb{C})$, $G = G_0 *_{C_i}$, with cyclic groups C_i, and which justify the following.

Theorem. *For each integer N, there is a compact polyhedron P_N in the hyperbolic 3-space $\mathbb{H}^3$ from which face identifications yield at least $2N$ different (up to homeomorphisms) 3-dimensional closed, orientable hyperbolic 3-manifolds with totally geodesic surfaces of genus $2N$ admitting an isometry group of the order $2N$.*

In our examples, the corresponding groups $G_0 = G_0(N) \subset PSL(2,\mathbb{C})$ have quotients

$$M(G_0) = [\mathbb{H}^3 \cup \Omega(G_0)]/G_0$$

under the G_0-action of the discontinuity set of G_0 in the closure of the hyperbolic space $\mathbb{H}^3$ with the following properties.

Firstly, each component of each lift of a boundary component is a Euclidean disc in $\mathbb{C}$.

* The first author was partially supported by The Ohio State University.

Secondly, the construction then consists of gluing totally geodesic surfaces in $M(G_0)$ which are incompressible and boundary-parallel in $M(G_0)$.

Moreover, since these totally geodesic surfaces admit a large number of isometries, this allows a large number of different gluings. They correspond to HNN-extensions of the group G_0 by hyperbolic screw-translations.

An easy modification of our construction gives a similar result for closed non-orientable manifolds.

At the end of 1988, when the Russian version of the paper was complete, A. Vesnin gave another algebraic approach (it will appear elsewhere) to the construction of hyperbolic 3-manifolds with the same volume based on group epimorphisms in the group $\mathbb{Z}_2 \times \mathbb{Z}_2 \times \mathbb{Z}_2$ (cf. **[VE]**). Also, as the first author recently was informed by A. D. Mednych, A. Kawauchi announced a topological approach to the construction of an example of similar manifolds with the same volume which are "imitations" of the torus $T^3 = S^1 \times S^1 \times S^1$ see **[K]**. Both these approaches do not give any fundamental polyhedron which is shared by manifolds with the same volume.

2. Hyperbolic polyhedra with great axial symmetries

To start the construction consider a quadrilateral in the hyperbolic plane $\mathbb{H}^2$ with vertices A, B, C, D such that their angles at vertices B and C are right and the angles at vertices A and D are equal to t, where $t < \pi/2$.

Given $t < \pi/2$, the set Q_t of such quadrilaterals $ABCD$ is described by a certain continuous parameter, namely the length $\ell = |BC|$ of the side BC of the quadrilateral — see Figure 1.

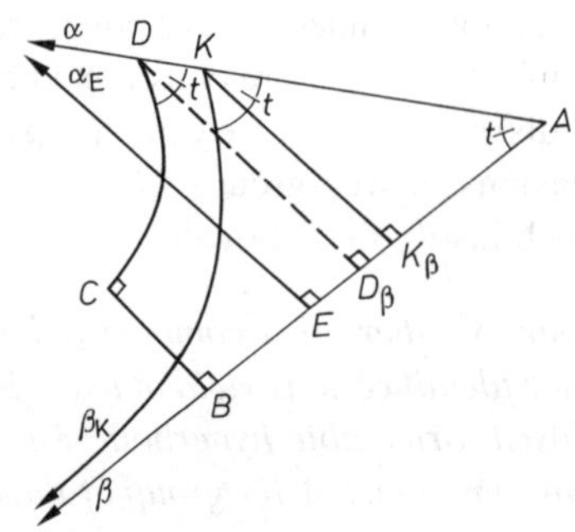

Figure 1

Let α and β be the hyperbolic rays in $\mathbb{H}^2$ from the point A which pass through points D and B respectively, and let $\alpha_E \subset \mathbb{H}^2$ be the ray parallel to α that orthogonally intersects the ray β. Then, for each quadrilateral $ABCD$ from Q_t, the intersection point $E = \alpha_E \cap \beta$ lies in the interval AB. Also, let β_K be the parallel to β in $\mathbb{H}^2$ that intersects α in a point K with the angle t. Then, for each quadrilateral $ABCD$ from the family Q_t, the intersection point K lies in the interval AD.

For the rays $\alpha_K \subset \alpha$ and $\beta_E \subset \beta$ starting from the points K and E, respectively, we see that:

1. Each point of the ray α_K distinct from K can be the vertex D of some quadrilateral $ABCD$ from Q_t;

2. Each point of the ray β_E distinct from the point E can be the vertex B of some quadrilateral from Q_t.

In addition, if K_β is the orthogonal projection to β of the point K, then the orthogonal projection D_β to β of the vertex D of each quadrilateral $(ABCD) \in Q_t$ lies in the interval $EK_\beta \subset \beta$ — see Figure 1.

Now let the plane $\mathbb{H}^2$ be embedded into the hyperbolic 3-space $\mathbb{H}^3$ as the plane ω, and let θ_A, θ_B, θ_C and θ_D be hyperbolic rays orthogonal to ω in a half-space $\omega_+ \subset \mathbb{H}^3$, $\partial\omega_+ = \omega$, where A, B, C and D are vertices of a quadrilateral from Q_t. Since they give a hyperbolic lines sheaf, for any quadrilateral from Q_t, there is a number $h = h(\ell)$ such that the hyperbolic plane $\omega_h \subset \omega_+$ orthogonal to θ_B with $d(B, \omega_h \cap \theta_B) = h$ intersects the ray θ_C in a point C_1 with the angle $s = \pi/6$. In other words, in the plane spanned by the rays θ_C and θ_B, we have a quadrilateral shown in Figure 2 (where $B_1 = \omega_h \cap \theta_B$), and (see **[B**, 7.17.1**]**) $\sinh h(\ell) = \sqrt{3}/\sinh \ell$.

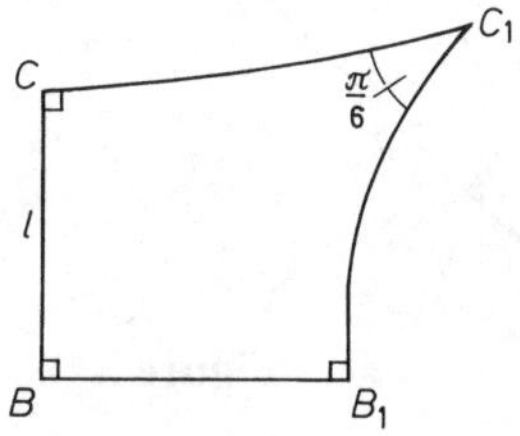

Figure 2

Denote by γ a hyperbolic line which is the line of intersection of the plane ω_h and the plane spanned by θ_C and θ_D. For different values of the parameter ℓ, $\ell = |BC|$, the line γ will either intersect the ray θ_D or is parallel with θ_D or hyperparallel with θ_D. Denote by ℓ_0 the value of ℓ such that we have the second case, i.e. γ is parallel with θ_D. For this parameter ℓ_0, the line of intersection δ of the plane $\omega_h = \omega_{h(\ell_0)}$ and the plane spanned by the rays θ_B and θ_A intersects the ray θ_A in a point A_1 — see Figure 3.

Below we will consider a subset $Q_{t,\ell_0} \subset Q_t$ such that the quadrilateral $ABCD \subset \omega$ has the side BC of length less than ℓ_0. In which case, we have the third alternative above, i.e., the line γ and the ray θ_D are hyperparallels. Note that the length $h = h(\ell)$ of the interval BB_1 in the ray θ_B will be increasing if the parameter ℓ, $\ell = |BC|$, will be decreasing. This follows from the consideration of the quadrilateral BB_1C_1C in Figure 2. Moreover, for sufficiently small value $\ell_1 < \ell_0$ of the parameter ℓ, there is a quadrilateral $ABCD \in Q_{t,\ell_0}$ such that the line δ and the ray θ_A are parallel. Now we restrict to the subset $Q_{t,\ell_1} \subset Q_{t,\ell_0} \subset Q_t$ of quadrilaterals $ABCD$ with $|BC| < \ell_1$. For these quadrilaterals, the line δ and the ray θ_A are hyperparallel.

For some $m > 0$, consider a point $A_2 \in \theta_A$ with the distance $d(A, A_2) = m$. Let τ_m be the plane orthogonal to θ_A which intersects the ray θ_A in the point A_2. Since the line δ and the ray θ_A are hyperparallel there is a number $m = m(\ell) > 0$ such that the planes ω_h and τ_m intersect with a dihedral angle $\pi/3$, i.e., the line of intersection of the

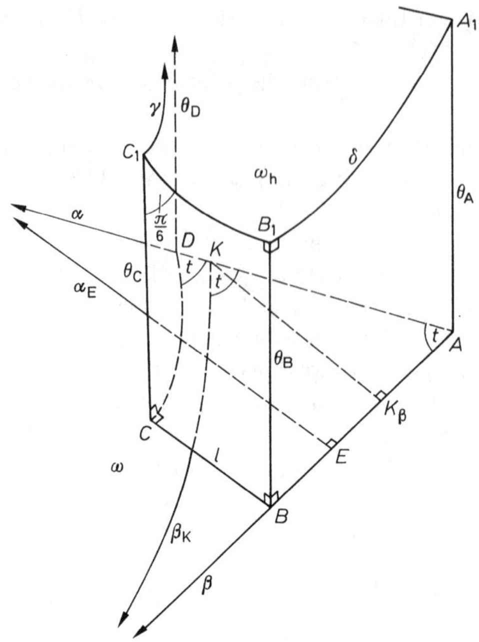

Figure 3

plane τ_m and the plane spanned by the rays θ_A and θ_B intersects the line δ in a point M with the angle $\pi/3$. Note that, for sufficiently small $\epsilon > 0$, the orthogonal projection M_β of the point M to β lies between the points A and K_β if $ABCD \in Q_{t,\ell_1}$ is such that $|BC| = \ell_1 - \epsilon$; see Figure 4. At the same time, the orthogonal projection D_β of the vertex D of the quadrilateral $ABCD \in Q_{t,\ell_1}$ lies in β between the points E and K_β; see Figure 1.

From our construction, it is easy to see that the distance between the points D_β and M_β in β tends to zero if the length $\ell = |BC|$ is decreasing. Therefore there exists a quadrilateral $ABCD \in Q_{t,\ell_1}$ with a common projection $D_\beta = M_\beta$ of the points D and M (denote this quadrilateral by q_{t,ℓ_2}, $\ell_2 < \ell_1 < \ell_0$). For this case $\ell = \ell_2$, the line ξ of intersection of the planes $\tau_{m(\ell_2)}$ and $\omega_{h(\ell_2)}$ (the edge of the dihedral angle with value $\pi/3$) and the line passing through points D and D_β lie in a common plane which contains the ray θ_D and the interval MM_β. Therefore the lines γ, ξ and θ_D lie in a hyperbolic lines sheaf and consequently there exists a plane τ which is orthogonal to γ, ξ, θ_D and the line ζ of intersection of the plane τ_m and the plane spanned by the rays θ_D and θ_A; see Figure 5.

Now consider the compact polyhedron $\Delta_t \subset \mathbb{H}^3$ which is the intersection of half-spaces bounded by the following planes in $\mathbb{H}^3$:

1) ω, ω_h, τ_m, τ;
2) the plane spanned by the rays θ_A and θ_B;
3) the plane spanned by the rays θ_B and θ_C;
4) the plane spanned by the rays θ_C and θ_D;

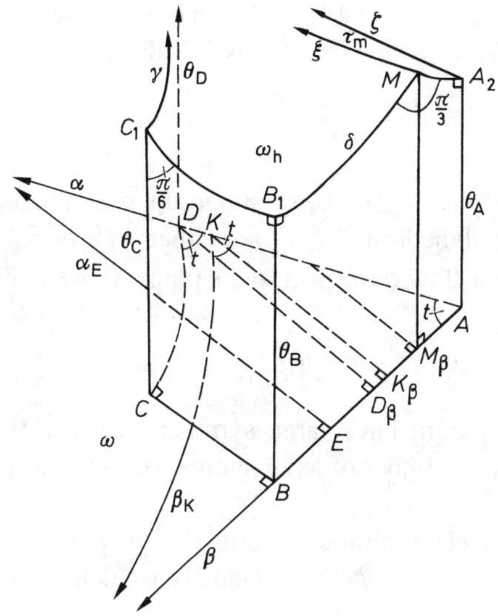

Figure 4

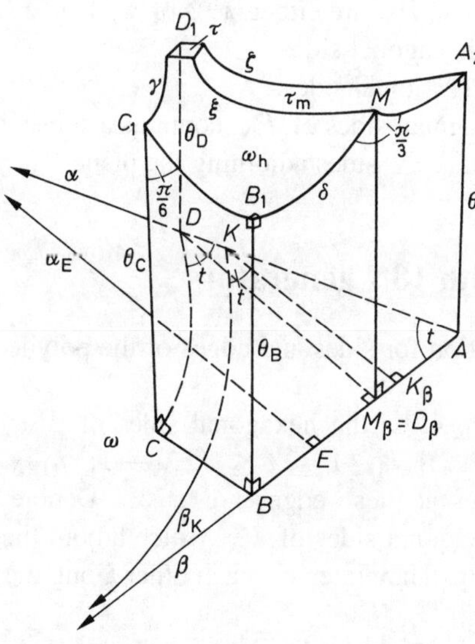

Figure 5

5) the plane spanned by the rays θ_D and θ_A

where $h = h(\ell_2), m = m(\ell_2)$ and the quadrilateral $ABCD$ in the plane ω is equal to q_{t,ℓ_2}.

The dihedral angles of the polyhedron Δ_t are either $\pi/2$ or $\pi/3$ or $\pi/6$ or t, $0 < t < \pi/2$. For the values π/n, $n \in \mathbb{Z}$, $n \geq 4$, of the parameter t, the polyhedra Δ_t are Coxeter polyhedra. Hence they define regular tesselations of the hyperbolic space

$\mathbb{H}^3$ (cf. **[A]**, **[MA]**). The union of all polyhedra of the tesselation with the common edge DD_1, $D_1 = \tau \cap \theta_D$, gives a new Coxeter polyhedron

$$\Delta^0_{2\pi/n} = \bigcup_{g \in G_D} g(\Delta_{\pi/n}), \qquad n \in \mathbb{Z}, \quad n \geq 4, \tag{2.1}$$

where G_D is a corresponding Coxeter subgroup which fixed the edge DD_1. The dihedral angles of the polyhedron $\Delta^0_{2\pi/n}$ are either $\pi/2$ or $\pi/3$ or $2\pi/n$, $n = 4, 5, \cdots$ Finally, for the reflection $\mathcal{T}$ with respect to the plane τ, we define the following Coxeter polyhedra

$$\Delta'_n = \Delta^0_{2\pi/n} \cup \mathcal{T}(\Delta^0_{2\pi/n}), \qquad n \in \mathbb{Z}, \quad n \geq 4. \tag{2.2}$$

These Coxeter polyhedra have large symmetry groups (see also Remark 4.2), in particular, there is a large number of axial symmetries with respect to the line containing the edge DD_1.

We will consider a special infinite subfamily of the polyhedra (2.2) which correspond to the values $n = 4N$; $N \in \mathbb{Z}$, $N \geq 1$. Namely we define

$$P_N = \Delta'_{4N} = \Delta^0_{\pi/2N} \cup \mathcal{T}(\Delta^0_{\pi/2N}), \qquad N \in \mathbb{Z}, \quad N \geq 1. \tag{2.3}$$

Each polyhedron P_N has the following properties.

(i) Dihedral angles of P_N are either $\pi/2$ or $\pi/3$ or $\pi/2N$.

(ii) P_N has $12N$ hexagonal sides.

(iii) P_N has two $12N$-gonal sides.

(iv) The set of pentagonal sides of P_N contains a subset of $12N$ sides adjoining the plane ω and a subset of $12N$ sides adjoining the plane $\mathcal{T}(\omega)$.

3. Web-groups with 18N generators

We now fix some notation for sides and edges of the polyhedron P_N from (2.3) — see Figure 6.

Let $A_1, A_2, \cdots, A_{12N}$ be the hexagonal sides of P_N which intersect each other along edges $a_i = A_i \cap A_{i+1}$, $1 \leq i \leq 12N - 1$; $a_{12N} = A_{12N} \cap A_1$. We note that the dihedral angles at these edges are $\pi/3$. Denote by $B_1, \cdots, B_{12N}$ and by $C_1, \cdots, C_{12N}$ the pentagonal sides of P_N which adjoin the opposite $12N$-gonal sides of P_N, correspondingly. They intersect each other along the following edges:

$$b_i = B_i \cap B_{i+1}, \quad c_i = C_i \cap C_{i+1}, \quad 1 \leq i \leq 2N - 1$$

and

$$b_{2N} = B_{12N} \cap B_1, \quad c_{2N} = C_{12N} \cap C_1.$$

The dihedral angles at these edges have values either $\pi/2$ or $\pi/2N$.

The compatibility of the enumerations above is such that every side A_i, $1 \leq i \leq 2N - 1$, intersects four pentagonal sides

$$B_i, \quad B_{i+1}, \quad C_i, \quad C_{i+1}$$

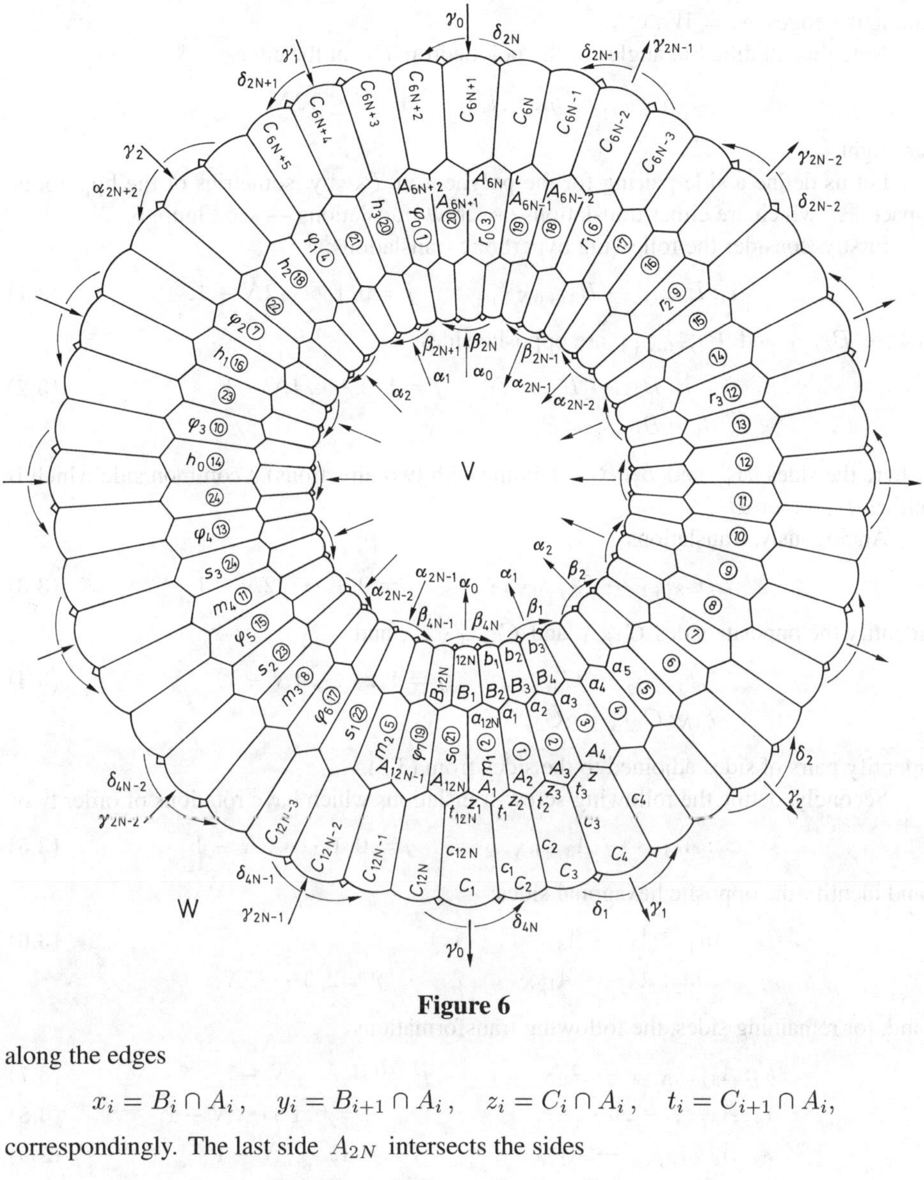

Figure 6

along the edges

$$x_i = B_i \cap A_i\,, \quad y_i = B_{i+1} \cap A_i\,, \quad z_i = C_i \cap A_i\,, \quad t_i = C_{i+1} \cap A_i,$$

correspondingly. The last side A_{2N} intersects the sides

$$B_{2N}\,, \quad B_1\,, \quad C_{2N}\,, \quad C_1$$

along the edges

$$x_{2N} = B_{2N} \cap A_{2N}\,, \quad y_{2N} = B_1 \cap A_{2N}\,, \quad z_{2N} = C_{2N} \cap A_{2N}\,, \quad t_{2N} = C_1 \cap A_{2N}\,,$$

correspondingly.

Finally, denote by V and W two opposite $12N$-gonal sides of P_N which lie in hyperbolic planes ω and $\mathcal{T}(\omega)$, correspondingly. Here V intersects the sides B_i,

$1 \le i \le 2N$, along the edges $v_i = V \cap B_i$, and W intersects the sides C_i, $1 \le i \le 12N$, along the edges $w_i = W \cap C_i$.

Note that all dihedral angles of the polyhedron P_N at the edges

$$v_i, w_i, x_i, y_i, z_i, t_i\,; \qquad 1 \le i \le 12N,$$

are right.

Let us define a side-pairing for the polyhedron P_N by isometries of the hyperbolic space $\mathbb{H}^3$ which are either translations or screw-translations — see Figure 6.

Firstly, consider the following hyperbolic translations:

$$\alpha_j: B_{3j+1} \to B_{3j+6N+1}\,, \qquad j = 0, 1, \cdots, 2N-1, \tag{3.1}$$

where B_{3j+1} and $B_{3j+6N+1}$ are opposite sides;

$$\begin{aligned} \beta_j: B_{3j} &\to B_{3j+2}\,, \qquad j = 1, 2, \cdots, 4N-1, \\ \beta_{4N}: B_{12N} &\to B_2, \end{aligned} \tag{3.2}$$

where the sides B_{3j} and $\beta_j(B_{3j})$ adjoin (with two directions) a common side which is either B_{3j+1} or B_1.

Analogously, translations

$$\gamma_j: C_{3j+1} \to C_{3j+6N+1}\,, \qquad j = 0, 1, \cdots, 2N-1 \tag{3.3}$$

identify the opposite sides C_{3j+1} and $C_{3j+6N+1}$, and

$$\begin{aligned} \delta_j: C_{3j} &\to C_{3j+2}\,, \qquad j = 1, 2, \cdots, 4N-1, \\ \delta_{4N}: C_{12N} &\to C_2 \end{aligned} \tag{3.4}$$

identify pairs of sides adjoined to the sides from (3.3).

Secondly, define the following screw-translations which have rotations of order two:

$$\varphi_j: A_{3j+2} \to A_{3j+6N+2}\,, \qquad j = 0, 1, \cdots, 2N-1 \tag{3.5}$$

and identify the opposite hexagonal sides;

$$\begin{aligned} m_1^{-1}: A_1 &\to A_3 \\ m_j: A_{3j} &\to A_{12N-3j+4}\,, \qquad j = 2, 3, \cdots, N; \end{aligned} \tag{3.6}$$

and, for remaining sides, the following transformations

$$h_j: A_{3j+3N+4} \to A_{9N-3j}\,, \qquad j = 0, 1, \cdots, N-1; \tag{3.7}$$

$$r_j: A_{3j+4} \to A_{6N-3j}\,, \qquad j = 0, 1, \cdots, N-1; \tag{3.8}$$

$$s_j: A_{6N+3j+4} \to A_{12N-3j}\,, \qquad j = 0, 1, \cdots, N-1 \tag{3.9}$$

complete the sides identification, except for $12N$-gonal sides V and W. These two sides can be identified by $2N$ different isometries of the hyperbolic space $\mathbb{H}^3$

$$T_j: V \to W, \qquad j = 0, 1, \cdots, 2N-1, \tag{3.10}$$

where each T_j, $j \ne 0$, is a screw-translation which is the composition of the hyperbolic translation T_0 and a rotation U_i with angle $\pi j/2N$ around the orthogonal to the sides V and ω line, i.e., around the symmetry axis DD_1 of the polyhedron P_N.

The transformations (3.1)–(3.9) generate a group G_N of isometries of $\mathbb{H}^3$. Let

$$G_N = \langle \alpha_j, \beta_i, \gamma_j, \delta_i, \varphi_j, m_k, h_\ell, r_\ell, s_\ell \rangle, \tag{3.11}$$

where $0 \le j \le 2N-1,\ 1 \le i \le 4N,\ 1 \le k \le N,\ 0 \le \ell \le N-1$.

Note that the generators of the group G_N identify sides of some hyperbolic polyhedron $P_N^0,\ P_N \subset P_N^0$, of infinite volume. We obtain this polyhedron $P_N^0 \subset \mathbb{H}^3$ as the intersection of half-spaces with the same set of boundary planes as for P_N except for the planes ω and $\mathcal{T}(\omega)$ (i.e., P_N^0 does not have sides V and W).

Lemma 3.1. *The group $G_N \subset PSL(2,\mathbb{C})$ from* (3.11) *is a geometrically finite web-group in the plane $\mathbb{C}$ for which any component of the discontinuity set is an Euclidean disc.*

Proof. Firstly, we note that the polyhedron $P_N^0 \subset \mathbb{H}^3$ is a fundamental polyhedron of the group G_N. We will have this fact after checking the conditions for the Poincaré-Aleksandrov theorem on the fundamental polyhedron (see **[A]**, **[AP2]**, **[MA]**), namely, the edges condition.

We will do this in the next section simultaneously with the proof that G_N is torsion-free.

The discontinuity set $\Omega(G_N)$ of the group G_N is the union of $g(\overline{P_N^0} \cap \partial\mathbb{H}^3),\ g \in G_N$, where $\overline{P_N^0} \cap \partial\mathbb{H}^3$ has two components F_1 and F_2. Denote by

$$\begin{aligned} \Gamma_1 &= \Gamma_1(N) = \langle \alpha_j, \beta_i : 0 \le j \le 2N-1, 1 \le i \le 4N \rangle \\ \Gamma_2 &= \Gamma_2(N) = \langle \gamma_j, \delta_i : 0 \le j \le 2N-1, 1 \le i \le 4N \rangle \end{aligned} \tag{3.12}$$

subgroups of the group G_N which are generated by side-pairings of the polygons F_1 and F_2, respectively. Since these sets of generators have invariant planes ω and $\mathcal{T}(\omega)$ respectively, the groups Γ_1 and Γ_2 are Fuchsian groups in the plane $\mathbb{C} = \partial\mathbb{H}^3$ with invariant Euclidean discs

$$\Omega_1 = \Omega_1(N) = \bigcup_{g \in \Gamma_1} g(F_1), \qquad \Omega_2 = \Omega_2(N) = \bigcup_{g \in \Gamma_2} g(F_2) \tag{3.13}$$

where $\partial\Omega_1 = \partial\omega,\ \partial\Omega_2 = \partial(\mathcal{T}(\omega))$.

Therefore each component Ω_i of the discontinuity set $\Omega(G_N) = \bigcup_{g \in G_N} g(F_1 \cup F_2)$ is some Euclidean disc $g_i(\Omega_j),\ g_i \in G_N,\ j = 1,2$, with the Fuchsian stabilizer $G_{\Omega_i} \subset G_N$ such that $G_{\Omega_i} = g_i \Gamma_j g_i^{-1}$.

The finite sidedness of the fundamental polyhedron P_N^0 shows the group G_N is geometrically finite. This completes the proof. □

Remark 3.2. The manifold

$$M(G_N) = [\mathbb{H}^3 \cup \Omega(G_N)]/G_N$$

is compact and has the minimal convex retract

$$M_{G_N} = [H_{G_N} \backslash \Lambda(G_N)]/G_N$$

which is a compact hyperbolic 3-manifold with totally geodesic boundary (see **[TH]**, **[AP1]**)

$$\partial M_{G_N} = (\omega/\Gamma_1) \cup (\mathcal{T}(\omega)/\Gamma_2) . \tag{3.14}$$

Each component ($S_1 = \omega/\Gamma_1$ or $S_2 = \mathcal{T}(\omega)/\Gamma_2$) of the boundary is a closed totally geodesic surface of genus $2N$, parallel to the boundary component (Ω_1/Γ_1 or Ω_2/Γ_2, respectively) of the manifold $M(G_N)$.

Lemma 3.3. *The totally geodesic surfaces S_1, S_2, $S_1 \cup S_2 = \partial M_{G_N}$, are isometric and have orientation preserving isometry groups of the order $2N$.*

Proof. The hyperbolic translation T_0 from (3.10) maps the fundamental polygon F_1 of the subgroup $\Gamma_1 \subset G_N$ to the fundamental polygon F_2 of the subgroup $\Gamma_2 \subset G_N$ and preserves the side identification. In other words,

$$\Gamma_2 = T_0 \Gamma_1 T_0^{-1} . \tag{3.15}$$

This shows that S_1 and S_2 are isometric.

Moreover, from the axial symmetry of the polyhedra P_N and P_N^0, we see that there are rotations of these polyhedra with angles $\pi j/2N$, $0 \le j \le 2N-1$, preserving the sides identifications of the sides of F_1 and F_2. Actually, such rotations permute the generators of each group Γ_i, $i = 1, 2$ from (3.12):

$$\alpha_j \to \alpha_{j'} , \quad \beta_K \to \beta_{k'} .$$

These rotations induce orientation preserving isometries of the totally geodesic surfaces S_1 and S_2. This completes the proof. □

4. Manifolds which share the polyhedron P_N

We now check the validity of the conditions of the Poincaré-Aleksandrov theorem (see **[A]**, **[AP2]**, **[MA]**). Simultaneously we will show that the web-group G_N from (3.11) and its HNN-extensions by screw-translations T_j from (3.10),

$$G_N^j = \langle G_N, T_j \rangle, 0 \le j \le 2N-1 , \tag{4.1}$$

are torsion free discrete subgroups of $PSL(2, \mathbb{C})$ with fundamental polyhedra P_N^0 and P_N correspondingly.

On the set of polyhedron edges, we have an equivalence relation defined by identification transformations (3.1)–(3.10). This gives a set of edge cycles in the boundary of the polyhedron. Given such a cycle, let us show that the sum of dihedral angles of P_N along edges of the cycle is 2π, i.e., the cycle is unessential.

First, we consider all $2N$ cycles of edges a_i, $i = 1, \cdots, 12N$, i.e., the following:

$$\begin{aligned} a_3 &= A_3 \cap A_4 \xrightarrow{m_1} a_1 \\ &= A_1 \cap A_2 \xrightarrow{\varphi_0} a_{6N+1} \\ &= A_{6N+1} \cap A_{6N+2} \xrightarrow{h_{N-1}} a_{6N+3} \\ &= A_{6N+4} \cap A_{6N+3} \xrightarrow{s_0} a_{12N-1} \\ &= A_{12N-1} \cap A_{12N} \xrightarrow{\varphi_{2N-1}^{-1}} a_{6N-1} \\ &= A_{6N} \cap A_{6N-1} \xrightarrow{r_0^{-1}} a_3 \\ &= A_3 \cap A_4, \end{aligned}$$

and for $j = 2, \cdots, N$:

$$\begin{aligned} a_{3j} &= A_{3j} \cap A_{3j+1} \xrightarrow{m_j} a_{12N-3j+4} \\ &= A_{12N-3j+5} \cap A_{12N-3j+4} \xrightarrow{\varphi_{2N-j+1}^{-1}} a_{6N-3j+4} \\ &= A_{6N-3j+4} \cap A_{6N-3j+5} \xrightarrow{h_{N-j}} a_{6N+3j} \\ &= A_{6N+3j+1} \cap A_{6N+3j} \xrightarrow{s_{j-1}} a_{12N-3j+2} \\ &= A_{12N-3j+2} \cap A_{12N-3j+3} \xrightarrow{\varphi_{2N-j}^{-1}} a_{6N-3j+2} \\ &= A_{6N-3j+3} \cap A_{6N-3j+2} \xrightarrow{r_{j-1}^{-1}} a_{3j} \\ &= A_{3j} \cap A_{3j+1}. \end{aligned}$$

The second set of a-cycles have the form $(j = 1, \cdots, N)$:

$$\begin{aligned} a_{3j+1} &= A_{3j+1} \cap A_{3j+2} \xrightarrow{r_{j-1}} a_{6N-3j+3} \\ &= A_{6N-3j+4} \cap A_{6N-3j+3} \xrightarrow{h_{N-j}} a_{6N+3j-1} \\ &= A_{6N+3j-1} \cap A_{6N+3j} \xrightarrow{\varphi_{j-1}^{-1}} a_{3j-1} \\ &= A_{3j} \cap A_{3j-1} \xrightarrow{m_j} a_{12N-3j+3} \\ &= A_{12N-3j+3} \cap A_{12N-3j+4} \xrightarrow{s_{j-1}^{-1}} a_{6N+3j+1} \\ &= A_{6N+3j+2} \cap A_{6N+3j+1} \xrightarrow{\varphi_j^{-1}} a_{3j+1} \\ &= A_{3j+1} \cap A_{3j+2}. \end{aligned}$$

Therefore, each of the a-cycles above contains 6 edges with dihedral angles $\pi/3$. Hence these cycles are unessential.

For edges x_i, y_i, z_i and t_i, $i = 1, \cdots, 12N$, we have the following cycles.

$$x_3 = A_3 \cap B_3 \xrightarrow{m_1} z_1 = C_1 \cap A_1 \xrightarrow{\gamma_0} t_{6N} = A_{6N} \cap C_{6N+1} \xrightarrow{r_0^{-1}}$$
$$\to y_4 = B_5 \cap A_4 \xrightarrow{\beta_1^{-1}} x_3 = A_3 \cap B_3,$$

and for $j = 2, \cdots, N$:

$$x_{3j} = A_{3j} \cap B_{3j} \xrightarrow{m_j} z_{12N-3j+4} = C_{12N-3j+4} \cap A_{12N-3j+4} \xrightarrow{\gamma_{2N-j+1}^{-1}}$$
$$\to t_{6N-3j+3} = A_{6N-3j+3} \cap C_{6N-3j+4} \xrightarrow{r_{j-1}^{-1}} y_{3j+1} = B_{3j+2} \cap A_{3j+1} \xrightarrow{\beta_j^{-1}}$$
$$\to x_{3j} = A_{3j} \cap B_{3j}.$$

Analogously for y_{3j} as the starting point:

$$y_3 = A_3 \cap B_4 \xrightarrow{m_1} t_1 = C_2 \cap A_1 \xrightarrow{\delta_{4N}^{-1}} z_{12N} = A_{12N} \cap C_{12N} \xrightarrow{s_0^{-1}} x_{6N+4} =$$
$$= B_{6N+4} \cap A_{6N+4} \xrightarrow{\alpha_1^{-1}} y_3 = A_3 \cap B_4,$$

and for $j = 2, \cdots, N$:

$$y_{3j} = A_{3j} \cap B_{3j+1} \xrightarrow{m_j} t_{12N-3j+4} = C_{12N-3j+5} \cap A_{12N-3j+4} \xrightarrow{\delta_{4N-j+1}^{-1}}$$
$$\to z_{12N-3j+3} = A_{12N-3j+3} \cap C_{12N-3j+3} \xrightarrow{s_{j-1}^{-1}} x_{6N+3j-1} =$$
$$= B_{6N+3j+1} \cap A_{6N+3j+1} \xrightarrow{\alpha_j^{-1}} y_{3j} = A_{3j} \cap B_{3j+1}.$$

For z_{3j} we have:

$$z_3 = A_3 \cap C_3 \xrightarrow{m_1} x_1 = B_1 \cap A_1 \xrightarrow{\alpha_0} y_{6N} = A_{6N} \cap B_{6N+1} \xrightarrow{r_0^{-1}}$$
$$\to t_4 = C_5 \cap A_4 \xrightarrow{\delta_1^{-1}} z_3 = A_3 \cap C_3,$$

and for $j = 2, \cdots, N$:

$$z_{3j} = A_{3j} \cap C_{3j} \xrightarrow{m_j} x_{12N-3j+4} = B_{12N-3j+4} \cap A_{12N-3j+4} \xrightarrow{\alpha_{2N-j+1}^{-1}}$$
$$\to y_{6N-3j+3} = A_{6N-3j+3} \cap B_{6N-3j+4} \xrightarrow{r_{j-1}^{-1}} t_{3j+1} = C_{3j+2} \cap A_{3j+1} \to$$
$$\xrightarrow{\delta_j^{-1}} z_{3j} = A_{3j} \cap C_{3j}.$$

Finally, for t_{3j} we have:

$$t_3 = A_3 \cap C_4 \xrightarrow{m_1} y_1 = B_2 \cap A_1 \xrightarrow{\beta_{4N}^{-1}} x_{12N} = A_{12N} \cap B_{12N} \xrightarrow{s_0^{-1}}$$
$$\to z_{6N+4} = C_{6N+4} \cap A_{6N+4} \xrightarrow{\gamma_1^{-1}} t_3 = A_3 \cap C_4,$$

and for $j = 2, \cdots, N$:

$$t_{3j} = A_{3j} \cap C_{3j+1} \xrightarrow{m_j} y_{12N-3j+4} = B_{12N-3j+5} \cap A_{12N-3j+4} \xrightarrow{\beta^{-1}_{4N-j+1}}$$
$$\to x_{12N-3j+3} = A_{12N-3j+3} \cap B_{12N-3j+3} \xrightarrow{s^{-1}_{j-1}} z_{6N+3j+1} =$$
$$= C_{6N+3j+1} \cap \cap A_{6N+3j+1} \xrightarrow{\gamma^{-1}_j} t_{3j} = A_{3j} \cap C_{3j+1}.$$

Now we start by r_j^{-1} and obtain for $j = 0,\ 1, \cdots, N-1$:

$$x_{6N-3j} = A_{6N-3j} \cap B_{6N-3j} \xrightarrow{r_j^{-1}} z_{3j+4} = C_{3j+4} \cap A_{3j+4} \xrightarrow{\gamma_{j+1}}$$
$$\to t_{6N+3j+3} = A_{6N+3j+3} \cap C_{6N+3j+4} \xrightarrow{h^{-1}_{N-j-1}} y_{6N-3j+1} =$$
$$= B_{6N-3j+2} \cap A_{6N-3j+1} \xrightarrow{\beta^{-1}_{2N-j}} x_{6N-3j} = A_{6N-3j} \cap B_{6N-3j},$$

and

$$z_{6N-3j} = A_{6N-3j} \cap C_{6N-3j} \xrightarrow{r_j^{-1}} x_{3j+4} = B_{3j+4} \cap A_{3j+4} \xrightarrow{\alpha_{j+1}}$$
$$\to y_{6N+3j+3} = A_{6N+3j+3} \cap B_{6N+3j+4} \xrightarrow{h^{-1}_{N-j-1}} t_{6N-3j+1} =$$
$$= C_{6N-3j+2} \cap A_{6N-3j+1} \xrightarrow{\delta^{-1}_{2N-j}} z_{6N-3j} = A_{6N-3j} \cap C_{6N-3j}.$$

For initial maps h_j^{-1} and for $j = 0,\ 1, \cdots, N-1$ we obtain:

$$x_{9N-3j} = A_{9N-3j} \cap B_{9N-3j} \xrightarrow{h_j^{-1}} z_{3N+3j+4} = C_{3N+3j+4} \cap A_{3N+3j+4} \to$$
$$\xrightarrow{\gamma_{N+j+1}} t_{9N+3j+3} = A_{9N+3j+3} \cap C_{9N+3j+4} \xrightarrow{s^{-1}_{N-j-1}} y_{9N-3j+1} =$$
$$= B_{9N-3j+2} \cap A_{9N-3j+1} \xrightarrow{\beta^{-1}_{3N-j}} x_{9N-3j} = A_{9N-3j} \cap B_{9N-3j},$$

and

$$z_{9N-3j} = A_{9N-3j} \cap C_{9N-3j} \xrightarrow{h_j^{-1}} x_{3N+3j+4} = B_{3N+3j+4} \cap A_{3N+3j+4} \to$$
$$\xrightarrow{\alpha_{N+j+1}} y_{9N+3j+3} = A_{9N+3j+3} \cap B_{9N+3j+4} \xrightarrow{s^{-1}_{N-j-1}} t_{9N-3j+1} =$$
$$= C_{9N-3j+2} \cap A_{9N-3j+1} \xrightarrow{\delta^{-1}_{3N-j}} z_{9N-3j} = A_{9N-3j} \cap C_{9N-3j}.$$

Finally, for initial identifications φ_j, we have 2 sets with $2N$ cycles:

$$x_2 = A_2 \cap B_2 \xrightarrow{\varphi_0} z_{6N+2} = C_{6N+2} \cap A_{6N+2} \xrightarrow{\delta^{-1}_{2N}} t_{6N-1} = A_{6N-1} \cap C_{6N} \to$$
$$\xrightarrow{\varphi_{2N-1}} y_{12N-1} = B_{12N} \cap A_{12N-1} \xrightarrow{\beta_{4N}} x_2 = A_2 \cap B_2,$$

and for $j = 1, \cdots, 2N-1$:

$$x_{3j+2} = A_{3j+2} \cap B_{3j+2} \xrightarrow{\varphi_j} z_{6N+3j+2} + C_{6N+3j+2} \cap A_{6N+3j+2} \rightarrow$$
$$\xrightarrow{\delta_{2N+j}^{-1}} t_{6N+3j-1} = A_{6N+3j-1} \cap C_{6N+3j} \xrightarrow{\varphi_{j-1}^{-1}} y_{3j-1} = B_{3j} \cap A_{3j-1} \rightarrow$$
$$\xrightarrow{\beta_j} x_{3j+2} = A_{3j+2} \cap B_{3j+2};$$

$$z_2 = A_2 \cap C_2 \xrightarrow{\varphi_0} x_{6N+2} = B_{6N+2} \cap A_{6+2} \xrightarrow{\beta_{2N}^{-1}} y_{6N-1} =$$
$$= A_{6N-1} \cap B_{6N} \xrightarrow{\varphi_{2N-1}} t_{12N-1} = C_{12N} \cap A_{12N-1} \xrightarrow{\delta_{4N}} z_2 = A_2 \cap C_2,$$

and for $j = 1, \cdots, 2N-1$:

$$z_{3j+2} = A_{3j+2} \cap C_{3j+2} \xrightarrow{\varphi_j} x_{6N+3j+2} = B_{6N+3j+2} \cap A_{6N+3j+2} \xrightarrow{\beta_{2N+j}^{-1}}$$
$$\rightarrow y_{6N+3j-1} = A_{6N+3j-1} \cap B_{6N+3j} \xrightarrow{\varphi_{j-1}^{-1}} t_{3j-1} = C_{3j} \cap A_{3j-1} \xrightarrow{\delta_j}$$
$$\rightarrow z_{3j+2} = A_{3j+2} \cap C_{3j+2}.$$

We see that each cycle above has four edges. This proves the unessentiality of all these cycles since each edge x_i or y_i or t_i or z_i is the edge of a right dihedral angle.

Let us consider cycles of edges b_i and c_i, $i = 1, \cdots, 12N$. From our construction of the polyhedron P_N in §2, we have that dihedral angles at edges

$$b_2, b_5, b_8, \cdots, b_{12N-1}; c_2, c_5, c_8, \cdots, c_{12N-1}$$

are equal $\pi/2N$. The dihedral angles at the other edges b_i and c_i are right.

The first case gives us two cycles with $4N$ edges:

$$b_2 = B_3 \cap B_2 \xrightarrow{\beta_1} b_5 = B_6 \cap B_5 \xrightarrow{\beta_2} b_8 = B_9 \cap B_8 \xrightarrow{\beta_3} \cdots$$
$$\xrightarrow{\beta_j} b_{3j+2} = B_{3j+3} \cap B_{3j+2} \xrightarrow{\beta_{j+1}} b_{3j+5} = B_{3j+6} \cap B_{3j+5} \xrightarrow{\beta_{j+2}} \cdots$$
$$\xrightarrow{\beta_{4N-1}} b_{12N-1} = B_{12N} \cap B_{12N-1} \xrightarrow{\beta_{4N}} b_2 = B_3 \cap B_2,$$

and

$$c_2 = C_3 \cap C_2 \xrightarrow{\delta_1} c_5 = C_6 \cap C_5 \xrightarrow{\delta_2} c_8 = C_9 \cap C_8 \xrightarrow{\delta_3} \cdots$$
$$\xrightarrow{\delta_j} c_{3j+2} = C_{3j+3} \cap C_{3j+2} \xrightarrow{\delta_{j+1}} c_{3j+5} = C_{3j+6} \cap C_{3j+5} \xrightarrow{\delta_{j+2}} \cdots$$
$$\xrightarrow{\delta_{4N-1}} c_{12N-1} = C_{12N} \cap C_{12N-1} \xrightarrow{\delta_{4N}} c_2 = C_3 \cap C_2.$$

This shows that these two cycles are unessential. The second case of edges with right angles give two sets of cycles. Namely, for edges b_i:

$$b_1 = B_1 \cap B_2 \xrightarrow{\alpha_0} b_{6N} = B_{6N} \cap B_{6N+1} \xrightarrow{\beta_{2N}} b_{6N+1} = B_{6N+1} \cap B_{6N+2} \xrightarrow{\alpha_0^{-1}}$$
$$\rightarrow b_{12N} = B_{12N} = B_{12N} \cap B_1 \xrightarrow{\beta_{4N}} b_1 = B_1 \cap B_2,$$

and for $j = 1, \cdots, 2N-1$:

$$b_{3j+1} = B_{3j+1} \cap B_{3j+2} \xrightarrow{\alpha_j} b_{6N+3j} = B_{6N+3j} \cap B_{6N+3j+1} \xrightarrow{\beta_{2N+j}}$$
$$\to b_{6N+3j+1} = B_{6N+3j+1} \cap B_{6N+3j+2} \xrightarrow{\alpha_j^{-1}} b_{3j} = B_{3j} \cap B_{3j+1} \xrightarrow{\beta_j}$$
$$\to b_{3j+1} = B_{3j+1} \cap B_{3j+2}.$$

Finally, for edges c_i, we obtain analogous $2N$ cycles each with 4 edges if we replace letters b, B, α and β above by letters c, C, γ and δ correspondingly.

This completes the proof of the unessentiality for every edge cycle in the boundary of the polyhedron P_N^0. Therefore we have the following

Lemma 4.1. *The web-group $G_N \subset PSL(2,\mathbb{C})$ is a torsion free discrete group with the fundamental polyhedron $P_N^0 \subset \mathbb{H}^3$.* □

Now we can complete the proof of the theorem stated in the Introduction. Namely, consider the manifold $M(G_N)$ and its minimal convex retract $M_{G_N} \subset M(G_N)$ (see Lemmas 3.1, 4.1 and Remark 3.2) which is a hyperbolic 3-manifold with two totally geodesic boundary components S_1, S_2, $S_1 \cup S_2 = \partial M_{G_N}$. Due to Lemma 3.3, these hyperbolic surfaces of genus $2N$ are isometric and have orientation preserving isometry groups of order $2N$. Hence we obtain the desired different (up to homeomorphisms) closed hyperbolic 3-manifolds $M_0, M_1, \cdots, M_{2N-1}$ from the theorem, if we glue the boundary components S_1 and S_2 of ∂M_{G_N} by different isometries $f_0, \cdots, f_{2N-1}$ which lift to screw-translations $T_0, \cdots, T_{2N-1}$ in $\mathbb{H}^3$ from (3.10).

In other words, we consider the closed hyperbolic 3-manifolds

$$M_i = \mathbb{H}^3/G_N^i, \; i = 0, \cdots, 2N-1, \tag{4.2}$$

where the torsion free discrete groups G_N^i from (4.1) are HNN-extensions of the group G_N by the screw-translations T_i.

Here all groups G_N^i are torsion free discrete groups and share the fundamental polyhedron $P_N \subset \mathbb{H}^3$. These facts directly follow from the second Maskit combination theorem (see **[MA]** or **[AP3]**) or, independently, from the direct consideration of cycles of edges v_i and w_i in the boundary of sides $V, W \subset 2P_N$. Namely, as in the case of the polyhedron P_N^0, we see that each such cycle contains exactly 4 edges with right dihedral angles, i.e., it is unessential.

For the proof that the manifolds M_i and M_j are non-homeomorphic, we suppose the contrary. In this case the Mostow rigidity theorem **[MO]** implies that there exists $T \in \mathrm{ISOM}(\mathbb{H}^3)$ such that

$$T(G_N *_{\langle T_i \rangle})T^{-1} = G_N *_{\langle T_j \rangle}, \qquad i \neq j.$$

and consequently, $TG_NT^{-1} = G_N$.

Conjugating the group G_N in $PSL(2,\mathbb{C})$ we may assume that the isomorphism $T_* : G_N \to G_N$ is the identity on the subgroup Γ_1. Hence $T|_{\Lambda(\Gamma_1)} = id$ in the limit circle $S^1 = \Lambda(\Gamma_1) = \partial\Omega_1$ due to a theorem of Tukia **[TU]** on isomorphisms of geometrically finite Kleinian groups. Therefore $T = id$ in $\mathbb{H}^3$. This contradicts the fact that the isomorphism

$$(T_i \cdot T_j^{-1})_*: \Gamma_2 \to \Gamma_2$$

is not the identity for $i \neq j$, i.e., this induces a non-identical homeomorphism $\Lambda(\Gamma_2) \to \Lambda(\Gamma_2)$ of the limit circle $\partial\Omega_2$. This completes the proof of the theorem.

Remark 4.2. For sufficiently large N, the constructed closed hyperbolic 3-manifolds $M_0, M_1, \cdots, M_{2N-1}$ are non-arithmetic.

In fact for each manifold M_i, its fundamental group G_N^i is a finite index subgroup of the symmetry group Γ_N of the tesselation in the hyperbolic 3-space determined by the compact polyhedron P_N.

On the other hand, this symmetry group (generated by reflections in sides of the polyhedron $\Delta_{\pi/4N}$) is non-arithmetic due to the Vinberg arithmeticity criterium **[V]**. Namely, the Gram matrix for the polyhedron $\Delta_{\pi/4N}$ has the following submatrix:

$$\begin{pmatrix} 1 & 0 & 0 & -a \\ 0 & 1 & -\cos\pi/4N & 0 \\ 0 & -\cos\pi/4N & 1 & 0 \\ -a & 0 & 0 & 1 \end{pmatrix}$$

which correspond to the edge DD_1 of the polyhedron $\Delta_{\pi/4N}$ (i.e., to 4 sides which intersect this edge). Here the number $-a$ is the cosine of the angle between normalized vectors which are orthogonal to planes τ and ω and are directed away from the polyhedron $\Delta_{\pi/4N}$ in the linear model of the hyperbolic 3-space — see **[V]**. More exactly, this number $-a$ is found from the fact that the hyperbolic distance $d = d(l_2, N)$ between planes ω and τ satisfy the equality:

$$\tanh d = \frac{\sinh(l_2).\sinh(l_2/2).\sin(\pi/4N).\sqrt{3+4(\sinh(l_2/2))^2}}{3(\cosh(l_2))^2.\sqrt{(\sinh(l_2/2))^2+(\cos\pi/4N)^2}}. \tag{4.3}$$

This formula follows from the consideration of the right angle pentagon side with vertexes D_1, D, C and C_1 (see figure 5) and from formulas (7.17.1) and (7.18.1) in **[B]** while the number l_2, $l_2 < l_1 < l_0$, appears from the consideration of the quadrilateral $ABCD$ and the pentagon ABB_1MA_2 and from our condition $D_\beta = M_\beta$ in the construction of the polyhedron $\Delta_{\pi/4N}$.

It is easy from (4.3) that the edge with arbitrary large order of symmetry has simultaneously arbitrary small hyperbolic length (for sufficiently large N).

Added in Proof. As Fritz Grunewald recently pointed out to the first author, a different (algebraic) approach to the construction of *arithmetic non-compact* hyperbolic 3-manifolds witht the same volume is possible (however, without the condition that the manifolds share a fundamental polyhedron).

Actually, this is related to the following result for the non-compact case (see **[GS]**):

Given a non-cocompact co-finite arithmetic group $G \subset \mathrm{PSL}(2,\mathbb{C})$ and an integer N, there exists a finite index subgroup $G_N \subset G$ with a surjective homomorphism

$$h: G_N \to F(N),$$

where $F(N)$ is the free group on N generators.

Namely, a consequence of this result gives us that the number of conjugacy classes of subgroups in G of large finite index N is commensurable with an exponential function of N.

References

[A] Aleksandrov, A. D., On filling of a space by polyhedra (Russian). Vestnik Leningrad Univ., Ser. Mat.-Fiz.-Him. 9 (1954), no. 2, 33–43.

[AP1] Apanasov, B. N., Geometrically finite hyperbolic structures on manifolds. Annals of Glob. Anal. and Geom. 1 (1983), no. 3, 1–22.

[AP2] Apanasov, B. N., The effect of dimension four in Aleksandrov's problem of filling a space by polyhedra. Annals of Glob. Anal. and Geom. 4 (1986), no. 2, 243–261.

[AP3] Apanasov, B. N., The Geometry of Discrete Groups in Space and Uniformization Problems, Kluwer Academic Publishers (Math. and its Appl., 40), Dordrecht, 1990, to appear (Extended English edition of "Discrete Transformations Groups and Manifold Structures" (Russian), Nauka, Novosibirsk, 1983).

[B] Beardon, A. F., The Geometry of Discrete Groups, Graduate texts in Math. 91 (Springer Verlag 1983).

[GS] F. J. Grunewald, J. Schwermer, Free non-abelian quotients of SL_2 over orders of imaginary quadratic number fields, J. Algebra 69 (1981), 298–304.

[K] Kawauchi, A., Imitations of (3,1)-dimensional manifold pairs. Sugaku Exposit. 2 (1989), no. 2, 141–156.

[MK] Makarov, V. S., Geometric methods for construction of discrete groups of the Lobachevsky space motions (Russian), in: Problems of Geometry, 15, Moscow, VINITI, 1983, 3–59.

[MA] Maskit B., "Kleinian groups" Berlin, Springer-Verlag, 1988.

[MO] Mostow, G. D., Quasi-conformal mappings in n-space and the rigidity of hyperbolic space forms. Publ. Math. IHES, 34 (1968), 53–104.

[TH] Thurston, W., "The geometry and topology of 3-manifolds" Mimeographed Lect. Notes, Princeton Univ., 1978/79.

[TU] Tukia, P., On isomorphisms of geometrically finite Möbius groups. Publ. Math. IHES 61 (1985), 171–274.

[V] Vesnin A. Yu., Three-dimensional hyperbolic manifolds of Löbel type. Siberian Math. J. 28 (1987), no. 5, 731–734.

[V] Vinberg, E.B. Discrete groups generated by reflections in Lobachevsky spaces. Math. USSR Sbornik 1 (1967), 429–444.

[W] Wielenberg, N., Hyperbolic 3-manifolds which share a fundamental polyhedron. Riemann Surfaces and Related Topics: Proc. 1978 Stony Brook Conference, ed. I. Kra and B. Maskit, Ann. of Math. Studies 97, Princeton Univ. Press, 1981, 505–513.

Department of Mathematics, University of Oklahoma, Norman, OK 73019

Institute of Mathematics, Acadamy of Sciences of Moldova, Kishinev, Moldova 277028

Link Complements and Integer Rings of Class Number Greater than One

Mark D. Baker

1. Introduction

Let $K_m = \mathbb{Q}(\sqrt{-m})$ be an imaginary quadratic number field, and let $\mathcal{O}_m$ be the integers of K_m. The *Bianchi modular group*, $\mathrm{PSL}(2, \mathcal{O}_\mathrm{m})$, is a discrete subgroup of $\mathrm{PSL}(2, \mathbb{C})$, the full group of orientation-preserving isometries of hyperbolic 3-space, $\mathbb{H}^3$.

If Γ is a torsion-free subgroup of finite index in a Bianchi group, then $\mathbb{H}^3/\Gamma$ is a finite volume hyperbolic 3-manifold. There are a number of examples where these manifolds are homeomorphic to link complements in S^3 (see **[B]**, **[H]**, **[Ri1]**, **[T]** and **[W]**). Moreover, all such examples involve the five cases $m = 1,\ 2,\ 3,\ 7$ and 11 for which $\mathcal{O}_m$ has a Euclidean algorithm, and in particular K_m has class number one.

Furthermore, results on the cohomology of $\mathrm{PSL}(2, \mathcal{O}_\mathrm{m})$ show that the only values of m for which Γ as above can give rise to link complements are from the following list, which arises from the solution to "The Cuspidal Cohomology Problem" (see **[V]**):

$$1, 2, 3, 5, 6, 7, 11, 15, 19, 23, 31, 39, 47, 71.$$

Concerning knot complements, the situation is much simpler. Riley (**[Ri1]**) showed that for the figure eight knot, k, the complement $S^3 \setminus k$ is homeomorphic to $\mathbb{H}^3/\Gamma$ where Γ is a subgroup of index 12 in $\mathrm{PSL}(2, \mathcal{O}_3)$, and this was followed by Reid's (**[Re]**) surprising result that k is the unique arithmetic knot.

In this paper we exhibit link complements homeomorphic to $\mathbb{H}^3/\Gamma$ for $\Gamma \subset \mathrm{PSL}(2, \mathcal{O}_{15})$ and $\Gamma \subset \mathrm{PSL}(2, \mathcal{O}_{23})$. K_{15} and K_{23} have class number 2 and 3 respectively.

2. Notation

Let $I = \langle 2, \omega \rangle$ where $\omega = (-1 + \sqrt{-15})/2$ and $J = \langle 2, \omega' \rangle$ where $\omega' = (-1 + \sqrt{-23})/2$ be prime ideals dividing 2 in $\mathcal{O}_{15}$ and $\mathcal{O}_{23}$ respectively. Let Γ_1 (respectively Γ_2) be the principal congruence subgroup of level I in $\mathrm{PSL}(2, \mathcal{O}_{15})$ (respectively level J in $\mathrm{PSL}(2, \mathcal{O}_{23})$. Now denote by ℓ_1 the six-circle link shown in figure 1 and ℓ_2 the nine-circle link in figure 2.

Note that $\{1, \omega\}$ and $\{1, \omega'\}$ are integral bases for $\mathcal{O}_{15}$ and $\mathcal{O}_{23}$ respectively. The main result is:

Theorem. (a) *$\mathbb{H}^3/\Gamma_1$ is homeomorphic to $S^3 \setminus \ell_1$.*
(b) *$\mathbb{H}^3/\Gamma_2$ is homeomorphic to $S^3 \setminus \ell_2$.*

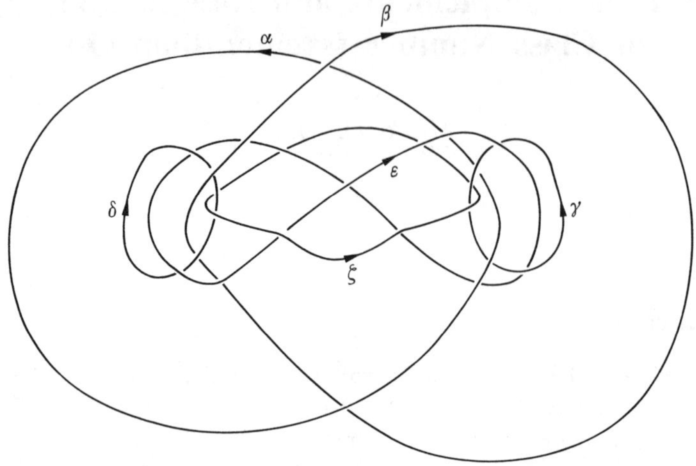

Figure 1

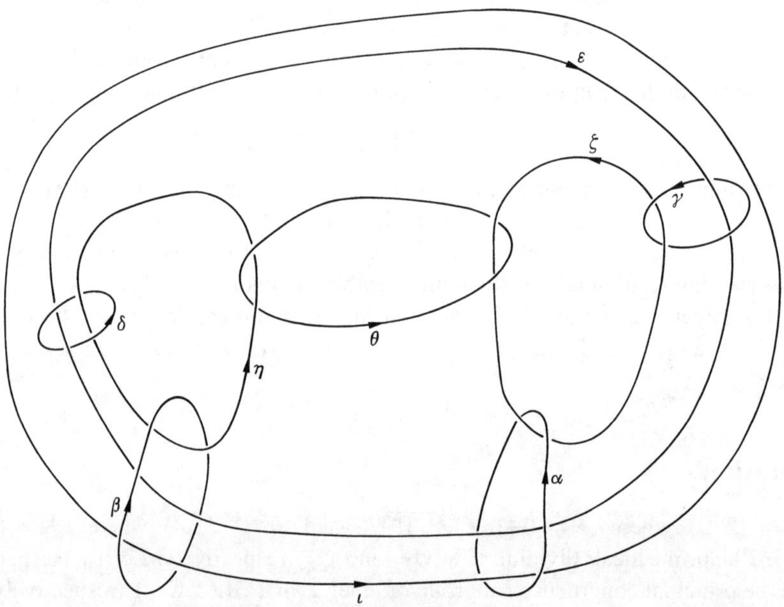

Figure 2

Since the n-fold cyclic cover of a link complement branched over an unknotted compoment of the link is again a link complement, this theorem allows us to construct infinitely many link complements corresponding to subgroups of $\mathrm{PSL}(2, \mathcal{O}_{15})$ and $\mathrm{PSL}(2, \mathcal{O}_{23})$.

3. Proof of the Theorem

By theorems 1 and 3 of **[Ri2]**, $\mathbb{H}^3/\Gamma_i$ is homeomorphic to $S^3 \setminus \ell_i$ $(i = 1, 2)$ if Γ_i is isomorphic to $\pi_1(S^3 \setminus \ell_i)$. We shall prove part (a) of the Theorem; part (b) is analogous, although somewhat more complicated.

3.1. Denote by D_1 the standard Ford fundamental region for the action of Γ_1 on $\mathbb{H}^3$ (see **[Ri1]** for details of this construction). Then $\mathbb{H}^3/\Gamma_1 \cong \overline{D}_1/\Gamma_1$. The projection onto $\mathbb{C}$ of $\overline{D}_1$ with face pairings is given in figure 3.

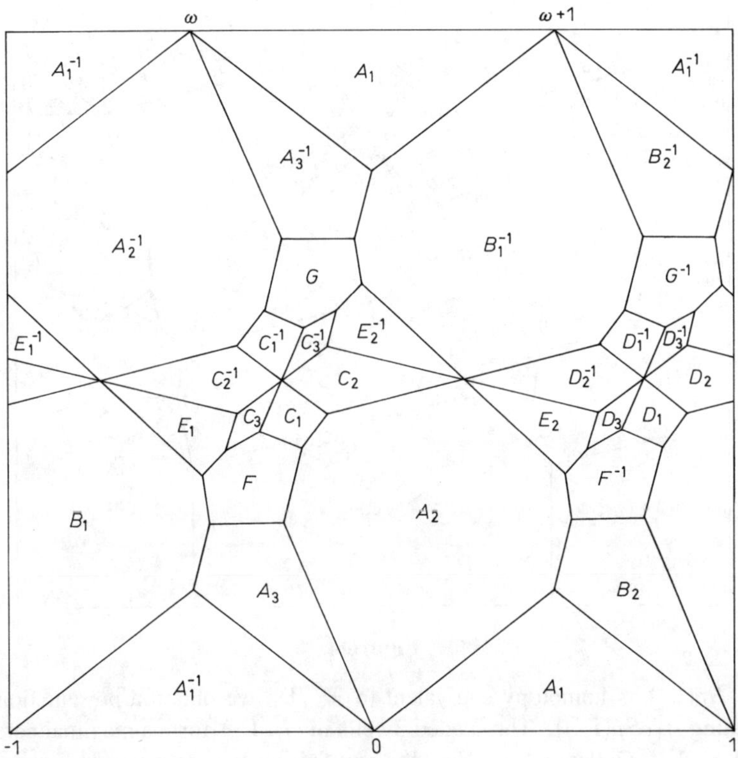

Figure 3

One can check that:

(i) The subgroup $\Gamma_1{}^{\infty} \subset \Gamma_1$ of elements fixing infinity is generated by $\{\left(\begin{smallmatrix}1 & 2\\ 0 & 1\end{smallmatrix}\right), \left(\begin{smallmatrix}1 & \omega\\ 0 & 1\end{smallmatrix}\right)\}$.

(ii) The matrices whose isometric hemispheres contribute to $\overline{D}_1$ are (up to conjugation by elements of the form $\left(\begin{smallmatrix}1 & b\\ 0 & 1\end{smallmatrix}\right)$ for $b \in \mathcal{O}_{15}$), M and M^{-1} for $M \in \{\left(\begin{smallmatrix}1 & 0\\ \omega & 1\end{smallmatrix}\right), \left(\begin{smallmatrix}1 & 0\\ 2 & 1\end{smallmatrix}\right), \left(\begin{smallmatrix}1 & 0\\ 2+\omega & 1\end{smallmatrix}\right), \left(\begin{smallmatrix}-1-\omega & \omega\\ -2+\omega & 3\end{smallmatrix}\right), \left(\begin{smallmatrix}5+\omega & 2-3\omega\\ 2-\omega & -3-2\omega\end{smallmatrix}\right), \left(\begin{smallmatrix}1+2\omega & 4+\omega\\ 4 & 1-2\omega\end{smallmatrix}\right), \left(\begin{smallmatrix}-5 & 2+2\omega\\ 2\omega & 3\end{smallmatrix}\right), \left(\begin{smallmatrix}-1+2\omega & 4\\ 4+\omega & -1-2\omega\end{smallmatrix}\right), \left(\begin{smallmatrix}-7 & 6+3\omega\\ -2+\omega & 5\end{smallmatrix}\right)\}$.

(iii) The points $\{0, \pm 1, \omega, \omega + 1, \frac{\omega}{2}, \frac{\omega \pm 1}{2}, \frac{\omega}{2} + 1\}$ in $\mathbb{C}$ and ∞ are the cusps of D_1.

3.2. The 2-complex, modulo face and edge identifications, pictured in figure 4 is a spine, $Sp(\Gamma_1)$, for $\mathbb{H}^3/\Gamma_1$. The 2-cells not lying over the cusps $(S_F, S_G, 1 - - - 8)$ are on the boundary of $\overline{D}_1$ and are pairwise identified by Γ_1 while the interiors of the 2-cells over the cusps $(T_A, T_B, T_C, T_D, {T_E}^1, {T_E}^2)$ are in the interior of $\overline{D}_1$. One sees that $\overline{D}_1/\Gamma_1$ retracts onto $Sp(\Gamma_1)$ along geodesic lines.

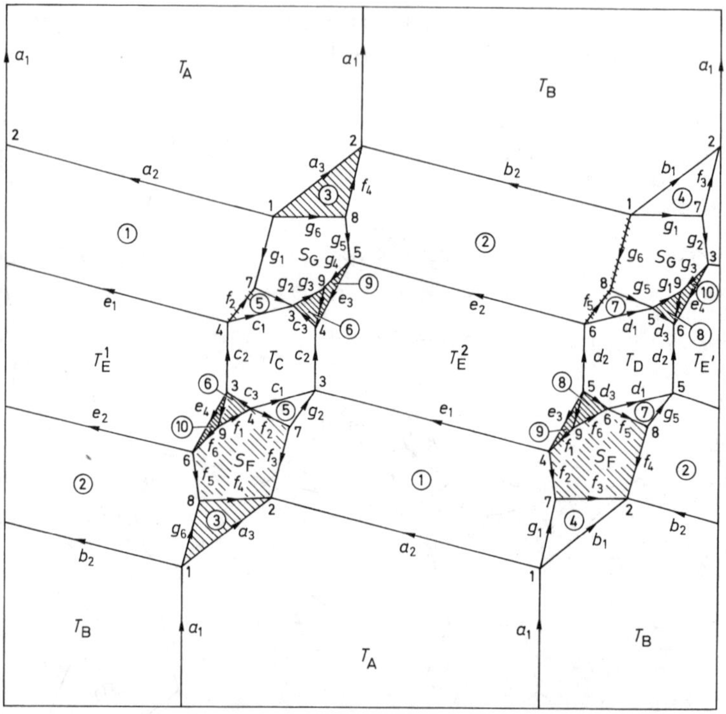

Figure 4

Since $Sp(\Gamma_1)$ is homotopy equivalent to $\mathbb{H}^3/\Gamma_1$ we obtain a presentation for Γ_1 by computing $\pi_1(Sp(\Gamma_1))$. The shaded 2-cells in $Sp(\Gamma_1)$ form a maximal contractible subcomplex, $\mathcal{K}_1$. Collapse $\mathcal{K}_1$. Now the remaining edges give the generators and the 2-cells give the relations. With this we have:

Proposition 1. *A presentation for* Γ_1 *is given by:*

Generators : $a_1,\ b_1,\ c_2,\ d_1,\ d_2,\ e_1.$

Relations :

$$a_1(b_1e_1{d_1}^{-1}b_1a_1){a_1}^{-1}(b_1e_1{d_1}^{-1}b_1a_1)^{-1} = 1;$$

$$b_1a_1({d_2}^{-1}e_1c_2{d_1}^{-1}a_1)(b_1a_1)^{-1}({d_2}^{-1}e_1c_2{d_1}^{-1}a_1)^{-1} = 1;$$

$$({b_1}^{-1}d_1)c_2({b_1}^{-1}d_1)^{-1}{c_2}^{-1} = 1;$$

$$d_1 d_2 {d_1}^{-1} {d_2}^{-1} = 1;$$

$$e_1 c_2 d_2 {e_1}^{-1} {c_2}^{-1} {d_2}^{-1} = 1. \qquad \square$$

3.3. Using figure 1, we next compute a Wirtinger presentation for $S^3 \setminus \ell_1$ (see for example **[Ro]**):

Generators : $\alpha,\ \beta,\ \gamma,\ \delta,\ \epsilon,\ \zeta$

Relations :

$$\alpha(\epsilon\gamma^{-1}\beta)\alpha^{-1}(\epsilon\gamma^{-1}\beta)^{-1} = 1;$$

$$\beta(\alpha\delta^{-1}\zeta\epsilon\zeta^{-1})\beta^{-1}(\alpha\delta^{-1}\zeta\epsilon\zeta^{-1})^{-1} = 1;$$

$$\gamma(\epsilon^{-1}\zeta^{-1}\alpha^{-1})\gamma^{-1}(\epsilon^{-1}\zeta^{-1}\alpha^{-1})^{-1} = 1;$$

$$\delta(\beta^{-1}\epsilon^{-1}\zeta^{-1})\delta^{-1}(\beta^{-1}\epsilon^{-1}\zeta^{-1})^{-1} = 1;$$

$$\epsilon(\gamma^{-1}\alpha\zeta\epsilon\beta\delta^{-1}\zeta)\epsilon^{-1}(\gamma^{-1}\alpha\zeta\epsilon\beta\delta^{-1}\zeta)^{-1} = 1.$$

3.4. Finally one checks that the following map on the generators determines an isomorphism between $\pi_1(S^3 \setminus \ell_1)$ and $\Gamma_1 \cong \pi_1(\mathbb{H}^3/\Gamma_1)$.

$$\alpha \to a_1, \beta \to b_1 a_1, \gamma \to d_1 c_2 {b_1}^{-1}, \delta \to {a_1}^{-1} d_1 d_2 a_1,$$

$$\epsilon \to b_1 e_1 c_2 {b_1}^{-1}, \zeta \to (b_1 e_1 c_2 {d_1}^{-1} a_1)^{-1}.$$

Remark. Part (b) of the Theorem is proved in exactly the same way. However the figures and presentations are considerably more complicated and will not be included here. They are available upon request.

References

[B] M. D. Baker, Link complements and imaginary quadratic number fields. To appear, Adv. in Math.

[H] A. Hatcher, Hyperbolic structures of arithmetic type on some link complements. J. London Math. Soc. (2) 27 (1983), 345-355.

[Re] A. W. Reid, Arithmeticity of knot complements. To appear J. London Math. Soc.

[Ri1] R. Riley, A quadratic parabolic group. Math. Proc. Camb. Phil. Soc. 77 (1975), 281-288.

[Ri2] R. Riley, Discrete parabolic representations of link groups. Matematika 22 (1975), 141-150.

[Ro] D. Rolfsen, Knots and Links. Publish or Perish, Berkeley, 1976.

[T] W. Thurston, Geometry and Topology of 3-manifolds. Mimeographed notes, Princeton University, 1978.

[V] K. Vogtmann, Rational homology of Bianchi groups. Math. Ann. 272 (1985) 399-419.

[W] N. Wielenberg, The structure of certain subgroups of the Picard group. Math. Proc. Camb. Phil. Soc. 84 (1978), 427-436.

I. R. M. A. R., Université de Rennes 1, Campus de Beaulieu, 35042 Rennes Cedex, France
Email: baker@FRCICB81.bitnet

Some Relations Between Spectral Geometry and Number Theory

Robert Brooks *

In his paper **[Mc]**, whose object was to show that the spectrum of the Laplacian of a Riemannian surface S determines the surface up to finitely many possibilities, Henry McKean proved the following result, which he called the "Riemann hypothesis for Riemann surfaces":

Theorem (McKean [Mc]). *If S is a hyperbolic Riemann surface, then the first eigenvalue $\lambda_1(S)$ of the Laplacian on S satisfies*

$$\lambda_1(S) \geq 1/4.$$

Here, the term "Riemann surface" denotes a compact, oriented surface with a constant curvature metric. The term "hyperbolic" means that the constant is equal to -1, so that S has genus > 1.

The number $1/4$ arises because it is the bottom $\lambda_0(\mathbb{H}^2)$ of the L^2-spectrum of the Laplacian of the hyperbolic plane. The content of McKean's theorem was thus to relate the first eigenvalue $\lambda_1(S)$ with the bottom of the spectrum of the universal cover $\widetilde{S} = \mathbb{H}^2$.

Actually, the term "proved" is used somewhat loosely here, because McKean's proof, to which we will return shortly, contained a fatal mistake. Indeed, it was observed shortly afterwards by Burt Randol that the result was indeed wrong, in the following strong sense:

Theorem (B. Randol [Ra]). *Let S be a hyperbolic Riemann surface. Then there exist arbitrarily large i-fold coverings S_i of S such that $\lambda_1(S_i) \to 0$ as $i \to \infty$.*

The examples of Randol are fairly easy to describe: Let γ be a simple closed geodesic on S which does not divide S into two pieces. Then we may open S up by cutting it along γ, make i copies of S for some large i, and have them "link hands" to form a circle S_i of surfaces as in Figure 1.

To see that $\lambda_1(S_i) \to 0$ as $i \to \infty$, we may construct test functions f_i on S_i in the following way: Assuming that i is even, divide S_i into two sets A and B of copies of S — for instance, the left half of S_i and the right half — so that A and B meet in two copies of γ. Then let f_i be $+1$ on A and -1 on B, and change in some standard way from $+1$ to -1 in neighborhoods of the two copies of γ. Substituting f_i into the

* Partially supported by the NSF

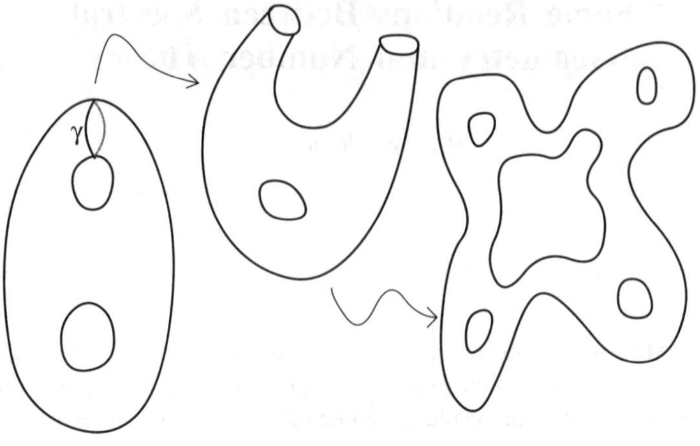

Figure 1

Rayleigh characterization of λ_1 given by

$$\lambda_1(S_i) = \inf_f \frac{\int_{S_i} ||\mathrm{grad}(f)||^2}{\int_{S_i} f^2},$$

where the inf is taken over all f satisfying $\int_{S_i} f = 0$, we see that $\int_{S_i} ||\mathrm{grad}(f_i)||^2$ is bounded independent of i, while $\int_{S_i} f_i^2$ grows linearly in i. It follows that $\lambda_1(S_i)$ tends to zero.

Here is another example to convince one that McKean's theorem is quite false: Let γ now be a closed geodesic on S which does divide S into two pieces. Then pinching γ smaller and smaller produces a family of Riemann surfaces with a long thin neck as in Figure 2. As the neck gets longer and thinner, one may easily produce test functions f_t whose Rayleigh quotients get smaller and smaller, proving that $\lambda_1(S_t) \to 0$ as $t \to \infty$.

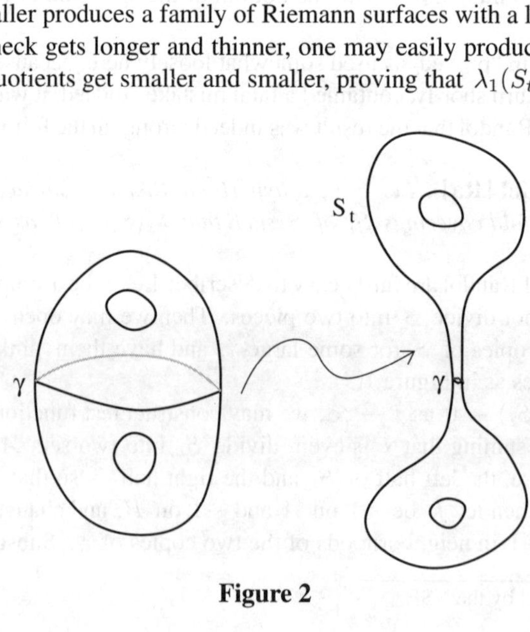

Figure 2

Thus, McKean's theorem is quite false even for surfaces of fixed genus.

Somewhat previous to this development, however, was a fascinating theorem of Selberg **[Se]**, who showed the following:

Theorem ([Se]). *Let* $\Gamma = \mathrm{PSL}(2, \mathbb{Z})$, *and*

$$\Gamma_n = \left\{ \begin{pmatrix} a & b \\ c & d \end{pmatrix} \in \Gamma : \begin{pmatrix} a & b \\ c & d \end{pmatrix} \equiv \pm \begin{pmatrix} 1 & 0 \\ 0 & 1 \end{pmatrix} \pmod n \right\}.$$

Then, for all n,

$$\lambda_1(\mathbb{H}^2/\Gamma_n) \geq \frac{3}{16}.$$

The Γ_n's are called the congruence subgroups of Γ. Note that $\mathbb{H}^2/\Gamma_n$ is a finite area (non-compact) surface, which covers the orbifold $\mathbb{H}^2/\Gamma$. The number $3/16$ enters in a kind of curious way, but for our purposes here, note that it is less than and vaguely in the same neighborhood as $1/4$. Indeed, Selberg raised the conjecture that one could improve $3/16$ to be $1/4$. Our interest in this theorem is that he obtains a lower bound for $\lambda_1(\mathbb{H}^2/\Gamma_n)$ independent of n, even though the surfaces $\mathbb{H}^2/\Gamma_n$ get larger and larger.

Selberg's proof makes crucial use of the Weil theorem on zeta functions of curves over finite fields (the "Riemann hypothesis for curves over finite fields"), although a weaker bound could be obtained using an elementary technique due to Davenport (**[Dav]**).

Nonetheless, in light of Randol's examples, it makes sense to ask the question: From a geometric point of view, what is responsible for Selberg's theorem? In other words, what qualitative features of the congruence surfaces distinguish them from the kinds of surfaces where McKean's theorem is quite false?

Our hope here is to capture for the geometer the essence of the number theory which is responsible for Selberg's $3/16$ theorem. Roughly speaking, our answer is twofold:
(i) congruence surfaces are short and fat;
(ii) congruence surfaces have interesting symmetries.

We will be more precise about this below, but, regarding the first point, let us remark that Randol's examples are all long and thin. Therefore, the problem in (i) is to quantify the notion of "short and fat," and of course to verify that the congruence surfaces are indeed short and fat. Regarding the second point, we emphasize that it is not enough for there to be a lot of symmetries — rather, we demand that they should be interesting. We must also argue that conditions (i) and (ii) lead to good lower bounds for λ_1.

The plan of this paper is as follows: in §1, we discuss in general terms how one can understand the relationship between the spectrum of the Laplacian of a manifold M and the geometry of M, in the case where one has bounds on the geometry of M. In §2, we specialize to the case of constant curvature -1, where the machinery of spherical functions becomes available to us. In §3, we show how the same picture can be carried over to the context of the Laplacian acting on graphs. This is a powerful technique, since analysis on graphs is fairly easy to carry out, and in general it is not difficult to transfer results about graphs to results about hyperbolic manifolds. We then give an example of this line of thought in §4, where we take ideas which emerge naturally from the graph-theoretic picture and translate them into hyperbolic geometry to get our desired explanation of Selberg's theorem.

We would like to thank the Department of Mathematics at UCLA for its gracious hospitality while this paper was written, and Ohio State University for its invitation to participate in its Special Year in Low-Dimensional Topology. We would also like to thank Jacob Iliadis and Peter Sarnak for their critical readings of this manuscript.

1. Spectral geometry in the presence of bounded geometry

A basic question is to understand the relationship between the spectrum of the Laplacian of a manifold M and the geometry of M. This is the question "Can one hear the shape of a drum?" raised by Mark Kac **[Kc]**, understood in its broadest sense.

In this section, we will review this question in the following light: suppose that M is compact and has constant curvature, or at least has bounded curvature. How do the low eigenvalues of M affect, and how are they affected by, the geometry of M?

A very successful approach to understanding the first eigenvalue $\lambda_1(M)$ is contained in the following theorem, due to Jeff Cheeger:

Theorem (Cheeger's inequality [Ch]).

$$\lambda_1(M) \geq \frac{1}{4} h^2 ,$$

where h is the Cheeger isoperimetric constant of M:

$$h(M) = \inf_S \frac{\operatorname{area}(S)}{\min(\operatorname{vol}(A), \operatorname{vol}(B))} ,$$

where S runs over hypersurfaces of M dividing M into two parts A and B.

Here, if M is n-dimensional, we denote by $\operatorname{vol}(A)$ the n-dimensional volume of A, and by $\operatorname{area}(S)$ the $(n-1)$-dimensional volume of S.

Cheeger's inequality has an interesting converse, due to Peter Buser:

Theorem (Buser [Bu]).

$$\lambda_1(M) \leq c_1 h(M) + c_2 h^2(M) ,$$

where c_1 and c_2 are positive constants depending only on a lower bound for the Ricci curvature of M.

In particular, if we assume constant curvature -1, then λ_1 is bounded above and below by h, so that λ_1 will be near zero if and only if the same is true of h. We remark that the constants in Buser's theorem are explicit, but not very sharp.

In this way, we can understand all the examples of the introduction.

Cheeger's constant exerts powerful control over the geometry of M, given at least some local control over M. For instance, let us consider the diameter. Then one has:

Theorem.

$$\operatorname{diam}(M) \leq C_1(h, r) \log \operatorname{vol}(M) + C_2(h, r)$$

where $C_1(h, r), C_2(h, r)$ are constants depending only on the isoperimetric constant h of M, and the injectivity radius r of M.

Proof. Let us pick a point x in M. Denote by $V(t,x)$ the volume of a metric ball of radius t about x.

If M has constant curvature -1 and r is the injectivity radius of M, then we may calculate $V(r,x)$ as the volume of a ball of radius r in hyperbolic space. If M has, say, negative curvature, then we may estimate $V(r,x)$ from below by the volume of a ball of radius r in Euclidean space. In the general case (no curvature assumptions at all!), a lower bound for $V(r/2,x)$ is given by Croke **[Cr]**.

In any case, the problem is now to estimate $V(t,x)$ for $t > r$. Here, we have

$$\frac{V'(t,x)}{V(t,x)} \geq h,$$

as long as $V(t,x) < (1/2)\,\mathrm{vol}(M)$, so that, after integrating, we obtain

$$V(t,x) \geq e^{h(t-r)} V(r,x),$$

up until $V(t,x) = (1/2)\,\mathrm{vol}(M)$. This will happen, therefore, by time t_0, where

$$t_0 = (1/h)(\log(\mathrm{vol}(M)) - \log(2) - \log(V(r))) + r.$$

It follows that, for any two points x and y, the balls of radius t_0 about x and y must intersect. Therefore, diam $(M) \leq 2t_0$, and the theorem is proved. □

Somewhat curiously, this estimate can be improved in some ways in higher dimensions. One has the following theorems of Gromov and Burger-Schroeder:

Theorem ([Gr]). *Let M be a manifold of dimension ≥ 4 whose curvatures are pinched between 0 and -1. Then*

$$\mathrm{diam}(M) \leq (\mathrm{const})\mathrm{vol}(M)^3.$$

In particular, the dependence of diam on h can be eliminated, at the expense of replacing the logarithmic dependence on volume with polynomial dependence.

Theorem ([BS]). *Let M be a rank 1 symmetric space of dimension ≥ 4. Then*

$$\mathrm{diam}(M) \leq \frac{(\mathrm{const}) \log(\mathrm{vol}(M)) + (\mathrm{const})}{\lambda_1}.$$

Here, the dependence of diam on h is replaced by dependence on λ_1. Also, (const) denotes a constant which depends only on the dimension of M.

Probably, the rank 1 assumption in the theorem of Burger and Schroeder can be weakened to curvature pinched between $-a^2$ and -1.

The dimension assumption $n \geq 4$ is necessary here. In dimension 2, one may shrink a non-separating curve to 0, increasing diameter but leaving h, and hence λ_1, bounded away from 0. In dimension 3, one may do hyperbolic Dehn surgery about a short closed geodesic, producing manifolds of small injectivity radius, and hence large diameter, but with volume bounded from above and λ_1 bounded from below.

An important step in the proofs of these theorems is to show that this cannot happen in higher dimensions. While there are certainly hyperbolic manifolds of higher dimensions which are non-compact but with finite volume, there do not exist hyperbolic manifolds in dimensions greater than 3 which have arbitrarily small geodesics but whose volumes remain bounded. Thus, for hyperbolic manifolds of dimensions greater than 3, an upper bound on volume contains implicitly a lower bound on the injectivity radius. To prove these theorems, one must make this implicit bound more explicit. See **[BS]** for details.

2. Spherical functions

Let $\mathbb{H}^n$ denote hyperbolic space of dimension n. If we fix a point $y \in \mathbb{H}^n$, and if $f(r)$ is a function of one variable $r \geq 0$, we may regard f as a function on $\mathbb{H}^n$ by setting

$$f(x) = f(\operatorname{dist}(x, y)).$$

We now claim:

Theorem. *For each λ, there is a unique function $S_\lambda(r) = S_\lambda^n(r)$ on $\mathbb{H}^n$ such that*
(i) $\Delta(S_\lambda) = \lambda \cdot S_\lambda$;
(ii) $S_\lambda(0) = 1, \qquad S'_\lambda(0) = 0.$

Proof. Let us write down the differential equation for S_λ. We have

$$\Delta(S_\lambda) = -\operatorname{div}(\operatorname{grad}(S_\lambda)).$$

Writing the hyperbolic metric in polar coordinates

$$ds^2 = dr^2 + (\sinh r)^2 d\theta^2$$

where $d\theta^2$ is the standard metric on the sphere S^{n-1}, we have

$$\operatorname{grad}(S_\lambda) = S'_\lambda \frac{\partial}{\partial r}$$

so that

$$\begin{aligned}\operatorname{div}(\operatorname{grad}(S_\lambda)) d\operatorname{vol} &= d\left[(\sinh(r))^{n-1} dr \wedge (d\theta)\left(S'_\lambda \frac{\partial}{\partial r}\right)\right] \\ &= \left(S''_\lambda + \frac{(n-1)\cosh(r)}{\sinh(r)} S'_\lambda\right) d\operatorname{vol}.\end{aligned}$$

Hence,

$$\operatorname{div}(\operatorname{grad}(S_\lambda)) = S''_\lambda + (n-1)\coth(r) S'_\lambda,$$

and so S_λ is the unique solution to the equation

$$S''_\lambda + [(n-1)\coth(r)] S'_\lambda + \lambda S_\lambda = 0$$

with initial conditions $S_\lambda(0) = 1$, $S'_\lambda(0) = 0$, as desired. □

Noting that $\coth(r) \to 1$ as $r \to \infty$, standard comparison arguments show that the solution to this equation will look like the solution to the equation

$$f'' + (n-1) \cdot f' + \lambda \cdot f = 0, \quad f'(0) = 0, f(0) = 1$$

for large r. This is then easily seen to be

$$f(r) = e^{-\frac{(n-1)}{2}r}[\cosh((\frac{\sqrt{(n-1)^2-4\lambda}}{2})r)$$

$$+\frac{(n-1)}{\sqrt{(n-1)^2-4\lambda}}\sinh((\frac{\sqrt{(n-1)^2-4\lambda}}{2})r)].$$

Notice that the solutions change character at $\lambda = (n-1)^2/4$. For $\lambda < (n-1)^2/4$, $S_\lambda(r)$ is a positive function which is decaying exponentially, while for $\lambda > (n-1)^2/4$, the solution oscillates with amplitude on the order of $e^{-\frac{(n-1)}{2}r}$. This change in behavior happens because $(n-1)^2/4$ is the bottom $\lambda_0(\mathbb{H}^n)$ of the spectrum of $\mathbb{H}^n$, see, for instance, **[Su]** for a discussion.

Standard comparison arguments also show that $S_\lambda(r)$ is a decreasing function of λ up to the first zero of $S_\lambda(r)$. In particular, $S_\lambda(r)$ is decreasing in λ for all r and all $\lambda < (n-1)^2/4$.

3. Spectral geometry of graphs

In understanding the global analysis of manifolds with bounded geometry, it is often helpful and interesting to model the problem at hand by a graph. The reasons for doing so are two-fold.

First of all, if M_0 is a given manifold, then a rich class of manifolds which share the same bounded geometry assumption is given by the covering spaces $\{M_i\}$ of M_0. After fixing generators $g_1, \ldots, g_k$ for $\pi_1(M_0)$, the coarse global geometry of M_i can be modeled on what might be called the Cayley graph Γ_i described as follows:

(i) the vertices of Γ_i are the cosets $\pi_1(M_0)/\pi_1(M_i)$;

(ii) two vertices are joined by an edge if and only if they differ by left-multiplication of a generator g_i.

If one can successfully translate the analytic problem at hand in terms of a problem about graphs, then one has a powerful technique available to answer it.

Furthermore, if $\pi_1(M_0)$ is rich enough (for instance, if it maps onto a free group on two generators), then the covering spaces of M_0 form a family which is as rich as possible in terms of the graph theory.

The second reason for doing this is that analysis on graphs is usually fairly easy. If one can understand the problem there, one can usually work one's way back to the geometric case with a fair amount of insight.

We apply these considerations to the kinds of questions considered in §1. If Γ is a graph, finite or infinite, one may define the Laplacian of Γ as follows: defining $L^2(\Gamma)$ to be the space of L^2-functions on the vertices of Γ, we may define

$$\Delta(f)(x) = \sum_{y\sim x}[f(x) - f(y)],$$

where "$y \sim x$" means that the vertices x and y are joined by an edge. Then Δ is a self-adjoint positive semidefinite operator on $L^2(\Gamma)$, and therefore has a spectrum.

Here are some sample theorems illustrating the connection between the Laplacian on graphs and on manifolds:

Theorem 1 ([B2]). *Let M' be an infinite covering of M, and let Γ' be the graph of $\pi_1(M)/\pi_1(M')$, relative to some choice of generators $g_1, \dots, g_h$ of $\pi_1(M)$.*
Then $\lambda_0(M') = 0$ if and only if $\lambda_0(\Gamma') = 0$.

In particular, if $M' = \widetilde{M}$ is the universal covering of M, then we have $\lambda_0(\widetilde{M}) = 0$ if and only if $\pi_1(M)$ is an amenable group.

Theorem 2 ([B5], [B1]). *Let $\{M_i\}$ be a family of finite coverings of M, and Γ_i the corresponding graphs.*
Then $\lambda_1(M_i) \to 0$ if and only if $\lambda_1(\Gamma_i) \to 0$.

Sketch of Proofs. The idea of the proofs is to make use of Cheeger's inequality. Indeed, there are analogues of the inequalities of Cheeger and Buser for graphs: if Γ is a finite graph, then set

$$h(\Gamma) = \inf_E \frac{\#(E)}{\min(\#(A), \#(B))}$$

where E runs over collections of edges such that $\Gamma - E$ has two components A and B. Then one has that

$$(\text{const})\, h^2 \leq \lambda_1(\Gamma) \leq (\text{const})\, h,$$

where (const) depends on the maximal number of edges meeting at a vertex.

To see that $\lambda_1(M_i) \leq (\text{const})\, h(\Gamma_i)$, we construct test functions f_i which are constant inside each fundamental domain, and which taper off in a standard way whenever a fundamental domain for A adjoins one for B. We then have

$$\int_{M_i} ||\text{grad}(f_i)||^2 \leq (\text{const})\#(E),$$

and

$$\int_{M_i} f_i^2 \geq (\text{const})\#(A).$$

This shows that $\lambda_0(\Gamma) = 0$ implies $\lambda_0(M') = 0$ (in Theorem 1) or that $\lambda_1(\Gamma_i) \to 0$ implies $\lambda_1(M_i) \to 0$ (in Theorem 2).

To obtain the opposite implication, suppose that $h(M_i) \to 0$. Then, one can show, using techniques of geometric measure theory, that there is a hypersurface S_i which realizes the isoperimetric ratio (in the case of Theorem 1, one must exercise a modicum of care to guarantee compactness). Standard variational arguments then show that the mean curvature of S_i is bounded by $h(M_i)$.

It then follows that a ball of fixed radius in S_i, which can therefore only pass through finitely many fundamental domains of M in M_i, must have a fixed amount of area. Denoting by $\mathcal{E}_i$ the collection of fundamental domains through which S_i passes, and $\mathcal{A}_i$, $\mathcal{B}_i$ the components of $M_i - \mathcal{E}_i$ we have the inequalities

$$\#(\mathcal{E}_i) \leq (\text{const})\text{area}(S_i)$$
$$\#(\mathcal{A}_i) \geq (\text{const})\text{volume}(A_i)$$

from which

$$h(\Gamma_i) \leq (\text{const})\, h(M_i)$$

is immediate.

This concludes the proofs of Theorems 1 and 2, except for the assertion $h(\Gamma) = 0$ if and only if $G = \pi_1(M)$ is amenable for Γ the Cayley graph of the group G. When properly translated, this is just the classical characterization of amenability due to Folner **[F]**. □

We should remark that there is an approach to Theorems 1 and 2, due to Marc Burger (**[Bur1]** and **[Bur2]**), which actually allows one to go further. If M is a compact manifold and M' a covering of M, let f be an eigenfunction of L^2-norm 1 of the Laplacian on M'. Denote by F the corresponding function on the graph of the covering whose value at any vertex is given by taking the average value of f over the corresponding copy of the fundamental domain of M in M'. In this way, one can see that all the low eigenvalues of M, not just the first one, are determined by the low eigenvalues of the graph. In effect, if the L^2-norm of F is large, then its Rayleigh quotient is controlled by the eigenvalues of the graph. On the other hand, if its L^2-norm is small, then in some fundamental domain, f has average value near zero, and so its eigenvalue is bounded from below by some large fraction of the first eigenvalue with Neumann conditions of the fundamental domain. See **[Bur1]** and **[Bur2]** for details.

In the case where the M_i's are normal coverings of M, so that $\pi_i = \pi_1(M)/\pi_1(M_i')$ is a group, the spectral properties of the graph Γ_i can be analyzed further. Indeed, we may then identify $L^2(\Gamma_i)$ with $L^2(\pi_i)$, which may be further decomposed into its irreducible components. To that end, it is worth remarking that there is an analogue of the Laplacian defined for unitary representations, which we call the representational Laplacian, given by the formula

$$\Delta_{\mathcal{H}}(X) = \sum_i (X - g_i(X))$$

for g_i fixed generators of a group, and X an element of $\mathcal{H}$, a unitary representation of Γ.

As before, $\Delta_{\mathcal{H}}$ is a self-adjoint, positive semi-definite operator, and so has a spectrum. The lowest eigenvalue $\lambda_0(\mathcal{H})$ has a special interpretation as the "Kazhdan distance" from $\mathcal{H}$ to the trivial representation, see **[B1]** for a discussion. One then has:

Definition. A group Γ has Kazhdan's Property T if there exists an $\varepsilon > 0$ with the following property:

For all unitary representations $\mathcal{H}$ which do not contain the trivial representation as a direct sum,

$$\lambda_0(\mathcal{H}) > \varepsilon.$$

It is then a theorem of Kazhdan **[Kz]** that if Γ is a discrete subgroup of a Lie group G with cofinite volume, then Γ has Property T if and only if the same is true for G (with Property T for Lie groups suitably defined).

Furthermore, the simple Lie groups with Property T are known — for instance, the Lie groups of non-compact type of rank > 1 have Property T, while the hyperbolic groups (e.g. $\mathrm{PSL}(2,\mathbb{R})$) do not, as is demonstrated by Randol's counterexamples to McKean's theorem.

Using these ideas, it is a simple matter to construct large Riemann surfaces S_i whose first eigenvalues $\lambda_1(S_i)$ are bounded away from 0 — for instance, let $\Gamma = \mathrm{PSL}(n,\mathbb{Z})$ for any $n \geq 3$, and let $\phi : \pi_1(S) \to \Gamma$ be any surjective homomorphism. If $\{\Gamma_i\}$ is a family of subgroups of finite index of Γ, with $[\Gamma : \Gamma_i] \to \infty$ as $i \to \infty$, then setting S_i to be the covering of S with $\pi_1(S_i) = \phi^{-1}(\Gamma_i)$, Property T says that

$$\lambda_0(L^2(\Gamma_i) \ominus (\text{constants})) > \epsilon$$

for some ϵ, where "$\ominus$" denotes the "direct minus" — that is,

$$[L^2(\Gamma_i) \ominus (\text{constants})] \oplus (\text{constants}) \simeq L^2(\Gamma_i).$$

Theorem 2 then says that $\lambda_1(S_i) > C$ for some positive constant C as $i \to \infty$.

If we now set $\Gamma = \mathrm{PSL}(2,\mathbb{Z})$, and Γ_p the congruence subgroups $\ker(\mathrm{PSL}(2,\mathbb{Z}) \to \mathrm{PSL}(2,\mathbb{Z}/p)$, Selberg's theorem (except for the constant $3/16$) is equivalent to the following assertion:

Assertion. *There exists a constant* $\epsilon > 0$ *such that, for all* p,

$$\lambda_1(\mathcal{H}_p) > \epsilon$$

for $\mathcal{H}_p = L^2(\mathrm{PSL}(2,\mathbb{Z}/p)) \ominus (\text{constants})$, *with fixed generators on* $\mathrm{PSL}(2,\mathbb{Z})$.

The left-hand side may be computed to some extent. To that end, we observe that the irreducible representations of $\mathrm{PSL}(2,\mathbb{Z}/p)$ for p a prime, are known and actually fairly understandable **[GGP]**. They come in two families, the discrete and continuous series (so named by analogy with the Lie Group case).

The dimensions of these representations are all at least $(p-1)/2$, which is approximately the cube root of the order of $\mathrm{PSL}(2,\mathbb{Z}/p)$. Since the size of an irreducible representation of a group G can be of dimension at most the square root of G, it is rather striking that we obtain only representations of dimension the cube root of G. We remark that the representations which enter into the Randol examples are all from the group $\mathbb{Z}/p$, and hence are only one-dimensional.

The high dimension of the irreducible representations would then seem to offer a striking contrast between these two cases. It remains to be seen that this contrast is reflected in the behavior of λ_1 — we will see that that is the case in the next section.

We remark that a detailed discussion of the Kazhdan constants for the irreducible representations of $\mathrm{PSL}(2,\mathbb{Z}/p)$ is contained in **[B1]**. A crucial role is played there by the Koosterman sums

$$S_\chi(a,b,p) = \sum_{y \not\equiv 0 (\mathrm{mod}\ p)} \zeta^{(ay+by^{-1})},$$

for $\zeta = e^{2\pi i/p}$ and χ an even multiplicative character $(\mathrm{mod})p$. These Kloosterman sums, which also enter into Selberg's proof of his theorem via the Weil estimate

$$|S_\chi(a,b,p)| < 2\sqrt{p},$$

enter here as matrix coefficients for the irreducible representation associated to χ.

4. Congruence subgroups

We may now complete our discussion in the introduction in the following way: Suppose S is a Riemann surface which has a group Γ of "interesting symmetries." From §3, we now know how to interpret the word "interesting" — the non-trivial irreducible representations of Γ should have large dimension relative to the order of Γ.

We then have the following:

Theorem ([B3]). *The k-th eigenvalue $\lambda_k(S)$ satisfies*

$$\left(\sqrt{1-4\lambda_k}\right) r(S) \le \log\left(\frac{\mathrm{vol}(S)}{k+1}\right) + C,$$

where $r(S)$ is the injectivity radius of S, C is a constant, and the inequality is vacuous if $\lambda_k \ge \frac{1}{4}$.

Sketch of Proof. Let f be a function in the span of all eigenfunctions of eigenvalue $\le \lambda < (1/4)$. We may lift f to a function in the hyperbolic plane $\mathbb{H}^2$ which we also denote by f.

If now y is an arbitrary point in $\mathbb{H}^2$, we may average f about y to obtain the function f^{av} defined by

$$f^{av}(r) = \frac{1}{\mathrm{length}(S(r))} \int_{S(r)} f(x)\, dx,$$

where $S(r) = \{x : \mathrm{dist}(x,y) = r\}$ is the hyperbolic circle about y of radius r.

If $f(y) > 0$, then one obtains the estimate

$$f^{av}(x) \ge f(y) S_\lambda(\mathrm{dist}(x,y)),$$

which follows from the fact that the S_λ 's are positive and decreasing in λ, for $\lambda < 1/4$.

If r is less than the injectivity radius of S, then we can carry out this calculation on S rather than in $\mathbb{H}^2$, so that we have

$$(\dagger)\ \int_S |f|^2 \int_{B(r,y)} |f(x)|^2 \ge \int_{B(r,y)} |f^{av}(x)|^2 \ge |f^{av}(y)|^2 \int_0^r |S_\lambda^2(r)|\,(\mathrm{length}(r))\, dr,$$

where the second inequality comes from the fact that averaging decreases L^2-norm, and where length (r) denotes the length of a circle of radius r in the hyperbolic plane.

We now have the following:

Lemma. *If $\mathcal{F}$ is a family of functions on S of dimension k, then there exists a function $f \in \mathcal{F}$ satisfying:*
(i) $\int_S |f|^2 = 1$;
(ii) *for some point $y \in S$,*

$$|f(y)|^2 \ge \frac{k}{\mathrm{vol}(S)}.$$

Proof. If $f_1, \dots, f_k$ is an orthonormal basis of $\mathcal{F}$, then $\int f_1^2 + \cdots + f_k^2 = k$. Hence there is a point y such that $f_1^2(y) + \cdots + f_k^2(y) \ge \frac{k}{\mathrm{vol}(S)}$.

If we now set

$$h = f_1(y) \cdot f_1 + \cdots + f_k(y) \cdot f_k,$$

we may verify that $\int_S h^2 = h(y)$, so that

$$f(x) = \frac{1}{\sqrt{h(y)}} h(x)$$

then satisfies the conclusions of the lemma. □

The proof of the theorem now follows from the observation that $1 = \int_S |f|^2 \geq \int_{B(r,y)} |f|^2$ and the inequality (†). □

We now apply these considerations to the congruence subgroups. To that end, let Γ be a cocompact arithmetic subgroup of $\mathrm{PSL}(2, \mathbb{R})$, and let Γ_p be the p-th congruence subgroup. Then $\Gamma/\Gamma_p \cong \mathrm{PSL}(2, \mathbb{Z}/p)$, and furthermore $S_p = \mathbb{H}^2/\Gamma_p$ is a finite covering of $S = \mathbb{H}^2/\Gamma$.

It is to S_p that we will apply the above considerations.

First of all, we have that

$$\mathrm{vol}(S_p) = \mathrm{vol}(S) \cdot \#(\mathrm{PSL}(2, \mathbb{Z}/p)) \sim (\mathrm{const})\, p^3,$$

and that the eigenfunctions of Δ are of one of two types:

Either

(i) they are invariant under $\mathrm{PSL}(2, \mathbb{Z}/p)$, and so descend to functions on S (and hence are bounded below by $\lambda_1(S)$);

or

(ii) they lie in eigenspaces of dimension at least $(p - 1/2)$.

We may now apply the above theorem once we know an estimate for the injectivity radius of S_p. But the injectivity radius is just 1/2 the length of the shortest closed geodesic, and for $\gamma \in \mathrm{PSL}(2, \mathbb{R})$, primitive in Γ, the length $\ell(\gamma)p$ of the geodesic fixed by γ is given by

$$2 \cosh \left(\frac{\ell(\gamma)}{2} \right) = tr(\gamma).$$

To compute $\mathrm{tr}\,(\gamma)$, we reduce (mod p) to obtain

$$\gamma \equiv \pm \begin{pmatrix} 1 & 0 \\ 0 & 1 \end{pmatrix} \pmod{p},$$

so that $\mathrm{tr}\,(\gamma) \equiv \pm 2 \pmod{p}$. Actually, it is an elementary computation that these conditions force $\mathrm{tr}\,(\gamma) \equiv \pm 2 \pmod{p^2}$. The case $\mathrm{tr}\,(\gamma) = \pm 2$ is ruled out by the assumption that Γ_p is cocompact, since $\mathrm{tr}\,(\gamma) = \pm 2$ implies that $\mathbb{H}^2/\Gamma_p$ has a cusp.

It follows that

$$\gamma(S_p) \sim \log(p^2).$$

Applying our estimate, we see that

$$(\sqrt{1 - 4\lambda_k}) 2 \log(p) \geq 3 \log(p) - \log(k) + (\mathrm{const}),$$

which gives a lower bound for λ_k as soon as $k >$ (const) $p^{1+\epsilon}$. For instance, setting $\epsilon = 1$, we get the bound

$$\lambda_{(\text{const})\, p^2}(S) \geq \frac{3}{16}.$$

This falls a little short of our desired goal, which would be

$$\lambda_{(\text{const})\, p} \geq (\text{const}),$$

because either $\lambda_1(S_p) = \lambda_1(S)$, or $\lambda_1(S_p) = \lambda_{\frac{p-1}{2}}(S_p)$.

This discussion was completed in an elegant manner by Sarnak and Xue **[SX]**, see also Huxley **[Hu]**. We will here only give a rough sketch of what is involved.

First of all, we observe that the assertion of our theorem can be recast in terms of the Selberg trace formula. Let us recall the framework of this formula briefly.

If $k(r)$ is a rapidly decreasing function of r, we may regard k as a function on hyperbolic space by the formula

$$k(x, y) = k(\operatorname{dist}(x, y)).$$

For Γ a cocompact subgroup of $\mathrm{PSL}(2, \mathbb{R})$, set

$$K_\Gamma(x, y) = \sum_{\gamma \in \Gamma} k(\gamma x, y).$$

Then, for $S = \mathbb{H}^2/\Gamma$,

$$\int_S K_\Gamma(x, x)dx = \sum_i \hat{k}(\gamma_j),$$

where $\hat{k}$ is the "Selberg transform" of k:

$$\hat{k}(\lambda) = \int_0^\infty k(r) S_\lambda(r) \ \operatorname{area}(r)\, dr.$$

The considerations of our theorem now follow from taking

$$\begin{aligned} k(r) &\equiv 1 \quad \text{for} \quad r < r(S), \\ &\equiv 0 \quad \text{for} \quad r > r(S). \end{aligned}$$

We are now in a position to push beyond the injectivity radius of S! By a careful choice of k, one can essentially trade off having a few short geodesics with the growth of balls of a larger radius to obtain better bounds on λ_1. This together with elementary bounds for integer points in regions is sufficient to obtain a proof of Selberg's theorem for cocompact arithmetic groups, with a constant slightly worse than $3/16$.

References

[B1] R. Brooks, Combinatorial problems in spectral geometry, in "Curvature and Topology of Riemannian Manifolds", Springer Lecture Notes 1201 (1988), 14–32.

[B2] R. Brooks, The fundamental group and the spectrum of the laplacian, Comm. Math. Helv. 56 (1981), 581–596.

[B3] R. Brooks, Injectivity radius and low eigenvalues of hyperbolic manifolds, J. Reine Ang. Math. 390 (1988), 117–129.

[B4] R. Brooks, On the angles between certain arithmetically defined subspaces of $\mathbb{C}^n$, Ann. Inst. Fourier XXXVII (1987), 175–185.

[B5] R. Brooks, The spectral geometry of a tower of coverings, J. Diff. Geom. 23 (1986), 97–107.

[B6] R. Brooks, The spectral geometry of k-regular graphs, to appear.

[Bur1] M. Burger, Petites valeurs propres du laplacien et topologie de Fell, Thèse, Lausanne, 1986

[Bur2] M. Burger, Spectre du laplacien, graphes, et topologie de Fell, Comm. Math. Helv. 63 (1988), 226–252

[BS] M. Burger and V. Schroeder, Volume, diameter, and the first eigenvalue of locally symmetric spaces of rank one, J. Diff. Geom. 26 (1989), 273–284.

[Bu] P. Buser, A note on the isoperimetric constant, Ann. Sci. Ec. Norm. Sup. 15 (1982), 213–230.

[Ch] J. Cheeger, A lower bound for the smallest eigenvalue of the laplacian, in Gunning, "Problems in Analysis" Annals of Math. Studies 31 (Princeton University Press, 1970), 195–199.

[Cr] C. Croke, Some isoperimetric inequalities and eigenvalue estimates, Ann. Sci. Ec. Norm. Sup. 13 (1980), 419–435.

[Dav] H. Davenport, On certain exponential sums, J. Reine Ang. Math. 169 (1933), 158–176.

[D] P. Diaconis, "Group Representations in Probability and Statistics", Institute of Mathematical Statistics Lecture Notes – Monograph Series 11 (Institute of Math. Statistics, Hayward, Ca, 1988).

[F] E. Folner, On groups with full Banach mean values, Math Scand. 3 (1955), 243–254.

[GGP] Gelfand, Graev, and Pyatetskii-Shapiro, "Representation Theory and Automorphic Functions", W. B. Saunders Ca, (1969).

[Gr] M. Gromov, Manifolds of negative curvature, J. Diff. Geom. 13 (1978), 223–230.

[Hu] M. N. Huxley, Exceptional eigenvalues and congruence subgroups, in Hejhal, Sarnak, and Terras, "The Selberg Trace Formula and Related Topics", Contemp. Math 53 (1986), 341–349.

[Kc] M. Kac, Can one hear the shape of a drum? Amer. Math Monthly 73 (1966), 1–23.

[Kz] D. Kazhdan, Connections of the dual space of a group with the structure of its closed subgroups, Functional Anal. Appl. (1968), 63–65.

[Mc] H.P. McKean, Selberg's trace formula as applied to a compact Riemann surface, Comm. Pure Appl. Math 25 (1972), 225–246.

[Ra] B. Randol, Small eigenvalues of the laplace operator on compact Riemann surfaces, Bull. AMS (1974), 990–1000.

[SX] P. Sarnak and X. Xue, rough notes.

[Se] A. Selberg, On the estimation of Fourier coefficients of modular forms, Proc. Symp. Pure Math, VIII (1965), 1–15.

[Su] D. Sullivan, Related aspects of positivity in Riemannian geometry, J. Diff. Geom. 25 (1987), 327–351.

Department of Mathematics, University of Southern California, Los Angeles, CA 90089-1113
Email: rbrooks@mtha.usc.edu

Normalizers and Centralizers of Elementary Abelian Subgroups of the Mapping Class Group

S. Allen Broughton

Abstract. Let $\bar{F}$ be a finite subgroup of Γ_σ, the mapping class group of a closed surface S of genus $\sigma \geq 2$, and let $N(\bar{F})$ be the normalizer of $\bar{F}$ in Γ_σ. The main result of this paper is to give a description of $N(\bar{F})$ as an extension of $\bar{F}$ by an explicitly defined subgroup of finite index in the mapping class group of an associated quotient surface S/F, punctured at the branch points. The quotient group $N(\bar{F})/\bar{F}$ fits into an exact sequence $N_B \to N(\bar{F})/\bar{F} \to N_S$, where N_B is of finite index in a braid group and N_S is of finite index in the mapping class group of S/F. A characterization of the image of $N(\bar{F})$ in $\mathrm{Aut}(\bar{F})$ is given so that the centralizer, $Z(\bar{F})$, is also determined. These results are then applied to the important case of elementary abelian p-subgroups of mapping class groups. Sample results include a classification of all elementary abelian subgroups with finite normalizers and a description of the normalizers of the rank 1 elementary abelian subgroups.

1. Introduction

Let $S = S_\sigma$ be a closed orientable surface of genus $\sigma \geq 2$ and let $\mathrm{Diff}^+(S)$ denote the group of orientation-preserving diffeomorphisms of S. The mapping class group of S is the group $\Gamma = \Gamma_\sigma = \Gamma(S) = \pi_0(\mathrm{Diff}^+(S))$. There are a number of reasons for studying the normalizers and centralizers of the elementary abelian subgroups of Γ, and more generally the partially ordered set $\mathcal{A}$ of elementary abelian subgroups of Γ. We shall just mention one application to the calculation of the Farrell cohomology of a mapping class group (cf. **[Brw**, Ch. X]). Since Γ has finite virtual cohomological dimension then, the p-component of the Farrell cohomology $\hat{H}^*(\Gamma)_{(p)}$ may be computed by calculating the Γ-equivariant Farrell cohomology $\hat{H}^*_\Gamma(\mathcal{A})_{(p)}$. In particular if all the elementary abelian subgroups in Γ have rank 1 then the equivariant cohomology has a decomposition

$$\hat{H}^*_\Gamma(\mathcal{A})_{(p)} \simeq \prod_{P \in \mathcal{P}} \hat{H}^*(N(P))_{(p)},$$

where $\mathcal{P}$ is a set of representatives of conjugacy classes of subgroups of order p in Γ and $N(P)$ is the normalizer of P.

The paper is partially expository, we supply some background and recall appropriate results from other sources, so that the presentation will be somewhat self-contained. The rest of the paper is organized as follows. In §2 we recall some terminology and results about mapping class groups and finite group actions on surfaces. In §3 we present results on the combinatorial enumeration of finite subgroups of mapping class groups,

Proposition 3.1, and characterizations of normalizers and centralizers, Theorem 3.2 and Corollary 3.3. Theorem 3.2 and Corollary 3.3 are the main new results, we include Proposition 3.1 for completeness. In §4 we apply the methods and results of §3 to the case of elementary abelian subgroups of Γ. There we present results such as necessary and sufficient conditions for elementary abelian subgroups to exist in the mapping class group of a given genus, a classification of all elementary abelian subgroups with finite normalizers, and a description of normalizers of rank 1 elementary abelian subgroups.

This paper grew out of discussions with conference participants Henry Glover and Fred Cohen. The author would like to thank them for their pointing out that more information on normalizers would be useful and that the author's methods in **[Br1]** and **[Br2]** might yield some answers.

2. Group actions, Fuchsian groups, mapping class groups

Group actions on surfaces. The theory of the finite subgroups of the mapping class group Γ may be effectively reformulated as the theory of topological equivalence classes of finite group actions on S.

Definition 2.1. The finite group G ***acts (smoothly, effectively and orientably) on*** the orientable surface S if there is a monomorphism $\epsilon : G \to \mathrm{Diff}^+(S)$. If ϵ' is another action on another surface S', then ϵ and ϵ' are ***topologically equivalent*** if there is an $\omega \in \mathrm{Aut}(G)$ and a diffeomorphism $h : S \to S'$ such that

$$\epsilon'(g) = h\epsilon(\omega(g))h^{-1}, \quad \textit{for all } g \in G. \tag{2.1}$$

Let $F = \epsilon(G)$ denote the group of diffeomorphisms of S determined by the G-action and $\overline{F}$ the image of F in Γ, under the natural map $\mathrm{Diff}^+(S) \to \Gamma$, $h \mapsto \overline{h}$. Because of the Nielsen Realization Theorem, proved by Kerckhoff **[K]**, every finite subgroup of Γ occurs as an $\overline{F}$. Furthermore, F acts faithfully on the homology of S, according to the Lefschetz Trace Theorem **[FK]**, so the map $F \to \overline{F}$ is always an isomorphism. When we re-interpret the finite subgroups of the mapping as finite group actions we get a more manageable though equivalent problem. The problems are equivalent because of the following proposition.

Proposition 2.1. *The assignment $\epsilon(G) = F \to \overline{F}$ defines a one-to-one correspondence of the topological equivalence classes of finite group actions on S with the conjugacy classes of finite subgroups of Γ.*

Proof. Clearly, under the given assignment equivalent actions on S yield conjugate subgroups of Γ, so the mapping of equivalence classes of actions to conjugacy classes of subgroups is well-defined. As noted above, this mapping is surjective because of the Nielsen Realization Theorem. The proof that the mapping is one-to-one will be given at the end of §3 after we have set up the appropriate notation. □

The quotient space $T = S/G$ is a Riemann surface of genus $\tau \leq \sigma$. The quotient map $q : S \to T$ is branched over t points $Q_1, \ldots, Q_t \in T$. For $P \in q^{-1}(Q_i)$ let $G_P = \{g \in G : gP = P\}$. The integer $n_i = |G_P| \geq 2$ is independent of $P \in q^{-1}(Q_i)$

and is called the ***branching order*** at Q_i. We call the $(t+1)$-tuple $(\tau{:}\, n_1, \ldots, n_t)$ the ***branching data*** of G acting on S. Normally, we order the $Q_1, \ldots, Q_t$, so that $n_1 \leq \ldots \leq n_t$, and say that the branching data is ***normally ordered*** in this case. Throughout the remainder of the paper, unless otherwise noted, we use Λ to denote the branching data $(\tau{:}\, n_1, \ldots, n_t)$ and assume that Λ is normally ordered. The branching data and the order of G are related by the Riemann-Hurwitz equation **[FK]**:

$$\frac{2\sigma - 2}{|G|} = 2\tau - 2 + \sum_{j=1}^{t}(1 - \frac{1}{n_j}). \tag{2.2}$$

Fuchsian groups and generating vectors. Finite group actions on S, in turn, can be reformulated in terms of Fuchsian groups and so-called surface-kernel epimorphisms. Let $\mathbb{H}$ be the upper half plane, the universal cover of S, and $\pi : \mathbb{H} \to S$ the covering projection. For any subgroup $E \subseteq \mathrm{Diff}^+(S)$ define the group

$$\tilde{E} = \{\tilde{h} \in \mathrm{Diff}^+(\mathbb{H}) : \pi \circ \tilde{h} = g \circ \pi, \text{ for some } g \in E\}. \tag{2.3}$$

If we give S a conformal structure by pulling back a conformal structure on T, and then pull the conformal structure on S back to $\mathbb{H}$, then $F = \epsilon(G)$ will act conformally on S, and $\tilde{F}$ and $K = \tilde{1} \simeq \pi_1(S)$ will be Fuchsian groups. The group K is torsion-free, $S \simeq \mathbb{H}/K$, and the sequence $K \to \tilde{F} \to F$ is exact. The map $\xi : \tilde{F} \to F$ of this sequence is induced by the action of $\tilde{F}$ on $\mathbb{H}/K$. Let $\eta : \tilde{F} \to G$ be the map $\epsilon^{-1} \circ \xi$. Such surjective maps from the Fuchsian group $\tilde{F}$ to G with torsion free kernel are called ***surface-kernel epimorphisms*** (c.f. **[H1]**, **[H2]**, **[L]**).

From this construction, we see that every Λ-action of G on S yields a surface-kernel epimorphism to G. Conversely suppose $K \to \tilde{F} \to G$ is a surface-kernel epimorphism. Then G acts on $\mathbb{H}/K$, $\mathbb{H}/K$ has genus σ by the Riemann-Hurwitz equation, and S and $\mathbb{H}/K$ are diffeomorphic because they have the same genus. Thus G has a Λ-action on S arising from η, and we see that G-actions on S are equivalent to surface-kernel epimorphisms of $\tilde{F} \to G$.

It is well known that $\tilde{F}$ has a presentation of the following sort (cf. eg. **[ZVG]**).

$$\begin{aligned}\tilde{F} = \langle &\alpha_1, \ldots, a_\tau, \beta_1, \ldots, \beta_\tau, \gamma_1, \ldots, \gamma_t : \\ &\textstyle\prod_{i=1}^{\tau}[\alpha_i, \beta_i]\prod_{j=1}^{t}\gamma_j = \gamma_1^{n_1} = \ldots \gamma_t^{n_t} = 1\rangle.\end{aligned} \tag{2.4}$$

Let Π_Λ denote the abstract group whose presentation is given by the right hand side above. If $t = 0$ then $\Pi_\Lambda \simeq \pi_1(S_\tau)$. Define the elements a_i, b_i, c_j of G by

$$\begin{aligned} a_i &= \eta(\alpha_i), \quad 1 \leq i \leq \tau, \\ b_i &= \eta(\beta_i), \quad 1 \leq i \leq \tau, \\ c_j &= \eta(\gamma_j), \quad 1 \leq j \leq t. \end{aligned} \tag{2.5}$$

These elements generate G,

$$\prod_{i=1}^{\tau}[a_i, b_i]\prod_{j=1}^{t} c_j = 1, \tag{2.6}$$

and

$$o(c_j) = n_j, \tag{2.7}$$

because of (2.4), and since $K = \ker(\eta)$ is torsion free.

Definition 2.2. Let $\Lambda = (\tau : n_1, \ldots, n_t)$. We call any $(2\tau + t)$-tuple of elements of G, $(a_1, \ldots, a_\tau, b_1, \ldots, b_\tau, c_1, \ldots, c_t)$, satisfying (2.6) and (2.7) a Λ-***vector***. Such a vector is called a ***generating*** Λ-***vector*** if $G = \langle a_1, \ldots a_\tau, b_1, \ldots b_\tau, c_1, \ldots, c_t \rangle$. We denote by $\mathrm{GV}(G, \Lambda)$ the set of generating Λ-vectors of G.

The formula (2.5) establishes a one-to-one correspondence between the set of generating Λ-vectors of G and the set of surface-kernel epimorphisms of $\tilde{F} \to G$ (see **[H1]** for proof that the homomorphism determined by a generating vector has torsion free kernel). The relation of topological equivalence of actions induces an equivalence relation on the set of generating vectors and, in turn, these equivalence classes give a concrete enumeration of the conjugacy classes of finite subgroups of the mapping class group. In **[Br2]** it is established that these equivalence classes may be recovered as the orbits of a subgroup of index 2 in $\mathrm{Aut}(G) \times \mathrm{Aut}(\tilde{F})$, acting on the set $\mathrm{GV}(G, \Lambda)$. A similar, much earlier result is given in **[L]**. Using the correspondence between generating vectors and surface-kernel epimorphisms above, we may define this action by the formula:

$$\eta \mapsto \omega \circ \eta \circ \delta, \ (\omega, \delta) \in \mathrm{Aut}(G) \times \mathrm{Aut}(\tilde{F}). \tag{2.8}$$

Remark 2.1. Note that the action of $\mathrm{Inn}(\tilde{F})$ shows up as $\mathrm{Inn}(G)$-action, so if G is abelian then the action of $\mathrm{Aut}(G) \times \mathrm{Aut}(\tilde{F})$ factors through $\mathrm{Aut}(G) \times \mathrm{Out}(\tilde{F})$.

Remark 2.2. Since $\mathrm{Aut}(G)$ and $\mathrm{Aut}(\tilde{F})$ have commuting actions on $\mathrm{GV}(G, \Lambda)$ then $\mathrm{Aut}(\tilde{F})$ permutes the $\mathrm{Aut}(G)$-classes of generating vectors. This action always factors through $\mathrm{Out}(\tilde{F})$ even for non-abelian G.

Remark 2.3. It follows from (2.6) that $(0 : n)$ and $(0 : m, n), m \neq n$ can never occur as branching data. It follows from (2.2) that $(0 : -)$, $(1 : -)$, $(0 : n, n)$, $(0 : 2, 2, n)$, $(0 : 2, 3, n)$, for $3 \leq n \leq 6$, $(0 : 2, 4, 4)$, $(0 : 3, 3, 3)$ and $(0 : 2, 2, 2, 2)$ cannot occur if $\sigma \geq 2$. According to **[Ma]**, if G is abelian then, the branching data cannot be $(\tau : n)$ and $\mathrm{l.c.m.}(n_1, \ldots, \hat{n}_j, \ldots, n_t) = \mathrm{l.c.m.}(n_1, \ldots, n_t)$, for $j = 1, \ldots, t$, where $\hat{n}_j$ means n_j is omitted.

Normalizers of Fuchsian groups. In our final proposition of this section we spell out the relationship between the normalizer of $\tilde{F}$ in $\mathrm{Diff}^+(\mathbb{H})$, $\mathrm{Aut}(\tilde{F})$ and the mapping class group of S/G. We will need to introduce some notation. First, let c be the relator $\prod_{i=1}^{\tau} [\alpha_i, \beta_i] \prod_{j=1}^{t} \gamma_j$ in the free group

$$\mathfrak{F} = \langle \alpha_1, \ldots, a_\tau, \beta_1, \ldots, \beta_\tau, \gamma_1, \ldots, \gamma_t \rangle. \tag{2.9}$$

According to **[ZVC**, Th. 5.8.2], each automorphism of $\tilde{F}$ is induced by an automorphism $\phi : \mathfrak{F} \to \mathfrak{F}$ such that $\phi(c), \phi(\gamma_1), \ldots, \phi(\gamma_t)$ are conjugate to the elements $c^\lambda, \gamma_{r_1}^\lambda, \ldots, \gamma_{r_t}^\lambda$, respectively, where $\lambda = \pm 1$ and $(r_1, \ldots, r_t)$ is a permutation of $(1, \ldots, t)$ satisfying $n_i = n_{r_i}$. An automorphism of $\tilde{F}$ is called an ***orientation-preserving automorphism*** if $\lambda = 1$ for the inducing automorphism ϕ above. Let

$\mathrm{Aut}^+(\tilde{F})$ denote the group of orientation-preserving automorphisms and $\mathrm{Out}^+(\tilde{F})$ its image in $\mathrm{Out}(\tilde{F})$.

Next, we introduce another universal covering. Let $T^\circ = T - \{Q_1, \dots, Q_t\}$, $S^\circ = q^{-1}(T^\circ)$ and $\mathbb{H}^\circ = \pi^{-1}(T^\circ)$. The map $q \circ \pi : \mathbb{H}^\circ \to T^\circ$ is a covering. Factor the universal cover $\rho : \mathbb{H} \to T^\circ$ as a composition $\rho = p \circ (\pi \circ q) : \mathbb{H} \to \mathbb{H}^\circ \to T^\circ$ where $p : \mathbb{H} \to \mathbb{H}^\circ$ is the universal cover of $\mathbb{H}^\circ$. The group of covering transformations of ρ is a Fuchsian group $L \simeq \pi_1(T^\circ)$ with presentation

$$L = \langle \alpha_1, \dots, a_\tau, \beta_1, \dots, \beta_\tau, \gamma_1, \dots, \gamma_t : \prod_{i=1}^{\tau} [\alpha_i, \beta_i] \prod_{j=1}^{t} \gamma_j = 1 \rangle . \tag{2.10}$$

Each element $\gamma_j \in L$ is a parabolic covering transformation corresponding to a loop about the puncture Q_j and the remaining generators are hyperbolic transformations corresponding to loops running around and through the handles of T°. From this interpretation, it is clear that the group of covering transformations of $p : \mathbb{H} \to \mathbb{H}^\circ$ is the normal subgroup M of L generated by $\{\gamma_1^{n_1}, \dots, \gamma_t^{n_t}\}$ and that the $\tilde{F}$-action on $\mathbb{H}^\circ$ is induced by the action of L/M on $\mathbb{H}/M$.

Finally, every diffeomorphism of T° has a unique extension to T, and therefore determines a permutation of the punctures $\{Q_1, \dots, Q_t\}$. Thus, we may define a subgroup, $\Gamma_\Lambda = \Gamma_\Lambda(T^\circ)$ of the mapping class group of T°, consisting of the mapping classes which preserve the branching order of punctures according to the branching data Λ. If h is an orientation-preserving diffeomorphism of T° preserving branching orders, then h lifts all the way up through ρ to an orientation-preserving map $\hat{h}$ on $\mathbb{H}$, normalizing L. The map $\hat{h}$ conjugates each γ_j to an L-conjugate of some γ_{r_j}, since h maps punctures to punctures. Furthermore, $n_j = n_{r_j}$ because h preserves branching orders, hence $\hat{h}$ normalizes M. Therefore, the map $\hat{h}$ induces a map $\tilde{h}$ on $\mathbb{H}^\circ \simeq \mathbb{H}/M$ and this map clearly normalizes $\tilde{F}$.

Proposition 2.2. *Let $\tilde{F}$, T, Λ and $\Gamma_\Lambda(T^\circ)$ be as above. Let $\mathcal{N}$ be the normalizer of $\tilde{F}$ in* Diff($\mathbb{H}$), *and let $\mathcal{N}^+ = \mathcal{N} \cap \mathrm{Diff}^+(\mathbb{H})$. Then we have the following.*

1) *Every automorphism of $\tilde{F}$ has a geometric realization, i.e., the automorphism is induced by conjugation by a suitable element $\tilde{h} \in \mathcal{N}$. In particular, this is true for $K = \tilde{1}$.*
2) *The automorphism in* 1) *lies in $\mathrm{Aut}^+(\tilde{F})$ if and only $\tilde{h} \in \mathcal{N}^+$.*
3) *The natural maps $\mathcal{N}^+ \to \mathrm{Out}^+(\tilde{F})$ and $\mathcal{N}^+ \to \Gamma_\Lambda(T^\circ)$ induce an isomorphism between $\mathrm{Out}^+(\tilde{F})$ and $\Gamma_\Lambda(T^\circ)$.*

Proof. 1) This is an extension of the Nielsen-Dehn theorem and is proven in **[ZVC**, Th. 5.8.3].

2) This follows from the construction of the homeomorphism $\tilde{h}$ in the proof of Theorem 5.8.3 of **[ZVC]**, though there is not a specific statement of this fact in the reference. The essential idea is that the relators lift to boundaries of polygons and discs in the universal cover and $\tilde{h}$ will reverse the orientation of these boundaries if and only if $\tilde{h}$ is orientation-reversing.

3) A proof of a version of 3) for fundamental groups of surfaces is given in **[ZVC**, Th. 5.15.3]. Our proof is a simple modification of that proof. It suffices to show that both maps in 3) are surjective and have the same kernel. From the discussion preceding the statement of the proposition it is clear $\mathcal{N} \to \Gamma_\Lambda(T^\circ)$ is surjective and that $\mathcal{N}^+ \to \mathrm{Out}^+(\tilde{F})$ is surjective, by 1). By lifting homotopies up through $\pi \circ q : \mathbb{H}^\circ \to T^\circ$, filling in punctures, and taking base points into consideration, it is easily shown that every element of the kernel of $\mathcal{N}^+ \to \Gamma_\Lambda(T^\circ)$ may be factored as product of an element of $\tilde{F}$ and a diffeomorphism which is $\tilde{F}$-equivariantly isotopic to the identity map. But, by the Baer theorem, **[ZVC**, Th. 5.14.1], homeomorphisms which induce the identity map on $\tilde{F}$ are $\tilde{F}$-equivariantly isotopic to the identity map. It follows immediately that both maps have the same kernel. □

Remark 2.4. The point of identifying mapping classes with automorphisms is that it translates topological equivalence of generating vectors into an algebraic equivalence, as well as easy construction of elements of $\mathrm{Aut}^+(\tilde{F})$ via automorphisms of $\mathcal{F}$.

Remark 2.5. In the next section we shall make use of the (trivial) observation that every group-theoretic construction for $\tilde{F}$ has an analogue for the isomorphic group Π_Λ.

3. Characterization of normalizers and centralizers

Classification of finite subgroups of Γ. The finite subgroups of a mapping class group may be classified by the following proposition.

Proposition 3.1. *Let $\mathrm{Aut}^+(\Pi_\Lambda)$ and its action on $\mathrm{GV}(G,\Lambda)$, be defined as in Remark 2.4. The conjugacy classes of finite subgroups of the mapping class group Γ_σ are in one-to-one correspondence with the $\mathrm{Aut}(G) \times \mathrm{Out}^+(\Pi_\Lambda)$-orbits in a finite number of $\mathrm{GV}(G,\Lambda)$. The G and Λ are chosen from the finite set of isomorphism classes of finite groups G and normally ordered branching data Λ such that the Riemann-Hurwitz equation (2.2) holds and $\mathrm{GV}(G,\Lambda)$ is not empty.*

Proof. The proposition is an immediate consequence of our discussion in §2 and Propositions 2.1 and 2.2. Reordering the branching data merely means relabeling the branch points on T, so no new actions arise from the generating vectors obtained by reordering branching data. □

Normalizers and centralizers. Now we give a characterization of normalizers and centralizers, the main result of this paper.

Theorem 3.2. *Let Γ be the mapping class group of a surface S and $\bar{F}$ a finite subgroup of Γ. Let $\Lambda = (\tau : n_1, \ldots, n_t)$ be the branching data and $\mathbf{x} \in \mathrm{GV}(G,\Lambda)$ the generating vector corresponding to $\bar{F}$, as guaranteed by Proposition 3.1. Let $\mathrm{Aut}^+(\Pi_\Lambda)$ and the action of $\mathrm{Aut}(G) \times \mathrm{Aut}^+(\Pi_\Lambda)$ on $\mathrm{GV}(G,\Lambda)$ be as defined in Remark 2.4 and let $\Sigma_{\mathbf{x}}$ denote the stabilizer of $\mathbf{x}$ under this action. Let Σ_G and Σ_Λ be the images of $\Sigma_{\mathbf{x}}$ in $\mathrm{Aut}(G)$ and $\Gamma_\Lambda \simeq \mathrm{Out}^+(\Pi_\Lambda)$, respectively, and let $N(\bar{F})$ and $Z(\bar{F})$ denote the normalizer and centralizer of $\bar{F}$ in Γ, respectively. Then, there are exact sequences*

$$\bar{F} \to N(\bar{F}) \to \Sigma_\Lambda \tag{3.1}$$

and

$$Z(\bar{F}) \to N(\bar{F}) \to \Sigma_G. \tag{3.2}$$

Proof. Let the groups F, $\tilde{F}$, K, G and $\mathcal{N}^+$ be as in §2 and let $\mathcal{M}^+ = \{\tilde{h} \in \mathcal{N}^+ : \tilde{h}K\tilde{h}^{-1} = K\}$. To prove (3.1) it suffices to show that there are maps

$$\zeta_1 : \mathcal{M}^+ \to N(\bar{F}), \quad \zeta_2 : \mathcal{M}^+ \to \Gamma_\Lambda,$$

such that

$$\ker(\zeta_1) \subset \ker(\zeta_2), \quad \ker(\zeta_2)/\ker(\zeta_1) \simeq \bar{F}, \tag{3.3.i}$$

and

$$\mathrm{im}(\zeta_1) = N(\bar{F}), \quad \mathrm{im}(\zeta_2) = \Sigma_\Lambda. \tag{3.3.ii}$$

Let us first define the maps and show that the images are as claimed in (3.3.ii). Let $\tilde{h} \in \mathcal{M}^+$, since $\tilde{h}$ normalizes K then $\tilde{h}$ induces a diffeomorphism h of $S = \mathbb{H}/K$. Since $\tilde{h}$ also normalizes $\tilde{F}$ then h normalizes F and hence $\bar{h}$ normalizes $\bar{F}$ in Γ. Define ζ_1 to be the map $\tilde{h} \mapsto \bar{h}$ just constructed. To show that ζ_1 maps onto $N(\bar{F})$ we argue as follows. The group $\tilde{F}$ induces a group F^* of automorphisms of K by conjugation. The map $\tilde{F} \to F^*$ is an isomorphism, since centralizers in $\mathrm{PSL}_2(\mathbb{R})$ are abelian. The map also takes K onto $K^* = \mathrm{Inn}(K)$, and both F^* and $\bar{F}$ have the same finite image in $\mathrm{Out}^+(K)$. Now let $\bar{h} \in \Gamma$ and let $\tilde{h} \in \mathrm{Diff}(\mathbb{H})$ be a geometric realization of $\bar{h}$ guaranteed by 1) of Proposition 2.2. If $\bar{h}$ normalizes $\bar{F}$ then, since F^* and $\bar{F}$ both have the same image in $\mathrm{Out}^+(K)$, $\tilde{h}$ must normalize the preimage of this group in $\mathrm{Aut}(K)$, namely F^*. Since $\tilde{F} \to F^*$ is an isomorphism then $\tilde{h}$ induces an outer automorphism of $\tilde{F}$. Using 1) of Proposition 2.2 again we get a possibly different $\tilde{h}'$, inducing the given automorphism on $\tilde{F}$. Now $\tilde{h}'$ normalizes $\tilde{F}$ and induces the same automorphism on F^* as does $\tilde{h}$, in particular both maps preserve the subgroup of K^* of F^*. It follows then that $\tilde{h}'$ normalizes K since $\tilde{F} \to F^*$ is an isomorphism. It further follows that $\bar{h} \in N(\bar{F})$ is the image of the element $\tilde{h}' \in \mathcal{M}^+$, proving the desired surjectivity.

Now let $\eta : \tilde{F} \to G$ be the surface-kernel epimorphism corresponding to $\mathbf{x}$. Consider the action of $\mathrm{Aut}(G) \times \mathcal{N}^+$ on $\mathrm{GV}(G, \Lambda)$ induced by the homomorphism $\mathcal{N}^+ \to \mathrm{Aut}(\tilde{F})$ and formula (2.8), where the δ of (2.8) is the automorphism induced by conjugation by $\tilde{h} \in \mathcal{N}^+$. There will be an $\omega \in \mathrm{Aut}(G)$ such that the pair $(\omega, \tilde{h})$ fixes $\mathbf{x}$ if and only if $\tilde{h}$ normalizes the kernel of η and in this case ω will be the unique automorphism of G such that

$$\omega(\eta(y)) = \eta(\tilde{h}y\tilde{h}^{-1}) \quad \text{for all} \quad y \in \tilde{F}. \tag{3.4}$$

But $\tilde{h}$ normalizes $\ker(\eta) = K$ if and only if $\tilde{h} \in \mathcal{M}^+$. Thus the map $\mathcal{N}^+ \to \mathrm{Aut}^+(\tilde{F}) \to \Gamma_\Lambda$ restricts to a surjective map $\zeta_2 : \mathcal{M}^+ \to \Sigma_\Lambda$.

It remains to show (3.3.i). By construction, any diffeomorphism in $\ker(\zeta_1)$ must be the product of a covering transformation of $\mathbb{H} \to S$, i.e., an element of K, and a diffeomorphism K-equivariantly homotopic to the identity. Next, by the uniqueness of ω in (3.4) and 3) of Proposition 2.2 it follows that projection $\mathrm{Aut}(G) \times \mathrm{Aut}^+(\Pi_\Lambda) \to$

$\mathrm{Aut}^+(\Pi_\Lambda)$ is one-to-one when restricted to the stabilizing subgroup $\Sigma_{\mathbf{x}}$. Therefore, $\ker(\zeta_2)$ will consist of elements of $\mathcal{N}^+$ which are products of elements of $\tilde{F}$ and a diffeomorphism which is $\tilde{F}$-equivariantly isotopic to the identity. Let D_K, $D_{\tilde{F}}$, denote the groups of diffeomorphisms which are, respectively, K-equivariantly, $\tilde{F}$-equivariantly isotopic to the identity. From the above $\ker(\zeta_1) = K \ltimes D_K$ and $\ker(\zeta_2) = \tilde{F} \ltimes D_{\tilde{F}}$. By **[ZVC**, Th. 5.16.2**]** $D_K = D_{\tilde{F}}$ and (3.3.i) holds,, hence (3.1) is proven.

The exactness of (3.2) follows from the uniqueness of the automorphism ω in (3.4). In (3.2) the natural action of $N(\bar{F})/Z(\bar{F})$ on $\bar{F}$ has been transferred to G by the inverse of the map $g \mapsto \epsilon(g) \mapsto \overline{\epsilon(g)}$. All is now proven. □

Braid and surface decomposition of normalizers. There is an exact sequence $\mathcal{B}_{\tau,t} \to \Gamma(T^\circ) \to \Gamma(T)$. The kernel $\mathcal{B}_{\tau,t}$ of this map is a central quotient of the full braid group on t strings in S_τ (cf. **[Bi**, Th. 4.3**]**). This homomorphism induces a structure on $N(\bar{F})/\bar{F} \simeq \Sigma_\Lambda \subset \Gamma_\Lambda \subset \Gamma(T^\circ)$. We call $\Sigma_\Lambda \cap \mathcal{B}_{\tau,t}$ the ***braid part*** of $N(\bar{F})$ and the image of Σ_Λ in $\Gamma(T)$ the ***surface part*** of $N(\bar{F})$.

The group G, and hence $\bar{F}$, also has a structure similar to the braid and surface structure described above. Let $\mathcal{C}$ be the subgroup of Π_Λ generated by the γ_j's in (2.4). Let $C = \langle c_1, \ldots, c_t \rangle$ be the image of $\mathcal{C}$ in G. We also call $G_B = C$ the ***braid part*** of G and $G_S = G/C$ the ***surface part*** of G. As C is normal then $G_S = G/C$ acts on S/C, and, as is easily shown, this action is fixed point free. Therefore S/C has genus $|G_S|\tau + 1$.

The relationship between the braid and surface structures on G and $N(\bar{F})$ is the following. The braid and surface structure of G is induced by the exact sequence $\mathcal{C} \to \Pi_\Lambda \to \Pi_\tau$. Since $\mathcal{C}$ is invariant under $\mathrm{Aut}^+(\Pi_\Lambda)$, this exact sequence also induces a homomorphism $\mathrm{Aut}^+(\Pi_\Lambda) \to \mathrm{Aut}^+(\Pi_\Lambda/\mathcal{C})$ and, hence, by Proposition 2.2 induces the sequence $\mathcal{B}_{\tau,t} \to \Gamma(T^\circ) \to \Gamma(T)$. But now, it is this sequence that induces the braid and surface structure on $N(\bar{F})$. The relations between the structures is particularly evident in the situation of Proposition 4.3.

In the corollary that follows we apply **[Bi**, §4.1**]** to our decomposition above. Theorem 4.3 of **[Bi]** does not apply to branching data of the form $(1\colon n)$, but we shall not need this for our applications to elementary abelian subgroups because of Remark 2.3.

Corollary 3.3. *Let notation be as in Theorem 3.2, let $\mathcal{B}_{\tau,t} \to \Gamma(T^\circ) \to \Gamma(T)$ be the exact sequence given above. Let N_B and N_S be the braid and surface parts of $N(\bar{F})$, respectively, and let G_B and G_S be the braid and surface parts of G, respectively. Then we have the following.*

1) *The sequence $N_B \to N(\bar{F})/\bar{F} \to N_S$ is exact.*
2) *For $\tau \geq 2$, $\mathcal{B}_{\tau,t}$ is the full braid group on t strings in S_τ. For $\tau = 1$, $t \geq 2$ and for $\tau = 0$, $t \geq 3$, $\mathcal{B}_{\tau,t}$ is the quotient of the full braid group by its centre. The braid part N_B is a subgroup of finite index in $\mathcal{B}_{\tau,t}$. If G is abelian then N_B contains the pure braid group (mod the centre if $\tau \leq 1$).*
3) *The generating Λ-vector $\mathbf{x}$ for the G-action projects to a generating $(\tau\colon -)$-vector $\mathbf{y}$ for the action of G_S on S/G_B. The surface part N_B of $N(\bar{F})$ is the subgroup of Γ_τ that maps the $\mathrm{Aut}'(G_S)$-class of $\mathbf{y}$ to itself, where $\mathrm{Aut}'(G_S)$ is the group of automorphisms of G_S that lift to G.*

Proof. The statements follow easily from the previous discussion, Theorem 4.3 in **[Bi]** and the following observations. The pure braid group contains the centre of the full braid group **[Bi**, p. 156**]**. The full mapping class group $\Gamma(T^\circ)$ acts on the finite set consisting of the disjoint union of $\mathrm{GV}(G,\Lambda')$ where the Λ' are obtained by arbitrarily reordering the indices n_j of Λ and are not normally ordered. We may obviously restrict this action to the subgroup $\mathcal{B}_{\tau,t}$. For general G, the braid part N_B is the subgroup of $\mathcal{B}_{\tau,t}$ that maps the $\mathrm{Aut}(G)$-class of the generating vector $\mathbf{x}$ back to the same class. If G is abelian then the action of $\mathcal{B}_{\tau,t}$ factors through the action of the symmetric group Σ_t acting on the branch points of T. However, the pure braid group is just the kernel of this action so N_B contains the pure braid group (mod the centre if $\tau \leq 1$). □

Conclusion of proof of Proposition 2.1. We need to show that if two finite subgroups F_1, F_2 of diffeomorphisms of a surface S have conjugate images in Γ then F_1 and F_2 are conjugate in $\mathrm{Diff}^+(S)$. Let $h \in \mathrm{Diff}^+(S)$ satisfy $h\bar{F}_1h^{-1} = \bar{F}_2$. Then hF_1h^{-1} and F_2 have the same image in Γ, so we may as well assume from the outset that F_1 and F_2 have the same image in Γ. For each $i = 1, 2$ carry out the following construction. Put a conformal structure on S/F_i, pull it back to $\mathbb{H}$ to get a holomorphic covering projection $\pi_i \to S$. Let K_i be the group of covering transformations of π_i and let $\tilde{F}_i \supset K_i$ be the Fuchsian group which is the lift of F_i (Formula (2.3)). Pick a base point x_0 on S and base point in $\mathbb{H}$ compatible with the covering projection π_i. This selection of base points determines an isomorphism $K_i \to \pi_1(S, x_0)$. This isomorphism and the inner action of $\tilde{F}_i$ on K_i gives rise to a monomorphism $\psi_i : \tilde{F}_i \to \mathrm{Aut}^+(\pi_1(S, x_0))$. Since $\bar{F}_1 = \bar{F}_2$ then $\psi_1(\tilde{F}_1) = \psi_2(\tilde{F}_2)$ holds as well as $\psi_1(K_1) = \psi_2(K_2)$. Therefore, $\psi = \psi_2^{-1} \circ \psi_1$ maps $\tilde{F}_1$ to $\tilde{F}_2$ and K_1 to K_2. This isomorphism may be geometrically realized by a diffeomorphism $\tilde{h}$ of $\mathbb{H}$ **[McB]**. Now $\tilde{h}$ induces a map h of $S = \mathbb{H}/K_1$ onto $S = \mathbb{H}/K_2$ which clearly conjugates $F_1 \simeq \tilde{F}_1/K_1$ onto $F_2 \simeq \tilde{F}_2/K_2$. All is now proven. □

4. Elementary abelian subgroups

For this section we replace the G of our previous discussion by P, an elementary abelian group isomorphic to $(\mathbb{Z}_p)^r$. Let all other notation be as above. If P acts on S_σ then the branching data Λ must satisfy $n_1 = \ldots = n_t = p$. In **[Br1]** it was shown that S_σ admits a Λ-action of P under the conditions of the following proposition.

Proposition 4.1. *Let p be a prime, $P = (\mathbb{Z}_p)^r$ and $\Lambda = (\tau{:}\, p, \ldots, p)$, where there are t branch points. If Λ is not an excluded type as in Remark 2.3 then S_σ admits a Λ-action of P if and only if the following hold:*

$$2\sigma - 2 = p^{r-1}[(2\tau - 2 + t)p - t], \tag{4.1.i}$$

$$r \leq 2\tau, \quad if\ t = 0, \tag{4.1.ii}$$

$$r < 2\tau + t, \quad if\ t > 1. \tag{4.1.iii}$$

Furthermore,

$$p \leq 2\sigma + 1. \tag{4.2}$$

Equation (4.2) has been known for a long time. See **[H2]** for a recent proof as well as necessary and sufficient arithmetic conditions on the genus of a surface in order for the surface to admit a $\mathbb{Z}_p$-action.

Remark 4.1. In **[Br1]** a table of the maximal ranks of elementary abelian subgroups for $\sigma \leq 40$ and $p \leq 81$ (because of (4.2)) is given. From the table in **[Br1]** it is apparent that for low genus there are fairly few elementary abelian subgroups with rank greater than 1 except for the prime 2. Every surface admits a least a $\mathbb{Z}_2 \times \mathbb{Z}_2$-action. As usual, the prime 2 presents a more difficult, special case.

Next let us identify the action of $\mathrm{Out}^+(\Pi_\Lambda)$ on $\mathrm{GV}((P,\Lambda)$.

Proposition 4.2. *Let p, P and Λ be as in Proposition 4.1. Then, the $\mathrm{Out}^+(\Pi_\Lambda)$-action on a typical generating vector $(a_1,\ldots,a_\tau,b_1,\ldots,b_\tau,c_1,\ldots,c_t)$ includes the actions listed below: In this list of actions we only write down the action for elements of the generating vector that are actually changed.*

Type I.1	$a_i \mapsto a_i + nb_i\,, \quad n \in \mathbb{Z}\,,$
Type I.2	$a_i \mapsto b_i\,,\; b_i \mapsto -a_i\,,$
Type I.3	$a_i \mapsto a_i + nb_i\,,\; b_{i+1} \mapsto b_{i+1} - nb_i\,, \quad n \in \mathbb{Z}\,,$
Type I.4	$a_i \mapsto a_{i+1}\,,\; a_{i+1} \mapsto a_i\,,\; b_i \mapsto b_{i+1}\,,\; b_{i+1} \mapsto b_i\,,$
Type II	$c_j \mapsto c_{j+1}\,,\; c_{j+1} \mapsto c_j\,,$
Type III.a	$a_i \mapsto a_i - nc_j\,, \quad n \in \mathbb{Z},$
Type III.b	$b_i \mapsto b_i - nc_j\,, \quad n \in \mathbb{Z}\,.$

Proof. Types I and III are given in **[H2]**. Those of type II can be obtained by observing that $\gamma_j \mapsto \gamma_{j+1}$, $\gamma_{j+1} \mapsto \gamma_{j+1}^{-1}\gamma_j\gamma_{j+1}$ defines an automorphism in $\mathrm{Out}(\Pi_\Lambda)$, because of Remark 2.4. All of these automorphisms may be constructed by computing the automorphism induced by a Dehn twist along appropriately chosen curves and then applying commutativity relations. □

Remark 4.2. All of these transformations are valid for actions of abelian groups, except that for Type II we have taken products of transformations that yield permutation preserving branching orders. For elementary abelian subgroups this action simply amounts to arbitrary permutations of the c_j's.

Now let us give a standard (though not unique) form for generating Λ-vectors of P, that is suggested by the braid and surface decomposition of normalizers . A similar form for generating vectors of cyclic groups was given in **[H2]**.

Proposition 4.3. *Let p, P and Λ be as in Proposition 4.1. Then, we may transform a given generating vector $(a_1,\ldots,a_\tau,b_1,\ldots,b_\tau,c_1,\ldots,c_t)$, by using the $\mathrm{Out}(\Pi_\Lambda)$-action alone, so that it satisfies the properties below. There are integers d, e and f satisfying*

$$0 \leq e \leq d \leq \tau, \quad 1 \leq f < t \quad \text{and} \quad 2d + e + f = r,$$

such that:

i) $a_1, \ldots, a_d, b_1, \ldots, b_e, c_1, \ldots, c_f$ *form a basis for* P,
ii) *all other* a_i *'s and* b_i *'s are zero,*
iii) $c_{f+1}, \ldots, c_t$ *are linear combinations of* $c_1, \ldots, c_f$,
iv) d, e *and* f *are invariants of the* P*-action.*

Proof. To prove i) – iii) we just apply the "symplectic" analogue of Gaussian elimination. Let f be the rank of the braid part C of P. Using a combination of transformations of Type II we may assume that $\{c_1, \ldots, c_f\}$ form a basis for C. Next, consider the pairs (a_i, b_i). If these two elements are not independent mod C then by a combination of transformations of Types I.1 and I.2 we may assume that $b_i \in C$. Next by a combination of transformations of Type III.b we may assume $b_i = 0$. If $a_i \in C$ also then we may also assume $a_i = 0$. Now suppose that b_i is a linear combination (mod C) of a_j 's and b_j 's for $j < i$. By using a combination of transformations of Type I.4, I.3, I.1 and then I.4 again we may subtract a multiple of b_j from b_i whilst fixing all of the a_j 's and b_j 's for $j < i$. By using these transformations in conjunction with transformations of Type I.2 we may also subtract multiples of the a_j 's from b_i, still fixing the a_j 's and b_j 's for $j < i$. Thus we may transform b_i into an element of C and then, using Type III transformations as above transform b_i into 0. Combining all of this with transformations of Type I.2 we may also expunge a_i if it is also a linear combination of a_j 's and b_j 's for $j < i$. We may now proceed to put the generating vector in normal form. Select, if possible, a pair (a_i, b_i) which is linearly independent mod C, and switch it with the pair (a_1, b_1) using transformations of type I.3. As above, we may zero out any a_i 's or b_i 's, for $i > 1$ which are linearly dependent (mod C) on a_1 and b_1. Now select another linearly independent pair, move it into the (a_2, b_2) position and zero out as above. Continue this process until no more new linearly independent pairs can be found. We may assume that among the remaining pairs the b_i 's are all zero. Pick one of the remaining non-zero a_i 's and move the pair $(a_i, 0)$ to the location immediately following the last linearly independent pair. Zero out the remaining a_i 's where possible. continue this operation until no non-zero a_i 's remain. When the above process is complete, the collection of all non-zero a_i 's and b_i 's that remain will form a linearly independent set (mod C).

In proving iv) note that we have already proven that f is invariant under $\mathrm{Out}^+(\Pi_\Lambda)$. Now we only need to show that d is invariant since $e = r - 2d - f$. The action is defined by a homomorphism $H_1(S_\tau; \mathbb{Z}_p) \to P/C$ and therefore determines a cohomology class $\xi \in H^1(S_\tau; P/C)$. If X represents the fundamental class in of $H_2(S_\tau; \mathbb{Z}_p)$ then $\xi \cup \xi \cap X$ is an element of $(P/C) \otimes (P/C)$ invariant under all homeomorphisms of S_τ. We may view $\xi \cup \xi \cap X$ as an element of $\mathrm{Hom}((P/C)^*, (P/C))$ where the * denotes the dual vector space. It is easy to calculate that the rank of this transformation is $2d$. □

Remark 4.3. From the proposition it follows that the real difficulty in enumerating conjugacy classes of elementary abelian subgroups is the case $\tau = 0$, the other actions can be easily derived from these. The rank 1 case has been completely enumerated by means of generating functions in **[L]** for $\tau = 0$ actions and in **[H2]** in the general case.

We finish with two examples.

Example 4.1. Elementary abelian subgroups with finite normalizers. It is well known that Γ_Λ is finite if and only if the corresponding Teichmüller space is zero dimensional. This dimension is $6\tau - 6 + 2t$ at least in the cases where $\sigma \geq 2$. Hence, the only possibilities are the so-called triangle groups $\lambda = (0{:}\, p, p, p), p > 3$, taking the excluded cases in Remark 2.3 into account. By Proposition 4.1 P has rank 1 or 2.

Most of the rank 1 case is worked out in **[Br2]**. The genus is given by $\sigma = (p-1)/2$ according to Formula (4.1.i). Every generating vector is determined by a triple of integers (k_1, k_2, k_3) with $1 \leq k_i \leq p-1$ and $k_3 \equiv -(k_1 + k_2) \bmod p$. There are $(p-1)(p-2)$ such triples. By Remarks 2.1 and 4.2, the $\mathrm{Aut}(P) \times \mathrm{Out}^+(\Pi_\Lambda)$ action will be effected by the action of $\mathbb{Z}_p^* \times \Sigma_3$, the first factor acting by multiplication and the second by permutations. The only automorphisms in $\mathbb{Z}_p^* \times \Sigma_3$ which have fixed points on $\mathrm{GV}(P, (0{:}\, p, p, p))$ are conjugate to $(c_1, c_2, c_3) \to (c_2, c_1, c_3)$ which fixes $(c_1, c_1, (p-2)c_1)$ and $(c_1, c_2, c_3) \to (ac_3, ac_1, ac_2)$, where $p \equiv 1 \bmod 3$, $a^3 \equiv 1 \bmod p$, and which fixes (c_1, ac_1, a^2c_1). The conjugacy classes of elementary abelian subgroups and their normalizers are:

$p \equiv 1 \bmod 3$	$p \not\equiv 1 \bmod 3$	Normalizer
$(1,1,p-2)$-class	$(1,1,p-2)$-class	$\mathbb{Z}_2 \times \mathbb{Z}_p$
$(1,a,a^2)$-class, $a^3 \equiv 1 \bmod p$, $a \leq p/2$		$\mathbb{Z}_3 \ltimes \mathbb{Z}_p$
$(p-7)/6$ other classes	$(p-5)/6$ other classes	$\mathbb{Z}_p$

In the rank 2 case σ will be $(p-1)(p-2)/2$. If (c_1, c_2, c_3) is any generating $(0{:}\, p, p, p)$-vector of $(\mathbb{Z}_p)^2$ then c_1 and c_2 must be linearly independent and $c_3 = -(c_1 + c_2)$. All triples of vectors are $\mathrm{Aut}(P)$-equivalent and so there is a single class of generating vectors. It now follows from Theorem 3.2 that the normalizer is isomorphic to $\Sigma_3 \ltimes (\mathbb{Z}_p)^2$ with the Σ_3 action given, in matrix terms, by

$$(1,2)(3) \to \begin{pmatrix} 0 & 1 \\ 1 & 0 \end{pmatrix}, \quad (1,2,3) \to \begin{pmatrix} 0 & -1 \\ 1 & -1 \end{pmatrix}.$$

Example 4.2. Rank 1 elementary abelian subgroups (cf. **[L]**). The action of a subgroup U of $\mathrm{Aut}(P)$ on P is generated by multiplication by an r-th root of unity, u, where $r = |U|$. Let u be a root of unity inducing the action of $N(P)/Z(P)$ on P. The action on generating vectors factors through $\mathrm{Aut}(P) \times \Sigma_t$, therefore, it follows that there are integers $k_1 \leq k_2 \leq \ldots \leq k_s$, such that $t = sr$ and that the given generating vector is equivalent to the vector

$$(k_1, uk_1, \ldots, u^{r-1}k_1, k_2, uk_2, \ldots, u^{r-1}k_2, \ldots, k_s, uk_s, \ldots, u^{r-1}k_s).$$

The k_j's are not necessarily unique, but we can make them unique by redefining k_j to be the smallest number in the range $1, \ldots, p-1$ equivalent mod p to one of the numbers $k_j, uk_j, \ldots, u^{r-1}k_j$. Assume that this has been done. Define $n_{i,j} = u^i k_j$ for $1 \leq i \leq r$, $1 \leq j \leq s$, and for $1 \leq \ell \leq p-1$ let e_ℓ be the number of $n_{i,j}$ that equal ℓ . Next, consider Σ_t as the symmetric group of the rectangular array of ordered pairs $\{(i,j) : 1 \leq i \leq r,\ 1 \leq j \leq s\}$. Let E be the subgroup of Σ_t isomorphic to $\mathbb{Z}_r \ltimes (\Sigma_{e_1} \times \cdots \times \Sigma_{e_{p-1}})$. The first factor is the cyclic group generated by the permutation $(i,j) \to (i+1, j) \bmod r$, this induces the permutation $u^i k_j \to u^{i+1} k_j$ on the $n_{i,j}$. The second factor is the stabilizer in Σ_t of the collection $\{n_{i,j} : 1 \leq i \leq r,\ 1 \leq j \leq s\}$, Σ_t acting on the indices of $n_{i,j}$. (If L denotes the stabilizer in Σ_s of $(k_1, \ldots, k_s)$ then

E is just the wreath product of L by $\mathbb{Z}_r$, acting as above.) The braid part of the $N(P)$ will be the preimage of E under the natural map $\mathcal{B}_{\tau,t} \to \Sigma_t$. The surface part of $N(P)$ will be Γ_τ. Both of these statements follow from Corollary 3.3 and Proposition 4.3.

References

[Bi] J. Birman, Braids, Links and Mapping Class Groups, Annals of Math. Studies, No. 82, Princeton U. Press (1974).

[Br1] S.A. Broughton, The equisymmetric stratification of the moduli space and the Krull dimension of mapping class groups, to appear in Topology and its Applications.

[Br2] S. A. Broughton, Classifying finite group actions on surfaces of small genus, to appear in Journal of Pure and Applied Algebra.

[Brw] K. Brown, Cohomology of Groups, Graduate Texts in Math. No. 87, Springer-Verlag, Berlin, New York (1982).

[FK] H. Farkas & I. Kra, Riemann Surfaces, Graduate Texts in Math., No. 71, Springer-Verlag, Berlin, New York (1980).

[H1] J. Harvey, Cyclic groups of automorphisms of compact Riemann surfaces, Quarterly J. of Math. (Oxford Ser. 2), 17 (1966), 86–97.

[H2] J. Harvey, On Branch Loci in Teichmüller space, Transactions Amer. Math. Soc., 153 (1971), 387–399.

[K] S. Kerckhoff, The Nielsen realization problem, Annals of Math., 117 (1983), 235–265.

[L] E.K. Lloyd, Riemann surface transformation groups, Journal of Combinatorial Theory, Ser A, Vol 13, (1972) 17–27.

[Ma] C. Maclachlan, Abelian groups of automorphisms of compact Riemann surfaces, Proc. London Math. Soc., Ser. 3, 15 (1965), 699–712.

[Mc] A.M. McBeath, The classification of non-euclidean crystallographic groups, Can. J. Math., 19 (1966), 1192–1205.

[ZVC] H. Zieschang, E. Vogt, H.D. Coldeway, Surfaces and Planar Discontinuous Groups, Lecture Notes in Math., No. 835, Springer–Verlag, Berlin, New York (1980).

Department of Mathematics, Cleveland State University, Cleveland, OH 44115
Email: r1076@csuohio.bitnet

Invariants of 3-Manifolds from Conformal Field Theory

Sylvain E. Cappell *,**, *Ronnie Lee** *and Edward Y. Miller**

0. Introduction

In 1989, Reshetikhin and Turaev **[RT1]**, **[RT2]** constructed a family of invariants of 3-manifolds based upon the theory of quantum groups of $U_q(sl(n))$. Subsequently, Kirby-Melvin **[KM]**, Ko-Smolinsky **[KS]**, and Lickorish **[L]**, provided different verifications of these invariants in certain special cases of $U_q(sl(2))$. In retrospect, the proofs discussed by these authors are all based upon a combinatorial approach to 3-manifold theory, known as the Kirby calculus. The object of this paper is to present these invariants from a different perspective. We follow the Witten framework of SU(2)-level k conformal field theory as developed by Atiyah-Hitchin-Lawrence-Segal in **[AHLS]** and Tsuchiya-Ueno-Yamada in **[TUY]**. Some modifications are made so that the vacuum state bundle can be extended to a compactification of a moduli space of Riemann surfaces. The invariants of 3-manifolds are then obtained from taking the inner product of two vacuum vectors associated to handle bodies in a Heegaard decomposition. In order to make this inner product independent of all the choices, a phase factor correction term has to be introduced and this is achieved by using the Shale-Weil cochain in the theory of metaplectic groups **[LV]**. This cochain has also arisen in Atiyah's study of framings of 3-manifolds **[A]**.

Starting from the axioms of Moore-Seiberg for rational conformal field theory, L. Crane had shown in **[C]** how invariants of a similar nature could be developed (see Theorem 3.3). A refinement in our approach is that we have a surgery formula, similar to the theory of Kirby-Melvin in **[KM]**, and this is obtained by comparing the above Shale-Weil cochain with the signature invariant of Sharpe in **[S1]**, **[S2]**, **[S3]**. Another approach to resolving these problems has also been announced by K. Walker.

From the present viewpoint of conformal field theory, it is unnecessary to limit the scope to SU(2)-level k theory. In a future paper, we plan to study the situation of other compact Lie groups as well as the generalization to equivariant settings.

It is a pleasure to acknowledge our debts to Crane, Greenspoon, Frankel, Lepowsky, Moore, Morava, Smolinsky, Zuckerman, and especially to Kirby who showed us a preliminary version of his manuscript with Melvin. Conversations with all of them have helped us in understanding this subject.

* Work partially supported by NSF

** John Simon Guggenheim Foundation Fellow

1. Conformal field theories

Roughly speaking, a 2-dimensional conformal field theory means an assignment to a Riemann surface Σ possibly with singularities of a finite dimensional vector space $Z(\Sigma)$, called the ***state space***. To a one-parameter family of deformations of Riemann surfaces $\delta = \{\Sigma_t;\, 0 \le t \le 1\}$ we have a corresponding continuous family of isomorphisms $Z(\delta)_t : Z(\Sigma_0) \to Z(\Sigma_t)$, $0 \le t \le 1$, called ***propagators***. Analogous to the axioms of topological quantum field theories **[AHLS**, pp. 3–16**]** these state spaces and propagators satisfy certain functorial properties which will be made precise in Theorem 1.1.

To experts there are many examples of conformal field theory: in general corresponding to a compact Lie group G and an appropriate class in $H^3(G;\mathbb{Z})$ there exists at least one such theory **[MS]**, **[P]**. In this paper we will concentrate on the case when $G = \mathrm{SU}(2)$ and the class is any positive multiple $k[G]$ of the generator$[G]$. This positive integer k is often referred to as the ***level.***

Given a Riemann surface Σ we denote by $R(\Sigma)$ the space

$$\mathrm{Hom}(\pi_1(\Sigma), \mathrm{SU}(2))/\,\mathrm{SU}(2)$$

of representations $\rho : \pi_1(\Sigma) \to \mathrm{SU}(2)$ modulo the equivalence relation defined by conjugation by elements of $\mathrm{SU}(2)$, i.e.

$$\rho \sim \rho' \ \text{ iff } \ \rho(x) = g \cdot \rho'(x) \cdot g^{-1} \ \text{ for some } \ g \in \mathrm{SU}(2)\,.$$

From the viewpoint of algebraic geometry, $R(\Sigma)$ admits an interpretation as the moduli space of semistable, rank 2, degree 0, holomorphic vector bundle $E(\rho)$ over Σ (see **[NS]**). With this interpretation, $R(\Sigma)$ becomes a projective variety together with a positive line bundle $\mathrm{Det}(\overline{\partial})$, given by the determinant line bundle of the Cauchy-Riemann operator $\overline{\partial} : \Omega^0(\Sigma;\, E(\rho)) \to \Omega^1(\Sigma;\, E(\rho))$. (For the definition of determinant line bundle and its properties see **[BF]**). Following Witten's suggestion, the ***state space*** $Z(\Sigma)$ in our theory is the space $\Gamma(R(\Sigma), \mathrm{Det}(\overline{\partial})^{\otimes k})$ of holomorphic sections of the kth-power line bundle $\mathrm{Det}(\overline{\partial})^{\otimes k}$ on $R(\Sigma)$.

Note that $R(\Sigma)$, as a topological space, depends only on the fundamental group $\pi_1(\Sigma)$, but $Z(\Sigma)$ depends heavily on the complex structure on Σ. As we vary the complex structure we vary the state space, and to clarify the situation we have:

Theorem 1.1. *Let $\mathcal{M}_{g,1}$ denote the moduli space of genus g Riemann surfaces with a base tangent vector at a base point. Then on $\mathcal{M}_{g,1}$ there exists a holomorphic vector bundle $\mathcal{Z} \to \mathcal{M}_{g,1}$ whose fiber $\mathcal{Z} \mid \Sigma$ over a point Σ can be canonically identified with $Z(\Sigma)$. Furthermore this bundle $\mathcal{Z} \to \mathcal{M}$ has a projective flat connection $\nabla(\mathcal{Z})$.*

This theorem is also known as the ***global existence of conformal field theory***[1)]; there are at least six different proofs by various authors: Axelrod-Pietra-Witten **[APW]**, Beilinson-Kazhdan **[BK]**, Beilinson-Schechtman **[BS]**, Hitchin **[H]**, Segal **[Se]**, Tsuchyia-Ueno-Yamada **[TUY]**. Given the abundance of literature on this topic, it is perhaps enough just to explain the term 'projective flat connection'.

1) In contrast, local existence can be obtained by just sewing Riemann surfaces and state spaces together.

Generally, a ***flat connection*** is a connection on a vector bundle which gives us a reduction of the structure group to a discrete group. A ***projective flat connection*** $\nabla(\mathcal{Z})$ is one defined on the projective bundle $P(\mathcal{Z})$ such that the structure group of $P(\mathcal{Z})$ is reduced to a projective representation

$$\Gamma_{g,1} = \pi_1(\mathcal{M}_{g,1}) \to \mathrm{PGL}(d), \quad d = \dim Z(\Sigma)$$

of the fundamental group $\pi_1(\mathcal{M}_{g,1})$ (the mapping class group $\Gamma_{g,1}$).

In fact, in the aforementioned proofs of Theorem 1.1, more than just the reduction of structure group can be concluded. In **[H**; 3.2.1 and 4.12] and **[TUY**; 5.33], the obstruction for lifting the projective connection $\nabla(\mathcal{Z})$ to one over $\mathcal{Z}$ is identified with a cohomology class in $H^1(\mathcal{M}_{g,1}, T\mathcal{M}_{g,1})$. We can kill this obstruction by considering a $\mathbb{C}^*$-bundle over $\mathcal{M}_{g,1}$. Let $\mathcal{L} \to \mathcal{M}_{g,1}$ denote the determinant line bundle associated to the Cauchy-Riemann operators $\bar{\partial} : \Omega^0(\Sigma) \to \Omega^1(\Sigma)$ and let $\pi : \mathring{\mathcal{L}} \to \mathcal{M}_{g,1}$ denote the $\mathbb{C}^*$-bundle on $\mathcal{M}_{g,1}$ obtained by removing the zero section from $\mathcal{L}$. Then, pulling back by π, there is a vector bundle $\pi^*\mathcal{Z}$ defined on $\mathring{\mathcal{L}}$. Since $H^2(\mathring{\mathcal{L}}; \mathbb{C}) = 0$, the obstruction for the projective connection $\nabla(\mathcal{Z})$ dies in $\pi^*\mathcal{Z}$ and so there exists a well-defined flat connection $\nabla(\pi^*\mathcal{Z})$ on $\pi^*\mathcal{Z}$. In **[BS**; p. 680], the following description of the monodromy representation

$$\tilde{\rho} : \widetilde{\Gamma}_{g,1} = \pi_1(\mathring{\mathcal{L}}) \to \mathrm{GL}(d), \quad d = \dim Z(\Sigma)$$

can be found. The fundamental group $\widetilde{\Gamma}_{g,1}$ is a central extension of $\Gamma_{g,1}$ by an infinite cyclic group $\mathbb{Z}(i)$,

$$1 \to \mathbb{Z}(i) \to \widetilde{\Gamma}_{g,1} \to \Gamma_{g,1} \to 1.$$

The restriction of $\tilde{\rho}$ to a generator i of this center acts on $Z(\Sigma)$ by multiplication by $\exp[2\pi\sqrt{-1} \cdot c]$ where the term $c = 3k/(k+2)$ is known as the ***central charge***. (In the general setting of compact Lie group G, the central charge $c = (\dim G)\, k/(k+h)$, and $h =$ dual Coxeter number).

Now consider a one-parameter deformation $\delta_t = \{\Sigma_t;\, 0 \le t \le 1\}$ of nonsingular Riemann surfaces. On the moduli space $\mathcal{M}_{g,1}$ it represents a parametrized path δ_t connecting the two Riemann surfaces Σ_0, Σ_1. Note that an element $\widetilde{\Sigma}$ in $\mathring{\mathcal{L}}$ can be interpreted as a Riemann surface Σ together with a nonzero class $[\widetilde{\Sigma}]$ in $\Lambda^g H^{1,0}(\Sigma)$. By choosing two such orientation classes $[\widetilde{\Sigma}_0]$, $[\widetilde{\Sigma}_1]$ in $\Lambda^g H^{1,0}(\Sigma_0)$, $\Lambda^g H^{1,0}(\Sigma_1)$, we obtain liftings of Σ_0, Σ_1 to $\widetilde{\Sigma}_0, \widetilde{\Sigma}_1$. Since the induced homomorphism $\pi_* : \widetilde{\Gamma}_{g,1} \to \Gamma_{g,1}$ between fundamental groups is onto, we can lift δ_t to a path $\gamma_t = \{\widetilde{\Sigma}_t;\, 0 \le t \le 1\}$ in $\mathring{\mathcal{L}}$ connecting $\widetilde{\Sigma}_0$ and $\widetilde{\Sigma}_1$. Using the above connection $\nabla(\pi^*\mathcal{Z})$, we can parallel translate one state space $Z(\widetilde{\Sigma}_0)$ to another $Z(\widetilde{\Sigma}_1)$ along the path γ_t, and hence obtain a continuous family of propagators $Z(\gamma)_t : Z(\widetilde{\Sigma}_0) \to Z(\widetilde{\Sigma}_t)$ between state spaces. Note that if the endpoints $\widetilde{\Sigma}_0$ and $\widetilde{\Sigma}_1$ are fixed,then because of the flatness of the connection $\nabla(\pi^*\mathcal{Z})$ the propagator $Z(\gamma_1) : Z(\widetilde{\Sigma}_0) \to Z(\widetilde{\Sigma}_1)$ depends only on the choice of the path class γ_t. A different lifting γ_t' of δ_t with the same endpoints results in a propagator $Z(\gamma_1')$ which can differ from $Z(\gamma_1)$ by some power of $\exp[2\pi\sqrt{-1}\cdot c]$. In addition, the projective connection $\nabla(\mathcal{Z})$ depends on certain choices of differential geometric data whereas $\nabla(\pi^*\mathcal{Z})$ is free of these choices and is completely determined by the monodromy representation ρ. Thus in the literature (e.g. **[MS]**) the monodromy representation really refers to $\nabla(\pi^*\mathcal{Z})$. In addition, this monodromy representation ρ can be obtained from solving the associated differential equations over various submoduli spaces in $\mathcal{M}_{g,1}$; because of Theorem 1.1 these subrepresentations can be pieced together to form ρ.

2. Factorization

The discussion in the previous paragraphs has to be extended to cover deformations of singular Riemann surfaces. By a ***stable Riemann surface***, we mean a Riemann surface Σ_0 with singularities modeled after the intersection, $z_1 z_2 = 0$, of two coordinate axes in the plane. Topologically we can view such a surface as obtained by shrinking a simple closed curve α_t (away from the basepoint) on a nonsingular surface Σ_t to a point.

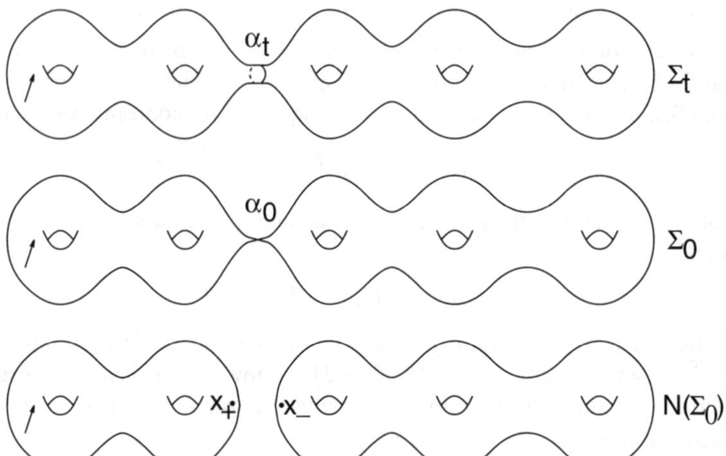

Figure 2.1

We can desingularize a stable Riemann surface Σ_0 by locally pulling apart the two sheets $z_1 = 0$ and $z_2 = 0$ near the singularity. The result is a nonsingular surface $N(\Sigma_0)$ with pairs of marked points x_+, x_-, called the ***normalization*** of Σ_0. To avoid trivial collapsing or duplication of the same operation, we impose on our stable Riemann surface Σ_0 the condition that the normalization $N(\Sigma_0)$ does not contain a component which is a 2-sphere with less than three marked points.

Theorem 2.2. *Let $\overline{\mathcal{M}}_{g,1}$ denote the moduli space of stable Riemann surfaces of genus g with a base tangent vector at a base point. Then there exists an extension of $\mathcal{Z} \to \mathcal{M}_{g,1}$ to a holomorphic vector bundle $\overline{\mathcal{Z}} \to \overline{\mathcal{M}}_{g,1}$, and an extension of the connection $\nabla(\mathcal{Z})$ to $\nabla(\overline{\mathcal{Z}})$ with logarithmic singularity along the subvariety $\overline{\mathcal{M}}_{g,1} - \mathcal{M}_{g,1}$.*

The proof of Theorem 2.2 can be found in **[TUY**; Thm. 5.33, 6.2.6], and a version of this is known as the ***Factorization Theorem***. As we move from a nonsingular Riemann surface Σ_t to a stable one, such as Σ_0 in Figure 2.1, the fiber $\overline{\mathcal{Z}} \mid \Sigma_0 = Z(\Sigma_0)$ admits the following decomposition ([**AHLS**, p. 71], [**TUY**, Thm. 6.2.6]):[2)]

$$Z(\Sigma) = \bigoplus_{1\le\lambda\le k} Z(\Sigma_+;\ {}^{\lambda}_{x_+}) \otimes Z(\Sigma_-;\ {}^{\lambda}_{x_-}).$$

Here $(\Sigma_\pm, x_\pm)$ stands for the two components of $N(\Sigma_0)$ together with their marked point $x_\pm$. Given a Riemann surface $(\Sigma;\ {}^{\lambda_1,\ldots,\lambda_n}_{x_1,\ldots,x_n})$ with n marked points $x_1, \ldots, x_n$, and n weights $\lambda_1, \ldots, \lambda_n$ on these marked points, the symbol $R(\Sigma;\ {}^{\lambda_1,\ldots,\lambda_n}_{x_1,\ldots,x_n})$ stands for the space of SU(2)-representations of $\pi_1(\Sigma - \{x_1, \ldots, x_n\})$ whose monodromy around x_i is given by the matrix

$$\begin{bmatrix} \exp(\frac{2\pi\sqrt{-1}}{k}(\lambda_i - 1)) & 0 \\ 0 & \exp(-\frac{2\pi\sqrt{-1}}{k}(\lambda_i - 1)) \end{bmatrix}.$$

On $R(\Sigma;\ {}^{\lambda_1,\ldots,\lambda_n}_{x_1,\ldots,x_n})$, there exists the determinant line bundle $\mathrm{Det}(\overline{\partial})$, and the space $\Gamma(R(\Sigma;\ {}^{\lambda_1,\ldots,\lambda_n}_{x_1,\ldots,x_n});\ \mathrm{Det}(\overline{\partial}^{\otimes k}))$ of sections of the kth-power line bundle $\mathrm{Det}(\overline{\partial})^{\otimes k}$ is denoted by $Z(\Sigma;\ {}^{\lambda_1,\ldots,\lambda_n}_{x_1,\ldots,x_n})$.

Now given a nonsingular Riemann surface Σ of genus g, by pinching $3g-3$ disjoint simple closed curves, we can reduce Σ to a stable Riemann surface Σ_0 whose normalization $N(\Sigma_0)$ consists of a union of disjoint 2-spheres $(S^2; x_1, x_2, x_3)$ with 3 marked points. Thus knowing the answer for $Z(S;\ {}^{\lambda_1,\ldots,\lambda_n}_{x_1,\ldots,x_n})$ [**AHLS**, p. 112] and applying repeatedly a decomposition formula slightly more general than the above

$$Z(\Sigma;\ {}^{\lambda_1,\ldots,\lambda_n}_{x_1,\ldots,x_n}) = \bigoplus_{1\le\lambda\le k} Z(\Sigma_+;\ {}^{\lambda_1,\ldots,\lambda_n,\lambda}_{x_1,\ldots,x_n,x_+}) \otimes Z(\Sigma_-;\ {}^{\lambda_1,\ldots,\lambda_n,\lambda}_{x_1,\ldots,x_n,x_-}),$$

we obtain a formula for $Z(\Sigma)$. See Figure 2.3.

Theorem 2.4. *Let V_i denote the irreducible* SU(2)*-representation with dimension i, and let V^i_{jk} denote the space of intertwiners* $\mathrm{Hom}_{\mathrm{SU}(2)}(V_j \otimes V_k, V_i)$*. Then there exists an isomorphism*

$$Z(\Sigma) \simeq \oplus V^1_{i_1 i_1} \otimes V^{i_1}_{i_1 j_1} \otimes V^{j_1}_{i_2 k_2} \otimes V^{k_2}_{i_2 j_2} \otimes \cdots \otimes V^{i_g}_{i_g 1},$$

where the indices $i_1, \ldots, i_g, i_1, \ldots, i_{g-1}, k_2, \ldots, k_{g-1}$ run through all integers between 1 and k [**MS**; p. 231].

3. Definition of invariants

To define our invariants, we need yet another moduli space. This is because the projective flat connection $\nabla(\mathcal{Z})$ blows up along $\overline{\mathcal{M}}_{g,1} - \mathcal{M}_{g,1}$. Recall, in Figure 2.1, the procedure of shrinking geodesics α_t to obtain a stable Riemann surface. Locally near a singular point, the deformation takes the form of a family of hyperboloids $z_1 z_2 = t$ parametrized

2) There are several different approaches to the state space $Z(\Sigma)$ in the literature, for example [**AGS2**], [**AHLS**], [**FK**], [**MS**], [**Se**]. Segal's approach in [**AHLS**, p. 71] depends on an interpretation of infinite dimensional Lie group in [**PS**, Prop. 8.11.5]; however this difficulty is avoided in [**TUY**, Thm. 6.2.6] where the state space $Z(\Sigma;\ {}^{\lambda_1,\ldots,\lambda_n}_{x_1,\ldots,x_n})$ is defined more directly.

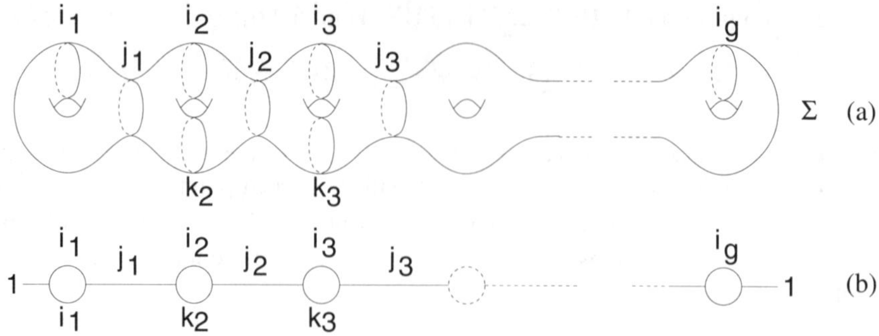

Figure 2.3 a,b

Here (a) represents a Riemann surface with $3g-3$ marked curves with integer weights $i_1, \ldots, i_g, j_1, \ldots, j_{g-1}, k_2, \ldots, k_{g-1}$.

The graph (b) is associated with the marked curves on the surface in an obvious manner, and each trivalent vertex in the graph represents a factor in the factorization in Theorem 2.4.

by t. The absolute value $|t|$ controls the length of the minimal geodesic α_t separating the hyperboloid $z_1 z_2 = t$, and $\arg t$ is an angle parameter controlling the rotation of the two sides of $\{z_1 z_2 = t\} - \alpha_t$. Hence, fixing $\arg t$ and letting $|t| \to 0$, produces in the limit a stable Riemann surface $z_1 z_2 = 0$ together with a prescribed isometry between two cone neighborhoods $z_1 = 0$ and $z_2 = 0$. Regarding this last information as part of the data in defining ***stable Riemann surface with local isometries at singularities***, we obtain a new moduli space $\widehat{\mathcal{M}}_{g,1}$. The following is a variation of the theme of Theorem 2.2.

Theorem 3.1. *The line bundle $\mathcal{L}$ and the vector bundle $\mathcal{Z}$ can be extended to bundles $\widehat{\mathcal{L}}$ and $\widehat{\mathcal{Z}}$ on $\widehat{\mathcal{M}}_{g,1}$. Similarly, the flat connecton $\nabla(\pi^*\mathcal{Z})$ can be extended to a flat connection $\nabla(\pi^*\widehat{\mathcal{Z}})$ on $\pi^*\widehat{\mathcal{Z}}$.*

Proof. From the interpretation of $\mathcal{M}_{g,1}$, $\overline{\mathcal{M}}_{g,1}$, $\widehat{\mathcal{M}}_{g,1}$ as moduli spaces, we have a commutative diagram

$$\begin{array}{ccc} \mathcal{M}_{g,1} & \hookrightarrow & \widehat{\mathcal{M}}_{g,1} \\ & \searrow & \downarrow \text{ forgetful map} \\ & & \overline{\mathcal{M}}_{g,1} \end{array}$$

Pulling back the vector bundles $\overline{\mathcal{Z}}$ and $\overline{\mathcal{L}}$ via the forgetful map, we obtain the definition of the bundles $\widehat{\mathcal{Z}}$ and $\widehat{\mathcal{L}}$ on $\widehat{\mathcal{M}}_{g,1}$.

The moduli space $\widehat{\mathcal{M}}_{g,1}$ can be regarded as the ***Borel-Serre enlargement***[3)] of $\mathcal{M}_{g,1}$. To construct $\widehat{\mathcal{M}}_{g,1}$, we can start with a normal slice N to a divisor D in $\overline{\mathcal{M}}_{g,1} - \mathcal{M}_{g,1}$. Then the deformation parameter t, discussed in the beginning of this section, represents

3) Strictly speaking, the Borel-Serre enlargement is defined only for locally symmetric spaces. The space $\widehat{\mathcal{M}}_{g,1}$ is an obvious generalization.

a local coordinate of N and $t = 0$ gives the equation of D. Consider the coordinate transformation

$$z \mapsto t = \exp(2\pi\sqrt{-1}z)$$

from the upper half space $\{z \mid \mathrm{Im}z > 0\}$ to the punctured disk $\{t \mid |t| < 1, t \neq 0\}$. Note that the limits of the points $z = x + \sqrt{-1}\,y$, as $y \to \infty$, in the upper half plane are sent to the origin $t = 0$. In the Borel-Serre enlargment we fill up these limits $x + \sqrt{-1}\infty$ by a line at infinity modulo the equivalence relation $x \sim x' \Leftrightarrow x \equiv x' \bmod \mathbb{Z}$ to get a circular boundary for $N - 0$. This is precisely the construction of $\widehat{\mathcal{M}}_{g,1}$: each vertical line $\{z = x_0 + \sqrt{-1}\,y \mid x_0 \text{ fixed}\}$ controls a ray in the t-coordinate with fixed $\arg t$ parameter.

Note that $\widehat{\mathcal{M}}_{g,1}$, as a manifold with corners, has the same homotopy type as $\mathcal{M}_{g,1}$. Hence the $\mathbb{C}^*$-bundle $(\widehat{\mathfrak{L}})^\circ$ has the same fundamental group as $\mathfrak{L}^\circ$.

From this the monodromy representation $\rho : \pi_1(\mathfrak{L}^\circ) \to \mathrm{GL}(d)$ can be extended to $\pi_1((\widehat{\mathfrak{L}})^\circ)$. Since the flat connection $\nabla(\pi^*\mathfrak{Z})$ is completely determined by the monodromy, it can be extended to a flat connection $\nabla(\pi^*\hat{\mathfrak{Z}})$ on $\pi^*\widehat{\mathfrak{Z}}$. This proves Theorem 3.1. □

Let H be a handle body $\#(D^2 \times S^1)$ with boundary isomorphic to Σ, $\partial H \simeq \Sigma$. By inserting $3g - 3$ disjoint disks $D_a^2 \subset H$, which cut the handle body into 3-balls, we obtain a deformation $\delta_t = \{\Sigma_t;\ 0 \leq t \leq 1\}$ of $\Sigma = \Sigma_1$ by shrinking the curves ∂D_a^2 to curves with smaller radii in D_a^2. At the end of this deformation Σ_0, we obtain a union of 2-spheres with 3 marked points, and filling in these 2-spheres by 3 balls we recover H. Note that, at each stage of the above deformation, Σ_t is a stable Riemann surface with local isometries at the singularities. In other words, we have a smooth path δ_t in $\widehat{\mathcal{M}}_{g,1}$ starting from Σ and terminating at Σ_0. Choosing orientations on $\Lambda^g H^{1,0}(\Sigma)$ and $\Lambda^g H^{1,0}(\Sigma_0)$, we obtain lifting of Σ, Σ_0 to points $\widetilde{\Sigma}$, $\widetilde{\Sigma}_0$, and in the same manner lifting the path δ_t to γ_t in $(\widehat{\mathfrak{L}})^\circ$. By parallel translation along γ_t via the connection $\nabla(\pi^*\widehat{\mathfrak{Z}})$, we obtain an isomorphism

$$Z(\gamma_1) : Z(\widetilde{\Sigma}) \simeq \oplus V^1_{i_1 i_1} \otimes V^{i_1}_{i_1 j_1} \otimes V^{j_1}_{i_2 k_2} \cdots \otimes V^{i_g}_{i_g 1} \tag{3.2}$$

(cf. Theorem 2.4).

From Theorem 2.4, the factor V^1_{11} is canonically isomorphic to $\mathrm{Hom}(\mathbb{C} \otimes \mathbb{C};\ \mathbb{C}) \simeq \mathbb{C}$ and hence a preferred generator v_1 in V^1_{11}. By setting the indices $i_1, \ldots, i_g, j_1, \ldots, j_{g-1}$, $k_2, \ldots, k_{g-1}$ all to one in (3.2), we obtain an element $v_1 \otimes v_1 \otimes \cdots \otimes v_1$ in the summand $V^1_{11} \otimes \cdots \otimes V^1_{11}$, and then pulling back by means of the isomorphism $Z(\gamma_1)$ we obtain an element v_H in $Z(\widetilde{\Sigma})$. This element v_H is referred to as the ***vacuum vector associated to the handle body*** H, $\partial H \simeq \Sigma$.

The dual vector bundle $\widehat{\mathfrak{Z}}^*$ of $\widehat{\mathfrak{Z}}$ and its pullback $\pi^*(\widehat{\mathfrak{Z}}^*)$ on $\widehat{\mathfrak{L}}^\circ$ have similar structures: projective flat connection $\nabla(\widehat{\mathfrak{Z}}^*)$, flat connection $\nabla(\pi^*\widehat{\mathfrak{Z}}^*)$, Factorization Theorem, and base vector $v_1 \otimes v_1 \otimes \cdots \otimes v_1$ **[TUY**; p. 461**]**. Hence, for a (dual) handle body H' with $\partial H' \simeq -\Sigma$, we can carry out the same procedure to produce an element $V^*_{H'}$ in $Z(\widetilde{\Sigma})^*$.

Theorem 3.3 (L. Crane). *The above vacuum vectors v_H and $v^*_{H'}$ are well-defined up to the multiplication of some power of* $\exp[2\pi\sqrt{-1} \cdot \frac{3k}{k+2}]$. *For a 3-manifold M, the* *absolute value* $|\langle v_H, v^*_{H'}\rangle|$ *of the product* $\langle v_H, v^*_{H'}\rangle$ *is a well-defined invariant where* H *and* H' *are the handle bodies in a Heegaard decomposition of* M, $M = H \cup_\Sigma H'$, $\partial H \simeq -\partial H' \simeq \Sigma$, *and* v_H, $v^*_{H'}$ *are the associated vacuum vectors.*

Proof. To prove Theorem 3.3, we observe that the above definition of v_H depends on the choices of $(3g-3)$ disjoint disks $D^2_a \subset H$, $a = 1, \ldots, 3g-3$. As indicated in the following figure, there are definitely different ways of inserting these disks:

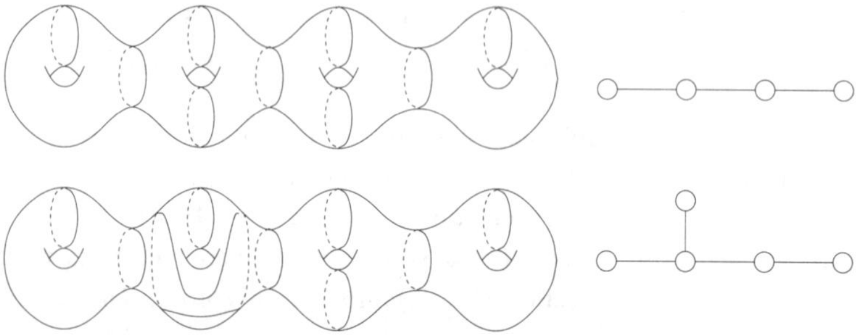

In **[C]**, Crane proved that different 2-disk decompositions can be obtained from a given one by applying the action of the group $\Gamma(H)$. Here $\Gamma(H)$ is a subgroup in $\Gamma_{g,1}$ consisting of diffeomorphisms of Σ which can be extended to H. This subgroup $\Gamma(H)$ is generated by the so-called Suzuki generators. From the axioms of Moore-Seiberg in **[MS]**, Crane verified that the vacuum vector v_H is preserved under the action of Suzuki generators and hence $\Gamma(H)$. As explained in **[MS]**, our monodromy representation ρ satisfies the axioms of Moore-Seiberg, and hence the proof of Theorem 3.3 follows. □

The ambiguity of the phase factor term in Theorem 3.3 can be eliminated. First of all, in the classical treatment of the moduli space $\mathcal{M}_{g,1}$ there is the period mapping $\sigma : \mathcal{M}_{g,1} \to \mathcal{A}_g$ which imbeds $\mathcal{M}_{g,1}$ into the moduli space $\mathcal{A}_g$ of abelian varieties of genus g. On $\mathcal{A}_g$, there is the 'theta' line bundle $\mathcal{W} \to \mathcal{A}_g$ whose sections on the universal covering space $\widetilde{\mathcal{A}}_g$ give rise to theta functions. The pullback of $\mathcal{W}$ under the period map $J : \mathcal{M}_{g,1} \to \mathcal{A}_g$ is by definition our line bundle $\mathcal{L}$. Thus, after removing the 0-section from $\mathcal{W}$, we have a $\mathbb{C}^*$-bundle $\mathring{\mathcal{W}} \to \mathcal{A}_g$ and a commutative diagram

$$\begin{array}{ccc} \mathring{\mathcal{L}} & \xrightarrow{p} & \mathring{\mathcal{W}} \\ \downarrow{\scriptstyle \pi} & & \downarrow{\scriptstyle \text{projection}} \\ \mathcal{M}_{g,1} & \xrightarrow{J} & \mathcal{A}_g \,. \end{array}$$

On the fundamental group level, this gives rise to homomorphisms between two central extensions:

$$\begin{array}{ccccccccc} 1 \longrightarrow & \mathbb{Z}(i) & \longrightarrow & \widetilde{\mathrm{Sp}}_{2g} & \longrightarrow & \mathrm{Sp}_{2g} & \longrightarrow 1 \\ & \uparrow{\scriptstyle \mathrm{id}} & & \uparrow{\scriptstyle p_*} & & \uparrow{\scriptstyle J_*} & \\ 1 \longrightarrow & \mathbb{Z}(i) & \longrightarrow & \widetilde{\Gamma}_{g,1} & \longrightarrow & \Gamma_{g,1} & \longrightarrow 1 \end{array} \tag{3.4}$$

where $\widetilde{\mathrm{Sp}}_{2g} = \pi_1(\mathring{\mathcal{W}})$ and the vertical arrows p_* and J_* are the homomorphisms induced by p and J.

In the study of theta functions, there is a famous 1-cochain

$$\Phi : \widetilde{\mathrm{Sp}}_{2g} \to \mathbb{Z}$$

known as ***Shale-Weil cochain*** in **[LV**; 1.7.8], **[T**; formula 2]. Originally constructed to correct certain ambiguity of projective Shale-Weil representation, this cochain Φ has the following properties.

(a) Φ is additive with respect to the action by the central elements i,
$\Phi(g_1 g_2) = \Phi(g_1) + \Phi(g_2)$ for $g_1 \in \mathbb{Z}(i)$, $g_2 \in \widetilde{\mathrm{Sp}}_{2g}$;(3.5)

(b) The restriction of Φ to the center is an isomorphism $\Phi(i) = 1$.

Pulling back Φ by means of p_*, we obtain a 1-cochain on $\widetilde{\Gamma}_{g,1}$,

$$\Psi : \widetilde{\Gamma}_{g,1} \to \mathbb{Z}, \quad \Psi(g) = \Phi(p_*(g)),$$

with similar properties.

Theorem 3.6. *Let $M = H \cup_\Sigma H'$ be a Heegaard decomposition of M, and let v_H, $v^*_{H'}$ be vacuum vectors defined by means of the deformation paths γ_H and $\gamma_{H'}$ in $(\widehat{\mathfrak{L}})^\circ$. Then*

$$I(M) = \exp[-2\pi\sqrt{-1} \cdot \frac{3k}{k+2} \Psi(\gamma_H \circ \gamma_{H'}^{-1})] \langle v_H, v^*_{H'} \rangle$$

is a well-defined topological invariant of M.

Proof. Note that if we choose different liftings $\widetilde{\gamma}_H, \widetilde{\gamma}_{H'}$ of deformations, we have different vacuum vectors $\tilde{v}_H$, $\tilde{v}^*_{H'}$. The resulting product $\langle \tilde{v}_H, \tilde{v}^*_{H'} \rangle$ differs from $\langle v_H, v^*_{H'} \rangle$ by a factor $\exp[2\pi\sqrt{-1} \cdot \frac{3k}{k+2} \cdot \ell]$ where ℓ is the number of rotations around the fiber of $\mathring{\mathfrak{L}}$ to bring the loop $\gamma_H \circ \gamma_{H'}^{-1}$ to $\tilde{\gamma}_H \circ \tilde{\gamma}_{H'}^{-1}$, i.e.

$$(\tilde{\gamma}_H \circ \tilde{\gamma}_{H'}^{-1}) = (i)^\ell \cdot (\gamma_H \circ \gamma_{H'}^{-1}).$$

Since $\Psi(\tilde{\gamma}_H \circ \tilde{\gamma}_{H'}^{-1}) = \ell + \Psi(\gamma_H \circ \gamma_{H'}^{-1})$, this discrepancy is corrected by the difference between $\exp[-2\pi\sqrt{-1} \cdot \frac{3k}{k+2} \cdot \Psi(\gamma_H \circ \gamma_{H'}^{-1})]$ and $\exp[-2\pi\sqrt{-1} \cdot \frac{3k}{k+2} \cdot \Psi(\tilde{\gamma}_H \circ \tilde{\gamma}_{H'}^{-1})]$. The rest of the proof of Theorem 3.6 follows from an argument similar to Theorem 3.3.□

Remark 3.7. In the above discussion, we have chosen a fixed complex structure on the common boundary Σ of the two handle bodies H and H'. If we vary this complex structure, the loop $\gamma_H \circ \gamma_{H'}^{-1}$ will be varied, however, the homotopy class of the loop remains unchanged and represents a fixed element in the mapping class group $\Gamma_{g,1}$.

4. Surgery formula

In **[KM]**, Kirby and Melvin described the Reshetikhin-Turaev invariant $\tau(M)$ in terms of the following formula:

$$\tau(M) = \alpha_L \cdot \sum_{\vec{\lambda}} [\vec{\lambda}] \cdot J_{L,\vec{\lambda}}. \tag{4.1}$$

Here M is obtained from S^3 by performing surgery on framed link L,

$\alpha_L = \left(\sqrt{\frac{2}{k+2}} \sin \frac{\pi}{k+2}\right)^n \left(\exp\left[-2\pi\sqrt{-1} \cdot \frac{3k}{k+2} \cdot \frac{\sigma}{8}\right]\right)$,

k = the fixed level,

n = the number of components of L,

σ = the signature of the linking matrix of L,

$\vec{\lambda} = (\lambda_1, \ldots, \lambda_n), 1 \leq \lambda_i \leq k$,

$[\vec{\lambda}] = \prod_{i=1}^n \left(\sin \frac{\pi \lambda_i}{k+2} \;/\; \sin \frac{\pi}{k+2}\right)$,

$J_{L,\vec{\lambda}}$ = invariant of framed colored link $(L, \vec{\lambda})$ defined in [**KM**; Thm. 4.4].

Similar formulas can also be found in [**L**]. As pointed out in [**KM**; 1.4], the above formula (4.1) differs from the definition of Reshetikhin-Turaev [**RT**; Thm. 1] by the factor $\exp[-2\pi\sqrt{-1} \cdot \frac{3k}{8(k+2)} b_1(M)]$ where $b_1(M)$ is the first Betti number of M. In view of these discrepancies, we will prove below:

Theorem 4.2. *Up to a phase factor which depends on the Betti number $b_1(M)$ and the linking pairing $\operatorname{Tor} H_1(M) \times \operatorname{Tor} H_1(M) \to \mathbb{Q}/\mathbb{Z}$, the invariants $I(M)$ and $\tau(M)$ are the same. In the case of integral homology spheres M, they are identical: $\tau(M) = I(M)$.*

To prove $I(M)$ satisfies a surgery formula, we have to relax the condition in Theorem 3.6 so that the formula:

$$I(M) = \exp[-2\pi\sqrt{-1} \cdot \frac{3k}{k+2} \cdot \Psi(\gamma_w \cdot \gamma_{w'}^{-1})]\langle v_w, v_{w'}^* \rangle \tag{4.3}$$

holds whenever we have a more general kind of decomposition of

$$M = W \cup W', \quad \text{where } \partial W \simeq \partial W' \simeq W \cap -W' \simeq \Sigma,$$

than just Heegaard decompositions. In such a decomposition, W and W' are obtained from attaching 1-, 2-, 3-handles to the separating surface Σ corresponding to the operations: connecting up different components; adding handles; shrinking simple closed curves to points; filling up 2-spheres by 3-balls. These operations give rise to paths δ_w, $\delta_{w'}$ from a base point to Σ in $\widehat{\mathcal{M}}_{g,1}$. After choosing some liftings $\gamma_w, \gamma_{w'}$ of $\delta_w, \delta_{w'}$, we obtain the definition of the terms $v_W, v_{W'}^*$ and $\Psi(\gamma_\omega \cdot \gamma_{\omega'}^{-1})$ in (4.1). In fact, in the proof of Theorem 3.6 and Theorem 3.3, we have similar constructions in order to prove the independence of Heegaard decomposition of M.

We now apply (4.3) to the situation when M is obtained by performing surgery on a framed link $L \subset S^3$. Let $W' = L \times D^2$ be a tubular neighborhood of L and $W = S^3 - L \times D^2$ be the complement of this neighborhood in S^3. Then M can be written as a union $W \cup W'$ where ∂W is glued to $\partial W'$ by the mapping

$$g : S^1 \times S^1 \to S^1 \times S^1, \qquad (z_1, z_2) \mapsto (z_2^{-1}, z_1),$$

on each of the torus components in $L \times S^1$. By Theorem 2.4 the state space $Z(S^1 \times S^1)$ is isomorphic to the sum $\oplus_{1 \leq i \leq k} \mathbb{C}(e_i)$ where each summand has a preferred basis element e_i. The vacuum vector $v_{D^2 \times S^1}^*$ associated to the solid torus (different from the tubular neighborhood $S^1 \times D^2 \subset L \times D^2$ by the sewing map g) can be expressed as the sum

$$v^*_{D^2\times S^1} = \sum_{1\le i\le k} \sqrt{\frac{2}{k+2}} \sin\left(\frac{\pi i}{k+2}\right) \cdot e_i .$$

Here the coefficient $\sqrt{\frac{2}{k+2}} \sin\left(\frac{\pi i}{k+2}\right)$ is the entry $[\rho(g)]_{1,i}$ in the matrix presentation of $\rho(g)$ in **[AGS1**; formula (36)]. Since the state space $Z(L \times S^1)$ is isomorphic to the tensor product n times of the component $Z(S^1 \times S^1)$, we have

$$Z(L \times S^1) \simeq \bigoplus_{\vec\lambda} \mathbb{C}(e_{\vec\lambda}), \quad \text{where} \quad e_{\vec\lambda} = e_{\lambda_1} \otimes \cdots \otimes e_{\lambda_n}, \quad 1 \le \lambda_i \le k,$$

and the vacuum vector $\tilde v^*_{w'}$ is the sum

$$\sum_{\vec\lambda} \sqrt{\frac{2}{k+2}} \sin\left(\frac{\pi\lambda_1}{k+2}\right) \cdot \sqrt{\frac{2}{k+2}} \sin\left(\frac{\pi\lambda_2}{k+2}\right) \cdots \sqrt{\frac{2}{k+2}} \sin\left(\frac{\pi\lambda_n}{k+2}\right) e_{\vec\lambda}$$
$$= \left(\sqrt{\frac{2}{k+2}} \sin\frac{\pi}{k+2}\right)^n \sum_{\vec\lambda} [\vec\lambda]\, e_{\vec\lambda} .$$

Putting these expressions into (4.3), we have

$$I(M) = \left(\sqrt{\frac{2}{k+2}} \sin\frac{\pi}{k+2}\right)^n \exp[-2\pi\sqrt{-1} \cdot \frac{3k}{k+2} \Psi(\gamma_\omega \cdot \gamma_{w'}^{-1})] \sum_{\vec\lambda} [\vec\lambda] \langle v_W, e_{\vec\lambda}\rangle .$$

Let $I(L, \vec\lambda)$ denote the term $\langle v_W, e_{\vec\lambda}\rangle$. Clearly $I(L, \vec\lambda)$ can be interpreted as an invariant of framed, colored link L; the term 'color' refers to the weight λ_i assigned to the ith-component of the link. To compute $I(L, \vec\lambda)$ we represent our link as a diagram of $2m$ braids B:

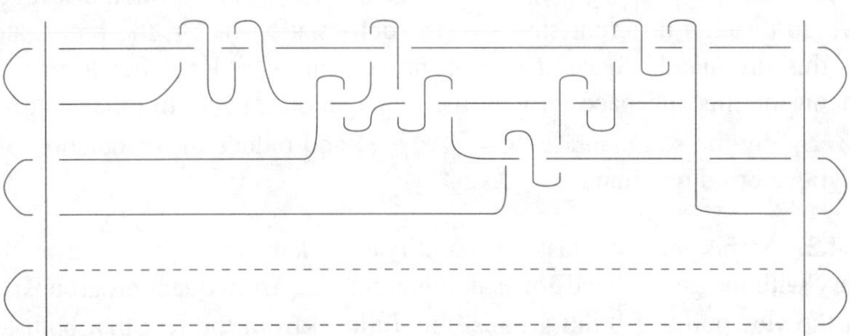

The top and bottom of the $(2j-1)^{st}$*-braid is connected up to the* $(2j)^{th}$*-braid by an arc. Such a configuration is also referred to as a plat in* **[B**, p. 192].

Our manifold $W = S^3 - L \times D^2$ is obtained by removing some thickened tubes around the braids B and the connecting arcs. Let P_t, $-\infty < t < \infty$, be the family of horizontal planes. At $t < -\epsilon$, the plane P_t is below the link; at $t = -\epsilon$ it touches the bottom connecting arcs at m minimum points; between $-\epsilon$ and $1+\epsilon$, it cuts the braids B in $2m$ points, at $t = 1+\epsilon$, it touches the top connecting arcs at m maximum points; finally at $t > 1+\epsilon$, the plane P_t is above the link L. For each t, $-\infty < t < \infty$, we denote by W_t the subspace in W obtained by removing all the points below P_t. Then $\{W_t\}_{-\infty<t<\infty}$ represents a family of deformations of W: starting from removing a small ball and gradually exhausting the entire manifold W except for a small ball. Accordingly their boundaries $\Sigma_t = \partial W_t$ represent a family of deformations of stable Riemann surfaces: starting from the disjoint union of $L \times S^1$ and S^2, then to its connected sum $\# L \times S^1$, and finally S^2 obtained by shrinking m disjoint curves to points.

Of course, the above deformations give us paths $\delta_t = \{\Sigma_t\}$ in $\widehat{\mathcal{M}}_{m,1}$. In this situation, there exists in fact a consistent choice of lifting of δ_t to paths γ_t in $(\widehat{\mathcal{L}})^\circ$. The reason is that the Riemann surfaces Σ_t obtained in the above paragraph lie in a $K(\pi,1)$-subspace $\widehat{\mathcal{M}}_{0,m+1}$ in $\widehat{\mathcal{M}}_{2m,1}$ whose fundamental group $\pi_1(\widehat{\mathcal{M}}_{0,m})$ is the framed braid group. It is not difficult to show that the induced homomorphism $H^2(\Gamma_{m,1}) \to H^2(\Gamma_{0,2m+1})$ on second cohomology is trivial and so the restriction $\widehat{\mathcal{L}} \mid \widehat{\mathcal{M}}_{0,2m+1}$ admits a trivialization.

Using the above liftings γ_t of δ_t, we can associate to the framed, colored braids a propagator $Z(\gamma) : Z(\Sigma^*_{-\epsilon'}) \to Z(\Sigma^*_{1+\epsilon'})$, $\epsilon' > \epsilon$. In turn, this propagator $Z(\gamma)(e_{\vec{\lambda}})$ gives us an element in $Z(\Sigma_{1+\epsilon'})^*$ and also a number $\langle v_0, Z(\gamma)(e_{\vec{\lambda}})\rangle$ by evaluating $Z(\gamma)(e_{\vec{\lambda}})$ on the vacuum vector v_0, $Z(\gamma)(e_{\vec{\lambda}})$ to the 3-ball. It is not difficult to show that $\langle v_0, Z(\gamma)(e_{\vec{\lambda}})\rangle$ is the same as $\langle v_W, e_{\vec{\lambda}}\rangle$ and hence the link invariant $I(L, \vec{\lambda})$.

The state space $Z(\Sigma_t)$, $-\epsilon < t < 1+\epsilon$, and the element $Z(\gamma_t)(e_{\vec{\lambda}})$ in this state space is recorded by the intersection $B \cap P_t$ of framed, colored braids B and the horizontal planes P_t. There are $2m$ such points $\{x_1, x_2, \ldots, x_{2m}\}$ with framing induced from the framing on L. In addition, there are colors $\lambda_1, \ldots, \lambda_{2m}$ (irreducible SU(2)-representations) assigned to these points: we assign λ_i (the representation V_{λ_i}) to be the same as the component of L at x_i when its orientation points upward, and otherwise assign $-\lambda_i$ (the dual of V_{λ_i}). Compactifying the plane P_t by adding a point at ∞, we obtain the data of a 2-sphere S^2_t with $2m+1$ framed, colored, points $(S^2_t;\ {}^{\lambda_1,\ldots,\lambda_{2m}\ 1}_{x_1,\ldots,x_{2m}\ \infty})$. The state space $Z(\Sigma_t)$ contains $Z(S^2_t;\ {}^{\lambda_1,\ldots,\lambda_{2m}\ 1}_{x_1,\ldots,x_{2m}\ \infty}) = [V_{\lambda_1} \otimes \cdots \otimes V_{\lambda_{2m}}]^{\mathrm{SU}(2)}$ as a direct summand, and as we go through different stages of the deformation of Σ_t, the propagator $Z(\gamma_t)$ preserves this summand. Since at $t = 0$, the element $Z(\gamma_0)(e_{\vec{\lambda}})$ lies in this subspace, it remains inside this subspace through the propagation $Z(\gamma_t)$. In other words, we can replace $Z(\Sigma_t)$ by the summand $Z(S^2_t;\ {}^{\lambda_1,\ldots,\lambda_{2m}}_{x_1,\ldots,x_{2m}})$ and reduce the calculation of $I(L, \vec{\lambda})$ to identifying a braid invariant.

Remark 4.3. At this point, our task is to identify the link invariant $I(L, \vec{\lambda})$ from conformal field theory with the generalized Jones polynomial $J_{L,\vec{\lambda}}$ from quantum groups.

From the viewpoint of braid groups, this follows from the fact that both level k – SU(2) conformal field theory and the theory of quantum group $U_q(\mathrm{SL}(2))$ give the same representation of braid groups. The proof of this last assertion can be found in **[AGS1]**, **[AGS2]**, and **[Ko]**. Instead of going through all the details, we will outline some of the main ideas based on the work of Fröhlich-King **[FK]** and Kirby-Melvin **[KM]**.

In **[FK]**, Fröhlich and King investigated these framed, colored braid invariants when all the colors are the same $\lambda_1 = \lambda_2 = \cdots = \lambda_{2m}$. Among other things, they proved in the case of constant color 2, $\lambda_1 = \lambda_2 = \cdots = \lambda_m = 2$ (the basic SU(2)-representation), the invariant $I(L,\vec{2})$ satisfies the skein relation. Up to a factor which can be fixed by requiring $I(\text{unknot},\vec{2}) = 1$, the invariant $I(L,\vec{2})$ can be identified with certain special values of HOMFLY polynomials as in Kirby-Melvin invariant [**KM**, Lemma 4.3], i.e.

$$I(L,\vec{2}) = J_{L,\vec{2}} =$$
$$= [2]\exp\left[\frac{3}{4}\cdot\frac{2\pi\sqrt{-1}}{k+2}\cdot(L\cdot L)\right] P_L(\sqrt{-1}\exp\frac{2\pi\sqrt{-1}}{k+2},\ 2\sin\frac{2\pi}{k+2}),$$
$$P_L(s,t) = \text{two variables HOMFLY polynomial for framed link } L.$$

Theorem 4.4. *For a framed, colored link L the invariant $I(L,\vec{\lambda})$ coincides with the Kirby-Melvin invariant $J_{L,\vec{\lambda}}$.*

Proof. In [**KM**; Thm. 4.4], Kirby and Melvin proved that their link invariant $J_{L,\vec{\lambda}}$ can be expressed as a weighted sum of $J_{L^{\vec{\mu}-2\vec{j}}}$,

$$J_{L,\vec{\lambda}} = \sum_{\vec{j}=0}^{\vec{\mu}/2}(-1)^{\vec{j}}\binom{\vec{\mu}-\vec{j}}{\vec{j}} J_{L^{\vec{\mu}-2\vec{j}}},$$

$$\text{where}\quad (-1)^{\vec{j}} = \prod_{i=1}^{n}(-1)^{j_i}\quad\text{and}\quad \binom{\vec{\mu}-\vec{j}}{\vec{j}} = \prod_{i=1}^{n}\binom{\mu_i-j_i}{j_i}.$$

Here the notation $L^{\vec{\mu}-2\vec{j}}$ stands for the link obtained from L by performing the cable knot construction $(\mu_i - 2j_i)$ times on the ith-component of L and $J_{L^{\vec{\mu}-2\vec{j}}} = J_{L^{\vec{\mu}-2\vec{j}},\vec{2}}$ is the framed link invariant with constant color 2. The key steps of their proofs are the following properties of $J_{L,\vec{\lambda}}$:

(a) If L' is a sublink of L obtained by removing some 1-colored components, then $J_{L,\vec{\lambda}} = J_{L',\vec{\lambda}\mid L'}$.

(b) If L' is obtained from L by splitting a component K of L into two parallel translations K_1, K_2 via the framing and if the color on all the other components remains the same and if $\lambda(K) = \lambda'(K_1) + \lambda'(K_2)$, then $J_{L,\vec{\lambda}} = J_{L',\vec{\lambda}'}$.

By interpreting our invariant $I(L,\vec{\lambda})$ in terms of braids, it is not difficult to show that these properties also hold for $I(L,\vec{\lambda})$. Hence, by the same argument as in [**KM**], we can express $I(L,\vec{\lambda})$ as a similar weighted sum of $I(L^{\vec{\mu}-2\vec{j}},\vec{2})$. The proof of Theorem 4.4 follows because at the constant $\vec{2}$ color both $I(L^{\vec{\mu}-2\vec{j}},\vec{2})$ and $J_{L^{\vec{\mu}-2\vec{j}}}$ are given by the same special values of the HOMFLY polynomial. □

5. Signature correction term

To complete the proof of Theorem 4.2, it remains to compare the two phase factor terms $\exp[-2\pi\sqrt{-1}\cdot\frac{3k}{(k+2)}\cdot\Psi(\gamma_H\circ\gamma_{H'}^{-1})]$ and $\exp[-2\pi\sqrt{-1}\cdot\frac{3k}{(k+2)}\cdot\frac{\sigma}{8}]$. Since $\Psi(\gamma_H\circ\gamma_{H'}^{-1})$ is the value of the 1-cochain $\Psi : \widetilde{\Gamma}_{g,1} \to \mathbb{Z}$ on $\gamma_H\circ\gamma_{H'}^{-1}$, the natural strategy is to interpret $\frac{\sigma}{8}$ in the same manner. Unfortunately, carried along with the procedure of framed surgery is the notion of framing, and therefore the term $\sigma/8$ does not quite fit into the setting of the moduli space $\mathcal{M}_{g,1}$.

The remedy is to introduce an additional ***even spin structure*** on the stable Riemann surfaces. For a nonsingular Riemann surface, a spin structure can be viewed as a quadratic refinement $\theta : H_1(\Sigma_g; \mathbb{Z}/2) \to \mathbb{Z}/2$ of its intersection pairing, $\langle x, y\rangle = \theta(x) + \theta(y) + \theta(x + y) \bmod 2$, and an even spin structure is a quadratic refinement with zero Arf invariant **[LMW**; p. 192]. Regarding θ as part of the data of a nonsingular Riemann surface, we obtain the definition of the moduli space $\mathcal{M}_{g,1}(\theta)$ of Riemann surfaces of genus g with a base tangent vector and a preferred even spin structure (Σ_g, θ). Note that given a nonsingular Riemann surface of genus g, there are exactly $2^{g-1}(2^g + 1)$ even spin structures. Hence the forgetful map $p : \mathcal{M}_{g,1}(\theta) \to \mathcal{M}_{g,1}$ is a $2^{g-1}(2^g + 1)$-sheeted covering map.

Since $\widehat{\mathcal{M}}_{g,1}$ has the same homotopy type as $\mathcal{M}_{g,1}$, we can extend the covering $p : \mathcal{M}_{g,1}(\theta) \to \mathcal{M}_{g,1}$ to $\widehat{p} : \widehat{\mathcal{M}}_{g,1}(\theta) \to \widehat{\mathcal{M}}_{g,1}$ over $\widehat{\mathcal{M}}_{g,1}$. In this enlargement $\widehat{\mathcal{M}}_{g,1}(\theta)$, not every boundary component is actually needed. Recall that a stable Riemann surface Σ_0 is obtained from a nonsingular one Σ by pinching some disjoint simple enclosed curves $\{\beta_a\}$ to points. If Σ has an even spin structure $\theta : H_1(\Sigma; \mathbb{Z}/2) \to \mathbb{Z}/2$ and if the curves β_α satisfy the equation $\theta(\beta_\alpha) = 0$ then on the homology $H_1(\Sigma_0; \mathbb{Z}/2)$ of the singular surface, $\Sigma_0 = \Sigma/\beta_\alpha \sim *$, there exists an induced quadratic refinement $\theta : H_1(\Sigma_0; \mathbb{Z}/2) \to \mathbb{Z}/2$. In our application, it is enough to consider the subspace in $\widehat{\mathcal{M}}_{g,1}(\theta)$ consisting of all the spin Riemann surfaces (Σ_0, θ) constructed in this manner. The reason is because we are primarily interested in the situation of spin handle body (H, θ) whose boundary $(\partial H, \theta \mid \partial H)$ is isomorphic to (Σ, θ). By inserting 2-disks $\{D_a\}$ and shrinking their boundaries $\partial D_a = \beta_a$ to points we obtain a deformation of Σ to stable Riemann surfaces. By definition, $\theta(\beta_a) = 0$, and so this deformation represents a path δ_t in this subspace of $\widehat{\mathcal{M}}_{g,1}(\theta)$, and from (Σ_g, θ) to some chosen base point.

Pulling back the bundle $\widehat{\mathcal{Z}}$ of states, the Hodge line bundle $\widehat{\mathcal{L}}$, the $\mathbb{C}^*$-bundle $(\widehat{\mathcal{L}})^\circ$ by means of $\widehat{p}$, we obtain corresponding objects: $\widehat{\mathcal{Z}}(\theta) = \widehat{p}^*(\widehat{\mathcal{Z}})$, $\widehat{\mathcal{Z}}(\theta) = \widehat{p}^*(\widehat{\mathcal{Z}})$, $\widehat{\mathcal{L}}(\theta) = \widehat{p}^*(\widehat{\mathcal{L}})$, $\widehat{\mathcal{L}}(\theta)^\circ = \widehat{p}^*((\widehat{\mathcal{L}})^\circ)$ defined on $\widehat{\mathcal{M}}_{g,1}(\theta)$. Given a 3-manifold (M, θ) with spin structure we can decompose (M, θ) into the union of two handle bodies with compatible spin structures (H, θ) and (H', θ'). Hence from the discussion in the previous paragraph, we have paths $\delta_H, \delta_{H'}$ on $\widehat{\mathcal{M}}_{g,1}(\theta)$ and also lifted paths $\gamma_H, \gamma_{H'}$ on $\widehat{\mathcal{L}}(\theta)^\circ$. Parallel translations along these paths $\gamma_H, \gamma_{H'}$ give us vacuum vectors $v_H, v^*_{H'}$ and hence their product $\langle v_H, v^*_{H'}\rangle$.

The classical period map $J : \mathcal{M}_{g,1} \to \mathcal{A}_g$ of $\mathcal{M}_{g,1}$ to the moduli space $\mathcal{A}_g$ of abelian varieties is covered by a mapping $J(\theta) : \mathcal{M}_{g,1}(\theta) \to \mathcal{A}_g(\theta)$ on their spin counterparts. On the level of fundamental groups, this induces a surjection $J(\theta)_* : \Gamma_{g,1}(\theta) \to \mathrm{Sp}_{2g}(\theta)$ of the spin mapping class group $\Gamma_{g,1}(\theta)$ onto the theta group $\mathrm{Sp}_{2g}(\theta)$. Note, in the theory of framed surgery of dimension $4\ell - 1$, we have to consider the group of automorphisms of a nonsingular, skew-symmetric pairing over $\mathbb{Z}^{2g}$, which preserves a fixed Arf invariant zero quadratic refinement. This is a subgroup of finite index in $\mathrm{Sp}_{2g}(\mathbb{Z})$ and is precisely the theta group $\mathrm{Sp}_{2g}(\theta)$. Analogous to (3.4), we have a diagram of central extensions

$$\begin{array}{ccccccccc} 1\longrightarrow & \mathbb{Z}(i) & \longrightarrow & \widetilde{\mathrm{Sp}}_{2g}(\theta) & \longrightarrow & \mathrm{Sp}_{2g}(\theta) & \longrightarrow 1 \\ & \uparrow & & \uparrow & & \uparrow & \\ 1\longrightarrow & \mathbb{Z}(i) & \longrightarrow & \widetilde{\Gamma}_{g,1} & \longrightarrow & \Gamma_{g,1}(\theta) & \longrightarrow 1 \end{array}$$

where $\widetilde{\Gamma}_{g,1}(\theta)$ is the fundamental group $\pi_1((\widehat{\mathfrak{L}}(\theta))^\circ)$ of the $\mathbb{C}^*$-bundle. Using the Shale-Weil cochain $\Phi : \widetilde{\mathrm{Sp}}_{2g}(\theta) \to \mathbb{Z}$, we obtain a correction term for $\langle v_H, v^*_{H'}\rangle$. In short, all the operations in the previous construction of $I(M)$ can be repeated in this spin setting.

The 3-sphere S^3 is endowed with a standard framing which can be extended to the 4-ball B^4, $\partial B^4 \simeq S^3$. Given a link $L \subset S^3$, this framing on S^3 induces a framed link structure $L \times D^2 \subset S^3$ on L. As we perform surgery on this framed link, we obtain a framed 3-manifold M at the end. Note that the data of a stable framing on a 2-dimensional manifold coincide with those of a spin structure. Thus the process of cutting S^3 open along $L \times D^2$ and sewing it back by a frame preserving diffeomorphism is the same as gluing by a spin preserving diffeomorphism. In other words, the previous discussion on the framed manifold M in §4 can be regarded as taking place in the spin moduli space $\widetilde{\mathfrak{M}}_{g,1}(\theta)$ and $\widehat{\mathfrak{L}}_{g,1}(\theta)^\circ$.

From the work of Sharpe **[S1]**, **[S2]**, **[S3]**, the above Shale-Weil cochain can be compared with an obstruction in surgery theory. First of all, there exists a universal central extension of the infinite theta group

$$1\longrightarrow\mathbb{Z}[\kappa]\longrightarrow \mathrm{St}\,\mathrm{Sp}(\theta)\longrightarrow \mathrm{Sp}(\theta)\longrightarrow 1 \tag{5.1}$$

where $\mathrm{Sp}(\theta)$ stands for the direct limit $\lim_{\to} \mathrm{Sp}_{2g}(\theta)$ and $\mathrm{St}\,\mathrm{Sp}(\theta)$ stands for the Steinberg symplectic group. The definition $\mathrm{St}\,\mathrm{Sp}(\theta)$ can be found in **[S2**; p. 456], and strictly speaking we have to use the elementary symplectic group $\mathrm{E}\,\mathrm{Sp}(\theta)$ but Sharpe's central extension can be extended to $\mathrm{Sp}(\theta)$. Let $\mathrm{TU}(\theta)$ denote the subgroup in $\mathrm{Sp}(\theta)$ consisting of the matrices

$$\begin{bmatrix} a & 0 \\ 0 & {}^t a^{-1}\end{bmatrix}\begin{bmatrix} I & b \\ 0 & I\end{bmatrix}, \quad \text{where } b = {}^t b, \text{ and } b_{ii} \equiv 0 \bmod 2.$$

Then $\mathrm{TU}(\theta)$ can be lifted to a subgroup in $\mathrm{St}\,\mathrm{Sp}(\theta)$, and the coset space $\mathrm{St}\,\mathrm{Sp}(\theta)/\,\mathrm{TU}(\theta)$ can be interpreted as the ***set of relative, 4ℓ-dimensional surgery data.*** By a relative, 4ℓ-dimensional surgery problem, we mean a degree 1, normal map $f : (N^{4\ell}; \partial_+ N, \partial_- N) \to (X \times I; X \times 1, X \times 0)$ from a simply connected manifold triple $(N; \partial_+ N, \partial_- N)$ to a corresponding Poincaré complex $(X \times I; X \times 1, X \times 0)$ which restricts to a homotopy equivalence on $\partial_+ N$, and the problem is to modify such a triple to a homotopy equivalence. In **[S1**; §6] Sharpe reduces the geometric problem to a highly connected situation where $\pi_i(f) = 0, 0 \le i \le 2\ell - 1$, and represents it algebraically by a pair (**[D]**,A). This pair is referred to as a ***surgery data***, and in **[S2**; Thm. 6.1], Sharpe showed that there exists a one-to-one correspondence between a coset space $\mathrm{St}\,\mathrm{Sp}(\theta)/\,\mathrm{TU}(\theta)$ and the set of all these surgery data (**[D]**,A).

A consequence of the above interpretation of $\mathrm{St}\mathrm{Sp}(\theta)$ is the following composite map:

$$\sigma : \mathrm{StSp}(\theta) \longrightarrow \mathrm{StSp}(\theta)/\mathrm{TU} \xrightarrow{\text{signature}} \mathbb{Z}$$

where the map on the right hand side is given by the signature obstruction of the surgery problem (**[D]**, A). From the additive properties of signatures, it is not difficult to show that, viewed as 1-cochain of $\mathrm{StSp}(\theta)$, the map σ has the following properties:

(a) σ is additive with respect to the action of the central element κ:

$$\sigma(\kappa \cdot g) = \sigma(\kappa) + \sigma(g);(5.2)$$

(b) $\sigma(\kappa) = 8$.

Lemma 5.3. *There exists a commutative diagram of central group extensions*

$$\begin{array}{ccccccccc} 1 \longrightarrow & Z[i] & \longrightarrow & \widehat{\mathrm{Sp}}(\theta) & \longrightarrow & \mathrm{Sp}(\theta) & \longrightarrow 1 \\ & \uparrow{\scriptstyle \times 8} & & \uparrow & & \uparrow{\scriptstyle \mathrm{id}} & \\ 1 \longrightarrow & \mathbb{Z}[\kappa] & \longrightarrow & \mathrm{StSp}(\theta) & \longrightarrow & \mathrm{Sp}(\theta) & \longrightarrow 1 \end{array}$$

where the left vertical map stands for the multiplication by 8.

Proof. The proof of Lemma 5.3 follows from the fact that the central extension $1 \to \mathbb{Z} \to \widetilde{\mathrm{Sp}}(\mathbb{Z}) \to \mathrm{Sp}(\mathbb{Z}) \to 1$ represents a generator of $H^2(\mathrm{B}\,\mathrm{Sp}(\mathbb{Z});\mathbb{Z})$, and so it can be identified with the universal central extension given by the Steinberg symplectic group $1 \to \mathbb{Z} \to \mathrm{StSp}(\mathbb{Z}) \to \mathrm{Sp}(\mathbb{Z}) \to 1$. This last extension is related to relative surgery problem of Poincaré complexes in the same way as the sequence $1 \to \mathbb{Z} \to \mathrm{StSp}(\theta) \to \mathrm{Sp}(\theta) \to 1$. The reason is that the aforementioned results of Sharpe can be extended to cover the algebraic surgery problems without using quadratic refinement, and so its counterpart $\mathrm{Sp}(\mathbb{Z})$.

Using the above interpretation, it follows that there exists a commutative diagram

$$\begin{array}{ccccccccc} 1 \longrightarrow & Z[i] & \longrightarrow & \widetilde{\mathrm{Sp}}(\theta) & \longrightarrow & \mathrm{Sp}(\theta) & \longrightarrow 1 \\ & \uparrow{\scriptstyle \times 8} & & \uparrow & & \uparrow{\scriptstyle \mathrm{incl}} & \\ 1 \longrightarrow & \mathbb{Z}[\kappa] & \longrightarrow & \mathrm{StSp}(\theta) & \longrightarrow & \mathrm{Sp}(\theta) & \longrightarrow 1 \end{array}$$

induced by forgetting the framing data on surgery. The left vertical arrow is the multiplicatin by 8 because a surgery representative of this element κ is the E_8-manifold of Kervaire-Milnor which has signature 8. This proves Lemma 5.3. □

On the moduli space $\widehat{\mathcal{A}}_g(\theta)$, the central extension

$$1 \longrightarrow \mathbb{Z}[\kappa] \longrightarrow \mathrm{StSp}_{2g}(\theta) \longrightarrow \mathrm{Sp}_{2g}(\theta) \longrightarrow 1$$

gives rise to a line bundle $\mathcal{U}$. An interpretation of Lemma 5.3 is that there exists an isomorphism of the eighth power $\mathcal{U}^{\otimes 8}$ of $\mathcal{U}$ onto $\widehat{\mathcal{W}}(\theta)$. In other words, let $\mathcal{N}$ denote the pullback of $\mathcal{U}$ by $\widehat{p} : \widehat{\mathcal{M}}_{g,1}(\theta) \to \widehat{\mathcal{A}}_g(\theta)$. Then $\mathcal{N}^{\otimes 8} \simeq \widehat{\mathcal{L}}(\theta)$. Suppose, as in §4, we are given a 3-manifold $M = W \cup W'$ obtained by performing surgery on framed link

$L \times D^2 \subset S^3$. Then, as discussed before, there exists a loop $\delta_W \cdot \delta_{W'}^{-1}$ of deformations in $\widehat{\mathcal{M}}_{g,1}(\theta)$ and to construct our invariant we lift this loop to $\widehat{\mathcal{L}}(\theta)^\circ$. An interpretation of Sharpe's result in this situation is that the surgery information provides us a lifting of $\delta_W \cdot \delta_{W'}^{-1}$, to $\mathcal{E}_W \cdot \mathcal{E}_{W'}^{-1}$ in $\mathring{\mathcal{N}}$. The same operations in constructing $I(M)$ can be repeated using the lifted paths $\mathcal{E}_W$ and $\mathcal{E}_{W'}$: pull back the bundle $\widehat{\mathcal{Z}}(\theta)$ of states to get a flat bundle on $\mathring{\mathcal{N}}$; parallel translate by the flat connection along $\mathcal{E}_W$ and $\mathcal{E}_{W'}$ to produce vacuum vectors U_W and $U_{W'}^*$; then form the product $\langle U_W, U_{W'}^* \rangle$. As for the correction term, we consider the 1-cochain defined by the composite

$$\frac{\sigma \circ \widehat{p}}{8} : \pi_1(\mathring{\mathcal{N}}) \xrightarrow{\mathring{\widehat{p}}} \pi_1(\mathring{\mathcal{U}}) = \mathrm{StSp}_{2g}(\theta) \xrightarrow{\sigma/8} \mathbb{Z}[\tfrac{1}{8}],$$

and multiply $\langle U_W, U_{W'}^* \rangle$ by the phase factor term

$$\exp[-2\pi\sqrt{-1} \cdot \frac{3k}{k+2} \cdot \frac{\sigma \cdot \widehat{p}}{8}(\mathcal{E}_W \cdot \mathcal{E}_{W'}^{-1})] \cdot \langle U_W, U_{W'}^* \rangle .$$

From the above discussion, this last formula is just another expression of our invariant $\tau(M)$.

Let us now consider Ψ also as a 1-cochain defined on $\pi_1(\mathring{\mathcal{N}})$:

$$\Psi : \pi_1(\mathring{\mathcal{N}}) \longrightarrow \mathrm{StSp}_{2g}(\theta) \longrightarrow \widetilde{\mathrm{Sp}}(\theta) \xrightarrow{\Phi} \mathbb{Z}.$$

In the case when M is an integral homology sphere, the loop $\mathcal{E}_W \cdot \mathcal{E}_{W'}^{-1}$ is sent to an element $\widehat{p}(\mathcal{E}_W \cdot \mathcal{E}_{W'}^{-1})$ in $\mathrm{Ker}[\mathrm{StSp}_{2g}(\theta) \to \mathrm{Sp}_{2g}(\theta)]$ or in other words some multiple of κ. By Lemma 5.3, the two cochains $\sigma/8$ and Φ agree on κ, and therefore the two correction terms and the two invariants $I(M)$, $\tau(M)$ become identical in this situation.

For a general 3-manifold, we observe that both the cochains $\sigma/8$ and Φ satisfy (3.5) and so their difference $\sigma/8 - \Phi$ descends to a cochain on $\mathrm{Sp}(\theta)$. On the other hand, the difference $\sigma/8 - \Phi$ is a topological invariant because both $I(M)$ and $\tau(M)$ are topological invariants for every k.

It follows that $\sigma/8 - \Phi$ is invariant under the left- and right-actions of the mapping class group $\Gamma(H)$ and $\Gamma(H')$ of handle bodies. Passing to the corresponding symplectic group, this means that the cochain $\sigma/8 - \Phi$ can be regarded as a map of the double coset space $\mathrm{TU} \backslash \mathrm{Sp} / \mathrm{TU}$.

From the work of Wall in **[Wa1**; p. 60] and **[Wa2**; Theorem 1], it is not difficult to see that there is a bijection between $\mathrm{TU} \backslash \mathrm{Sp} / \mathrm{TU}$ and the set of algebraic data derived from the first Betti number $b_1(M)$ and the linking pairing

$$\mathrm{Tor}\ H_1(M) \times \mathrm{Tor}\ H_1(M) \to \mathbb{Q}/\mathbb{Z}$$

of M. This completes the proof of Theorem 4.2.

References

[AGS1] L. Alvarez-Gaume, C. Gomez, and G. Sierra, Quantum group interpretation of some conformal field theories, Phys. Lett. B 220 (1989), 142–152.

[AGS2] L. Alvarez-Gaume, C. Gomez, and G. Sierra, Duality and quantum groups, Nuclear Phys. B330 (1990), 347–398.

[A] M. F. Atiyah, On framings of 3-manifolds, Topology 29 (1990), 1–7.

[AHLS] M. F. Atiyah, N. Hitchin, R. Lawrence and G. Segal, Oxford seminar on Jones-Witten theory, Michaelmas term, 1988.

[APW] S. Axelrod, S. D. Pietra, and E. Witten, Geometric quantization of Chern Simons gauge theory, Institute for Advanced Study preprint (1989).

[BK] A. A. Beilinson and D. Kazhdan, Lectures given by Kazhdan at Yale University and Courant Institute (1989).

[BS] A. A. Beilinson and V. V. Schechtman, Determinant line bundles and Virasoro algebras, Comm. Math. Phys. 118 (1980), 651–701.

[B] J. S. Birman, Braids, links and mapping class groups, Ann. of Math. Studies 82 (1974), Princeton University Press and University of Tokyo Press, Princeton, New Jersey.

[BF] J.-M. Bismut and D. S. Freed, The analysis of elliptic families I: Metrics and connections on determinant bundles, Comm. Math. Phys. 106 (1986), 159–176.

[C] L. Crane, 2-D physics and 3-D topology, to appear in Comm. Math. Phys.

[FK] J. Fröhlich and C. King, The Chern-Simons theory and knot polynomials, Comm. Math. Phys. 126 (1989), 167–199.

[H] N. J. Hitchin, Flat connections and geometric quantization, Comm. Math. Phys. 131 (1990), 347–380.

[KM] R. Kirby and P. Melvin, On the 3-manifold invariants of Witten and Reshetikhin-Turaev for $s\ell(2;\mathbb{C})$, U. C. Berkeley, preprint (1990).

[Ko] T. Kohno, Linear representations of braid groups and classical Yang-Baxter equations, Contemporary Math. Vol. 78 (1988), 339–363.

[KS] K. H. Ko and L. Smolinsky, A combinatorial matrix in 3-manifold theory, Louisiana State University preprint (1990).

[L] W. B. R. Lickorish, Invariants for 3-manifolds from the combinatorics of the Jones polynomial, Cambridge University preprint (1989).

[LMW] R. Lee, E. Miller and S. H. Weintraub, Rochlin invariants, theta functions and the holonomy of some determinant line bundles, J. Reine angew. Math. 392 (1988), 187–218.

[LV] G. Lion and M. Vergne, The Weil Representation, Maslov Index and Theta Series, Prog. Math. 6 (1980).

[MS] G. Moore and N. Seiberg, Classical and quantum conformal field theory, Comm. Math. Phys. 123 (1989), 177–254.

[NS] M. S. Narasimhan and C. S. Seshadri, Holomorphic vector bundles on a compact Riemann surface, Math. Ann. 155 (1964), 69–80.

[P] D. M. Pickrell, to appear.

[PS] A. Presley and G. Segal, Loop Groups, Oxford University Press (1986).

[RT1] N. Ya. Reshetikhin and V. G. Turaev, Ribbon graphs and their invariants derived from quantum groups, Comm. Math. Phys. 127 (1990).

[RT2] N. Ya. Reshetikhin and V. G. Turaev, Invariant of 3-manifolds via link polynomials and quantum groups, M.S.R.I. preprint (1989).

[Se] G. Segal, Two-dimensional conformal field theories and modular functors, in: IXth International Conference on Mathematical Physics, eds. B.Simmon, A. Truman, and I. M. Davies, Adem Hilger (1989), 22–37.

[S1] R. W. Sharpe, On the structure of unitary Steinberg group, Ann. of Math. Vol. 96 (1972), 444–479.

[S2] R. W. Sharpe, Surgery on compact manifolds: the bounded even dimensional case, Ann. of Math. 98 (1973), 187–209.

[S3] R. W. Sharpe, Surgery and unitary K_2, in: An algebraic K-theory III, Hermitian K-theory and Geometric Applications, LNM 343 (Springer Verlag, 1973), 464–471.

[T] V. G. Turaev, A cocycle for the symplectic first Chern class and the Maslov index, Functional Anal. Appl. 18 (1984), 35–39.

[TUY] A. Tsuchiya, K. Ueno and Y. Yamada, Conformal field theory on universal family of stable curves with gauge symmetries, Advanced Studies in Pure Math.19 (1989), 456–566.

[Wa1] C. T. C. Wall, Surgery on Compact Manifolds, Academic Press, 1970.

[Wa2] C. T. C. Wall, Quadratic forms on finite groups and related topics, Topology 2 (1964), 281–298.

[W] E. Witten, Quantum field theory and Jones polynomial, Comm. Math. Phys. 121 (1989), 351–399.

Courant Institute, New York, NY 10012

Yale University, New Haven, CT 06520

Polytechnic University of New York, Brooklyn, NY 11201

An Undetected Slope in a Knot Manifold

Darryl Cooper and Darren Long

0. Introduction

The work of Culler and Shalen **[CS]** using representation varieties to construct incompressible surfaces via degeneration has been seminal; it is still the only general technique available. However, it raises a natural question: Which surfaces are so obtained?

In order to formalize this question, we proceed as follows. Suppose that M is a knot complement, that is to say, a compact 3-manifold with a single torus boundary component. Choose coordinates α, β for $\pi_1(\partial M)$; an embedded incompressible, ∂-incompressible surface $(F, \partial F)$ gives rise to a collection of parallel simple closed loops in ∂M. Choose one such and write the homology class it represents as $\alpha^a \beta^b$; then to $(F, \partial F)$ we associate the rational number a/b (possibly $\infty = 1/0$) which we call the ***boundary slope*** of $(F, \partial F)$. A theorem of Hatcher **[H]** implies that M has only a finite number of boundary slopes.

There are several natural notions in this context, which we now define. We will say that a surface $(F, \partial F)$ is ***detected*** if there is a sequence $\{\rho_n\}$ of representations of the fundamental group of M into $\mathrm{SL}(2, \mathbb{C})$, which lies on a curve inside the character variety of $\pi_1(M)$ and converges to an ideal point, so that if one looks at the action of $\pi_1(M)$ on the tree coming from this sequence the group $\pi_1(F)$ is conjugate into an edge stabiliser.

A boundary slope a/b is ***detected*** if there is a sequence of representations $\{\rho_n\}$ as above with:

(a) The traces $\mathrm{tr}(\rho_n(\alpha^a \beta^b))$ remain bounded as $n \to \infty$.

(b) There is some homology class γ in $\pi_1(\partial M)$ for which $\mathrm{tr}(\rho_n(\gamma))$ does not remain bounded as $n \to \infty$.

The results of **[CS]** imply that many slopes are detected. However, as an example, a fibre of a fibration over S^1 cannot be detected by irreducible representations, but the slope of a fibre may be detected because of the presence of another, non-fibre surface which also has this slope. This happens with the knot 8_{20}. A fibre can be detected if we allow abelian representations.

In this note we give an example which shows that for knots in rational homology spheres it is possible for rather complicated surfaces to be undetected:

Theorem 1.5. *There is a knot complement in a rational homology sphere containing a boundary slope of* $1/6$ *which is not detected.*

Our example is also surprising in the sense that although this slope is not detected by representations into $\mathrm{SL}(2, \mathbb{C})$, it is detected by representations into $\mathrm{PSL}(2, \mathbb{C})$, if one uses some appropriate notions of degeneration; for example that of **[B]**.

Before describing the example, we make some remarks of a general nature. Suppose that a 3-manifold M has two boundary components each of which is a torus. Then there are maps between character varieties

$$i_j : X(M) \longrightarrow X(T_j), \quad j = 1, 2$$

induced by restriction to the boundary. There are now two phenomena we wish to understand. Firstly, the image of one or both of the i_j's may not have complex dimension 2. It seems to be a hard and interesting problem to say when this does or does not happen, but we shall not explore this question here. The second problem is that even if the image is two dimensional, there may be curves which are missed out of the image. Accordingly, if $\dim(i_j(X(M))) = \dim(X(T_j)) = 2$, any component of $i_j(X(M)) \setminus X(T_j)$ of complex dimension 1 we call a ***forbidden curve***.

We observe for example that for a Brunnian link it is always the case that there are two forbidden curves given by $\mathrm{tr}(\mu) = \pm 2$. The special property of the Whitehead link which we exploit is that it has a third forbidden curve coming from the presence of an immersed punctured Klein bottle.

Here is an informal description of the example. Let W be the complement in S^3 of the Whitehead link. By consideration of the representations of the fundamental group $\pi_1(W)$ into $\mathrm{SL}(2,\mathbb{C})$ we find that the simple closed curve C of slope 2 on one of the boundary tori has the property that there is a curve of representations of $\pi_1(W)$ which send C to $+\mathrm{Id}$ but only finitely many which send C to $-\mathrm{Id}$; that is to say, $\mathrm{tr}(C) = -2$ gives a forbidden curve.

The loop D of slope 6 on the trefoil has the property that every irreducible representation of its fundamental group must send D to $-\mathrm{Id}$; so that if we glue these manifolds together in a way which sends C to D, then the possibilities for representations of the resulting manifold are severely limited.

The authors thank the referee for many comments which helped to improve the clarity of this paper.

1. The example

In this section we outline the construction. Of necessity, this involves some calculations; some of these are relegated to an appendix.

Our example is built from the complement of the Whitehead link in S^3; we denote this manifold by W. It has two torus boundary components which can be exchanged by a symmetry. Fix some labelling once and for all of these tori as T_0 (i.e. "outer") and T_i ("inner"). The (meridian, longitude) pairs we denote by (μ_0, λ_0) and (μ_i, λ_i) respectively.

Following **[CCGLS]**, one can compute the $\mathrm{SL}(2,\mathbb{C})$ "eigenvalue variety" of $\pi_1(W)$ (see the appendix), we denote this by $E(\pi_1(W))$. This is defined quite generally (see the appendix), but one can give an especially clear heuristic description for the Whitehead link, since it is a two bridge link. For one sees easily that $\pi_1(W)$ is generated by μ_0 and μ_i. Then given an irreducible representation ρ, we may conjugate so that the generators map

$$\mu_i \mapsto \begin{pmatrix} m_i & 1 \\ 0 & 1/m_i \end{pmatrix} \quad \text{and} \quad \mu_0 \mapsto \begin{pmatrix} m_0 & 0 \\ t & 1/m_0 \end{pmatrix}.$$

The group relation now imposes on m_0, m_i and t a single polynomial constraint $F(m_i, m_0, t) = 0$ which describes an affine algebraic set. (Of course this description is a mild simplification since there is more than one way of doing the preliminary conjugation.)

There are two projection mappings $i_0: E(\pi_1(W)) \to E(\pi_1(T_0))$ and $i_i: E(\pi_1(W)) \to E(\pi_1(T_i))$ which come from restricting a representation to the boundary. The image of these maps describes which eigenvalue pairs are possible on the boundary tori. We use (m_0, ℓ_0) etc. as notation for the eigenvalue pairs.

Note that the image of either mapping cannot possibly fill up all of $E(\pi_1(T_0)) = \mathbb{C}^2$, since the Whitehead link is a pure link so that restricting $m_0 = \pm 1$ forces $\ell_0 = \pm 1$. Thus the image $i_0(E(\pi_1(W)))$ misses out most of the curves $m_0 = \pm 1$. Since we are dealing with eigenvalues, the image must also avoid the curves $m_0 = 0$ and $\ell_0 = 0$. We shall show explicitly in this example that there is another curve which is (largely) missed out; calculation of this curve is our next task.

In order to find what eigenvalues are possible on the boundary tori, we use elimination theory **[M]**; details are deferred to the appendix. The result is that we form polynomials in the eigenvalue triples (m_i, m_0, ℓ_i) and (m_i, m_0, ℓ_0). We call these the outer and inner polynomials denoting them $O(m_i, m_0, \ell_i)$ and $I(m_i, m_0, \ell_i)$ respectively.

As explained in the appendix, the crucial property of these polynomials is that given any triple of nonzero complex numbers (m_i, m_0, ℓ_0) with $O(m_i, m_0, \ell_0) = 0$ (or a triple (m_i, m_0, ℓ_i) with $I(m_i, m_0, \ell_i) = 0$), there is an irreducible representation of $\pi_1(W)$ whose restriction to the inner boundary torus has eigenvalue m_i for the meridian, and on the outer boundary torus has eigenvalues (m_0, ℓ_0) for the meridian-longitude pair.

Lemma 1.1. *If a representation of the Whitehead link complement has $m_0^2.\ell_0 = -1$, then m_0 is ± 1 or $\pm i$.*

Proof. The loop $C = \mu_0^2 \lambda_0$ can be expressed in terms of the generators μ_i, μ_0 of the fundamental group of the Whitehead link as a conjugate of

$$\mu_0(\mu_i \mu_0^{-1} \mu_i)\mu_0(\mu_i \mu_0^{-1} \mu_i)^{-1}.$$

Putting r $= (\mu_i \mu_0^{-1} \mu_i)$, we recognise this as the relator in the Klein bottle group. (It follows that there is an immersion of a punctured Klein bottle into the Whitehead link with boundary $\mu_0^2 \lambda_0$.)

Now suppose that we have a representation ρ of the Whitehead link for which $\mathrm{tr}(\rho(\mu_0^2 \lambda_0)) = -2$. If $\rho(\mu_0^2 \lambda_0)$ is parabolic, then it follows that ρ is parabolic on all of T_0. Alternatively if $\mu_0^2 \lambda_0 = -I$, then $\rho(r \mu_0 r^{-1}) = -\rho(\mu_0^{-1})$, and it follows that $\mathrm{tr}(\rho(\mu_0)) = 0$. We have shown that if $\mathrm{tr}(\rho(\mu_0^2 \lambda_0)) = -2$ then $\mathrm{tr}(\rho(\mu_0)) = 0,\ \pm 2$, but $\mu_0, \mu_0^2 \lambda_0$ is a basis of $\pi_1(T_0)$, from which it follows that there are only finitely many possibilities for the character of ρ restricted to T_0. □

Remark. An alternative proof comes from the fact that if one puts $\ell_0 = -1/m_0^2$ in $O(m_i, m_0, \ell_0)$ then the resulting polynomial has only the given roots together with the root $m_i = 0$, which is forbidden for a representation.

This is the special feature of the Whitehead link complement which we are about to exploit; a representation of $\pi_1(W)$ having $m_0^2\ell_0 = -1$ has only finitely many possibilities for m_0. (Remarkably, this fails to be true for the curve $m_0^2\ell_0 = +1$.) It would be interesting to have other hyperbolic examples.

We now return to the matter in hand:

Lemma 1.2. *Let F denote an incompressible, ∂-incompressible surface in the Whitehead manifold which is detected in* $\mathrm{SL}(2,\mathbb{C})$. *If the slope of F on the inner torus is $1/6$, the only slopes possible on the outer torus are 2, -24, 26. All these slopes occur.*

Proof. We recall from **[CCGLS]** that information concerning boundary slopes may be obtained from the Newton polygon. We briefly sketch the salient details. Given a polynomial f(x,y) we define its ***Newton polygon*** to be the convex hull of the set in the plane defined by: $\{(r,s)|x^r y^s$ has nonzero coefficient in $f(x,y)\}$.

Roughly, the edges of the Newton polygon describe ways of going to infinity in the affine variety f(x,y)=0 which keep some ratio $x^\alpha y^\beta$ bounded. Let us briefly recall how this works. For notational simplicity, suppose that the Newton polygon has a slope of zero and that all coefficients lie below this slope. Then we may write $f(x,y) = y^\beta P_\beta(x) + \sum_{r=o}^{\beta-1} y^r P_r(x)$ where the $P_j(x)$ are polynomials in x. Factoring out y^β we see that for very large y the term which dominates the behaviour of the equation $f(x,y)=0$ is $P_\beta(x)$ and it follows that we may obtain points on the curve $f(x,y)=0$ provided we choose x to be very close to a root of $P_\beta(x)$. In other words, there is a way of going to infinity along the curve defined by $f(x,y)$ which keeps x bounded, i.e., a slope of zero.

The precise result is the following:

Theorem [CCGLS]. *Suppose that a knot has eigenvalue variety defined by the polynomial $f(\lambda,\mu)$. Form a convex body in the $\lambda\mu$-plane by taking the convex hull of the points $\{(\lambda,\mu)|\lambda^r\mu^s$ has non zero coefficient in $f(\lambda,\mu)\}$. Then the slopes of faces of this convex body are boundary slopes of the knot.*

The above theorem has a natural interpretation for any curve of representations, even in the context of links; namely it describes what happens when one restricts a curve of representations to a boundary component. The reason for this is the following: If one projects such a curve into the $\mathbb{C}^2$ of possible eigenvalues associated with a boundary component, the complex dimension of the image can be at most one. If this image is a point, then the sequence is constant on the boundary component. If not, then (after possibly taking Zariski closure) we see a complex curve in $\mathbb{C}^2$. It is a general fact **[Ha**, Proposition 1.13] that a curve in $\mathbb{C}^2$ is the zero set of a single polynomial in two variables and therefore has associated to it a Newton polygon. Ideal points on this curve give rise to splittings of the group of the link and the boundary slopes on the given component give slopes of surfaces which can arise.

To prove 1.2, we seek to understand degenerations with the property that the class $\mu_i \cdot \lambda_i^6$ has bounded trace. If we have any curve of degenerations then the behaviour for the eigenvalues of the inner boundary component is coded by the polyonomial $I(m_i, m_0, \ell_i)$. The possibilities for the eigenvalues can be understood by replacing m_i by C/ℓ_i^6, where C is any function bounded away from 0 and ∞. Making this substitution in $I(m_i, m_0, \ell_i)$; this yields a polynomial which has Newton polygon given in Figure 1a.

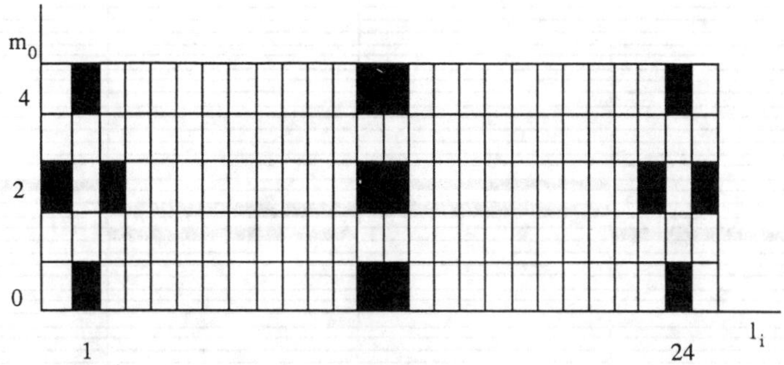

Figure 1a

A nonzero term $m_0^r \ell_i^s$ in the polynomial is indicated by a black rectangle in the s, r entry. Notice that coefficients of the extremal vertices are as indicated in Figure 1b.

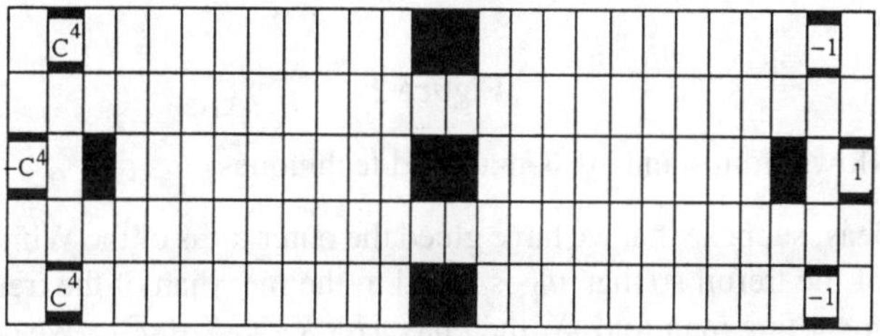

Figure 1b

The slopes in this picture are ± 2 and 0. The 0 slope corresponds to the case when ℓ_i is bounded away from ∞, that is to say, the slope on the inner torus is 0 which we have excluded in our hypothesis.

Analysis of the slopes of ± 2 now follows along exactly the lines of the first paragraph of this proof. Consider the slope of 2 coming from the top left of the Newton polygon. Since C is bounded away from ∞, two terms which dominate are $-C^4 m_0^2(1 - m_0^2 \ell_i)$ so that we may go to infinity on the curve in a way which has $m_0^2 \ell_i$ converging to 1. Similarly for the other slopes and we see that in all cases we have m_0^2 converges to $\ell_i^{\pm 1}$.

There are two cases, and making the substitutions $m_i = -1/\ell_i^6$ and $\ell_i = m_0^{\pm 2}$ in $O(m_i, m_0, \ell_0)$ we see that in either case, we obtain a polynomial which has Newton polygon indicated in Figure 2.

It follows we obtain the candidate slopes of 2, 26 and -24. Further, from the remarks made above all these boundary slopes do arise in W, completing the proof. □

Lemma 1.3. *Suppose that $(F_j, \partial F_j)$ is an incompressible, ∂-incompressible surface in a manifold $(M_j, \partial M_j)$ for $j = 1, 2$. Then gluing a component of ∂M_1 to a component of ∂M_2 so that ∂F_1 is glued to ∂F_2 yields an incompressible, ∂-incompressible surface.*

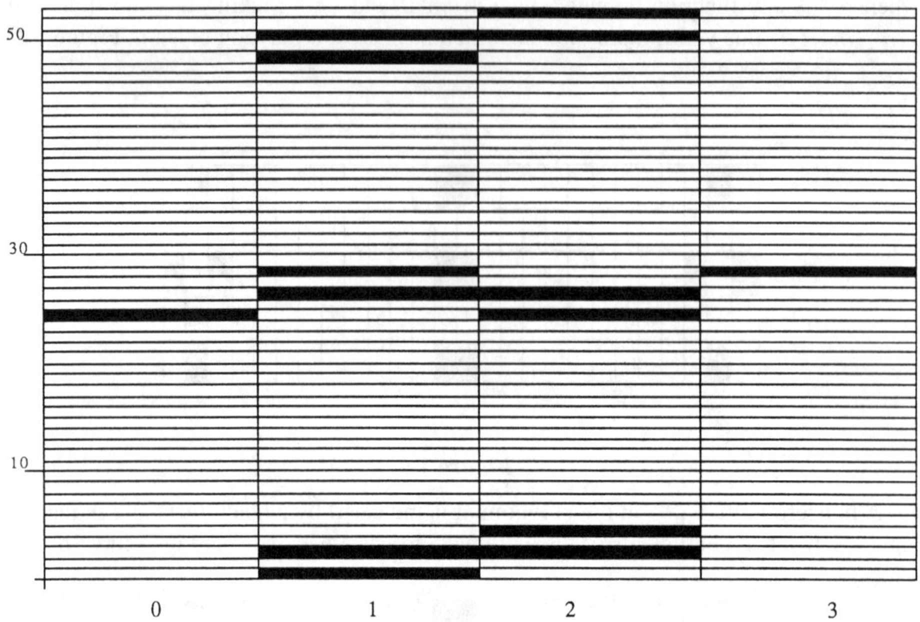

Figure 2

Proof. This follows from standard 3-manifold techniques. □

To fix our ideas, suppose that we have glued the outer torus of the Whitehead manifold to the exterior of the trefoil so that μ_0 is glued to the meridian of the trefoil μ_T and λ_0 to $\mu_T^4\lambda_T$. The resulting manifold X then has $H_1(X) = \mathbb{Z} \oplus \mathbb{Z}_4$. We continue to refer to the boundary torus of X as the inner torus, etc.

In this gluing we see that the loop of slope 2 on the outer torus is glued to the loop of slope 6 in the trefoil. The incompressible annulus of the trefoil has slope 6, and it follows that:

Corollary 1.4. *The manifold X contains an incompressible, ∂-incompressible surface U, of slope $1/6$.* □

Our main claim is:

Theorem 1.5. *The slope $1/6$ is not detected by degeneration of representations into* $\mathrm{SL}(2,\mathbb{C})$.

Proof. The crucial subclaim here is:

Subclaim. *The only degenerations of the Whitehead link group which are bounded on the outer torus produce slopes 0 or 4 on the inner torus.*

Proof of subclaim. Since the representation is bounded on the outer torus, m_0 is bounded away from 0 and ∞. We can therefore plot a Newton polygon for $I(m_i, m_0, \ell_i)$; see Figure 3.

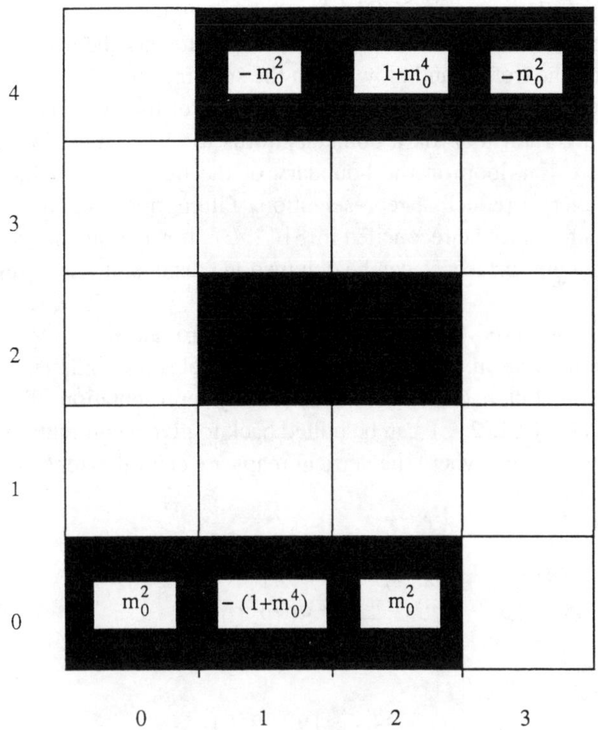

Figure 3

As above, since m_0 is bounded away from 0 and ∞ the only slopes possible on the inner torus are 0 and 4. This proves the subclaim. □

Suppose that this slope is detected, that is to say, there is a degeneration ρ_n of $\pi_1(X)$, so that the representations are blowing up on the boundary torus of X. There are two possibilities:

Case A. The representations ρ_n restrict to abelian representations of the trefoil complement. For such representations, the longitude of the trefoil has eigenvalue 1. We claim that this cannot be the only loop on the splitting torus with bounded trace, since the longitude of the trefoil is glued to the loop on the outer torus $\mu_0^{-4}\lambda_0$, and -4 was not in the list of possibilities of Lemma 1.2. Therefore the degeneration is bounded on the outer torus. But the subclaim implies that this produces slopes of 0 or 4 on the inner torus, a contradiction.

Case B. The representations are nonabelian on the trefoil. Then a calculation shows that the curve of slope 6 on the trefoil is mapped to Id by every such representation. Since this curve is glued to the curve of slope 2 on the outer torus, we see via Lemma 1.1 that only finitely many representations are possible on the splitting torus; whence the ρ_n are bounded and we again obtain a slope of 0 or 4 on the outer torus.

In either case, we obtain a contradication, completing the proof. □

Remark. Using the techniques of **[CCGLS]**, one can compute the A-polynomial for this knot; its Newton polygon has slopes 0, 4 and ± 1; these latter arising from representations which are abelian on the trefoil, and blow up on the splitting torus.

What has happened here is that most representations of the Whitehead link complement which carry a certain loop on a boundary torus to $\pm\mathrm{Id}$, must in fact carry it to $+\mathrm{Id}$. Further, there is a loop on the boundary of the trefoil complement which has to go to $-\mathrm{Id}$ for every irreducible representation. Gluing these curves together gives representations which cannot be reconciled in $\mathrm{SL}(2,\mathbb{C})$; however it shows that (with appropriate definitions) the surface U can be obtained via degenerations of representations into $\mathrm{PSL}(2,\mathbb{C})$.

We have already observed that $H^2(X;\mathbb{Z}_2)$ is nonzero and these representations do not lift. We outline the reason for this — the following elegant argument was shown to us by Andrew Casson. Let $\rho\colon G \to \mathrm{PSL}(2,\mathbb{C})$ be any representation. Then the natural projection $\mathrm{SL}(2,\mathbb{C}) \to \mathrm{PSL}(2,\mathbb{C})$ can be pulled back to give a commutative diagram of groups and homomorphisms, where the vertical maps are central extensions:

$$\begin{array}{ccc} \mathbb{Z}_2 & & \mathbb{Z}_2 \\ \downarrow & & \downarrow \\ G' & \xrightarrow{\rho'} & \mathrm{SL}(2,\mathbb{C}) \\ \downarrow & & \downarrow \\ G & \xrightarrow{\rho} & \mathrm{PSL}(2,\mathbb{C}) \ . \end{array}$$

Now the action of G on $\mathbb{Z}_2$ is trivial, and so the possibilities for groups G' correspond to elements of $H^2(G;\mathbb{Z}_2)$. If G is the fundamental group of a Haken 3-manifold M, then $H^2(M;\mathbb{Z}_2) \cong H^2(G;\mathbb{Z}_2)$. By Poincaré duality one sees that if the boundary of M is a torus and $H_1(M;\mathbb{Z}_2) \cong \mathbb{Z}_2$ then $H^2(M;\mathbb{Z}_2) \cong 0$. Thus $G' \cong \mathbb{Z}_2$, and so $\rho'|G$ is a lift of ρ to $\mathrm{SL}(2,\mathbb{C})$.

In particular, this shows that for knot complements in S^3 representations into $\mathrm{PSL}(2,\mathbb{C})$ always lift (compare **[C]**). Thus an incompressible surface in a knot complement is detected by $\mathrm{SL}(2,\mathbb{C})$ if and only if it is detected by $\mathrm{PSL}(2,\mathbb{C})$.

Concluding Remarks. This paper grew out of an attempt to show that every knot has an irreducible representation into $\mathrm{SL}(2,\mathbb{C})$, the idea being that one would try to patch together representations across incompressible tori. In the notation established above, this means that one needs to understand the image $i_0(E(\pi_1(W)))$. This involves two parts; one first needs to show that $i_0(E(\pi_1(W)))$ has complex dimension two, then to understand the forbidden curves. Both of these problems appear to be hard. One means of attack is given by:

Conjecture. *Suppose that M is a compact hyperbolic 3-manifold with boundary a torus and that the interior of M admits a complete hyperbolic structure ρ. Suppose that α is any non-trivial element of the fundamental group of M. Is it true that $\mathrm{tr}(\rho\alpha)$ is not constant on the component of the representation space containing ρ?*

An affirmative answer to this conjecture would mean that the projection of the character variety of a 2-component hyperbolic link N to the character variety of one of its boundary components T say always has complex dimension 2. The reason is that by **[T]** the component of the character variety of N containing the complete representation has complex dimension 2. Suppose that the restriction of this character variety to T has dimension 1, then there is a hyperbolic Dehn filling of T using a representation ρ of N. This gives a 3-manifold M with a single torus boundary component T' say, and since ρ may be varied keeping $\rho|T$ fixed, it follows that the trace of the core of the Dehn filling does not vary as ρ is varied near the complete representation of N.

Finally, we mention that although the most useful notion for a slope is that it be detected, there is a weaker notion: One may say that a slope is ***weakly detected*** if it is the boundary of some detected surface. The result of **[CS]** show that detected slopes are weakly detected, however the annulus in a connected sum shows that the converse may be false.

2. Appendix: Some calculations

With respect to the meridians shown below (Fig. 4), the fundamental group of the complement of the Whitehead link is presented as

$$< \mu_i, \mu_0 \mid \mu_i^{-1}\mu_0\mu_i\mu_0^{-1}\mu_i\mu_0\mu_i^{-1}\mu_0^{-1} = \mu_i\mu_0^{-1}\mu_i^{-1}\mu_0\mu_i^{-1}\mu_0^{-1}\mu_i\mu_0 > .$$

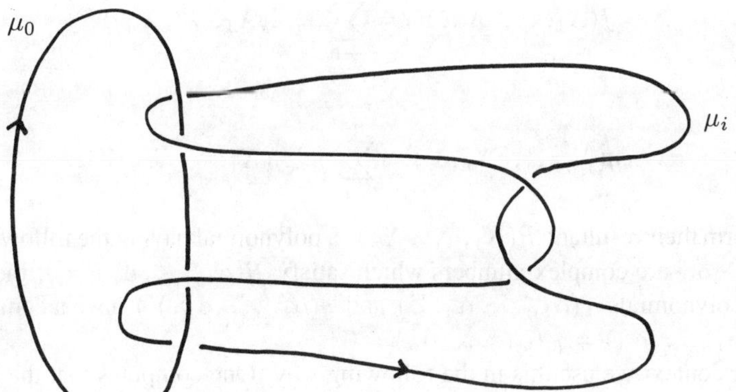

Figure 4

We are interested in representations up to conjugacy, so one easily finds that it is sufficient to consider representations

$$\mu_i \mapsto \begin{pmatrix} m_i & 1 \\ 0 & 1/m_i \end{pmatrix} \quad \text{and} \quad \mu_o \mapsto \begin{pmatrix} m_o & 0 \\ t & 1/m_o \end{pmatrix}.$$

Then the condition that we obtain a representation is the single polynomial

$$F(m_i, m_0, t) = -(m_i m_0) + m_i^3 m_0 + m_i m_0^3 - m_i^3 m_0^3 + \\ +(m_i^2 - m_i^4 + m_0^2 - 4m_i^2 m_0^2 + m_i^4 m_0^2 - m_0^4 + m_i^2 m_0^4)t + \\ +(-(m_i m_0) + 2m_i^3 m_0 + 2m_i m_0^3 - m_i^3 m_0^3)t^2 - m_i^2 m_0^2 t^3 = 0.$$

The theory of **[CS]** usually uses traces and the character variety, but this involves a loss of information in the sign of the slopes when considering the Newton polygon.

Briefly, one can form an eigenvalue variety as follows. Let $X(\pi_1(W))$ be a component of the affine algebraic set of characters. Then there is a projection map $X(\pi_1(W)) \to X(\pi_1(\partial W))$, coming from the restriction map, the target being some subvariety of $X(\pi_1(T_0)) \times X(\pi_1(T_i))$.

Considering the relevant component of the representation variety $R(\pi_1(W))$, we may form by projection an eigenvalue variety for the boundary $E(\pi_1(\partial W))$. Then there is a map $E(\pi_1(\partial W)) \to X(\pi_1(\partial W))$ which is generically two to one. Then the eigenvalue variety $E(\pi_1(W))$ comes from forming the pull back in the diagram

$$\begin{array}{ccc} & & E(\pi_1(\partial W)) \\ & & \downarrow \\ X(\pi_1(W)) & \longrightarrow & X(\pi_1(\partial W)) \end{array} .$$

We are interested in finding what eigenvalues are possible on the boundary. To do this one uses elimination theory **[M]** for which we briefly recall some details. Given polynomials

$$f(X_1, \ldots, X_n, Y) = \sum_{i=0}^{n} p_{n-i}(X_1, \cdots, X_n) Y^i$$

and

$$g(X_1, \ldots, X_n, Y) = \sum_{i=0}^{m} q_{m-i}(X_1, \cdots, X_n) Y^i,$$

we can form their resultant $R(X_1, \ldots, X_n)$, a polynomial having the following property. If $\alpha_1, \ldots, \alpha_n$ are complex numbers which satisfy $R(\alpha_1, \ldots, \alpha_n) = 0$, then either:
(i) the polynomials $f(\alpha_1, \ldots, \alpha_n, Y)$ and $g(\alpha_1, \ldots, \alpha_n, Y)$ have a common root, or
(ii) $p_0(\alpha_1, \ldots, \alpha_n) = q_0(\alpha_1, \ldots, \alpha_n) = 0$.

In our context we use this in the following way. One computes that the longitude on the outside torus is the word

$$\mu_0^{-1} \mu_i^{-1} \mu_0 \mu_i \mu_0^{-1} \mu_i \mu_0 \mu_i^{-1},$$

and one computes using the above generators a rational function $Q(m_0, m_i, t)$ which is the $(1,1)$ entry of this matrix; that is to say, if we set ℓ_0 to be the eigenvalue of the longitude on the outer torus, we have $\ell_0 = Q(m_0, m_i, t)$, which may be re-written as a polynomial function $g(m_0, m_i, t, \ell_0) = 0$. We then eliminate the variable t between this polynomial and $F(m_i, m_0, t)$.

This yields:

$$O(m_i, m_0, \ell_0) = \ell_0(-1+m_0)(1+m_0)(1+\ell_0 m_0^2)+ \\ +m_i^4\ell_0(-1+mo)(1+m_0)(1+\ell_0 m_0^2)+ \\ +m_i^2(1+\ell_0^2+2\ell_0 m_0^2-2\ell_0^2 m_0^2-\ell_0 m_0^4-\ell_0^3 m_0^4).$$

In general, elimination theory involves loss of information in the affine context because of possibility (ii). However, in our case the highest power of t occurring in F is a cube and its coefficient is $m_i^2 m_0^2$ so that we see that given any triple of nonzero complex numbers m_i, m_0, ℓ_0 which satisfy $O(m_i, m_0, \ell_0) = 0$, (i) applies and we may find a representation of $\pi_1(W)$ which gives this eigenvalue triple on the boundary.

References

[B] M. Bestvina, Degenerations of the hyperbolic space, Duke J. Math 56 (1988), 143–161.

[CCGLS] D. Cooper, M. Culler, H. Gillet, D. D. Long and P. B. Shalen, Plane curves associated to character varieties of 3-manifolds, Preprint.

[C] M. Culler, Lifting representations to covering groups. Adv. in Math. 59 (1986), 64–70.

[CS] M. Culler and P. B. Shalen, Varieties of group representations and splittings of 3-manifolds, Annals of Math. 117 (1983), 109–146.

[Ha] R. Hartshorne, "Algebraic Geometry", Graduate Text Mathematics 52, Springer Verlag, 3rd. printing 1983.

[H] A. Hatcher, On the boundary curves of incompressible surfaces. Pacific J. of Math. 99 (1982) 373–377.

[HT] A. Hatcher and W.P. Thurston, Incompressible surfaces in 2-bridge knot complements, Inv. Math. 79 (1985) 225–246.

[M] D. Mumford, "Algebraic Geometry I: Complex Projective Varieties", Grundlehren 221, Springer-Verlag, 1976.

[T] W. Thurston, "The geometry and topology of 3-manifolds", Princeton notes (1978).

Department of Mathematics, University of California, Santa Barbara, CA 93106
Email: cooper%henri@hub.ucsb.edu

Department of Mathematics, University of California, Santa Barbara, CA 93106
Email: long%henri@hub.ucsb.edu

Heegaard Splittings of the Three-Ball

*Charles Frohman**

0. Introduction

This paper investigates the topology of a class of Heegaard splittings of the three-ball. These Heegaard splittings have an easy topological classification. This is in contrast to the fact that arbitrary Heegaard splittings of the three-ball cannot possibly have a simple classification. Inspired by this we include an added condition in the definition of a Heegaard splitting and we call Heegaard splittings not satisfying this condition naive Heegaard splittings. The added condition is that the splitting surface carry the fundamental groups of the two three-manifolds obtained by cutting along the splitting surface. More or less this is equivalent to requiring that the Heegaard splitting can be studied using Haken's lemma. There does not exist much literature on Heegaard splittings of manifolds with boundary. What little there is only considers Heegaard surfaces that do not have boundary such as in **[C-G]** and **[B-O]**. The naive notion of Heegaard splitting is implicit in works such as **[M]**. The author believes that it is fair to give credit to Bonahon, Casson, Gordon, and Otal for giving the proper notion of a Heegaard splitting with boundary. The arguments in this paper form a good lesson on the systematic use of Haken's lemma in the study of Heegaard splittings.

As an application we give a short proof of a theorem of Meeks and Yau. Let M be a three-manifold and let F and F' be surfaces embedded in M, then F and F' are topologically equivalent if there exists a homeomorphism $h: M \to M$ so that $h(F) = F'$. If h can be chosen with $h|_{\partial M} = Id$ then F and F' are topologically equivalent relative to ∂M.

Theorem 3.1. *(Meeks and Yau) If F and F' are two connected properly embedded minimal surfaces of finite type in $\mathbb{R}^3$ and if F and F' have the same topological type, then F and F' are topologically equivalent as subsets of $\mathbb{R}^3$.*

Finally at the end of the last section we ask two questions which we deem to be worthy of future study. We would like to thank Bill Meeks for explaining his work with Yau. Also we would like to thank Richard Skora for suggesting the use of graphs to describe the topology of our splittings on the boundary.

* Work partially supported by NSF grant DMS 900293

1. Preliminaries

We work in the topologically locally flat category. Unless otherwise specified, all manifolds will be connected. A three-manifold M is said to be irreducible if every sphere S embedded in M bounds a ball. A compact orientable irreducible three-manifold H with connected boundary ∂H is a handlebody if the inclusion map $i: \partial H \to H$ induces a surjection $\pi_1(\partial H) \to \pi_1(H)$. This implies that H has a handle structure having only zero handles and one handles.

Let M be a three-manifold. A naive Heegaard splitting of M is a properly embedded surface F so that there exist handlebodies H_1 and H_2 contained in M with $H_1 \cup H_2 = M$ and $H_1 \cap H_2 = F$. The surface F is a Heegaard splitting if there exist submanifolds C_1 and C_2 of M with $C_1 \cup C_2 = M$ and $C_1 \cap C_2 = F$ having the properties: 1) the C_i are irreducible and 2) the inclusions $\pi_1(F) \to \pi_1(C_i)$ are surjections. In the language of Casson-Gordon or Bonahon-Otal, the manifolds C_i are compression bodies. As is proved in **[C-G]** and **[B-O]**, a form of Haken's lemma can be used to study such Heegaard splittings.

Generalized Haken's Lemma. *Let F be a Heegaard splitting of M.*

i) If there exists a sphere embedded in M that does not bound a ball, then there exists a sphere in M that does not bound a ball, intersecting M in a single closed curve of transverse intersection.

ii) Let D be a properly embedded disk in M so that ∂D is a disjoint from ∂F. Further suppose that ∂D is not null homotopic in the complement of F. There exists a disk D' so that $\partial D' = \partial D$ and the intersection of D' with F is a single circle of transverse intersection. □

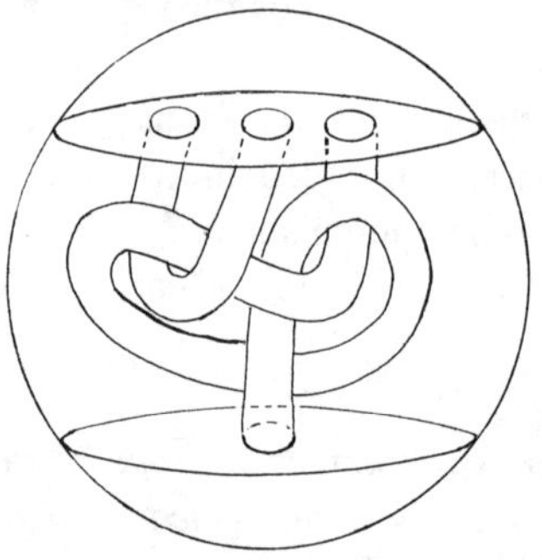

Figure 1a

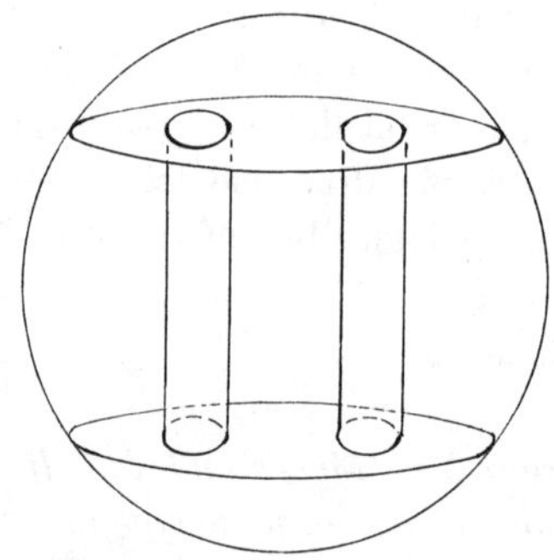

Figure 1b

In Figure 1a a naive Heegaard splitting of the three-ball is pictured. In Figure 1b a Heegaard splitting of the three-ball is shown. The problem with the naive definition of Heegaard splitting is that it does not rule out knotting.

Figure 2 is a genus one Heegaard splitting of the three-ball with one boundary component. It is an easy exercise to show that up to topological equivalence this is the only such Heegaard splitting of the three-ball. There is an operation on Heegaard splittings called stabilization. Let F be a Heegaard splitting of M. Let B be a ball contained in M so

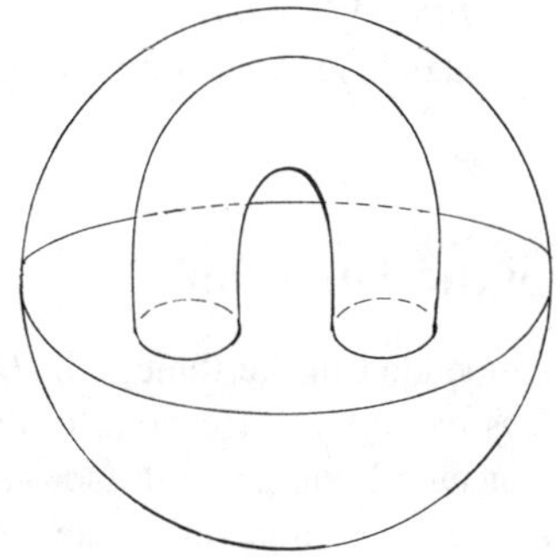

Figure 2

that $B \cap F$ is a disk. Replace the part of F inside the ball by a genus one Heegaard splitting of B having boundary $\partial B \cap F$. This operation is well defined up to topological equivalence. A Heegaard splitting is reducible if there exists a sphere S intersecting F in a single circle c so that the circle c does not bound a disk in F. If the manifold M is irreducible, it follows from a theorem of Waldhausen **[W1]** that F is the result of stabilizing a Heegaard splitting of lower genus. The result of Waldhausen being used, states that a Heegaard splitting of the ball with one boundary component is the result of repeatedly stabilizing a properly embedded disk. Hence two Heegaard splittings of the ball having the same boundary (a single circle) are topologically equivalent relative to the boundary of the ball.

Let $F \subset M$ be a properly embedded surface in a three-manifold. A compressing disk for F is an embedded disk D with the properties that $D \cap F = \partial D$, and ∂D does not bound a disk in F. Let $D \times [0,1]$ be a collaring of D in M so that $D \times [0,1] \cap F = \partial D \times [0,1]$. Let F' be the surface $(F - \partial D \times [0,1]) \cup D \times \{0,1\}$. The surface F' is the result of 1-surgery on F along D. Let $p \in \operatorname{int} D$, the arc $p \times [0,1]$ is called a cocore of the 1-surgery. Let k be an embedded arc in M such that $k \cap F = \partial k$. Let $k \times D$ be a collaring of k so that $k \times D \cap F = \partial k \times D$. Let F'' be the surface $(F - \partial k \times D) \cup k \times \partial D$. The surface F'' is the result of 0-surgery on F along k. Let $p \in \operatorname{int} k$ be a point. The disk $p \times D$ is called a cocore of the 0-surgery. Topologically 1-surgery and 0-surgery are inverses of one another. The result of 0-surgery along the cocore of a 1-surgery on F is topologically equivalent in M to F. The result of 1-surgery along the cocore of a 0-surgery on F is topologically equivalent in M to F.

If F is a properly embedded surface in M and F has a compressing disk, then we say that F is compressible. If F has no compressing disks, then F is incompressible. A boundary compression is a disk D embedded in M so that the boundary of D consists of an arc a in F and arc b in ∂M. Furthermore, the arc a should be all of the intersection of F and D, and the arc a should not cut a disk out of F. If such a disk exists, then F is boundary compressible. If no such arc exists, then F is boundary incompressible. There is an operation on F completely analogous to surgery only using a boundary compression. If C is an irreducible three-manifold and F is a surface in ∂C that carries the fundamental group of C, then the only incompressible boundary incompressible surfaces S in C with ∂S contained in F are either boundary parallel or disks. To prove the generalized Haken's lemma, combine this with the proof of Haken's lemma given in **[J]**.

2. Heegaard Splittings of the Three-Ball

Let F be a surface properly embedded in the three-ball B. The graph $\Gamma(F)$ is constructed as follows. The vertices of $\Gamma(F)$ correspond to the connected components of $\partial B - F$. Two vertices are connected by an edge if and only if the closures of the corresponding components of $\partial B - F$ have nonempty intersection. The graph $\Gamma(F)$ is a tree. The distance between two vertices v and v' in $\Gamma(F)$ is the least number of edges of $\Gamma(F)$ that must be transversed in a path from v to v'. The diameter of $\Gamma(F)$ is the maximum distance between two points in $\Gamma(F)$. We say two graphs are equivalent if there is a $1-1$ correspondence between their vertices that corresponds pairs of vertices that are connected by an edge to pairs of vertices connected by an edge. The goal of this section is to prove:

Theorem 2.1. *Let F and F' be two Heegaard splittings of the three-ball B. Then F and F' are topological equivalent if they have the same genus and $\Gamma(F)$ is equivalent to $\Gamma(F')$. Furthermore, if F and F' have the same boundary and the same genus, then they are equivalent relative to ∂B.*

The proof will be in two steps. The first step is to show that if F is a Heegaard splitting of the three-ball of positive genus, then F is obtained by repeatedly stabilizing a planar surface. The second step is to show that planar Heegaard splittings F of the three-ball are classified by the equivalence class of $\Gamma(F)$. Using Haken's lemma this can be reduced to the case when F is a disk, annulus or pair of pants. These three-cases can be worked quite easily.

Proposition 2.2. *Let F be a Heegaard splitting of the three-ball. If the genus of F is positive, then F is the result of repeatedly stabilizing a planar surface.*

Proof. Every surface F has an intersection pairing

$$H_1(F) \otimes H_1(F) \to Z\,.$$

The genus of the surface is one half the maximum dimension of a subspace of $H_1(F)$ on which the intersection form is nondegenerate. Let C_1 and C_2 be the submanifolds of B given in the definition of a Heegaard splitting. If F has positive genus one of the subspaces $\ker\{H_1(F) \to H_1(C_i)\}$ is not contained in the null space of the intersection form, otherwise the dimension of $H_1(B)$ would be positive. Without loss of generality, we may assume that there exists $x \in \ker\{H_1(F) \to H_1(C_i)\}$ so that x is not in the null space of the intersection form. The class x can be represented by a simple closed curve c on F, and there exists a properly embedded surface G in C_1 with $\partial G = c$. Since $\pi_1(F) \to \pi_1(C_1)$ is a surjection the set $\pi_1(C_1, F)$ consists only of the trivial arc. Consequently, the only incompressible boundary incompressible surfaces in C_1 with boundary in F are disks. Hence, there is a sequence of surgeries and disk modifications on S so that the resulting surface G' is a family of disks. This process does not alter the homology class of ∂G in F. Therefore, there is a properly embedded disk D in C_1 with ∂D a nonseparating curve. Let s' be a simple closed curve in C_1 that intersects D in a single point of transverse intersection so that there exists an annulus A embedded in C_1 whose boundary is the union of s and s' . We claim that F is a Heegaard splitting of $B - N(s')$ where $N(s')$ is a regular neighborhood of s' that has been chosen small enough that it is contained in the interior of C_1. To check that F is a Heegaard splitting of $B - N(s')$ we must prove that $\pi_1(F) \to \pi_1(C_1 - N(s'))$ is a surjection. The annulus A can easily be chosen so that it intersects D in a single arc of transverse intersection. Let W be a regular neighborhood of $A \cup D$ in C_1. We show the configuration in Figure 3. Notice that the frontier of W in C_1 is a disk D'. The fundamental group of $C_1 - N(s')$ is the free product of the fundamental group of $W - N(s')$ and the complement of W. The fundamental group of F is the amalgamated product of the fundamental group of $W \cap F$ and the complement of W along $\partial(W \cap F)$. From Figure 3 it is clear that the fundamental group of $W \cap F$ surjects onto the fundamental group of $W - N(s')$. By hypothesis the fundamental group of the complement of W in F surjects onto the fundamental group of the complement of W. This confirms that F is a Heegaard splitting of $B - N(s')$.

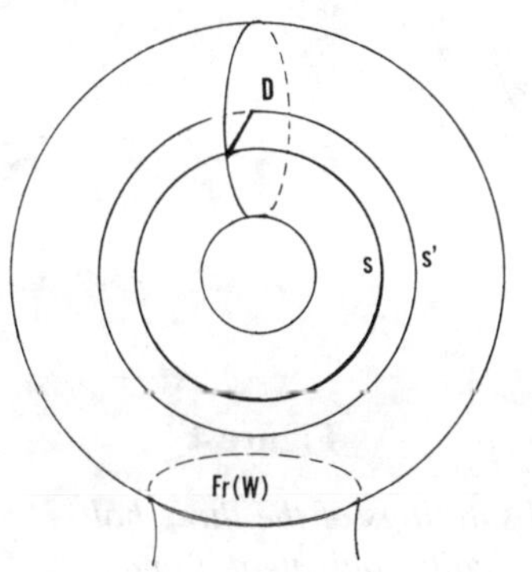

Figure 3

Notice that ∂B is a sphere in $B - N(s')$ that does not bound a ball. By Haken's lemma there exists a sphere S not bounding a ball in $B - N(s')$ intersecting F in a single simple closed curve of transverse intersection. The part of F lying inside the ball that S bounds in B must have positive genus, otherwise, C_1 or C_2 is not irreducible. By Waldhausen's theorem F is the result of stabilizing a lower genus surface. An induction on genus shows that F is the result of repeatedly stablizing a planar Heegaard splitting.□

In **[F]** we give a very similar proof of the following reducibility criterion that was conjectured by Bob Craggs. As an exercise the reader might give his own proof. It is easy to see that if F is a Heegaard splitting of a closed manifold M then there exists an embedded graph X so that F is the boundary of a regular neighborhood of X. Suppose that M is irreducible. If there exists a ball B in M with the property that $B \cap X$ contains a cycle of the graph X then F is reducible.

A circle c embedded in a surface F is boundary parallel if there exists an annulus A embedded in F with ∂A consisting of c and a boundary component of F. A surface F is irreducible if the only circles embedded in F are boundary parallel. It is easy to see that a surface is irreducible if and only if it is a disk, annulus or pair of pants. If F is a disk or annulus, then up to topological equivalence there is only one naive Heegaard splitting of the ball where F is the splitting surface. Furthermore, if there are two such with the same boundary, then they are topologically equivalent relative to the boundary of the ball. Hence we will concentrate on the pair of pants.

Let P denote the pair of pants. The boundary of P consists of three circles. It is easy to check that if P is properly embedded in the ball, then $\Gamma(P)$ is either a line segment or a star as pictured in Figure 4.

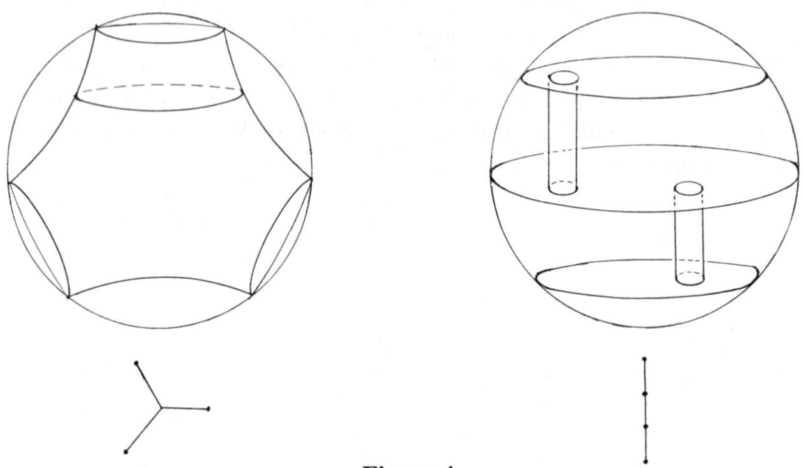

Figure 4

Lemma 2.3. *Two Heegaard splittings of the three-ball B where the splitting surface is a pair of pants P are topologically equivalent if and only if they have the same graph. If two Heegaard splittings of the ball by a pair of pants have the same boundary, then they are topologically equivalent relative to ∂B.*

Proof. Let C_1 and C_2 be the two three-manifolds obtained from B by cutting along P. In the case that $\Gamma(P)$ is a star then without loss of generality we can assume ∂C_1 consists of the union of P and a pair of pants P' lying in ∂B, and ∂C_2 is a sphere. Since ∂C_1 is a surface of genus two, $\dim H_1(C_1) \geq 2$. On the other hand the inclusion of P into C_1 induces a surjection of fundamental groups, so that $\dim H_1(C_1) \leq 2$. Since the fundamental groups of P and C_1 are free it can be concluded that $\pi_1(P) \to \pi_1(C_1)$ is an isomorphism. It now follows from a theorem of Waldhausen **[W2]** that C_1 is homeomorphic to $P \times [0,1]$ where the restriction of the homeomorphism to $P \times \{0\}$ is the inclusion of P into C_1. On the other hand, since the ball is irreducible, C_2 is a ball. The theorem now follows in the case that $\Gamma(P)$ is a star.

Now assume that $\Gamma(P)$ is a line segment. The closures of the complement of ∂P in ∂B consist of two disks and two annuli. Push offs of the two disks into the ball yield two compressing disks for P. The result of 1-surgery along these is three-disks, D_1, D_2 and D_3. Without loss of generality, we may assume that there exist four balls B_1, B_2, B_3 and B_4 embedded in B so that the union of the B_i is B, and the intersection of B_i with B_{i+1} is D_i. Let k_1 and k_2 be the cocores of the surgeries (see Figure 5). Since P is connected we may assume that one component of the boundary of k_i is in D_i and the other is in D_{i+1}. Since the result of boring out along the k_i is a Heegaard splitting the boundary of a collaring of k_i must carry the fundamental group of $B_{i+1} - k_i$. By Papakyriakopoulos' unknotting lemma the arcs k_i are unknotted. From these observations it is now clear how to construct a homeomorphism between any two of these.

The second part of the theorem follows from a careful rereading of the proofs already given with a new hypothesis. □

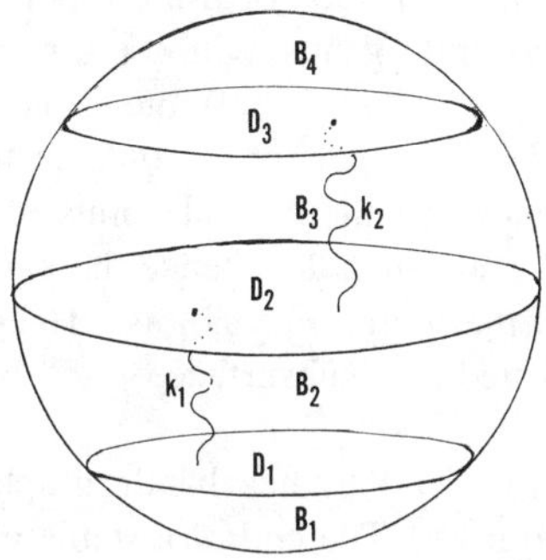

Figure 5

Proof of Theorem 2.1. Let F be a Heegaard splitting of the three-ball. If F has positive genus, then by Proposition 2.2 it is the result of stabilizing a planar Heegaard splitting. Hence we may assume that F is planar. Let $\Gamma(F)$ be the graph of F. We will assume that $\Gamma(F)$ has more than four vertices. If the valence of a vertex is greater than three, there exists a disjoint family of simple closed curves cutting the offending component of the complement of $\partial B - \partial F$ into a collection of pairs of pants. Using Haken's lemma we can find a family of disks embedded in the ball having these circles as boundary so that each disk intersects F in a single circle of transverse intersection. We call such disks reducing disks. Cutting along this disk we get a family of Heegaard splittings of the ball so that the corresponding graphs have no vertices of valence greater than three. If Theorem 2.1 is true for such graphs, then it is true in general, by just gluing the resulting homeomorphisms together. If $\Gamma(F)$ has more than one valence three vertex, then we can choose a circle that lies in a component of $\partial B - F$ that bounds a reducing disk, so that the result of cutting leaves two graphs, each with fewer valence three vertices. Hence we can assume that $\Gamma(F)$ has at most one valence three-vertex. If $\Gamma(F)$ has a valence three-vertex we can choose curves inside the component of $\partial B - \partial F$ that are parallel to each boundary component and bound reducing disks. The result of cutting leaves a star and some line segments. If any line segment has length greater than three, then we can choose reducing disks to cut it into line segments of length less than or equal to three. We have now reduced the proof of Theorem 2.1 to Lemma 2.3.

3. Applications

By a minimal surface in $\mathbb{R}^3$ we mean a connected properly embedded surface F without boundary that has mean curvature zero. This implies that the gaussian curvature of F is nonpositive. Let $d_p : \mathbb{R}^3 \to \mathbb{R}$ be distance from the point p. For a generic point p the function d_p restricted to F will be a morse function. Since F has nonpositive gaussian curvature, this morse function can have only index zero and index one critical points. In **[F-M]** it is proved that if F is a minimal surface in $\mathbb{R}^3$ and H_1 and H_2 are the two three-manifolds obtained by cutting $\mathbb{R}^3$ along F, then the inclusions $F \to H_i$ induce surjections on the fundamental groups. This is a property special to minimal surfaces. For instance, in **[M]** an embedding of a surface of negative gaussian curvature is given that does not have this property.

Suppose that F is a minimal surface of finite topological type. This means that F has finitely many ends and finite genus. Choose a generic d_p. There exists R so that the boundary of $B_R = \{d \mid d_p(x) \leq R\}$ intersects F transversely and $F \cap B_R$ is homotopy equivalent to F. The part of F outside the B_R is homeomorphic to $\partial B_R \cap F \times [0, \infty)$, this means that the critical points of d_p outside of B_R come in cancelling pairs of 0-handles and 1-handles, hence the part of F lying outside the ball is topologically standard. Furthermore, $F \cap B_R$ is a Heegaard splitting of B_R. From this one sees that F is classified as a subsurface of $\mathbb{R}^3$ by the graph of this Heegaard splitting.

A theorem of Meeks and Yau **[M-Y]** states that the graph of $F \cap B_R$ is a line segment. The idea of the proof is as follows. The end of a stable minimal surface is asymptotic to a catenoid or a plane. In either case there is a well defined tangent plane at infinity. That is, the tangent planes to the minimal surface tend towards one tangent direction as you head to infinity. Two such stable ends are disjoint if and only if the corresponding tangent planes are the same. If F is a minimal surface in $\mathbb{R}^3$, then it is possible to find a family of simple closed curves on F that separate any two ends of F and bound a stable minimal surface in the complement of F. The result now follows from studying these minimal surfaces. These results prove:

Theorem 3.1. [Meeks–Yau]. *Two minimal surfaces of finite type in $\mathbb{R}^3$ are topologically equivalent if and only if they have the same topological type.*

It can also be seen that the considerations above allow us to prove the following theorem about nonpositively curved surfaces in $\mathbb{R}^3$. We say that a surface F is good if it carries the fundamental groups of the result cutting $\mathbb{R}^3$ along F. If F has finite type we define the graph of F at infinity to be the graph associated with the intersection of F with a ball that is so large that outside the ball F is just a product of its intersection with the boundary of the ball and the positive reals.

Theorem 3.2. *Two good nonpositively curved surfaces in $\mathbb{R}^3$ of finite type are topologically equivalent if and only if they have the same topological type and the same graph at infinity.*

This raises the question of the topology of good nonpositively curved surfaces that are not of finite type. The following question is inspired by the work of Meeks and Yau on minimal surfaces. The idea is to pick a good system of balls that exhaust $\mathbb{R}^3$, and look at the graphs of the surface inside those balls.

Question 3.3. Is there an $\mathbb{R}$-tree at infinity that allows a classification of good nonpositively curved surfaces in $\mathbb{R}^3$ to be given?

Along another line of inquiry, suppose that H is a handlebody, and F is a closed Heegaard splitting of H. It is a result of Bonahon and Otal that such Heegaard splittings of M are classified by their genus. This can be proved quite easily from our viewpoint by choosing a family of meridian disks for the handlebody that are reducing disks for F.

Question 3.4. Is there a finiteness theorem for Heegaard splittings of handlebodies that parallels the classification of Heegaard splittings of the three-ball given here. Can such a theorem be used to give finiteness theorems for Heegaard splittings of arbitrary irreducible three-manifolds?

The idea would be to find a class of Heegaard splittings of handlebodies that admit a finiteness theorem based on genus and the number of boundary components. Then try to prove finiteness theorems for Heegaard splittings of three-manifolds by putting any Heegaard splitting of a given genus in good position with a chosen Heegaard splitting.

References

[B-O] F. Bonahon, J-P Otal, Scindement de Heegaard des éspaces lenticulaires, Ann. Sci. Ec. Norm. Sup. IV Ser. 16, (1983), 451–466.

[C-G] A. Casson, C. McA.Gordon, On reducing Heegaard splittings of three-manifolds, Top. and Appl., (1987), 275–284.

[F] C. Frohman, The topological uniqueness of triply periodic minimal surfaces in R^3, Jour. of Diff. Geo. 31 (1990), 277–284.

[F-M] C. Frohman, W. Meeks, The topological uniqueness of Heegaard splittings and minimal surfaces with one end in R^3, Preprint.

[J] W. Jaco, "Lectures on three-manifold topology" C.B.M.S. regional conference series in mathematics 43, Amer. Math. Soc., 1977.

[M] W. Meeks, The topological uniqueness of minimal surfaces in three dimensional Euclidian space, Topology 20 (1979), 389–410.

[M-Y] W. Meeks, S. T. Yau, The topology of minimal surfaces of finite type, Preprint.

[W1] F. Waldhausen, Heegaard Zerlegungen der Drei-Sphäre, Topology 7 (1967) 195–203.

[W2] F. Waldhausen, On irreducible three-manifolds which are sufficiently large, Annals of Math. 87 (1968), 56–88.

University of Iowa, Department of Mathematics, Iowa City, IA 52242

On the Borromean Orbifolds: Geometry and Arithmetic

Hugh M. Hilden, María Teresa Lozano**
*José María Montesinos-Amilibia**

Introduction

This paper is a continuation of earlier work by the authors on universal knots, links, and groups. (**[HLM1]**, **[HLM2]**, **[HLMW]**). A knot or link in S^3 is ***universal*** if every closed oriented 3-manifold occurs as a branched covering of S^3 with branch set the knot or link. A finitely generated group of hyperbolic isometries, or Kleinian group, U, is called ***universal*** if the orbit space of the action of U on $\mathbb{H}^3$ is homeomorphic to S^3 and if every closed oriented 3-manifold M^3 occurs as the orbit space of a finite index subgroup $G(M^3)$ of U.

The concept of universal group grew naturally out of the concept of universal link. If the image of the axes of rotation of U is a knot or link k in S^3, the natural map $\mathbb{H}^3/G(M^3) \to \mathbb{H}^3/U = S^3$ is a branched covering with branch set k, so the knot or link k is universal. The group U acting on $\mathbb{H}^3$ endows its quotient S^3 with the structure of hyperbolic orbifold. The singular set is the knot or link k.

In **[HMLW]**, it was shown that universal groups existed and in **[HML2]** we studied in detail a particularly simple example of a universal group, the universal group of the Borromean rings (see Figure 1). (That is, the link k is the Borromean rings.)

Walter Neumann and Alan Reid pointed out to the authors that the universal group of the Borromean rings, which we studied in **[HLM2]**, (the group $B(4,4,4)$ in notation we shall shortly introduce), is in fact an 'arithmetic' group, as follows from results of Vinberg (**[Vi]**). Thus this group has 'more structure' (**[HLM2]**).

It follows directly from this observation, and the results of **[HLMW]**, that every closed oriented 3-manifold has the structure of an 'arithmetic orbifold'. So an immediate question arises as to how rare a flower is an arithmetic orbifold in the garden of hyperbolic orbifolds.

Evidence that arithmetic orbifolds are rare in dimension two was obtained by Takeuchi (**[T3]**) who produced the complete, but finite list of arithmetic groups among the infinitude of triangle groups. In fact he showed that for fixed signature σ there are only finitely many conjugacy classes of arithmetic Fuchsian groups of signature σ (**[T4]**).

Further evidence can be found in Reid's thesis (**[R]**) where arithmetic Kleinian groups and their Fuchsian subgroups are studied. Also, Borel (**[B]**) showed that the set of volumes of arithmetic hyperbolic 3-orbifolds is a discrete subset of $\mathbb{R}$.

* This research was supported by grants PB85-0336 and PB89-0105. The first author expresses his appreciation to the DGICYT for financial support while he was in Spain.

In this paper we produce a three parameter family $\widehat{B}(m,n,p)$, $3 \leqq m,n,p \leqq \infty$, of hyperbolic orbifolds with singular set the Borromean rings. (The 'angle' at the three components is $2\pi/m,\ 2\pi/n,\ 2\pi/p$.) In this family exactly eleven members are arithmetic: $(3,3,3)$, $(4,4,4)$, $(6,6,6)$, $(3,4,4)$, $(3,6,6)$, (∞,∞,∞), $(3,\infty,\infty)$, $(4,\infty,\infty)$, $(3,3,\infty)$, $(3,4,\infty)$, $(4,4,\infty)$.

Analogously, in another paper, **[HLM3]**, which relies on **[HKM]**, we classify the arithmetic figure eight orbifolds. (There are exactly six.)

The organization of the paper is as follows:

The first section deals with hyperbolic geometry preliminaries. The relationship between the Beltrami-Klein model of $\mathbb{H}^3$ and the Poincaré model of $\mathbb{H}^3$ is made via the 'intermediate' model used in **[HKM]** which facilitates the computation of isometries.

In the second section the basic definition of arithmetic groups and arithmetically definable groups are explained. Certain results (widely dispersed in the literature) are collected and set down.

In the third section we construct the orbifolds $\widehat{B}(m,n,p)$ in the Klein model of hyperbolic 3-space, and their Kleinian groups $B_{m,n,p} \leqq \mathrm{SL}(2,\mathbb{C})$.

In the fourth section, using the characterization of Vinberg (**[Vi]**) of arithmetic subgroups of $O(F_o,\mathbb{R})$ generated by reflections we obtain the list of arithmetic triples.

Finally, in the last section, we compute the quaternion algebras for the eleven arithmetic cases of Section 4, and we compute the orders in the quaternion algebras that determine their arithmeticity. We also compile some arithmetic information about the field $\mathbb{Q}(\alpha)$, where $\alpha^4+\alpha^2-1=0$, and about the quaternion algebra $\left(\frac{-1,-1}{\mathbb{Q}(\alpha)}\right)$, the relevant field and algebra for the universal arithmetic orbifold $\widehat{B}(4,4,4)$. In the appendix we use Reid's criterion to show non-arithmeticity in the cases $m,n,p \notin \{3,4,6,\infty\}$.

§1. Preliminaries on hyperbolic geometry

1. The points of hyperbolic 3-space $\mathbb{H}^3$ in the standard Beltrami-Klein model are the interior points of the unit ball centered at the origin in $\mathbb{R}^3$. In homogeneous coordinates, $(x:y:z:t)$, for $\mathbb{R}P^3$, the equation of S^2 is

$$x^2+y^2+z^2-t^2=0\,. \tag{1.1}$$

2. The connection between this model of $\mathbb{H}^3$, and the Poincaré half space model is made via the stereographic projection of S^2 on its equatorial plane $\mathbb{C}P^1$ from the point $(0:0:1:1)$.

The connection between the two models is clarified if we use a model which we shall call the ***intermediate model***. (Compare **[F]** and **[HKM]**.) We choose new coordinates for $\mathbb{C}\mathrm{P}^3$:

$$\theta_1 = x+iy,\quad \theta_2 = z-t,\quad \theta_3 = z+t,\quad \theta_4 = -x+iy\,. \tag{1.2}$$

The equation of the sphere S^2 becomes

$$\det\begin{vmatrix}\theta_1 & \theta_2\\ \theta_3 & \theta_4\end{vmatrix} = 0\,.$$

The intermediate model of $\mathbb{H}^3$ consists of the points of $\mathbb{R}P^3$ interior to the sphere

$$\{(\theta_1:\theta_2:\theta_3:\theta_4)\,|\,\theta_1\theta_4-\theta_2\theta_3=0,\quad \theta_2=\overline{\theta}_2,\quad \theta_3=\overline{\theta}_3,\quad \theta_1=-\overline{\theta}_4\}.$$

The points of $\mathbb{H}^3$, which we shall represent by matrices, satisfy

$$\det\begin{vmatrix}\theta_1 & \theta_2\\ \theta_3 & \theta_4\end{vmatrix}>0;\quad \theta_2=\overline{\theta}_2,\quad \theta_3=\overline{\theta}_3,\quad \theta_1=-\overline{\theta}_4.$$

We call this model intermediate because the point

$$\begin{vmatrix}\theta_1 & \theta_2\\ \theta_3 & \theta_4\end{vmatrix};\quad \theta_2=\overline{\theta}_2,\quad \theta_3=\overline{\theta}_3,\quad \theta_1=-\overline{\theta}_4,$$

also represents a circle in $\mathbb{C}\mathrm{P}^1$. In fact the equation

$$z=\frac{\theta_1\overline{z}+\theta_2}{\theta_3\overline{z}+\theta_4}\tag{1.3}$$

is the equation of a circle whose squared radius is $\det\begin{vmatrix}\theta_1 & \theta_2\\ \theta_3 & \theta_4\end{vmatrix}\Big/\theta_3^2$. Thus the circles of radius zero in $\mathbb{C}\mathrm{P}^1$ (the 'points' at $\mathbb{C}\mathrm{P}^1$) are represented by the points $\begin{bmatrix}\theta_1 & \theta_2\\ \theta_3 & \theta_4\end{bmatrix}$ with determinant zero.

The function which assigns to a point $z\in\mathbb{C}P^1$ the circle of radius zero and center z, represented by the point $\begin{bmatrix}\theta_1 & \theta_2\\ \theta_3 & \theta_4\end{bmatrix}$ of the intermediate model, is the stereographic projection of $\mathbb{C}\mathrm{P}^1$ on S^2 from the south pole of S^2. Moreover, we know that the stereographic projection assigns to a circle of radius different from zero (it can be real or purely imaginary) a circle of S^2. Now if the circle taken in $\mathbb{C}\mathrm{P}^1$ is that of equation (1.3) *its stereographic image is a circle of S^2 whose pole is* $\begin{bmatrix}\theta_1 & \theta_2\\ \theta_3 & \theta_4\end{bmatrix}$.

3. An immediate corollary to this observation is a convenient procedure for finding the equation in the Poincaré model of the reflection in a plane P in the Beltrami-Klein model. One proceeds as follows:

(i) Calculate the point $P^{\perp}$, the pole of P with respect to S^2, in coordinates $\begin{pmatrix}\theta_1 & \theta_2\\ \theta_3 & \theta_4\end{pmatrix}$ of the intermediate model using the change of coordinates (1.2);

(ii) The equation of reflection in P in the Poincaré model is

$$z'=\frac{\theta_1\overline{z}+\theta_2}{\theta_3\overline{z}+\theta_4}.$$

4. In the following section we mention, in passing, another representation of $\mathbb{H}^3$, known as the hyperboloid model. This is one of the leaves of the hyperboloid determined by the equation $f=-1$, where f is a quadratic form on $\mathbb{R}^4$ of signature $(3,1)$. This defines a structure of Minkowski space in $\mathbb{R}^4$. The metric induced in one of the leaves of $f=-1$ is Riemannian and defines the hyperboloid model of $\mathbb{H}^3$.

Letting F represent the matrix of f, we denote by $O(F,\mathbb{R})$ the group of matrices, M, in $\mathrm{GL}(4,\mathbb{R})$ such that $M^TFM=F$. The index two subgroup leaving invariant each leaf of the hyperboloid $f=-1$ is the group $\mathrm{Iso}\,\mathbb{H}^3$ of isometries of $\mathbb{H}^3$.

§2. Arithmetically Defined Groups: Preliminaries ([R], [R1])

1. Motivation and history. Let G be a locally compact topological group with a fixed left invariant Haar measure. A subgroup $\Gamma \leqq G$ is a ***lattice*** if it is discrete and if the homogeneous space G/Γ has finite measure. We say Γ has ***finite co-volume***. The group Γ is ***co-compact*** if G/Γ is compact.

As an example consider $G=\mathrm{SL}(n,\mathbb{R})$ and $\Gamma=\mathrm{SL}(n,\mathbb{Z})$. The group G is a semi-simple Lie group and Γ is a lattice in G. (It is true, but not at all obvious, that Γ has finite co-volume.) The lattice Γ is ***non co-compact***.

As a second example let $G=\mathbb{R}^n$ and $\Gamma_0=\mathbb{Z}^n$. In this example every lattice Γ in G is 'arithmetic', that is there is an automorphism $\pi:\mathbb{R}^n\to\mathbb{R}^n$ such that $\pi(\Gamma)=\mathbb{Z}^n$.

This second example serves as motivation to study a two part problem. Given G:

1. Define when a lattice $\Gamma \leqq G$ is 'arithmetic'.
2. Determine whether every lattice is 'arithmetic'.

For the first part of the problem we shall need to apply 'arithmetic methods' to do the study of G. It turns out that the most convenient class of groups to work with is the class of *connected semi-simple Lie groups with trivial center*, because a group G of this type is always a *linear algebraic group* defined over $\mathbb{Q}$.

Of course, the two Lie groups in our examples have non trivial centers, but the study of the general case reduces to the study of the trivial center case via Lie theory.

For these groups there is a general construction of arithmetic lattices due to Borel and Harish-Chandra that we shall describe later. (See **[Z**, pp. 1–5**]** or **[BH]**.)

The second part of the problem has the following solution. If the $\mathbb{R}$-rank of G is greater than one (the dimension of a maximal torus is greater than one) every lattice is arithmetic. This is a theorem of Margulis.

But the case where the $\mathbb{R}$-rank of G is one, the case that we are interested in, is different. In fact if G is $\mathrm{Iso}^+\,\mathbb{H}^3$, the group of direct isometries of hyperbolic three space, non arithmetic lattices exist. (Makarov, Vinberg.)

Vinberg considered $\mathrm{Iso}^+\,\mathbb{H}^3$ as a *real* simple connected Lie group with trivial center. He adapted the Borel and Harish-Chandra definition of arithmeticity of a lattice and characterized the lattices generated by reflections. He obtained non arithmetic lattices. (His arithmeticity criteria is given in terms of the Gram matrix of a polytope that is the fundamental domain of Γ.)

But Vinberg did not notice that $\mathrm{Iso}^+\,\mathbb{H}^3$ is also a complex Lie group (isomorphic to $\mathrm{PSL}(2,\mathbb{C})$) (see the appendix to the translation of **[Vi]**). Serre observed that if $\mathrm{Iso}^+\,\mathbb{H}^3$ is considered as a complex Lie group there are other possibilities for constructing arithmetic lattices via the Borel and Harish-Chandra method. In fact there are more arithmetically definable subgroups than those considered by Vinberg. This follows simply from the fact that $\mathbb{C}$ contains more algebraic number fields than $\mathbb{R}$.

Therefore for the moment we shall distinguish, following Vinberg, between ***real-arithmetic*** and ***complex-arithmetic*** groups.

The study of real-arithmetic groups probably began with Fricke and Klein, the complex-arithmetic groups with Bianchi. A characterization of the arithmetic subgroups of $\mathrm{Iso}^+ H^2$ was given by Takeuchi **[T2]**. This characterization was generalized to $\mathrm{Iso}^+ \mathbb{H}^3$ by Reid in his thesis **[R]**.

Our interest in these groups arises from the study of a particular subgroup of $\mathrm{Iso}^+ \mathbb{H}^3$, the group $B(4,4,4)$. (The notation will be explained later.) This group has the following *universal* property:

For every closed orientable three manifold M^3 there exists $\Gamma(M^3)$ of finite index $\leqq B(4,4,4)$ such that $\mathbb{H}^3/\Gamma(M^3) = M^3$.

It turns out that $B(4,4,4)$ is 'arithmetic'. We suspect that this fact will have interesting consequences in the theory of three manifolds. Therefore we shall study the arithmeticity of $B(4,4,4)$, as well as that of a family of groups $B(m,n,p)$ naturally associated with $B(4,4,4)$, in a very concrete way.

2. Linear algebraic groups; arithmetic subgroups.

Definition. A ***linear algebraic group defined over*** k, where k is a finite extension of $\mathbb{Q}$, is a subgroup G of $\mathrm{GL}(n,\mathbb{C})$ defined as follows.

$$G = \{(a_{ij}) \in \mathrm{GL}(n,\mathbb{C}) \mid p(a_{ij}, \det(a_{ij})^{-1}) = 0 \text{ for every } p \text{ in } I\} \tag{1}$$

Here I is an ideal in the polynomial ring $k[X_{11},\cdots,X_{nn},Y]$. Note that I is finitely generated so that it suffices to consider a finite set of polynomials in (1). Note also that the entries of a linear algebraic group defined over k need not lie in k. $\mathrm{SL}(n,\mathbb{C})$ is a linear algebraic group defined over $\mathbb{Q}$.

The significance of this definition is that G is at the same time a group and an affine subvariety of $\mathrm{GL}(n,\mathbb{C})$ defined over k. In effect we identify $\mathrm{GL}(n,\mathbb{C})$ with the affine subvariety V of $\mathbb{C}^{n^2+1}$ given by the equation

$$\det(X_{ij})Y = 1.$$

Here the matrix (a_{ij}) is identified with the point $(a_{11},\cdots,a_{nn},\det(a_{ij})^{-1})$. The coordinate ring of the variety V is

$$\mathbb{C}[a_{11},\cdots,a_{nn},\det(a_{ij})^{-1}].$$

For every ideal $I \subset k[a_{11},\cdots,a_{nn},\det(a_{ij})^{-1}]$ we obtain an affine subvariety of $\mathrm{GL}(n,\mathbb{C})$ defined over k (see **[Re]**). This affine subvariety is sometimes a group.

If B is a subring of $\mathbb{C}$ we define

$$\mathrm{GL}(n,B) = \{(a_{ij}) \in \mathrm{GL}(n,\mathbb{C}) \mid a_{ij} \in B, \det(a_{ij}) \text{ is a unit of } B\}.$$

We shall denote $G \cap \mathrm{GL}(n,B)$ by G_B. Note that $G_{\mathbb{R}}$ and $G_{\mathbb{C}}$ make sense because a field is a ring and that $G_{\mathbb{C}} = G$.

For every embedding σ of the algebraic number field k in $\mathbb{C}$ we define the linear algebraic group G^σ as follows:

$$G^\sigma = \{(a_{ij}) \in \mathrm{GL}(n,\mathbb{C}) \mid p^\sigma(a_{ij}, \det(a_{ij})^{-1}) = 0 \text{ for every } p \in I\}.$$

Here p^σ is obtained from p by applying σ to the coefficients of p.

Definition. Let G be a linear algebraic group defined over k. A subgroup $\Gamma \leqq G$ is ***arithmetic*** if Γ is commensurable with $G_{O(k)}$ ($\Gamma \cap G_{O(k)}$ has finite index in both Γ and $G_{O(k)}$). Here $O(k)$ is the ring of algebraic integers of k.

In the definition we do not insist that Γ is a lattice. Under certain conditions it can be proven that an arithmetic subgroup of a linear algebraic group G defined over k is a lattice.

Theorem 2.1 (Borel and Harish-Chandra). *Let G be a semi-simple Lie group which is a linear algebraic group defined over k. Let Γ be an arithmetic subgroup of G. Then Γ is finitely generated and moreover*

1. *Γ is a lattice in $G_{\mathbb{R}}$ if and only if $\Gamma \leqq G_{\mathbb{R}}$ and*
   (i) *k is totally real;*
   (ii) *$G^\sigma_{\mathbb{R}}$ is compact for every $\sigma : k \to \mathbb{R}$ such that $\sigma \neq$ identity. (In this case we say $G_{\mathbb{R}}$ is **admissible**.)*
2. *Γ is a lattice in G if and only if $\Gamma \leqq G$ and*
   (i) *k has exactly one complex place (one pair of complex conjugate embeddings);*
   (ii) *$G^\sigma_{\mathbb{R}}$ is compact for every $\sigma : k \to \mathbb{R}$ such that $\sigma \neq$ id or complex conjugation. (In this case we say G is **admissible**.)* □

Example 1. Let F be a symmetric matrix with entries in $k \subset \mathbb{R}$ and equivalent over $\mathbb{R}$ to $(1,1,1,-1)$. Define

$$\mathrm{SO}(F,\mathbb{C}) = \{(a_{ij}) \in \mathrm{GL}(n,\mathbb{C}) \mid (a_{ij})^t F(a_{ij}) = F, \det(a_{ij}) = 1\}.$$

Then $\mathrm{SO}(F,\mathbb{C})$ is a Lie group with center equal to $\pm I$. The group $\mathrm{Ad}\,\mathrm{SO}(F,\mathbb{C})$ of inner automorphisms of $\mathrm{SO}(F,\mathbb{C})$ is a connected simple Lie group with trivial center. It is also a linear algebraic group defined over k, although it is not easy to see why. It can be shown that every automorphism of $\mathrm{SO}(F,\mathbb{C})$ is inner and that a linear map from $\mathbb{C}^{n^2}$ to itself leaves $\mathrm{SO}(F,\mathbb{C})$ invariant and restricts to an automorphism of $\mathrm{SO}(F,\mathbb{C})$ if and only if a finite number of polynomial conditions are satisfied.

The group $\mathrm{Ad}\,\mathrm{SO}(F_0,\mathbb{C})_{\mathbb{R}} = \mathrm{Ad}\,\mathrm{SO}(F_0,\mathbb{R})$ is identified with $\mathrm{Iso}^+\,\mathbb{H}^3$ where F_0 is the diagonal matrix $(1,1,1,-1)$. This is a linear algebraic group defined over $\mathbb{Q}$. It is clear that $\mathrm{Ad}\,\mathrm{SO}(F,\mathbb{R})$ is isomorphic to $\mathrm{Ad}\,\mathrm{SO}(F_0,\mathbb{R}) = \mathrm{Iso}^+\,\mathbb{H}^3$. The isomorphism is of a special type that we shall now consider.

Definition. A homomorphism $f : H \to G$ between two linear algebraic groups defined over k is said to be defined over ℓ, where ℓ is a finite extension of k, if in the expression

$$f(h_{k\ell}) = (g_{ij})$$

the entry g_{ij} is expressable as a polynomial function of the variables $h_{11}, \cdots, h_{nn}$, $(\det(h_{ij}))^{-1}$ with coefficients in ℓ.

If f is an isomorphism and both f and f^{-1} are defined over ℓ for some finite extension ℓ then H is said to be a ***k-form of** G*.

Example 1 (continued). Since F is $\mathbb{R}$-equivalent to F_0, there exists T such that $T^t F_0 T = F$. The entries of T lie in a finite extension ℓ of k. It follows that $\operatorname{Ad}\mathrm{SO}(F,\mathbb{C})$ is a k-form of $\operatorname{Ad}\mathrm{SO}(F_0,\mathbb{C})$ induced by

$$f : \mathrm{SO}(F,\mathbb{C}) \to \mathrm{SO}(F_0,\mathbb{C}),$$
$$M \longmapsto TMT^{-1}.$$

Vinberg **[Vi]** proved that the k-forms of $\operatorname{Ad}\mathrm{SO}(F_0,\mathbb{C})$ are precisely the $\operatorname{Ad}\mathrm{SO}(F,\mathbb{C})$.

Remark. In this example we observed that the group $\operatorname{Ad}\mathrm{SO}(F,\mathbb{C})$ is isomorphic to $\mathrm{SO}(F,\mathbb{C})/\pm I$, the usual definition of $\mathrm{Iso}^+\mathbb{H}^3$. We needed to consider $\operatorname{Ad}\mathrm{SO}(F,\mathbb{C})$ in order to establish that $\mathrm{Iso}^+\mathbb{H}^3$ has the structure of a linear algebraic group defined over k.

The arithmetic subgroups of $\mathrm{SO}(F,\mathbb{C})$ are those commensurable with $\mathrm{SO}(F,\mathbb{C})_{O(k)}$.

Example 2. Let D be a generalized quaternion algebra over the field k. This means that as a k-vector space

$$D = \{x_1 + x_2 i + x_3 j + x_4 ij \mid x_1, x_2, x_3, x_4 \in k\}.$$

And there are two non zero elements a, b in k such that multiplication is determined by the rules $i^2 = a$, $j^2 = b$, $ij = -ji$ and multiplication is to be bilinear over k. The algebra D turns out to be an associative central simple algebra and we denote it by

$$D = \Big(\frac{a,b}{k}\Big).$$

The algebra D has a 'reduced norm' and 'reduced trace' given by the formulas

$$N(x_1 + x_2 i + x_3 j + x_4 ij) = x_1^2 - a x_2^2 - b x_3^2 + ab x_4^2,$$
$$T(x_1 + x_2 i + x_3 j + x_4 ij) = 2x_1.$$

The multiplicative group of invertible elements $D^\star$ of D can be made to operate on D by conjugation $x \longmapsto \{y \mapsto xyx^{-1}\}$. Conjugation preserves both the reduced norm and reduced trace and therefore leaves the three dimensional vector space of elements of trace zero invariant. The reduced norm when restricted to this subspace is a quadratic form represented by the symmetric matrix F with respect to the basis $\{i, j, ij\}$ where

$$F = \begin{bmatrix} -a & 0 & 0 \\ 0 & -b & 0 \\ 0 & 0 & ab \end{bmatrix}.$$

The action by conjugation is trivial only for non zero elements of the center $k^\star$ of $D^\star$. Thus we obtain an isomorphism

$$D^\star/k^\star \to \mathrm{SO}(F,\mathbb{C})_k.$$

In the special case where $k = \mathbb{C}$, $D = M(2, \mathbb{C})$ (there is only the one quaternion algebra over $\mathbb{C}$), $D^\star = \mathrm{GL}(2, \mathbb{C})$ and $D^\star/k^\star = \mathrm{PSL}(2, \mathbb{C})$. In the case where $a = b = -1$ $\mathrm{SO}(F, \mathbb{C}) = \mathrm{SO}(3, \mathbb{C})$, in the usual notation, and we have shown that $\mathrm{PSL}(2, \mathbb{C})$ is a linear algebraic group defined over $\mathbb{Q}$.

It is known that the k-forms of $\mathrm{SO}(3, \mathbb{C})$ are exactly the linear algebraic groups $\mathrm{SO}(3, F)$ defined over k.

The group $\mathrm{PSL}(2, \mathbb{C})$ is also isomorphic to $\mathrm{Iso}^+ \mathbb{H}^3$. Thus $\mathrm{Iso}^+ \mathbb{H}^3$ has the structure of a complex Lie group as well as that of a real Lie group. It is connected and has trivial center.

The group $\mathrm{SL}(2, \mathbb{C})$, the universal cover of $\mathrm{PSL}(2, \mathbb{C})$, is easily seen to be a complex Lie group and a linear algebraic group defined over $\mathbb{Q}$. Its k-forms are the universal covers of the k-forms of $\mathrm{PSL}(2, \mathbb{C})$ and are the groups of elements of norm 1 in the algebra $D \otimes_k \mathbb{C}$, denoted D^1.

The isomorphism $D \otimes_k \mathbb{C} \cong M(2, \mathbb{C})$ restricts to the isomorphism $D^1 \cong \mathrm{SL}(2, \mathbb{C})$ needed to define the k-form.

3. Arithmetically definable groups. Given a connected semi-simple Lie group with trivial center $\widetilde{G}$ there are two constructions of arithmetic lattices considered by Borel and Harish-Chandra.

First construction (real arithmetic lattices). For this construction $\widetilde{G}$ must be a real Lie group with the structure of a linear algebraic group defined over $\mathbb{Q}$.

First we choose a Lie group isomorphism $\varphi : \widetilde{G} \to G_{\mathbb{R}}$. (There may be many to choose from. Here G is a linear algebraic group defined over $\mathbb{Q}$.)

Next we choose a totally real field k. (Since $\mathbb{Q} \leq k$, G is defined over k also.)

Thirdly we choose a k-form of G. That is, a linear algebraic group H defined over k and an isomorphism $\lambda : G_{\mathbb{R}} \to H_{\mathbb{R}}$ defined over some finite extension ℓ of k. (Again, there may be many choices of H and λ.)

However, there is a condition that must be satisfied by H and k. For any $\sigma \neq id$, $\sigma : k \to \mathbb{R}$, $H^\sigma_{\mathbb{R}}$ must be compact.

Fourthly we choose any subgroup of $H_{\mathbb{R}}$ commensurable with $H_{O(k)}$, call it $\widehat{\Gamma}$.

Finally, we choose any group, call it Γ, conjugate in $\widetilde{G}$ to $\varphi^{-1}\lambda^{-1}\widehat{\Gamma}$.

The group Γ is a lattice of $\widetilde{G}$ by Theorem 2.1. Any such group (there are typically many because of the five choices) is called a ***real arithmetic lattice***.

Second Construction (complex arithmetic lattices). For this construction $\widetilde{G}$ must be a complex Lie group with the structure of a linear algebraic group defined over $\mathbb{Q}$.

First we choose a Lie group isomorphism $\varphi : \widetilde{G} \to G$. (Here $G = G_{\mathbb{C}}$ is a linear algebraic group defined over $\mathbb{Q}$. There may be many isomorphisms to choose from.)

Next we choose a number field k with exactly one complex place. (One pair of complex conjugate embeddings of k in $\mathbb{C}$. The rest, if any, are real. Since $\mathbb{Q} \leq k$, G is defined over k also.)

Thirdly we choose a k-form of G. That is, a linear algebraic group defined over k and an isomorphism $\lambda : G \to H$ defined over some finite extension ℓ of k. (Again, there are many choices of H and λ.)

However, there is a condition that must be satisfied by H and k. For any *real* embedding $\sigma : k \to \mathbb{R}$, $H^{\sigma}_{\mathbb{R}}$ must be compact.

Fourthly we choose any subgroup of $H = H_{\mathbb{C}}$ commensurable with $H_{O(k)}$, call it $\widehat{\Gamma}$.

Finally, we choose any group, call it Γ, conjugate in $\widetilde{G}$ to $\varphi^{-1}\lambda^{-1}\Gamma$.

The group Γ is a lattice of $\widetilde{G}$ by Theorem 2.1. Any such group is called a ***complex arithmetic lattice***.

Thus $\Gamma \leqq G$ is a real arithmetic lattice if it can be constructed via the first construction and a complex arithmetic lattice if it can be constructed via the second.

The reader may wonder what happens if, for example, k has two complex places or the 'condition' isn't satisfied. There is a theory for these groups too, expounded in **[BH]**, but it appears to us to have no three dimensional topological application, so we don't deal with it in this paper.

4. The special case $G_{\mathbb{R}} = \mathrm{Ad}\,\mathrm{SO}(F_0, \mathbb{R}) = \mathrm{Iso}^+\,\mathbb{H}^3$. Specializing the previous definition to the case of the real connected centerless Lie group $\mathrm{Ad}\,\mathrm{SO}(F_0, \mathbb{R}) = \mathrm{Iso}^+\,\mathbb{H}^3$ we shall say that $\Gamma \leqq \mathrm{Iso}^+\,\mathbb{H}^3$ is ***real arithmetic*** if, following Vinberg (**[Vi]**),

1. There exists a finite extension k of $\mathbb{Q}$, and a 4×4 symmetric matrix F with entries in k which is ***admissible***, i.e.:
   (i) k is totally real
   (ii) F is equivalent to F_0 over $\mathbb{R}$
   (iii) F^{σ} is *definite* for every non identity embedding $\sigma : k \to \mathbb{R}$;
2. There exists an isomorphism $\lambda : \mathrm{Ad}\,\mathrm{SO}(F_0, \mathbb{R}) \to \mathrm{Ad}\,\mathrm{SO}(F, \mathbb{R})$ such that $\lambda(\Gamma)$ is an arithmetic subgroup of $\mathrm{Ad}\,\mathrm{SO}(F, \mathbb{C})_{\mathbb{R}}$. (This means $\lambda(\Gamma)$ is commensurable with $\mathrm{Ad}\,\mathrm{SO}(F, \mathbb{C})_{O(k)}$).

We shall say that $\Gamma \leqq \mathrm{Iso}\,\mathbb{H}^3$ is real arithmetic if $\Gamma \cap \mathrm{Iso}^+\,\mathbb{H}^3$ is real arithmetic as a subgroup of $\mathrm{Iso}^+\,\mathbb{H}^3$.

(*Remark.* The isomorphism λ need not coincide with the isomorphism defined over $\ell \leqq k$ that defines $\mathrm{Ad}\,\mathrm{SO}(F, \mathbb{C})$ as a k-form of $\mathrm{Ad}\,\mathrm{SO}(F_0, \mathbb{C})$.)

This definition reduces to the previous one because $\mathrm{Ad}\,\mathrm{SO}(F; \mathbb{R}) \cong \mathrm{Ad}\,\mathrm{SO}(4)$ is compact and, as was explained before, the k-forms of $\mathrm{Ad}\,\mathrm{SO}(F_0, \mathbb{C})$ are precisely the $\mathrm{Ad}\,\mathrm{SO}(F, \mathbb{C})$. Thus a real arithmetic subgroup $\Gamma \leqq \mathrm{Iso}\,\mathbb{H}^3$ is a lattice.

At this point, it is convenient to introduce another definition of Vinberg (**[Vi]**). A lattice Γ of $\mathrm{Ad}\,\mathrm{SO}(F_0, \mathbb{R}) = \mathrm{Iso}^+\,\mathbb{H}^3$ (or of $\mathrm{Iso}\,\mathbb{H}^3$) is ***quasi-arithmetic*** if there exists an isomorphism

$$\lambda : \mathrm{Ad}\,\mathrm{SO}(F_0, \mathbb{R}) \to \mathrm{Ad}\,\mathrm{SO}(F, \mathbb{R})$$

where

1. F is admissible;
2. $\lambda(\Gamma)$ has a finite index subgroup contained in $\mathrm{Ad}\,\mathrm{SO}(f, \mathbb{C})_k$.

Note that in order for a *quasi-arithmetic* group Γ to be real arithmetic it is necessary and sufficient that Γ is commensurable with $\operatorname{Ad}\mathrm{SO}(F,\mathbb{C})_{O(k)}$.

The following can be proven. Let $\langle x,y\rangle$ be defined as equal to xF_0y^t. Then a lattice $\Gamma \leqq \operatorname{Iso}\mathbb{H}^3 (= \operatorname{Ad}O(F,\mathbb{R}))$ is quasi-arithmetic if there is a basis $\{e_1,e_2,e_3,e_4\}$ of $\mathbb{R}^4$ such that the matrix $F = (\langle e_i,e_j\rangle)$ satisfies:

(i) F has entries in the totally real number field k;

(ii) F^σ is definite for every $\sigma : k \to \mathbb{R}$, $\sigma \neq id$;

(iii) There exists $\Gamma_1 \leqq \Gamma$, of finite index, whose elements, written as matrices with respect to the basis $\{e_1,\cdots,e_4\}$ have their entries in k.

The lattice Γ is real arithmetic if also

(iv) There exists $\Gamma_1 \leqq \Gamma$ if finite index, whose elements written as matrices with respect to the basis $\{e_1,\cdots,e_4\}$ have their entries in $O(k)$.

A real arithmetic lattice is *co-compact* if and only if it satisfies one of the following conditions:

(i) the field k is not $\mathbb{Q}$;

(ii) the field k is $\mathbb{Q}$, but the form F doesn't represent zero over $\mathbb{Q}$.

5. The special case $G = \mathrm{PSL}(2,\mathbb{C}) = \operatorname{Iso}^+\mathbb{H}^3$. Now we specialize the definition of complex arithmetic groups to the case $G = \mathrm{PSL}(2,\mathbb{C}) = \operatorname{Iso}^+\mathbb{H}^3$.

The quaternion algebra $D = (\frac{a,b}{k})$ is ***admissible*** if

(i) k has exactly one complex place (pair of complex conjugate embeddings);

(ii) D is *ramified* at every real place. ($D \otimes_{k^\sigma} \mathbb{R} = \mathbb{H}$, where $\mathbb{H}$ is the Hamilton quaternion algebra. Here k^σ is the real image of k under the embedding σ.)

Note that D defines a k-form for $\mathrm{PSL}(2,\mathbb{C}) = \mathrm{SO}(F,\mathbb{C})$ where

$$F = \begin{pmatrix} -a & & \\ & -b & \\ & & ab \end{pmatrix}.$$

To say that D is admissible is to imply that $\mathrm{SO}(F^\sigma,\mathbb{C})_{\mathbb{R}} = \mathrm{SO}(3,\mathbb{R})$ which is compact or that $D^1 \otimes_{k^\sigma} \mathbb{R} = S^3$ is compact also.

Definition (Compare Borel **[B]**). $\Gamma \leq \mathrm{PSL}(2,\mathbb{C}) = \operatorname{Iso}^+\mathbb{H}^3$ is ***complex arithmetic*** if there exists an admissible quaternion algebra $D = (\frac{a,b}{k})$ and an isomorphism λ : $\mathrm{PSL}(2,\mathbb{C}) \to \mathrm{SO}(F,\mathbb{C})$ such that $\lambda(\Gamma)$ is an arithmetic subgroup of $\mathrm{SO}(F,\mathbb{C})$ (a subgroup commensurable with $\mathrm{SO}(F,\mathbb{C})_{O(k)}$).

Definition (Compare **[V]**, **[R]**). A discrete subgroup $\widetilde{\Gamma} \leqq \mathrm{SL}(2,\mathbb{C})$ is ***complex arithmetic*** if there exists an admissible quaternion algebra $D = (\frac{a,b}{k})$ and an order θ of D, such that $\widetilde{\Gamma}$ is commensurable with $f(\theta^1)$ where $\theta^1 = \theta \cap D^1$ and

$$f : D^1 \to D^1 \otimes \mathbb{C} = \mathrm{SL}(2,\mathbb{C}) \quad \text{is the canonical embedding.}$$

If $\widetilde{\Gamma}$ is of finite index in $f(\theta^1)$ we say that $\widetilde{\Gamma}$ is derived from a quaternion algebra.

Remark. As we noted earlier, $D^1 \otimes \mathbb{C}$ can be studied as a linear algebraic group defined over k. Its arithmetic subgroups are precisely those which like $\widetilde{\Gamma}$ are commensurable with $f(\theta^1)$, where θ is an order.

Proposition. $\widetilde{\Gamma} \leqq \mathrm{SL}(2,\mathbb{C})$ *is complex arithmetic if and only if* $\Gamma \leqq \mathrm{PSL}(2,\mathbb{C})$ *is complex arithmetic where* $\widetilde{\Gamma}$ *is the inverse image of* Γ *in the map* $\mathrm{SL}(2,\mathbb{C}) \to \mathrm{PSL}(2,\mathbb{C})$.

The proof follows from the fact that the image of $\theta^1 = \theta \cap D^1$ is commensurable with $\mathrm{SO}(F,\mathbb{C})_{O(k)}$ under the map μ

$$\begin{aligned} \mu : D^1 &\to \mathrm{SO}(F,\mathbb{C})_k \\ g &\longmapsto \{d \mapsto gdg^{-1}\}. \end{aligned} \qquad \square$$

An important consequence of this proposition is that it is usually unnecessary to distinguish between Γ and $\widetilde{\Gamma}$.

We deduce from Theorem 2.1 that Γ is complex arithmetic and a lattice. This lattice is co-compact if and only if D is a division algebra. If this is not the case, then k has no real place. In effect $D \otimes_k K$ is not a division algebra for any extension K of k.

The relationship between real- and complex-arithmetic groups is given by the following theorem of Reid (**[R]**):

Theorem (Reid). *The set of complex-arithmetic subgroups of* $\mathrm{Iso}^+ \mathbb{H}^3$ *that contain Fuchsian subgroups (lattices in* $\mathrm{Iso}^+ H^2$*) coincides with the set of real-arithmetic subgroups of* $\mathrm{Iso}^+ \mathbb{H}^3$.

The Russian school (Makarov, Vinberg, Nikulin, etc.) works with groups generated by reflections and they study real arithmeticity. The following proposition in combination with Reid's theorem above, shows that for reflection groups, real arithmeticity is no more special than complex arithmeticity. The groups studied in this paper are commensurable with groups generated by reflections, so from here on we simply speak about 'arithmetic subgroups' of $\mathrm{Iso}^+ \mathbb{H}^3$.

Proposition. *Let* $\Gamma \leq \mathrm{Iso}\, \mathbb{H}^3$ *be a group generated by reflections in the faces of a polytope. Then* Γ *contains Fuchsian subgroups (we follow the conventions of* **[Vi]** *about polytopes).*

Proof. The reflection planes in Γ decompose $\mathbb{H}^3$ in polytopes. Each of them is a fundamental domain for Γ. Take a plane r containing a face of one polytope. Then the stabilizer in Γ of r contains a Fuchsian subgroup G. In fact, r is tessellated by copies of faces of a polytope. Taking a copy for each face of the polytope we obtain a union of finitely many faces in r. This union is a fundamental domain for G. $\square$

Corollary. *Every complex-arithmetic subgroup* $\Gamma \leq \mathrm{Iso}\, \mathbb{H}^3$ *which is commensurable with a group generated by reflections is real-arithmetic.* $\square$

6. Characterization of arithmetic subgroups of $\mathrm{Iso}^+\mathbb{H}^3$**.** Reid in his thesis **[R]**, following work of Takeuchi in the case $\mathrm{SL}(2,\mathbb{R})$, characterized subgroups of $\mathrm{SL}(2,\mathbb{C})$ which are arithmetic.

Theorem 2.2 (Reid). *Let* $\Gamma \leqq \mathrm{SL}(2,\mathbb{C})$ *be a subgroup of finite co-volume. If* Γ *is derived from a quaternion algebra then*

(1) $K = \mathbb{Q}(\mathrm{tr}\, g : g \in \Gamma)$ *is a number field with one complex place;*
(2) $\mathrm{tr}(g)$ *is an algebraic integer, for every* $g \in \Gamma$*;*
(3) *For every embedding* $\sigma : K \to \mathbb{R}$, $\sigma(\mathrm{tr}(\Gamma)) \subset [-2, 2]$.

Conversely if Γ *satisfies*

(4) $K = \mathbb{Q}(\mathrm{tr}(g) : g \in \Gamma)$ *is a number field not totally real;*
(5) *(same as (2) above);*
(6) *For every embedding* $\sigma : K \to \mathbb{C}, \sigma \neq id$ *or conjugation,* $\sigma(\mathrm{tr}(\Gamma))$ *is bounded, then* Γ *is derived from a quaternion algebra.* □

Theorem 2.3 (Reid). *A subgroup* $\Gamma \leqq \mathrm{SL}(2,\mathbb{C})$ *is arithmetic if and only if* Γ^2 *is derived from a quaternion algebra, where* Γ^2 *is the group generated by squares of elements of* Γ. □

Vinberg, in **[Vi]**, characterized the subgroups of $\mathrm{Iso}\,\mathbb{H}^3$ generated by reflections in the faces of a polyhedron Π in $\mathbb{H}^3$ that are arithmetic, in terms of the Gram matrix of Π.

Suppose that Π is contained in the hyperboloid model of $\mathbb{H}^3$ (see 1.4). We take vectors $e_1, \ldots, e_n$ in the Minkowski space $\mathbb{R}^4$ that are orthogonal to the faces of Π. We normalize the vectors e_i such that $f(e_i, e_i) = 1$. The Gram matrix of Π is then $G = [a_{ij}]$ where $a_{ij} = f(e_i, e_j)$.

Let $b_{i_1 \cdots 1_m}$ be the cyclic products, i.e., the numbers $a_{i_1 i_2} a_{i_2 i_3} \ldots a_{i_m i_1}$, for any ordered sequence of indices of the set $\{1, \ldots, n\}$, $m \geq 1$.

Theorem 2.4 (Vinberg). *Let* Γ *be a discrete subgroup of* $\mathrm{Iso}\,\mathbb{H}^3$ *of finite co-volume, and generated by reflections in the faces of* Π. *Let* $\widetilde{K}$ *be the number field generated by the entries* a_{ij} *of the Gram matrix, and let* K *be the subfield generated by all the cyclic products. Let* D *be the determinant of an arbitrary principal submatrix of* G *of order 4 that is adjacent to a principal submatrix* C^+ *of order 3, where* C^+ *is the Gram matrix of a simplicial angle of* Π. *Then* Γ *is quasi-arithmetic if and only if*

(i) $\widetilde{K}$ *is a totally real number field;*
(ii) D *is a primitive element of the field* K*;*
(iii) *All the conjugates of* D *are positive.*

Alternatively Γ *is quasi-arithmetic if and only if* $\widetilde{K}$ *is totally real and for every embedding* $\sigma : \widetilde{K} \to \mathbb{R}$ *which is not the identity on* K *the matrix* $G^\sigma = [\sigma(a_{ij})]$ *is positive semidefinite.*

Γ *is arithmetic if and only if* Γ *is quasi-arithmetic and*

(iv) *the numbers* $2a_{ij}$ *are algebraic integers.* □

§3. The geometric description of the Borromean hyperbolic orbifolds $\widehat{B}(m,n,p)$

Let $f = \frac{x^2}{A^2} + \frac{y^2}{B^2} + \frac{z^2}{C^2} - t^2$, with $A,\ B,\ C$ real positive, be a quadratic form on $\mathbb{R}P^3$ defined in homogeneous co-ordinates $(x:y:z:t)$. The component $E(A,B,C)$ of $\mathbb{R}P^3 - (f=0)$, that contains the origin $(0:0:0:1)$ is the interior of an ellipsoid with longitudinal semiaxes $A,\ B,\ C$ and serves as a Beltrami-Klein model for hyperbolic 3-space $\mathbb{H}^3$.

Consider the polyhedron $P(\widehat{a},\widehat{b},\widehat{c})$ of Figure 1. Only the positive octant is shown in Figure 1. The rest is obtained by reflections in the coordinate planes. We shall call $P(\widehat{a},\widehat{b},\widehat{c})$ a ***pyritohedron***.

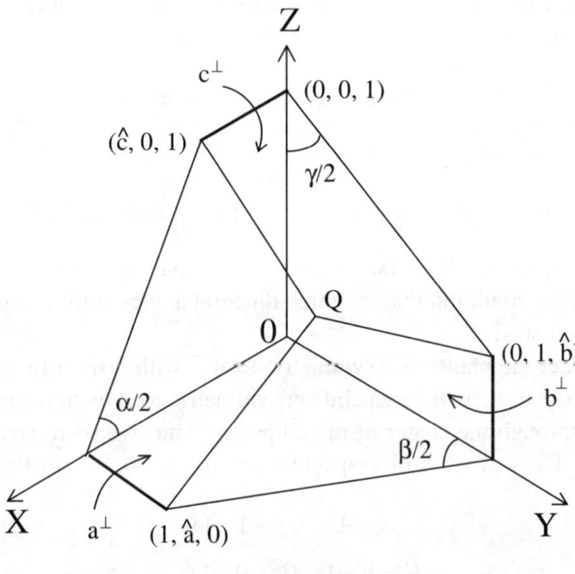

Figure 1

We think of the pyritohedron $P(\widehat{a},\widehat{b},\widehat{c})$ as being contained in the model $E(A,B,C)$ of $\mathbb{H}^3$. We shall find conditions on $A,\ B,\ C,\ \widehat{a},\ \widehat{b},\ \widehat{c}$ such that the dihedral angles at the three edges converging at the point Q are right angles.

Thus we consider the projective equations of the planes $a^\perp,\ b^\perp,\ c^\perp$:

$$a^\perp = \{(x:y:z:t) \mid x + (1-\widehat{c})z - t = 0\},$$

$$b^\perp = \{(x:y:z:t) \mid y + (1-\widehat{a})x - t = 0\},$$

$$c^\perp = \{(x:y:z:t) \mid z + (1-\widehat{b})y - t = 0\}.$$

The poles of these planes, $a,\ b,\ c$ with respect to the quadric $f=0$ are

$$a=(A^2:0:(1-\widehat{c})C^2:1),$$
$$b=((1-\widehat{a})A^2:B^2:0:1),$$
$$c=(0:(1-\widehat{b})B^2:C^2:1).$$

Since we wish to have $\cos(a^\perp,b^\perp)=\cos(b^\perp,c^\perp)=\cos(c^\perp,a^\perp)=0$ we must have

$$aFb^t=bFc^t=cFa^t=0$$

where F is the matrix of the form f. This immediately gives rise to the equations

$$(1-\widehat{a})A^2=(1-\widehat{b})B^2=(1-\widehat{c})C^2=1.$$

Thus we see that the dihedral angles at the edges of $P(\widehat{a},\widehat{b},\widehat{c})$ concurrent with Q are right angles if and only if

$$\widehat{a}=1-A^{-2};\quad \widehat{b}=1-B^{-2};\quad \widehat{c}=1-C^{-2}.$$

In this case the poles of $a^\perp,b^\perp,c^\perp$ are

$$a=(A^2:0:1:1);\quad b=(1:B^2:0:1);\quad c=(0:1:C^2:1).$$

Finally, given angles $\alpha,\ \beta,\ \gamma$ (see Figure 1) we would like to determine $A,\ B,\ C$ in such a way that the condition that the three dihedral angles converging at the point Q be right angles is satisfied.

Thus we consider the plane $a'^\perp$ symmetric to $a^\perp$ with respect to $y=0$. Its pole, a', with respect to $f=0$, will be Euclidean symmetric to a with respect to $y=0$ as this plane passes through the center of the ellipsoid. Thus we shall have for the planes symmetric to $a^\perp$, $b^\perp$, and $c^\perp$ with respect to $y=0,\ z=0,\ x=0$ the poles

$$a'=(A^2:0:-1:1),$$
$$b'=(-1:B^2:0:1),$$
$$c'=(0:-1:C^2:1).$$

Then

$$-\cos\alpha=\frac{aFa'^t}{\sqrt{aFa^t}\sqrt{a'Fa'^t}}=\frac{aFa'^t}{aFa^t}=\frac{A^2-C^{-2}-1}{A^2+C^{-2}-1}=\frac{1+C^2(1-A^2)}{1-C^2(1-A^2)}.$$

From this we derive the equations:

$$C^2(A^2-1)(1+\cos\alpha)=1-\cos\alpha$$
$$C^2(A^2-1)=\tan^2(\frac{\alpha}{2}).$$

Therefore we have, given $\alpha,\ \beta,\ \gamma$,

$$C^2(A^2-1)=\tan^2(\frac{\alpha}{2}),$$

$$A^2(B^2-1)=\tan^2(\frac{\beta}{2}),$$

$$B^2(C^2-1)=\tan^2(\frac{\gamma}{2}).$$

We deduce also that $\alpha,\ \beta,\ \gamma$ can take values in the internal $[0,\pi)$. We sum up what has been demonstrated so far in the following proposition.

Proposition 3.1. *The pyritohedron $P(\widehat{a},\widehat{b},\widehat{c})$ contained in the ellipsoidal model, $E(A,B,C)$, of $\mathbb{H}^3$ has hyperbolic right angles for the dihedral angles at the edges concurrent at Q if and only if:*

$$\widehat{a}=1-A^{-2},\quad \widehat{b}=1-B^{-2},\quad \widehat{c}=1-C^{-2}.$$

Moreover, the values of the hyperbolic angles α,β,γ satisfy

$$\tan^2(\frac{\alpha}{2})=C^2(A^2-1),\ \tan^2(\frac{\beta}{2})=A^2(B^2-1),\ \tan^2(\frac{\gamma}{2})=B^2(C^2-1).$$

The angles $\alpha,\ \beta,\ \gamma$ can take any value in the interval $[0,\pi)$. □

Next, we shall obtain generators for a discrete subgroup $B(m,n,p)$ of $\mathrm{Iso}^+\mathbb{H}^3$ such that the quotient $\widehat{B}(m,n,p):=\mathbb{H}^3/B(m,n,p)$ is the ***Borromean orbifold*** with underlying topological space S^3 and singular locus the Borromean rings and such that the isotropy groups of the components of the rings are cyclic of orders m,n,p (see Figure 2) where $m>2,\ n>2,\ p>2$

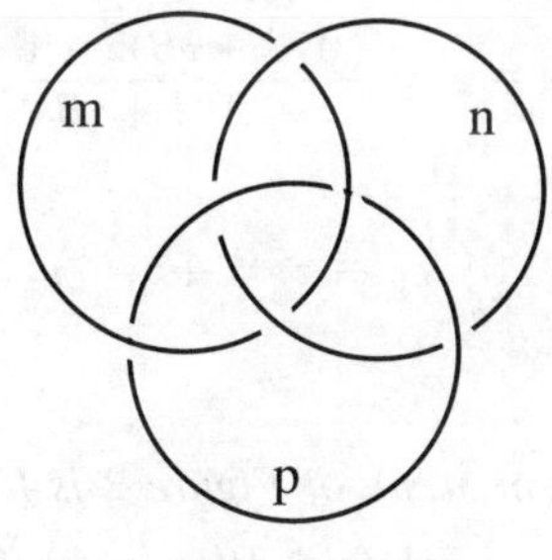

Figure 2

It is well known (see for example **[Th]**, **[HLMW]**) that $B(m,n,p)$ is generated by $g_a,\ g_b,\ g_c$, which are, by definition, rotations of $\frac{2\pi}{m},\ \frac{2\pi}{n},\ \frac{2\pi}{p}$ in the axes $\langle(1:0:0:1),$ $(1:\widehat{a}:0:1)\rangle$, $\langle(0:1:0:1),\ (0:1:\widehat{b}:1)\rangle$, $\langle(0:0:1:1),\ (\widehat{c}:0:1:1)\rangle$ respectively.

Our desire is to determine the group $B(m,n,p)$ as a subgroup of $\mathrm{PSL}(2,\mathbb{C})$ in the Poincaré half-space model. To achieve this it is convenient to represent $g_a,\ g_b,\ g_c$ as products of hyperbolic reflections. In effect letting $r_a,\ r_b,\ r_c$ be reflections in the planes $a^{\perp},\ b^{\perp},\ c^{\perp}$ and $r_x,\ r_y,\ r_z$ reflections in the planes $yz,\ xz,\ xy$ we have

$$g_a = r_a r_z\,, \quad g_b = r_b r_x\,, \quad g_c = r_c r_y\,.$$

As we have indicated in Section 2, it is easy to obtain a reflection in the half-space model when one knows the pole of the plane of reflection in the Beltrami-Klein model. We pass from the Beltrami-Klein model to the intermediate model via the two following co-ordinate changes:

$$X = \frac{x}{A}e, \quad Y = \frac{y}{B}, \quad Z = \frac{z}{C}, \quad T = t;$$
$$\theta_1 = X + iY\,, \quad \theta_2 = Z - T\,, \quad \theta_3 = Z + T\,, \quad \theta_4 = -X + iY\,.$$

These changes carry f to the form

$$\det \begin{bmatrix} \theta_1 & \theta_2 \\ \theta_3 & \theta_4 \end{bmatrix}$$

which defines the intermediate model.

In the intermediate model the planes of reflection of $r_a,\ r_b,\ r_c,\ r_x,\ r_y,\ r_z$ are determined by the coordinates of their poles

$$a = \begin{pmatrix} A & C^{-1}-1 \\ C^{-1}+1 & -A \end{pmatrix}, \quad b = \begin{pmatrix} A^{-1}+iB & -1 \\ 1 & -A^{-1}+iB \end{pmatrix}, \quad c = \begin{pmatrix} iB^{-1} & C-1 \\ C+1 & iB^{-1} \end{pmatrix};$$

$$(1:0:0:0) = \begin{pmatrix} 1 & 0 \\ 0 & -1 \end{pmatrix}, \quad (0:1:0:0) = \begin{pmatrix} i & 0 \\ 0 & i \end{pmatrix}, \quad (0:0:1:0) = \begin{pmatrix} 0 & 1 \\ 1 & 0 \end{pmatrix}.$$

Then, in the Poincaré half-space model these reflections correspond to the following antihomographies of $\mathbb{C}P^1$.

$$r_a \equiv z' = \frac{A\bar{z} + C^{-1} - 1}{(C^{-1}+1)\bar{z} - A}, \quad r_b \equiv z' = \frac{(A^{-1}+iB)\bar{z} - 1}{\bar{z} - A^{-1} + iB}, \quad r_c \equiv z' = \frac{iB^{-1}\bar{z} + C - 1}{(C+1)\bar{z} + iB^{-1}};$$

$$r_x \equiv z' = -\bar{z}, \quad r_y \equiv z' = +\bar{z}, \quad r_z \equiv z' = \frac{1}{\bar{z}}.$$

Thus we have:

Theorem 3.2. *The orbifold $\widehat{B}(m,n,p)$ of Figure 2 is hyperbolic for $3 \leqq m,n,p \leqq \infty$. The group $B(m,n,p)$ of the orbifold $\widehat{B}(m,n,p)$ is generated by the following homographies of $\mathbb{C}P^1$ which act on the Poincaré half-space model:*

$$g_a \equiv z' = \frac{(C^{-1}-1)z + A}{-Az + C^{-1} - 1}, \quad g_b \equiv z' = \frac{(A^{-1}+iB)z + 1}{z + A^{-1} - iB}, \quad g_c \equiv z' = \frac{iB^{-1}z + C - 1}{(C+1)z + iB^{-1}},$$

where $A,\ B,\ C$ satisfy

$$C^2(A^2-1) = \tan^2(\frac{\pi}{m}), \quad A^2(B^2-1) = \tan^2(\frac{\pi}{n}), \quad B^2(C^2-1) = \tan^2(\frac{\pi}{p}).$$

The homographies g_a, g_b, g_c *are images of the following elements of* $\mathrm{SL}(2,\mathbb{C})$*:*

$$h_a = \cos\frac{\pi}{m}\begin{pmatrix} 1-C & AC \\ -AC & 1+C \end{pmatrix}, \quad h_b = \cos\frac{\pi}{n}\begin{pmatrix} 1+iAB & A \\ A & 1-iAB \end{pmatrix},$$

$$h_c = \cos\frac{\pi}{p}\begin{pmatrix} 1 & iB(1-C) \\ -iB(1+C) & 1 \end{pmatrix},$$

which generate a subgroup of $\mathrm{SL}(2,\mathbb{C})$ *that we shall denote by* $B_{m,n,p}$*, and that projects to* $B(m,n,p)$*.* □

Remark 1. The orbifold $\widehat{B}(\infty,\infty,\infty)$ should be viewed as the exterior of the Borromean rings. It is an open hyperbolic manifold, as can be deduced from our construction. In effect for $\alpha=\beta=\gamma=0$, one deduces that $A=B=C=1$; that the ellipsoid is the sphere of radius one with center the origin. Moreover $\widehat{a}=\widehat{b}=\widehat{c}=0$. The pyritohedron degenerates to a rhombododecahedron inscribed in the sphere. This is exactly Thurston's construction in **[Th]** to give the complement of the Borromean rings a complete hyperbolic structure.

Remark 2. $\widehat{B}(2,2,2)$ is Euclidean (see Thurston (**[Th]**) for example). This can be deduced from our result on setting $A=B=C$ so that f can be written $f=x^2+y^2+z^2-A^2t^2$. As $\alpha=\beta=\gamma$ tends to π, $\tan^2(\frac{\alpha}{2})$ tends to infinity and A, B, C tend to infinity. The form f represents the sphere of infinite radius and the geometry induced in $\mathbb{R}^3$ tends to be Euclidean. (The pyritohedron becomes small relative to the sphere as the sphere grows. In the limit its geometry is Euclidean.) The pyritohedron tends toward a regular cube as $\widehat{a}=\widehat{b}=\widehat{c}$ tend to one. This is the construction, using a cube, that Thurston made for $\widehat{B}(2,2,2)$. (Compare **[Ho**, 4.19**]**.)

§4. The groups $B_{m,n,p}$ that are arithmetic

Since the groups $B(m,n,p)$ of the orbifolds $\widehat{B}(m,n,p)$ are subgroups of $O(F,\mathbb{C})$ where

$$F = \begin{bmatrix} A^{-2} & 0 & 0 & 0 \\ 0 & B^{-2} & 0 & 0 \\ 0 & 0 & C^{-2} & 0 \\ 0 & 0 & 0 & -1 \end{bmatrix}$$

(compare 1.4), we can investigate which ones are arithmetic and which ones are quasi-arithmetic. We shall see that the quasi-arithmetic groups are exactly those for which m, n, and p belong to $\{3,4,6,\infty\}$ and the arithmetic groups are exactly those corresponding to the eleven unordered triples $(6,6,6)$, $(4,4,4)$, $(3,3,3)$, $(4,4,3)$, $(6,6,3)$, (∞,∞,∞), $(\infty,\infty,3)$, $(\infty,\infty,4)$, $(\infty,3,3)$, $(\infty,4,3)$, $(\infty,4,4)$.

As our group $B_{m,n,p}$ is not generated by reflections we cannot directly apply Theorem 2.4 of Vinberg. But if we denote by $R(m,n,p)$ the group generated by reflections in the faces of the ***octant*** Π of Figure 1 we have:

Lemma 4.1. *The group $B(m,n,p)$ is an index eight subgroup of $R(m,n,p)$. Then $B(m,n,p)$ is arithmetic (resp. quasi-arithmetic) if and only if $R(m,n,p)$ is arithmetic (resp. quasi-arithmetic).* □

Now we use Theorem 2.4 of Vinberg. Let $d,\ e,\ f$ be the poles of the coordinate planes $yz,\ xz,\ xy$. The following normalized vectors are orthogonal to the faces of the octant Π:

$$\begin{aligned} a &= \rho(A^2,0,1,1), & \rho &= -\cos\frac{\pi}{m}C; \\ b &= \mu(1,B^2,0,1), & \mu &= -\cos\frac{\pi}{n}A; \\ c &= \nu(0,1,C^2,1), & \nu &= -\cos\frac{\pi}{p}B; \\ d &= \alpha(-1,0,0,0), & \alpha &= A; \\ e &= \beta(0,-1,0,0), & \beta &= B; \\ f &= \gamma(0,0,-1,0), & \gamma &= C. \end{aligned}$$

Therefore the Gram matrix $G=[a_{ij}]$ of Π is:

$$G = \begin{bmatrix} 1 & 0 & 0 & -CA\cos\frac{\pi}{m} & 0 & -\cos\frac{\pi}{m} \\ 0 & 1 & 0 & -\cos\frac{\pi}{n} & -AB\cos\frac{\pi}{n} & 0 \\ 0 & 0 & 1 & 0 & -\cos\frac{\pi}{p} & -BC\cos\frac{\pi}{p} \\ & & & 1 & 0 & 0 \\ & \bigstar & & 0 & 1 & 0 \\ & & & 0 & 0 & 1 \end{bmatrix}.$$

Therefore

$$\widetilde{K} = \mathbb{Q}(\{a_{ij}\}) = \mathbb{Q}(\cos\frac{\pi}{m},\cos\frac{\pi}{n},\cos\frac{\pi}{p},CA,AB,BC),$$
$$K = \mathbb{Q}(\{bi_1,\dots,i_m\}) = \mathbb{Q}(A^2,B^2,C^2).$$

This last is a consequence of the fact that the field K is generated by cyclic products, the cyclic products comprise the diagonal elements (all one), the squares of the entries of G, $(\cos^2\frac{\pi}{m},\cos^2\frac{\pi}{n},\cos^2\frac{\pi}{p},C^2A^2,C^2B^2,A^2B^2)$ and the element $a_{14}a_{42}a_{25}a_{53}a_{36}a_{61} = A^2B^2C^2\cos^2\frac{\pi}{m}\cos^2\frac{\pi}{n}\cos^2\frac{\pi}{p}$, and the elements $\cos^2\frac{\pi}{m}$, $\cos^2\frac{\pi}{n}$, and $\cos^2\frac{\pi}{p}$ belong to $\mathbb{Q}(A^2,B^2,C^2)$.

Lemma 4.1.1. *There is an integer j prime to the positive integer ℓ and belonging to the interval $(1,\ell/2)$ if and only if $\ell \neq 2,\ 3,\ 4,\ 6,\ \infty$.* □

Proposition 4.2. *If one of $m,\ n,\ p$ is not equal to $3,\ 4,\ 6$, or ∞, then the group $R(m,n,p)$ is not quasi-arithmetic.*

Proof. Suppose one of the numbers $m,\ n,\ p$ is not $3,\ 4,\ 6$ or ∞, let it be m, and let j be the integer given by Lemma 4.1.1, prime to m and in the interval $(1,m/2)$.

Using the fact that the arithmetic series $\{j+im \mid i \geqq 0\}$ contains infinitely many primes (Dirichlet's Theorem), we choose one, $q = j+im$, which also is prime to n and p. Let k' and ℓ' be (the unique) integers such that:

$$q \equiv k' \bmod n, \quad |k'| \in [1, n/2),$$
$$q \equiv \ell' \bmod p, \quad |\ell'| \in [1, p/2),$$

and let us denote $|k'|$ and $|\ell'|$ by k and ℓ respectively.

Note that $\cos\frac{2\pi}{m}q = \cos\frac{2\pi}{m}j$, $\cos\frac{2\pi}{n}q = \cos\frac{2\pi}{n}k$, and $\cos\frac{2\pi}{p}q = \cos\frac{2\pi}{p}\ell$. Next, we apply the non trivial automorphism

$$\theta : \cos\frac{2\pi}{c} \longmapsto \cos\frac{2\pi}{c}q$$

to the field $\mathbb{Q}(\cos^2\frac{\pi}{m}, \cos^2\frac{\pi}{n}, \cos^2\frac{\pi}{p}) \leqq \mathbb{Q}(\cos\frac{2\pi}{c})$ where c is the least common multiple of m, n, and p. This induces one or more non trivial embeddings of $\mathbb{Q}(A^2, B^2, C^2)$ in $\mathbb{R}$. These embeddings are found via the equations

$$\begin{aligned} \widehat{C}^2(\widehat{A}^2 - 1) &= \tan^2\frac{\pi}{m}j, \\ \widehat{A}^2(\widehat{B}^2 - 1) &= \tan^2\frac{\pi}{n}k, \\ \widehat{B}^2(\widehat{C}^2 - 1) &= \tan^2\frac{\pi}{p}\ell, \end{aligned} \tag{4.1}$$

which give rise to embeddings $\{A^2 \to \widehat{A}^2,\ B^2 \to \widehat{B}^2,\ C^2 \to \widehat{C}^2\}$ where $\widehat{A}^2$, $\widehat{B}^2$, $\widehat{C}^2$ are solutions of 4.1. Since

$$0 < \frac{\pi}{m}j < \frac{\pi}{2}, \quad 0 < \frac{\pi}{n}k < \frac{\pi}{2}, \quad 0 < \frac{\pi}{p}\ell < \frac{\pi}{2},$$

Proposition 3.1 guarantees us that there exist solutions with $\widehat{A}^2$, $\widehat{B}^2$, $\widehat{C}^2$ ***positive*** numbers. The embedding σ thus obtained of K in $\mathbb{R}$ is non trivial and the quadratic form

$$\frac{x^2}{\widehat{A}^2} + \frac{y^2}{\widehat{B}^2} + \frac{z^2}{\widehat{C}^2} - t^2$$

is of type (3.1).

We extend $\sigma : K \to \mathbb{R}$ to $\widetilde{\sigma} : \widetilde{K} \to \mathbb{R}$ and form the matrix $\widetilde{\sigma}(G)$. This matrix and the $\widehat{G}$ corresponding to the octant of the pyritohedron $P(\widehat{a}, \widehat{b}, \widehat{c})$ given by Proposition 3.1 have *equal* corresponding cyclic products.

Therefore the type of the matrix $\widetilde{\sigma}(G)$ coincides with the type of the matrix $\widehat{G}$. Since this one is indefinite, we deduce that $\widetilde{\sigma}(G)$ is indefinite. It follows then from Theorem 2.4 of Vinberg that $R(m, n, p)$ is not quasi-arithmetic. □

We shall study the cases $R(m, n, p)$, m, n, p in $\{3, 4, 6, \infty\}$ one at a time. We shall take as principal submatrix C^+ of order 3 (see 2.4) the first box of G. We shall take as D:

$$D = \det \begin{bmatrix} 1 & 0 & 0 & -AC\cos\frac{\pi}{m} \\ 0 & 1 & 0 & -\cos\frac{\pi}{n} \\ 0 & 0 & 1 & 0 \\ -AC\cos\frac{\pi}{n} & -\cos\frac{\pi}{n} & 0 & 1 \end{bmatrix} = 1 - A^2C^2\cos^2\frac{\pi}{n} - \cos^2\frac{\pi}{n}.$$

We shall prove that:

(i) $\widetilde{K} = \mathbb{Q}(\cos\frac{\pi}{m}, \cos\frac{\pi}{n}, \cos\frac{\pi}{p}, AC, BC, AB)$ is totally real;
(ii) D is a primitive element of $K = \mathbb{Q}(A^2, B^2, C^2)$;
(iii) the conjugates of D are positive in all cases.

This implies that all the $R(m,n,p)$ with m, n, p in $\{3,4,6,\infty\}$ are quasi-arithmetic. Finally we shall show

(iv) $2a_{ij}$ is an algebraic integer for every a_{ij} in G.

When this is also the case the group $R(m,n,p)$ is arithmetic.

In the sequel we shall calculate A, B, C by means of the following formulas. Let α, β, γ denote $\tan^2\frac{\pi}{m}$, $\tan^2\frac{\pi}{n}$, $\tan^2\frac{\pi}{p}$ respectively. Then

$$\alpha = \tan^2\frac{\pi}{m} = C^2(A^2-1),$$
$$\beta = \tan^2\frac{\pi}{n} = A^2(B^2-1),$$
$$\gamma = \tan^2\frac{\pi}{p} = B^2(C^2-1).$$

We obtain

$$A^2 = 1 + \cfrac{\alpha}{1 + \cfrac{\gamma}{1 + \cfrac{\beta}{A^2}}}.$$

Then:

$$A = \sqrt{\frac{\alpha+\gamma+1-\beta+\sqrt{\Delta}}{2(\gamma+1)}}.$$

Analogously

$$B = \sqrt{\frac{\alpha+\beta+1-\gamma+\sqrt{\Delta}}{2(\alpha+1)}},$$
$$C = \sqrt{\frac{\beta+\gamma+1-\alpha+\sqrt{\Delta}}{2(\beta+1)}},$$

where $\Delta = \alpha^2+\beta^2+\gamma^2+2(\alpha\beta+\alpha\gamma+\beta\gamma)+4\alpha\beta\gamma+2(\alpha+\beta+\gamma)+1$.

CASE (6, 6, 6). Here $\cos\frac{\pi}{6} = \frac{\sqrt{3}}{2}$; $\alpha = \beta = \gamma = \frac{1}{3}$; $\sqrt{\Delta} = \frac{4}{9}\sqrt{21}$, $A^2 = B^2 = C^2 = \frac{3+\sqrt{21}}{6}$; $D = -\frac{\sqrt{21}}{8}$, $K = \mathbb{Q}(\sqrt{21}) = \mathbb{Q}(D)$, $\sigma(D) = \frac{\sqrt{21}}{8} > 0$. $\widetilde{K} = \mathbb{Q}(\sqrt{3}, \sqrt{21})$ is totally real. Thus $R(6,6,6)$ is quasi-arithmetic.

Since $2\cos\frac{\pi}{m}CA = \sqrt{3}\cdot\frac{3+\sqrt{21}}{6} = \frac{\sqrt{3}+\sqrt{7}}{2} = x$ and x satisfies $x^4 - 5x + 1 = 0$ we see that $R(6,6,6)$ is arithmetic.

CASE (4, 4, 4). Here $\cos\frac{\pi}{4} = \frac{\sqrt{2}}{2}$; $\alpha = \beta = \gamma = 1$; $\sqrt{\Delta} = 2\sqrt{5}$, $A^2 = B^2 = C^2 = \frac{1+\sqrt{5}}{2}$; $D = \frac{1-\sqrt{5}}{4}$, $\sigma(D) = \frac{1+\sqrt{5}}{4} > 0$, $K = \mathbb{Q}(\sqrt{5}) = \mathbb{Q}(D)$. $\widetilde{K} = \mathbb{Q}(\sqrt{5}, \sqrt{2})$ is totally real. Thus $R(4,4,4)$ is quasi-arithmetic.

Since $2\cos\frac{\pi}{m}AC = \sqrt{2}\frac{1+\sqrt{5}}{2}$ is an algebraic integer, $R(4,4,4)$ is arithmetic.

CASE (3, 3, 3). Here $\cos\frac{\pi}{3} = \frac{1}{2}$; $\alpha = \beta = \gamma = 3$; $\sqrt{\Delta} = 4\sqrt{13}$, $A^2 = B^2 = C^2 = \frac{1+\sqrt{13}}{2}$; $D = \frac{-1-\sqrt{13}}{8}$, $\sigma(D) = \frac{-1+\sqrt{13}}{8} > 0$, $K = \mathbb{Q}(\sqrt{13}) = \mathbb{Q}(D)$. $\widetilde{K} = \mathbb{Q}(\sqrt{13})$ is totally real. Thus $R(3,3,3)$ is quasi-arithmetic.

Since $2\cos\frac{\pi}{m}CA = \frac{1+\sqrt{13}}{2}$ is an algebraic integer, $R(3,3,3)$ is arithmetic.

CASE (6, 6, 3). Here $\cos\frac{\pi}{6} = \frac{\sqrt{3}}{2}$, $\cos\frac{\pi}{3} = \frac{1}{2}$; $\alpha = \frac{1}{3}$, $\beta = \frac{1}{3}$, $\gamma = 3$; $\sqrt{\Delta} = \frac{4}{3}\sqrt{13}$, $A^2 = \frac{3+\sqrt{13}}{6}$, $B^2 = \frac{-1+\sqrt{13}}{2}$, $C^2 = \frac{3+\sqrt{13}}{2}$; $D = -\frac{3}{8}(3+\sqrt{13})$, $\sigma(D) = \frac{-3}{8}(3-\sqrt{13}) > 0$, $K = \mathbb{Q}(\sqrt{13}) = \mathbb{Q}(D.)$. $\widetilde{K} = \mathbb{Q}(\sqrt{13})$ is totally real. Thus $R(6,6,3)$ is quasi-arithmetic.

Since $2\cos\frac{\pi}{6}AC = \sqrt{\frac{11+3\sqrt{13}}{2}}$, $2\cos\frac{\pi}{6}AB = \sqrt{\frac{5+\sqrt{13}}{2}}$, $2\cos\frac{\pi}{3}BC = \sqrt{\frac{5+\sqrt{13}}{2}}$ are algebraic integers $R(6,6,3)$ is arithmetic.

CASE (6, 6, 4). Here $\cos\frac{\pi}{6} = \frac{\sqrt{3}}{2}$, $\cos\frac{\pi}{4} = \frac{\sqrt{2}}{2}$; $\alpha = \frac{1}{3}$, $\beta = \frac{1}{3}$, $\gamma = 1$; $\sqrt{\Delta} = \frac{2}{3}\sqrt{17}$, $A^2 = \frac{3+\sqrt{17}}{6}$, $B^2 = \frac{1+\sqrt{17}}{4}$, $C^2 = \frac{3+\sqrt{17}}{4}$; $D = \frac{-3}{16}(3+\sqrt{17})$, $\sigma(D) = \frac{-3}{16}(3-\sqrt{17}) > 0$, $K = \mathbb{Q}(\sqrt{17}) = \mathbb{Q}(D)$. $\widetilde{K}$ is totally real so $R(6,6,4)$ is quasi-arithmetic. But since $2\cos\frac{\pi}{m}AC = \sqrt{\frac{13+3\sqrt{17}}{4}}$ is not an algebraic integer, $R(6,6,4)$ is not arithmetic.

CASE (6, 4, 4). Here $\cos\frac{\pi}{4} = \frac{\sqrt{2}}{2}$, $\cos\frac{\pi}{6} = \frac{\sqrt{3}}{2}$; $\alpha = 1$, $\beta = 1$, $\gamma = \frac{1}{3}$; $\sqrt{\Delta} = \frac{4}{3}\sqrt{7}$, $A^2 = \frac{1+\sqrt{7}}{2}$, $B^2 = \frac{2+\sqrt{7}}{3}$, $C^2 = \frac{1+\sqrt{7}}{3}$; $D = -\frac{1}{6}(1+\sqrt{7})$, $\sigma(D) = \frac{-1}{6}(1-\sqrt{7}) > 0$, $K = \mathbb{Q}(\sqrt{7}) = \mathbb{Q}(D)$. Then $\widetilde{K}$ is totally real. But since $2\cos\frac{\pi}{m}AC = \sqrt{\frac{8+2\sqrt{7}}{3}}$ is not an algebraic integer it follows that $R(6,4,4)$ is quasi-arithmetic but not arithmetic.

CASE (6, 3, 3). Here $\cos\frac{\pi}{3} = \frac{1}{2}$, $\cos\frac{\pi}{6} = \frac{\sqrt{3}}{2}$; $\alpha = 3$, $\beta = 3$, $\gamma = \frac{1}{3}$; $\sqrt{\Delta} = \frac{4}{3}\sqrt{37}$, $A^2 = \frac{1+\sqrt{37}}{2}$, $B^2 = \frac{5+\sqrt{37}}{6}$, $C^2 = \frac{1+\sqrt{37}}{6}$; $D = \frac{-1}{24}(1+\sqrt{37})$, $\sigma(D) > 0$, $K = \mathbb{Q}(\sqrt{37}) = \mathbb{Q}(\sqrt{D})$. $\widetilde{K}$ is totally real so $R(6,3,3)$ is quasi-arithmetic.

But since $2\cos\frac{\pi}{m}AC = \sqrt{\frac{19+\sqrt{37}}{6}}$ is not an algebraic integer, $R(6,3,3)$ is not arithmetic.

CASE $(4,\ 4,\ 3)$. Here $\cos\frac{\pi}{4} = \frac{\sqrt{2}}{2}$, $\cos\frac{\pi}{3} = \frac{1}{2}$; $\alpha = 1$, $\beta = 1$, $\gamma = 3$; $\sqrt{\Delta} = 4\sqrt{3}$, $A^2 = \frac{1+\sqrt{3}}{2}$, $B^2 = \sqrt{3}$, $C^2 = 1+\sqrt{3}$; $D = -\frac{1}{2}(1+\sqrt{3})$, $\sigma(D) = -\frac{1}{2}(1-\sqrt{3}) > 0$, $K = \mathbb{Q}(\sqrt{3}) = \mathbb{Q}(D)$. $\widetilde{K}$ is totally real.

Since also, $2\cos\frac{\pi}{m}AC = \sqrt{4+2\sqrt{3}}$, $2\cos\frac{\pi}{n}AB = \sqrt{3+\sqrt{3}}$, $2\cos\frac{\pi}{p}BC = \sqrt{3+\sqrt{3}}$ are all algebraic integers, $R(4,4,3)$ is arithmetic.

CASE $(6,\ 4,\ 3)$. Here $\cos\frac{\pi}{6} = \frac{\sqrt{3}}{2}$, $\cos\frac{\pi}{4} = \frac{\sqrt{2}}{2}$, $\cos\frac{\pi}{3} = \frac{1}{2}$; $\alpha = \frac{1}{3}$, $\beta = 1$, $\gamma = 3$; $\sqrt{\Delta} = \frac{2}{3}\sqrt{73}$, $A^2 = \frac{5+\sqrt{73}}{12}$, $B^2 = \frac{-1+\sqrt{73}}{4}$, $C^2 = \frac{7+\sqrt{73}}{6}$; $D = \frac{-5-\sqrt{73}}{8}$, $\sigma(D) = \frac{-5+\sqrt{73}}{8} > 0$, $K = \mathbb{Q}(\sqrt{73}) = \mathbb{Q}(D)$. $\widetilde{K}$ is totally real; and $R(6,4,3)$ is quasi-arithmetic.

But since $2\cos\frac{\pi}{n}AB = \sqrt{\frac{17+\sqrt{73}}{6}}$ is not an algebraic integer we deduce that $R(6,4,3)$ is not arithmetic.

CASE $(4,3,3)$. Here $\cos\frac{\pi}{4} = \frac{\sqrt{2}}{2}$, $\cos\frac{\pi}{3} = \frac{1}{2}$; $\alpha = 1$, $\beta = 3$, $\gamma = 3$; $\sqrt{\Delta} = 10$, $A^2 = \frac{3}{2}$, $B^2 = 3$, $C^2 = 2$; $D = -\frac{3}{4}$. Therefore $K = \mathbb{Q}$ and $\widetilde{K}$ is totally real so $R(4,3,3)$ is quasi-arithmetic.

But since $2\cos\frac{\pi}{m}AB = \frac{3\sqrt{2}}{2}$ is not an algebraic integer, $R(4,3,3)$ is not arithmetic. The cases with $m = \infty$ are proven analogously and we obtain

Theorem 4.3. *The groups $R(m,n,p)$, $B(m,n,p)$ are quasi-arithmetic exactly when m, n, and p belong to $\{3,4,6,\infty\}$. Among these the arithmetic ones are the triples $(6,6,6)$, $(4,4,4)$, $(3,3,3)$, $(4,4,3)$, $(6,6,3)$, (∞,∞,∞), $(\infty,\infty,3)$, $(\infty,\infty,4)$, $(\infty,3,3)$, $(\infty,4,3)$, $(\infty,4,4)$.* □

§5. Trace computation lemmas

We want to compute the fields $K_1 = \mathbb{Q}(\mathrm{tr}(g) \mid g \in B_{m,n,p})$, $K_2 = \mathbb{Q}(\mathrm{tr}(g) \mid g \in B^2_{m,n,p})$. For this we utilize the following lemmas:

Lemma 5.1 (**[Vo]**, Compare **[Ma]**). *Let Γ be a finitely generated subgroup of $\mathrm{SL}(2,\mathbb{C})$. Let $\{\delta_1,\ldots,\delta_r\}$ be a set of generators of Γ. Let $t_{i_1\ldots i_s} = \mathrm{tr}(\delta_{i_1}\ldots\delta_{i_s})$, $1 \leqq s \leqq r$, $1 \leqq i_1 < i_2 < \ldots < i_s \leqq r$. Then, $\mathrm{tr}(\gamma)$, for $\gamma \in \Gamma$, is a polynomial with rational coefficients in the traces $t_{i_1\ldots i_s}$, where $1 \leqq s \leqq 3$.*

The proof makes use of the formulas

$$\mathrm{tr}(\alpha\beta) + \mathrm{tr}(\alpha^{-1}\beta) = \mathrm{tr}(\alpha)\cdot\mathrm{tr}(\beta)$$
$$\mathrm{tr}(\alpha\beta) = \mathrm{tr}(\beta\alpha)$$

and an induction argument on the length of a word in the generators. The case $s \leqq r$, which is all we need in this paper, is fairly easy to prove. The case $1 \leqq s \leqq 3$ is more difficult. □

Lemma 5.2. *Let* $g_1, \cdots, g_n \in \mathrm{SL}(2, \mathbb{C})$ *and suppose* $\mathrm{tr}(g_i) \neq 0$ *for* $i = 1, \cdots, k$. *Then*

$$\Pi_{k=1}^n g_k = (\Pi_{k=1}^n(\mathrm{tr}(g_k))^{-1})\Pi_{k=1}^n(\mathrm{id} + g_k^2),$$
$$(\Pi_{k=1}^n g_k)^2 = (\Pi_{k=1}^n(\mathrm{tr}(g_k))^2)^{-1}(\Pi_{k=1}^n(\mathrm{id} + g_k^2))^2.$$

Proof. The first formula follows immediately from the Cayley-Hamilton theorem. The second formula is just the square of the first formula. □

Lemma 5.3. *Let* $\Gamma \subset \mathrm{SL}(2, \mathbb{C})$ *be generated by* $g_1, \cdots, g_n$. *Let* $\Gamma^{(2)}$ *be the group generated by squares of elements of* Γ. *Suppose* $\mathrm{tr}(g_i) \neq 0$ *for* $i = 1, \cdots, n$. *Let* $\Gamma^{\mathrm{SQ}} = \langle g_1^2, \cdots, g_n^2 \rangle$ *be the subgroup of* Γ *generated by squares of the* g_i, $i = 1, \cdots, n$. *Then* k^2, *the trace field of* $\Gamma^{(2)}$, *equals the trace field of* Γ^{SQ}.

Proof. From the formula $g_i^2 - \mathrm{tr}(g_i)g_i + id = 0$, it follows that $(\mathrm{tr}(g_i))^2 = \mathrm{tr}(g_k^2) + 2$. Substituting this expression in the formula given by Lemma 5.2, (in the case of a *product* of squares of elements of Γ, not just one), and taking the trace, we see that the trace of an element of $\Gamma^{(2)}$ belongs to the trace field of Γ^{SQ}. On the other hand $\Gamma^{\mathrm{SQ}} \subset \Gamma^{(2)}$. □

Lemmas 5.2 and 5.3, although fairly trivial, can simplify the computation of the trace field of $\Gamma^{(2)}$ considerably. The group generated by $g_1^2, \ldots, g_n^2$ has the same number of generators as Γ. The index of $\Gamma^{(2)}$ in Γ is, typically, 2^n, and many more than n generators may be needed to generate $\Gamma^{(2)}$. Roughly speaking, the number of traces that need to be computed in a group with n generators is proportional to n factorial.

Thus to compute $K_1 = \mathbb{Q}(\mathrm{tr}(g) \mid g \in B_{m,n,p})$ and $K_2 = \mathbb{Q}(\mathrm{tr}(h) \mid h \in B_{m,n,p}^2) = \mathbb{Q}(\mathrm{tr}(k) \mid k \in B_{m,n,p}^{\mathrm{SQ}})$, we calculate the traces of h_a, h_b, h_c, $h_a h_b$, $h_a h_c$, $h_b h_c$, $h_a h_b h_c$ which we denote by t_a, t_b, t_{ab}, t_{ac}, t_{bc}, t_{abc}, and the traces of h_a^2, h_b^2, h_c^2, $h_a^2 h_b^2$, $h_a^2 h_c^2$, $h_b^2 h_c^2$, $h_a^2 h_b^2 h_c^2$, which we denote by T_a, T_b, T_c, T_{ab}, T_{ac}, T_{bc}, T_{abc}. We obtain the following. (The generators h_a, h_b, h_c can be found in the statement of Theorem 3.2.)

$$t_a = 2\cos\frac{\pi}{m}, \quad t_b = 2\cos\frac{\pi}{n}, \quad t_c = 2\cos\frac{\pi}{p},$$

$$t_{ab} = 2\cos\frac{\pi}{m}\cos\frac{\pi}{n}(1 - iABC), \quad t_{ac} = 2\cos\frac{\pi}{m}\cos\frac{\pi}{p}(1 - iABC),$$

$$t_{bc} = 2\cos\frac{\pi}{n}\cos\frac{\pi}{p}(1 - iABC), \quad t_{abc} = 2\cos\frac{\pi}{m}\cos\frac{\pi}{n}\cos\frac{\pi}{p}(1 - iABC)^2,$$

$$T_a = 4\cos^2\frac{\pi}{m} - 2, \quad T_b = 4\cos^2\frac{\pi}{n} - 2, \quad T_c = 4\cos^2\frac{\pi}{p} - 2,$$

$$T_{ab} = \frac{1}{2}t_a^2 t_b^2(1 - iABC) - t_a^2 - t_b^2 + 2, \text{ etc. } \ldots,$$

$$T_{abc} = 2\left[(\frac{t_a^2}{2} - 1)(\frac{t_b^2}{2} - 1)(\frac{t_c^2}{2} - 1) + \frac{t_a^2}{2}\frac{t_b^2}{2}\frac{t_c^2}{2}(iABC - A^2B^2C^2)\right].$$

Then the following proposition is an immediate consequence of the previous calculation.

Proposition 5.4.

$$K_1 = \mathbb{Q}(\cos\frac{\pi}{m}, \cos\frac{\pi}{n}, \cos\frac{\pi}{p}, iABC)$$

$$K_2 = \mathbb{Q}(\cos^2\frac{\pi}{m}, \cos^2\frac{\pi}{n}, \cos^2\frac{\pi}{p}, iABC)$$

□

The field $K_2 = \mathbb{Q}(\cos^2\frac{\pi}{m}, \cos^2\frac{\pi}{n}, \cos^2\frac{\pi}{p}, iABC)$ is an extension of the field

$$k = \mathbb{Q}(\cos^2\frac{\pi}{m}, \cos^2\frac{\pi}{n}, \cos^2\frac{\pi}{p}) = \mathbb{Q}(\cos\frac{2\pi}{m}, \cos\frac{2\pi}{n}, \cos\frac{2\pi}{p})$$
$$= \mathbb{Q}(\cos\frac{2\pi}{c})$$

where c is the least common multiple of m, n and p.

§6. Computation of the quaternion algebras and the special case $B_{4,4,4}$

In this section we compute the quaternion algebras for the eleven arithmetic cases of Section 4. We also compute the orders in the quaternion algebras that define their arithmeticity. We go into considerably more detail in the one case $B_{4,4,4}$.

It was shown (**[HLMW]**, **[HLM2]**) that this particular group is ***universal***. That is, given any closed oriented 3-manifold M^3 there is a finite index subgroup $G(M^3)$ of $B_{4,4,4}$ such that M^3 is homomorphic to the orbit space $\mathbb{H}^3/G(M^3)$, and because of its potential application to the study of 3-manifolds, we wish to explain its arithmetic and algebraic properties in more detail.

Theorem 6.4 of this section, in the case $B_{4,4,4}$ is implicit in **[R]**. We repeat it here explicitly to make it more accessible.

Also the number theoretic properties of the field $\mathbb{Q}(iA)$ are in the literature. We make no claim to originality in this section; the justification for Lemma 6.8 through Theorem 6.16 is again that of accessibility.

In Section 3, Theorem 3.2, generators for $B_{m,n,p}$ in $\mathrm{SL}(2,\mathbb{C})$ are given by

$$\begin{aligned} h_a &= \cos\frac{\pi}{m}\begin{pmatrix} 1-C & AC \\ -AC & 1+C \end{pmatrix}, \\ h_b &= \cos\frac{\pi}{n}\begin{pmatrix} 1+iAB & A \\ A & 1-iAB \end{pmatrix}, \\ h_c &= \cos\frac{\pi}{p}\begin{pmatrix} 1 & iB(1-C) \\ -iB(1-C) & 1 \end{pmatrix}. \end{aligned} \tag{6.1}$$

It follows immediately from Proposition 5.4 that the trace field of $B^2_{m,n,p}$ is $\mathbb{Q}(iABC)$. (In the eleven cases we are considering now, m, n, and p belong to $\{3,4,6,\infty\}$. So the squares of the cosines appearing in Proposition 5.4 are rational.)

Following Reid (**[R]**) we construct two interesting geometric objects from $B_{m,n,p}$. Let O_R denote the ring of integers of the field $\mathbb{Q}(iABC)$ and define

$$\begin{aligned} H &= \left\{ \sum_{i=1}^{n} k_i g_i \mid g_i \in B^2_{m,n,p},\ k_i \in \mathbb{Q}(iABC) \right\}, \\ O &= \left\{ \sum_{i=1}^{n} r_i g_i \mid g_i \in B^2_{m,n,p},\ r_i \in O_R \right\}. \end{aligned} \tag{6.2}$$

From the formulas (6.1), using the equation $X^2 = (\mathrm{tr}(X))X - id$, valid for X in $\mathrm{SL}(2,\mathbb{C})$, we obtain the following expressions:

$$\begin{aligned} h_a^2 &= (2\cos^2\frac{\pi}{m} - 1)id + 2\cos^2\frac{\pi}{m}\begin{bmatrix} -C & AC \\ -AC & C \end{bmatrix}, \\ h_b^2 &= (2\cos^2\frac{\pi}{n} - 1)id + 2\cos^2\frac{\pi}{n}\begin{bmatrix} iAB & A \\ A & -iAB \end{bmatrix}, \\ h_c^2 &= (2\cos^2\frac{\pi}{p} - 1)id + 2\cos^2\frac{\pi}{p}\begin{bmatrix} 0 & iB - iBC \\ -iB - iBC & 0 \end{bmatrix}. \end{aligned} \tag{6.3}$$

Now define matrices $\widehat{I}$ and $\widehat{J}$:

$$\widehat{I} = \begin{bmatrix} -C & 0 \\ 0 & C \end{bmatrix}, \qquad \widehat{J} = \begin{bmatrix} 0 & iB \\ -iB & 0 \end{bmatrix}. \tag{6.4}$$

Let $\widehat{K} = \widehat{I}\widehat{J}$ and observe that $\widehat{I}\widehat{J} = -\widehat{J}\widehat{I}$. The following set of formulas is obtained by straightforward manipulation of (6.3). (We wish to show that $\widehat{I}$ and $\widehat{J}$ belong to H.)

$$\begin{aligned} \frac{h_a^2 - (2\cos^2\frac{\pi}{m} - 1)id}{2\cos^2\frac{\pi}{m}} &= \widehat{I} + \left(\frac{-iABC}{B^2}\right)\widehat{J}, \\ \frac{h_b^2 - (2\cos^2\frac{\pi}{n} - 1)id}{2\cos^2\frac{\pi}{n}} &= \left(\frac{-iABC}{C^2}\right)\widehat{I} + \left(\frac{iABC}{B^2C^2}\right)\widehat{K}, \\ \frac{h_c^2 - (2\cos^2\frac{\pi}{p} - 1)id}{2\cos^2\frac{\pi}{p}} &= \widehat{J} + \widehat{K}. \end{aligned} \tag{6.5}$$

Let $\alpha,\ \beta,\ \gamma$ represent $\tan^2(\frac{\pi}{m}),\ \tan^2(\frac{\pi}{n}),\ \tan^2(\frac{\pi}{p})$ respectively. These are rational numbers in the eleven cases we consider, where $m,\ n,\ p \in \{3, 4, 6, \infty\}$.

According to Theorem 3.2 the numbers $A,\ B,\ C$ satisfy the following equations:

$$\begin{aligned} C^2A^2 - C^2 &= \alpha, \\ A^2B^2 - A^2 &= \beta, \\ B^2C^2 - B^2 &= \gamma. \end{aligned} \tag{6.6}$$

If we multiply the first of the equations in (6.6) by B^2 and add it to the third we can solve for B^2 in terms of $A^2B^2C^2$. We can similarly solve for A^2 and C^2 obtaining

$$\begin{aligned} A^2B^2C^2 - \gamma &= (\alpha + 1)B^2, \\ A^2B^2C^2 - \alpha &= (\beta + 1)C^2, \\ A^2B^2C^2 - \beta &= (\gamma + 1)A^2. \end{aligned} \tag{6.7}$$

In particular we note that $A^2,\ B^2$, and C^2 belong to the trace field $\mathbb{Q}(iABC)$ in the eleven cases.

The equations (6.5) can be considered a single matrix equation of form $\vec{v} = M\vec{w}$. We see by inspection that all the entries of the matrix M lie in the field $\mathbb{Q}(iABC)$ and we compute that the determinant of M equals $A^2 - iAB^{-1}C^{-1}$, which is not zero as $A,\ B$, and C are positive real numbers. Consequently the system is invertible and $\widehat{I}$, $\widehat{J},\ \widehat{K}$ can be written as linear combinations of $1,\ h_a^2,\ h_b^2,\ h_c^2$ with coefficients in the field $\mathbb{Q}(iABC)$.

Theorem 6.1. *The algebra H is the quaternion algebra $\left(\frac{C^2, B^2}{\mathbb{Q}(iABC)}\right)$.*

Proof. We need to show that H is a vector space of dimension four. Let $\{1^\star, h_a^\star, h_b^\star, h_c^\star\}$ be the dual basis of $\{1, h_a^2, h_b^2, h_c^2\}$ with respect to the trace form on $M(2,\mathbb{C})$ (i.e., the non degenerate bilinear form $(X,Y) \longmapsto \operatorname{tr} XY$). Then given $g \in B^2_{m,n,p},\ g = a_1 1^\star + a_2 h_a^\star + a_3 h_b^\star + a_4 h_c^\star$, with the coefficients a_i in $\mathbb{C}$. In fact, the a_i belong to O_R. To see that, for example, a_3 belongs to O_R multiply both sides of this equation by h_b^2 and take the trace. We see that $a_3 = \operatorname{tr}(h_b^2 g)$. But we saw in Section 4 that in each of the eleven cases the traces of the elements of $B^2_{m,n,p}$ were algebraic integers in O_R. Thus H, which is a vector space over $\mathbb{Q}(iABC)$, is contained in the $\mathbb{Q}(iABC)$-span of 1^*, h_a^*, h_b^*, h_c^*, a vector space of dimension four. So dimension of H is four. We may use 1, $\widehat{I}$, $\widehat{J}$, $\widehat{K}$ as a new basis for the quaternion algebra H. Since $\widehat{I}\widehat{J} = -\widehat{J}\widehat{I}$ and $\widehat{I}^2 = C^2 \cdot 1$ and $\widehat{J}^2 = B^2 \cdot 1$, H is the quaternion algebra

$$\left(\frac{C^2, B^2}{\mathbb{Q}(iABC)}\right). \qquad \square$$

Suppose $\varphi\colon \mathbb{Q}(iABC) \to \mathbb{R}$ is a real embedding of the field $\mathbb{Q}(iABC)$. Let $\widehat{A^2}$, $\widehat{B^2}$, $\widehat{C^2}$ be the images of the elements A^2, B^2, C^2 under the embedding φ.

Then $-A^2B^2C^2$, the square of $iABC$, is sent to a positive real number and $\widehat{A^2}\widehat{B^2}\widehat{C^2}$ is negative.

The numbers $C^2(A^2-1)$, $A^2(B^2-1)$, $B^2(C^2-1)$ are all positive rational numbers so that $\widehat{C^2}(\widehat{A^2}-1)$, $\widehat{A^2}(\widehat{B^2}-1)$, and $\widehat{B^2}(\widehat{C^2}-1)$ are also positive rational.

Since $\widehat{A^2}\widehat{B^2}\widehat{C^2}$ is negative either all three factors are negative or one is negative and the other two are positive. But if, say, $\widehat{A^2}$ and $\widehat{B^2}$ are positive then so is $\widehat{B^2}\widehat{C^2}$ since $\widehat{B^2}(\widehat{C^2}-1)$ is positive. So we see that all three are negative.

It follows that the quaternion algebra $\left(\frac{C^2,B^2}{\mathbb{Q}(iABC)}\right) \otimes_\varphi \mathbb{R} \cong \left(\frac{\widehat{C^2},\widehat{B^2}}{\mathbb{R}}\right)$ is the quaternion algebra of Hamilton and not $M(2,\mathbb{R})$.

Since in all eleven cases the field $\mathbb{Q}(iABC)$ has exactly one complex place we have the following.

Theorem 6.2. *The field $\mathbb{Q}(iABC)$ has exactly one complex place and the quaternion algebra $H = \left(\frac{C^2,A^2}{\mathbb{Q}(iABC)}\right)$ is ramified at every real place.* $\square$

The next step is to show that the ring O of (6.2) is an order.

Theorem 6.3. *The ring O is an order in the eleven cases of Theorem 4.2.*

Proof. To show that O is an order it suffices to show that O is a ring, (clear from its definition), that O contains O_R, (obvious), that O contains a basis for H over $\mathbb{Q}(iABC)$, $(\{1, h_a^2, h_b^2, h_c^2\})$, and that O is of finite type as an O_R module.

But $O \subset O_R 1^\star + O_R h_a^\star + O_r h_b^\star + O_R h_c^\star$ by the same argument as in the proof of Theorem 6.1. Moreover, the elements $1^\star$, $h_a^\star$, $h_b^\star$, $h_c^\star$ can be written as linear combinations of 1, h_a^2, h_b^2, h_c^2 with coefficients in $\mathbb{Q}(iABC)$. There is a rational integer m such that all the coefficients in these linear combinations lie in the ring O_R, after they have first been multiplied by m. Thus $m(O_R 1^\star + O_R h_a^\star + O_r h_b^\star + O_R h_c^\star) \subset O$, and it follows that O is of finite type. $\square$

Corollary 6.3.1. *In the eleven cases, the group $B^2_{m,n,p}$ is derived from a quaternion algebra and therefore the group $B_{m,n,p}$ is arithmetic.* $\square$

We wish to study the special case $B_{4,4,4}$ further. In this case $A = B = C$ and we have seen that A^2 belongs to the field $\mathbb{Q}(iABC) = \mathbb{Q}(iA^3)$. Thus iA belongs to $\mathbb{Q}(iA^3)$ and $\mathbb{Q}(iA^3) = \mathbb{Q}(iA)$.

In general, for quaternion algebras $(\frac{a,b}{k}) \cong (\frac{ax^2,by^2}{k})$ for any non zero x, y in k. Therefore $(\frac{C^2,B^2}{\mathbb{Q}(iA)}) = (\frac{A^2,A^2}{\mathbb{Q}(iA)}) = (\frac{-(iA)^2,-(iA)^2}{\mathbb{Q}(iA)}) = (\frac{-1,-1}{\mathbb{Q}(iA)})$.

Theorem 6.4. *The group $B^2_{4,4,4}$ is derived from the quaternion algebra $\left(\frac{-1,-1}{\mathbb{Q}(iA)}\right)$ where iA is a complex root of the equation $x^4 + x^2 - 1 = 0$.* □

(Remark: We can show that in all eleven cases the quaternion algebra is $\left(\frac{-1,-1}{\mathbb{Q}(iABC)}\right)$. The argument is tedious and it goes case by case so we don't include it. We don't know whether or not this follows from some more general argument, nor do we understand its geometric significance.)

Next we gather some facts about the quaternion algebra $\left(\frac{-1,-1}{\mathbb{Q}(iA)}\right)$.

There are three archimedean valuations of the field $\mathbb{Q}(iA)$, a complex valuation arising from the identity map $\mathbb{Q}(iA) \to \mathbb{C}$, and two real valuations, arising from the maps determined by sending iA to the two real roots of $x^4 + x^2 - 1 = 0$.

A quaternion algebra H over a field k is said to be ***ramified*** at the valuation v if the algebra $H \otimes_k k_v$ is a division algebra over k_v, the completion of k at v. For $H = (\frac{-1,-1}{\mathbb{Q}(iA)})$ the two algebras corresponding to the real valuations are $(\frac{-1,-1}{\mathbb{R}})$, which equal the Hamilton quaternions.

The element $x \cdot 1 + yI + zJ$ has the norm $x^2 + y^2 + z^2$ in $(\frac{-1,-1}{k})$. Consequently the algebra $(\frac{-1,-1}{k}) \otimes_k k_v$ is unramified if $x^2 + y^2 + z^2 = 0$ has a nontrivial solution in k_v. We shall show this equation has a nontrival solution in the completion of $\mathbb{Q}(iA)$ with respect to any non-archimedean valuation.

Lemma 6.5. *The equation $x^2 + y^2 + z^2 = 0$ has a nontrivial solution in $\mathbb{Q}_p$ for any $p \neq 2$.*

Proof. If -1 is a square in $\mathbb{Q}_p$, say $-1 = x^2$, then $x^2 + 1^2 = 0$ works, so assume that -1 is not a square in $\mathbb{Q}_p$.

By Hensel's lemma, there is a nontrivial solution of $x^2 + y^2 + z^2 = 0$ in $\mathbb{Z}_p$, with at least one of x, y, z a unit, if and only if there is a nontrivial solution of $\overline{x}^2 + \overline{y}^2 + \overline{z}^2 = \overline{0}$ in the residue class field $\mathbb{Z}$ (mod p). Let $B = \{\overline{x}^2 \in \mathbb{Z}(\text{mod} p) \mid \overline{x} \neq 0\}$ and let $C = \{-\overline{z}^2 \in \mathbb{Z}(\text{mod} p) \mid \overline{z} \neq 0\}$. Then $\sharp B = \sharp C = \frac{1}{2}(p-1)$, and $B \cap C = \emptyset$ (or -1 would be a square). Also let $B + B = \{\overline{x}^2 + \overline{y}^2 \mid \overline{x} \neq 0, \overline{y} \neq 0\}$. Then $\sharp(B+B) \geqq \frac{1}{2}(p-1)$ (or -1 would be a square). If $(B+B) \cap C \neq \emptyset$ we are done, as then the congruence $\overline{x}^2 + \overline{y}^2 + \overline{z}^2$ has a nontrivial solution. If $(B+B) \cap C = \emptyset$, then $B + B = B$, as there are exactly $p-1$ non zero elements of $\mathbb{Z}$ (mod p). But this is absurd as B would then contain $\overline{1}$, $\overline{1} + \overline{1} = \overline{2}$, $\overline{1} + \overline{2} = \overline{3}$, … . □

Corollary 6.5.1. *The algebra* $\left(\frac{-1,-1}{k_v}\right)$ *is not a division algebra if* k_v *is the completion of* $\mathbb{Q}(iA)$ *with respect to the valuation* v *where* v *is an extension of the* p*-adic valuation with* $p \neq 2$.

Proof. k_v contains $\mathbb{Q}_p$. □

We note that $x^2 + y^2 + z^2 = 0$ has no nontrivial solutions in $\mathbb{Q}_2$ because the square of a unit is congruent to zero mod eight.

The 2-adic valuation on $\mathbb{Q}$ has a unique extension to the field $\mathbb{Q}(\sqrt{5})$ because $x^2 - 5$ is irreducible over $\mathbb{Q}_2$. The extension is given by $|a + b\sqrt{5}| = (|a^2 - 5b^2|_2)^{\frac{1}{2}}$.

Lemma 6.6. $x^2 + y^2 + z^2 = 0$ *has a nontrivial solution in* $\mathbb{Q}(\sqrt{5})_v$ *where* v *is the unique extension of the* 2*-adic valuation in* $\mathbb{Q}$.

Proof. $$[\frac{1}{2}(\sqrt{5}-1)]^2 + [\frac{1}{2}(\sqrt{5}+1)]^2 + [\sqrt{5}]^2 = 8 \equiv 0 (\mathrm{mod}\, 8)$$

Since $|\sqrt{5}|_v = 1$, and $(|2\sqrt{5}|_v)^2 = \frac{1}{4}$, and $|[\sqrt{5}]^2 + [\frac{1}{2}(\sqrt{5}-1)]^2 + [\frac{1}{2}(\sqrt{5}+1)]^2|_v = \frac{1}{8} < \frac{1}{4}$, by Hensel's lemma the equation $z^2 + [\frac{1}{2}(\sqrt{5}-1)]^2 + [\frac{1}{2}(\sqrt{5}+1)]^2 = 0$ has a nontrivial solution in $\mathbb{Q}(\sqrt{5})_v$. □

Corollary 6.6.1. $(\frac{-1,-1}{k_v})$ *is unramified for* v *an extension of the* 2*-adic valuation on* $\mathbb{Q}$ *to the field* $\mathbb{Q}(iA)$. □

We summarize the preceding in a theorem.

Theorem 6.7. *The quaternion algebra* $\left(\frac{-1,-1}{\mathbb{Q}(iA)}\right)$ *is ramified exactly at its two real valuations.* □

The ring of algebraic integers O_R of the field $\mathbb{Q}(iA)$ is of interest to us and we wish to explicitly exhibit a $\mathbb{Z}$-basis for O_R.

Lemma 6.8. *Let* $\mu = \frac{1}{2}(\sqrt{5}+1)$. *A* $\mathbb{Z}$*-basis for the integers of* $\mathbb{Q}(\sqrt{5})$ *is* $\{1, \mu\}$.

Proof. A number in a quadratic extension E/F is integral over F if and only if both its norm and trace are integers of F.

Since $t = \mathrm{tr}_{\mathbb{Q}(\sqrt{5})/\mathbb{Q}}(a + b\mu) = 2a + b$ and $N = N_{\mathbb{Q}(\sqrt{5})/\mathbb{Q}}(a + b\mu) = a^2 + ab - b^2$ and $t^2 - 4N = 5b^2$, we see that, if $a + b\mu$ is an integer, then $b \in Z$ and $2a \in Z$. If $a = \frac{m}{2}$ then $2N = 2(\frac{m^2}{4} + \frac{m}{2}b - b^2) \in \mathbb{Z}$ which implies $\frac{1}{2}m^2 \in \mathbb{Z}$, i.e. m is even. Thus $a,\ b \in \mathbb{Z}$. □

The field $\mathbb{Q}(iA)$ is obtained by adjoining $\sqrt{-\mu}$ to $\mathbb{Q}(\sqrt{5})$ because $\sqrt{-\mu} = i\sqrt{\mu}$, and since $A^4 - A^2 - 1 = 0$ we see that $A^2 = \mu$.

Lemma 6.9. *If* $a + b\sqrt{-\mu}$ *is an integer of* $\mathbb{Q}(iA)$ *over* $\mathbb{Q}(\sqrt{5})$ *with* $a,\ b \in \mathbb{Q}(\sqrt{5})$, *then* a *and* b *are integers of* $\mathbb{Q}(\sqrt{5})$.

Proof. Let $t = \mathrm{tr}_{\mathbb{Q}(iA)/\mathbb{Q}(\sqrt{5})}(a+b\sqrt{-\mu}) = 2a$ and $N = N_{\mathbb{Q}(iA)/\mathbb{Q}(\sqrt{5})}(a+b\sqrt{-\mu}) = a^2+\mu b^2$. Then t and N are integers of $\mathbb{Q}(\sqrt{5})$ as is $t^2-4N = 4\mu b^2$. But μ is a unit of $\mathbb{Q}(\sqrt{5})$, $(\mu^{-1} = \mu - 1)$, so $4b^2$ is an integer of $\mathbb{Q}(\sqrt{5})$ as is $2b$.

Let $2a = \ell + m\mu$ and $2b = n + p\mu$ with $\ell, m, n, p \in Z$. Using $\mu^2 = 1+\mu$ we compute

$$N = \frac{1}{4}\left[(\ell^2+m^2+2np+p^2)+(2m\ell+m^2+n^2+2np+2p^2)\mu\right].$$

Therefore

(i) $\ell^2+m^2+2np+p^2 \equiv 0 \pmod 4$;

(ii) $2m\ell+m^2+n^2+2np+2p^2 \equiv 0 (\mathrm{mod}\, 4)$.

It's clear from (ii) that m and n have the same parity. If both are even then (i) implies $\ell^2+p^2 \equiv 0 (\mathrm{mod}\, 4)$ which is only possible with ℓ and p even, so all four integers are even.

On the other hand, if m and n are both odd, (i) cannot be satisfied with p even, so p must be odd. If m, n, and p are odd then (i) implies ℓ is even. With these values for ℓ, m, n, and p equation (ii) becomes

$$m^2+n^2+2np+2p^2 \equiv 0 \pmod 4,$$

which isn't satisfied for m, n, p are odd. Thus we have shown that equations (i) and (ii) imply all four integers ℓ, m, n, p are even, which implies a, b are integers in $\mathbb{Q}(\sqrt{5})$. □

Theorem 6.10. *The set* $\{1, iA, (iA)^2, (iA)^3\}$ *is a* $\mathbb{Z}$*-basis for* O_R.

Proof. This follows immediately from the preceding lemma, together with the fact that x in $\mathbb{Q}(iA)$ is integral over $\mathbb{Q}$ if it is integral over $\mathbb{Q}(\sqrt{5})$. □

The ring of integers of an algebraic number field is not necessarily a unique factorization domain, or a principal ideal domain. The standard example seems to be $\mathbb{Q}(\sqrt{-5})$ where 6 has the factorizations $2\cdot 3$ and $(1+\sqrt{-5})(1-\sqrt{-5})$.

However the ring O_R is in fact a Euclidean domain, which implies it is both a unique factorization domain and a principal ideal domain. This has been known to algebraic number theorists for a long time, but we shall present a proof anyway, because an explicit algorithm might prove useful, for example in decision problems about 3-manifolds.

Before describing the algorithm, we establish some elementary propositions using calculus.

Proposition 6.11. *Let* $f(x,y) = x^2+xy-y^2$. *Then* $|f(x,y)| \leqq \frac{245}{256}$ *on the rectangle* $|x|, |y| \leqq 7/8$.

Proof. (Calculus). The gradient vanishes only at $(0,0)$ where $f=0$, so the maximum and minimum are attained on the boundary. Since $f(a,b) = f(-a,-b)$ and $f(-a,b) = -f(b,a)$ we only need to check one side. We see that $g(y) = f(\frac{7}{8}, y) = \frac{5}{4}(\frac{7}{8})^2 - (y - \frac{7}{16})^2$ and $\frac{5}{4}(\frac{7}{8})^2 = \frac{245}{256}$. □

Proposition 6.12. *Let*

$$f(x,y) = (x-1+y)^2 + y^2,$$
$$g(x,y) = (x-1+y)^2 - (x-1)^2,$$
$$h(x,y) = (x-1+y)^2 - y^2.$$

Then $0 \leqq f(x,y) \leqq \frac{1}{2}$ *and* $\frac{-3}{4} \leqq g(x,y),\ h(x,y) \leqq \frac{1}{4}$ *for* (x,y) *in the triangle* $x+y \geqq \frac{1}{2},\ 0 \leqq x,\ y \leqq \frac{1}{2}$.

Proof. (Calculus). The gradients of $f,\ g,$ and h do not vanish inside the triangle so its enough to check boundary values, where, in effect, $f,\ g,$ and h become functions of one variable. It is an easy exercise to find their maxima. □

Proposition 6.13. *Let* $Z = (a+b\mu)+(c+d\mu)\sqrt{-\mu} \in \mathbb{Q}(iA)$ *with* $a,b,c,d \in \mathbb{Q}$. *Then*

$$N_{\mathbb{Q}(iA)/\mathbb{Q}(\sqrt{5})}(Z) = (a^2+b^2+(c+d)^2-c^2)+((a+b)^2-a^2+(c+d)^2+d^2).$$

Proof. This is precisely the same computation that was done in Lemma 6.8.

To show that $\mathbb{Q}(iA)$ has a Euclidean algorithm we must describe how to find an integer X in $\mathbb{Q}(iA)$ given a number x in $\mathbb{Q}(iA)$ such that $|N_{\mathbb{Q}(iA)/\mathbb{Q}}(X-x)| \leqq \delta < 1$.

This is the same as showing how to choose integers $A,B,C,D \in Z$, given $a,b,c,d \in \mathbb{Q}$, such that $|N_{\mathbb{Q}(iA)/\mathbb{Q}}[(A-a)+(B-b)\mu+\{(C-c)+(D-d)\}\sqrt{-\mu}]| \leqq \delta$. We will describe how to do this in the rest of this section. □

Let $\underline{A}$ be the greatest integer less than or equal to a and let $\bar{A}$ be the smallest integer greater than or equal to a, and similarly define $\underline{B},\ \underline{C},\ \underline{D},\ \bar{B},\ \bar{C},\ \bar{D}$.

We shall always choose A to be either $\underline{A}$ or $\bar{A}$ and similarly for $B,\ C,\ D$. Let $\widehat{A}$ be either $\underline{A}$ or $\bar{A}$ whichever is nearest to a with $\widehat{A} = \underline{A}$ in case of a tie.

We begin by choosing C and D.

Choose $D = \widehat{D}$ once and for all. Let $y = d - D$ and let $x = c - \widehat{C}$. If x and y have opposite signs or if x and y have the same sign but $|x+y| \leqq \frac{1}{2}$ choose $C = \widehat{C}$. If x and y have opposite signs and $|x+y| > \frac{1}{2}$ choose C to be the other nearest integer, not $\widehat{C}$.

Proposition 6.14.

$$0 \leqq (c-C+d-D)^2 + (d-D)^2 \leqq \frac{1}{2},$$
$$\tfrac{-3}{4} \leqq (c-C+d-D)^2 - (c-C)^2 \leqq \frac{1}{4}.$$ □

Proof. In the case where $y = d-D$ and $x = c-C$ have opposite signs or $|x = y| \leqq \frac{1}{2}$ this is clear since $|x|, |y| \leqq \frac{1}{2}$.

In the case where $x = d-D$ and $y = c-C$ have opposite signs and $|x+y| > \frac{1}{2}$, this follows from Proposition 6.11 since $(c-C+d-D)^2 = (1-x+y)^2$. □

Now we choose A and B.

Again set $x = a - \widehat{A}$ and $y = b - \widehat{B}$. Again, if either x and y have opposite signs, or x and y have the same sign and $|x + y| \leqq \frac{1}{2}$, choose $A = \widehat{A}$ and $B = \widehat{B}$. The inequalities $x^2 + y^2 \leqq \frac{1}{2}$ and $\frac{-1}{4} \leqq (x + y)^2 - x^2$ are satisfied.

Suppose x and y have the same sign but $|x + y| > \frac{1}{2}$. Observe that the triangle $0 \leqq x,\ y \leqq \frac{1}{2},\ x + y \geqq \frac{1}{2}$ is contained in the union of the two discs $x^2 + (1 - y)^2 \leqq \frac{5}{8}$, $(1 - x)^2 + y^2 \leqq \frac{5}{8}$. If the first inequality is satisfied, set $A = \widehat{A}$ and B equals the second nearest integer to b. If the first inequality is not satisfied, the second will be. Then set $B = \widehat{B}$ and A equals the 'second nearest' integer.

Proposition 6.15.
$$0 \leqq (a - A)^2 + (b - B)^2 \leqq \frac{5}{8},$$
$$\tfrac{-3}{4} \leqq (a - A + b - B)^2 - (a - A)^2 \leqq \frac{1}{4}.$$

Proof. The first inequality follows easily from the choices of A and B and the second follows from Proposition 6.11. □

Theorem 6.16.
$$\left| N_{\mathbb{Q}(iA)/\mathbb{Q}} \left[(a - A) + (b - B)\mu + \{(c - C) + (d - D\mu)\} \sqrt{-\mu} \right] \right| \leqq \frac{245}{256}.$$

The ring O_R is a Euclidean domain, unique factorization domain, and principal ideal domain.

Proof. The inequality follows directly from the formula for the norm (Proposition 6.12), the inequalities in Propositions 6.14 and 6.15, from
$$N_{\mathbb{Q}(iA)/\mathbb{Q}} = N_{\mathbb{Q}(\sqrt{5})/\mathbb{Q}} \cdot N_{\mathbb{Q}(iA)/\mathbb{Q}(\sqrt{5})},$$
and Proposition 6.10.

The second statement follows from the fact that a Euclidean domain is both a principal ideal domain and unique factorization domain. □

Appendix. The non arithmeticity properties using Reid's criterion

For its potential geometric significance we shall prove the following proposition using Reid's criterion.

Proposition. *If one of the indices $m,\ n,\ p$ is not equal to $3,\ 4,\ 6,$ or ∞, the field $K_2 = \mathbb{Q}(\mathrm{tr}(g) \mid g \in B^2_{m,n,p})$ has more than one complex place. In this case $B_{m,n,p}$ is not arithmetic.*

Proof. Suppose that one of $m,\ n,\ p$, say it is m, is not equal to $3,\ 4,\ 6$, or ∞. Let j be an integer prime to m and in the interval $(1, m/2)$.

Using the fact that the arithmetic series $\{j + im \mid i \geqq 0\}$ contains infinitely many primes (Dirichlet's Theorem), we choose one, $q = j + im$, which also is prime to n and p. Let k' and ℓ' be (the unique) integers such that:

$$q \equiv k' \bmod n, \quad |k'| \in [1, n/2),$$
$$q \equiv \ell' \bmod p, \quad |\ell'| \in [1, p/2),$$

and let us denote $|k'|$ and $|\ell'|$ by k and ℓ respectively.

Note that $\cos\frac{2\pi}{m}q = \cos\frac{2\pi}{m}j$, $\cos\frac{2\pi}{n}q = \cos\frac{2\pi}{n}k$, and $\cos\frac{2\pi}{p}q = \cos\frac{2\pi}{p}\ell$. Let c be the least common multiple of m, n, and p.

Next, apply the non trivial automorphism θ_q, which sends $\cos\frac{2\pi}{c}$ to $\cos\frac{2\pi q}{c}$, to the field k, using the integer q that we have just defined. This automorphism induces one or more non trivial embeddings of the field $K_2 = k(iABC)$ in $\mathbb{C}$ or in $\mathbb{R}$. These embeddings are found as follows.

The numbers A, B, and C satisfy the following system of equations

$$\begin{aligned} C^2(A^2-1) &= \tan^2\frac{\pi}{m}, \\ A^2(B^2-1) &= \tan^2\frac{\pi}{n}, \\ B^2(C^2-1) &= \tan^2\frac{\pi}{p}. \end{aligned} \tag{A.1}$$

The number $A^2B^2C^2$ satisfies the following equation, which is found by eliminating the terms A^2B^2, A^2C^2, and B^2C^2 from (A.1) and multiplying the three resulting equations together:

$$\begin{aligned} (A^2B^2C^2 - \tan^2\frac{\pi}{m})(A^2B^2C^2 - \tan^2\frac{\pi}{n})(A^2B^2C^2 - \tan^2\frac{\pi}{p}) = \\ A^2B^2C^2(1+\tan^2\frac{\pi}{m})(1+\tan^2\frac{\pi}{n})(1+\tan^2\frac{\pi}{p}). \end{aligned} \tag{A.2}$$

Apply the automorphism/embedding θ_q to the elements on the right hand side of (A.2) obtaining a new set of equations with the right hand sides replaced by $\tan^2\frac{\pi}{m}j$, $\tan^2\frac{\pi}{n}k$, $\tan^2\frac{\pi}{p}\ell$.

Because $0 < \frac{\pi}{m}j,\ \frac{\pi}{n}k,\ \frac{\pi}{p}\ell < \frac{\pi}{2}$, Proposition 3.1 guarantees us that there are solutions of the new equations, $\widehat{A}$, $\widehat{B}$, $\widehat{C}$, in which the numbers $\widehat{A}^2$, $\widehat{B}^2$, $\widehat{C}^2$ and therefore $\widehat{A}^2\widehat{B}^2\widehat{C}^2$ are real and positive.

The number $i\widehat{A}\widehat{B}\widehat{C}$ has negative, non zero square and so is pure imaginary. It solves a sixth degree equation (obtained by applying θ_q to the coefficients of (A.2)) with coefficients in k. Consequently the embedding θ_q of k extends to a necessarily complex embedding of $k(iABC)$ in $\mathbb{C}$. Since $k(iABC) = K_2$, the field K_2 has more than one complex place. This proves the proposition. □

References

[Bi] L. Bianchi, Sui gruppi di sostituzioni lineari con coefficienti appartenenti a corpi quadratici immaginari, Math. Ann. 40 (1892), 332–412.

[B] A. Borel, Commensurability classes and volumes of hyperbolic 3-manifolds, Ann. Scuola Norm. Sup. Pisa 8 (1981), 1–33.

[B1] A. Borel, Density and maximality of arithmetic subgroups, J. Reine Angew. Math. 224 (1966), 78–89.

[BH] A. Borel, Harish-Chandra, Arithmetic subgroups of algebraic groups, Ann. of Math. 75 (1962), 485–535.

[F] W. Fenchel, "Elementary Geometry in Hyperbolic Space", Stud. Math. 11, Walter de Gruyter, Berlin (1989).

[HKM] H. Helling, A. C. Kim, J. L. Mennicke, On Fibonacci groups, (to appear).

[HLM1] H. M. Hilden, M. T. Lozano, J. M. Montesinos, On knots that are universal, Topology 24 (1985), 499–504.

[HLM2] H. M. Hilden, M. T. Lozano, J. M. Montesinos, On the universal group of the Borromean rings, in "Proceedings of the 1987 Siegen conference on Differential Topology", ed. U. Koschorke. LNM 1350, Springer Verlag (1988), 1-13.

[HLM3] H. M. Hilden, M. T. Lozano, J. M. Montesinos, The arithmeticity of the figure eight knot orbifolds, (these Proceedings).

[HLMW] H. M. Hilden, M.T. Lozano, J. M. Montesinos, W. C. Whitten, On universal groups and three manifolds, Inventiones mathematicae 87 (1987), 441–456.

[Ho] C. D. Hodgson, Degeneration and regeneration of geometric structures on 3-manifolds, Ph.D. Thesis, Princeton (1986).

[M] Maclachlan C., A. Reid, Commensurability classes of arithmetic Kleinian groups and their Fuchsian subgroups, Math. Proc. Camb. Phil. Soc. 102 (1987), 251–257.

[Ma] W. Magnus, Rings of Fricke characters and automorphism groups of free groups, Math. Z. 170 (1980), 91–103.

[R] A. W. Reid, Arithmetic Kleinian groups and their Fuchsian subgroups, Ph.D. Thesis (1987), University of Aberdeen (part is in **[M]**).

[R1] A. W. Reid, Arithmetic Fuchsian groups: a survey, Master of Science Th., Aberdeen (1985).

[Re] N. Reid, "Undergraduate algebraic geometry", London Math. Society Student Texts 12, Cambridge University Press (1988).

[S] H. P. F. Swinnerton-Dyer, Arithmetic groups, in "Discrete groups and automorphic functions", ed. W. J. Harvey, London: Academic Press (1977), 307–401.

[T1] K. Takeuchi, On some discrete subgroups of $\mathrm{SL}(2,\mathbb{R})$, J. Fac. Sci. Univ. Tokyo Sec 1, 16 (1969), 97–100.

[T2] K. Takeuchi, A characterization of arithmetic Fuchsian groups, J. Math Soc. Japan 27 (1975), 600–612.

[T3] K. Takeuchi, Arithmetic triangle groups, J. Math. Soc. Japan 29 (1977), 91–106.

[T4] K. Takeuchi, Arithmetic Fuchsian groups of signature (1; e), J. Math. Soc. Japan 35 (1983).

[Th] W. Thurston, The geometry and topology of 3-manifolds, Lecture Notes, Princeton University 1978.

[V] M. F. Vigneras, "Arithmetique des Algèbres de Quaternions", LNM 800, Springer Verlag (1980).

[Vi] E. B. Vinberg, Discrete groups generated by reflections in Lobachevskii space, Math Sb. 114 (1967), 471–488.

[Vo] H. Vogt, Sur les invariants fondamentaux des équations differentielle linéaire du second ordre, Ann. Sci. Ecole Norm Sup. (3) 6 Suppl. 3-72 (1889) Thèse. Paris.

[Z] Robert J. Zimmer, "Ergodic theory and semi simple groups", Birkhauser (1984).

Department of Mathematics, University of Hawaii at Manoa, Honolulu, HI 96822

Department of Mathematics, University of Zaragoza, 50009 Zaragoza, Spain

Department of Mathematics, Universidad Complutense, 28040 Madrid, Spain

The Arithmeticity of the Figure Eight Knot Orbifolds

*Hugh M. Hilden, María Teresa Lozano, Jose María Montesinos-Amilibia**

The Fibonacci groups are abstract groups defined by the presentations:

$$F(2,m) = \langle X_1, \ldots, X_m : X_i X_{i+1} = X_{i+2} \quad \text{(indices modulo } m)\rangle$$

In the article **[HKM]**, the authors state that these groups are "the link between certain objects in 3-dimensional topology, in 3-dimensional hyperbolic geometry, in the theory of discontinuous transformation groups; in rank one Lie groups". In fact for $m = 2n$, there is a closed orientable 3-manifold M_n such that:

(i) The fundamental group $\pi_1(M_n) = F(2,2n)$.

(ii) The 3-manifold M_n is hyperbolic for $n > 3$. That is to say, $F(2,2n)$ acts as a discrete group of isometries on hyperbolic 3-space, $\mathbb{H}^3$, with quotient space M_n.

(iii) The group $F(2,2n)$, as a subgroup of $\mathrm{SL}(2,\mathbb{C})$, is arithmetic for the values $n = 4, 5, 6, 8, 12$, or ∞.

In this note, which can be considered as a sequel to **[HKM]**, we shall show that in addition, the group $F(2,2n)$ is related to knot theory.

In effect, aficionados of knot theory know that the figure eight knot is a fibered knot whose fiber is a punctured torus. It's monodromy is given by a matrix related to the Fibonacci sequence. Thus the following theorem should come as no surprise.

Theorem 1. *Let K_n be the n-fold cyclic branched covering of the figure eight knot. Then $\pi_1(K_n) = F(2,2n)$.*

Proof. The group of the figure eight knot K has the following presentation:

$$\pi_1(S^3 - K) = \langle x, A, C : xAx^{-1} = ACA, xCx^{-1} = CA\rangle \qquad (1)$$

In this presentation x is a meridian of K, and A and C are a basis for the fundamental group of a Seifert surface for K; a free group. (We shall prove this later).

To obtain K_n we cut S^3 along a Seifert surface F of K (a punctured torus) obtaining

$$h^3 = F \times [0,1]/\sim$$

where $(x,t) \sim (x,s)$ for $x \in \partial F$ and $s,t \in [0,1]$.

* This research was supported by grants PB85-0336 and PB89-0105. The first author expresses his appreciation to the DGICYT for financial support while he was in Spain.

To recover S^3 we use the diffeomorphism φ of F (the monodromy). Here φ is isotopic to an Anosov diffeomorphism that satisfies

$$\begin{aligned}\varphi_*(A) &= ACA \\ \varphi_*(C) &= CA\end{aligned} \tag{2}$$

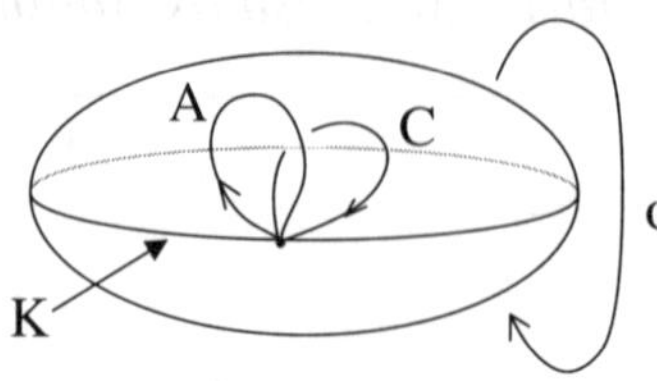

Figure 1

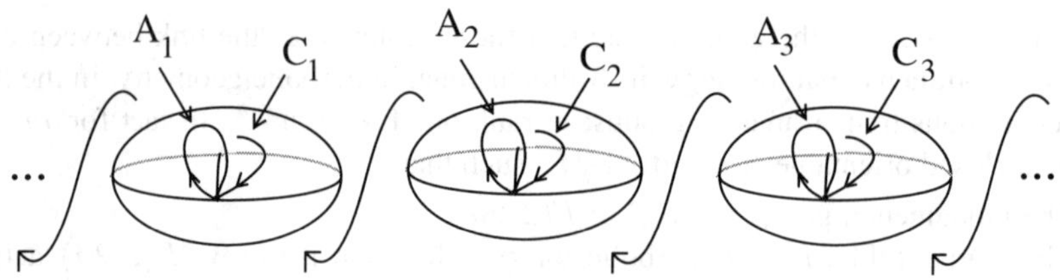

Figure 2

Moreover φ is the identity on ∂F (See Figure 1).

Now take n copies of h^3 and paste them together cyclically as illustrated in Figure 2, obtaining K_n.

Therefore $\pi_1(K_n)$ has generators $A_1, \dots, A_n, C_1, \dots, C_n$, and relations $A_1C_1A_1 = A_2, \dots, A_nC_nA_n = A_1$; $C_1A_1 = C_2, \dots, C_nA_n = C_1$.

They can be replaced by: $C_1A_1 = C_2, \dots, C_nA_n = C_1; A_1C_2 = A_2, \dots, A_nC_1 = A_1$, which correspond exactly to the Fibonacci relations: $X_1X_2 = X_3, X_2X_3 = X_4, \dots$, etc.

To finish the proof of the theorem we need to verify (1). It is well known that the figure eight knot, being fibered, is the binding of an open book with fibre F and monodromy

$$\varphi = \begin{bmatrix} 2 & 1 \\ 1 & 1 \end{bmatrix}$$

acting on $\widehat{F} = \mathbb{R}^2/\mathbb{Z}^2$. (See Figure 3.)

We can isotope φ slightly in a neighborhood of $(0,0)$ so that φ is the identity near $(0,0)$. The action of φ on $\pi_1 F$ is illustrated in Figure 4. (The base point for $\pi_1 F$ is not moved by φ. It is on the boundary of the punctured hole.)

Thus (1) has been proven. Another way to see (1) is to use the relations in the knot group derived from the knot projection in Figure 5.

Thus:

$$\pi_1(S^3 - K) = \langle x, y : x^{-1}yxy^{-1}xyx^{-1}y^{-1}xy^{-1} \rangle . \tag{3}$$

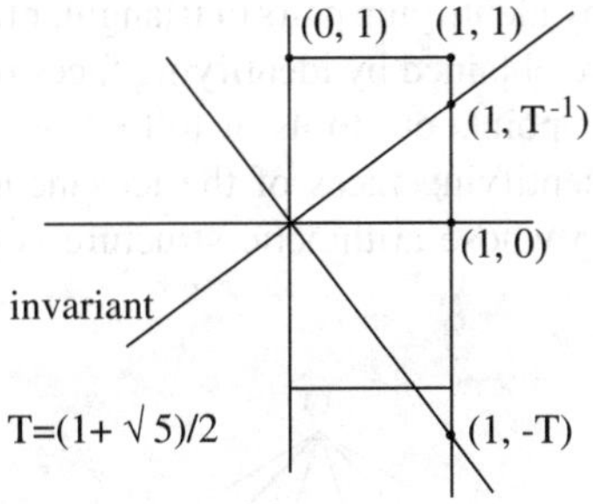

Figure 3

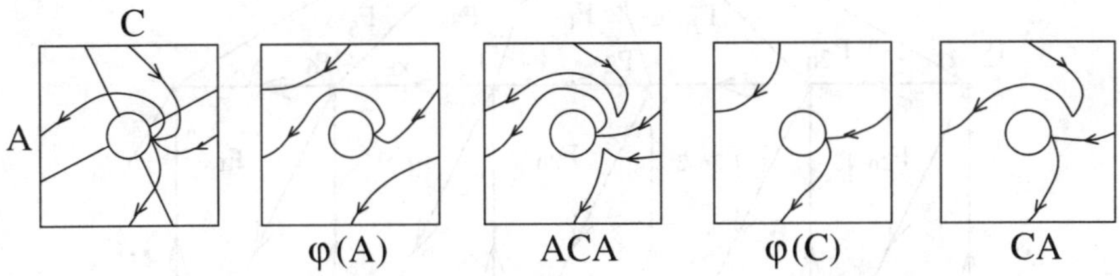

Figure 4

Figure 5

Substituting $a = yx^{-1}$ and $c = x^{-1}a^{-1}xa$ we obtain:

$$\begin{aligned}\pi_1(S^3 - K) &= \langle x, a, c : c^{-1}a^{-1}xax^{-1}a^{-1}, x^{-1}a^{-1}xac^{-1}\rangle \\ &= \langle x, a, c : xax^{-1} = aca, xac^{-1}x^{-1} = a\rangle \\ &= \langle x, a, c : xax^{-1} = aca, xc^{-1}x^{-1} = a^{-1}c^{-1}\rangle \\ &= \langle x, a, c : xax^{-1} = aca, xcx^{-1} = ca\rangle . \qquad \square\end{aligned}$$

Therefore the manifolds K_n and M_n (of **[HMK]**) have identical fundamental groups. The following theorem is not surprising. (This theorem has been obtained independently by J. Howie.)

Theorem 2. *The manifolds M_n and K_n are diffeomorphic; that is, M_n is the n-fold branched cyclic covering of the figure eight knot.*

Proof. We shall describe M_n. Figure 6 presents a generalized icosahedron. There is a central band of $2n$ triangular faces bounded above and below by n-pyramids. The

3-manifold M_n is obtained by identifying pairs of triangular faces as indicated in Figure 6. In particular M_3 and M_5 are obtained by identifying faces in a cube and an icosahedron respectively. (As the referee points out to us, it follows from results of Best **[B]**, that, since M_5 is obtained by identifying faces of the icosahedron, the group $F(2,10)$ is a subgroup of another group whose arithmetic structure is given in **[M]**, **[Bo]** and so $F(2,10)$ is arithmetic.)

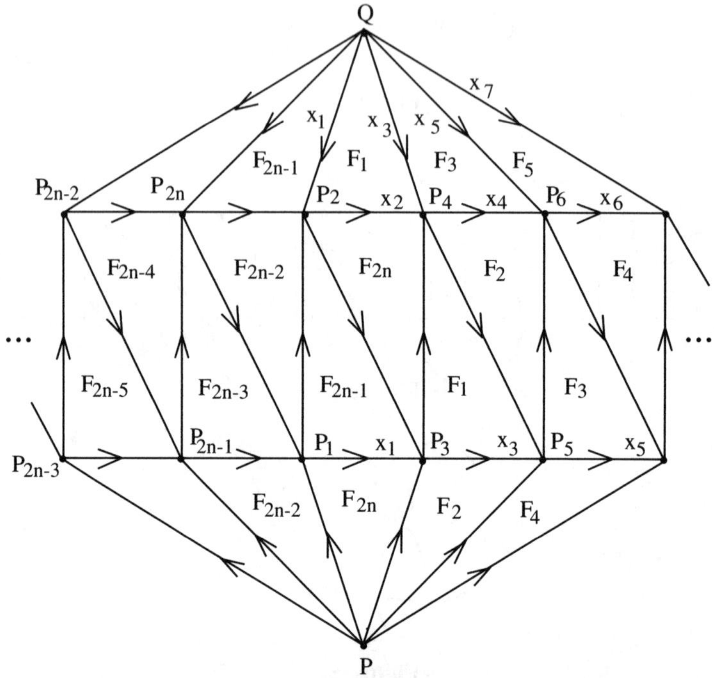

Figure 6

Rotation by $2\pi/n$ about the axis PQ defines an action of C_n, the cyclic group of order n, on M_n. The quotient of the action is S^3 and the image of the axis PQ is the knot k. The isotropy group of a point of M_n not on PQ is trivial.

The quotient M_n/C_n is obtained by taking a fundamental domain for the action of C_n on M_n and making identifications. See Figure 7.

A Heegaard diagram for this quotient appears in Figure 8.

The grey line in Figure 8 is the axis PQ. It lies below the diagram, inside the ball whose boundary is being identified along the disc pairs F_1, $\overline{F}_1$, and F_2, $\overline{F}_2$. Cancelling handles we obtain S^3 and the knot k. (See Figure 9 a, b, and c.) □

Since M_n is hyperbolic for $n \geq 4$ and Euclidean for $n = 3$, as Helling, Kim and Mennicke have beautifully shown in **[HKM]**, we have the following Corollary, where (K, n) is the orbifold with underlying space S^3, singular set K and isotropy group cyclic of order n.

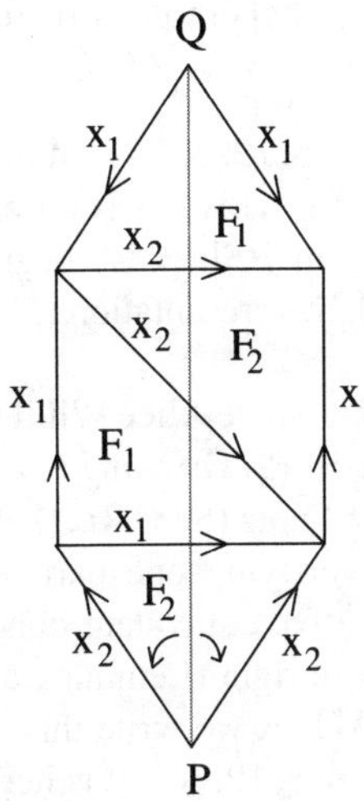

Figure 7

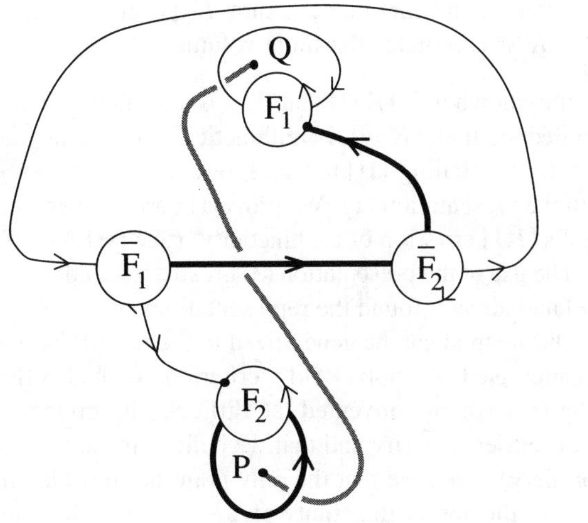

Figure 8

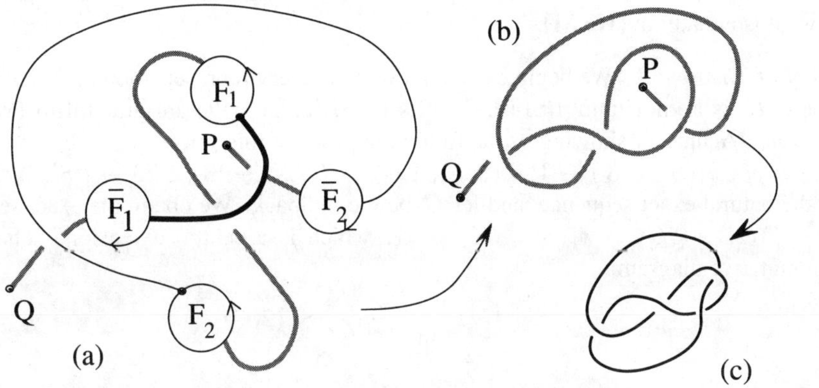

Figure 9

Corollary. *The orbifold* (K, n) *is Euclidean for* $n = 3$, *hyperbolic for* $n > 3$, *and spherical for* $n < 3$. □

For $n = 2$, M_n is the lens space $L(5, 2)$. For $n > 4$ this Corollary is already due to Thurston **[T]** who uses his method of completing the hyperbolic structure obtained by identifying two ideal tetrahedra. (Compare **[Ho**, 6.29.2, page 146, lines 14-21**]**.)

Theorem 3. *The group of the orbifold* (K, n) *is arithmetic exactly for* $n = 4, 5, 6, 8, 12, \infty$.

Remarks. Riley **[Ri1]** was the first to show that (K, ∞) is arithmetic. Later, Reid **[R]** proved that (K, ∞) is the only arithmetic knot, i.e., the only knot K for which (K, ∞) is arithmetic. Here (K, ∞) should be interpreted as the space $S^3 - K$. The symbol

∞ serves to indicate that the knot K is "at infinity" or that the hyperbolic structure on $S^3 - K$ is complete with finite volume.

It is shown in **[HKM]** that M_n is arithmetic for $n = 4, 5, 6, 8, 12, \infty$. From this one can deduce that (K, n) is arithmetic for the same values of n. This has been proved recently by Helling **[H]** using representations of $\pi_1(S^3 - K)$ in $\mathrm{PSL}(2, \mathbb{C})$ beginning with the presentation (1). We prove Theorem 3 beginning with the presentation (3) using Reid's **[R1]** criterion of arithmeticity. (See **[HLM]**.)

The use of the presentation (3) greatly reduced the work for us since Alice Whittemore **[W]** had already found the representations of $\pi_1(S^3 - K)$ in $\mathrm{SL}(2, \mathbb{C})$ using (3).

Our method can be generalized to the rest of the two bridge knots (See **[Ri2]**). When we attempted to apply Reid's criterion we had difficulties verifying one part of that criterion. We circumvented the difficulty by giving an alternative equivalent condition that is easier to verify and that we believe has interest in its own right (Lemma 1 and its Corollary). We note that the only point not resolved in **[HKM]**, as we write this, is the proof of the non-arithmeticity of M_n for the values $n \neq 4, 5, 6, 8, 12, \infty$. It is here that Reid's criterion turned out to be very useful (see Corollary 1.1 below.) It is likely that Reid's criterion can be used in the context of **[HKM]** to demonstrate non-arithmeticity in the remaining cases. Therefore the reader should consider Theorem 3 as an alternative to what is already in **[HKM]**.

Proof of Theorem 3. We begin by remarking that every representation of $\pi_1(S^3 - K)$, where K is a knot, into $\mathrm{PSL}(2, \mathbb{C})$ lifts to $\mathrm{SL}(2, \mathbb{C})$. We are grateful to Francisco Gonzalez-Acuña for showing us the following proof of this.

Let $\omega : \pi_1(S^3 - K) \to \mathrm{PSL}(2, \mathbb{C})$, let $1 \to \mathbb{Z}/2\mathbb{Z} \to \mathrm{SL}(2, \mathbb{Z}) \to \mathrm{PSL}(2, \mathbb{C}) \to 1$ be the natural exact sequence, and let G be the pullback. We obtain the exact sequence $1 \to \mathbb{Z}/2\mathbb{Z} \to G \to \pi_1(S^3 - K) \to 1$, which is a central extension. There is a commutative diagram:

$$\begin{array}{ccc} G & \xrightarrow{\omega_1} & \mathrm{SL}(2, \mathbb{C}) \\ \downarrow{\scriptstyle p_1} & & \downarrow{\scriptstyle p} \\ \pi_1(S^3 - K) & \xrightarrow{\omega} & \mathrm{PSL}(2, \mathbb{C}) \end{array}$$

But $\mathbb{Z}/2\mathbb{Z}$ central extensions of $\pi_1(S^3 - K)$ are classified by $H^2(\pi_1(S^3 - K); \mathbb{Z}/2\mathbb{Z})$ which is the same as $H^2(S^3 - K; \mathbb{Z}/2\mathbb{Z})$ because $S^3 - K$ is a $K(\pi, 1)$ space, by the asphericity of knots. Since $H^2(S^3 - K; \mathbb{Z}/2\mathbb{Z}) = 0$, the central extension $1 \to \mathbb{Z}/2\mathbb{Z} \to G \to \pi_1(S^3 - K) \to 1$ splits and there is a map $s : \pi_1(S^3 - K) \to G$ with $p_1 s$ equal to the identity. The map $\omega_1 s$ is the lift of ω that we wanted.

Now consider the presentation (3) of $\pi_1(S^3 - K)$ and let $\omega : \pi_1(S^3 - K) \to \mathrm{SL}(2, \mathbb{C})$ be a representation. We may write $w(x) = A$, $w(y) = B$ and the relation of the presentation (3) gives

$$A^{-1}BAB^{-1}ABA^{-1}B^{-1}AB^{-1} = I. \tag{4}$$

The representations of G in $\mathrm{SL}(2, \mathbb{C})$ were obtained in **[W]**. We can summarize the results as follows:

If $w(G)$ is not cyclic, and if we denote the traces of A and B by α and the trace of AB by β then

$$\beta = \frac{1}{2}\left(1+\alpha^2 \pm \sqrt{(\alpha^2-1)(\alpha^2-5)}\right) \tag{5}$$

and up to conjugation $w(G)$ is generated by

$$\begin{aligned} A &= \begin{bmatrix} \lambda & 0 \\ 0 & \lambda^{-1} \end{bmatrix} \quad , \quad \lambda = \frac{1}{2}\left(\alpha \pm \sqrt{\alpha^2-4}\right) \quad , \\ B &= \begin{bmatrix} \mu & 1 \\ \mu(\alpha-\mu)-1 & \alpha-\mu \end{bmatrix}, \quad \mu = \frac{\lambda\beta-\alpha}{\lambda^2-1} \quad . \end{aligned} \tag{6}$$

Formula 5 is obtained by taking traces in the following expression which is a consequence of (4).

$$AB^{-1}A^{-1}BAB^{-1} = BAB^{-1}A^{-1}$$

We make use of the following formulas in the computation:

(i) $t(A) = t(A^{-1})$
(ii) $t(XAX^{-1}) = t(A)$
(iii) $t(AB) = t(A)t(B) - t(AB^{-1})$.

Thus we obtain $t(AB^{-1})t(A^{-1}BAB^{-1}) - t(B^{-1}A) = t(BAB^{-1}A^{-1})$. Since $t(AB^{-1}) = \alpha^2 - \beta$, $t(BAB^{-1}A^{-1}) = t(A^{-1}BAB^{-1}) = \beta^2 - \beta\alpha^2 + 2\alpha^2 - 2$ we finally obtain $\beta - \alpha^2 + (\beta^2 - \beta\alpha^2 + 2\alpha^2 - 2)(\alpha^2 - \beta - 1) = 0$; $(\beta^2 - (1+\alpha^2)\beta + 2\alpha^2 - 1)(\alpha^2 - \beta - 2) = 0$.

If α is real and $\alpha^2 - \beta - 2 = 0$ then β is real. If α is real and $\beta^2 - (1+\alpha^2)\beta + 2\alpha^2 - 1 = 0$ then β is non-real if and only if $\alpha \in (1, \sqrt{5})$.

Thus we conclude that if α is real and positive and $w(G)$ is discrete and of finite covolume then (see **[W]**):

(i) $\alpha \in (1, \sqrt{5})$
(ii) w lifts to $\mathrm{SL}(2,\mathbb{C})$ (and we continue to call its lift w)
(iii) $\beta = \frac{1}{2}(1+\alpha^2 \pm \sqrt{(\alpha^2-1)(\alpha^2-5)})$
(iv) the representation exists for these values of α and β and, up to conjugation, w is defined by (6).

The hyperbolic structures of the (K, n) correspond to the discrete representations w_n such that A and B are elliptic elements of order n in $\mathrm{PSL}(2,\mathbb{C})$; that is $A^n = B^n = -I$. Then, by conjugating,

$$A = \begin{bmatrix} \lambda & 0 \\ 0 & \lambda^{-1} \end{bmatrix} \quad , \quad \lambda^n = -1 .$$

Thus $\lambda = e^{\pi i/n}$ and $\alpha = \operatorname{trace} A = 2\cos\pi/n$ is real and positive. As $4\cos^2\pi/n$ belongs to $(1,5)$ for all $n > 3$, we conclude that for $n > 3$ there exists a representation w that is given by (6).

Note that for each such n there are exactly two complex conjugate representations. These representations necessarily correspond to the hyperbolic structures of (K, n),

$n > 3$, which we know exist from the results of Thurston and Hodgson (or from **[HKM]**) mentioned earlier. Therefore the representations w_n given by (6) are discrete.

The two complex conjugate representations produce the same hyperbolic structure in (K, n) if we assume (S^3, K) is not oriented.

Let Γ_n be $w_n(G)$, let K_n be the field $\mathbb{Q}(t(g) \mid g \in \Gamma_n)$, let $\Gamma_n^{(2)}$ be the subgroup of Γ_n generated by squares of elements of Γ_n, and let L_n be the field $\mathbb{Q}(t(h) \mid h \in \Gamma_n^{(2)})$. By Lemma 4.1.2 of **[HLM]** we have

$$K_n = \mathbb{Q}(t(A), t(B), t(AB)) = \mathbb{Q}(\alpha, \beta)$$
$$L_n = \mathbb{Q}(t(A^2), t(B^2), t(A^2B^2)) = \mathbb{Q}(\alpha^2, \beta).$$

In effect $t(A^2) = t(A)^2 - 2 = \alpha^2 - 2 = t(B^2)$ and $t(A^2B^2) = t(A^2B)t(B) - t(A^2) = \beta\alpha^2 - 2\alpha^2 + 2$.

Therefore $L_n = Q(\cos^2 \pi/n, \theta_n)$ where $\theta_n^2 = (4\cos^2 \pi/n - 1)(4\cos^2 \pi/n - 5)$.

Proposition 1. *L_n has exactly one complex place if and only if $n = 4, 5, 6, 8, 12, \infty$.*

Proof. The embeddings of L_n in $\mathbb{C}$ are obtained by taking the automorphisms of the normal field $k_n = Q(\cos 2\pi/n)$ and extending them to L_n. (Note that $\cos 2\pi/n = 2\cos^2 \pi/n - 1$.) If j is an integer prime to n there is an automorphism taking $\cos 2\pi/n$ to $\cos 2\pi j/n$.

The image of θ_n under this automorphism is

$$\theta_n' = \pm\sqrt{(4\cos^2 \pi j/m - 1)(4\cos^2 \pi j/n - 5)}.$$

If j also belongs to the interval $(1, n/3)$ then $4\cos^2 \pi j/n$ belongs to $(1, 5)$ and θ_n' is non real. Since for every $n > 3$, $n \neq 4, 5, 6, 8, 12, \infty$ there exists a j prime to n and in the interval $(1, n/3)$, for these values of n the field L_n has more than one complex place.

It remains to be seen that for $n = 4, 5, 6, 8, 12, \infty$ the field L_n has exactly one complex place. This can be shown by direct inspection:

$$L_4 = \mathbb{Q}\left(\sqrt{-3}\right), \quad \theta_4 = \sqrt{-3}, \quad \cos \pi/4 = \frac{\sqrt{2}}{2},$$

$$L_5 = \mathbb{Q}\left(\sqrt{\frac{-1 - 3\sqrt{5}}{2}}\right), \quad \theta_5 = \sqrt{\frac{-1 - 3\sqrt{5}}{2}}, \quad \cos \frac{\pi}{5} = \frac{1 + \sqrt{5}}{4},$$

$$L_6 = \mathbb{Q}\left(\sqrt{-1}\right), \quad \theta_6 = 2\sqrt{-1}, \quad \cos \frac{\pi}{6} = \frac{\sqrt{3}}{2},$$

$$L_8 = \mathbb{Q}\left(\sqrt{-1 - 2\sqrt{2}}\right), \quad \theta_8 = \sqrt{-1 - 2\sqrt{2}}, \quad \cos \frac{\pi}{8} = \frac{\sqrt{2 + \sqrt{2}}}{2},$$

$$L_{12} = \mathbb{Q}\left(\sqrt{-2\sqrt{3}}\right), \quad \theta_{12} = \sqrt{-2\sqrt{3}}, \quad \cos \frac{\pi}{12} = \frac{\sqrt{2 + \sqrt{3}}}{2},$$

$$L_\infty = \mathbb{Q}\left(\sqrt{-3}\right), \quad \theta_\infty = \sqrt{-3}, \quad \cos\frac{\pi}{\infty} = 1. \qquad \square$$

All these fields have exactly one complex place. We note that $L_4,\ L_6$, and L_∞ have no real places and $L_5,\ L_8,\ L_{12}$ have two real places.

Corollary 1.1. *The only orbifolds that can be arithmetic are* (K, n) *where* $n = 4, 5, 6, 8, 12, \infty$.

Proof. Reid's criterion **[R1]** shows that a necessary condition for Γ_n to be arithmetic is that the field L_n has one complex place. □

We recall that Reid **[R1]** showed that $\Gamma \leqq \mathrm{SL}(2, \mathbb{C})$, of finite covolume, is arithmetic if $\Gamma^{(2)}$ satisfies the following conditions.

(i) The trace field L of $\Gamma^{(2)}$ is a finite extension of $\mathbb{Q}$ with exactly one complex place.
(ii) $t(g)$ is an algebraic integer for all $g \in \Gamma^{(2)}$.
(iii) Let $T = \{t(g) \mid g \in \Gamma^{(2)}\}$. Then $\sigma(T)$ is bounded for every real embedding $\sigma : L \to \mathbb{R}$.

Note. Condition (iii) is difficult to verify. (See **[HLM]** for one case in which it was proven.) We have substituted an equivalent condition which is easier to use in the present case.

Proposition 2. *The number* $t(g)$ *is an algebraic integer for every* $g \in \Gamma_n$.

Proof. It is enough to show that α and β are algebraic integers because they generate the ring generated by all the traces. $\alpha = 2\cos\pi/n$ is an algebraic integer and β satisfies $\beta^2 - (1+\alpha^2)\beta + 2\alpha^2 - 1 = 0$ which is monic with algebraic integer coefficients. □

Corollary 2.1. *The number* $t(h)$ *is an algebraic integer for every* $h \in \Gamma_n^{(2)}$. *Therefore* (K, n) *is arithmetic for* $n = 4, 6, \infty$.

Proof. We have shown that condition (i) is satisfied and condition (ii) follows immediately from Proposition 2 above. In the $n = 4, 6, \infty$ case condition (iii) is empty as L_n has no real places. □

To finish the proof of Theorem 3 it is enough to show that $\Gamma_n^{(2)}$ satisfies condition (iii) of Reid for $n = 5, 8, 12$. As we stated before, we shall do this by substituting another equivalent condition for condition (iii).

Lemma 1. *Let* Γ *be a subgroup of* $\mathrm{SL}(2, \mathbb{C})$ *that has elements*

$$\gamma_0 = \begin{bmatrix} \lambda & 0 \\ 0 & \lambda^{-1} \end{bmatrix}, \quad \lambda^2 \neq 1, \quad \gamma_1 = \begin{bmatrix} a & 1 \\ c & d \end{bmatrix}, \quad c \neq 0.$$

Let K *be the field* $Q(t(\gamma) \mid \gamma \in \Gamma)$. *Suppose that* K *is a finite extension of* Q. *Then:*

(i) *The algebra* $A(\Gamma) = \{\sum a_i\gamma_i \mid a_i \in K, \gamma_i \in \Gamma\} \leqq M(2,\mathbb{C})$ *is a quaternion algebra. In fact*

$$A(\Gamma) = \begin{cases} \left(\dfrac{\alpha^2/4-1, c}{K}\right), & \textit{where } \alpha = \lambda+\lambda^{-1}, \textit{ if } \lambda \notin K, \\ M(2,K), & \textit{if } \lambda \in K. \end{cases}$$

(ii) *If* $t(\gamma)$ *is an algebraic integer for every* $\gamma \in \Gamma$, *then* $O(\Gamma) = \{\sum b_i\gamma_i \mid b_i \in O(K), \gamma_i \in \Gamma\}$, *where* $O(K)$ *is the ring of integers of* K, *is an order of the algebra* $A(\Gamma)$.

Corollary 3.1. *With the same hypotheses as in Lemma 1, together with* $\lambda \notin K$, *if* $\varphi : K \to \mathbb{R}$ *is a real embedding such that* $\varphi(\alpha)^2 < 4$, *then:*
(i) $A(\Gamma) \otimes_{\varphi(K)} \mathbb{R}$ *is the Hamilton quaternions if and only if* $\varphi(c) < 0$.
(ii) *Let* $T = \{t(\gamma) \mid \gamma \in \Gamma\}$. *Then* $\varphi(T)$ *is bounded if and only if* $\varphi(c) < 0$.

Notes.
1) If Γ is discrete and co-compact then K is a number field (finite extension of $\mathbb{Q}$ [**M**, Theorem 4.1]).
2) If Γ has torsion elements, then by conjugating Γ, γ_0 and γ_1 can be chosen as above with $\lambda = e^{\pi i/n}$ such that $\alpha = 2\cos\pi/n$. Then every $\varphi : K \to \mathbb{R}$ restricted to $\mathbb{Q}(\alpha^2)$ is an automorphism, and $\cos^2\pi/n$ is sent to $\cos^2\pi j/n$ for some j. We deduce that if $\lambda \notin K, \varphi(\alpha)^2 < 4$ for all φ.
3) If Γ is discrete and of finite covolume there exist γ_0 and γ_1 as in the statement of Lemma 1.

Proof of Lemma 1. (Compare [**Ta**], [**R1**], [**HLM**].) The matrices $\{1, \gamma_0, \gamma_1, \gamma_2 = \gamma_0\gamma_1\}$ are linearly independent over $\mathbb{C}$ and thus a basis for $M(2,\mathbb{C})$. So the K-linear closure of $\{1, \gamma_0, \gamma_1, \gamma_2\}$ is a four dimensional K-vector space contained in $A(\Gamma)$. Let $T : M(2,\mathbb{C}) \times M(2,\mathbb{C}) \to \mathbb{C}$ be the trace form defined by $T(A,B) = t(AB)$. Then T is a non-degenerate bilinear form on $M(2,\mathbb{C})$ and a dual basis, $\{1^*, \gamma_0^*, \gamma_1^*, \gamma_2^*\}$ of $M(2,\mathbb{C})$, is defined. If $\gamma \in \Gamma$ we have $\gamma = x_0 1^* + x_1\gamma_0^* + x_2\gamma_1^* + x_3\gamma_2^*$ with the x coefficients in $\mathbb{C}$. But $t(\gamma\gamma_i) = x_i$ so that x_i in fact belongs to K as $t(\gamma\gamma_i)$ belongs to K. Thus we see that

$$\langle 1, \gamma_0, \gamma_1, \gamma_2\rangle_K \leqq A(\Gamma) \leqq \langle 1^*, \gamma_0^*, \gamma_1^*, \gamma_2^*\rangle_K.$$

Since both ends are four dimensional vector spaces we must have equality and $A(\Gamma)$ is a four dimensional vector space over K. Evidently $A(\Gamma)$ is a central algebra since a matrix that commutes with the basis $\{1, \gamma_0, \gamma_1, \gamma_2\}$ commutes with $M(2,\mathbb{C})$.

Arguing as in [**Ta**] or [**R1**] we deduce that if $\lambda \notin K$ then
(i) $c \in K$,
(ii) $A(\Gamma) = \left\{\begin{bmatrix} A & B \\ cB' & A' \end{bmatrix} \mid A, B \in K(\lambda)\right\}$ where A' is the conjugate of $A \in K(\lambda)$ over K.

(Let x' be the conjugate of x in $K(\lambda)$ over K. First verify that $\lambda' = \lambda^{-1}$ because $\lambda + \lambda^{-1} \in K$. Then solve for a and d in terms of $t(\gamma_1)$, $t(\gamma_2)$ and λ. Compute a'

and verify that it is d. Check that the basis vectors all have the form of (ii). That c belongs to K follows from $a' = d$ and determinant $\gamma_1 = 1$.)

According to Vignéras **[V**, p. 1**]** a quaternion algebra H over K is a central algebra of dimension four over K such that there exists a two dimensional subalgebra L and an invertible element c of K such that $H = L + Lu$ where $u^2 = c$, $um = \overline{m}u$ for all m in L, where $m \to \overline{m}$ is the non-trivial K-automorphism of L. Thus, in our case we can set

$$L = \left\{ \begin{bmatrix} A & 0 \\ 0 & A' \end{bmatrix} \mid A \in K(\lambda) \right\} \quad u = \begin{bmatrix} 0 & 1 \\ c & 0 \end{bmatrix}, \quad c \in K, \quad c \neq 0.$$

and check $H = L + Lu$, $u^2 = c$, and $um = \overline{m}u$ for m one of the basis vectors.

Let $\alpha = \lambda + \lambda^{-1}$ and define

$$i = \begin{bmatrix} \lambda - \alpha/2 & 0 \\ 0 & \lambda^{-1} - \alpha/2 \end{bmatrix}, \quad j = \begin{bmatrix} 0 & 1 \\ c & 0 \end{bmatrix}.$$

Then $i^2 = \alpha^2/4 - 1$, $j^2 = c$, and $ij = -ji$ so that $A(\Gamma) = \left(\dfrac{\alpha^2/4 - 1, c}{K}\right)$. This completes the proof of (i) of Lemma 1.

If $t(\gamma) \in O(K)$ for all $\gamma \in \Gamma$, the same proof as above shows that there exists an $m \in \mathbb{Z}$ such that

$$\langle m1^*, m\gamma_0^*, m\gamma_1^*, m\gamma_2^* \rangle_{O(K)} \leqq O(\Gamma) \leqq \langle 1^*, \gamma_0^*, \gamma_1^*, \gamma_2^* \rangle_{O(K)}.$$

It follows from this that $O(\Gamma)$ is a finitely generated $O(K)$ module. Since $O(\Gamma)$ is a subring of $A(\Gamma)$ that contains a basis of $A(\Gamma)$, $O(\Gamma)$ is an order. □

We shall apply Corollary 3.1 to the group $\Gamma_n^{(2)}$, $n = 5, 8, 12$. We take for the matrices γ_0, γ_1 conjugates of A^2 and B^2 (See (6)).

$$\gamma_0 = A^2 = \begin{bmatrix} \lambda^2 & 0 \\ 0 & \lambda^{-2} \end{bmatrix}, \quad \lambda^2 = e^{2\pi i/n}$$

$$\gamma_1 = \begin{bmatrix} \star & 1 \\ c & \star \end{bmatrix}, \quad c = \alpha^2(\mu(\alpha - \mu) - 1) \in L_n$$

We must show, to complete Theorem 3, that $\varphi(c) = \varphi(\alpha^2)\varphi(\mu(\alpha - \mu) - 1) < 0$ for every real embedding $\varphi : L_n \to \mathbb{R}$ and $n = 5, 8, 12$.

But we have

$$\alpha^2 = \begin{cases} \frac{3+\sqrt{5}}{2} & \text{n = 5} \\ 2 + \sqrt{2} & \text{n = 8} \\ 2 + \sqrt{3} & \text{n = 12} \end{cases}.$$

Since the embeddings $\varphi : L_n \to \mathbb{R}$ extend the embeddings $\sqrt{5} \to -\sqrt{5}$ or $\sqrt{2} \to -\sqrt{2}$ or $\sqrt{3} \to -\sqrt{3}$ it is clear that $\varphi(\alpha^2) > 0$.

Let $\delta = \mu(\alpha - \mu) - 1$ where $\mu = \dfrac{\lambda\beta - \alpha}{\lambda^2 - 1}$, $\beta = \dfrac{1}{2}(1 + \alpha^2 \pm \theta_n)$, $\alpha = 2\cos\pi/n$, $\theta_n = \sqrt{(\alpha^2-1)(\alpha^2-5)}$, and $\lambda = \dfrac{1}{2}(\alpha + \sqrt{\alpha^2 - 4})$. It remains to be shown that $\varphi(\delta) < 0$, for all $\varphi : L_n \to \mathbb{R}$, $n = 5, 8, 12$. The results of the laborious calculations are as follows:

The case n = 8. The field L_8 is $\mathbb{Q}(\theta_8)$ where $\theta_8 = \sqrt{-1 - 2\sqrt{2}}$. The real embeddings $\varphi_\pm : L_8 \to \mathbb{R}$ are determined by $\varphi_\pm(\theta_8) = \pm\sqrt{-1 + 2\sqrt{2}}$ and we obtain

$$\varphi_+(\delta) = \frac{-2 - 3\sqrt{2} - \sqrt{-2 + 4\sqrt{2}}}{4(1+\sqrt{2})} \approx -0.448422$$

$$\varphi_-(\delta) = \frac{-2 - 3\sqrt{2} + \sqrt{-2 + 4\sqrt{2}}}{4(1+\sqrt{2})} \approx -0.844470.$$

The case n = 12. The field L_{12} is $\mathbb{Q}(\theta_{12})$, where $\theta_{12} = \sqrt{-2\sqrt{3}}$. The real embeddings $\theta_\pm : L_{12} \to \mathbb{R}$ are determined by $\varphi_\pm(\theta_{12}) = \pm\sqrt{2\sqrt{3}}$ and we obtain

$$\varphi_+(\delta) = -\frac{3 + \sqrt{3} - \sqrt[4]{12}}{4 + 2\sqrt{3}} \approx -0.384619$$

$$\varphi_-(\delta) = -\frac{3 + \sqrt{3} + \sqrt[4]{12}}{4 + 2\sqrt{3}} \approx -0.883329.$$

The case n = 5. The field L_5 is $\mathbb{Q}(\theta_5)$, where $\theta_5 = \sqrt{\dfrac{-1 - 3\sqrt{5}}{2}}$. The real embeddings $\varphi_\pm : L_5 \to \mathbb{R}$ are given by $\varphi_\pm(\theta_5) = \pm\sqrt{\dfrac{-1 + 3\sqrt{5}}{2}}$ and we obtain

$$\varphi_+(\delta) \approx -0.404726$$

$$\varphi_-(\delta) \approx -0.871667.$$

This completes the proof of Theorem 3. □

Because of its ease of application, we shall now give another criterion for arithmeticity. This criterion is very similar to Reid's **[R1]** criterion.

Corollary 3.2. *Let* $\Gamma \leqq \mathrm{SL}(2, \mathbb{C})$ *be discrete and co-compact. If* Γ *is derived from a quaternion algebra then:*

(i) *The trace of* γ *is an algebraic integer for every* $\gamma \in \Gamma$.

(ii) *The field* $K = \mathbb{Q}(t(\gamma) \mid \gamma \in \Gamma)$ *has exactly one complex place.*

For every pair $\gamma_0 = \begin{bmatrix} \lambda & 0 \\ 0 & \lambda^{-1} \end{bmatrix}$, $\lambda^2 \neq 1$, $\gamma_1 = \begin{bmatrix} a & 1 \\ c & d \end{bmatrix}$, $c \neq 0$ *with* γ_0, $\gamma_1 \in \Gamma$ *we have:*

(iii) $\lambda \notin K$.
(iv) $\varphi(\lambda+\lambda^{-1})^2 < 4$ *and* $\varphi(c) < 0$ *for every embedding* $\varphi : K \to \mathbb{R}$.

Conversely, if there is a pair $\gamma_0,\ \gamma_1$ *in* Γ *for which (i), (ii), (iii), and (iv) are satisfied then* Γ *is derived from a quaternion algebra.*

If $\Gamma \leqq \mathrm{SL}(2,\mathbb{C})$ *is discrete and of finite covolume but not co-compact, then there is a conjugate of* Γ *derived from a quaternion algebra if and only if*
(i) *Trace* γ *is an algebraic integer for all* $\gamma \in \Gamma$,
(ii) $Q(t(\gamma) \mid \gamma \in \Gamma) = \mathbb{Q}(\sqrt{-D})$ *where* D *is a positive squarefree integer.*

Note. Recall (**[R1]**), that $\Gamma \leqq \mathrm{SL}(2,\mathbb{C})$ is arithmetic if and only if $\Gamma^{(2)}$ is derived from a quaternion algebra.

Proof. Let Γ be co-compact and derived from a quaternion algebra and suppose we have chosen $\gamma_0,\ \gamma_1$ as above. Then $\Gamma \leqq O^1 \leqq A$ where O^1 is the set of elements of norm one in an order O of the quaternion algebra A over L. Then $t(\Gamma) \leqq t(O^1)$ consists of algebraic integers. Moreover $K \leqq L$ can only admit one complex place.

On the other hand, since Γ is discrete and co-compact, K is not real. Thus (ii) is satisfied. If $\lambda \in K$ then $A(\Gamma) = M(2,K)$ by Lemma 1. Since $A(\Gamma) \leqq A,\ L$ has no real places as A would have to be ramified at them. Thus K has no real places.

Therefore $\Gamma \leqq \mathrm{SL}(2,\mathbb{Q}(\sqrt{-D}))$ and Γ would not be co-compact so that $\lambda \notin K$. If this occurs, Lemma 1 implies that

$$A(\Gamma) \otimes_{\varphi(K)} \mathbb{R} = \left(\frac{(\varphi(\alpha^2)-1)/4, \varphi(c)}{\mathbb{R}}\right) \leqq A \otimes_{\varphi(L)} \mathbb{R} = H$$

for every $\psi : L \to \mathbb{R}$ such that $\psi \mid K = \varphi \neq$ identity. We deduce that $\varphi(\alpha^2) < 4$ and $\varphi(c) < 0$ from Lemma 1.

The converse is a consequence of Lemma 1. The case in which Γ is not co-compact is left to the reader. □

Final note. It is interesting to see what occurs with Alice Whittemore's representation w when n takes values $\leqq 3$. The results can be compared with those of Hodgson **[Ho**, §6**]**. The formulas (6) give, for $\mu = 3$

$$A = \begin{bmatrix} \frac{1+i\sqrt{3}}{2} & 0 \\ 0 & \frac{1-i\sqrt{3}}{2} \end{bmatrix}, \quad B' = \begin{bmatrix} \frac{3-i\sqrt{3}}{6} & 1 \\ -2/3 & \frac{3+i\sqrt{3}}{6} \end{bmatrix}.$$

These can be conjugated to A and

$$B = \begin{bmatrix} \frac{3-i\sqrt{3}}{6} & \frac{\sqrt{6}}{3} \\ -\frac{\sqrt{6}}{3} & \frac{3+i\sqrt{3}}{6} \end{bmatrix}.$$

We note that A and B belong to $\mathrm{SU}(2) = \left\{\begin{bmatrix} a & b \\ -\bar{b} & \bar{a} \end{bmatrix}\right\} \cong S^3$. Therefore $\{\pm A, \pm B\}$ belongs to $\mathrm{PSU}(2) \cong \mathrm{SO}(3)$, and we see now that $\{\pm A, \pm B\}$ generate the holonomy of the Euclidean orbifold $(K, 3)$.

In effect, after stereographic projection, $\{\pm A, \pm B\}$ are $120°$ rotations in coplanar axes which pass through the origin in $\mathbb{R}^3$ and which form an angle β such that $\cos\beta = 1/3$ (see Figure 10a).

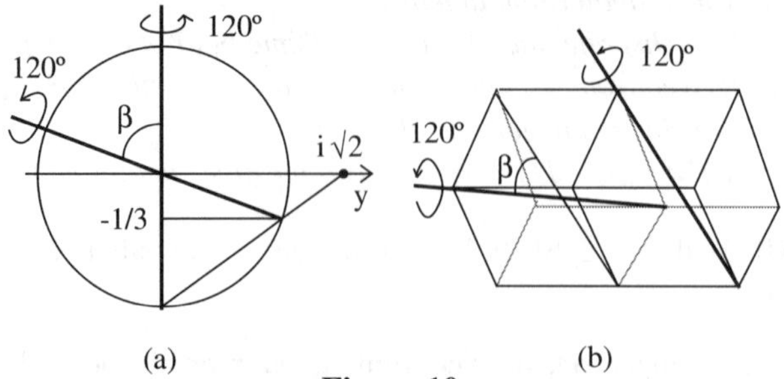

Figure 10

But the holonomy group of $(K, 3)$ is the group of direct isometries of a tetrahedron. It is generated by $120°$ rotations in the axes of Figure 10b (see **[T]** or **[Ho]**).

For $n = 2$, the formulas (6) give, up to conjugation

$$A = \begin{bmatrix} i & 0 \\ 0 & -i \end{bmatrix} \quad , \quad B = \begin{bmatrix} -i\frac{1+\sqrt{5}}{4} & \frac{\sqrt{10-2\sqrt{5}}}{4} \\ \frac{-\sqrt{10-2\sqrt{5}}}{4} & i\frac{1+\sqrt{5}}{4} \end{bmatrix}.$$

Here also $A, B \in \mathrm{SU}(2)$. The stereographic projections of A, B define $180°$ rotations in axes that pass through the origin in $\mathbb{R}^3$ and form an angle of $\pi/5$ (see Figure 11).

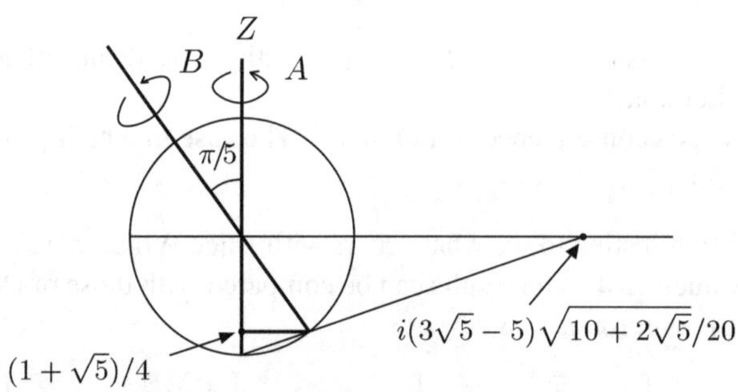

Figure 11

We see thus that $\{\pm A, \pm B\} \leq \mathrm{PSU}(2) \cong \mathrm{SO}(3)$ generates the dihedral group D_{10}. This group is the "action at ∞" of the orbifold group of $(K, 2)$. In effect, this group acts on S^3 and, via the Hopf map $S^3 \to S^2 = \mathbb{CP}^1$, acts in $\mathbb{CP}^1$. It is easy to see that this last action is that of D_{10}. (It's enough to remember that the double branched cover of S^3 branched over K is the lens space $L(5, 2)$.)

References

[B] L. A. Best, On torsion free subgroups of $\mathrm{PSL}(2,\mathbb{C})$ with compact orbit space, Can. J. Math. 23 (1971), 451–460.

[Bo] A. Borel, Commensurability classes and volumes of hyperbolic 3-manifolds, Ann. Scuola Norm. Sup. Pisa 8 (1981), 1–33.

[H] H. Helling, Private communication.

[HKM] H. Helling, A. C. Kim, J. L. Mennicke, On Fibonacci Groups, preprint.

[HLM] H. Hilden, M. T. Lozano and J. M. Montessinos, On the Borromean orbifolds: Geometry and Arithmetic, these Proceedings.

[Ho] C. D. Hodgson, Degeneration and regeneration of geometric structures on three-manifolds, Ph.D. Thesis, Princeton (1986).

[M] A. M. Macbeath, Commensurability of co-compact three-dimensional hyperbolic groups, Duke Math. Journal 50 (1983), 1245–1253.

[MR] C. Maclachlan and A. Reid, The arithmetic structure of tetrahedral groups of hyperbolic isometries, Matematika 36 (1989), 221–240.

[R] A.W. Reid, Arithmeticity of Knot Complements, to appear in Journal of the London Math. Soc.

[R1] A. W. Reid, Arithmetic Kleinian groups and their Fuchsian subgroups, Ph.D. Thesis (1987) Aberdeen; part of it is in C. Maclachlan and A. Reid: Math. Proc. Camb. Phil. Soc. 102 (1987) 251-257.

[Ri1] R. Riley, A quadratic parabolic group, Math. Proc. Cambridge Phil. Soc. 77 (1975), 281–288.

[Ri2] R. Riley, Non abelian representations of 2-bridge knot groups, Quart. J. Math Oxford 35 (1984), 191–208.

[T] W. Thurston, The Geometry and Topology of 3-Manifolds, mimeographed lecture notes, Princeton University, 1978.

[Ta] K. Takeuchi, A characterization of arithmetic Fuchsian groups, J. Math. Soc. Japan 27 (1975), 600–612.

[V] M. F. Vignéras, Arithmétique des Algébres de Quaternions, LNM 800, Springer Verlag (1980).

[W] A. Whittemore, On representations of the group of Listing's knot by subgroups of $\mathrm{SL}(2,\mathbb{C})$, Proc. Amer. Math. Soc. 40 (1973), 378–382.

Department of Mathematics, University of Hawaii at Manoa, Honolulu, HI 96822

Department of Mathematics, University of Zaragoza, 50009 Zaragoza, Spain

Department of Mathematics, Universidad Complutense, 28040 Madrid, Spain

Deduction of Andreev's Theorem from Rivin's Characterization of Convex Hyperbolic Polyhedra

Craig D. Hodgson

Abstract. This paper gives a derivation of Andreev's theorem characterizing compact polyhedra in 3-dimensional hyperbolic space $\mathbb{H}^3$ with dihedral angles $\leq \pi/2$, based on the work of Rivin which characterizes convex hyperbolic polyhedra in terms of a generalized Gauss map.

Introduction

The work of H. S. M. Coxeter **[Co1]** and F. Lannér **[La]** produced an elegant algebraic characterization of the possible sets of dihedral angles of N-dimensional simplices in spaces of constant curvature.

In his 1970 paper, E. M. Andreev **[An1]** characterized the remaining compact convex polyhedra in 3-dimensional hyperbolic space $\mathbb{H}^3$ with dihedral angles $\leq \pi/2$. These results of Andreev also provide a complete characterization of 3-dimensional hyperbolic reflection groups. Andreev's work was later put in a broader context by W. P. Thurston **[Th]**, who also noted some applications to conformal geometry via the theory of circle packings. Thurston's generalization of Andreev's Theorem also plays a significant part in the proof of Thurston's Uniformization Theorem for 3-manifolds (see **[Mo]**).

In his 1986 Ph.D. dissertation **[R]**, I. Rivin obtained a new characterization the shapes of convex hyperbolic polyhedra (with arbitrary dihedral angles $\leq \pi$) in terms of a generalized Gauss map. This paper gives a brief introduction to this work, and illustrates the use of Rivin's results by giving a new derivation of Andreev's theorem.

Acknowledgements. The derivation of Andreev's theorem presented below is a considerable simplification of Rivin's argument given in **[R]**. The author would like to thank Igor Rivin, Brian Bowditch, and the members of the Topology/Geometry Seminar at Melbourne University for many useful discussions related to this work.

1 Outline of Rivin's Work

This section gives a brief account of the characterization of compact convex polyhedra in hyperbolic 3-space obtained by Igor Rivin in **[R]**. The reader is referred to the paper **[RH]** for a detailed treatment of this work.

To explain Rivin's main result, first consider the Gauss map in Euclidean 3-dimensional space $\mathbb{E}^3$. The ***Gauss Map*** G of a compact convex polyhedron P in $\mathbb{E}^3$ is a *set-valued* function from P to the unit sphere $\mathbb{S}^2$, which assigns to each point p the set of outward unit normals to support planes to P at p. Thus, the whole of a face f of P is mapped under G to a single point – the outward unit normal to f. An edge e of P is mapped to a geodesic segment $G(e)$ on $\mathbb{S}^2$, whose length is easily seen to be the exterior dihedral angle at e. A vertex v of P is mapped by G to a spherical polygon $G(v)$, whose sides are the images under G of edges incident to v, and whose angles are easily seen to be the angles complementary to the planar angles of the faces incident to v. That is, $G(e_1)$ and $G(e_2)$ meet at angle $\pi - \alpha$ whenever e_1 and e_2 meet at angle α. The polygon $G(v)$ is also called the ***spherical polar*** of the link of v in P. (The ***link*** of a vertex is the intersection of a small sphere centered at v with P, rescaled, so that the radius is 1).

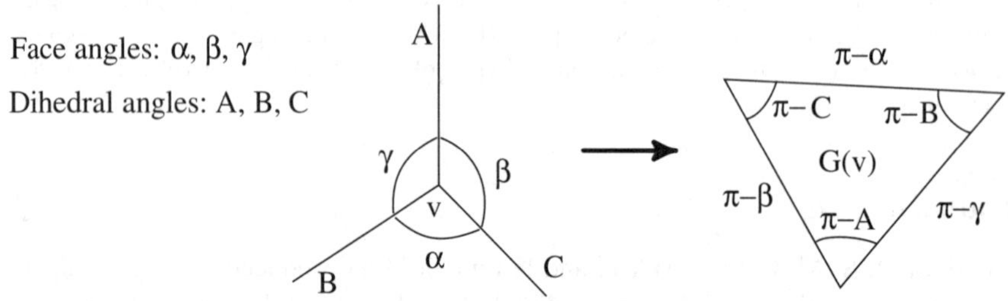

Figure 1. Gaussian image of a trivalent vertex.

Collecting the above observations together, it is seen that $G(P)$ is combinatorially Poincaré dual to P, while metrically it is the unit sphere $\mathbb{S}^2$.

A similar construction can be applied to a convex polyhedron P in $\mathbb{H}^3$. Associate to each vertex v of P a spherical polygon $G(v)$ spherically polar to the link of v in P (as in Figure 1). Glue the resulting polygons together into a closed surface, using the rule that $G(v_1)$ and $G(v_2)$ are identified isometrically whenever v_1 and v_2 share an edge.

The resulting path metric space $G(P)$ is clearly topologically $\mathbb{S}^2$ and the complex is still Poincaré dual to P. Metrically, however, it is no longer the round sphere. To see this, consider $G(f)$ which is the single common point of the spherical polygons $G(v_i)$, where v_i are the vertices of f. The angle of $G(v_i)$ incident to $G(f)$ is the exterior angle of f at v_i, and so by the Gauss-Bonnet Theorem, the sum of these angles is $2\pi + \text{area}(f) \neq 2\pi$. Thus $G(f)$ is a ***cone-like singularity***, or a ***cone point***, with ***cone angle*** greater than 2π. (A cone angle equal to 2π corresponds to a smooth point.)

This analogue of the Gauss map turns out to have rather remarkable properties, which are studied in detail in **[R]** and **[RH]**. The main result is a precise characterization of those surfaces that can arise as $G(P)$ for a compact convex polyhedron P in $\mathbb{H}^3$:

Theorem 1.1 (Rivin's characterization of compact convex polyhedra). *A path metric space (M, g) homeomorphic to $\mathbb{S}^2$ can arise as the Gaussian image $G(P)$ of a compact convex polyhedron P in $\mathbb{H}^3$ if and only if the following conditions hold:*

(a) *The metric g is piecewise spherical, with constant curvature 1 away from a finite collection of cone points c_i.*

(b) *The cone angles at c_i are greater than 2π.*

(c) *The lengths of closed geodesics of (M, g) are all strictly greater than 2π.*

Furthermore, the metric of $G(P)$ determines the hyperbolic polyhedron P uniquely (up to congruence).

Remarks. In condition (c) above, the word "closed geodesic" can be interpreted as either a closed "local" geodesic (i.e. the image of a circle under a locally distance minimizing mapping), or as a closed "global" geodesic (i.e. the image of a circle under an isometric mapping). The theorem holds with either interpretation of "geodesic". In a space with a piecewise spherical metric, any geodesic looks like spherical geodesic (arc of great circle) away from the cone points, and subtends an angle $\geq \pi$ on either side if it passes through a cone point.

The necessity of (a) and (b) is immediately apparent from the above discussion of G. The necessity of (c) follows from a hyperbolic version of Fenchel's Theorem ("the total geodesic curvature of a closed curve in $\mathbb{H}^3$ is greater than 2π"). The proof of uniqueness follows closely the argument used by Cauchy in the proof of his celebrated Rigidity Theorem for convex polyhedra in $\mathbb{E}^3$ (see **[Ca]**, **[Al]**). The proof of sufficiency of conditions (a)–(c) is based upon a "continuity" method, as used to great effect by Aleksandrov in his work on convex polyhedra (see **[Al]**).

A detailed proof of Theorem 1.1 can be found in **[RH]**. Similar results for the case of ideal hyperbolic polyhedra and applications are described in **[HRS]**.

2 Andreev's Theorem

The compact convex hyperbolic polyhedra with dihedral angles $\leq \pi/2$ were characterized by E.M. Andreev **[An1]**. This section gives a derivation of Andreev's Theorem from Rivin's Characterization Theorem 1.1. First we recall Andreev's Theorem in its original form.

Theorem 2.1 (Andreev). *Let P be a compact convex polyhedron in $\mathbb{H}^3$ with faces F_i and dihedral angles $\alpha_{ij} \leq \pi/2$ between faces F_i and F_j. Call a circular sequence of k edge-adjacent faces of P such that no three of these faces have a common point, a **k-prismatic element**. Then P has trivalent vertices, and*

1. $0 < \alpha_{ij} \leq \pi/2$.
2. *If $F_i \cap F_j \cap F_k$ is a vertex then $\alpha_{ij} + \alpha_{jk} + \alpha_{ki} > \pi$.*
3. *If F_i, F_j, F_k form a 3-prismatic element, then $\alpha_{ij} + \alpha_{jk} + \alpha_{ki} < \pi$.*
4. *If F_h, F_i, F_j, F_k form a 4-prismatic element, then $\alpha_{hi} + \alpha_{ij} + \alpha_{jk} + \alpha_{kh} < 2\pi$.*
5. *If F_s is a quadrilateral with sides in cyclic order α_{is}, α_{js}, α_{ks}, α_{ls}, then $\alpha_{is}+\alpha_{ks}+\alpha_{ij}+\alpha_{jk}+\alpha_{kl}+\alpha_{li} < 3\pi$ and $\alpha_{js}+\alpha_{ls}+\alpha_{ij}+\alpha_{jk}+\alpha_{kl}+\alpha_{li} < 3\pi$. (In other words, the assignment of dihedral angles shown in Figure 2 does not occur.)*

Moreover, these conditions are sufficient for an abstract polyhedron P with ≥ 5 vertices to be realizable as a compact convex polyhedron in $\mathbb{H}^3$ with dihedral angles α_{ij}.

Remark. An *abstract polyhedron* is a cell decomposition of the 2-sphere which can be realized by a convex polyhedron in 3-dimensional Euclidean (or hyperbolic) space. By a theorem of Steinitz, this is equivalent to requiring that the 1-skeleton of the cell decomposition is a 3-connected graph, i.e. removing any two vertices and the incident edges cannot disconnect the graph.

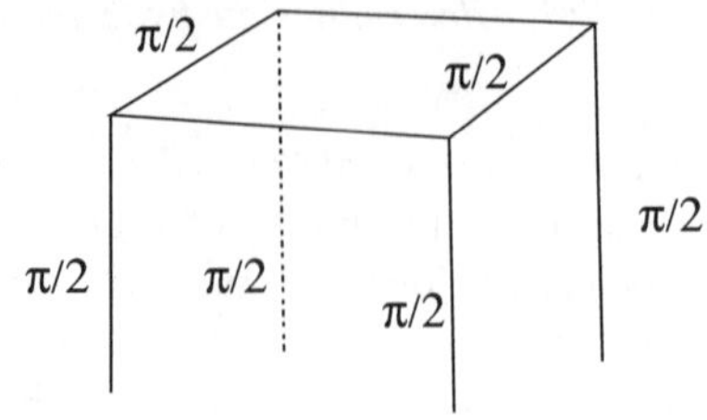

Figure 2. A forbidden quadrilateral.

Remarks. If P has 4 vertices, it is a simplex. In this case the above conditions are not sufficient for realizability. However, the possible dihedral angles of simplices in hyperbolic space of any dimension are easily characterized via linear algebra. (See Coxeter **[Co2]**, Thurston **[Th]** or Vinberg **[V]**.)

As Andreev points out, condition 5 follows from 1–4 except when P is a triangular prism.

Convention. Throughout the remainder of this section P will denote a convex hyperbolic polyhedron as in Andreev's Theorem, and we also assume

0. *P is not a simplex.*

We begin by reformulating Andreev's conditions 1–5 as conditions on the Gaussian image $P^* = G(P)$ of P. First, note that P^* is topologically a 2-sphere and the intrinsic metric on P^* is completely determined by the combinatorics of P and its dihedral angles: P^* has a natural triangulation by spherical triangles (polars of the links of vertices of P) with edge lengths $\pi - \alpha_{ij}$. These triangles have edge lengths in $[\pi/2, \pi)$ by Andreev's condition 1, and have perimeter $< 2\pi$ by condition 2, so are convex.

The following geometric facts about such triangles will prove useful.

Lemma 2.2 (Spherical triangles). *Let T be a convex spherical triangle with edge lengths in the interval $[\pi/2, \pi)$. Then*

(i) *each angle is no smaller than the opposite side,*
(ii) *each angle lies in the interval $[\pi/2, \pi]$,*
(iii) *the distance from each vertex v of T to a point p on the opposite edge is at least $\pi/2$. If equality occurs then at least one edge at v has length $\pi/2$; and if p is not a vertex, both edges at v have length $\pi/2$.*

Proof. (i) Let T have angles A, B, C and opposite sides of lengths a, b and c. It suffices to show that $A \geq a$. By the cosine law

$$\cos A = \frac{\cos a - \cos b \cos c}{\sin b \sin c}.$$

When $b = c = \pi/2$ have $A = a$. To prove the lemma we now calculate

$$\frac{\partial \cos A}{\partial b} = \frac{1}{\sin^2 b \sin c}(\cos c - \cos a \cos b) \leq 0$$

since $\cos c \leq 0$, $\cos a \cos b \geq 0$, $\sin c \geq 0$. Likewise

$$\frac{\partial \cos A}{\partial c} \leq 0.$$

It follows that $A \geq \alpha$ whenever $b \geq \pi/2$ and $c \geq \pi/2$, with equality only when $b = c = \pi/2$. This completes the proof of (i).

Part (ii) follows immediately from the cosine law, or from (i).

To prove (iii), place one vertex A at the north pole of $\mathbb{S}^2$. Then the other vertices B, C lie in the southern hemisphere; hence the geodesic BC (with length $< \pi$) must lie entirely in the southern hemisphere. □

This result has the following useful consequence.

Lemma 2.3 (Geodesic edge cycles). *A closed curve in the 1-skeleton Γ of P^* is a geodesic unless it contains a vertex with a single triangle on one side. Hence, a 3-cycle in Γ is a geodesic unless it bounds a triangle, and a 4-cycle in Γ is a geodesic unless it bounds a square made up of two triangles.*

Proof. Curves in Γ consist of smooth geodesic arcs joining cone points of P^*. Such curves are geodesics provided that the angle on each side of the curve at every cone point is $\geq \pi$. Thus the first statement follows immediately from part (i) of Lemma 2.2. The other statements follow by a simple Euler characteristic argument. □

When P is not a simplex, it follows from Lemma 2.3 that the k-prismatic elements of P with $k = 3$ or 4 are dual to the closed geodesics in P^* made up of k edges in the natural triangulation. So in our terminology, Andreev's conditions 0–5 imply:

0*. P^* is not a simplex.

1*. P^* is a topological 2-sphere with a piecewise spherical path metric, and can be triangulated by convex spherical triangles with all edge lengths in the interval $[\pi/2, \pi)$.

2*. The lengths of geodesics of P^* comprised of three or four edges of the triangulation are greater than 2π.

3*. P^* does not contain the configuration shown in Figure 3.

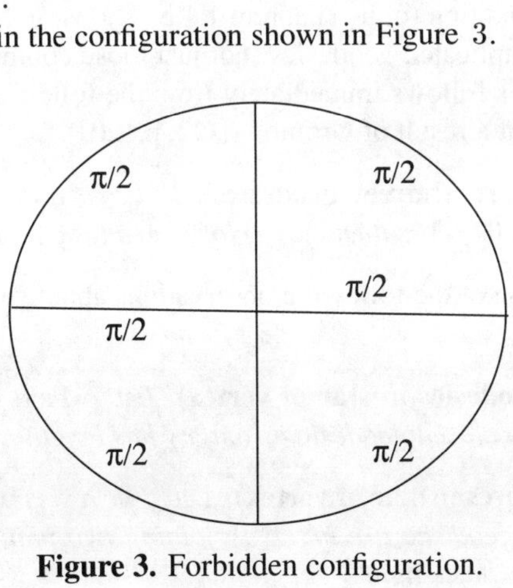

Figure 3. Forbidden configuration.

In the remainder of this section, we show that the hypotheses of Theorem 1.1 follow from the seemingly weaker conditions 0^*–3^*.

First we note that the combinatorial type of P is uniquely determined by the intrinsic metric on P^*. This follows from the following result:

Proposition 2.4 (Combinatorial type determined). *P^* has at most one geodesic triangulation whose vertex set is the set of cone points, with the lengths of all edges lying in the interval $[\pi/2, \pi)$.*

Proof. Let T and T' be triangulations satisfying the hypothesis of the theorem. If e is an edge of T' which is not an edge of T, then e must leave one triangle of T and enter another. Since all edge lengths of the triangles are $\geq \pi/2$ it follows from Lemma 2.2(iii) that $\mathrm{length}(e) \geq \pi$. □

We will call this triangulation of P^* the "Andreev triangulation" and call its edges "Andreev edges". This triangulation will be used throughout the following argument. For each vertex v, let $st(v)$ denote the star of v in the triangulation, (i.e. the union of all triangles containing v), let $ost(v)$ denote the interior of $st(v)$, and $lk(v) = \partial st(v)$ the link of v in the triangulation.

Next, we show that the metric on P^* satisfies condition (b) of Theorem 1.1.

Proposition 2.5 (Andreev cone points negatively curved). *A metric space P^* satisfying the dual Andreev conditions 0^*–3^* has all cone angles greater than 2π.*

Proof. Let v be a vertex of P^*. We first show that the link $lk(v)$ of v has length $> 2\pi$. Since each edge has length $\geq \pi/2$, this is clear if $lk(v)$ is a cycle of > 4 edges. If $lk(v)$ consists of 3 edges then it forms a geodesic by Lemma 2.3 (since P^* is not a simplex). Hence it has length $> 2\pi$ by condition 2^*. If $lk(v)$ consists of 4 edges, then it has length $> 2\pi$ and the cone angle at v is $> 2\pi$ unless all angles at v are exactly $\pi/2$ and each edge in $lk(v)$ has length exactly $\pi/2$. This is ruled out by condition 3^*. From Lemma 2.2(i) it now follows that $\text{cone angle at } v \geq \mathrm{length}(lk(v)) > 2\pi$. □

Finally we verify condition (c) in Theorem 1.1 by showing that Andreev's conditions imply that the more complicated geodesics (not just those comprised of Andreev edges) are longer than 2π. This follows immediately from the following proposition, which is a more precise version of a result of Gromov ([**G2**, p.120]).

Proposition 2.6 (No short Andreev geodesics). *If P^* is a metric space satisfying the dual Andreev conditions 0^*–3^*, then every closed geodesic in P^* has length $> 2\pi$.*

Proof. We will make use of the following observations about geodesics contained in the star of a vertex in P^*.

Lemma 2.7 (Closed geodesics in star of vertex). *Let v be a vertex of P^* with cone angle $> 2\pi$. Then every closed geodesic in $ost(v)$ has length $> 2\pi$.*

Lemma 2.8 (Geodesic arcs in star of vertex). *Let v be a vertex of P^* with cone angle $> 2\pi$, and let γ be a geodesic arc in $st(v)$ joining two points of $\partial st(v)$, such that $\gamma \cap ost(v) \neq \emptyset$. Then γ has length $\geq \pi$.*

Further, if γ has length π, then both endpoints of γ are at distance exactly $\pi/2$ from v and either:

(i) *γ passes through v, or*

(ii) *there is a geodesic γ' of length π contained in $\partial st(v)$ such that $\gamma \cup \gamma'$ bounds a lune (spherical bigon) in $st(v) \setminus v$.*

Proofs. Lemmas 2.7 and 2.8 are well-known. Proofs are sketched by Gromov in **[G2]**, and details can be found in **[CD**, Lemma 8.7, Theorem 8.4**]**. □

Now let γ be a closed geodesic of length $\leq 2\pi$ in P^*, and assume γ is not contained in the 1-skeleton Γ of the Andreev triangulation. To prove Proposition 2.6, we show that there must also be a closed geodesic γ' of length $\leq 2\pi$ contained in Γ.

Since γ cannot lie entirely in the interior of a convex spherical triangle, the set $V(\gamma)$ of vertices of Andreev triangles whose interior meets γ contains at least 4 vertices. If every pair of vertices in $V(\gamma)$ is joined by an edge of Γ, then Γ contains the complete graph on these vertices, hence $V(\gamma)$ consists of exactly 4 vertices (since Γ embeds in $\mathbb{S}^2$). Since P^* is not a simplex, it follows easily that γ lies entirely in the link of one vertex v, hence has length $> 2\pi$ by Lemma 2.7. (Compare Figure 4.)

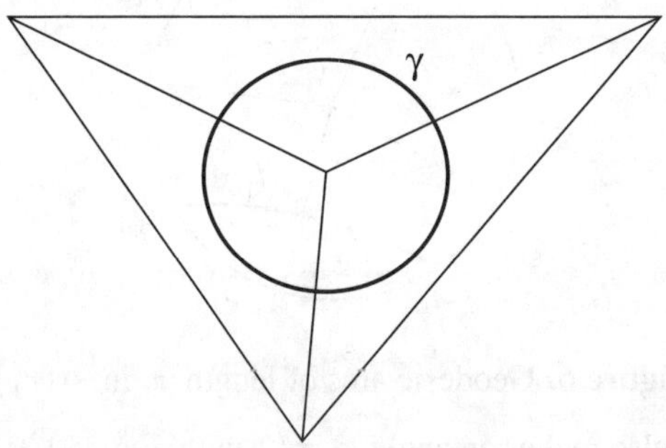

Figure 4. Geodesic contained in star of vertex.

Otherwise, there are vertices $v_1, v_2 \in V(\gamma)$ which are not joined by an edge. Then $ost(v_1)$ and $ost(v_2)$ are disjoint open stars meeting γ. By Lemma 2.8 each of $\gamma_1 = \gamma \cap st(v_1)$ and $\gamma_2 = \gamma \cap st(v_2)$ contains an arc of length $\geq \pi$, hence γ has length at least 2π. If γ has length 2π, then $\gamma \subset st(v_1) \cup st(v_2)$ and γ_1, γ_2 are both arcs of length exactly π. Let p and q be the endpoints of γ_1 (and of γ_2).

Case 1: Assume that v_1 and v_2 both lie on γ. Then we can replace γ by a curve γ' in Γ of length 2π which passes through v_1 and v_2 by Lemma 2.2(iii). Further, by Lemma 2.3, we can also arrange that γ' is a geodesic, unless v_1 or v_2 is a trivalent vertex of Γ. (This can occur, for example, if P^* is a triangular biprism as in Figure 5.) However, this situation is ruled out by condition 3^*.

Case 2: If p and q are both vertices of Γ, then the previous argument applies with p and q replacing v_1 and v_2. (Note that $ost(p)$, $ost(q)$ are disjoint since p and q are at distance π apart, so are not joined by an edge of Γ.)

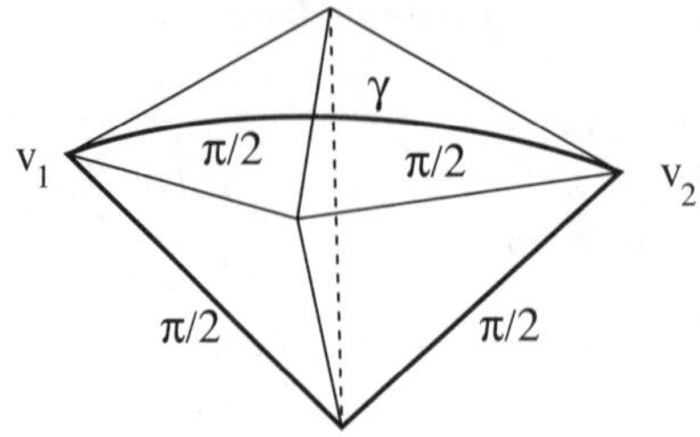

Figure 5. Geodesic of length 2π.

Case 3: Otherwise we have $v_1 \notin \gamma$ and p not a vertex of Γ (after relabelling v_1, v_2 and p, q if necessary). In this case, Lemma 2.8(ii) and Lemma 2.3(iii) show that p lies in an edge e at distance $\pi/2$ from v_1. Since γ_1 doesn't pass through v_1, there is an arc γ_1' of length π in $\partial st(v_1)$ with endpoints p and q, and the angle between γ_1, γ_1' is $< \pi/2$ at p (hence also at q). (See Figure 6.)

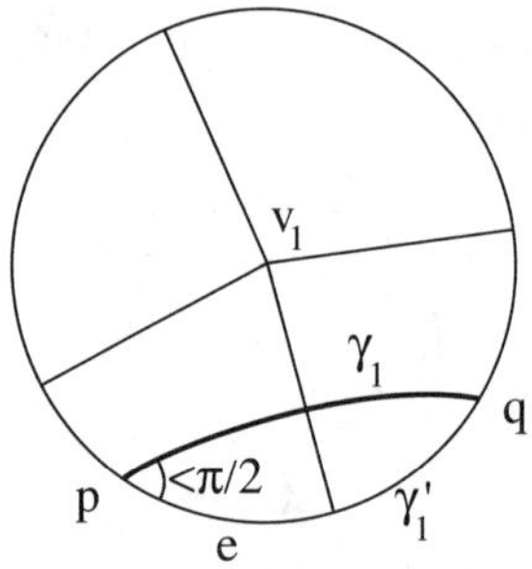

Figure 6. Geodesic arcs of length π in $st(v_1)$.

It follows that γ_2 also makes an angle $< \pi/2$ with $\partial st(v_2)$ at p. Hence γ_2 doesn't pass through v_2, and there is an arc γ_2' of length π in $\partial st(v_2)$ with endpoints p and q. Again, the angle between γ_2, γ_2' is $< \pi/2$ at p (hence also at q). In this case $\gamma' = \gamma_1' \cup \gamma_2'$ is a closed geodesic in Γ of length 2π. (The angle conditions guarantee that γ' does not back-track at p or q.)

This completes the proof of Proposition 2.6. □

Andreev's Theorem 2.1 now follows immediately from the Characterization Theorem 1.1 and Propositions 2.4, 2.5 and 2.6. □

Remark. A similar argument can be used to prove Andreev's characterization of finite volume hyperbolic polyhedra **[An2]**, based on the characterization of finite volume convex hyperbolic polyhedra recently obtained by Rivin (private communication). (See **[HRS]** for the case of ideal polyhedra.)

Bibliography

[Al] A. D. Aleksandrov, Convex Polyhedra, GITTL, Moscow, 1950 (Russian); German trans. Akademie Verlag, Berlin, 1958. (MR 12, 732; 19, 1192.)

[An1] E. M. Andreev, On convex polyhedra in Lobachevskii space, Math. USSR, Sbornik, 10 (1970), 413–440.

[An2] E. M. Andreev, On Convex Polyhedra of finite volume in Lobachevskii Space, Math. USSR, Sbornik, 12 (1970), 255–259.

[Ca] A. L. Cauchy, Sur les polygones et polyèdres, 2nd memoir, J. Ecole Polytechnique, 19 (1813), 87–98.

[CD] R. Charney and M. Davis, Singular metrics of nonpositive curvature on branched covers of Riemannian manifolds, Ohio State University Mathematical Research Institute Preprint 90-19, 1990.

[Co1] H. S. M. Coxeter, The polytopes with regular-prismatic vertex figures, part 2, Proc. London Math. Soc. (2), 34 (1932), 126–189.

[Co2] H. S. M. Coxeter, Regular Polytopes, 3rd edition, Dover, New York, 1973.

[G2] M. L. Gromov, Hyperbolic groups, In: S. M. Gersten, editor, Essays in group theory, pages 75–263. Springer, 1987, M.S.R.I. Publ. 8.

[HRS] C. D. Hodgson, I. Rivin, and W. D. Smith, A characterization of convex ideal polyhedra in hyperbolic 3-space and of convex polyhedra inscribed in the sphere, to appear in Bull. Amer. Math. Soc.

[La] F. Lannér, On complexes with transitive groups of automorphisms, Comm. Sem. Math. Univ. Lund 11 (1950) 1–71.

[Mo] J. W. Morgan, On Thurston's uniformization theorem for three-dimensional manifolds, In: J. W. Morgan and H. Bass, editors, The Smith Conjecture, pages 37–125. Academic Press, 1984.

[R] I. Rivin, On geometry of convex polyhedra in hyperbolic 3-space, Ph.D. thesis, Princeton, 1986.

[RH] I. Rivin and C. D. Hodgson, A characterization of compact convex polyhedra in hyperbolic 3-space, preprint, 1991.

[Th] W. P. Thurston, Geometry and topology of 3-manifolds, Revised Lecture Notes, Princeton Univ. Math. Dept., 1982.

[V] E. B. Vinberg, Hyperbolic reflection groups, Russian Math. Surveys, 40 (1985), 31–75.

Mathematics Department, University of Melbourne, Parkville, Victoria 3052, Australia
Email: cdh@mundoe.maths.mu.oz.au

Surgeries on the Whitehead Link Yield Geometrically Similar Manifolds

Craig D. Hodgson, G. Robert Meyerhoff, and Jeffrey R. Weeks

Abstract Dehn filling on one boundary component of the Whitehead link complement yields an interesting family of one-cusp hyperbolic 3-manifolds. The family contains infinitely many pairs of manifolds with the same volume, homology groups, Chern-Simons invariant, cusp volume, cusp shape, and horoball packing. Furthermore, each closed hyperbolic manifold obtained by Dehn filling the cusp of one member of a pair corresponds naturally to a closed manifold obtained by Dehn filling the cusp of the other, and corresponding closed manifolds have the same volume and Chern-Simons invariant (mod $1/2$). Even though the cusped manifolds are typically not knot complements, one can compute their Alexander polynomials to show that the members of each pair are nonhomeomorphic. The Alexander polynomial also resolves the question of which such manifolds fiber over the circle.

Consider the family of manifolds obtained by Dehn filling one boundary component of the Whitehead link complement (see Figure 1). With six exceptions these manifolds are hyperbolic; that is, they admit complete metrics which are locally isometric to hyperbolic 3-space. Throughout this paper we assume each of these manifolds is carrying such a metric, which is unique by **[Mostow]** and **[Prasad]**. Figure 2 shows what the hyperbolic structure looks like in a neighborhood of the remaining boundary component. The boundary component itself has been discarded. The remainder of the neighborhood is topologically a torus cross a half-line; geometrically, the torus shrinks exponentially fast as you move upward. Such a neighborhood of the (deleted) boundary component is called a cusp. Even though the cusp is noncompact, the metric is complete and the volume is finite. We let $WL(m,l)$ denote the one-cusp manifold obtained by performing an (m,l) Dehn filling on one boundary component of the left-handed Whitehead link complement and leaving the other component open.

We compute the volumes of these manifolds using the method of **[Milnor]**. The results appear in Figure 3a, where the volume of each $WL(m,l)$ is plotted at the point (m,l). Warning: Figure 3a also includes the volumes of the $WL(m,l)$ with m and l not relatively prime. These spaces are orbifolds, not manifolds. They are included to help bring out the symmetry in the set of volumes. Note that the same volume may occur several times in Figure 3a. In fact, the volumes are symmetric under the action of the dihedral group of order eight. To show this symmetry more clearly, the volumes have been replotted in Figure 3b in the following geometrically natural way. Consider the hyperbolic structure on the original Whitehead link complement (no Dehn filling), consider the Euclidean torus cross-section of the cusp which is used for Dehn filling (cf. Figure 2), and, finally, consider the lift of an (m,l)-curve to the universal cover of the torus, that is, to the Euclidean plane. With basepoints fixed, plot $\mathrm{vol}(WL(m,l))$ at the terminal end of

Figure 1

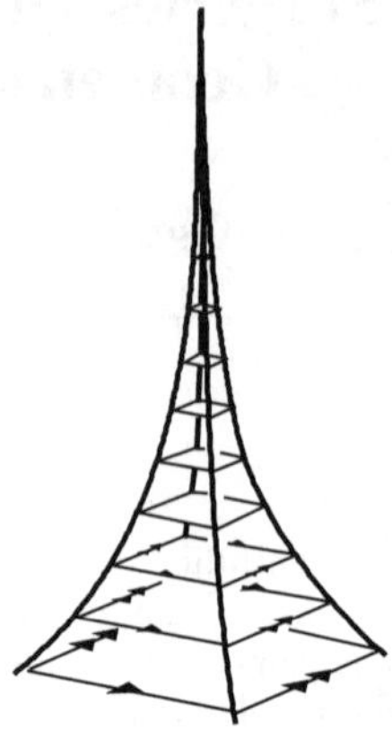

Figure 2

```
                                  |
 3.49 3.51 3.52 3.53 3.54 3.55 3.56 3.57 3.58 3.59 3.60 3.60 3.61
                                  |
                                  |
 3.38 3.39 3.42 3.44 3.46 3.49 3.51 3.53 3.54 3.56 3.57 3.58 3.59
                                  |
                                  |
 3.14 3.15 3.18 3.22 3.28 3.33 3.38 3.43 3.46 3.49 3.52 3.54 3.55
                                  |
                                  |
 2.67 2.57 2.54 2.57 2.67 2.83 3.01 3.16 3.28 3.36 3.42 3.46 3.49
                                  |
                                  |
 2.67 2.03  *   0.00 0.00 0.00  *   2.03 2.67 2.99 3.18 3.30 3.38
                                  |
                                  |
-3.14-2.92-2.54-1.78-0.00--+----+----+---0.00-1.78-2.54-2.92-3.14-
                                  |
                                  |
 3.38 3.30 3.18 2.99 2.67 2.03  *   0.00 0.00 0.00  *   2.03 2.67
                                  |
                                  |
 3.49 3.46 3.42 3.36 3.28 3.16 3.01 2.83 2.67 2.57 2.54 2.57 2.67
                                  |
                                  |
 3.55 3.54 3.52 3.49 3.46 3.43 3.38 3.33 3.28 3.22 3.18 3.15 3.14
                                  |
                                  |
 3.59 3.58 3.57 3.56 3.54 3.53 3.51 3.49 3.46 3.44 3.42 3.39 3.38
                                  |
                                  |
 3.61 3.60 3.60 3.59 3.58 3.57 3.56 3.55 3.54 3.53 3.52 3.51 3.49
                                  |
```

Figure 3a

3.38	3.33	3.28	3.22	3.18	3.15	3.14	3.15	3.18	3.22	3.28	3.33	3.38
3.33	3.26	3.16	3.07	2.99	2.94	2.92	2.94	2.99	3.07	3.16	3.26	3.33
3.28	3.16	3.01	2.83	2.67	2.57	2.54	2.57	2.67	2.83	3.01	3.16	3.28
3.22	3.07	2.83	2.44	2.03	1.83	1.78	1.83	2.03	2.44	2.83	3.07	3.22
3.18	2.99	2.67	2.03	*	0.00	0.00	0.00	*	2.03	2.67	2.99	3.18
3.15	2.94	2.57	1.83	0.00				0.00	1.83	2.57	2.94	3.15
3.14	2.92	2.54	1.78	0.00		+		0.00	1.78	2.54	2.92	3.14
3.15	2.94	2.57	1.83	0.00				0.00	1.83	2.57	2.94	3.15
3.18	2.99	2.67	2.03	*	0.00	0.00	0.00	*	2.03	2.67	2.99	3.18
3.22	3.07	2.83	2.44	2.03	1.83	1.78	1.83	2.03	2.44	2.83	3.07	3.22
3.28	3.16	3.01	2.83	2.67	2.57	2.54	2.57	2.67	2.83	3.01	3.16	3.28
3.33	3.26	3.16	3.07	2.99	2.94	2.92	2.94	2.99	3.07	3.16	3.26	3.33
3.38	3.33	3.28	3.22	3.18	3.15	3.14	3.15	3.18	3.22	3.28	3.33	3.38

Figure 3b

the lift. Figure 3b includes half-integer values of l; these spaces are not manifolds and are included only to bring out the symmetry in the space of volumes. A rigorous proof that corresponding volumes are equal is easily obtained using the method of the proof of Proposition 1 below, but here we will focus on the question of whether manifolds of equal volume are homeomorphic. Because the components of the Whitehead link are homologically unlinked, it is straightforward to see that $H_1(WL(m,l)) = \mathbb{Z} \oplus \mathbb{Z}/m$. Thus, for example, even though $WL(4,1)$ and $WL(8,-1)$ have the same volume, they cannot be homeomorphic because the first homology of one is $\mathbb{Z} \oplus \mathbb{Z}/4$, while that of the other is $\mathbb{Z} \oplus \mathbb{Z}/8$. A more interesting pair of examples is $WL(4,1)$ and $WL(4,-3)$. These examples, and others like them, are the focus of the remainder of this article. We will see that all their easily computable geometric invariants are identical, yet their Alexander polynomials are different.

We first give a brief exposition of the aforementioned geometric invariants, using $WL(4,1)$ and $WL(4,-3)$ as examples. We then formalize our claims as Propositions 1 through 3, and offer simple proofs.

All invariants—numerical, algebraic, and graphical—were computed using the computer program "snap pea". The program is available free of charge from Jeff Weeks. Please send requests to Jeff Weeks via e-mail at jweeks@midd.bitnet or weeks@poincare.geom.umn.edu, or via paper mail at Mathematics Department, Middlebury College, Middlebury, VT 05753 or Geometry Supercomputer Project, 1200 Washington Avenue South, Minneapolis, MN 55415. At present the graphical parts of the program (horoball packings, Dirichlet domains) require a Macintosh computer, but the algebraic and numerical invariants (including some not mentioned here, such as the fundamental group) can be computed on other computers. Work is underway to transfer the graphics to as large a variety of computers as possible.

We thank the Geometry Supercomputer Project for its continuing support of our work.

For more on the manifolds $WL(m,l)$ we recommend **[Neu-Reid]**.

Invariant #1: Volume. This is the most common and most useful invariant for distinguishing hyperbolic 3-manifolds. $WL(4,1)$ and $WL(4,-3)$ caught our interest precisely because their volumes are equal: $\mathrm{vol}(WL(4,1)) = \mathrm{vol}(WL(4,-3)) = 3.17729327\ldots$.

Invariant #2: Homology. The first homology group is another useful invariant for distinguishing 3-manifolds. Informally, the homology group reflects a manifold's topology while the volume reflects its geometry. $H_1(WL(4,1)) = H_1(WL(4,-3)) = \mathbb{Z} \oplus \mathbb{Z}/4$.

Invariant #3: Cusp volume. The cusp volume of a one-cusp hyperbolic 3-manifold is defined as follows. A horospherical cross-section of the cusp (e.g. one of the tori shown in Figure 2) divides the manifold into a compact piece and a noncompact piece. Push the cross-section in the direction of the compact piece until it first bumps into itself. The resulting noncompact piece is called a maximal cusp and its volume is called the cusp volume of the manifold. The cusp volume can be used to distinguish manifolds. For example, the complement of the 5_2 knot and that of a related 12-crossing pretzel knot (see Figure 4) both have volume $2.82812208\ldots$ and first homology $\mathbb{Z}$, yet we know they are nonhomeomorphic because the cusp volume of the former is $1.97346351\ldots$ while that of the latter is $2.01273328\ldots$. The cusp volume of both $WL(4,1)$ and $WL(4,-3)$ is $1.95940979\ldots$

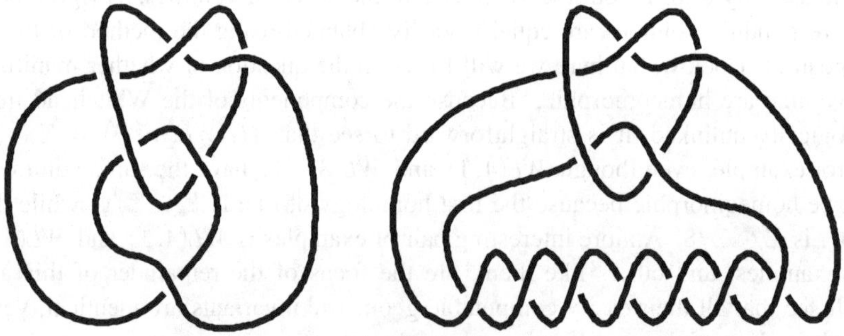

Figure 4

Invariant #4: Chern-Simons invariant. The Chern-Simons invariant is defined mod $1/2$ for oriented cusped hyperbolic 3-manifolds (see **[Meyerhoff 0]**), and is computed using the formula of **[Neumann]**. We don't yet have a good intuitive understanding of just what the Chern-Simons invariant measures; **[Meyerhoff 1]** surveys what is known. With standard orientations, $CS(WL(4,1)) = -CS(WL(4,-3)) = -0.04906304\ldots$ (mod $1/2$). The Chern-Simons invariant changes sign when a manifold's orientation is reversed. Thus, the fact that $CS(WL(4,1))$ and $CS(WL(4,-3))$ have opposite signs suggests that $WL(4,1)$ and $WL(4,-3)$ are in some sense oppositely oriented. The proof of Proposition 1 will make this idea more precise. In the meantime we reverse the orientation on $WL(4,1)$ to correspond to that of $WL(4,-3)$.

Invariant #5: Horoball packing. Imagine painting the boundary torus of the maximal cusp defined above with a special paint that allows you to see out of the cusp but does not allow you to see back in. The horoball packing is the image you see when you position yourself "at infinity" and peer down the cusp into the manifold. As in any Riemannian manifold, the image you see in the manifold is the same as the image you would see in the universal cover, and subjectively you feel as if you are seeing the universal cover. Thus the horoballs you see appear to form a symmetrical packing.

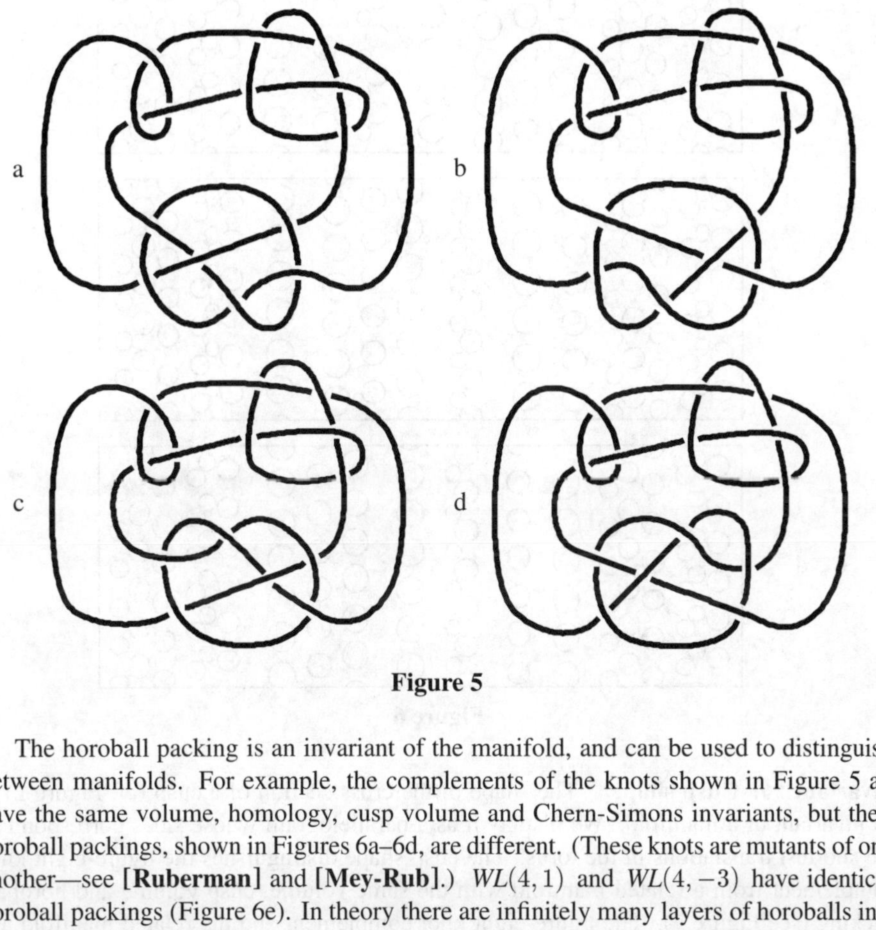

Figure 5

The horoball packing is an invariant of the manifold, and can be used to distinguish between manifolds. For example, the complements of the knots shown in Figure 5 all have the same volume, homology, cusp volume and Chern-Simons invariants, but their horoball packings, shown in Figures 6a–6d, are different. (These knots are mutants of one another—see **[Ruberman]** and **[Mey-Rub]**.) $WL(4,1)$ and $WL(4,-3)$ have identical horoball packings (Figure 6e). In theory there are infinitely many layers of horoballs in a horoball packing, but in practice, of course, the computer draws only those greater than a certain minimum size.

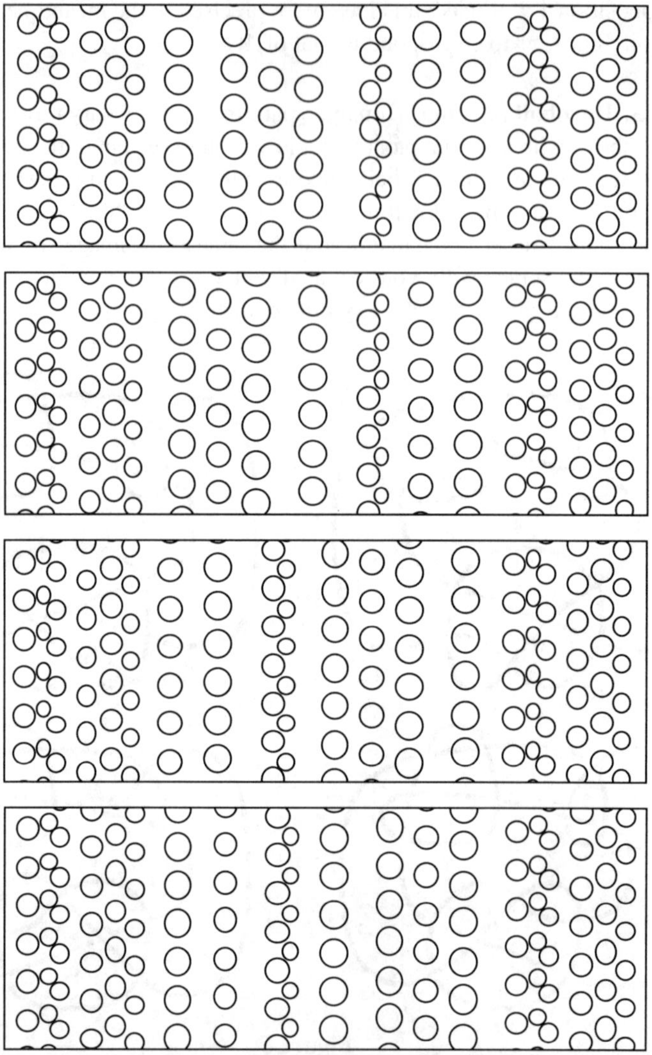

a

b

c

d

Figure 6

Invariant #6: Cusp shape. The shape of the cross-section of a cusp (cf. Figure 2) is an invariant of a manifold. We display it as a parallelogram whose sides correspond to the shortest translations in the torus. The cusp shape distinguishes the figure-eight knot complement from a related manifold with the same volume, cusp volume and horoball packing (see Figure 7). The figure-eight knot complement and the related manifold are punctured torus bundles whose monodromies differ by a $180°$ rotation. The cusps of $WL(4,1)$ and $WL(4,-3)$ are identical.

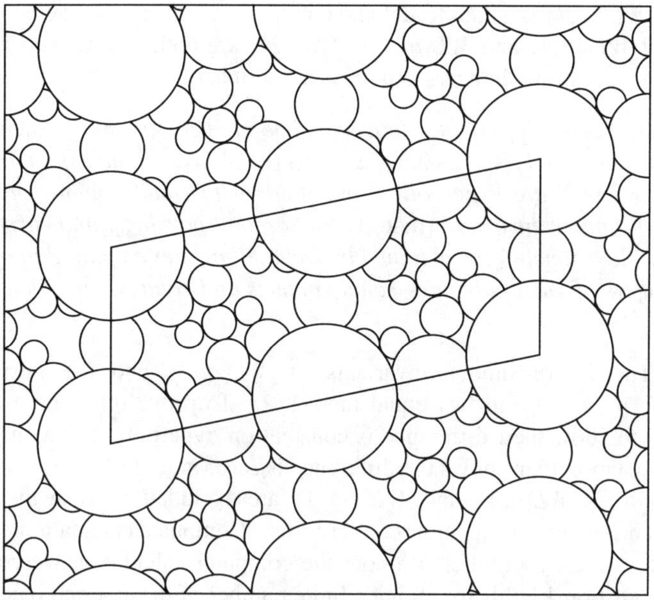

Figure 6e

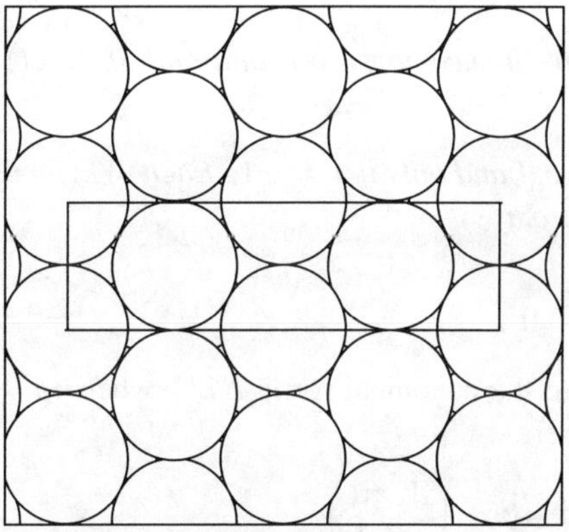

Figure 7a

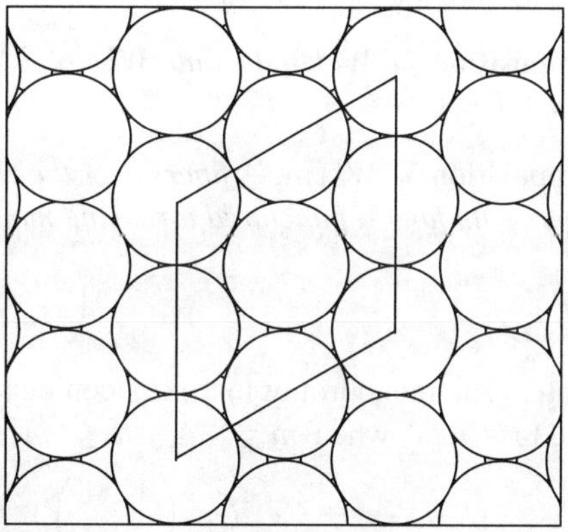

Figure 7b

Invariant #7: Dehn fillings. One can distinguish cusped manifolds by examining invariants of the closed manifolds obtained by Dehn filling the cusp. As an example, consider the figure-eight knot complement and the related manifold mentioned above: the smallest closed hyperbolic 3-manifold obtainable by Dehn surgery on the figure-eight knot has volume $0.98136882\ldots$, while the smallest for the related manifold has volume $0.94270736\ldots$. The closed manifolds obtained from $WL(4,1)$ correspond naturally to those obtained from $WL(4,-3)$, and corresponding manifolds have equal volumes and Chern-Simons invariants (mod $1/2$).

We are now ready to state and prove the assertions made in the abstract.

Note that $WL(m,l)$ is a manifold if and only if m and l are relatively prime integers; it follows that $WL(m,l)$ and $WL(m,-l-(m/2))$ are both manifolds if and only if m and l are relatively prime integers and m is a multiple of 4.

Proposition 1. *For each pair of relatively prime integers m and l such that m is a multiple of 4 and $(m,l) \notin \{\pm(0,1), \pm(4,-1)\}$, $WL(m,l)$ and the mirror image of $WL(m,-l-(m/2))$ are hyperbolic 3-manifolds with equal volume, homology, cusp volume, Chern-Simons invariant* (mod $1/4$), *horoball packings, and cusp shapes. Furthermore, the closed manifolds obtained by Dehn filling can be paired in a natural way, and the members of each pair have equal volumes and equal Chern-Simons invariants* (mod $1/2$).

Notes. (1) The Chern-Simons invariants of $WL(m,l)$ and the mirror image of $WL(m,-l-(m/2))$ are in fact equal mod $1/2$. To prove this, one uses the results of **[Yoshida]** to show their difference is constant on hyperbolic Dehn surgery space; a computer calculation then shows the difference is zero (mod $1/2$).
(2) The manifolds $WL(0,1)$ and $WL(-4,1)$ are excluded because they are not hyperbolic: each contains an incompressible torus. Computer calculations show that all remaining cases are hyperbolic. Without the computer calculation we could conclude only that Proposition 1 holds for all but a finite number of the asserted pairs.
(3) A computer calculation of length spectra suggests that $WL(m,l)$ and $WL(m,-l-(m/2))$ are not isospectral.

Proposition 2. *$WL(m,l)$ and $WL(m',l')$ are homeomorphic if and only if $(m,l) = \pm(m',l')$.*

Proposition 3. *$WL(m,l)$ fibers over the circle if and only if $l = \pm 1$. Each $WL(m,\pm 1)$ fibers with fiber a punctured torus and monodromy*

$$\begin{pmatrix} m+2 & 1 \\ -1 & 0 \end{pmatrix}.$$

Note. The monodromy matrix is conjugate to the canonical form RL^m when $m \geq 0$ and to $-RL^n$ when $m = -4-n \leq -4$, where

$$R = \begin{pmatrix} 1 & 1 \\ 0 & 1 \end{pmatrix} \quad \text{and} \quad L = \begin{pmatrix} 1 & 0 \\ 1 & 1 \end{pmatrix}.$$

In the remaining cases $(-4 < m < 0)$ the monodromy matrix has finite order and therefore defines a Seifert fibered space.

Proof of Proposition 1. Let M_0 denote $WL(m,l)$, and M_1 denote the mirror image of $WL(m,-l-(m/2))$. We've already observed that $H_1(M_0) = H_1(M_1) = \mathbb{Z} \oplus \mathbb{Z}/m$. The equality of all the remaining invariants follows from the fact that M_0 and M_1 have a common double cover. See Figure 8, but ignore the p's and q's for the time being. The hypothesis that m is even implies that the coverings are unbranched (more generally, a k-fold covering will be unbranched if and only if k divides m). The hyperbolic structures on each of M_0 and M_1 lift to the common double cover, yet Mostow's rigidity theorem says that this hyperbolic structure must be unique. Thus, for example, vol(doublecover) $= 2\,\mathrm{vol}(M_0) = 2\,\mathrm{vol}(M_1)$, which implies $\mathrm{vol}(M_0) = \mathrm{vol}(M_1)$. Similar reasoning shows that the cusp volumes and the Chern-Simons invariants are also equal. However, because the Chern-Simons invariant of the double cover is defined only mod $1/2$, this proof shows only that the Chern-Simons invariants of M_0 and M_1 are equal mod $1/4$. The horoball packings for M_0 and M_1 are the same because each is the same as the horoball packing for the double cover. (The definition of the horoball packing for the double cover is similar to that for the one-cusp manifolds, but with the additional requirement that the two "maximal cusps" have disjoint interiors and equal volumes.) Similarly, the cusp shapes are the same because each is the same as the shape of a cusp of the double cover. Finally, consider the closed manifold obtained by (p, q) Dehn filling on the cusp of M_0. Figure 8 shows that this manifold shares a double cover with the closed manifold obtained by $(p + 4q, q)$ Dehn filling on the cusp of M_1. When these manifolds are hyperbolic—which is true in all but finitely many cases—we may conclude that their volumes and their mod $1/2$ Chern-Simons invariants are equal. □

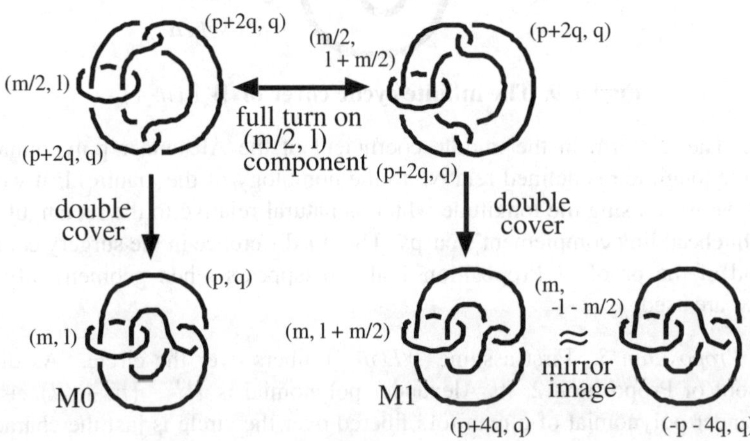

Figure 8

Proof of Proposition 2. The Alexander polynomial of *WL*(m, l) is $lt^2-(m+2l)t+l$, and is unique up to multiplication by $\pm t^n$. From here the proposition follows easily. Because *WL*(m, l) is not a knot complement when $m \neq 1$, a few words are in order about the definition and computation of the Alexander polynomial. The Alexander polynomial is traditionally defined in terms of the infinite cyclic cover of the complement M of a knot; the infinite cyclic cover is the covering space corresponding to the kernel of the map $\pi_1(M) \to H_1(M)$, with $H_1(M) \cong \mathbb{Z}$. In the case of $M =$ *WL*(m, l), $H_1(M) \cong \mathbb{Z} \oplus \mathbb{Z}/m$, so we must mod out by torsion to get a map from $\pi_1(M)$ onto $\mathbb{Z}$; that is, we define the infinite cyclic cover as the covering space corresponding to the kernel of the composed map $\pi_1(M) \to H_1(M) \to H_1(M)/\text{torsion} \cong \mathbb{Z}$. Furthermore, this is the only homomorphism from $\pi_1(M)$ onto $\mathbb{Z}$, up to an isomorphism of $\mathbb{Z}$. (Proof: any homomorphism from $\pi_1(M)$ onto $\mathbb{Z}$ factors through $H_1(M)$, and any homomorphism from $H_1(M)$ onto $\mathbb{Z}$ factors through $H_1(M)/$ torsion.) Informally speaking, this means we can't go wrong when we go to draw the infinite cyclic cover of M. We interchange the components of the Whitehead link, draw the infinite cyclic cover (Figure 9) and use the method of Section 7C of **[Rolfsen]** to immediately write down the Alexander polynomial. This method shows the geometrical origin of the coefficients of the Alexander polynomial. □

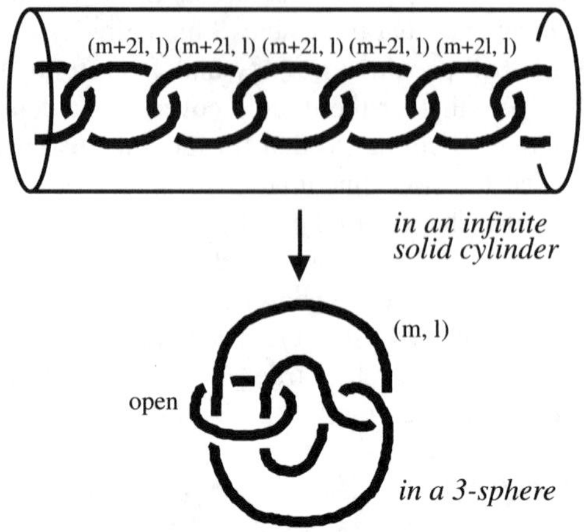

Figure 9. The infinite cyclic cover of $WL(m, l)$

Remark. The $2l$ term in the middle coefficient of the Alexander polynomial arises because the longitude is defined relative to the homology of the manifold; it would not appear if we were using the longitude which is natural relative to the rectangular shape of the Whitehead link complement's cusps. The $4q$ difference in the surgery coefficients at the end of the proof of Proposition 1 also disappears when geometrically natural longitudes are used.

Proof of Proposition 3. First assume *WL*(m, l) fibers over the circle. As discussed in the proof of Proposition 2, its Alexander polynomial is $lt^2 - (m + 2l)t + l$. But the Alexander polynomial of a manifold fibered over the circle is just the characteristic polynomial of the map induced by the monodromy on the first homology of the fiber, and is therefore monic (cf. the proof on p. 326 of **[Rolfsen]**, which goes through even for manifolds which aren't knot complements). Thus $l = \pm 1$.

Now assume $l = \pm 1$. Figure 10 shows how to fiber *WL*$(m, 1)$ over the circle. Figure 10 is best thought of as a movie showing the fiber cycling through the manifold, or more precisely as six frames from such a film loop. At time $t = 0$ the fiber is passing through the surgered-in solid torus. The fiber can be isotoped through the solid torus if and only if the image of a meridian on the boundary of the Whitehead link complement is identified with some longitude on the boundary of the solid torus. This occurs precisely when $l = \pm 1$. The monodromy matrix may be computed from Figure 10 by drawing a meridian and longitude on a fiber and following them as the fiber cycles once around the manifold. The value of m determines how much twist the fiber picks up as it passes through the surgered-in solid torus. □

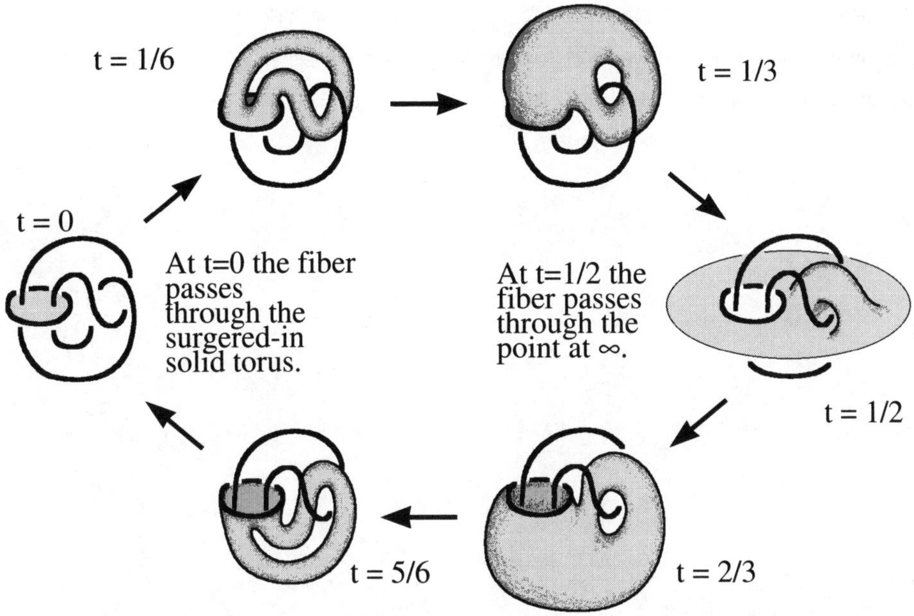

Figure 10

Bibliography

[Meyerhoff 0] R. Meyerhoff, Density of the Chern-Simons invariant for hyperbolic 3-manifolds, in: Low-dimensional Topology and Kleinian Groups, edited by D. B. A. Epstein, Cambridge Univ. Press, 1986.

[Meyerhoff 1] R. Meyerhoff, Geometric invariants of 3-manifolds, to appear in the Mathematical Intelligencer.

[Mey-Rub] R. Meyerhoff and D. Ruberman, Mutation and the eta invariant, J. Diff. Geo. 31 (1990), 101–130.

[Milnor] J. Milnor, Hyperbolic geometry: the first 150 years, Bulletin of the A.M.S. 6 (1982), 9–24.

[Mostow] G. D. Mostow, Strong Rigidity of Locally Symmetric Spaces, Ann. Math. Studies, No. 78, Princeton Univ. Press, Princeton, NJ, 1973.

[Neumann] W. D. Neumann, Combinatorics of triangulations and the Chern Simons invariant of hyperbolic 3-manifolds, these Proceedings.

[Neu-Reid] W. D. Neumann and A. W. Reid, Arithmetic of hyperbolic 3-manifolds, these Proceedings.

[Prasad] G. Prasad, Strong rigidity of $\mathbb{Q}$-rank 1 lattices, Invent. Math. 21 (1973), 255–286.

[Rolfsen] Dale Rolfsen, Knots and Links, Publish or Perish, 1976.

[Ruberman] D. Ruberman, Mutations and Volumes of Links in S^3, Invent. Math. 90 (1987), 189–215.

[Yoshida] T. Yoshida, The eta invariant of hyperbolic 3-manifolds, Invent. Math. 81 (1985), 473–514.

Department of Mathematics, University of Melbourne, Parkville, Victoria 3052, Australia

Department of Mathematics, University of Southern California, Los Angeles, CA 90089

Department of Mathematics, Middlebury College, Middlebury, VT 05753

Fundamental Group of the Complement of Affine Plane Curves

Shulim Kaliman

1. Introduction

W. Fulton and P. Deligne **[F]**, **[D]** proved that the complement in $\mathbb{P}^2$ of an algebraic curve with nodes as its only singularities had abelian fundamental group (this statement was known as Zariski's conjecture). Their result yields a corollary.

Theorem. *Let E be a line in $\mathbb{P}^2$, $\mathbb{C}^2 \cong \mathbb{P}^2 - E$, and $\{\Gamma_i\}_{i=1}^n$ be plane algebraic irreducible curves such that the curve $\Gamma = \bigcup_{i=1}^n \Gamma_i$ has only node singularities. Suppose that Γ intersects E transversally. Then $\pi_1(\mathbb{C}^2 - \Gamma_i) = \mathbb{Z}$ and $\pi_1(\mathbb{C}^2 - \Gamma) = \prod_{i=1}^n \pi_1(\mathbb{C}^2 - \Gamma_i) = \mathbb{Z}^n$.*

One can ask when the group $\pi_1(\mathbb{C}^2 - \Gamma)$ coincides with the direct product $\prod_{i=1}^n \pi_1(\mathbb{C}^2 - \Gamma_i)$ if the groups $\pi_1(\mathbb{C}^2 - \Gamma_i)$ are not abelian. We shall prove the following fact.

Main Theorem. *Let $\Gamma_1, \ldots, \Gamma_n$ be algebraic plane curves in $\mathbb{P}^2$, let $E \subset \mathbb{P}^2$ be a line, and $\mathbb{C}^2 \cong \mathbb{P}^2 - E$. Suppose that for every pair $i \neq j$ the curves Γ_i and $\bigcup_{k \neq i} \Gamma_k$ intersect transversally and $\Gamma_i \cap \Gamma_j \cap E = \emptyset$. Then $\pi_1(\mathbb{C}^2 - \bigcup_{i=1}^n \Gamma_i) = \prod_{i=1}^n \pi_1(\mathbb{C}^2 - \Gamma_i)$.*

The Main Theorem and the well-known Zariski theorem imply the following corollary.

Corollary. *Let $D_1, \ldots, D_n$ be algebraic divisors in $\mathbb{P}^m$, let $H \subset \mathbb{P}^m$ be a hyperplane, and let $\mathbb{C}^m \cong \mathbb{P}^m - H$. Suppose that for some algebraic variety Q of codimension 3 and for every $i \neq j$ the set $D_i \cap D_j \cap H - Q$ is empty, and D_i intersects $\bigcup_{k \neq i} D_k$ transversally outside Q. Then $\pi_1(\mathbb{C}^m - \bigcup_{i=1}^n D_i) = \prod_{i=1}^n \pi_1(\mathbb{C}^m - D_i)$.* □

2. Preliminaries

We shall keep the above notations. Since we do not suppose that Γ_i is irreducible, we can restrict ourselves to the case of two curves Γ_1 and Γ_2. Let W be the group of linear automorphisms of $\mathbb{C}^2$. One can consider W as an affine algebraic submanifold $\{(a, b, c, d, e, f) \in \mathbb{C}^6 \mid ad - bc \neq 0\}$ in $\mathbb{C}^6$ so that each point (a, b, c, d, e, f) corresponds to the automorphism $\alpha\colon (x, y) \to (ax + by + e, cx + dy + f)$. Put $R_k = \Gamma_k \cap \mathbb{C}^2$, $R^\alpha = \alpha(R_2)$, and let Γ^α be the closure of R^α in $\mathbb{P}^2$.

Lemma 1. *There exists a proper algebraic subvariety* $S \subset W$ *such that for every* $\alpha \in W - S$
(i) *the curve* Γ^α *intersects* Γ_1 *transversally;*
(ii) *the set* $\Gamma_1 \cap \Gamma^\alpha \cap E$ *is empty.*

Proof. Obviously the points of W for which either (i) or (ii) does not hold, generate an algebraic subvariety S of W. On the other hand for every $\alpha \in W$ that is close enough to the identity element of W the condition $\Gamma_1 \cap \Gamma_2 \cap E = \emptyset$ implies that $\Gamma_1 \cap \Gamma^\alpha \cap E = \emptyset$, and the curves R_1 and R^α have the same number of intersection points as the curves R_1 and R_2. This means that (i) and (ii) hold for such an automorphism. Thus $W - S$ is non-empty and open in the Zariski topology. □

Lemma 2. *If* $\alpha \in W - S$, *then the curve* $R_1 \cup R^\alpha$ *is isotopically equivalent to the curve* $R = R_1 \cup R_2$ *in* $\mathbb{C}^2$. *In particular, the group* $\pi_1(\mathbb{C}^2 - (R_1 \cup R^\alpha))$ *is isomorphic to the group* $\pi_1(\mathbb{C}^2 - R)$.

Proof. Fix $\beta \in W - S$. Denote the intersection points of R_1 and R^β by $\{a_i \mid i = 1, \ldots, m\}$. Let M_β be the set of lines in $\mathbb{C}^6 \supset W$ such that each line $G \in M_\beta$ is given by a linear mapping $t \to \gamma_t (t \in \mathbb{C})$ with $\gamma_0 = \beta$, and for $\alpha_t = \gamma_t \circ \beta^{-1}$ and every $i = 1, \ldots, m$ the curve $\{\alpha_t(a_i) \mid t \in \mathbb{C}\}$ is not tangent to R^β at the point a_i. We denote the rest of lines that contain β by N_β. If $G = \{\gamma_t = \alpha_t \circ \beta \mid t \in \mathbb{C}\} \in M_\beta$, then one can choose a small disc D_i in R^β such that D_i contains a_i and for $D_i^t = \alpha_t(D_i)$

$$D_i^t \cap D_i^{t'} = \emptyset, \tag{1}$$

when $t \neq t'$ and $|t|$, $|t'|$ are small. Put $\alpha = \alpha_{t_0}$ (here $|t_0| < \delta$ for a certain small $\delta > 0$). Without loss of generality, one may suppose that (1) holds for for $|t|, |t'| < 4$, $\gamma_t \in W - S$ for $|t| < 4$, and for every i and $|t| < 4$ the curve R_1 intersects D_i^t at one point b_i^t only, and the curve $\alpha(R_1)$ also intersects D_i^t at one point c_i^t only. Let $\phi: D_i \to \Delta = \{z \in \mathbb{C} \mid |z| < 1\}$ be a diffeomorphism such that $\phi(a_i) = 0$, and let $g_{u,v}: \Delta \to \Delta$ be a diffeomorphism such that $g_{u,v}(z) = z$ for $|z| > 1/2$, $g_{u,v}(u) = v$ (where $|u|$ and $|v|$ are small enough), and $g_{u,v}$ depends on u and v continuously. Put $u_i^t = \phi \circ \alpha_t^{-1}(b_i^t)$, $v_i^t = \phi \circ \alpha_t^{-1}(c_i^t)$, and $U_i = \bigcup_{|t|<2} D_i^t$. According to (1) the set U_i is a neighborhood of a_i. Define a diffeomorphism h_i^α of U_i in the following way: if $p \in D_i^t$, then $h_i^\alpha(p) = \alpha_t \circ \phi^{-1} \circ g_{c_i^t, b_i^t} \circ \phi \circ \alpha_t^{-1}(p)$. Clearly $h_i^\alpha(\alpha(R_1) \cap U_i) = R_1 \cap U_i$ and $h_i^\alpha(\alpha_t(R^\beta) \cap U_i) \subset \alpha_t(R^\beta)$. Let $\tilde{\alpha}_t$ be the extension of α_t to $\mathbb{P}^2$ and $U = \bigcup_{|t|<2} \Gamma^{\gamma_t}$. For every $|t| < 2$ we define the embedding $H_t: U \to \mathbb{P}^2$ that coincides with $\tilde{\alpha}_t$ outside $\bigcup_{i=1}^m U_i$ and with $H_t = h_i^{\alpha_t} \circ \alpha_t$ on U_i. Then $H_t(\Gamma^\beta) = \Gamma^{\gamma_t}$ and the vector field dH_t/dt is tangent to $E = \mathbb{P}^2 - \mathbb{C}^2$ (since $d\tilde{\alpha}_t/dt$ is tangent to E) and to R_1. Using a partition of unity, one can obtain a vector field $w(t)$ on $\mathbb{P}^2$ such that $w(t)$ is tangent to E and to R_1, $w(t)|_{U'} = dH_t/dt$ for $U' = \bigcup_{|t|<1} \Gamma^{\gamma_t}$, and $w(t) \equiv 0$ outside a certain neighborhood of U'. The phase flow associated with this field yields an isotopical equivalence between $R \cup R^\beta$ and $R \cup R^{\gamma_{t_0}}$ for $|t_0| < \delta$. Let T_G^β be the subset of $G \cap W - S$ such that $s \in T_G^\beta$, if for an intersection point a of R_1 and R^{γ_s} the curve $\{\gamma_t \circ \gamma_s^{-1}(a) \mid t \in \mathbb{C}\}$ is tangent to R^{γ_s} at the point a (since T_G^β is an algebraic subvariety, it is finite). Using the above arguments, it is easy to show that $R_1 \cup R^\beta$ is isotopically equivalent to $R \cup R^{\gamma_t}$ for $\gamma_t \notin T_G^\beta \cup S$. Put $V_\beta = \{\gamma \in W - S \mid \gamma \in T_G^\beta$ for a line $G \in M_\beta$, or γ belongs to a line $G' \in N_\beta\}$. We have proved that $R \cup R^\beta$ is isotopically equivalent to $R \cup R^\gamma$ for $\gamma \in W - (V_\beta \cup S)$. It remains to note that the set $\bigcap_{\beta \in M-S} V_\beta$ is empty. Indeed, if $\tau \in V_\beta - S$, then $W - (S \cup V_\beta \cup V_\tau) \neq \emptyset$, since $V_\beta \cup V_\tau \cup S$ is a proper algebraic subvariety of W. Hence $R \cup R^\tau$ is isotopically equivalent to $R \cup R^\beta$. This yields the desired conclusion. □

3. Proof of the Main Theorem

Let (x, y) be a coordinate system in $\mathbb{C}^2$, and let R_1, R_2, R^α be the same as above.

Lemma 3. *Suppose that the restriction of the projection $\psi\colon \mathbb{C}^2 \to \mathbb{C}$ given by the formula $\psi(x, y) = x$ to an affine algebraic curve Q is a proper mapping $\phi\colon Q \to \mathbb{C}$. Then there exists a bidisc $B = \{(x, y) \mid |x| < r, |y| < q\}$ such that the embedding $\rho\colon B - Q \to \mathbb{C}^2 - Q$ generates an isomorphism between the fundamental groups $\pi_1(B - Q)$ and $\pi_1(\mathbb{C}^2 - Q)$.*

Proof. Choose r so that the projections of all the branch points of ϕ belong to $\Delta_r = \{x \mid |x| < r\}$. Let $B' = \psi^{-1}(\overline{\Delta}_r) - Q$. We can take q such that $B' - B = \overline{\Delta}_r \times \{y \mid |y| > q\}$. Clearly, $B - Q$ is a deformation retract of B'. On the other hand, since Δ_r contains the projections of all the branch points of ϕ, the manifold $G = \mathbb{C}^2 - (Q \cup B')$ can be considered as a bundle over $\{x \mid |x| > r\}$, whose fiber is a punctured complex line. Thus G can be contracted to the set $\psi^{-1}(\partial\Delta_r) - Q$ in $\mathbb{C}^2 - Q$, i.e., B' is a deformation retract of $\mathbb{C}^2 - Q$. Hence $B - Q$ is also a deformation retract of $\mathbb{C}^2 - Q$. This implies the desired conclusion. □

Let a generic line intersect R_k at n_k different points. Without loss of generality, one may suppose that $l = \{x = 0\}$ is a generic line for R_1 and R_2. In particular, this means that $\psi|_{R_k}\colon R_k \to \mathbb{C}$ is a proper mapping, where ψ is the same as in the preceding lemma. Let $l \cap R_1 = \{(0, u_1), \ldots, (0, u_{n_1})\}$ and $\operatorname{Re} u_i < \operatorname{Re} u_j$ for $i < j$.

Lemma 4. *Let W, S, R^α be as in Lemma 1. Then there exists an automorphism $\alpha \in W - S$ such that the set $l \cap R^\alpha = \{(0, v_1), \ldots, (0, v_{n_2})\}$ satisfies*

(1) $\operatorname{Re} v_i << \operatorname{Re} v_j$ *for $i < j$;*

(2) $\operatorname{Re}(v_1 - u_{n_1})$ *is a sufficiently large positive number.*

Proof. It is easy to choose an automorphism $\alpha(x, y) = (x, ay + b)$, where b is positive and $a \in \mathbb{C}$, so that for the set $l \cap R^\alpha$ the conditions (1) and (2) hold. Moving α a little, if necessary, one may suppose that $\alpha \in W - S$. □

From now on we shall suppose that R_2 itself satisfies (1) and (2), in other words we can take $\alpha = id$ in the preceding lemma. Let w be a point in the line l with a sufficiently large $\operatorname{Im} w$ and let $\nu_1, \ldots, \nu_{n_1}, \mu_1, \ldots, \mu_{n_2}$ be a system of loops such that each of them consists of an interval with w as one of the ends and a small circle about a point of the set $\{u_1, \ldots, u_{n_1}, v_1, \ldots, v_{n_2}\}$ (see Fig. 1). Let $R = R_1 \cup R_2$ and $\rho\colon l - R \to \mathbb{C}^2 - R$ be the natural embedding. Then the loops $\{\rho(\nu_i)\}$ generate the subgroup G_1 of $\pi_1(\mathbb{C}^2 - R)$ and the loops $\{\rho(\mu_i)\}$ generate a subgroup G_2.

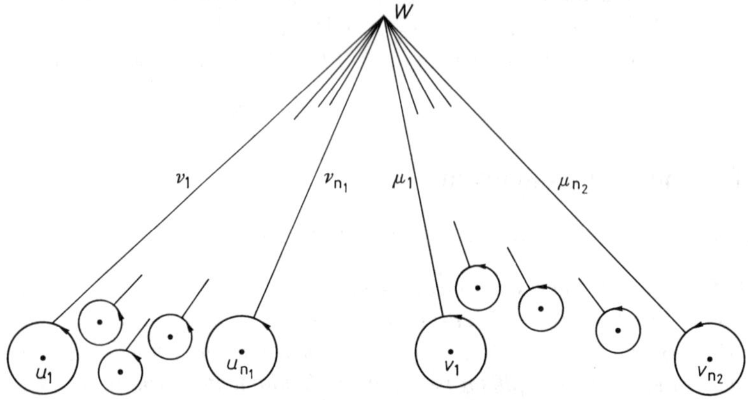

Figure 1

Lemma 5. *The subgroups G_1 and G_2 generate the fundamental group $\pi_1(\mathbb{C}^2 - R)$. Moreover, if $q_k: \mathbb{C}^2 - R \to \mathbb{C}^2 - R_k$ is a natural embedding, then it generates an isomorphism between G_k and $\pi_1(\mathbb{C}^2 - R_k)$.*

Proof. Since l is a generic line, by the well-known Zariski theorem the embedding $\rho: l - R \to \mathbb{C}^2 - R$ generates an epimorphism from $\pi_1(l - R)$ to $\pi_1(\mathbb{C}^2 - R)$. Moreover, by the same theorem, the loops $\{\rho(\nu_i)\}$ generate $\pi_1(\mathbb{C}^2 - R_1)$ and the loops $\{\rho(\mu_i)\}$ generate $\pi_1(\mathbb{C}^2 - R_2)$. Hence G_1 and G_2 generate $\pi_1(\mathbb{C}^2 - R)$, and the image $q_{k*}(G_k)$ coincides with $\pi_1(\mathbb{C}^2 - R_k)$ (where q_{k*} is the homomorphism induced by q_k). Now it remains to check that the restriction of q_{k*} to G_k has trivial kernel. Let a bidisc B be as in Lemma 3 for the curve R_1 instead of Q, and A be an analogous bidisc for R_2. Put $\alpha(x, y) = (x, y + d)$, where d is a sufficiently large positive number. Clearly, the assumptions of Lemma 3 hold for R^α. Replacing R_2 by R^α, if necessary, we may suppose that $A \cap B = \emptyset$, $A \cap R_1 = \emptyset$, and $B \cap R_2 = \emptyset$. Let $w_1 \in l \cap B$ and $w_2 \in l \cap A$. Change the system of loops in $l - R$ (see Fig. 2). Instead of the loop ν_i we consider a loop, which can be described as the union of two intervals ($[w, w_1]$ and the second interval η_i that connects w_1 and a point close to u_i) and a small circle about u_i. In the same way we change the loops $\{\mu_i\}$. Consider the loop ν_i^0, which can be described as the union of η_i and a small circle about u_i. Suppose that $[\nu_i]$ is the element of $\pi_1(\mathbb{C}^2 - R, \rho(w))$ corresponding to the loop ν_i and the element $[\nu_i^0]$ of $\pi_1(\mathbb{C}^2 - R, \rho(w_1))$ corresponds to ν_i^0. Let $[\gamma] = \prod[\nu_{i_k}]$ be an element of $\pi_1(\mathbb{C}^2 - R, \rho(w))$ that belongs to the kernel of the homomorphism q_{1*}. This implies that $[\gamma^0] = \prod[\nu_i^0]$ must be the identity element of the group $\pi_1(\mathbb{C}^2 - R, \rho(w_1))$. But $[\gamma^0]$ corresponds to a loop that belongs to $l \cap B$. By Lemma 3, $[\gamma^0]$ is the identity element of $\pi_1(B - R, \rho(w_1))$, and thus $[\gamma]$ is the identity element of $\pi_1(\mathbb{C}^2 - R, \rho(w))$. The same arguments show that q_{2*} also has trivial kernel. □

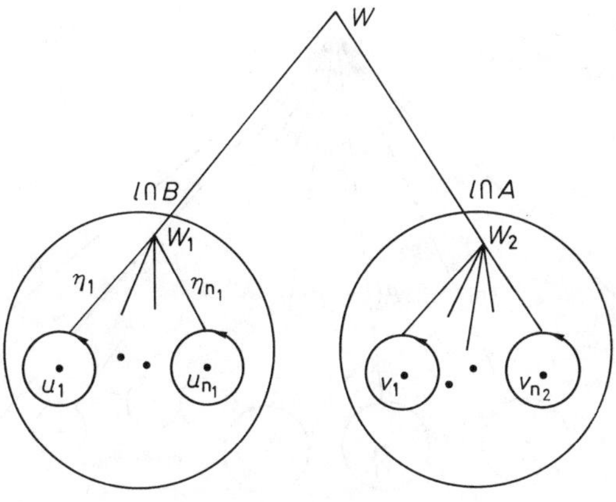

Figure 2

Lemma 6. *Let G_1 and G_2 be as above. Then every pair of elements $g_1 \in G_1$ and $g_2 \in G_2$ is commutative.*

Proof. Put $\alpha_t(x, y) = (x, y - t(v_1 - u_{n_1}))$. Since S is a proper algebraic submanifold, there may be only several points $0 < t_1, \ldots, t_k < 1$ in the interval $[0, 1]$ such that $\alpha_{t_i} \notin W - S$. Moving α_t a little in small neighborhoods of these points, one can suppose that α_t ($t \in [0, 1]$) is a continuous curve in $W - S$. Denote the y-coordinate of the point $\alpha_t(0, v_i)$ by v_i^t. We shall choose a system of loops $\nu_1, \ldots, \nu_{n_1}, \mu_1^t, \ldots, \mu_{n_2}^t$ for the set $\{u_1, \ldots, u_{n_1}, v_1^t, \ldots, v_{n_2}^t\}$ in the same way we did for the set $\{u_1, \ldots, u_{n_1}, v_1, \ldots, v_{n_2}\}$ (see Fig. 3). By Lemma 2 R and $R_1 \cup R^\lambda$ are isotopically equivalent if $\lambda \in W - S$. Thus there exists a proper isomorphism $f_\lambda\colon \pi_1(\mathbb{C}^2 - R) \to \pi_1(\mathbb{C}^2 - (R_1 \cup R^\lambda))$. If γ is a loop in $l - (R_1 \cup R^\lambda)$ that contains the point w, then we denote the corresponding element of the fundamental group $\pi_1(\mathbb{C}^2 - (R_1 \cup R^\lambda), \tilde{w})$ by $[\gamma]$ (here the point $\tilde{w} = (0, w) \in \mathbb{C}^2$). If $\gamma \subset l - (R_1 \cup R^\lambda \cup R^\theta)$, then $[\gamma]$ may be used as the element of $\pi_1(\mathbb{C}^2 - (R_1 \cup R^\theta), \tilde{w})$ or as the element of $\pi_1(\mathbb{C}^2 - (R_1 \cup R^\lambda, \tilde{w})$. In order to avoid misunderstanding in this case we shall write $[\gamma]_\theta$ or $[\gamma]_\lambda$ to distinguish these elements when it is necessary. Clearly, $f_{\alpha_t}([\nu_i]_{\alpha_0}) = [\nu_i]_{\alpha_t}$ and $f_t([\mu_i]) = [\mu_i^t]$ (see Fig. 3). When ϵ is small enough, $v_1^{1-\epsilon}$ is close to u_{n_1}, and we can change the loops ν_{n_1} and $\mu_1^{1-\epsilon}$ in the way shown in Fig. 4. If we delete the interval $[w, w_0]$, we obtain the corresponding loops ν' and μ', which have the common point w_0 and belong to a small neighborhood of u_{n_1}. Since $\alpha_1 \in W - S$, the curve R_1 intersects R^{α_1} transversally at the point $(0, u_{n_1}) \in \mathbb{C}^2$. Hence for a sufficiently small ϵ and a certain small neighborhood V of $(0, u_{n_1})$ in $\mathbb{C}^2$ the set $M = V - (R_1 \cup R^{\alpha_{1-\epsilon}})$ is homeomorphic to $\mathbb{C}^2 - \{xy = 0\}$. Therefore $\pi_1(M) = Z \oplus Z$. In particular, the natural embedding of the loops ν' and μ' in M generates commutative elements in $\pi_1(M)$. This means that elements $[\nu_{n_1}]$ and $[\mu_1^{1-\epsilon}]$ are commutative in $\pi_1(\mathbb{C}^2 - (R_1 \cup R^{\alpha_{1-\epsilon}}))$. But $[\nu_{n_1}]_{\alpha_{1-\epsilon}} = f_{\alpha_{1-\epsilon}}([\nu_{n_1}]_{\alpha_0})$ and $[\mu_1^{1-\epsilon}] = f_{\alpha_{1-\epsilon}}([\mu_1])$. Hence $[\nu_{n_1}]$ and $[\mu_1]$ are commutative in $\pi_1(\mathbb{C}^2 - R)$. Since the element $[\mu_1^{1-\epsilon}]$ coincides with $[\nu_{n_1}]^{-1}[\mu_1^{1-\epsilon}][\nu_{n_1}]$, the loop $\overline{\mu}_1$ (see Fig. 5) generates the element $[\mu_1^{1-\epsilon}]$. Let η be a curve that connects the points u_{n_1-1} and $v_1^{1-\epsilon}$ and consists of a small arc near the point u_{n_1} and an interval (see Fig. 6). Let $c\colon [0, 1] \to \eta$ be a homeomorphism

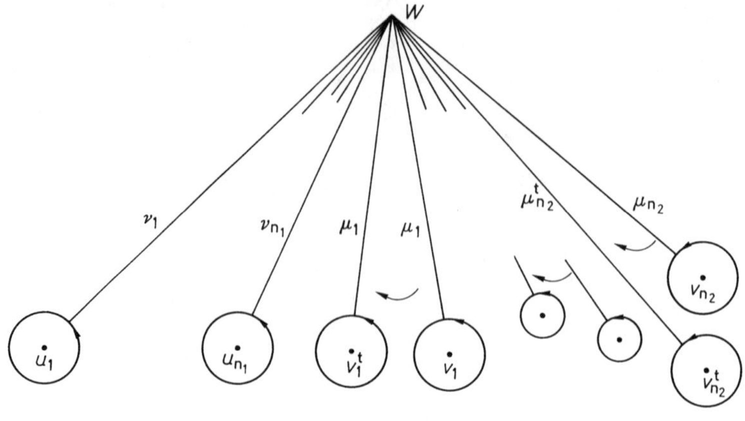

Figure 3

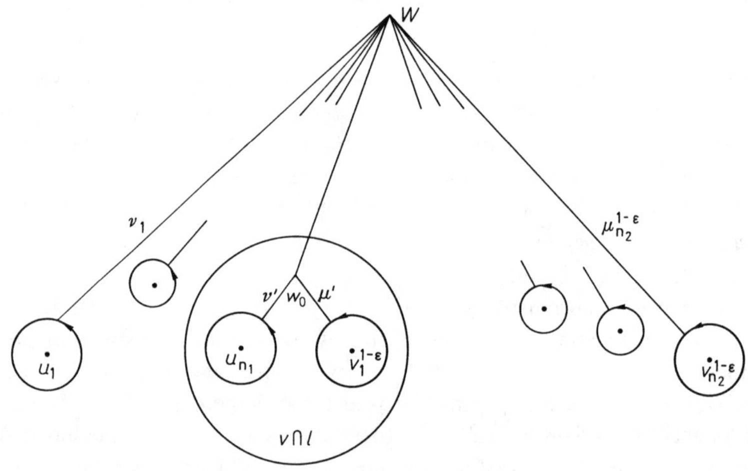

Figure 4

with $c(0) = v_1^{1-\epsilon}$ and $c(1) = u_{n_1-1}$. Put $\tau_t(x,y) = (x, y - t(v_1^{1-\epsilon} - c(t)))$ and $\beta_t = \tau_t \circ \alpha_{1-\epsilon}$. Moving β_t a little, if necessary, we may suppose that $\beta_t \in W - S$ for every $t \in [0,1]$. Then this family of automorphisms transforms Fig. 5 to Fig. 7 isotopically. The point $\tilde{v}_i^{1-\delta}$ in Fig. 7 coincides with the y-coordinate of $\beta_{1-\delta}(0, v_i)$ (here δ is a sufficiently small number) and the loops $\{\beta_{1-\delta}(\mu_i)\}$ do not intersect the loops $\{\nu_i\}$ due to condition (1) of Lemma 4. Repeating the above arguments we see that the pair $[\nu_{n_1-1}]$ and $[\mu_1]$ is commutative, and Fig. 7 can be replaced by Fig. 8, where the loop $\hat{\mu}$ generates again the element $[\beta_{1-\delta}(\mu_1)] = f_{\beta_{1-\delta}}([\mu_1])$. In the same way it is easy to prove that every pair $[\nu_i]$ and $[\mu_1]$ is commutative. Then, repeating this procedure, we can show that every pair $[\nu_i]$ and $[\mu_2]$ is commutative, and so on. □

Let id_k be the identity element of G_k, and let id be the identity element of $\pi_1(\mathbb{C}^2 - R)$. By Lemma 6 there exists a homomorphism $F: G_1 \times G_2 \to \pi_1(\mathbb{C}^2 - R)$ such that

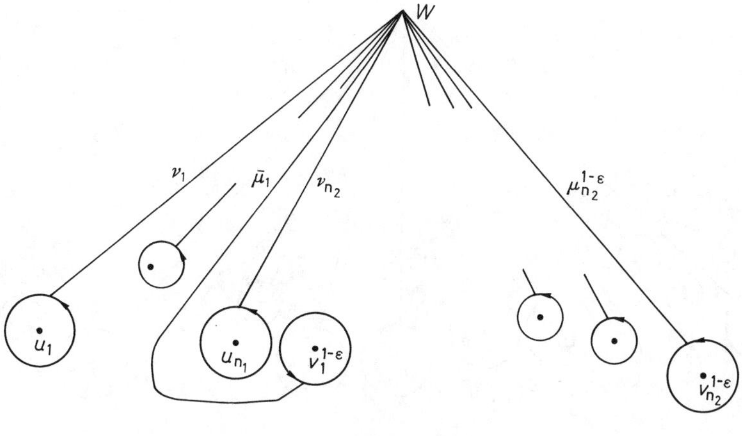

Figure 5

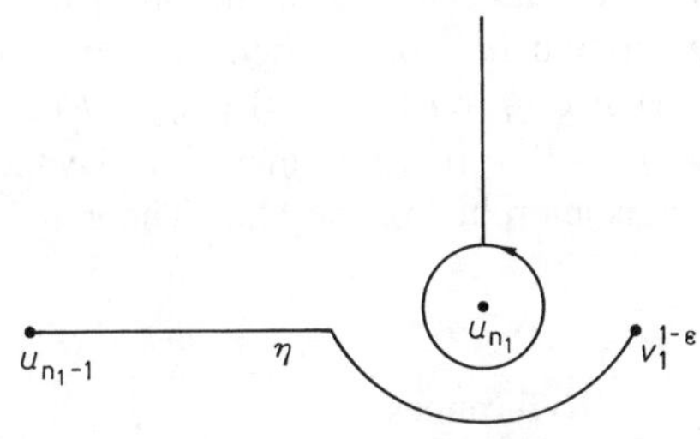

Figure 6

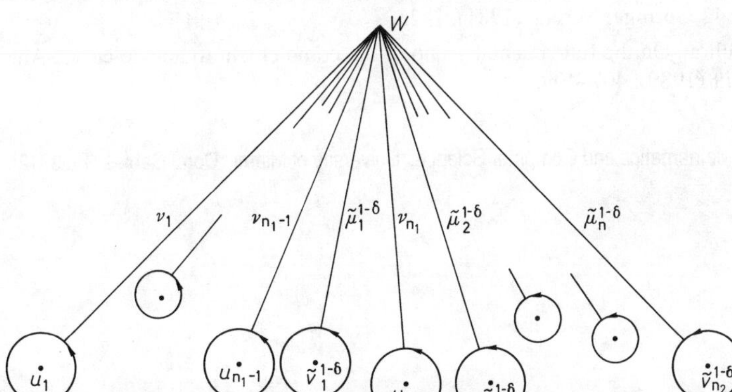

Figure 7

$F(g_1 \times id_2) = g_1$ for every $g_1 \in G_1$ and $F(id_1 \times g_2) = g_2$ for every $g_2 \in G_2$. By Lemma 5, F is an epimorphism, and in order to complete the proof of the Main Theorem we must show that F has trivial kernel. Suppose that $F(g_1 \times g_2) = id$. Let

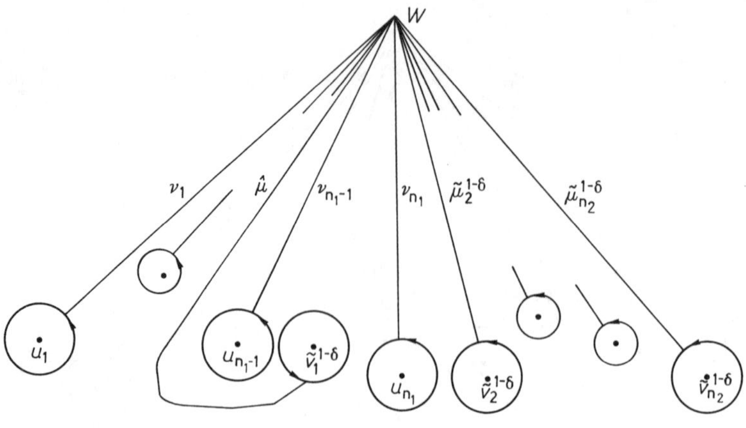

Figure 8

$q_{k*}: \pi_1(\mathbb{C}^2 - R) \to \pi_1(\mathbb{C}^2 - R_k)$ be the same as in the proof of Lemma 5. Since $G_k \cong \pi_1(\mathbb{C}^2 - R_k)$ we keep the notation id_k for the identity element of $\pi_1(\mathbb{C}^2 - R_k)$. Clearly, $q_{1*} \circ F(id_1 \times g_2) = id_1$. Hence $q_{1*} \circ F(g_1 \times g_2) = q_{1*} \circ F(g_1 \times id_2) = g_1$. On the other hand $q_{1*} \circ F(g_1 \times g_2) = q_{1*}(id) = id_1$, i.e., $g_1 = id_1$. By the same arguments $g_2 = id_2$. This means that F is an isomorphism. The Main Theorem is proved. □

References

[D] P. Deligne, Le groupe fondamental du complément d'une courbe plane n'ayant que des points doubles ordinates est abélian, Sem. Bourbaki, 1979/1980, Lecture Notes in Math. vol. 842, Springer-Verlag, (1981), 1–25.

[F] W. Fulton, On the fundamental group of the complement to a node curve, Ann. of Math. (2) 111 (1980), 407–409.

Department of Mathematics and Computer Science, University of Miami, Coral Gables, FL 33124

Totally Geodesic Surfaces in Hyperbolic Link Complements

William Menasco and Alan W. Reid

1. Introduction

The question of whether a 3-manifold contains a closed (embedded or immersed) incompressible surface is of considerable importance in the study of 3-manifolds. In this article we shall be mainly concerned with closed incompressible surfaces in hyperbolic link complements, i.e., link complements $S^3 \setminus L$ which are diffeomorphic to $\mathbb{H}^3/\Gamma$, where Γ is a torsion-free Kleinian group. More precisely we focus on the geometry of closed embedded incompressible surfaces in hyperbolic link complements and in particular on the existence of closed embedded totally geodesic surfaces in such link complements.

Although much work has been done on determining which link complements in S^3 contain closed embedded incompressible surfaces and in describing criteria to detect this (see for example **[LP]**, **[Me1]**, **[Me2]** and **[O]**), the emphasis has been topological rather than geometric. In particular when the link complement is hyperbolic it is of interest to understand how the surface group acts on the sphere-at-infinity of hyperbolic space (see §2.1); however this seems hard to describe, although partial results in the form of algebraic criteria are given in **[R2]**.

The main results of this paper are Theorem 1 and its corollaries, and Theorem 4 and its corollaries. In particular, Theorem 1 determines a relationship between the geometry of a closed embedded incompressible surface in a hyperbolic link complement and properties of the link projection. Although the proof is, for the most part elementary, the result provides a striking connection between the combinatorics of the link projection and the hyperbolic geometry—for example an easy consequence is Corollary 1 which shows that if a hyperbolic link complement contains a closed embedded totally geodesic surface, then the link is non-alternating.

Theorem 4 presents a strange dichotomy between the nature of certain embedded incompressible separating surfaces in hyperbolic 3-manifolds and the question of when a 2-generator subgroup of the fundamental group of a hyperbolic 3-manifold is free. In particular one consequence of Theorem 4 is that a 2-generator hyperbolic 3-manifold of finite volume cannot contain an embedded orientable totally geodesic surface (see Corollary 3). A special case of Corollary 3 is Corollary 4 which deals with the case that the 2-generator hyperbolic 3-manifold is a knot complement.

What will be apparent in the subsequent discussions is the surprising role of annuli in the study of closed embedded incompressible surfaces in hyperbolic 3-manifolds. For example in hyperbolic link complements accidental parabolic elements occur naturally and frequently in closed embedded incompressible surfaces. Indeed, to construct an example of a link complement that contains a closed embedded incompressible surface without an accidental parabolic (see §2.3 for an example) requires some effort.

It is important that the surface be embedded in the previous discussions, since it is known that many alternating hyperbolic link complements contain closed immersed totally geodesic surfaces, e.g., any alternating link which is arithmetic (cf. **[M]**). In addition in **[AR]** and **[ALR]** it is shown that many alternating link complements contain closed immersed quasi-Fuchsian surfaces.

Acknowledgements. The first author acknowledges the support of the NSF for this research. The second author wishes to thank C. Adams and D. Long for useful conversations and correspondence, in particular D. Long for help with §3. The second author is also especially grateful to W. Neumann for countless helpful and illuminating discussions.

2. Existence of closed embedded totally geodesic surfaces in hyperbolic link complements

2.1. In this section we prove the main result which restricts the geometric type of closed embedded incompressible surfaces in alternating link complements. In addition we show how the proof of this result may be applied in more generality. Before stating and proving these results we review some necessary material, our references for which are **[Ma]** and **[Mo]**.

A Kleinian group Γ acts by linear-fractional transformation on $\mathbb{C} \cup \infty$ (the sphere-at-infinity of hyperbolic space) and is called *quasi-Fuchsian* if its limit set is a Jordan curve and no element of Γ interchanges its complementary components. Γ is *Fuchsian* when the Jordan curve is a geometric circle.

Now if M is a hyperbolic 3-manifold and $S \subset M$ an incompressible surface, then S is *totally geodesic* if, under a faithful discrete representation of $\pi_1(M)$, the induced representation of $\pi_1(S)$ is Fuchsian.

Let S be a surface and $\rho: \pi_1 S \to \mathrm{PSL}(2, \mathbb{C})$ a faithful discrete representation. $\rho(\pi_1(S))$ (resp. S) is said to contain an *accidental parabolic* element if there is an essential, non-boundary loop α on S such that $\rho([\alpha])$ is a parabolic element.

Recall that a Kleinian group Γ is *geometrically finite*, if there exists a finite sided convex fundamental polyhedron in $\mathbb{H}^3$ for Γ. Otherwise, Γ is called *geometrically infinite*.

Theorem 1. *Let L be an alternating link or a closed 3-braid such that $S^3 \setminus L$ is hyperbolic. Suppose $S \subset S^3 \setminus L$ is a closed embedded incompressible surface. Then, S is neither quasi-Fuchsian nor geometrically infinite. In particular S is not totally geodesic.*

Proof. Let us at the outset fix some notation:

Let Γ_L denote a faithful discrete representation of $\pi_1(S^3 \setminus L)$, which by the Mostow Rigidity Theorem is unique up to conjugacy in $\mathrm{PSL}(2, \mathbb{C})$ and F the induced representation of $\pi_1(S)$ in Γ_L.

That S cannot be geometrically infinite follows from the work of Bonahon **[B]** and Thurston **[T]** which implies that a geometrically infinite surface in a hyperbolic

3-manifold is virtually the fiber of a fibration over the circle and this cannot be the case here as S is closed.

The key result required to complete the proof is the following "Meridian Lemma", proved for alternating links in **[Me1]** and for closed 3-braids in **[LP]**, and stated here only for hyperbolic links.

Meridian Lemma. *Let L be a hyperbolic link which is alternating or a closed 3-braid. If Σ is a closed embedded incompressible surface in $S^3 \setminus L$, then Σ contains a circle isotopic in $S^3 \setminus L$ to a meridian of L.*

Thus, in particular, S will contain a circle isotopic to a meridian of L. In terms of the hyperbolic structure, this implies the existence of an accidental parabolic element in F—recall, that under a faithful discrete representation meridians of the link map to parabolic elements, and as S is closed, clearly this parabolic element is accidental. However, a quasi-Fuchsian group cannot contain an accidental parabolic element (cf. **[Ma**, Proposition D17]).

This is easy to see, if S is totally geodesic since the universal cover in this case is a genuine hyperbolic plane and the existence of parabolic elements in F implies there are points in the universal cover moved an arbitrarily small distance. This corresponds to the surface S having a cusp, which is absurd as S is closed. □

From Theorem 1 we get the following criterion for links to be non-alternating:

Corollary 1. *Let L be a hyperbolic link. If $S^3 \setminus L$ contains a closed embedded totally geodesic surface then L is non-alternating.* □

Remark 1. The existence of accidental parabolic elements is a subtle phenomena and indeed there is evidence in the literature which indicates a belief that if a surface was not quasi-Fuchsian then it was geometrically infinite. This is indeed the case when the surface group contains no parabolic elements (cf. **[Mo**, Corollary 9.2]). However, in the presence of accidental parabolic elements the surface group can be either geometrically finite or geometrically infinite; this latter case cannot occur for a closed incompressible surface in a hyperbolic 3-manifold (see the argument in the proof of Theorem 1). In §4, to illustrate this point and convince the reader we give explicit examples of geometrically finite surface groups with accidental parabolic elements which arise in certain arithmetic link complements.

2.2. A common method for constructing closed embedded incompressible surfaces in hyperbolic link complements in S^3 is that of "peripheral tubing" (see figure 1)—that is start with an embedded incompressible surface with boundary, some of whose boundary components are meridians of the link and tube together some of these meridional boundary components (see **[Me1]**, **[Me2]**, **[LP]** and **[O]** for examples).

Clearly the operation of peripheral tubing yields a loop on the surface which is a meridian of the link. Hence arguing as in §2.1 we have:

Theorem 2. *Let L be a hyperbolic link and $S \subset S^3 \setminus L$ a closed surface obtained by peripheral tubing. Then either S is compressible or S is geometrically finite and contains accidental parabolic elements.*

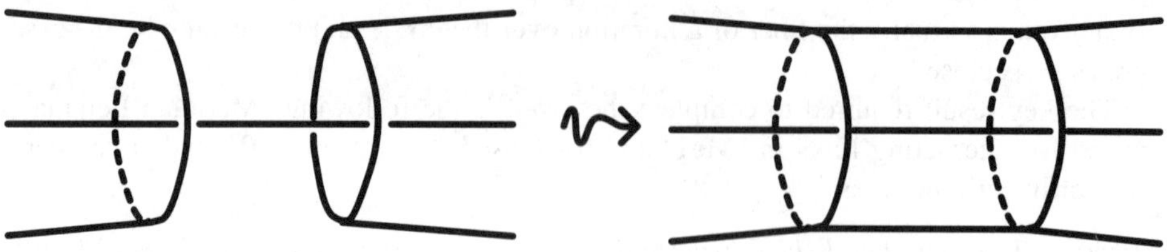

Figure 1

2.3. The natural question that arises from 2.1 and 2.2 is whether there exist any hyperbolic link complements containing a closed embedded totally geodesic surface. There is an obvious candidate construction, given by the next lemma, for constructing examples of this phenomena. It is then a matter of exhibiting an explicit example.

Lemma 1. *Let* $M = \mathbb{H}^3/\Gamma$ *be a hyperbolic 3-manifold with boundary, admitting an orientation-reversing involution* σ *such that the fixed-point set of* σ *is 2-dimensional and* σ *acts fixed-point freely on the boundary of* M*. Then* M *contains a closed embedded totally geodesic surface.*

Proof. It is well-known that $Fix(\sigma)$ is embedded and totally geodesic (see for example **[L**, Theorem 15]). The assumption that σ acts fixed-point freely on the boundary of M implies that $Fix(\sigma)$ is closed. □

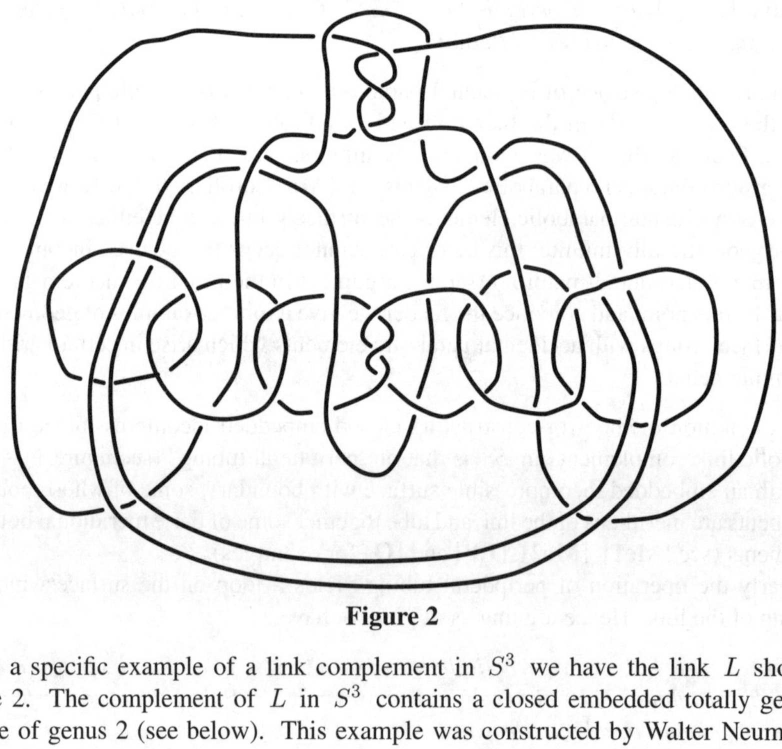

Figure 2

As a specific example of a link complement in S^3 we have the link L shown in Figure 2. The complement of L in S^3 contains a closed embedded totally geodesic surface of genus 2 (see below). This example was constructed by Walter Neumann to whom we are grateful. The link was shown to be hyperbolic using Jeff Weeks' program "Snappea". The volume is approximately $49.9159398784423\ldots$ and the triangulation found by "Snappea" contains 54 simplices. Some discussion of how this example was constructed is in order.

The starting point is the link 2_1^2, denoted herein by H, and shown in Figure 3.

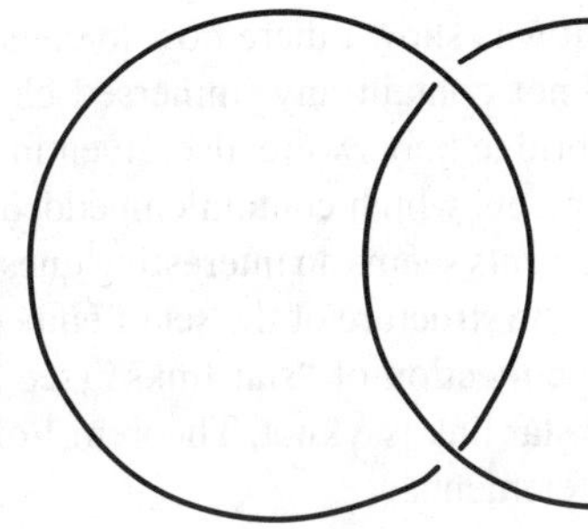

Figure 3

Now $S^3\backslash H$ is homeomorphic to $T^2 \times (0,1)$, and there is an obvious involution $\sigma: T^2 \times (0,1) \to T^2 \times (0,1)$ defined by $\sigma(x,t) = (x, 1-t)$. The fix-point set of this involution is clearly the torus occurring as the level $t = 1/2$. We shall denote this torus by T. Notice that the involution σ can be thought of as inversion in T.

Now take two copies of $S^3\backslash H$ and take the connected sum S^3 with S^3 along the copies of T to give the picture in Figure 4.

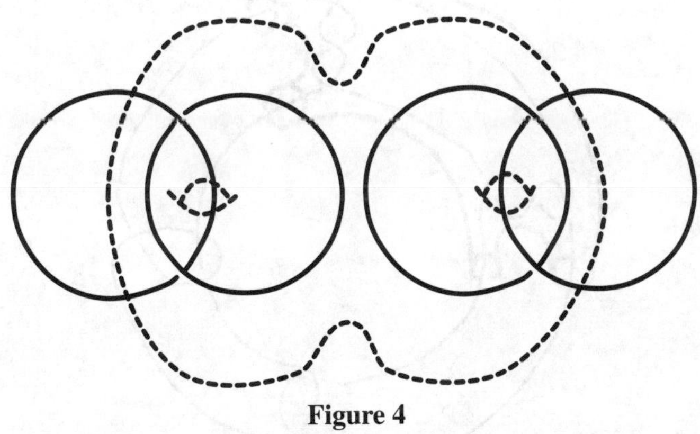

Figure 4

The involution σ extends to give an involution of this new link complement with fix-point set the genus 2 surface $T\sharp T$ (shown in Figure 4 as a broken line).

To complete the construction we want to construct a hyperbolic link which is respected by the involution σ. This was done by adding pairs of components to eliminate any obvious obstructions to hyperbolicity; i.e., essential 2-spheres, annuli etc., and then giving the link L to "Snappea" to look for a hyperbolic structure.

Remark 2. The question of existence of closed embedded totally geodesic surfaces in hyperbolic knot complements seems considerably more subtle—for one thing, Lemma 1 does not apply. Moreover, the only example of a hyperbolic knot complement known to the authors that contains even an immersed closed totally geodesic surface is the figure-eight knot. Furthermore, it follows from **[M]** that there are infinitely many such surfaces in this case. These arise from the arithmetic nature of this knot, however as shown in **[R1]** this knot is the only arithmetic knot.

Negative results about existence of even immersed closed totally geodesic surfaces were obtained in **[R2]**, in particular it was shown there how to construct infinitely many twist knots whose complements do not contain any immersed closed totally geodesic surfaces (other examples of 2 and 3 bridge knots were also given in **[R2]**).

Indeed, to construct knot complements which contain embedded incompressible surfaces without accidental parabolic elements seems an interesting question. A large class of examples of link complements where the structure of the set of embedded incompressible surfaces has been investigated is the collection of "star links" (see **[O]** for a definition) and in particular in the case when the star link is a knot, Theorem 1 of **[O]** always implies the existence of accidental parabolic elements.

In the examples above the accidental parabolic elements have arisen from meridians of link components, however this is not always the case.

Example. Let K be an alternating star link and K_0 the augmented alternating link (in the notation of **[A]**) formed from K as shown in Figure 6 (which shows the case when K is the star link $K(1/3, 1/3, 2/3, 1/2)$ shown in Figure 5 in the notation of **[O]**).

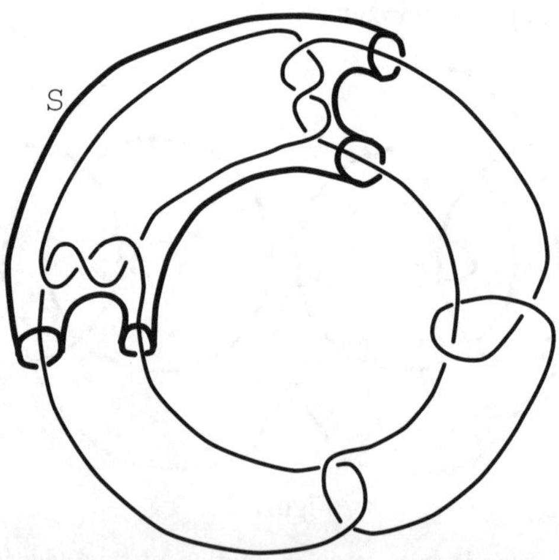

Figure 5

Now let S be an "obvious" 4-punctured sphere in $S^3 \setminus K$ (for example in the case of $K(1/3, 1/3, 2/3, 1/2)$ take S as shown in Figure 5); this will be incompressible so long as S bounds two tangles on each side (cf. **[O**, Corollary 2.14**]**). Clearly in $S^3 \setminus K_0$ this 4-punctured sphere contains a loop isotopic in $S^3 \setminus K_0$ to the added unknotted component. Tubing S as described by **[O**, Theorem 2**]** then gives an example of a closed incompressible surface.

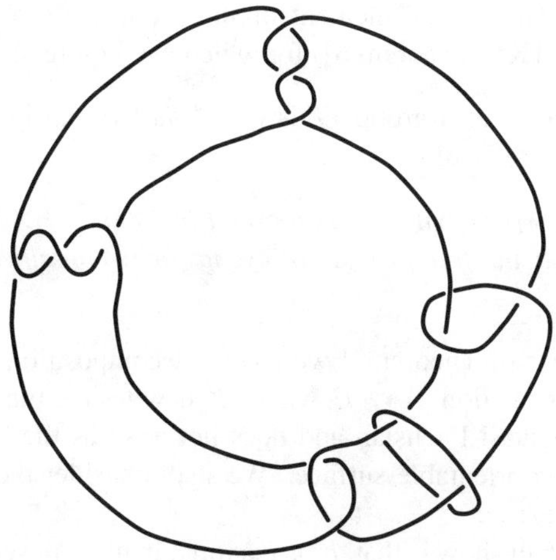

Figure 6

2.4. Although the example of §2.3 exhibited surfaces with accidental parabolics which were not meridians, the accidental parabolic was still embedded. The natural question is whether this is always the case. In this direction, the following result is a direct consequence of the proof of the Annulus Theorem (see for example **[J]**, Theorem 8.13 pp. 154–155).

Theorem 3. *Let M be a hyperbolic 3-manifold of finite volume and $S \subset M$ a closed embedded incompressible surface which contains an accidental parabolic element. Then S contains an embedded accidental parabolic element.* □

3. Two-generator hyperbolic 3-manifolds

3.1. In this section we prove a result that puts limitations on the nature of separating embedded incompressible surfaces in certain hyperbolic 3-manifolds. This has a particularly succinct statement when the fundamental group is 2-generator (in the sequel we refer to such manifolds as *2-generator hyperbolic 3-manifolds*). This provides further evidence on the paucity of closed embedded totally geodesic surfaces in hyperbolic knot complements. Before stating the theorem we require a definition:

Definition. Let M be a hyperbolic 3-manifold of finite volume and S an embedded incompressible separating quasi-Fuchsian surface. An *obstructing annulus* for S is an essential properly embedded annulus in M_0 where M_0 is M cut open along S.

Theorem 4. *Let M be a hyperbolic 3-manifold of finite volume and $S \subset M$ an embedded orientable incompressible separating quasi-Fuchsian surface with no obstructing annulus. Then either any non-abelian 2-generator subgroup of $\pi_1(M)$ is free or S is the boundary of a regular neighbourhood of a non-orientable embedded incompressible surface. In particular, in this latter case, S is not totally geodesic.*

Proof of Theorem 4. The crucial element of the proof is the following Theorem of Karrass and Solitar (cf. [**KS**, Theorem 6]) for which we require additional notation.

Notation. A subgroup U of a group G is called *malnormal* in G if for all $g \in G$, $g \neq 1$ we have $gUg^{-1} \cap U = 1$.

Theorem. *Let $G = A *_U B$ with U malnormal in A and B. Then any 2-generator subgroup of G is either the free-product of cyclics or is contained in a conjugate of a factor.* □

Under the hypothesis of Theorem 4 we have a decomposition of $G = \pi_1(M)$ as a free product with amalgamation $A *_F B$ where F denotes the fundamental group of S.

Suppose that S is quasi-Fuchsian and does not arise as the boundary of a regular neighbourhood of a non-orientable surface. We shall consider the Theorem of Karrass and Solitar in this context.

The first thing we shall show is that F is maximal in its limit set; that is, if J denotes the limit set of F then $Stab(J, G) = F$. Thus suppose that $Stab(J, G) \neq F$. Now as $Stab(J, G)$ is also quasi-Fuchsian (this is a slight abuse of notation as $Stab(J, G)$ may contain orientation-reversing isometries), the index of F in $Stab(J, G)$ is finite. When S is closed it follows from [**S**, Theorems 1.3 and 1.4] that the index of F in $Stab(J, G)$ is 2 and $Stab(J, G)$ is the fundamental group of a non-orientable surface. However this corresponds to precisely the situation that S is the boundary of a regular neighbourhood of a non-orientable surface. Thus F is maximal in its limit set. In the non-closed case, since as above, the index of F in $Stab(J, G)$ is finite, the proofs of Theorems 1.3 and 1.4 of [**S**] may be modified to obtain the same conclusion as above.

We now claim that F is malnormal in A and B. For if there exists $x \neq 1$ in $A \setminus F$ or $B \setminus F$ such that $F' = x^{-1}Fx$ has non-trivial intersection with F, this implies the existence of an element of $\pi_1(M)$ whose fix-point(s) lie on J and J'. If the element is loxodromic, then the hypothesis that S has no obstructing annulus implies that the element stabilizes J. To see this observe that the translate of J by the loxodromic element is forced by the "no obstructing annulus" hypothesis to intersect J transversely—the only other intersection pattern is that J and its translate touch at two points and this is exactly the picture one sees in the case that S has an obstructing annulus. Now the transverse intersection pattern corresponds to S having a self-intersection, which is clearly false. Thus $xJ = J$ and by the first part of the proof it follows that $x \in F$, which was assumed not to be the case. If the element is parabolic then we may find a parabolic element p of F such that $x^{-1}px$ and p have the same fixed-point. Thus x and p commute. However in the fundamental groups of the pieces into which S separates M the parabolic subgroup to which x and p belong has rank 1. Without loss we can assume p is primitive in the above discussion, so that x is a power of p. But by assumption x is not an element of F.

Consequently G now satisfies the hypothesis of the Theorem of Karrass and Solitar. Therefore if H is a non-abelian 2-generator subgroup of G which is not free then H is conjugate into A or B. However, a general result of Jaco-Shalen on 2-generator subgroups of Haken 3-manifold groups (cf. [**JS**]) implies that H, if not free, must be of finite index in G, so H cannot be conjugate into A or B. Thus H is free.

Finally, S cannot be totally geodesic, for if it were, the relevent non-orientable surface would also be totally geodesic with the same limit set and hence S would not embed. □

The proof of Theorem 4 shows that a totally geodesic surface has no obstructing annulus. Thus Theorem 4 above and its proof have the following corollaries for 2-generator hyperbolic 3-manifolds:

Corollary 3. *Let M be a 2-generator hyperbolic 3-manifold of finite volume. If S is an orientable embedded separating incompressible surface then S is not totally geodesic.* □

It is conjectured that a 2-generator knot complement has tunnel number one. Recall a knot has tunnel number one if upon attaching a 1-cell to the knot the complement of a regular neighbourhood of the result is a genus 2 handlebody. Since any closed embedded surface in S^3 is orientable we have as a special case of Corollary 3,

Corollary 4. *The complement of a hyperbolic tunnel number one knot cannot contain a closed embedded totally geodesic surface.* □

Remark 3. Within the class of 2-generator knot complements are the 2-bridge knots and embedded incompressible surfaces in the complement of 2-bridge knots were classified in **[HT]**. In particular it was shown there that there are no closed embedded incompressible surfaces. A reasonable question to ask is whether this is also true for a tunnel number one knot.

Remark 4. Notice that an elementary corollary of the Theorem of Karrass and Solitar is that 2-generator knots are prime. This was proved in **[N]**, but the central lemma there, Lemma 3, seems to have a gap in its proof. Its proof works when the assumptions are strengthened to those of the Theorem of Karrass and Solitar, but then Norwood's lemma is a consequence of their Theorem.

Remark 5. There cannot be an analogous result for 2-generator hyperbolic 3-manifolds whose fundamental group is an HNN-extension, since for example the Whitehead Link group is 2-generator and the complement of the Whitehead Link in S^3 contains a non-separating embedded totally geodesic thrice-punctured sphere.

3.2. In view of the contents of §2.1–3.1, a reasonable conjecture would seem to be:

Conjecture 1. Let K be a hyperbolic knot, then $S^3 \setminus K$ does not contain a closed embedded totally geodesic surface.

4. Geometrically finite surface groups with accidental parabolics

In this section, for the readers convenience, we give some explicit examples of geometrically finite surfaces with accidental parabolic elements which arise in certain cusped arithmetic hyperbolic 3-manifolds, and in particular in some arithmetic link complements. These surfaces are obtained using a combination theorem of Maskit (cf. **[Ma**, Chap. 7**]**). We first set up some notation.

O_d will denote the ring of integers in the quadratic imaginary number field $\mathbb{Q}(\sqrt{-d})$ and we shall assume in what follows that $d \neq 1, 2, 3$ (to apply the First Combination Theorem of Maskit—see below). Let $x_1 = \left(\begin{smallmatrix} 0 & -1 \\ 1 & 0 \end{smallmatrix}\right)$ and $x_2 = \left(\begin{smallmatrix} 0 & -1 \\ 1 & -1 \end{smallmatrix}\right)$. Then $\langle x_1, x_2 \rangle$ is the modular group, which we shall denote by M. Let h be the matrix $\left(\begin{smallmatrix} 1 & \sqrt{-d} \\ 0 & 1 \end{smallmatrix}\right)$, let $x_3 = hx_1h^{-1}$, $x_4 = hx_2h^{-1}$ and $N = hMh^{-1}$. Notice that since h commutes with $t := x_1x_2$, N contains the translation t. Let F_d be the group generated by M and N. F_d is discrete since it is a subgroup of the group $\mathrm{PSL}(2, O_d)$. With this notation we will prove:

Theorem 5. *F_d is a geometrically finite Kleinian group which contains accidental parabolic elements and is isomorphic to a Fuchsian group of signature* $(0; 2, 2, 3, 3; 0)$.

Proof. To prove the theorem we use the First Combination Theorem of Maskit (cf. [**Ma**, Chap. 7 Theorem C2]) which requires some terminology.

Definition. Let G and H be Kleinian groups which act discontinuously somewhere in $\mathbb{C} \cup \infty$ with $G \cap H = J$ (non-trivial). An *interactive pair* (X, Y) for G and H consists of two J-invariant non-empty disjoint sets X and Y, such that every element of $G \setminus J$ maps X into Y and every element of $H \setminus J$ maps Y into X.

In addition the interactive pair (X, Y) is called *proper* if either there is a point of X that is not H-equivalent to any point of Y, or there is a point of Y that is not G-equivalent to any point of X.

Let $T = \langle t \rangle$, then as remarked above $T \subset M \cap N$, indeed $T = M \cap N$. Define $X_1 = \{z | \mathrm{Im} z > 1\}$ and $X_2 = \{z | \mathrm{Im} z < 1\}$. We claim that (X_1, X_2) is a proper interactive pair for M and N.

To see this; certainly X_1 and X_2 are T-invariant. Also observe that every element $\left(\begin{smallmatrix} a & b \\ c & d \end{smallmatrix}\right)$ of $M \setminus T$ has $|c| \geq 1$, hence every element of $M \setminus T$ maps X_1 into X_2. Similarly it is easy to see that every element of $N \setminus T$ maps X_2 into X_1. Moreover since the real axis is a subset of X_2 which is stabilized by M this exhibits points of X_2 which are not M-equivalent to any point of X_1. Hence (X_1, X_2) is a proper interactive pair for M an N (this fails in the cases of $d = 1, 2, 3$).

It now follows from [**Ma**, Theorem A10] that F_d is the free product of M and N with T amalgamated. The orbifolds obtained by quotienting the respective hyperbolic planes stabilized by M and N are both 2-spheres with two cone points of order 2 and 3 and one puncture. The amalgamation corresponds to the 2-sphere with four cone points of orders 2, 2, 3, 3 obtained by identifying these two orbifolds along their boundaries. Evidently F_d cannot be quasi-Fuchsian as the limit set of F_d contains the limit sets of M and N. It is also clear by construction that F_d contains accidental parabolic elements.

To prove that F_d is geometrically finite it is a matter of checking the remaining hypotheses of the First Combination Theorem, for this implies that F_d is geometrically finite if and only if M and N are—and this latter statement is true since M and N are Fuchsian. We leave the reader to check that the remaining hypotheses of [**Ma**, Chap. 7, Theorem C2] are satisfied. □

If Γ is a torsion-free subgroup of finite index in $\mathrm{PSL}(2, O_d)$ then $\Gamma \cap F_d$ gives the advertised surface group. In particular we can take Γ to be the fundamental group of a link complement in S^3 (see for example **[H]**).

Remark 6. In the case of $d \neq 3$ mod 4, the groups F_d were also considered in **[FF]** where it was shown that $\mathrm{PSL}(2, O_d)$ decomposes as a free product with amalgamation along F_d. The question of geometrical finiteness was not discussed in **[FF]**.

References

[A] C. Adams, Augmented alternating link complements are hyperbolic. In "Low-dimensional Topology and Kleinian Groups", editor, D. B. A. Epstein, 115–130, London Math. Soc. Lecture Notes Series 112, Cambridge Univ. Press, 1987.

[AR] I. R. Aitchison and J. H. Rubinstein, An introduction to polyhedral metrics of non-positive curvature on 3-manifolds. Preprint.

[ALR] I. R. Aitchison, E. Lumsden and J. H. Rubinstein, Polyhedral metrics on alternating link complements, in preparation.

[B] F. Bonahon, Bouts des varietes hyperboliques de dimension 3. Ann. of Math. 124 (1986), 71–158.

[FF] B. Fine and C. Frohman, Some amalgam structures for Bianchi groups. Proc. A.M.S. 102 (1988), 221–229.

[H] A. Hatcher, Hyperbolic structures of arithmetic type on some link complements. J. London Math. Soc. 27 (1983), 345–355.

[HT] A. Hatcher and W. P. Thurston, Incompressible surfaces in 2-bridge knot complements. Invent. Math. 79 (1985), 225–246.

[J] W. Jaco, Lectures on Three-Manifold Topology. C.B.M.S 43 , 1980.

[JS] W. Jaco and P. B. Shalen, Seifert fibered spaces in 3-manifolds. Memoirs of the A.M.S. 220 (1979).

[KS] A. Karrass and D. Solitar, The free product of two groups with a malnormal amalgamated subgroup. Canadian J. Math. 23 (1971), 933–959.

[L] H. B. Lawson, Lectures on Minimal Submanifolds. Publish or Perish, Berkeley, 1980.

[LP] M. T. Lozano and J. H. Przytycki, Incompressible surfaces in the exterior of a closed 3-braid I, surfaces with horizontal boundary components. Math. Proc. Camb. Phil. Soc. 98 (1985), 275–299.

[M] C. Maclachlan, Fuchsian subgroups of the groups $\mathrm{PSL}(2, O_d)$. In "Low-dimensional Topology and Kleinian Groups", editor, D. B. A. Epstein, 305–311, London Math. Soc. Lecture Series, 112, Cambridge Univ. Press, 1987.

[Ma] B. Maskit, Kleinian Groups. Grundlehren der mathematischen Wissenschaften 287, Springer-Verlag 1988.

[Me1] W. Menasco, Closed incompressible surfaces in alternating knot and link complements. Topology 23 (1984), 37–44.

[Me2] W. Menasco, Determining incompressibility of surfaces in alternating knot and link complements. Pacific J. Math. 117 (1985), 353–370.

[Mo] J. W. Morgan, On Thurston's Uniformization Theorem for three-dimensional manifolds. In "The Smith Conjecture", editors J. W. Morgan and H. Bass, 37–125. Academic Press, 1984.

[N] F. H. Norwood, Every two-generator knot is prime. Proc. A.M.S. 86 (1982), 143–147.

[O] U. Oertel, Closed incompressible surfaces in complements of star links. Pacific J. Math. 111 (1984), 209–230.

[R1] A. W. Reid, Arithmeticity of knot complements. To appear, J. London Math. Soc.

[R2] A. W. Reid, Totally geodesic surfaces in hyperbolic 3-manifolds. To appear, Proc. Edinburgh Math. Soc.

[S] G. P. Scott, On sufficiently large 3-manifolds. Quart. J. Math. Oxford (2) 23 (1972), 159–172.

[T] W. P. Thurston, The Geometry and Topology of 3-Manifolds. Mimeographed lecture notes, Princeton Univ., 1977.

Department of Mathematics, S.U.N.Y. at Buffalo, Buffalo, NY 14214
Email: MTHTWWM@UBVMSC.bitnet

Department of Mathematics, The Ohio State University, Columbus, OH 43210
Mathematical Sciences Research Institute, Berkelely, CA 94720
Email: reid@mps.ohio-state.edu

Examples of Quasi-Geodesic Flows on Hyperbolic 3-Manifolds

*Lee Mosher**

Let M be a hyperbolic 3-manifold, $\phi : M \times \mathbb{R} \to M$ a flow on M with no stationary points. An interesting problem is to study the relationship between the hyperbolic geometry of M and the topology of ϕ. More specifically, is ϕ a *quasi-geodesic* flow? This means that in the universal cover $\widetilde{M} \approx \mathbb{H}^3$, each flow line of the lifted flow $\widetilde{\phi}$ is a quasi-geodesic, i.e., the flow line has bounded Hausdorff distance from some true hyperbolic geodesic.

One kind of example occurs when M fibers over the circle with fiber A and pseudo-Anosov monodromy map $f: A \to A$, where ϕ is taken to be the suspension flow of f. Cannon and Thurston, in a remarkable paper **[CT]**, deduce several fascinating relationships between the hyperbolic geometry of M and the topology of the fibering. In particular, they show that ϕ is quasi-geodesic.

Are there other examples of quasi-geodesic flows on hyperbolic 3-manifolds, besides pseudo-Anosov suspension flows? We shall show that this is indeed the case. Gabai has described a new method for producing interesting flows that are not suspension flows **[G2]**. We shall look at some simple examples of Gabai's flows, and prove that they are quasi-geodesic. Gabai's flows are always transverse to a finite depth foliation of M (see **[G1]** for the definition). Our examples were inspired by examining the simplest possible finite depth foliations that fail to be fibrations over S^1.

Our method of proof is quite different from that of Cannon and Thurston. When ϕ is a pseudo-Anosov suspension flow, they use the pseudo-Anosov data to construct a metric d_{CT} on $\widetilde{M}$ for which it is obvious that each flow line is a geodesic. Then they show that d_{CT} is equivariantly quasi-isometric to the hyperbolic metric, so each flow line is a hyperbolic quasi-geodesic.

Our flow ϕ will be transverse to an incompressible compact surface $A \subset M$ which is not a fiber of any fibration. ϕ will have a single periodic orbit γ missing A; every other orbit of ϕ, periodic or non-periodic, will intersect A. The key fact used (Theorem 2.2) is due to Thurston, which says that $\pi_1(A)$, thought of as a subgroup of $\pi_1(M) \subset \mathrm{Isom}_+(\mathbb{H}^3)$, is a quasi-fuchsian surface group. From this theorem, it follows that each lift of A to $\widetilde{M} \approx \mathbb{H}^3$ is the interior of a properly embedded disc in the closed ball $\bar{\mathbb{H}}^3 = \mathbb{H}^3 \cup S^2_\infty$. Moreover, one can show that these discs satisfy a nice uniform separation property in $\mathbb{H}^3$. Using uniform separation, any flow segment of $\widetilde{\phi}$ which crosses these discs with sufficient frequency can be shown to make uniform progress away from its previous locations, which characterizes quasi-geodesic behavior. Any

* This work was done with the aid of an NSF Postdoctoral Research Fellowship, Grant #DMS-8705932.

other flow segment of $\widetilde{\phi}$, which stays away from the discs for a long stretch of time, can be shown to stay close to a lift of γ, and quasi-geodesic behavior follows there as well.

Besides whatever intrinsic curiosity these examples might arouse, quasi-geodesic flows have applications in 3-manifold topology. Thurston has defined a natural norm x on $H_2(M;\mathbb{R})$, having the property that the unit ball B_x is a polyhedron. According to the results of **[Mos]**, certain types of quasi-geodesic flows on M can be used to compute information about x, as follows.

Suppose ϕ is a flow on M. Fried defines in **[Fr1]** a natural cone of homology directions $D_\phi \subset H_1(M;\mathbb{R})$. Roughly speaking, a homology direction is any class that can be approximated by a normalized homology class of a long, almost closed orbit of ϕ. When ϕ has a Markov partition, e.g. when ϕ is pseudo-Anosov, Fried has shown how to compute this cone explicitly in terms of the symbolic dynamics of ϕ. His computations show that D_ϕ is finite-sided. Let $D_\phi^\perp \subset H_2(M;\mathbb{R})$ denote the dual cone, with respect to the intersection pairing. Let $\chi_\phi \subset H^2(M;\mathbb{R})$ denote the Euler class of the oriented plane bundle normal to the tangent vector field of ϕ. By the Universal Coefficients Theorem, χ_ϕ can be regarded as a linear functional on $H_2(M;\mathbb{R})$.

Theorem. Flows represent faces of B_x **[Mos]**. *Let ϕ be a pseudo-Anosov flow on M. Suppose also that ϕ is quasi-geodesic, and that ϕ has no dynamically parallel orbits. Then*

$$\sigma = D_\phi^\perp \cap \chi_\phi^{-1}(-1)$$

is a face of B_x.

To say that ϕ is pseudo-Anosov means that it has the same local behavior as the suspension flow of a pseudo-Anosov surface homeomorphism, although ϕ need not be a suspension flow itself.

To say that ϕ has no dynamically parallel orbits means that there is no annulus A, immersed in M, such that $\mathrm{int}(A)$ is embedded and transverse to ϕ, and each boundary component of A is mapped as a covering space to a periodic orbit of ϕ.

When ϕ is the suspension flow of a pseudo-Anosov surface homeomorphism, the conclusions of this theorem follow from work of Fried **[Fr2]**, without assuming any extra hypotheses. The examples constructed in this paper are used in **[Mos]**, in conjunction with the above theorem, to give new computations of faces of B_x.

Gabai's construction of flows transverse to finite depth foliations **[G2]** actually produces flows which are pseudo-Anosov. Together with his construction of finite depth foliations **[G1]**, this shows that pseudo-Anosov flows exist in great abundance on 3-manifolds. As a natural generalization to the examples described here, we ask:

Question. Given a pseudo-Anosov flow ϕ transverse to a finite depth foliation, is ϕ quasi-geodesic?

Sergio Fenley, in this thesis **[Fe]**, was able to generalize the results of **[CT]** to arbitrary depth one foliations. Thus, by blending the Cannon-Thurston-Fenley techniques with the methods of this paper, the above question is very likely to be true for all depth one foliations.

I conjecture that the above question is true in general. The key to the proof in the special case of depth 1 would be a result which follows from work of Thurston and Bonahon, that characterizes those incompressible surfaces in M which are quasi-fuchsian. To answer the above question affirmatively, one needs more delicate arguments which take place in certain 3-dimensional submanifolds of M; this is the subject of work in progress. If the question turns out to be true, then the theorem *Flows represent faces of* B_x will become much more widely applicable, although dynamically parallel orbits remain problematical.

Contents

1. The examples

1.1. Motivation from finite depth foliations. We begin with a motivational example, which requires a familiarity with finite depth foliations (see **[G1]** or **[CC]**). Consider a depth one foliation $\mathcal{F}$ with one compact leaf A. This means that the foliation $\mathcal{F} \,\big|\, M \quad A$ is a fibration over the circle. According to the theory of finite depth foliations, the fiber is a non-compact surface S with finitely many ends, and each end spirals into A. Let $g : S \to S$ denote the monodromy map, which is well-defined up to isotopy. Each end of S can be classified as either an attracting or a repelling end, depending on whether g moves points towards or away from the end.

The absolute simplest possible behavior occurs when g is a cocompact, free, properly discontinuous $\mathbb{Z}$-action. An example is the translation of a ladder surface depicted in diagram 1, having exactly one attracting and one repelling end. But in this case, it easily follows that M cut open along A is homeomorphic to $A \times I$, so that A is a fiber of a fibration over S^1.

To avoid such trivial behavior, we do not allow g to be isotopic to a translation. This can be guaranteed by requiring that the maximal compact invariant set $\Omega(g)$ is non-empty, over the entire isotopy class of g. The simplest kind of non-trivial $\Omega(g)$ is a single hyperbolic fixed point p, which we assume. Recall that a hyperbolic fixed point is smoothly modelled on the action of the map $h : \mathbb{R}^2 \to \mathbb{R}^2$ defined by $h(x, y) = (\lambda x, \lambda^{-1} y)$, where $\lambda > 1$, having a fixed point at the origin.

The restrictions on $\Omega(g)$ imply that p has an interesting invariant neighborhood under the action of g. Take the neighborhood N of the origin in $\mathbb{R}^2$ defined by the inequalities

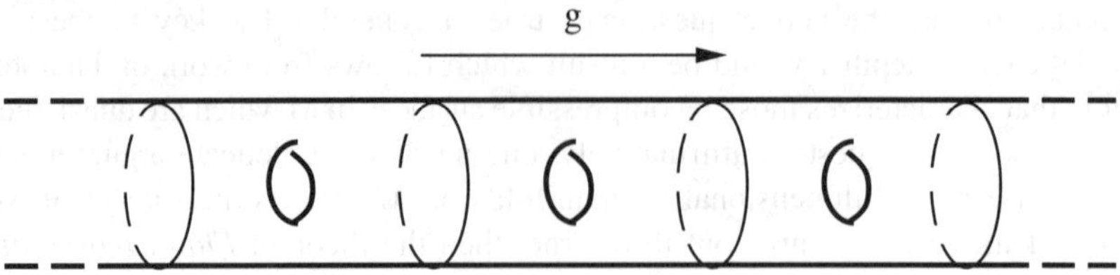

Diagram 1. A translation of a ladder surface

$-1 \leq x \cdot y \leq 1$. N is invariant under the action of the map h. There is a proper map $i : N \to S$ which is a homeomorphism onto its image, taking each of the two ends $y = \pm\infty$ of N to a repelling end of S, and taking each of the two ends $x = \pm\infty$ of N to an attracting end of S, such that i conjugates h to g. g acts on the invariant surface $S - \operatorname{int}(i(N))$ by a translation, with four boundary components which are invariant lines of g, glued to the four boundary components of $i(N)$.

Now consider the suspension flow of g, which is a flow ϕ_1 defined on $M - A$. Let M_1 denote the compact manifold obtained by putting a boundary on $M - A$. The boundary has two components, each homeomorphic to A, denoted $\partial_- M_1$ and $\partial_+ M_1$. The map g can be adjusted near the ends of S so that, after reparameterizing near the ends of $M - A$, ϕ_1 extends continuously to a flow on M_1 which points inward along $\partial_- M_1$ and outward along $\partial_+ M_1$. In more precise terminology, ϕ_1 is only a "semi-flow", which means that for a given $x \in M_1$, $\phi_1(x, t)$ is defined for some closed, connected set of values of t, not necessarily all of $\mathbb{R}$. The maximal invariant set of ϕ_1 is the set

$$\{x \mid \phi_1(x,t) \text{ is defined for all } t \in \mathbb{R}\}.$$

By construction, the maximal invariant set forms a single hyperbolic periodic orbit of ϕ_1, denoted C. The flow ϕ on M is induced from ϕ_1 by gluing $\partial_- M_1$ and $\partial_+ M_1$ back together.

The salient features of our motivational example are:

a) ϕ is transverse to a compact surface A.

b) There is a unique orbit C of ϕ that does not intersect A, and C is hyperbolic.

There is an additional feature which we shall henceforth ignore, namely that C is homologically non-trivial in M_1, which follows because C has algebraic intersection number 1 with S.

Gabai is able to prove that after performing a preliminary isotopy of the gluing map $\partial_+ M_1 \mapsto \partial_- M_1$, the flow ϕ is pseudo-Anosov. However, we shall not make use of the pseudo-Anosov property in this paper.

1.2. The actual example. To begin our actual example, we shall describe all possible flows satisfying properties a) and b), subject to the important restrictions that M is irreducible and A is incompressible. We shall also introduce conditions that guarantee M is atoroidal, which by Thurston's uniformization theorem for 3-manifolds **[Mor]** implies M is hyperbolic. These conditions are also designed to facilitate the proof of the quasi-geodesic property.

Every (untwisted) periodic hyperbolic orbit of a 3-dimensional flow has a canonical regular neighborhood, on which the flow is conjugate to the flow on a round handle. A round handle is a solid torus equipped with a particular flow, and is an invention of Dan Asimov **[A]**. Round handles were first used by Asimov to construct interesting flows on 3-manifolds. We begin with the definition of a round handle.

Notation. Given a flow ψ on a space X, $x \in X$, and $t \in \mathbb{R}$, we often write $x \cdot t$ as a shorthand for $\psi(x,t)$. Sometimes this notation will be used even when no specific symbol is introduced for the flow itself, as in the next paragraph.

First define a flow on $\mathbb{R}^3$ where

$$(x,y,z) \cdot t = (x \cdot \lambda^t, y \cdot \lambda^{-t}, z+t), \quad \lambda > 1.$$

Now take the restriction of this flow to the infinite solid cylinder $I \times I \times \mathbb{R}$, and mod out by a single translation in the z-direction. The result is a semi-flow ϕ_0 on a solid torus T with the following properties. ∂T decomposes into a disjoint union of two subsurfaces, $\partial_- T \cup \partial_+ T$, each of which is itself a disjoint union of two annuli. The common boundary of these two subsurfaces forms a curve system $\sigma(T)$ consisting of four parallel closed curves on ∂T, and each component of $\sigma(T)$ is a generator of $\pi_1(T) \approx \mathbb{Z}$. ϕ_0 points inward along $\partial_-(T)$, outward along $\partial_+(T)$, and is externally tangent along the curve $\sigma(T)$. External tangency along a curve is illustrated in diagram 2a). The maximal closed invariant set of ϕ_0 is a single untwisted hyperbolic periodic orbit, which is the quotient of the z-axis; we denote this orbit by $\mathrm{Core}(T)$, $C(T)$, or just C. The solid torus T, equipped with the flow ϕ_0, is called a round handle.

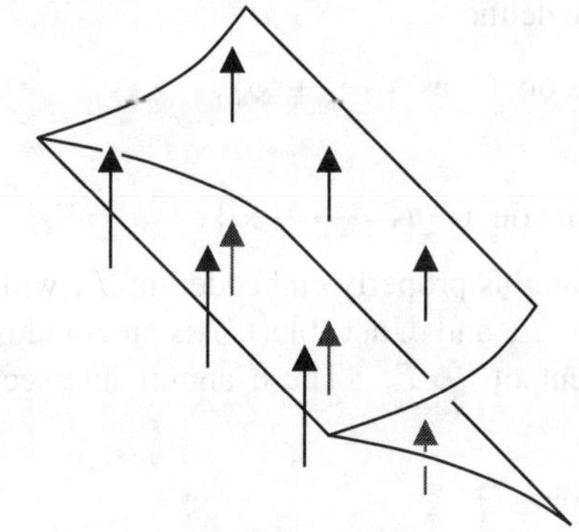

a) Curve tangency

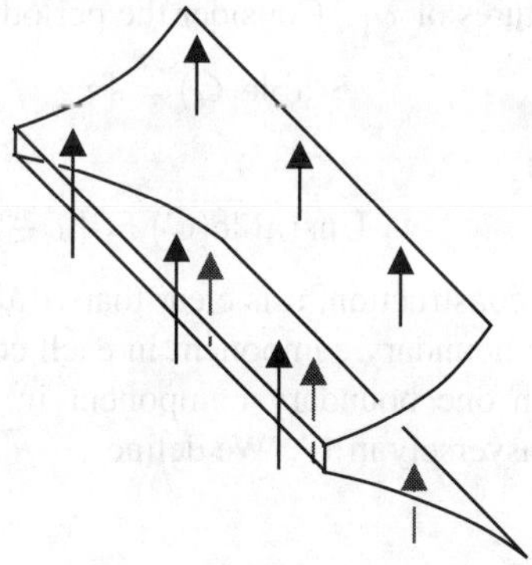

b) Annulus tangency

Diagram 2

We shall need a slight variation of this construction, that occurs quite commonly in the literature. By shaving off a tubular neighborhood of $\sigma(T)$, (still using T to denote the new object) we can replace each component of $\sigma(T)$ with an annulus in ∂T that is tangent to the flow, such that the flow lines leave T at each boundary component of this annulus. This is illustrated in diagram 2b). We refer to this collection of annuli as $A(T)$, and we write $R_\pm(T) = \partial_\pm(T) - \mathrm{int}(A(T))$. With this notation, T is a "sutured manifold" in the terminology of Gabai, and the components of $\sigma(T)$ are called the "sutures", as in **[G1]**. Notice that the flow on $A(T) \approx \sigma(T) \times I$ is isotopic to a product flow in the I direction.

Remark. In the motivational example described in §1.1, the suspension flow of g restricted to $i(N)$ yields a "peeled round handle", i.e. a round handle T with $R_\pm(T)$ removed, and with the flow near $R_\pm(T)$ slowed down so that it takes a flow line infinitely long to reach $R_\pm(T)$.

Let S be a compact oriented surface with four boundary components, such that no component of S is a disc. Put the product flow on $S \times I$. Let M_1 be obtained from $S \times I$ and T by gluing $\partial S \times I$ homeomorphically to $A(T)$ in such a way that the flows match up, up to reparameterization. We also require that M_1 be orientable, which is easily arranged. Let ϕ_1 be the induced flow on M_1. The maximal closed invariant set of ϕ_1 consists of the single periodic orbit $\text{Core}(T)$. Since no component of S is a disc, it follows easily that M_1 is irreducible and ∂M_1 is incompressible.

Finally, choose a homeomorphic gluing map $f : \partial_+(M_1) \to \partial_-(M_1)$. Let M be the 3-manifold obtained from M_1 by gluing. Let ϕ be the flow on M induced by ϕ_1. Let $A \subset M$ be the surface obtained as the quotient of $\partial_-(M_1)$ and $\partial_+(M_1)$. Clearly M is irreducible, and A is incompressible.

Now we formulate some extra conditions that are designed to insure that M is atoroidal, as well as to facilitate the proof of the quasi-geodesic property. First we require that the surface S has no component which is an annulus. This guarantees that M_1 is atoroidal, and ∂M_1 has no torus component.

In order to formulate the remaining condition, we must examine some of the dynamical features of ϕ_1. Consider the periodic orbit C. We define

$$\text{Stable}(C) = \{x \in M_1 \mid x \cdot t \text{ limits on } C \text{ as } t \to +\infty\},$$

and

$$\text{Unstable}(C) = \{x \in M_1 \mid x \cdot t \text{ limits on } C \text{ as } t \to -\infty\}.$$

By construction, it is clear that $\text{Stable}(C)$ is an annulus properly embedded in T, with one boundary component in each component of $\partial_- T$; and $\text{Unstable}(C)$ is an annulus with one boundary component in each component of $\partial_+ T$. These annuli intersect transversely in C. We define

$$\lambda_- = \partial_- M_1 \cap \text{Stable}(C),$$

and

$$\lambda_+ = \partial_+ M_1 \cap \text{Unstable}(C).$$

Each of these is a curve system with two components.

Our final condition is:

(∗) The two curve systems $f(\lambda_+)$ and λ_- intersect efficiently and bind the surface $\partial_- M_1$.

Efficient intersection means that no arc of $f(\lambda_+)$ is path homotopic to an arc of λ_-. It is easy to see that an arbitrary map f can be isotoped to satisfy this condition. Binding means that for any homotopically non-trivial closed curve $\gamma \subset \partial_- M_1$, every curve γ' homotopic to γ intersects at least one of $f(\lambda_+)$ or λ_-. Binding together with efficient intersection guarantees that each component of $M - \big(f(\lambda_+) \cup \lambda_-\big)$ is an even sided polygon with ≥ 4 sides, whose sides lie alternately in $f(\lambda_+)$ and λ_-.

Proposition 1.1. *M is atoroidal.*

Proof. Suppose there were an incompressible torus K. Since M_1 is atoroidal and its boundary has no torus component, then any torus isotopic to K must intersect A. By isotoping K, we can assume that

$$K \cap A = c_1 \cup c_2 \cup \ldots \cup c_N ,$$

where $c_1, c_2, \ldots, c_N$ is a family of simple closed curves, non-trivial in K, which are indexed by $\mathbb{Z}/N$ and written in circular order around K. Let R_n be the annulus in K separating c_n from c_{n+1}, for $n \in \mathbb{Z}/N$. Picture R_n as a properly embedded annulus in M_1. By a further isotopy of K, we can assume that each R_n is not ∂-parallel in M_1. Up to isotopy, there are two different types of non-∂-parallel annuli in M_1:

I) A product annulus in $S \times I$, i.e. $c \times I$ for some circle c in S.

II) An annulus in T connecting distinct boundary components of either $R_-(T)$ or $R_+(T)$.

Consider two adjacent annuli R_{n-1} and R_n, for some $n \in \mathbb{Z}/N$. These annuli share c_n as a boundary component. We show that this is impossible. After cutting open along A, there are curves c_n^+ in $\partial_+(M_1)$ and c_n^- in $\partial_-(M_1)$ that are the traces of c_n. By considering the two possible types of annuli described above, notice that by isotoping c_n^+, we can make it disjoint from λ_+. Similarly, we can make c_n^- disjoint from λ_-. But since $f(c_n^+) = c_n^-$, this contradicts the choice of f. □

It is possible to formulate weaker conditions on f that are necessary and sufficient for M to be atoroidal, but we will avoid this in order to make our examples simple.

In order to guarantee that ϕ not be a suspension flow, it is sufficient to show that ϕ has a homologically trivial periodic orbit. The only possible candidate is C, since every other periodic orbit has non-zero intersection number with A. There are two ways to arrange this.

The first way is for C to be homologically trivial in M_1, which happens if and only if some component of S has a single boundary component. This, however, rules out the motivational examples that arise from depth one foliations.

The other way is to arrange that there are oriented curves $\gamma_\pm \subset \partial_\pm M_1$ such that $f(\gamma_+) = -\gamma_-$, and $\gamma_+ + \gamma_- + C$ is a homologically trivial cycle in M_1. The way to proceed is as follows. First choose curves $\gamma_\pm$ so that $\gamma_+ + \gamma_- + C$ is homologically trivial. Then choose a preliminary gluing map f' so that $f'(\gamma_+) = -\gamma_-$. Unfortunately, f' may fail to satisfy property $(*)$. To fix this, choose an automorphism h of $\partial_+ M_1$ that acts trivially on first homology but mixes up the surface sufficiently. For example, there are pseudo-Anosov maps acting trivially on first homology. Finally, let $f = f' \circ h^N$, for some large value of N. Then C will be homologically trivial in M, and f will satisfy property $(*)$.

Remark. By the results of [**Mos**, section 6] the tangent vector field $T\phi$ is not even homotopic to the tangent vector field of a suspension flow.

2. Proof that ϕ is quasi-geodesic

To start the argument, we note that $\mathrm{Core}(T)$ is homotopically non-trivial in M, and so each connected lift is a quasi-geodesic.

Roughly speaking, the argument then proceeds like this. A quasi-geodesic is a curve that makes uniform progress away from its previous locations. Given an arbitrary flow line l of ϕ, as we follow along l for a while one of two things happens: either l spirals around very close to $\mathrm{Core}(T)$ or l crosses A many times. When l spirals around very close to the quasi-geodesic $\mathrm{Core}(T)$, it must spiral around many times, and so makes uniform progress in that part. And when l crosses A many times, it must make uniform progress because in the universal cover each component of A is a disc bounded by a quasi-circle, and these discs are (in some sense) uniformly separated from each other.

Now we develop a more rigorous argument. Let K be a meridian disc of the round handle T, transverse to the flow on T. For concreteness, we fix K to be the image, under the quotient map $I \times I \times \mathbb{R}$ to T, of $I \times I \times 0$. Let P be the period of the core of T.

Lemma 2.1. *There exists a number $Q > 0$ with the following property: if $l = x \cdot [0, t]$ is any flow segment with $t > Q$, and if $l \cap A = \emptyset$, then $l \subset T$, and l crosses K exactly n or $n+1$ times where $n \cdot P < t \leq (n+1) \cdot P$, with crossings only counted in* $\mathrm{int}(l)$.□

The number Q can be thought of as an upper bound to the return time to A, over all points outside of T.

We need some properties of connected lifts of A. Henceforth, the term "lift" will be used as a shorthand for "connected lift". It is evident that A is not a fiber of any fibration of M, because M_1, which is obtained by cutting M open along A, is not a product. This can be verified by showing explicitly that the inclusion induced map $\pi_1(\partial_-(M_1)) \to \pi_1(M_1)$ is not an isomorphism.

We shall be working throughout with the Poincaré disc model for $\mathbb{H}^3$, in which $\mathbb{H}^3$ as a set is identified with the open unit ball in $\mathbb{R}^3$, $\bar{\mathbb{H}}^3$ is the closed unit ball with the usual topology, and $\partial\bar{\mathbb{H}}^3 = S^2_\infty$ is the unit 2-sphere in $\mathbb{R}^3$, known as the sphere at infinity.

We shall need the following theorem:

Theorem 2.2. *Let M be a closed hyperbolic manifold, $A \subset M$ an incompressible surface. Then one of the following possibilities holds:*

i) *A is a fiber of a fibration of M, with base space either the interval orbifold I or the circle T^1.*

ii) *Under the inclusion $\pi_1(A) \subset \pi_1(M) \subset \mathrm{Isom}_+(\mathbb{H}^3)$, $\pi_1(A)$ is a quasi-fuchsian group.*

In conclusion ii), $\pi_1(A)$ is identified with the stabilizing subgroup of any connected lift $\widetilde{A} \subset \widetilde{M} \approx \mathbb{H}^3$ of A. We remind the reader that $\pi_1(A)$ is quasi-fuchsian if there is a quasi-conformal homeomorphism $h : S^2_\infty \to S^2_\infty$ such that $h^{-1} \circ \pi_1(A) \circ h$ is a

Fuchsian group. Thus, the inclusion $\mathbb{H}^2 \approx \widetilde{A} \subset \mathbb{H}^3$ is a quasi-isometry, and it extends to a properly embedded disc $\mathrm{cl}(\widetilde{A}) \subset \overline{\mathbb{H}}^3$, whose boundary is a quasi-circle.

This theorem is a consequence of the results of **[T]**, together with Bonahon's results of **[B]**, all applied to the covering space of M corresponding to $\pi_1(A)$. The theorem is not explicitly stated in **[T]**, although it can be pieced together from scattered results from chapters 8 and 9 of **[T]**, together with Bonahon's theorem. In particular, we refer the reader to Propositions 8.3.4, 8.4.3, 9.2.2 of **[T]**, and Theorem A of **[B]**. □

Let $\widetilde{A}_1$, $\widetilde{A}_2$ be any two lifts of A. Choose points $x_n \in \widetilde{A}_n$, and a path $\widetilde{\rho}$ from x_1 to x_2. Let ρ be the projection of $\widetilde{\rho}$ into M. Define $N(\widetilde{A}_1, \widetilde{A}_2)$ to be the algebraic intersection number with A of any closed curve obtained by closing up ρ using a path staying entirely in A; clearly this is well-defined. If, furthermore, ρ can be chosen to be a flow segment, then we say that $\widetilde{A}_2$ is flow accessible from $\widetilde{A}_1$. More generally, given any two subsets $B,\ C \subset \mathbb{H}^3$, we say that C is flow accesible from B if there is a flow segment $x \cdot [b, c]$ such that $x \cdot b \in B$ and $x \cdot c \in C$.

Proposition 2.3. *Suppose that* $\widetilde{A}_2$ *is flow accessible from* $\widetilde{A}_1$. *If* $N(\widetilde{A}_1, \widetilde{A}_2) \geq 2$, *then*

$$\mathrm{cl}(\widetilde{A}_1) \cap \mathrm{cl}(\widetilde{A}_2) = \emptyset.$$

Proof. We need a standard lemma about Kleinian groups.

Lemma 2.4. *Let* Γ *be a cocompact, discrete, properly discontinuous group of hyperbolic isometries. Let* $G,\ H$ *be two quasi-fuchsian subgroups of* Γ. *Let* $K = G \cap H$. *Let* $\Lambda(Q)$ *denote the limit set of a group* Q. *Then*

$$\Lambda(G) \cap \Lambda(H) = \Lambda(K).$$

Proof. Suppose that $z \in \Lambda(G) \cap \Lambda(H)$. Let $D,\ E$ be proper, quasi-isometric images of $\mathbb{H}^2$ in $\mathbb{H}^3$ on which $G,\ H$ respectively act as universal covering groups of closed hyperbolic surfaces. Choose half-geodesics $l \subset D \approx \mathbb{H}^2$ and $m \subset E \approx \mathbb{H}^2$ such that l, m both converge to z, and note that l and m are quasi-geodesics in $\mathbb{H}^3$. Choose base points $\xi \in D$ and $\eta \in E$. For some sufficiently large number $R > 0$, we can choose sequences $x_n \in l,\ y_n \in m,\ g_n \in G$, and $h_n \in H$, such that the following conditions hold:

$$x_n \to z \text{ and } y_n \to z \text{ as } n \to \infty;$$

$$d(x_n, y_n) < R;$$

$$d(g_n(\xi), x_n) < R \text{ and } d(h_n(\eta), y_n) < R.$$

Hence $d\big((h_n^{-1} \circ g_n)(\xi), \eta\big)$ is uniformly bounded for all n, and so the sequence $h_n^{-1} \circ g_n$ takes on only finitely many values in the group Γ. By passing to a subsequence, we can assume that the entire sequence is constant, i.e. that

$$h_0^{-1} \circ g_0 = h_n^{-1} \circ g_n \text{ for all } n.$$

It follows that

$$\vartheta_n = h_n \circ h_0^{-1} = g_n \circ g_0^{-1}$$

is a sequence of elements of the group $K = G \cap H$. Moreover, the sequence ϑ_n operating on the point $g_0(\xi)$ limits on z, so $z \in \Lambda(K)$. This finishes the proof that $\Lambda(G) \cap \Lambda(H) \subset \Lambda(K)$.

The other inclusion is obvious. □

Proof of 2.3 (continued). Choose a base point $\widetilde{x}$ for $\mathbb{H}^3$, so that $\pi_1(M)$ is canonically identified with a group of hyperbolic isometries. The stabilizer subgroups of $\widetilde{A}_1$ and $\widetilde{A}_2$ are identified with quasi-fuchsian subgroups G_1 and G_2 of $\pi_1(M)$ by Theorem 2.2; incidently, these subgroups are conjugate. According to the above lemma, we need only show that $G_1 \cap G_2 = \{1\}$.

The subgroups G_1 and G_2 can be characterized in the following manner. Let $\widetilde{\rho}$ be a flow segment from $\widetilde{A}_1$ to $\widetilde{A}_2$, and for convenience choose $\widetilde{x}$ to be the endpoint of $\widetilde{\rho}$ in $\widetilde{A}_1$. Let $\widetilde{y}$ be the other endpoint of $\widetilde{\rho}$ in $\widetilde{A}_2$. Let ρ, x, y be the images in M. Then G_1 can be identified with $\pi_1(A, x)$, and G_2 can be identified with $\rho * \pi_1(A, y) * \rho^{-1}$, where "$*$" denotes the path concatenation operator.

Suppose there is a non-trivial element in $G_1 \cap G_2$. Then it follows easily that there is an immersed annulus $R : S^1 \times I \to M$ such that $S^1 \times 0$ and $S^1 \times 1$ go to non-trivial curves in A, and $R(\xi \times I)$ is identical with the flow segment ρ, where $\xi \in S^1$ is some base point. By transversality, we can jiggle R so that $R^{-1}(A)$ consists of the two boundary components, a single core curve $C = S^1 \times \frac{1}{2}$ of the annulus, several homotopically trivial closed curves disjoint from $\xi \times I$, and several properly embedded arcs disjoint from $\xi \times I$.

Recall that M_1 is obtained by gluing T to $S \times I$, and M is obtained from M_1 by identifying $\partial_+(M_1)$ and $\partial_-(M_1)$ by a gluing homeomorphism f. Since A is incompressible, we can homotop R relative to $(S^1 \times \partial I) \cup C$, so as to eliminate the arcs and trivial curves; and then we can perform a similar homotopy to arrange that $\text{image}(R) \cap T = \emptyset$. By a further homotopy relative to $S^1 \times \partial I$, we can arrange so that R restricts to maps $R_1 : S^1 \times [0, \frac{1}{2}] \to S \times I$ and $R_2 : S^1 \times [\frac{1}{2}, 1] \to S \times I$ that are immersed product annuli. But this clearly contradicts condition $(*)$ on f, because $R_1(S^1 \times \frac{1}{2})$ is a non-trivial curve in $\partial_+(M_1)$ missing λ_+, which is taken by f to the curve $R_2(S^1 \times \frac{1}{2})$, non-trivial in $\partial_-(M_1)$ and missing λ_-. □

Let $\mathcal{F}$ be the collection of all flow segments with both endpoints on A, parameterized by flow time. We make $\mathcal{F}$, and its subspaces to be defined, into a topological space by imposing the compact open topology. Let $\mathcal{F}^1$ consist of all $\alpha \in \mathcal{F}$ such that $|\text{int}(\alpha) \cap A| = 1$. Define maps $k_i : \mathcal{F}^1 \to A$, for $i \in \{0, 1, 2\}$, where $k_i(\alpha)$ is the i-th point of intersection of α with A.

To complement Proposition 2.3, we shall make a study of the structure of $\mathcal{F}^1$. For each $n = 0, 1, 2, 3, \ldots$, let λ_+^n consist of all $x \in A$ such that the flow line going backwards from x intersects A less than n times, and define λ_-^n similarly using forward time. Our definition gives $\lambda_-^0 = \lambda_+^0 = \emptyset$. Clearly λ_-^1 and λ_+^1 are each curve systems consisting of a pair of simple closed curves. In fact, λ_-^1 is the image under the quotient map $M_1 \to M$ of the curve system $\lambda_- \subset \partial_-(M_1)$, and λ_+^1 is the image of $\lambda_+ \subset \partial_+(M_1)$. The hypothesis on the gluing map $f : \partial_+(M_1) \to \partial_-(M_1)$ implies that λ_-^1 and λ_+^1 bind A.

In addition, the reader should have no trouble verifying that

$$\lambda^1_+ \subset \lambda^2_+,$$

and that λ^2_+ is a lamination on A obtained in the following manner. Let k be the closure of a component of $\lambda^1_- - \lambda^1_+$, an arc properly embedded in the closure of a component of $A - \lambda^1_+$. Each component of λ^1_+ inherits an orientation from $\mathrm{Core}(T)$, by isotopy through the annulus $\mathrm{Unstable}(T) \subset T$. Let k' be obtained from k by spinning the endpoints of k around λ^1_+ infinitely in the direction opposite to the preferred orientation on λ^1_+. This spinning operation can be performed simultaneously on all components of $\lambda^1_- - \lambda^1_+$, and the result is the lamination λ^2_+. From this description, it is clear that the only compact leaves of λ^2_+ are those contained in λ^1_+, and that each non-compact leaf spirals into a compact leaf. Also, there are only finitely many non-compact leaves. From this it follows that $M - \lambda^2_+$ has only finitely many components.

By a similar argument, one can prove the analogous facts about

$$\lambda^1_- \subset \lambda^2_-,$$

except that components of $\lambda^1_+ - \lambda^1_-$ are spun around components of λ^1_- in a direction that agrees with the preferred orientation that λ^1_- inherits from $\mathrm{Core}(T)$.

The following proposition is an easy consequence of this discussion:

Proposition 2.5. *For each $i \in \{0,1,2\}$, $k_i : \mathcal{F}^1 \to A$ is a homeomorphism onto its image, and the image consists exactly of $A - (\lambda^i_+ \cup \lambda^{2-i}_-)$.* □

Corollary 2.6. *$\mathcal{F}^1$ has finitely many connected components, in $1-1$ correspondence under the map k_i with the components of $A - (\lambda^i_+ \cup \lambda^{2-i}_-)$.* □

Lemma 2.7. *Let $l = x \cdot \mathbb{R}$ be any flow line of ϕ, $\widetilde{l}$ a lift of l in $\mathbb{H}^3$. Then the ends of $\widetilde{l}$ converge to distinct points on the sphere at infinity.*

Proof. If l has an end that is eventually disjoint from A, then that end must eventually stay in the round handle forever, and therefore is asymptotic to $\mathrm{Core}(T)$. Thus, the associated end of $\widetilde{l}$ converges to an endpoint of the appropriate lift of $\mathrm{Core}(T)$.

Suppose l has an end that crosses A infinitely many times. The corresponding end of $\widetilde{l}$ thus crosses an infinite sequence $\widetilde{A}_n$ of distinct lifts of A, and clearly $\widetilde{l}$ crosses each $\widetilde{A}_n$ exactly once. According to the result of Thurston quoted above, each lift of A extends to a properly embedded disc in $\bar{\mathbb{H}}^3$, whose boundary is a quasi-circle. Since only finitely many lifts of A can intersect any compact subset of $\mathbb{H}^3$, it follows by equivariance of the action of $\pi_1(M)$ that the Euclidean diameter of $\widetilde{A}_n$ approaches 0 as $n \to \infty$. This shows that the end of $\widetilde{l}$ under consideration limits on a unique point in $\partial\bar{\mathbb{H}}^3$.

Let $x, y \in \partial\bar{\mathbb{H}}^3$ be the two endpoints of $\widetilde{l}$. We must show that $x \neq y$. There are several cases, depending on the cardinality of $l \cap A$.

If l crosses A zero times, then l is the closed non-trivial loop $\mathrm{Core}(T)$, so $\widetilde{l}$ is a quasi-geodesic and $x \neq y$.

If l crosses A one time, then there is a pair of non-trivial simple closed curves $C_1 \subset \lambda^1_-$, $C_2 \subset \lambda^1_+$, a lift $\widetilde{A}$ of A, and lifts $\widetilde{C}_1$, $\widetilde{C}_2$ of C_1, C_2 in $\widetilde{A}$, such that the two ends of $\widetilde{l}$ converge, respectively, to ends of $\widetilde{C}_1$ and $\widetilde{C}_2$. In the hyperbolic structure on $\widetilde{A} \approx \mathbb{H}^2$, the endpoints of $\widetilde{C}_1$ and $\widetilde{C}_2$ are disjoint in $\partial\bar{\mathbb{H}}^2$, since C_1, C_2 are non-isotopic. Since $\mathbb{H}^2 \approx \widetilde{A} \subset \mathbb{H}^3$ is a quasi-disc, it follows that the ends of $\widetilde{l}$ are disjoint in $\partial\bar{\mathbb{H}}^3$.

If l crosses A two or more times, then there are lifts $\widetilde{A}_1$, $\widetilde{A}_2$, with $\widetilde{A}_2$ flow accessible from $\widetilde{A}_1$ and $N(\widetilde{A}_1, \widetilde{A}_2) \geq 2$, such that if R is the open annulus separating $\widetilde{A}_1$ and $\widetilde{A}_2$ (whose existence is guaranteed by Proposition 2.3), then the endpoints of l are in distinct components of $\partial\bar{\mathbb{H}}^3 - R$. □

Since $\partial\widetilde{l}$ consists of two distinct points, there is a geodesic g such that $\partial g = \partial\widetilde{l}$. The following lemma can be used to simplify the verification that $\widetilde{l}$ is a quasi-geodesic.

Lemma 2.8. *Suppose that m is a proper embedding of $\mathbb{R}$ in $\mathbb{H}^3$, g is a geodesic, the ends of m converge respectively to the ends of g, and there exists $K > 0$ such that each point x of m is within distance K of some point of g. Then m is a K-quasi-geodesic.*

Proof. We must show that each point y of g is within distance K of a point of m. For each $x \in m$, let $\rho(x) \in g$ denote the point of g closest to x. Clearly ρ is continuous, and the hypotheses imply that ρ is a proper map, taking the ends of m respectively to the ends of g. Thus, ρ is onto. Since $d(x, \rho(x)) < K$ for all x, the lemma follows. □

We shall use this without comment in what follows.

Choose a point $\widetilde{x}$ in $\widetilde{l}$. We must exhibit an upper bound K for the distance from $\widetilde{x}$ to g, that is independent of the choice of $\widetilde{x}$. Incidently, the bound K will also be independent of $\widetilde{l}$, which shows that ϕ is a "uniformly" quasi-geodesic flow. The key to finding the bound K is the following lemma, which allows us to pin down the location of some point of g, to within a bounded distance.

Lemma 2.9. *Let Q_0, Q_1 be disjoint compact subsets of $\partial\bar{\mathbb{H}}^3$. Then there exists a compact subset $W \subset \mathbb{H}^3$ such that for each geodesic g having one endpoint in Q_0 and the other endpoint in Q_1, $g \cap W \neq \emptyset$.*

Proof. The proof is a simple convexity argument.

Choose a continuous function $\theta : \bar{\mathbb{H}}^3 \to [0,1]$ such that $Q_0 \subset \theta^{-1}(0)$ and $Q_1 \subset \theta^{-1}(1)$. Let $X = \theta^{-1}(\frac{1}{2})$, and let $\partial X = X \cap \partial\bar{\mathbb{H}}^3$. Note that ∂X is a compact subset of $\partial\bar{\mathbb{H}}^3$ disjoint from Q_0 and Q_1. Thus, ∂X can be covered by finitely many closed round discs $D_i \subset \partial\bar{\mathbb{H}}^3$ such that each D_i is disjoint from Q_0 and Q_1. Let P_i be the totally geodesic plane in $\mathbb{H}^3$ bounded by ∂D_i, and let $H_i \subset \bar{\mathbb{H}}^3$ be the closed topological ball with boundary $D_i \cup P_i$. If g is any geodesic with one boundary point in Q_0 and the other in Q_1, it follows by convexity that $g \cap H_i = \emptyset$ for each i. Since $X \cap g \neq \emptyset$ for each such g, then clearly $W = \mathrm{cl}(X - \cup_i H_i)$ intersects g. Also, W is a compact subset of $\bar{\mathbb{H}}^3$, and by construction it is disjoint from $\partial\bar{\mathbb{H}}^3$. □

In order to apply this lemma, we must consider several cases. We adopt the notation from before Lemma 2.9. Let x be the projection of $\widetilde{x}$ in M.

Case 1. Suppose there exist numbers a, b such that

$$a \leq 0 \leq b,$$

$$x \cdot [a,b] \subset T,$$

$$x \cdot a \in \partial_-(T),$$

$$x \cdot b \in \partial_+(T),$$

$$b - a \geq \vartheta,$$

where ϑ is a constant to be determined later.

Fix a lift $\widetilde{T}$ of T, and identify $\pi_1(T) \approx \mathbb{Z}$ with the subgroup of $\pi_1(M)$ leaving $\widetilde{T}$ invariant. By equivariance, we can assume that $\widetilde{x} \in \widetilde{T}$. Let $\widetilde{C}$ be the lift of $\mathrm{Core}(T)$ with $\widetilde{C} \subset \widetilde{T}$. Let $\widetilde{A}_0$ be the lift of A containing $\widetilde{x} \cdot a$, and let $\widetilde{A}_1$ be the lift containing $\widetilde{x} \cdot b$. We write x_a, x_b as shorthand for $\widetilde{x} \cdot a$ and $\widetilde{x} \cdot b$, and X as shorthand for the flow segment $\widetilde{x} \cdot [a,b]$.

This is the case where the flow segment X stays close to $\mathrm{Core}(T)$, so in some sense it is obvious that X is quasi-geodesic. However, we are trying to prove that X stays close to the specific geodesic g, so there is a certain amount of technical work to do in order to pin down the location of g.

Clearly there are points y_a, $y_b \in \widetilde{C}$, and a constant K_1 independent of x, a, b, such that

$$d(x_a, y_a) \leq K_1,$$

$$d(x_b, y_b) \leq K_1,$$

and X is within K_1 of the segment $[y_a, y_b]$ of $\widetilde{C}$. We shall show that there is another constant K_2, independent of choices, and points z_a, $z_b \in g$, such that $d(x_a, z_a), d(x_b, z_b) \leq K_2$. Since $\widetilde{C}$ is itself a quasi-geodesic, it will follow that there is a constant K, independent of choices, such that X is within K of the segment $[z_a, z_b]$ of g. Thus, $\widetilde{x} \in X$ is within distance K of g.

The component of $\partial_-(\widetilde{T})$ containing x_a is a surface $P_0 \subset \widetilde{A}_0$. P_0 supports a cocompact action by the group $\pi_1(T)$, whose quotient is the component of $\partial_-(T)$ containing x_a. Divide P_0 into fundamental domains, and let D_0 be the fundamental domain containing x_a. Note that there are at most finitely many lifts $\widetilde{A}'$ of A, such that D_0 is flow accessible from $\widetilde{A}'$, and $N(\widetilde{A}', D_0) = 1$; this is a consequence of Corollary 2.6. Each $\widetilde{A}'$ has the property that $\widetilde{A}_1$ is flow accessible from it, and $N(\widetilde{A}', \widetilde{A}_1) = 2$. Thus, $\mathrm{cl}(\widetilde{A}')$ is always disjoint from $\partial\widetilde{C}$ by Proposition 2.3, since $\partial\widetilde{C} \subset \mathrm{cl}(\widetilde{A}_1)$. Let $Q(\widetilde{A}')$ be that closed disc in $\partial\mathbb{H}^3$, bounded by the curve $\partial(\mathrm{cl}(\widetilde{A}'))$, which is disjoint from $\partial\widetilde{C}$. Let Q_0 be the union of $Q(\widetilde{A}')$, over the finitely many choices for $\widetilde{A}'$. Q_0 is a compact set disjoint from $\partial\widetilde{C}$. If we follow $\widetilde{l}$ backwards from x_a, the endpoint must lie in Q_0. This is true because $\widetilde{l}$ either crosses some $\widetilde{A}'$, or is asymptotic to a geodesic in some $\widetilde{A}'$.

In a similar manner, the component of $\partial_+(\widetilde{T})$ containing x_b is a surface $P_1 \subset \widetilde{A}_1$, acted upon by $\pi_1(T) \approx \mathbb{Z}$, and has a fundamental domain D_1 with $x_b \in D_1$. From D_1 we can construct a compact set Q_1, disjoint from $\partial\widetilde{C}$, such that if we follow the flow line forwards from $\widetilde{x}_b$, the endpoint must be in Q_1. The construction of Q_0 (resp. Q_1)

depends only on the fundamental domain D_0 (resp. D_1), and the choice is equivariant with respect to the action of $\pi_1(T)$.

Once and for all, divide $\partial\bar{\mathbb{H}}^3 - \partial\widetilde{C}$ into annular fundamental domains for the action of $\pi_1(T)$. From the construction of Q_0 and Q_1, together with Lemma 2.1, it follows that ϑ can be chosen sufficiently large so that Q_0 and Q_1 are separated by at least one annular fundamental domain for the action of $\pi_1(T)$. We remark that ϑ is independent of all choices made in this argument, in particular ϑ is independent of the choice of $\widetilde{l}$ and the point $\widetilde{x} \in \widetilde{l}$.

Let $R_0 \subset \partial\bar{\mathbb{H}}^3 - \partial\widetilde{C}$ be the fundamental domain for the action of $\pi_1(T)$ lying closest to Q_0, subject to the condition that R_0 separates Q_0 from Q_1. Let Q_i' be the component of $\partial\bar{\mathbb{H}}^3 - R_0$ containing Q_i. g has its endpoints in Q_0', Q_1' respectively. Applying Lemma 2.9, it follows that there is a constant K_2 such that every geodesic connecting Q_0' to Q_1' comes within K_2 of each point in the compact set D_0. In particular, there is a point $z_a \in g$ within distance K_2 of x_a. Since Q_0' and Q_1' depend only on D_0, and since the dependence is natural with respect to the action of $\pi_1(T)$, it follows that K_2 is independent of all choices. A similar argument shows that K_2 can also be chosen so that some point z_b of g comes within K_2 of x_b.

Case 2. Suppose that there exists numbers a, b, c such that

$$a \leq 0 \leq c,$$

$$a \leq b \leq c,$$

$$x \cdot [a, c] \cap A = \{x \cdot a, x \cdot b, x \cdot c\},$$

$$b - a \leq \vartheta, \qquad c - b \leq \vartheta,$$

where ϑ is as in Case 1.

The argument for this case is similar to, but simpler than, the argument for case 1. Fix once and for all a lift $\widetilde{A}_0$ of A and a fundamental domain D for $\widetilde{A}_0$. By equivariance, we can assume that $\widetilde{x} \cdot b \in D$. Thus, $\widetilde{x}$ is contained in the compact set $D \times [-\vartheta, \vartheta]$.

By Corollary 2.6 there are only finitely many pairs of lifts A_i^-, A_i^+ such that $\widetilde{x} \cdot a \in A_i^-$ and $\widetilde{x} \cdot c \in A_i^+$. Moreover, since A_i^+ is flow accessible from A_i^- and $N(A_i^-, A_i^+) = 2$, it follows from Proposition 2.3 that $\partial A_i^- \cap \partial A_i^+ = \emptyset$. Thus, there are disjoint closed discs $Q_i^-, Q_i^+ \in \partial\bar{\mathbb{H}}^3$ such that $\partial Q_i^- = \partial A_i^-$ and $\partial Q_i^+ = \partial A_i^+$. It is clear that the flow line $\widetilde{l}$ has one endpoint in Q_i^- and the other in Q_i^+, so the same is true of the geodesic g. From Lemma 2.9 it follows that there is a number K_i such that any geodesic connecting Q_i^- and Q_i^+ comes within distance K_i of the compact set $D \cdot [-\vartheta, \vartheta]$. In particular, some point of g comes within distance K_i of the point $\widetilde{x}$. Now choose K to be an upper bound for the finite set of numbers K_i.

Case 3. Suppose neither Case 1 nor Case 2 holds. Then there is some number a with $|a| \leq 2\vartheta$ such that $y = x \cdot a$ satisfies the hypothesis of Case 1. By Case 1, $\widetilde{y} = \widetilde{x} \cdot a$ is within a bounded distance K' of the geodesic g. By equivariance of the action of $\pi_1(M)$ on $\mathbb{H}^3$, together with compactness of M, there is a number K'' such that for each flow segment X of $\widetilde{\phi}$ whose flow time is bounded by 2ϑ, $\operatorname{diam}(X) \leq K''$. Thus, $\widetilde{x}$ is within distance $K = K' + K''$ of g.

Bibliography

[A] D. Asimov, "Round handles and non-singular More-Smale flows", Annals of Math. 102 (1979), 41-54.

[B] F. Bonahon, "Bouts des variétés hyperboliques de dimension 3", Annals of Math. 124 (1986), 71–154.

[CC] J. C. Cantwell & L. W. Conlon, "Poincaré-Bendixon theory for leaves of codimension one", Trans. Amer. Math. Soc. 265 (1981) 181–209.

[CT] James W. Cannon & William P. Thurston, "Group invariant Peano curves", preprint.

[Fe] Sérgio Roberto Fenley, "Depth one foliations in hyperbolic 3-manifolds", Princeton University disseration, October 1989.

[Fr1] D. Fried, "The geometry of cross sections to flows", Topology, vol. 21 no. 4 (1982), pp. 353–371.

[Fr2] D. Fried. "Fibrations over S^1 with pseudo-Anosov monodromy". Exposé 14 of "Travaux de Thurston sur les surfaces". Astérisque 66-67 (1979), 251–266.

[G1] D. Gabai, "Foliations and the Topology of 3-manifolds", J. Diff. Geom. 18 (1983), pp. 445–503.

[G2] D. Gabai, "Laminations transverse to finite depth foliations", to appear.

[Mor] J. Morgan, "On Thurston's uniformization theorem for 3-dimensional manifolds", in The Smith Conjecture, ed. by J. Morgan and H. Bass, Academic Press, 1984, pp. 37–125.

[Mos] L. Mosher, "Dynamical systems and the homology norm of a 3-manifold", preprint.

[T] W. P. Thurston, "The geometry and topology of three-manifolds", Princeton University Press, preprint.

Lee Mosher, Department of Mathematics and Computer Science, Rutgers University, Newark, NJ 07102
Email: dpsbb@cunyvm.bitnet

Combinatorics of Triangulations and the Chern-Simons Invariant for Hyperbolic 3-Manifolds

Walter D. Neumann

1. Introduction

In this paper we prove some results on combinatorics of triangulations of 3-dimensional pseudo-manifolds, improving on results of **[NZ]**, and apply them to obtain a simplicial formula for the Chern-Simons invariant of an ideally triangulated hyperbolic 3-manifold. Combining this with **[MN]** gives a simplicial formula for the η invariant also.

In effect, the main ingredient in the formula is the sum of the "Rogers dilogarithm" of the complex parameters of the ideal tetrahedra of the triangulation, but the choice of the appropriate branch of the Rogers dilogarithm for each simplex involves unexpected combinatorics (cf. Remark 4 below for this interpretation of the formula).

The combinatorial part of this paper (Sects. 4–6) is self-contained and of independent interest. For instance, T. Yoshida **[Y2]** has used these combinatorics (in the version of **[NZ]**) to study character varieties and boundary slopes in the spirit of Culler-Shalen **[CS]**.

In the remainder of this Introduction we summarize the application to the Chern-Simons invariant. All manifolds in this paper are assumed to be oriented.

If M is a complete hyperbolic 3-manifold which is compact, then its Chern-Simons invariant $\mathrm{CS}(M)$ is well-defined modulo $2\pi^2$. If M is non-compact then Bob Meyerhoff has shown in **[M]** that there is still a natural definition of $\mathrm{CS}(M)$ which is well-defined modulo π^2. Let $\mathcal{V}(M) = \mathrm{Vol}(M) + i\,\mathrm{CS}(M)$, which is well-defined modulo $i2\pi^2\mathbb{Z}$ or $i\pi^2\mathbb{Z}$.

A formula for $\mathcal{V}(M')$ mod $i\pi^2\mathbb{Z}$, as M' varies over the hyperbolic Dehn surgeries on M, was conjectured in **[NZ]** and proved by T. Yoshida in **[Y1]**. This formula is of theoretical interest but is not practical for actually computing $\mathcal{V}(M')$. It is not hard to reverse the derivation in **[NZ]** to obtain a computable formula in terms of an ideal triangulation of M. However, the resulting formula involves an unknown constant which depends on the combinatorics of the triangulation of M and which seems hard to determine in general. Using a result of Dupont **[D]** we can find a version of the formula in which this constant is at least a rational multiple of $i\pi^2$ (Theorem 1 below). Using a more careful analysis of the relevant combinatorics we are able to give a version (Theorem 2) in which the constant is conjecturally in $(i\pi^2/6)\mathbb{Z}$ and is thus determined up to a six-fold ambiguity (since it lives in $\mathbb{C}$ mod $i\pi^2\mathbb{Z}$).

Suppose M has an ideal triangulation which subdivides it into n ideal tetrahedra

$$M = \Delta_1 \cup \ldots \cup \Delta_n .$$

Choose an edge of the j-th tetrahedron and let z_j^0 be the complex parameter which then describes this tetrahedron (we are following the notation of **[NZ]** which may be consulted for more details — see also Sect. 2). Let

$$\mathcal{Z}^0 = \begin{pmatrix} \log z_1^0 \\ \vdots \\ \log z_n^0 \\ \log(1-z_1^0) \\ \vdots \\ \log(1-z_n^0) \end{pmatrix}.$$

Recall from **[NZ]** (see also Sect. 2) that if M has h cusps then the z_j^0 are determined by so-called consistency and cusp relations which can be written in the form

$$U\mathcal{Z}^0 = \pi i \mathbf{d},$$

where U is a certain integral $(n+2h)\times 2n$-matrix and

$$\mathbf{d} = \begin{pmatrix} d_1 \\ \vdots \\ d_{n+2h} \end{pmatrix}$$

is some integral vector. The equation

$$U\mathbf{c} = \mathbf{d} \qquad (*)$$

has a solution $\mathbf{c} = \mathcal{Z}^0/\pi i \in \mathbb{C}^{2n}$. Since U is an integral matrix, $(*)$ also has solutions

$$\mathbf{c} = \begin{pmatrix} c'_1 \\ \vdots \\ c'_n \\ c''_1 \\ \vdots \\ c''_n \end{pmatrix} \in \mathbb{Q}^{2n}.$$

We shall see that solutions $\mathbf{c}$ can be found in $\mathbb{Z}^{2n}$, in fact even in a certain sublattice of $\mathbb{Z}^{2n}$.

Let M' be the result of a hyperbolic Dehn surgery on M obtained by deforming the parameters z_j^0 to new values z_j (cf. e.g., **[NZ]**). Topologically M' differs from M in that a new closed geodesic γ_j has been added at the j-th cusp for some $j \in \{1, \dots, h\}$. Let λ_j be the complex number which has real part equal to the length of this geodesic and imaginary part equal to its torsion (the latter is only well-defined modulo 2π). If no geodesic has been added at the j-th cusp we put $\lambda_j = 0$.

Let $\mathcal{R}(z)$ be the "Rogers dilogarithm function," which is related to the usual dilogarithm Li_2 by

$$\begin{aligned} \mathcal{R}(z) &= \frac{1}{2}\log(z)\log(1-z) + \mathrm{Li}_2(z) \\ &= \frac{1}{2}\log(z)\log(1-z) - \int_0^z \frac{\log(1-t)}{t}\,dt. \end{aligned}$$

Theorem 1. *Given any solution* $\mathbf{c} \in \mathbb{C}^{2n}$ *to* $(*)$, *there exists a constant* $\alpha = \alpha(\mathbf{c})$ *such that if* M' *is the result of hyperbolic Dehn surgery on* M, *then*

$$\mathcal{V}(M') = \alpha - \frac{\pi}{2}\sum_{j=1}^{h}\lambda_j - i\sum_{\nu=1}^{n}\Big(\mathcal{R}(z_\nu) - \frac{i\pi}{2}(c'_\nu \log(1-z_\nu) - c''_\nu \log(z_\nu))\Big).$$

Moreover, if $\mathbf{c} \in \mathbb{R}^{2n}$ *then* α *is pure imaginary and if* $\mathbf{c} \in \mathbb{Q}^{2n}$ *then* $\alpha \in i\pi^2\mathbb{Q}$.

In Sect. 2 we will define a "parity condition" on $\mathbf{c} \in \mathbb{Z}^{2n}$ which depends on the combinatorics of our situation.

Theorem 2. *There exist solutions* $\mathbf{c}$ *to* $(*)$ *in* $\mathbb{Z}^{2n}$ *satisfying the parity condition. For such* $\mathbf{c}$, *the constant* α *of Theorem* 1 *is well-defined (i.e., independent of* $\mathbf{c}$*) modulo* $i\pi^2/2$ *(and is conjecturally an integer multiple of* $i\pi^2/6$*).*

The formula of Theorem 1 is a fairly easy application of the methods in **[NZ]**; we prove it in Sect. 3, along with the rationality statement for α and the second sentence of Theorem 2. The existence of $\mathbf{c}$ as in Theorem 2 needs the combinatorial results of Sects. 4–6 and is proved in Sect. 6.

Remarks. 1. If one drops the parity condition in Theorem 2 then α is well-defined modulo $i\pi^2/4$ instead of $i\pi^2/2$ and is conjecturally an integral multiple of $i\pi^2/12$.

2. Denote $Sol_0 = \{\mathbf{c} \in \mathbb{Z}^{2n} : U\mathbf{c} = \mathbf{d}\}$ and $Sol = \{\mathbf{c} \in Sol_0 : \mathbf{c}$ satisfies the parity condition$\}$. It follows from Theorem 4.2 that $Sol_0/Sol \cong H^1(M^*;\mathbb{Z}/2) = Hom(\Gamma/P, \mathbb{Z}/2)$, where M^* is the end compactification of M and P is the subgroup of $\Gamma = \pi_1 M$ generated by all parabolic elements of Γ. In particular, $Sol = Sol_0$ if Γ/P has odd order.

3. Our formula has been implemented in Jeff Weeks' program "snap pea" by R. Meyerhoff and C. Hodgson, using $\mathbf{c}$ in Sol_0 rather than Sol (see **[HMW]** for a brief description of this program). There is therefore now a large accumulation of experimental evidence for the $i\pi^2/12$ conjecture of Remark 1 above. In cases when $Sol = Sol_0$ the computations appear to confirm the stronger $i\pi^2/6$ conjecture, but I have not checked systematically.

4. Note that $\mathcal{R}(z)$ is a multivalued function of z. It has singularities at 0 and 1 and is single-valued on the universal cover of $\mathbb{C} - \{0,1\}$. We only need $\mathcal{R}(z)$ (mod π^2), and in fact $\mathcal{R}(z)$ (mod $2\pi^2$) is well-defined on the universal abelian cover X of $\mathbb{C} - \{0,1\}$ ($\mathcal{R}$ itself is well-defined on the universal nilpotent cover). A point of X is determined by a point z of a fundamental domain for the cover and an integer pair $(c', c'') \in \mathbb{Z}^2 = \pi_1(\mathbb{C} - \{0,1\})$. The value of $\mathcal{R}$ mod $2\pi^2$ at this point is then

$$\mathcal{R}(z) + i\pi(c' \log(1-z) - c'' \log(z)),$$

where $\mathcal{R}(z)$ is the standard branch on the fundamental domain. Thus, when $\mathbf{c}$ is integral, the summand

$$\mathcal{R}(z_\nu) - \frac{i\pi}{2}(c'_\nu \log(1-z_\nu) - c''_\nu \log(z_\nu)) \pmod{\pi^2}$$

in our formula can be thought of as representing a branch at z_ν of one of the four functions

$$\mathcal{R}(z), \quad \mathcal{R}(z)+\frac{i\pi}{2}\log(z), \quad \mathcal{R}(z)+\frac{i\pi}{2}\log(1-z), \quad \text{or}$$

$$\mathcal{R}(z)+\frac{i\pi}{2}\left(\log(1-z)+\log(z)\right) \qquad \pmod{\pi^2},$$

according to the parity of c'_ν and c''_ν. Each of the first three of these functions can be naturally associated to the choice of an edge of the ideal tetrahedron with parameter z, since choosing a different edge replaces z by $1-1/z$ or $1/(1-z)$ and $\mathcal{R}(z)$ satisfies the functional equations

$$\mathcal{R}(1-\frac{1}{z}) = \mathcal{R}(z)+\frac{i\pi}{2}\log(z)-\frac{\pi^2}{6}$$

$$\mathcal{R}(\frac{1}{1-z}) = \mathcal{R}(z)+\frac{i\pi}{2}\log(1-z)+\frac{\pi^2}{6}.$$

The last of the above four functions is mapped to a different branch of itself, up to a constant, by the transformation $z \mapsto (1-1/z)$.

5. We believe there should be a topological interpretation of an integer vector $\mathbf{c}$ as in Theorem 2: it should be associated with some kind of extra structure on M. Remark 4 and Section 2 both give some support for this belief.

Acknowledgements. The support of the N.S.F. for this work is gratefully acknowledged. Don Zagier encouraged me to finally write up this work, despite the lack of proof of the conjecture of Theorem 2.

2. Consistency and cusp conditions, and the conditions on c

An ideal hyperbolic structure on a tetrahedron Δ is specified by assigning a complex number with positive imaginary part to each edge, these numbers being related as in Fig. 1. The geometric meaning of these parameters is that the Euclidean triangle cut off by a horosphere section at a vertex is similar to the Euclidean triangle pictured in Fig. 1 with vertices 0, 1, and z in the complex plane; the parameter z is only well-defined after choosing an edge of Δ, and parameters for different edges are related as illustrated. Instead of these parameters z, $1/(1-z)$, and $(z-1)/z$, it will be convenient to use their natural logarithms

$$\log(z), \quad -\log(1-z), \quad \log(1-z)-\log(z)+\pi i$$

(unless otherwise stated, log is always the standard branch on $\mathbb{C}-(-\infty,0]$). We call these the ***log-parameters*** of the tetrahedron.

Given a tetrahedron Δ, we shall choose a labelling of its edges $e_1, e_2, \ldots, e_6$ as in Fig. 2. Thus e_j and e_{j+3} are opposite edges for $j=1,2,3$.

Let M be a hyperbolic 3-manifold which is ideally triangulated as

$$M=\Delta_1\cup\ldots\cup\Delta_n. \tag{2.1}$$

Choose a labelling e_j, $j=1,\ldots,6$, of the edges of each Δ_ν as above. Let z_ν be the parameter for Δ_ν with respect to the edge e_1 of Δ_ν. We do not assume the hyperbolic

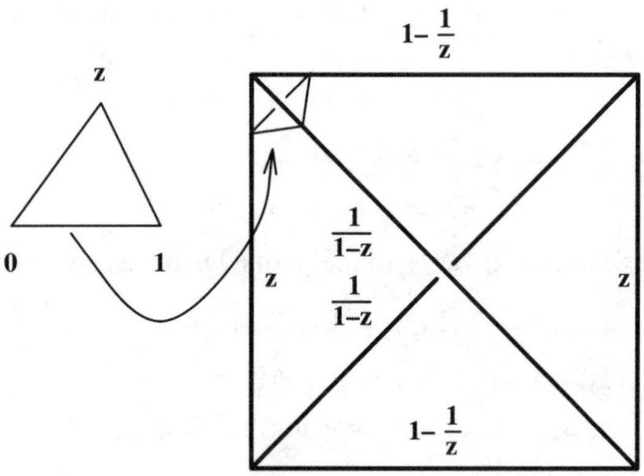

Figure 1

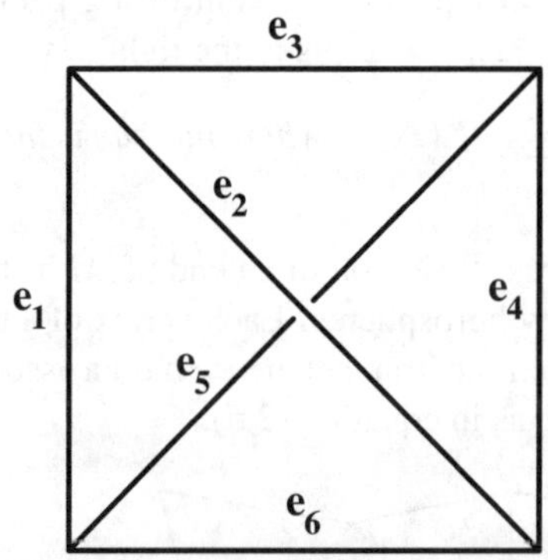

Figure 2

structure on M is complete, but we assume it results by deforming from a complete structure on M with tetrahedral parameters z_ν^0 say.

Consistency conditions. Each edge of each tetrahedron of the triangulation of M has an associated log-parameter. The condition that the ideal tetrahedra fit together around an edge E of the triangulation is equivalent to the condition that the log-parameters sum to $2\pi i$ around this edge. We call this the ***consistency condition at the edge*** E. It has the form

$$\sum_{\nu=1}^{n} r'_\nu(E)\log(z_\nu) + r''_\nu(E)\log(1-z_\nu) = d(E)\pi i \tag{2.2}$$

for some integer $d(E)$ and integers $r'_\nu(E)$ and $r''_\nu(E)$, $(\nu = 1, \ldots, n)$.

We can describe the integer $d(E)$ as follows. Associate an integer $\epsilon_i(\Delta_\nu)$ to the i-th edge of Δ_ν as follows:

$$\begin{aligned}\epsilon_1(\Delta_\nu) = \epsilon_4(\Delta_\nu) = 0,\\ \epsilon_2(\Delta_\nu) = \epsilon_5(\Delta_\nu) = 0,\\ \epsilon_3(\Delta_\nu) = \epsilon_6(\Delta_\nu) = 1.\end{aligned} \tag{2.3}$$

Associate a complex number $\zeta_j(\Delta_\nu)$ to the j-th edge of Δ_ν by:

$$\begin{aligned}\zeta_1(\Delta_\nu) = \zeta_4(\Delta_\nu) &= \log(z_\nu),\\ \zeta_2(\Delta_\nu) = \zeta_5(\Delta_\nu) &= -\log(1-z_\nu),\\ \zeta_3(\Delta_\nu) = \zeta_6(\Delta_\nu) &= -\log(z_\nu) + \log(1-z_\nu).\end{aligned} \tag{2.4}$$

The log-parameters of Δ_ν are then

$$\mu_j(\Delta_\nu) = \zeta_j(\Delta_\nu) + \epsilon_j(\Delta_\nu)\pi i, \quad j = 1, \dots, 6. \tag{2.5}$$

Equation (2.2) is the result of equating the sum of log-parameters around the edge E with $2\pi i$ and moving the $\epsilon_j(\Delta_\nu)\pi i$ terms to the right of the equation. Thus

Lemma 2.1. $d(E) = 2 - \sum_\kappa \epsilon_{j_\kappa}(\Delta_{\nu_\kappa})$ *where the sum is the sum of the* $\epsilon_j(\Delta_\nu)$ *around the edge* E. □

Cusp conditions. The torus T section of an end of M is triangulated by triangles cut off the vertices of the Δ_ν by horospheres. Each vertex of a triangle of this triangulation determines some edge of a tetrahedron, and hence has an associated log-parameter, which can be written $\mu = \zeta + \epsilon\pi i$ as in equation (2.5).

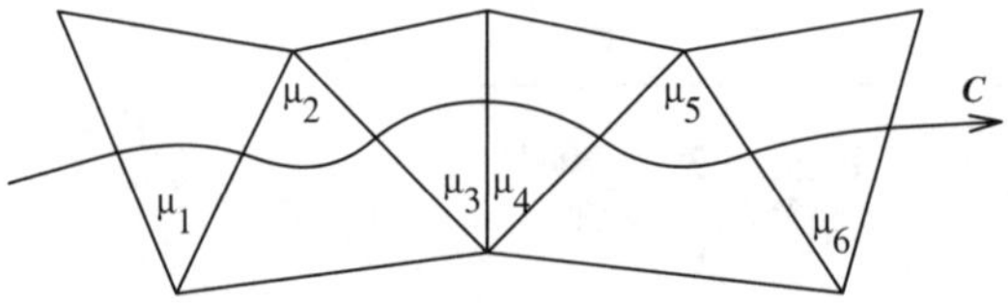

$\mu(C) = -\mu_1 + \mu_2 - \mu_3 - \mu_4 + \mu_5 - \mu_6 + \dots$

Figure 3

Let C be an essential simple closed curve in T which is in general position with respect to the 1-skeleton of the triangulation of T and, moreover, "has no back-tracking," in the sense that it never departs a 2-simplex across the same edge by which it entered. Then as C passes through a 2-simplex it determines a vertex of the 2-simplex, the vertex between the entering and departing edges, and a sign $+$ or $-$ according as it goes past this vertex counter-clockwise or clockwise. Let $\mu(C) = \sum \pm\mu_j$ be the sum over the vertices that C passes of the corresponding log-parameter μ_j with sign $\pm$ as above (Fig. 3). Clearly,

$$\mu(C) = \sum_{\nu=1}^{n} r'_\nu(C)\log(z_\nu) + r''_\nu(C)\log(1-z_\nu) - d(C)\pi i \tag{2.6}$$

for some integer $d(C)$ and integers $r'_\nu(C)$ and $r''_\nu(C)$, $(\nu = 1, \dots, n)$.

It is not hard to see (cf. Sect. 6 or **[NZ]**) that, given the consistency conditions, $\mu(C)$ only depends on the homology class of C in T and μ gives a homomorphism from $H_1(T)$ to $\mathbb{C}$. In fact, $\mu(C) = 2\log\lambda(H(C))$, where $\lambda(H(C))$ is an eigenvalue of the holonomy $H(C)$ of C.

The ***cusp condition*** corresponding to C is the condition that C have parabolic holonomy, that is, $\mu(C) = 0$, which by (2.6) can be written

$$\sum_{\nu=1}^{n} r'_\nu(C)\log(z_\nu) + r''_\nu(C)\log(1-z_\nu) = d(C)\pi i. \tag{2.7}$$

Just as for Lemma 2.1 one sees:

Lemma 2.2. *$d(C) = -\sum \pm\epsilon_j$, where the ϵ_j are the ϵ's (cf. equation (2.3)) at the vertices that C passes, with signs $\pm$ given as above.* □

Consistency and cusp conditions. For each edge E of our triangulation we have the consistency condition (2.2). At each cusp of M choose two curves C as above representing a basis of the first homology of the cusp and consider the corresponding cusp conditions (2.7). The complete hyperbolic structure on M —the one with tetrahedral parameters z_ν^0 —is determined by all of these consistency and cusp conditions together. This is a system of linear equations with integral coefficients, which can be written

$$U\mathcal{Z}^0 = \pi i\mathbf{d},$$

as described in the Introduction, with

$$\mathcal{Z}^0 = \begin{pmatrix} \log z_1^0 \\ \vdots \\ \log z_n^0 \\ \log(1-z_1^0) \\ \vdots \\ \log(1-z_n^0) \end{pmatrix}.$$

Now suppose we are given a $2n$-vector $\mathbf{c} = (c'_1, \ldots, c'_n, c''_1, \ldots, c''_n)^{\mathrm{t}}$. Assign numbers $\eta_i(\Delta_\nu)$ to the edges of Δ_ν as follows:

$$\begin{aligned} \eta_1(\Delta_\nu) &= \eta_4(\Delta_\nu) = c'_\nu, \\ \eta_2(\Delta_\nu) &= \eta_5(\Delta_\nu) = -c''_\nu, \\ \eta_4(\Delta_\nu) &= \eta_6(\Delta_\nu) = -c'_\nu + c''_\nu + 1. \end{aligned} \tag{2.8}$$

Proposition 2.3. *The equation $U\mathbf{c} = \mathbf{d}$ (equation $(*)$ of the Introduction) is equivalent to the following conditions:*

C1. *the sum of the η's around any edge E of the triangulation is* 2;

C2. *for any path C as above in a torus section of an end of M, the signed sum of the η's over the vertices adjacent to the path C is zero.*

Proof. Replace $\log(z_\nu^0)$ by c'_ν and $\log(1-z_\nu^0)$ by c''_ν in the proofs of Lemmas 2.1 and 2.2. □

The parity conditions. We consider an integral vector $\mathbf{c} \in \mathbb{Z}^{2n}$. Thus, the $\eta_j(\Delta_\nu)$ defined above are integers. Let C be any closed path in M which is in general position with respect to the 2-skeleton of our triangulation. We also assume C "has no back-tracking" in the sense that it never departs a 3-simplex across the same face by which it entered. Then as C passes through a 3-simplex it determines an edge of the 3-simplex — the edge common to the entering and departing faces — and hence an integer η as above. We call the modulo-2 sum of these η's along C the ***parity of* c *along*** C.

The ***parity condition on*** $\mathbf{c}$ is the condition:

C3. *The parity of* $\mathbf{c}$ *along* C *is even for all* C *as above.*

The combinatorial part of Theorem 2 is re-formulated and made precise in part (i) of the following theorem, which will be proved in Sect. 6.

Theorem 2.4.

(i) *There exists* $\mathbf{c} \in \mathbb{Z}^{2n}$ *satisfying the above conditions* C1, C2, *and* C3.

(ii) *Any two elements* $\mathbf{c}_1$ *and* $\mathbf{c}_2$ *of* $\mathbb{Z}^{2n}$ *satisfying conditions* C1, C2, *and* C3, *differ by an integral linear combination of the vectors*

$$J_{2n}\mathbf{r}(E) = (r_1''(E), \dots, r_n''(E), -r_1'(E), \dots, -r_n'(E))^{\mathrm{t}},$$

where the $\mathbf{r}(E)$ *are the coefficient vectors of the consistency conditions and*

$$J_{2n} = \begin{pmatrix} 0 & I_n \\ -I_n & 0 \end{pmatrix}.$$

Remark. The parity condition C3 implies that C1 holds modulo 2. Given that C1 holds modulo 2, the parity of $\mathbf{c}$ along C is easily seen to depend only on the homology class of C and to give a homomorphism $H_1(M;\mathbb{Z}/2)\to\mathbb{Z}/2$. If C2 holds modulo 2, this homomorphism vanishes on $H_1(\mathcal{E};\mathbb{Z}/2)$, where $\mathcal{E}$ is the union of the ends of M. Thus, one then need only check the parity condition on a set of representatives for generators of $H_1(M;\mathbb{Z}/2)/\operatorname{Im} H_1(\mathcal{E};\mathbb{Z}/2) = H_1(M^*;\mathbb{Z}/2)$, where M^* is the end compactification of M.

3. The formula for the Chern-Simons invariant

In this section we prove Theorems 1 and 2 of the Introduction, assuming Theorem 2.4, which will be proved in Sect. 6.

Proof of Theorem 1. Let M be triangulated, as in the previous section, into ideal tetrahedra with parameters z_ν. We do not assume M has the complete hyperbolic structure. Under certain conditions that we recall below, the metric completion M' of this structure on M is a hyperbolic manifold, in which case it is a Dehn filling of M in the topological sense. To review the conditions we need some notation.

Number the ends of M by $j = 1, \dots, h$ say. Choose a specific oriented basis $\mathbf{l}_j$, $\mathbf{m}_j$ for the homology of the j-th end of M for each j, and let

$$u_j = \mu(\mathbf{m}_j), \quad v_j = \mu(\mathbf{l}_j), \quad (j = 1, \dots, h),$$

where $\mu(\mathbf{m}_j)$ means $\mu(C)$ for some curve C representing $\mathbf{m}_j$, and the same for $\mu(\mathbf{l}_j)$.

If $u_j = v_j = 0$ then M is still complete at the j-th end, that is, it has a cusp there. If coprime integers p_j and q_j exist with

$$p_j u_j + q_j v_j = 2\pi i,$$

then M' is a hyperbolic manifold near the j-th end of M, obtained by adding a geodesic γ_j to M there in such a way that it is topologically a (p_j, q_j)-Dehn filling, that is, a Dehn filling which kills the homology class $p_j \mathbf{m}_j + q_j \mathbf{l}_j$. In all other cases M' is not a hyperbolic manifold at the j-th end of M.

Remark on orientations. In **[NZ]** $\{\mathbf{m}_j, \mathbf{l}_j\}$ was an oriented basis rather than $\{\mathbf{l}_j, \mathbf{m}_j\}$. For the justification of the different convention used here, see the "Note on orientations" in section 6 of **[NR]** (these Proceedings) or in **[MN]**.

Assume now that M' is a hyperbolic manifold. Let λ_j be as in the Introduction: $\lambda_j = 0$ if the j-th end of M is complete, and otherwise λ_j is the "complex length" of the added geodesic γ_j, that is, its length plus i times its torsion.

Lemma 3.1. $\operatorname{Re}\lambda_j = -\frac{1}{2\pi}\operatorname{Im}(v_j \overline{u_j})$.

Proof. This is Proposition 4.3 of **[NZ]**, except that u_j and v_j have been interchanged, in keeping with the above orientation convention. □

Following equation (2.6), we can write

$$u_j = \sum_{\nu=1}^{n} \left(r'_\nu(\mathbf{m}_j)\log(z_\nu) + r''_\nu(\mathbf{m}_j)\log(1-z_\nu)\right) - d(\mathbf{m}_j)\pi i$$

$$v_j = \sum_{\nu=1}^{n} \left(r'_\nu(\mathbf{l}_j)\log(z_\nu) + r''_\nu(\mathbf{l}_j)\log(1-z_\nu)\right) - d(\mathbf{l}_j)\pi i.$$

Given a solution $\mathbf{c} = (c'_1, \ldots, c'_n, c''_1, \ldots, c''_n)^{\mathrm{t}}$ to the equation $U\mathbf{c} = \mathbf{d}$, we can rewrite this

$$\begin{aligned} u_j &= \sum_{\nu=1}^{n} \left(r'_\nu(\mathbf{m}_j)(\log(z_\nu) - c'_\nu \pi i) + r''_\nu(\mathbf{m}_j)(\log(1-z_\nu) - c''_\nu \pi i)\right) \\ v_j &= \sum_{\nu=1}^{n} \left(r'_\nu(\mathbf{l}_j)(\log(z_\nu) - c'_\nu \pi i) + r''_\nu(\mathbf{l}_j)(\log(1-z_\nu) - c''_\nu \pi i)\right). \end{aligned} \tag{3.1}$$

As in **[NZ]**, we write the matrix U as

$$U = \begin{pmatrix} M \\ L \\ R \end{pmatrix} = \begin{pmatrix} C \\ R \end{pmatrix},$$

where the rows of R are the coefficients of the consistency conditions and

$$L = \begin{pmatrix} r_1'(\mathbf{l}_1) & \dots & r_n'(\mathbf{l}_1) & r_1''(\mathbf{l}_1) & \dots & r_n''(\mathbf{l}_1) \\ \vdots & & \vdots & \vdots & & \vdots \\ r_1'(\mathbf{l}_h) & \dots & r_n'(\mathbf{l}_h) & r_1''(\mathbf{l}_h) & \dots & r_n''(\mathbf{l}_h) \end{pmatrix},$$

$$M = \begin{pmatrix} r_1'(\mathbf{m}_1) & \dots & r_n'(\mathbf{m}_1) & r_1''(\mathbf{m}_1) & \dots & r_n''(\mathbf{m}_1) \\ \vdots & & \vdots & \vdots & & \vdots \\ r_1'(\mathbf{m}_h) & \dots & r_n'(\mathbf{m}_h) & r_1''(\mathbf{m}_h) & \dots & r_n''(\mathbf{m}_h) \end{pmatrix}.$$

We also write

$$\mathcal{Z} = \begin{pmatrix} \log z_1 \\ \vdots \\ \log z_n \\ \log(1-z_1) \\ \vdots \\ \log(1-z_n) \end{pmatrix}.$$

Then equations (3.1) can be written in matrix form:

$$\begin{pmatrix} u_1 \\ \vdots \\ u_n \\ v_1 \\ \vdots \\ v_n \end{pmatrix} = C(\mathcal{Z} - \mathbf{c}\pi i), \tag{3.2}$$

and the consistency relations can be written

$$R(\mathcal{Z} - \mathbf{c}\pi i) = 0. \tag{3.3}$$

We recall the main combinatorial lemma of **[NZ]**. Let J_{2m} denote the $2m \times 2m$ matrix

$$J_{2m} = \begin{pmatrix} 0 & I \\ -I & 0 \end{pmatrix}.$$

Lemma 3.2. *Given* $\mathbf{x}$ *and* $\mathbf{y}$ *in* $\mathbb{C}^{2n}$ *satisfying* $R\mathbf{x} = R\mathbf{y} = 0$, *we have*

$$\mathbf{x}^{\mathrm{t}} J_{2n} \bar{\mathbf{y}} = \frac{1}{2} \mathbf{x}^{\mathrm{t}} C^{\mathrm{t}} J_{2h} C \bar{\mathbf{y}}.$$

Proof. This is Corollary 2.4 of **[NZ]**, modified in accordance with our different orientation convention. It also follows easily from Theorem 4.1 below. □

We can apply this lemma with $\mathbf{x} = \mathbf{y} = \mathcal{Z} - \mathbf{c}\pi i$ to get

$$\begin{aligned}\sum_{j=1}^{h} \operatorname{Im}(v_j \bar{u}_j) &= \frac{i}{2}(u_1, \ldots, v_h) J_{2h} \begin{pmatrix} \bar{u}_1 \\ \vdots \\ \bar{v}_h \end{pmatrix} \\ &= \frac{i}{2}(\mathcal{Z} - \mathbf{c}\pi i)^{\mathrm{t}} C^{\mathrm{t}} J_{2h} C(\bar{\mathcal{Z}} + \bar{\mathbf{c}}\pi i) \\ &= i(\mathcal{Z} - \mathbf{c}\pi i)^{\mathrm{t}} J_{2n}(\bar{\mathcal{Z}} + \bar{\mathbf{c}}\pi i) \\ &= -2\sum_{\nu=1}^{n} \operatorname{Im}\big(\log z_\nu \log \overline{1 - z_\nu}\big) \\ &\qquad + 2\pi \sum_{\nu=1}^{n} \operatorname{Re}\big(\bar{c}'_\nu \log(1 - z_\nu) - \bar{c}''_\nu \log z_\nu\big) \\ &\qquad + i\pi^2 \mathbf{c}^{\mathrm{t}} J_{2n} \bar{\mathbf{c}}.\end{aligned}$$

By Lemma 3.1 this gives

$$\begin{aligned}\frac{\pi}{2} \operatorname{Re}\Big(\sum_{j=1}^{h} \lambda_j\Big) &= \alpha_0 + \frac{1}{2}\sum_{\nu=1}^{n} \operatorname{Im}\big(\log z_\nu \log \overline{1 - z_\nu}\big) \\ &\qquad - \frac{1}{2}\pi \operatorname{Re}\Big(\sum_{\nu=1}^{n} \big(\bar{c}'_\nu \log(1 - z_\nu) - \bar{c}''_\nu \log z_\nu\big)\Big),\end{aligned}$$

with $\alpha_0 = -(i/4)\pi^2 \mathbf{c}^{\mathrm{t}} J_{2n} \bar{\mathbf{c}}$, which vanishes if $\mathbf{c}$ is real. Combining this with the elementary formula

$$\frac{1}{2} \operatorname{Im}\big(\log z \log \overline{1 - z}\big) = \frac{1}{2} \operatorname{Im}\big(\log z \log(1 - z)\big) - \log|z| \arg(1 - z),$$

we get

$$\begin{aligned}\frac{\pi}{2} \operatorname{Re}\Big(\sum_{j=1}^{h} \lambda_j\Big) &= \alpha_0 + \frac{1}{2}\sum_{\nu=1}^{n} \operatorname{Im}\big(\log z_\nu \log(1 - z_\nu)\big) - \sum_{\nu=1}^{n} \log|z_\nu| \arg(1 - z_\nu) \\ &\qquad - \frac{1}{2}\pi \operatorname{Re}\Big(\sum_{\nu=1}^{n} \big(\bar{c}'_\nu \log(1 - z_\nu) - \bar{c}''_\nu \log z_\nu\big)\Big). \qquad (3.4)\end{aligned}$$

The function $\log|z| \arg(1-z) + \operatorname{Im} \operatorname{Li}_2(z)$ is called the "Bloch-Wigner dilogarithm." It is the volume of the ideal tetrahedron with parameter z (cf. **[NZ]**, for example). Thus

$$\operatorname{Vol}(M') = \sum_{\nu=1}^{n} \big(\log|z_\nu| \arg(1 - z_\nu) + \operatorname{Im} \operatorname{Li}_2(z_\nu)\big). \qquad (3.5)$$

Adding (3.4) to (3.5) gives

$$\begin{aligned}\mathrm{Vol}(M')+\frac{\pi}{2}\,\mathrm{Re}\Big(\sum_{j=1}^{h}\lambda_j\Big) &= \alpha_0+\frac{1}{2}\sum_{\nu=1}^{n}\mathrm{Im}\big(\log z_\nu\log(1-z_\nu)\big)+\sum_{\nu=1}^{n}\mathrm{Im}\,\mathrm{Li}_2(z_\nu)\\ &\quad -\frac{1}{2}\pi\,\mathrm{Re}\Big(\sum_{\nu=1}^{n}(\overline{c'_\nu}\log(1-z_\nu)-\overline{c''_\nu}\log z_\nu)\Big)\\ &= \alpha_0+\sum_{\nu=1}^{n}\mathrm{Im}\,\mathcal{R}(z_\nu)-\frac{1}{2}\pi\,\mathrm{Re}\Big(\sum_{\nu=1}^{n}(\overline{c'_\nu}\log(1-z_\nu)-\overline{c''_\nu}\log z_\nu)\Big),\end{aligned}$$

which can be written

$$\begin{aligned}&\mathrm{Re}\Big(\mathcal{V}(M')+\frac{\pi}{2}\sum_{j=1}^{h}\lambda_j\Big)=\\ &\alpha_0+\mathrm{Re}\Big(-i\sum_{\nu=1}^{n}\big(\mathcal{R}(z_\nu)-\frac{i\pi}{2}(\overline{c'_\nu}\log(1-z_\nu)-\overline{c''_\nu}\log z_\nu)\big)\Big).\end{aligned}\tag{3.6}$$

This confirms the real part of the formula of Theorem 1. But T. Yoshida shows in **[Y1]**, confirming a conjecture of **[NZ]**, that $\mathcal{V}(M')+\frac{\pi}{2}\sum_{j=1}^{h}\lambda_j$ is a complex analytic function. Since a complex analytic function is determined up to a constant by its real part, the formula of Theorem 1 is proved.

To complete the proof of Theorem 1 we must show that the constant α is in $i\pi^2\mathbb{Q}$ if $\mathbf{c}$ is rational. We use work of Dupont **[D]** (that builds on work of Bloch and Wigner and Dupont and Sah), which we now quote.

Let $\mathcal{P}_\mathbb{C}$ be the "Bloch group", generated by symbols $\{z\}$ with $z\in\mathbb{C}-\{0,1\}$ subject to the relations

$$\sum_{i=0}^{4}\{[a_0:\ldots:\widehat{a_i}:\ldots:a_4]\}=0\quad\text{for distinct points } a_0,\ldots,a_4\in\mathbb{C}-\infty\,,$$

where $[z_0:z_1:z_2:z_3]$ means the cross-ratio $(z_0-z_2)(z_1-z_3)/(z_0-z_3)(z_1-z_2)$.

There is a commutative diagram with exact rows

$$\begin{array}{ccc} H_3(\mathrm{SL}(2,\mathbb{C})^\delta;\mathbb{Z}) & \xrightarrow{\sigma} & \mathcal{P}_\mathbb{C}\\ \big\downarrow{\scriptstyle c} & & \big\downarrow{\scriptstyle \rho}\\ \mathbb{C}/\mathbb{Q} & \xrightarrow{1\wedge\mathrm{id}} & \Lambda^2_\mathbb{Q}\mathbb{C},\end{array}$$

where the map of importance to us is ρ, given by the formula

$$\rho(\{z\})=\frac{\log z}{2\pi i}\wedge\frac{\log(1-z)}{2\pi i}+1\wedge\frac{\mathcal{R}(z)}{2\pi^2}\,.\tag{3.7}$$

Dupont shows that the map c is $2\widehat{C_2}$, where $\widehat{C_2}$ is the Cheeger-Chern-Simons class associated to the Chern polynomial c_2. Our hyperbolic manifold M' represents an element of $H_3(\mathrm{SL}(2,\mathbb{C})^\delta;\mathbb{Z})$ and in our terminology his result is that

$$\sum_{\nu=1}^{n}\rho(\{z_\nu\}) = 1\wedge 2\widehat{C_2}(M) = 1\wedge \frac{i}{2\pi^2}\mathcal{V}(M'). \tag{3.8}$$

This equation will be our starting point.

(Dupont used $\Lambda^2_{\mathbb{Z}}\mathbb{C}$ rather than $\Lambda^2_{\mathbb{Q}}\mathbb{C}$. They are the same thing: if s and t are integers and $z,w\in\mathbb{C}$ then in $\Lambda^2_{\mathbb{Z}}\mathbb{C}$ we have $(s/t)z\wedge w = (s/t)z\wedge t(1/t)w = sz\wedge(1/t)w = z\wedge(s/t)w$. For us $\Lambda^2_{\mathbb{Q}}\mathbb{C}$ is more convenient.)

Now suppose M' results from M by (p_j,q_j)-Dehn surgery at the j-th end. Then, as described at the start of this section,

$$p_j\frac{u_j}{2\pi i} + q_j\frac{v_j}{2\pi i} = 1.$$

Choose integers r_j and s_j with $p_js_j - q_jr_j = 1$. Then in **[NZ]** it is shown that

$$r_j\frac{u_j}{2\pi i} + s_j\frac{v_j}{2\pi i} = \frac{\lambda_j}{2\pi i}.$$

(**[NZ]** had a different sign due to the differing orientation convention.) Hence, taking wedge product of these two equations,

$$\frac{u_j}{2\pi i}\wedge\frac{v_j}{2\pi i} = 1\wedge\frac{\lambda_j}{2\pi i}. \tag{3.9}$$

For $\mathbf{w} = (w_1,\ldots,w_{2m})$ and $\mathbf{w}' = (w'_1,\ldots,w'_{2m})$ in $\mathbb{C}^{2m}$ we define $\mathbf{w}\wedge\mathbf{w}' = (1/2)\sum_{j=1}^{m}(w_j\wedge w'_{m+j} - w_{m+j}\wedge w'_j)\in\Lambda^2_{\mathbb{Q}}\mathbb{C}$. It is a formal observation that whenever rational matrices R and C satisfy Lemma 3.2, they will satisfy:

$$R\mathbf{x} = R\mathbf{y} = 0 \quad\Longrightarrow\quad \mathbf{x}\wedge\mathbf{y} = \frac{1}{2}C\mathbf{x}\wedge C\mathbf{y}, \qquad \text{for } \mathbf{x},\mathbf{y}\in\mathbb{C}^{2n}. \tag{3.10}$$

($\mathbb{C}$ could be replaced by any $\mathbb{Q}$-vector space for this.)

Now suppose we have a rational solution $\mathbf{c}$ to $U\mathbf{c} = \mathbf{d}$. We shall denote $\mathcal{Z} - \mathbf{c}\pi i$ by $\mathcal{Z}_0$. Using, in turn, (3.8) and (3.7), (3.10), and (3.9),

$$
\begin{aligned}
1 \wedge \frac{i}{2\pi^2}\mathcal{V}(M') &= \frac{\mathcal{Z}}{2\pi i} \wedge \frac{\mathcal{Z}}{2\pi i} + \sum_{\nu=1}^{n} 1 \wedge \frac{\mathcal{R}(z_\nu)}{2\pi^2} \\
&= \frac{\mathcal{Z}_0}{2\pi i} \wedge \frac{\mathcal{Z}_0}{2\pi i} + \frac{1}{2}\left(\frac{\mathcal{Z}}{2\pi i} \wedge \mathbf{c} + \mathbf{c} \wedge \frac{\mathcal{Z}}{2\pi i}\right) - \frac{1}{4}\mathbf{c} \wedge \mathbf{c} + \sum_{\nu=1}^{n} 1 \wedge \frac{\mathcal{R}(z_\nu)}{2\pi^2} \\
&= \frac{1}{2}\sum_{j=1}^{h} \frac{u_j}{2\pi i} \wedge \frac{v_j}{2\pi i} + \frac{1}{2}\sum_{\nu=1}^{n}\left(c'_\nu \wedge \frac{\log 1 - z_\nu}{2\pi i} - c''_\nu \wedge \frac{\log z_\nu}{2\pi i}\right) \\
&\quad + \sum_{\nu=1}^{n} 1 \wedge \frac{\mathcal{R}(z_\nu)}{2\pi^2} \\
&= \frac{1}{2}\sum_{j=1}^{h} 1 \wedge \frac{\lambda_j}{2\pi i} + \frac{1}{2}\sum_{\nu=1}^{n}\left(1 \wedge c'_\nu \frac{\log 1 - z_\nu}{2\pi i} - 1 \wedge c''_\nu \frac{\log z_\nu}{2\pi i}\right) \\
&\quad + \sum_{\nu=1}^{n} 1 \wedge \frac{\mathcal{R}(z_\nu)}{2\pi^2}.
\end{aligned}
$$

Hence

$$
\begin{aligned}
\frac{i}{2\pi^2}\mathcal{V}(M') &= \frac{1}{2}\sum_{j=1}^{h} \frac{\lambda_j}{2\pi i} + \frac{1}{2}\sum_{\nu=1}^{n}\left(c'_\nu \frac{\log 1 - z_\nu}{2\pi i} - c''_\nu \frac{\log z_\nu}{2\pi i}\right) \\
&\quad + \sum_{\nu=1}^{n} \frac{\mathcal{R}(z_\nu)}{2\pi^2} \quad (\mathrm{mod}\ \mathbb{Q}),
\end{aligned}
$$

which is the desired rationality result. □

Proof of Theorem 2. We assume Theorem 2.4. Suppose we have two $\mathbf{c}$'s as in Theorem 2, that is, satisfying the conditions C1, C2, and C3 of Section 2 (cf. Theorem 2.4). Denote their difference by $\mathbf{s}$. Replacing one of these $\mathbf{c}$'s by the other in the formula of Theorem 1 changes the formula by $(\pi/2)\mathbf{s}^t J_{2n}\mathcal{Z}$. To prove Theorem 2 we must show this is a multiple of $i\pi^2/2$. But by Theorem 2.4, $\mathbf{s}^t J_{2n}$ is a linear combination of the rows of the coefficient matrix R of the consistency condition, so the consistency condition (equation (2.2)) implies that $\mathbf{s}^t J_{2n}\mathcal{Z}$ is πi times an integer. □

Remark. It is easy to check that the formula for $\mathcal{V}(M')$ of Theorem 2 is invariant modulo $i\pi^2/6$ under "cocycle moves" to change the triangulation: replace two ideal simplices $\langle V_0, V_1, V_2, V_3\rangle$ and $\langle V'_0, V_1, V_2, V_3\rangle$ which have a common face by three— $\langle V_0, V'_0, V_1, V_2\rangle$, $\langle V_0, V'_0, V_2, V_3\rangle$, and $\langle V_0, V'_0, V_3, V_1\rangle$.

4. Combinatorics of 3-cycles

By an ***n*-cycle with boundary** will be meant any n-complex K obtained from a finite disjoint union of closed n-simplices by iteratively gluing together pairs of free $(n-1)$-faces by simplicial identification maps (a "free" face is one that has not yet been glued) in such a way that any open q-simplex still embeds after the identification. If no free faces remain K will be called a ***closed n*-cycle** or just an ***n*-cycle** (these are also called "n-dimensional normal pseudo-manifolds" in the literature). Alternatively, 1-cycles are compact quasi-simplicial 1-manifolds and n-cycles can then be defined inductively for $n \geq 2$ as finite quasi-simplicial complexes whose vertex links are connected $(n-1)$-cycles. (A "quasi-simplicial complex" is a finite CW-complex built from simplices such that the attaching map $\partial\Delta^q \to K^{q-1}$ for each q-simplex is simplicial and is injective on each open $(q-k)$-face of Δ^q. The second barycentric subdivision of such a complex is a simplicial complex.) We will not distinguish notationally the complex K and its underlying topological space. The complement $K - K^{n-3}$ of the $(n-3)$-skeleton of an n-cycle K is a manifold, and by an ***orientation*** of K we mean an orientation of this manifold.

For $n \leq 2$ an n-cycle is an n-manifold. A 3-cycle is topologically a manifold except for finitely many singular points where the local structure is that of a cone on a closed connected surface.

To an oriented 3-simplex Δ we shall associate a 2-dimensional bilinear space J_Δ over $\mathbb{Z}$ as follows. As a $\mathbb{Z}$-module J_Δ is generated by the six edges $e_1, \ldots, e_6$ of Δ (see Fig. 2) with the relations:

$$e_i - e_{i+3} = 0 \quad \text{for } i = 1, 2, 3.$$

$$e_1 + e_2 + e_3 = 0.$$

Thus, opposite edges of Δ represent the same element of J_Δ, so J_Δ has three "geometric" generators, and the sum of these three generators is zero. The bilinear form on J_Δ is the non-singular skew-symmetric form given by

$$\langle e_1, e_2\rangle = \langle e_2, e_3\rangle = \langle e_3, e_1\rangle = -\langle e_2, e_1\rangle = -\langle e_3, e_2\rangle = -\langle e_1, e_3\rangle = 1.$$

Now suppose K is an oriented 3-cycle. For each i let C_i be the free $\mathbb{Z}$-module on the ***unoriented*** i-simplices of K. Let J be the direct sum $\coprod J_\Delta$, summed over the oriented 3-simplices of K. There is a unique reasonable way of defining natural homomorphisms

$$\alpha: C_0 \longrightarrow C_1$$

and

$$\beta: C_1 \longrightarrow J.$$

Namely, α takes a vertex to the sum of the incident edges (with an edge counted twice if both endpoints are at the given vertex). The J_Δ component of β takes an edge E of K to the sum of those edges e_i in the edge set $\{e_1, e_2, \ldots, e_6\}$ of Δ which are identified with E in K.

The natural basis of C_i gives an identification of C_i with its dual space and the bilinear form on J gives an identification of J with its dual space. With respect to these identifications, the dual map

$$\alpha^*: C_1 \longrightarrow C_0$$

is easily seen to map an edge E of K to the sum of its endpoints, and the dual map

$$\beta^*: J \longrightarrow C_1$$

can be described as follows. To each 3-simplex Δ of K we have a map $j = j_\Delta$ of the edge set $\{e_1, e_2, \ldots, e_6\}$ of Δ to the set of edges of K: put $j(e_i)$ equal to the edge that e_i is identified with in K. For e_i in J_Δ we have

$$\beta^*(e_i) = j(e_{i+1}) - j(e_{i+2}) + j(e_{i+4}) - j(e_{i+5}) \quad \text{(indices mod 6).}$$

This is shown pictorially in Fig. 4.

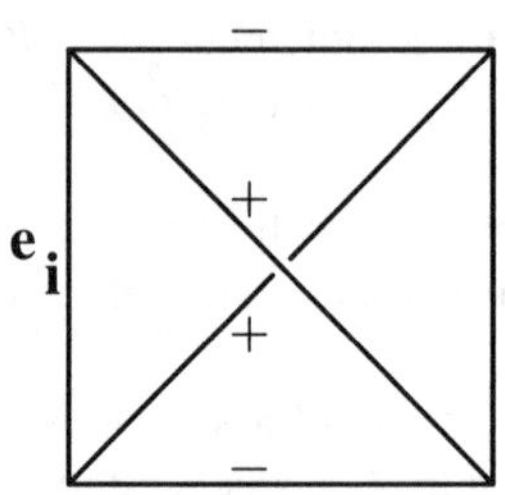

Figure 4

We shall show that $\operatorname{Im}\beta \subseteq \operatorname{Ker}\beta^*$. Since $\operatorname{Ker}\beta^* = (\operatorname{Im}\beta)^\perp$, the form on J then induces a form on $\operatorname{Ker}\beta^* / \operatorname{Im}\beta$ which is non-degenerate on $(\operatorname{Ker}\beta^* / \operatorname{Im}\beta)/\{\text{Torsion}\}$. We shall denote this form also by $\langle\, ,\, \rangle$.

Let K_0 be the result of removing a small open cone neighborhood of each vertex V of K, so ∂K_0 is the disjoint union of the links L_V of the vertices of K.

Theorem 4.1. *The sequence*

$$\mathfrak{J}: \quad 0 \longrightarrow C_0 \xrightarrow{\alpha} C_1 \xrightarrow{\beta} J \xrightarrow{\beta^*} C_1 \xrightarrow{\alpha^*} C_0 \longrightarrow 0$$

is a chain complex. Tensored with $\mathbb{Z}[\frac{1}{2}]$, it is exact except in the middle, where its homology is the first homology of ∂K_0:

$$(\operatorname{Ker}\beta^* / \operatorname{Im}\beta) \otimes \mathbb{Z}[\frac{1}{2}] = H_1(\partial K_0; \mathbb{Z}[\frac{1}{2}]) = \coprod_{V \in K^0} H_1(L_V; \mathbb{Z}[\frac{1}{2}]).$$

Moreover, the bilinear form $\langle\, ,\, \rangle$ on $(\operatorname{Ker}\beta^* / \operatorname{Im}\beta) \otimes \mathbb{Z}[\frac{1}{2}]$ is twice the intersection form on $H_1(\partial K_0; \mathbb{Z}[\frac{1}{2}])$.

Note that only vertices at which K is not a manifold contribute their homology in this theorem. Thus the chain complex $\mathfrak{J}$ computes the "local homology" of the singularities of K.

Remark. In [**NZ**, Theorem 2.2 to Proposition 2.5] combinatorial results equivalent to the above Theorem were proved under the extra assumption that the complement of the vertices, $K - K^0$, carries a complete hyperbolic structure of finite volume.

We will need the homology of $\mathcal{J}$ without tensoring with $\mathbb{Z}[\frac{1}{2}]$, but the result is more technical. We number homology groups of $\mathcal{J}$ as follows:

$$\begin{array}{ccccccccccccc} 0 & \longrightarrow & C_0 & \xrightarrow{\alpha} & C_1 & \xrightarrow{\beta} & J & \xrightarrow{\beta^*} & C_1 & \xrightarrow{\alpha^*} & C_0 & \longrightarrow & 0. \\ & & H_5(\mathcal{J}) & & H_4(\mathcal{J}) & & H_3(\mathcal{J}) & & H_2(\mathcal{J}) & & H_1(\mathcal{J}) & & \end{array}$$

Theorem 4.2. *The homology groups* $H_i(\mathcal{J})$ *are*

$$H_5(\mathcal{J}) = 0, \quad H_4(\mathcal{J}) = \mathbb{Z}/2, \quad H_1(\mathcal{J}) = \mathbb{Z}/2,$$
$$H_3(\mathcal{J}) = \mathcal{H} \oplus H^1(K;\mathbb{Z}/2), \quad H_2(\mathcal{J}) = H_1(K;\mathbb{Z}/2),$$

where $\mathcal{H} = \operatorname{Ker}(H_1(\partial K_0;\mathbb{Z}) \to H_1(K_0;\mathbb{Z}/2))$. *The isomorphism* $H_3(\mathcal{J})/\{\text{Torsion}\} \to \mathcal{H}$ *is an isometry for the form on* $H_3(\mathcal{J}) = \operatorname{Ker}\beta^*/\operatorname{Im}\beta$ *and the intersection form on* $\mathcal{H} \subseteq H_1(\partial K_0) = \coprod H_1(L_V)$.

Remark. The isomorphism of $H_2(\mathcal{J})$ with $H_1(K;\mathbb{Z}/2)$ is given by the obvious map— an element of $\operatorname{Ker}\alpha^*$ represents a modulo-2 simplicial 1-cycle in K. The torsion of $H_3(\mathcal{J})$ then follows by the universal coefficient theorem. The map of $H_3(\mathcal{J})/\{\text{Torsion}\}$ to $\mathcal{H}$ is less obvious, and is pictured in Fig. 9 below. In Theorem 5.1 we give a more direct computation of $H_3(\mathcal{J})$ in terms of cohomology, which we need later, but its proof uses the above theorem.

Proof. We shall work over $\mathbb{Z}$. The version over $\mathbb{Z}[\frac{1}{2}]$ follows by the same proof, or directly from the result over $\mathbb{Z}$.

To show that $\mathcal{J}$ is a chain complex we must show $\beta \circ \alpha = 0$, $\beta^* \circ \beta = 0$, and $\alpha^* \circ \beta^* = 0$. The first and third of these equations are dual to each other and hence equivalent, and the third equation is clear from Fig. 4. For the second equation note that, for an edge E of K, $\beta^*\beta(E)$ is a sum of contributions $\pm E'$ over edges E' which are adjacent to E in some 3-simplex Δ of K. Each such E' spans with E a 2-simplex face of Δ and contributes also with opposite sign to $\beta^*\beta(E)$ for the 3-simplex on the other side of this face. Thus $\beta^*\beta(E)$ is zero.

We must next discuss the homology groups $H_i(\mathcal{J})$.

$H_1(\mathcal{J})$ is $\operatorname{Coker}(\alpha^*)$ and we must show it is $\mathbb{Z}/2$. If $\epsilon\colon C_0 \to \mathbb{Z}/2$ is the map which takes an element to the mod-2 sum of its coefficients, then certainly $\operatorname{Im}(\alpha^*) \subseteq \operatorname{Ker}(\epsilon)$. For any 2-simplex F of K, let E_1, E_2, and E_3 be the three edges of F (which may not be distinct, since K is just quasi-simplicial). Then $\alpha^*(E_1 + E_2 - E_3) = 2V$, where V is the common vertex of E_1 and E_2 in F, and $\alpha^*(E_2 - E_3) = V - V'$, where V' is the vertex at the other end of E_1. Since $\operatorname{Ker}(\epsilon)$ is generated by elements of these two types, $\operatorname{Im}(\alpha^*) = \operatorname{Ker}(\epsilon)$, as was to be proved.

Any element of $\operatorname{Ker}(\alpha^*)$, taken modulo 2, is a simplicial mod-2 1-cycle in K. Thus we have a map $\operatorname{Ker}(\alpha^*) \to H_1(K;\mathbb{Z}/2)$. This map vanishes on $\operatorname{Im}(\beta^*)$ (see Fig. 4), so it induces a map $H_2(\mathcal{J}) \to H_1(K;\mathbb{Z}/2)$. We claim this is an isomorphism.

A closed simplicial path in K represents an element of $H_1(K;\mathbb{Z}/2)$ (the mod-2 sum of the 1-simplices along the path) and $H_1(K;\mathbb{Z}/2)$ is generated by such elements. If the path has odd length, it can be modified to have even length by replacing some 1-simplex E by the sum of the other two edges of some 2-simplex containing E. Once it has even length, the alternating sum of the 1-simplices along the path is an element of $\mathrm{Ker}(\alpha^*)$ which represents the given element of $H_1(K;\mathbb{Z}/2)$. In particular, the map $H_2(\mathfrak{J}) \to H_1(K;\mathbb{Z}/2)$ is onto.

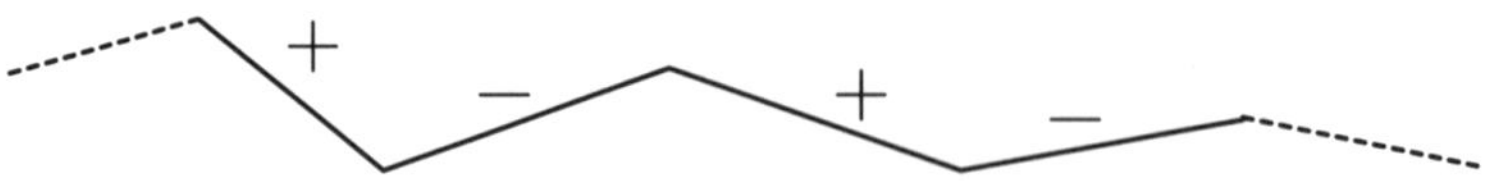

Figure 5

$H_2(\mathfrak{J})$ is generated by such alternating sums along paths, so we can draw elements of $H_2(\mathfrak{J})$ as in Fig. 5. Alternating paths as in Fig. 4 give the relations. Now if F and F' are 2-simplices with a common edge E, and E_1 and E_2 (respectively E_1' and E_2') are the other two edges of F (respectively F'), then any occurrence of $E_1 - E_2$ in an alternating path can be replaced by $E_1' - E_2'$ without changing the represented element of $H_2(\mathfrak{J})$ (see Fig. 6a; there is an orientation of E determined by $E_1 - E_2$ which should agree with the one determined by $E_1' - E_2'$). Indeed, if F and F' represent adjacent vertices in the link of E then this uses a single application of a relation of the type in Fig. 4, so in general it follows from the connectedness of the link of E. We therefore denote $E_1 - E_2$ by E^{+-} or E^{-+} (this assumes an implicit orientation of E) and draw it as in Fig. 6b.

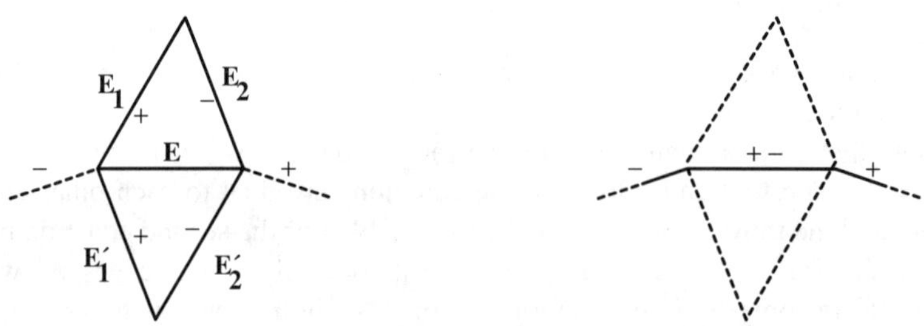

Figure 6 (a) and (b)

In a similar way, the connectedness of the link at a vertex easily implies that the configuration of Fig. 7a represents 0 in $H_2(\mathfrak{J})$, so an E followed by a $-E'$ is equivalent to E^{+-} followed by E'^{+-} (Fig. 7b).

Thus any alternating path is equivalent to the corresponding path of $^{+-}$ edges, or, doing the conversion in the opposite direction around the path, also to the corresponding path of $^{-+}$ edges. In particular, it is equivalent to its own negative, so the signs on the path are irrelevant to the represented element in $H_2(\mathfrak{J})$. Moreover, the boundary of a 2-simplex represents 0 (since $E_1^{+-} + E_2^{+-} + E_3^{+-} \sim E_1^{+-} + E_2 - E_3 = 0$), so $H_2(\mathfrak{J})$ is $H_1(K;\mathbb{Z}/2)$, as claimed.

The computation of $H_4(\mathfrak{J})$ and $H_5(\mathfrak{J})$, as well as the torsion in $H_3(\mathfrak{J})$, now follows by standard duality arguments (the universal coefficient theorem for cohomology), since

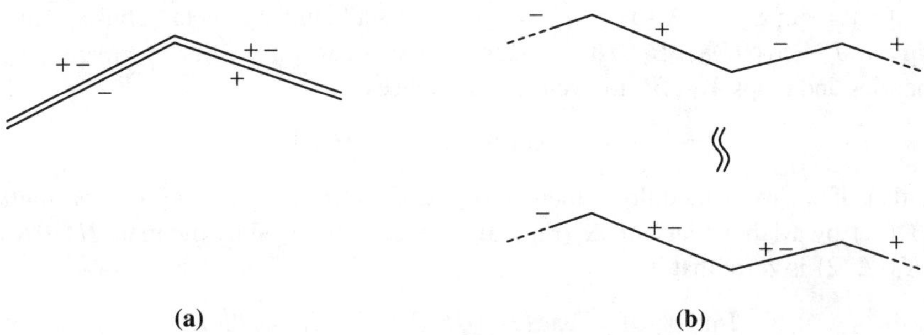

Figure 7

our chain complex is self-dual. It remains to compute the free part of $H_3(\mathcal{J})$. To do so we will need to define maps $\gamma\colon H_3(\mathcal{J}) \to \mathcal{H}$ and $\delta\colon H_1(\partial K_0) \to H_3(\mathcal{J})$.

Recall that K_0 denotes the result of removing a small open cone neighborhood of each vertex in K. K_0 can be constructed by gluing truncated tetrahedra (Fig. 8).

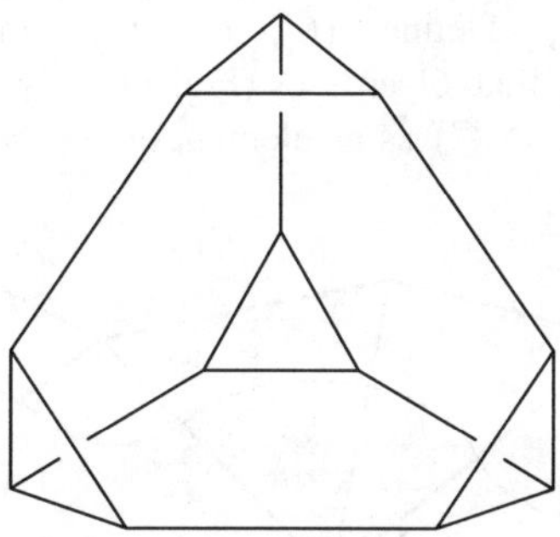

Figure 8

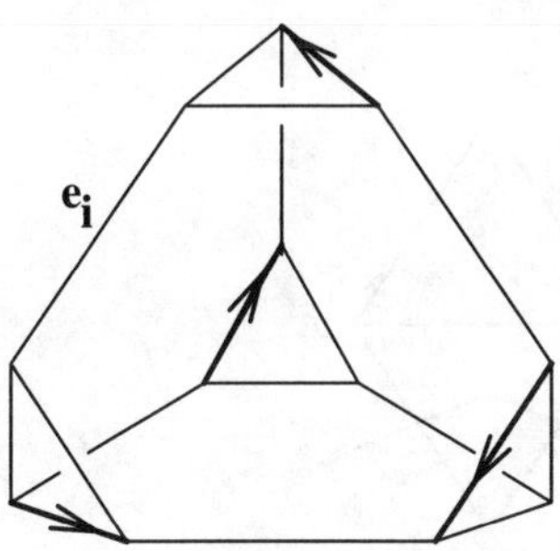

Figure 9

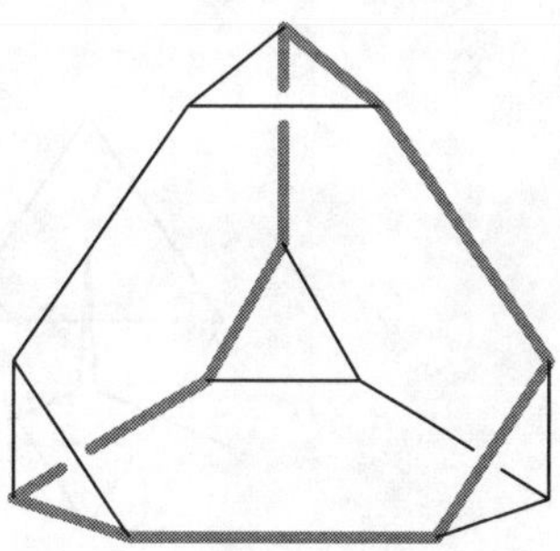

Figure 10

Definitionof $\gamma\colon H_3(\mathcal{J}) \to \mathcal{H}$. The triangular faces of the truncated tetrahedra give a quasi-simplicial triangulation of ∂K_0. Let $S_i(\partial K_0), Z_i(\partial K_0)$, and $B_i(\partial K_0)$ be the groups of simplicial chains, cycles, and boundaries for this triangulation. Let Δ be a 3-simplex of K. To an edge e_i of this simplex we associate a simplicial 1-chain $\gamma_0(e_i) \in S_1(\partial K_0)$ as in Fig. 9. For opposite edges e_i and e_{i+3} $(1 \leq i \leq 3)$ of Δ one has $\gamma_0(e_i) = \gamma_0(e_{i+3})$. Also, $\gamma_0(e_1 + e_2 + e_3)$ is a boundary cycle. Thus γ_0 induces a map $\gamma_1 \colon J \to S_1(\partial K_0)/B_1(\partial K_0)$. Moreover, it is easy to see that γ_1 maps $\operatorname{Im}\beta$ to boundaries and maps $\operatorname{Ker}\beta^*$ to cycles, so it induces a map

$$\gamma \colon H_3(\mathfrak{J}) = \operatorname{Ker}\beta^* / \operatorname{Im}\beta \to H_1(\partial K_0).$$

Note that, if we work modulo 2, then $\gamma_0(e_i)$ differs from the geometric representative for $\beta^*(e_i)$ by a 1-boundary in Δ (Fig. 10). Thus γ followed by the map $H_1(\partial K_0) \to H_1(K_0; \mathbb{Z}/2)$ is zero; that is

$$\operatorname{Im}\gamma \subseteq \mathcal{H} = \operatorname{Ker}(H_1(\partial K_0) \to H_1(K_0; \mathbb{Z}/2)).$$

Definitionof $\delta \colon H_1(\partial K_0) \longrightarrow H_3(\mathfrak{J})$. We use the dual cell decomposition of the quasi-triangulation of ∂K_0 to compute $H_1(\partial K_0)$. A simple cellular path C in the dual cell decomposition of ∂K_0 determines a vertex of each 2-simplex it passes through—the vertex common to the two edges of the 2-simplex that it crosses—and a sign $+$ or $-$ according as C goes counterclockwise or clockwise around this vertex. A vertex of a 2-simplex of ∂K_0 corresponds to an edge of a 3-simplex Δ of K, and hence to an element e of $J = \coprod J_\Delta$. Define $\delta(C)$ to be the signed sum of these elements e over all 2-simplices of ∂K_0 that C crosses (Fig. 11). By inspection, $\beta^*\delta(C) = 0$, so $\delta(C) \in \operatorname{Ker}\beta^*$. We consider $\delta(C)$ as an element of $\operatorname{Ker}\beta^* / \operatorname{Im}\beta = H_3(\mathfrak{J})$.

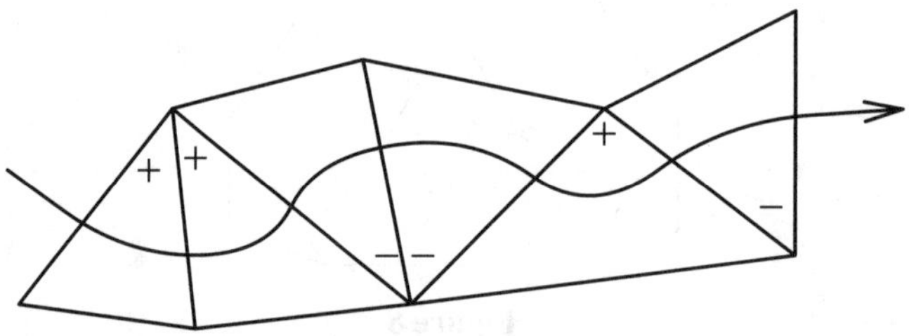

Figure 11

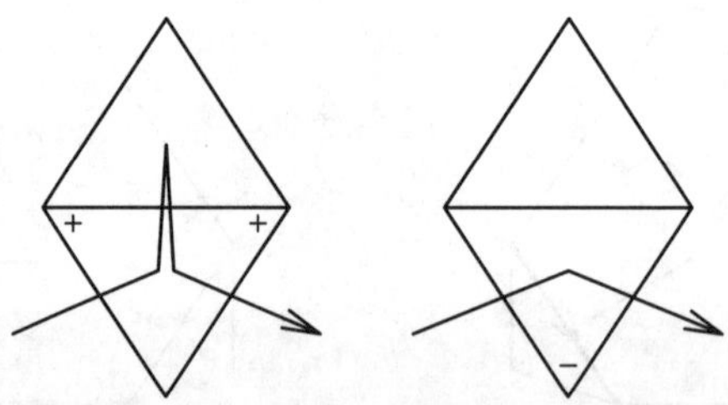

Figure 12

The definition of δ extends in the obvious fashion to arbitrary closed paths in the dual cell complex of ∂K_0 (the contribution where a path back-tracks, i.e., enters a 2-simplex and immediately departs across the same edge is zero). Eliminating back-tracking in a closed path C does not change the value of $\delta(C)$, see Fig. 12. If the path C is the boundary path of a 2-cell, then $\delta(C)$ is β of the corresponding edge of K, so it is zero in $H_3(\mathfrak{J})$. It follows that $\delta(C) \in H_3(\mathfrak{J})$ only depends on the homology class of C, so δ gives a well-defined homomorphism from $H_1(\partial K_0)$ to $H_3(\mathfrak{J})$.

Lemma 4.3.
(i) $\gamma \circ \delta\colon H_1(\partial K_0) \to H_1(\partial K_0)$ *is multiplication by* 2.
(ii) γ *and* δ *are isomorphisms after tensoring with* $\mathbb{Z}[\frac{1}{2}]$.
(iii) $\langle \delta x, \delta y\rangle = 2x \cdot y$, *where* $x \cdot y$ *means intersection form in* $H_1(\partial K_0)$.
(iv) $\gamma a \cdot \gamma b = 2\langle a, b\rangle$ *for* $a, b \in H_3(\mathfrak{J})$.
(v) $x \cdot \gamma a = \langle \delta x, a\rangle$ *for* $a \in H_3(\mathfrak{J})$ *and* $x \in H_1(\partial K_0)$, *that is,* $\gamma\colon H_3(\mathfrak{J})/\{\text{Torsion}\} \to H_1(\partial K_0)$ *and* $\delta\colon H_1(\partial K_0) \to H_3(\mathfrak{J})/\{\text{Torsion}\}$ *are adjoint maps with respect to the forms on* $H_1(\partial K_0)$ *and* $H_3(\mathfrak{J})/\{\text{Torsion}\}$.

Proof. For (i) it suffices to show that $\gamma\delta(C)$ is homologous to $2C$ for a simple closed cellular path C. Now (see Fig. 13), $\gamma\delta(C)$ will consist of contributions "near" C and contributions "far from" C.

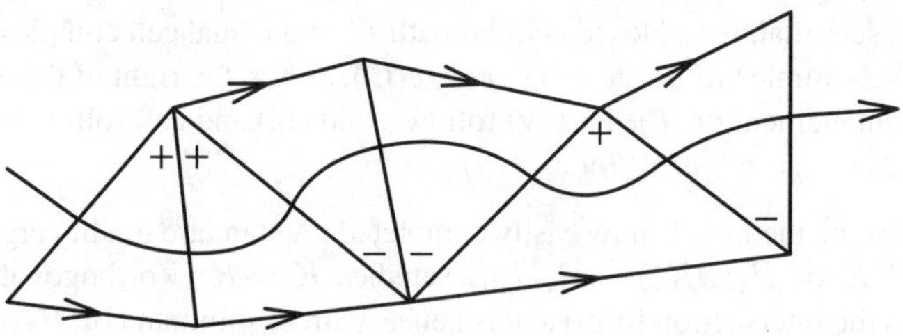

Figure 13a. The "near" contributions.

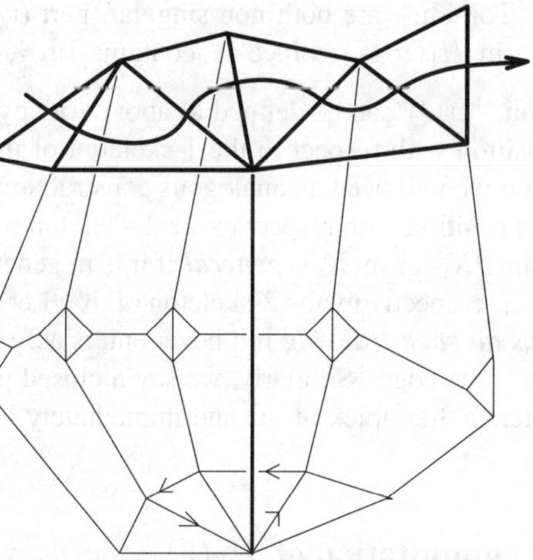

Figure 13b. The "far" contributions.

Contributions of the various types may overlap, but this does not affect the argument. By inspection one sees that the "near" contribution is homologous to $2C$ (it is the boundary of a regular neighborhood of C), while each "far" contribution is located at the far end of an edge of K which starts at a vertex of a 2-simplex of ∂K_0 through which

C passes, and is a null-homologous loop of length equal to the number of 2-simplices of ∂K_0 that C passes through as it passes that vertex. Thus (i) holds.

Since we already know that $H_3(\mathcal{J})$ has only even torsion, to deduce (ii) from (i) it suffices to show that the ranks of $H_3(\mathcal{J})$ and $H_1(\partial K_0)$ agree. But the rank of $H_3(\mathcal{J})$ equals the Euler characteristic of our chain complex, which is $n_0 - n_1 + 2n_3 - n_1 + n_0 = 2(n_0 - n_1 + n_3)$, where n_i is the number of i-simplices of K. Since $2n_3 = n_2$ (every 2-simplex is on two 3-simplices and every 3-simplex has four 2-faces), $2(n_0 - n_1 + n_3) = 2(n_0 - n_1 + n_2 - n_3) = 2\chi(K)$. The desired equation $2\chi(K) = \operatorname{rank} H_1(\partial K_0)$ now follows from the fact that the Euler characteristic of a compact 3-manifold is half the Euler characteristic of its boundary.

Note that (iii) through (v) are mutually equivalent, given (i) and (ii). (iii) is proved in **[NZ**, Section 3**]** (in **[NZ]** ∂K_0 consists of tori, but this is not used in the proof) but for completeness we give a proof here. Let S_1 and S_1' be the groups of simplicial 1-chains of ∂K_0 and 1-chains for the dual cell complex of ∂K_0. Intersection number defines a bilinear form $S_1' \times S_1 \to \mathbb{Z}$, which induces the usual intersection form on $H_1(\partial K_0)$. By inspection one sees that, for a closed cellular path C in the dual cell complex of ∂K_0 and an edge e of a 3-simplex of K, $C \cdot \gamma_0(e) = \langle \delta(C), e \rangle$ (on the right of this equation e is interpreted as an element of J). Part (v) follows, and (iii) and (iv) follow from (v)—for instance $x \cdot (2y) = x \cdot \gamma\delta(y) = \langle \delta(x), \delta(y) \rangle$. □

The proof of the theorem is now easily completed. A standard duality argument shows that the kernel K of $H_1(\partial K_0) \to H_1(K_0)$ satisfies $K = K^\perp$ (orthogonal complement with respect to the intersection form) and is hence a direct summand of $H_1(\partial K_0)$ of rank $h = (\operatorname{rank} H_1(\partial K_0))/2$. It follows that $\mathcal{H} = \operatorname{Ker}(H_1(\partial K_0) \to H_1(K; \mathbb{Z}/2))$ has index 2^h in $H_1(\partial K_0)$. On the other hand, since the intersection form on $H_1(\partial K_0)$ and the form $\langle\, ,\, \rangle$ on $H_3(\mathcal{J})/\{\text{Torsion}\}$ are both non-singular, part (iv) of the lemma implies that $\operatorname{Im}\gamma$ has index 2^h in $H_1(\partial K_0)$. Since $\mathcal{H}$ contains $\operatorname{Im}\gamma$, they are equal. □

Remark and Definition. $\delta(C)$ can be defined as above for any closed path C in ∂K_0 which is in general position with respect to the 1-skeleton of the quasi-triangulation of ∂K_0. In the next section we will need an analogous construction for closed curves C in K which are in general position with respect to the 2-skeleton of K. We shall therefore simply say that a path in ∂K_0 or in K is ***general*** if it is in general position with respect to the 1-skeleton of ∂K_0, respectively the 2-skeleton of K. Let C be such a closed path in ∂K_0. We say C ***has no back-tracking*** if it never enters a 2-simplex and immediately leaves again across the same edge. Similarly, we say a closed path in K ***has no back-tracking*** if it never enters a 3-simplex of K and immediately leaves it again across the same face.

5. Cohomological computation of $H_3(\mathcal{J})$

The main result of this section is Theorem 5.1, which gives an explicit computation of $H_3(\mathcal{J})$ in terms of two maps $\gamma': H_3(\mathcal{J}) \to H^1(\partial K_0; \mathbb{Z})$ and $\gamma_2': H_3(\mathcal{J} \otimes \mathbb{Z}/2) \to H^1(K_0; \mathbb{Z}/2)$. We must first define these maps.

$\gamma': H_3(\mathcal{J}) \to H^1(\partial K_0; \mathbb{Z})$ is just the Poincaré dual of the map γ of the previous section, that is $\gamma' = PD \circ \gamma$, where $PD: H_1(\partial K_0) \to H^1(\partial K_0)$ is the Poincaré duality isomorphism. We can also express it as the dual of the map $\delta\colon H_1(\partial K_0)\to H_3(\mathfrak{J})$ as follows:

$$\gamma'(a)(c) = \langle a, \delta(c)\rangle \quad \text{for } c \in H_1(\partial K_0). \tag{5.1}$$

Indeed, $\gamma'(a)(c) = PD(\gamma(a))(c) = \gamma(a)\cdot c = \langle a, \delta(c)\rangle$, where the last equality is by Lemma 4.3.

In a similar way we shall describe $\gamma_2'\colon H_3(\mathfrak{J}\otimes\mathbb{Z}/2)\to H^1(K_0;\mathbb{Z}/2)$ as the dual of a map $\delta_2\colon H_1(K_0;\mathbb{Z}/2)\to H_3(\mathfrak{J}\otimes\mathbb{Z}/2)$:

$$\gamma_2'(a)(c) = \langle a, \delta_2(c)\rangle \quad \text{for } c \in H_1(K_0;\mathbb{Z}/2), \tag{5.2}$$

where δ_2 is defined as follows.

If $c\in H_1(K_0;\mathbb{Z}/2)$ is represented by a general closed path C in K_0, then each time C passes through a 3-simplex Δ from one face to another, it determines an edge e of Δ, namely the edge common to the two faces, and hence an element, also denoted e, of $J\otimes\mathbb{Z}/2$. We let $\delta_2(c)$ be the class in $H_3(\mathfrak{J}\otimes\mathbb{Z}/2)$ of the sum of these elements e. The proof that this sum is indeed a cycle, i.e., in $\operatorname{Ker}(\beta^*\otimes\mathbb{Z}/2)$, and that modulo $\operatorname{Im}(\beta\otimes\mathbb{Z}/2)$ it only depends on the homology class of C, is entirely analogous to the corresponding proof for δ.

Let $\iota\colon H_3(\mathfrak{J})\to H_3(\mathfrak{J}\otimes\mathbb{Z}/2)$ be the natural map. We shall be interested in the composition

$$\gamma_2'\iota\colon\ H_3(\mathfrak{J})\xrightarrow{\iota} H_3(\mathfrak{J}\otimes\mathbb{Z}/2)\xrightarrow{\gamma_2'} H^1(K_0;\mathbb{Z}/2).$$

Theorem 5.1. *The following diagram is a pullback diagram:*

$$\begin{array}{ccc} H_3(\mathfrak{J}) & \xrightarrow{\gamma'} & H^1(\partial K_0;\mathbb{Z}) \\ \big\downarrow{\scriptstyle \gamma_2'\iota} & & \big\downarrow{\scriptstyle r} \\ H^1(K_0;\mathbb{Z}/2) & \xrightarrow{i^*} & H^1(\partial K_0;\mathbb{Z}/2), \end{array}$$

where i^* is induced by $i\colon\partial K_0\to K_0$ and r is reduction modulo 2. Equivalently, the following sequence is exact:

$$0\to H_3(\mathfrak{J})\xrightarrow{(\gamma',\gamma_2'\iota)} H^1(\partial K_0;\mathbb{Z})\oplus H^1(K_0;\mathbb{Z}/2)\xrightarrow{r-i^*} H^1(\partial K_0;\mathbb{Z}/2)\to 0.$$

Proof. The commutativity of the diagram is immediate from (5.1) and (5.2) and the fact that if C is a cellular curve on ∂K_0 and C_2 is the result of pushing C inside K_0, then $\delta_2(C_2)=\delta(C) \pmod 2$.

The rest of the proof will take several steps.

Step 1. *$\gamma_2'\colon H_3(\mathfrak{J}\otimes\mathbb{Z}/2)\to H^1(K_0;\mathbb{Z}/2)$ is surjective.*

Until further notice coefficient group $\mathbb{Z}/2$ is understood.

Step 1 is most easily seen by translating to homology via the Poincaré duality isomorphism $H^1(K_0)\to H_2(K_0,\partial K_0)$. Note that $H_2(K_0,\partial K_0)=H_2(K)$ by the exact sequence for the pair $(K, K-K_0)$ and excision. We shall construct a map $\gamma_2\colon H_3(\mathfrak{J}\otimes\mathbb{Z}/2)\to H_2(K)$ and show it is the Poincaré dual of $\gamma_2'\colon H_3(\mathfrak{J}\otimes\mathbb{Z}/2)\to H^1(K_0)$.

Let, for the moment, S_2, Z_2, and B_2 denote the groups of simplicial 2-chains, 2-cycles, and 2-boundaries of K with coefficients $\mathbb{Z}/2$. For an edge e of a 3-simplex

Δ of K, let $\gamma_2(e) \in S_2$ be the sum of the two 2-simplices represented by the two faces of Δ that meet in e. This clearly induces a map $\gamma_2\colon J \to S_2/B_2$. Moreover, it is geometrically clear that this map maps $\operatorname{Ker}(\beta^* \otimes \mathbb{Z}/2)$ to cycles and $\operatorname{Im}(\beta \otimes \mathbb{Z}/2)$ to zero, and hence induces a map

$$\gamma_2\colon H_3(\mathfrak{J} \otimes \mathbb{Z}/2) \to H_2(K).$$

Now for an edge e of a 3-simplex of k and a closed curve in K_0, we have by inspection, $\gamma_2(e) \cdot C = \langle e, \delta_2(C)\rangle \pmod 2$. Thus, for $a \in J$ we have $\gamma_2(a) \cdot C = \langle a, \delta_2(C)\rangle = \gamma_2'(a)(C)$, whence γ_2' is indeed the Poincaré dual of γ_2.

Now note that any modulo-2 simplicial 2-cycle $Z \in Z_2$ is a sum of an even number of 2-simplices, since $\partial Z = 0$ and the boundary of a single 2-simplex is the sum of an odd number of 1-simplices. Call two 2-simplices of K "adjacent" if they lie on a common 3-simplex and call their sum an "adjacent pair". Any sum of two 2-simplices can be rewritten modulo 2 as a sum of adjacent pairs (connect the two 2-simplices by a path in K and form adjacent pairs from the successive 2-simplices that the path meets). Thus any $Z \in Z_2$ is a sum of adjacent pairs. Each of these adjacent pairs determines an edge of a 3-simplex, hence an element e of $J \otimes \mathbb{Z}/2$. Let a be the sum of these elements e. Clearly $Z = \gamma_2(a)$, and the fact that $\partial Z = 0$ translates directly to $a \in \operatorname{Ker}(\beta^* \otimes \mathbb{Z}/2)$. Thus $\gamma_2\colon H_3(\mathfrak{J} \otimes \mathbb{Z}/2) \to H_2(K)$ is surjective. Hence γ_2' is surjective.

Step 2. *The following sequence is exact (coefficients $\mathbb{Z}/2$):*

$$0 \to H_1(K_0) \xrightarrow{\delta_2} H_3(\mathfrak{J} \otimes \mathbb{Z}/2) \xrightarrow{\gamma_2'} H^1(K_0) \to 0.$$

We have just shown the surjectivity of γ_2' and the injectivity of δ_2 follows from this and equation (5.2), since $\delta_2(c) = 0$ implies $\gamma_2'(a)(c) = 0$ for all a, hence $x(c) = 0$ for all $x \in H^1(K_0)$, hence $c = 0$.

The equation $\gamma_2'\delta_2 = 0$ is equivalent, by the definition of γ_2', to the condition that $\langle \delta_2(C), \delta_2(C')\rangle = 0 \pmod 2$ for any two closed curves C and C' in K_0. We can assume each curve is general, and then it is easy to see that $\langle \delta_2(C), \delta_2(C')\rangle \pmod 2$ counts the number of instances of C and C' passing through the same 2-face of a 3-simplex of K. But for each such instance, they pass through the same face viewed as a face of the adjacent 3-simplex, so the total number is zero modulo 2.

We now know $\operatorname{Im}\delta_2 \subseteq \operatorname{Ker}\gamma_2'$, and to show this inclusion is an equality it suffices to show that $\dim H_3(\mathfrak{J}\otimes\mathbb{Z}/2) = \dim H_1(K_0) + \dim H^1(K_0)$, that is, $\dim H_3(\mathfrak{J}\otimes\mathbb{Z}/2) = 2\dim H_1(K_0)$. But Theorem 4.2 and the universal coefficient theorem imply that $H_3(\mathfrak{J}\otimes \mathbb{Z}/2)$ has dimension $\dim H_1(\partial K_0) + 2\dim H_1(K)$, so we must show $\dim H_1(K_0) = (1/2)\dim H_1(\partial K_0) + \dim H_1(K)$. The long exact sequence for $(K, K - K_0)$ and excision show $H_2(K, K_0) = H_1(\partial K_0)$, and inserting this in the long exact sequence for (K, K_0) gives an exact sequence

$$H_1(\partial K_0) \xrightarrow{i_*} H_1(K_0) \xrightarrow{j_*} H_1(K) \to 0,$$

where the maps are the natural maps. Since $\dim \operatorname{Ker}(i_*) = (1/2)\dim H_1(\partial K_0)$ by Poincaré duality, the desired equality follows.

Note that the above sequence induces a short exact sequence

$$0 \to \mathcal{K} \xrightarrow{i_*} H_1(K_0) \xrightarrow{j_*} H_1(K) \to 0,$$

with $\mathcal{K} = H_1(\partial K_0)/\operatorname{Ker}(i_*)$.

Step 3. (*Coefficients are still* $\mathbb{Z}/2$.) *There is a commutative diagram with exact rows and columns:*

$$\begin{array}{ccccccccc}
 & & 0 & & 0 & & & & \\
 & & \downarrow & & \downarrow & & & & \\
0 & \to & \mathcal{K} & \xrightarrow{\delta^{(2)}} & H_3(\mathcal{J}) \otimes \mathbb{Z}/2 & \xrightarrow{\gamma_2' \iota} & H^1(K_0) & \to & 0 \\
 & & \downarrow{\scriptstyle i_*} & & \downarrow{\scriptstyle \iota} & & \downarrow{\scriptstyle =} & & \\
0 & \to & H_1(K_0) & \xrightarrow{\delta_2} & H_3(\mathcal{J} \otimes \mathbb{Z}/2) & \xrightarrow{\gamma_2'} & H^1(K_0) & \to & 0 \\
 & & \downarrow{\scriptstyle j_*} & & \downarrow{\scriptstyle \kappa} & & & & \\
 & & H_1(K) & \xrightarrow{=} & H_1(K) & & & & \\
 & & \downarrow & & \downarrow & & & & \\
 & & 0 & & 0 & & & &
\end{array}$$

Here the middle vertical sequence is the universal coefficient theorem for the chain complex $\mathcal{J}$ and the middle horizontal sequence is the sequence of Step 2. We must describe the top horizontal sequence and prove commutativity of the diagram; exactness of the top sequence then follows by a diagram chase.

The map $\delta^{(2)}$ is induced by the map $\delta \otimes \mathbb{Z}/2\colon H_1(\partial K_0) \to H_3(\mathcal{J}) \otimes \mathbb{Z}/2$: indeed, $\iota(\delta \otimes \mathbb{Z}/2) = \delta_2 i_*$ (this was pointed out at the beginning of the proof of Theorem 5.1), so $\delta \otimes \mathbb{Z}/2$ has kernel $\operatorname{Ker}(i_*)$, so $\delta^{(2)}$ can be defined to make the upper left square commute.

It remains to show commutativity of the lower square. For $a \in H_3(\mathcal{J} \otimes \mathbb{Z}/2)$ one computes $\kappa(a)$ as follows. Represent a by an element $A \in J$. Then $\beta^*(A)$ is 0 modulo 2. The desired element $\kappa(a) \in H_2(\mathcal{J}) = H_1(K)$ is represented by $(1/2)\beta^*(A)$.

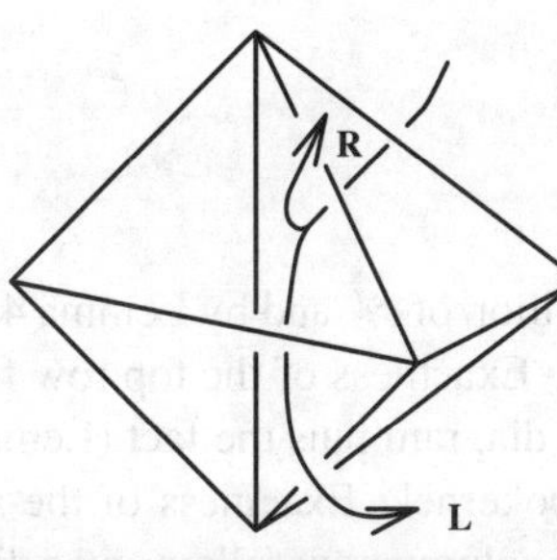

Figure 14

Figure 15

Consider a general path C in K_0. As C passes through a 3-simplex Δ it determines an edge e of Δ, as previously described, and $\delta_2(C)$ is represented by the sum of the elements of J corresponding to these e. Call the edge E of K corresponding to e the "axis" of C passing through Δ. In the next 3-simplex C may continue around

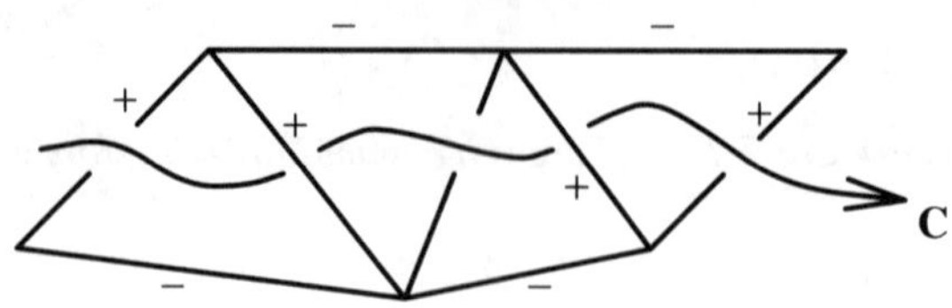

Figure 16

the same axis, turn left, or turn right (see Fig. 14). We can eliminate left turns from the path C without changing its homology class, as illustrated in Fig. 15. Thus, without loss of generality, C has no left turns. The successive axes of C then form a path in K homologous to C. Moreover, each pair of successive axes determines a face of a 3-simplex in K and hence an additional edge (the third edge of that face). By inspection, one sees that $\kappa\delta_2(C)$ is represented by the sum of the axes minus the sum of these additional edges (Fig. 16), and this is homologous to C modulo 2, as desired.

Step 4. (*Omitted coefficients are still* $\mathbb{Z}/2$.) *There is a commutative diagram with exact rows and columns:*

$$\begin{array}{ccccccccc}
 & & 0 & & 0 & & & & \\
 & & \downarrow & & \downarrow & & & & \\
0 & \to & H_1(\partial K_0;\mathbb{Z}) & \xrightarrow{\delta} & H_3(\mathfrak{J}) & \xrightarrow{\gamma_2'\iota} & H^1(K_0) & \to & 0 \\
 & & \downarrow{\scriptstyle \cdot 2} & & \downarrow{\scriptstyle (\gamma',\gamma_2'\iota)} & & \downarrow{\scriptstyle =} & & \\
0 & \to & H_1(\partial K_0;\mathbb{Z}) & \xrightarrow{(PD,0)} & H^1(\partial K_0;\mathbb{Z})\oplus H^1(K_0) & \xrightarrow{pr_2} & H^1(K_0) & \to & 0 \\
 & & \downarrow & & \downarrow{\scriptstyle r-i^*} & & & & \\
 & & H_1(\partial K_0) & \xrightarrow{PD} & H^1(\partial K_0) & & & & \\
 & & \downarrow & & \downarrow & & & & \\
 & & 0 & & 0 & & & &
\end{array}$$

The commutativity of the top left square is by definition of γ' and by Lemma 4.3, and the commutativity of the rest of the diagram is trivial. Exactness of the top row follows directly from exactness of the top row of the previous diagram plus the fact (Lemma 4.3 and Theorem 4.2) that δ is injective with 2-torsion cokernel. Exactness of the second row and first column is trivial. Exactness of the middle column now follows by a diagram chase, completing the proof of Step 4 and of Theorem 5.1. □

6. Proof of Theorem 2.4

Let $\overline{J}_\Delta$ be defined like J_Δ but without the relation $e_1 + e_2 + e_3 = 0$; that is, $\overline{J}_\Delta$ is generated by the six edges $e_1, \dots, e_6$ of Δ with the relations $e_i = e_{i+3}$ for $i = 1, 2, 3$. We give $\overline{J}_\Delta$ the standard bilinear form $(e_i, e_j) = \delta_{ij}$ for $i, j \in \{1, 2, 3\}$. Let $\overline{J}$ be the orthogonal sum of the $\overline{J}_\Delta$. The map $\beta^*: J \to C_2$ can be factored as

$$\beta^*: J \xrightarrow{\beta_1} \overline{J} \xrightarrow{\beta_2} C_2,$$

with β_1 and β_2 defined on the Δ-component by

$$\left.\begin{aligned} \beta_1(e_i) &= e_{i+1} - e_{i+2} \\ \beta_2(e_i) &= j(e_i) + j(e_{i+3}) \end{aligned}\right\} \text{for } i = 1, 2, 3. \tag{6.1}$$

Note that $\beta_1: J \to \overline{J}$ and the natural projection $p: \overline{J} \to J$ are adjoint maps with respect to the forms on J and $\overline{J}$; that is,

$$(\beta_1(a), x) = \langle a, p(x) \rangle \quad \text{for } a \in J \text{ and } x \in \overline{J}. \tag{6.2}$$

Let C be a general closed path in ∂K_0 with no back-tracking. As described in Sect. 4, as C passes through a 2-simplex of ∂K_0 it determines an edge e of the corresponding 3-simplex Δ and a sign $\pm$, and hence an element—which we call $\pm e$—of $\overline{J}$. Let $\overline{\delta}(e)$ be the sum of all these $\pm e$. Thus the map $\delta: H_1(\partial K_0) \to H_3$ of Sects. 4 and 5 is given by

$$\delta([C]) = [p\overline{\delta}(C)]. \tag{6.3}$$

For $x \in \overline{J}$ define an integer-valued map $\overline{\gamma}(x)$ on the set of closed non-back-tracking paths on ∂K_0 by

$$\overline{\gamma}(x)(C) = (x, \overline{\delta}(C)). \tag{6.4}$$

(Note that $\overline{\gamma}(x)(C)$ will usually not just depend on the homology class of C.) The map γ' of the previous section is given by

$$\gamma'(a) = \overline{\gamma}(\beta_1(a)) \quad \text{for } a \in J, \tag{6.5}$$

since, by (6.4), (6.2), (6.3), and the definition of γ', $\overline{\gamma}(\beta_1(a))(C) = (\beta_1(a), \overline{\delta}(C)) = \langle a, \delta(C) \rangle = \gamma'(a)(C)$.

Similarly, if C is a general closed path with no back-tracking in K, then as C passes through a 3-simplex Δ of K it determines an edge e of Δ (the edge common to the two faces that it passes through), and we define $\overline{\delta}_2(C) \in \overline{J} \otimes \mathbb{Z}/2$ to be the modulo 2 sum of these e. For $x \in \overline{J}$ define a $\mathbb{Z}/2$-valued map $\overline{\gamma}_2(x)$ on the set of general closed non-back-tracking paths in K by

$$\overline{\gamma}_2(x)(C) = (x, \overline{\delta}_2(C)) \pmod 2. \tag{6.6}$$

As for $\overline{\gamma}$ one sees that the map γ_2' of the previous section is given by

$$\gamma_2'(a) = \overline{\gamma}_2\beta_1(a) \quad \text{for } a \in J. \tag{6.7}$$

Lemma 6.1. *Suppose every component of ∂K_0 is a torus. Then there exists an element $\eta \in \overline{J}$ satisfying:*
1. *the $\overline{J}_\Delta$ component of η has coefficient sum 1 for each 3-simplex Δ,*
2. *$\beta_2(\eta) = 2\sum E$, twice the sum of all the edges of K,*
3. *$\overline{\gamma}(\eta)(C) = 0$ for every general essential simple closed curve C in ∂K_0 with no back-tracking,*
4. *$\overline{\gamma}_2(\eta)(C) = 0$ modulo 2 for every general closed curve in K_0 with no back-tracking.*

Moreover, any such η is unique up to $\operatorname{Im}\beta_1\beta$.

Proof. Choose any element x of $\overline{J}$ which satisfies condition 1. The contribution at a vertex V of K to $\alpha^*(\beta_2(x) - 2\sum E)$ is $n_2(V) - 2n_0(V)$, where $n_i(V)$ is the number of i-simplices in the link L_V of vertex V. But $n_2(V) - 2n_0(V) = -2\chi(L_V)$, which is zero since L_V is a torus. Thus $\beta_2(x) - 2\sum E$ is in $\operatorname{Ker}\alpha^*$.

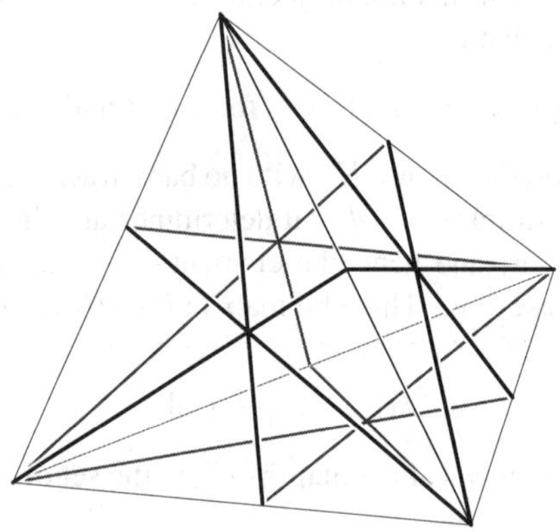

Figure 17

We claim that $\beta_2(x) - 2\sum E$ is in $\operatorname{Im}\beta^*$, that is, it represents zero in $H_2(\mathfrak{J}) = H_1(K;\mathbb{Z}/2)$. Indeed, let x_Δ be the $\overline{J}_\Delta$-component of x and consider $\beta_2(x_\Delta)$ as a modulo-2 1-chain in $\Delta \subseteq K$. The boundary of this 1-chain is the sum of the vertices of Δ. The same is true for the 1-chain S_Δ pictured in Fig. 17 (the edges of the tetrahedron are not part of the chain). Hence $\beta_2(x_\Delta) - S_\Delta$ is a 1-cycle in $\Delta \subseteq K$, hence a 1-boundary. Thus, as a modulo-2 1-cycle, $\beta_2(x)$ differs from $\sum S_\Delta$ by a boundary. Since $\sum S_\Delta$ is identically zero modulo 2, $\beta_2(x) - 2\sum E$ is a modulo-2 boundary, as claimed.

Now choose $a \in J$ with $\beta^*(a) = \beta_2(x) - 2\sum E$. Then $y = x - \beta_1(a)$ satisfies conditions 1 and 2 of the lemma.

Now it is not hard to verify, given that y satisfies conditions 1 and 2 of the lemma, that for a general essential simple closed curve C in ∂K_0 without back-tracking, $\overline{\gamma}(y)(C)$ only depends on the homology class of C. Moreover, if one restricts to curves of this type (or more generally, to general curves C without back-tracking that are isotopic through immersed curves to essential simple closed curves, i.e., C should have no "winding"), the map $C \mapsto \overline{\gamma}(y)(C)$ defines a homomorphism $H^1(\partial K_0) \to \mathbb{Z}$, and hence an element c of $H^1(\partial K_0)$. Similarly, evaluation of $\overline{\gamma}_2$ on general closed curves in K_0 without back-tracking defines an element $c_2 \in H^1(K_0;\mathbb{Z}/2)$. Moreover, $r(c) = i^*(c_2)$. Thus, by Theorem 5.1, there exists a $w \in H_3(\mathfrak{J})$ with $(\gamma'(w), \gamma_2'\iota(w)) = (c, c_2)$. Let $a \in J$ be a representative for w. Then $\eta = y - \beta_1(a)$ satisfies conditions 1, 2, 3, and 4.

Given two elements η satisfying the conditions, their difference ζ will satisfy:
$1'$. The $\overline{J}_\Delta$ of ζ has coefficient sum 0 for each 3-simplex Δ,
$2'$. $\beta_2(\zeta) = 0$,
and conditions 3 and 4 of the lemma. Condition $1'$ is equivalent to $\zeta \in \operatorname{Im}\beta_1$, say $\zeta = \beta_1(c)$. Condition $2'$ then says that $c \in \operatorname{Ker}\beta^*$, so c is a cycle for $H_3(\mathfrak{J})$. Conditions 3 and 4 say that c represents 0 in $H_3(\mathfrak{J})$, so c is in $\operatorname{Im}\beta$. □

Note that for any η satisfying condition 1 of the lemma one can solve equations (2.8) for c'_ν and c''_ν. Theorem 2.4 is thus just a restatement of the above lemma, so it is proved. □

References

[CS] M. Culler and P. B. Shalen, Varieties of group representations and splittings of 3-manifolds, Ann. of Math. 117 (1983), 109–146

[D] Johan L. Dupont, The dilogarithm as a characteristic class for flat bundles, Proc. of the Northwestern conference on cohomology of groups (Evanston, Ill., 1985), J. Pure Appl. Algebra 44 (1987), 137–164.

[HMW] Craig D. Hodgson, G. Robert Meyerhoff, and Jeffrey R. Weeks, Surgeries on the Whitehead link yield geometrically similar manifolds, These Proceedings.

[MN] R. Meyerhoff and Walter D. Neumann, An asymptotic formula for the eta invariants of hyperbolic 3-manifolds, to appear in Comment. Math. Helvetici (1991).

[M] R. Meyerhoff, The Chern-Simons invariant for hyperbolic 3-manifolds, Thesis, Princeton 1981.

[NR] W. D. Neumann and Alan W. Reid, Arithmetic of hyperbolic 3-manifolds, These Proceedings.

[NZ] Walter D. Neumann and Don Zagier, Volumes of hyperbolic three-manifolds, Topology 24 (1985), 307–332.

[Y1] Tomoyoshi Yoshida, The η-invariant of hyperbolic 3-manifolds, Invent. Math. 81 (1985), 473–514

[Y2] Tomoyoshi Yoshida, On ideal points of deformation curves of hyperbolic 3-manifolds with one cusp, To appear in Topology

Department of Mathematics, The Ohio State University, Columbus, OH 43210
Email: neumann@mps.ohio-state.edu

Arithmetic of Hyperbolic Manifolds

Walter D. Neumann and Alan W. Reid

1. Introduction

By a hyperbolic 3-manifold we mean a complete orientable hyperbolic 3-manifold of finite volume, that is a quotient $\mathbb{H}^3/\Gamma$ with $\Gamma \subset \mathrm{PSL}_2\mathbb{C}$ a discrete subgroup of finite covolume (here briefly "a Kleinian group").

Among hyperbolic 3-manifolds, the arithmetic ones form an interesting, and in many ways more tractable, subclass. The tractability comes from the availability of arithmetic tools and invariants. For example, an arithmetic manifold $M = \mathbb{H}^3/\Gamma$ is determined up to commensurability by its defining field k (a number field with exactly one complex place) and quaternion algebra A (which is ramified at all real places of k). Any such pair (k, A) determines a unique commensurability class of arithmetic hyperbolic 3-manifolds.

One aim of this paper is to try to extend arithmetic considerations to more general hyperbolic 3-manifolds. For example, a commensurability invariant pair $(k(M), A(M))$ consisting of a non-totally-real number field and a quaternion algebra over it is defined for any $M = \mathbb{H}^3/\Gamma$ (Sects. 2 and 3; we also write $(k(\Gamma), A(\Gamma))$), but it fails to be a complete commensurability invariant of M—non-commensurable M can have the same $A(M)$ (see Sect. 10 for examples). Nevertheless, $k(M)$ and $A(M)$ do contain useful information—for instance (Theorem 3.2 and Proposition 3.3) ramification of $A(M)$ at a finite prime forces subgroups of Γ to have non-trivial abelianizations, $A(M)$ gives a good amphicheirality test (Proposition 3.4), and $k(\Gamma)$ composes under amalgamation of Kleinian groups along a non-elementary Kleinian group (Theorem 2.8 and **[NR1]**) and is therefore a mutation invariant.

The trace field $\mathbb{Q}(\mathrm{tr}\,\Gamma)$ of Γ is not a commensurability invariant and the field $k(\Gamma)$ is in fact the smallest field among the trace fields of finite index subgroups of Γ. We call it the ***invariant trace field*** of Γ. We show that the trace field $\mathbb{Q}(\mathrm{tr}\,\Gamma)$ is a Galois $(\mathbb{Z}/2)^m$-extension of the invariant trace field $k(\Gamma)$ and there is a precise Galois relationship between subgroups of Γ and their trace fields, the largest subgroup of Γ with trace field $k(\Gamma)$ being normal with quotient $(\mathbb{Z}/2)^m$ (Theorem 2.2).

It turns out that the existence of parabolic elements in Γ facilitates many arithmetic questions. For example, in this case $A(\Gamma)$ is just the matrix algebra $\mathrm{M}_2(k(\Gamma))$. Moreover, $k(\Gamma)$ equals the field generated by the tetrahedral parameters of the ideal tetrahedra of any ideal triangulation of $\mathbb{H}^3/\Gamma$ (Theorem 2.4). If one conjugates Γ so that three parabolic fixed points are at 0, 1, and ∞ in $\mathbb{C} \cup \{\infty\} = \partial\overline{\mathbb{H}}^3$, then $k(\Gamma)$ is also the field generated by all parabolic fixed points (Lemma 2.5).

As another example, an arithmetic orbifold with cusps cannot have geodesics shorter than 0.431277313 (cf. Theorem 4.6 and Corollary 4.7); a corresponding result for compact orbifolds is conjectured (with a bound of 0.09174218) but depends on Lehmer's conjecture of number theory.

We illustrate the greater tractability of non-compact manifolds also by doing some explicit computations for a family of manifolds related to Dehn surgeries on one component of the Whitehead link. Before describing this we need some terminology.

The most striking dichotomy in the arithmetic of hyperbolic groups is a result of Margulis (cf. **[Z**, Ch. 6**]**). Define the "commensurator of Γ" to be

$$\mathrm{Comm}(\Gamma) = \{g \in \mathrm{Isom}(\mathbb{H}^3) : [\Gamma{:}\Gamma \cap g^{-1}\Gamma g] < \infty\},$$

and let $\mathrm{Comm}^+(\Gamma)$ be its orientation preserving subgroup. Margulis shows that $\mathrm{Comm}^+(\Gamma)$ is discrete (and hence contains Γ with finite index) if and only if Γ is non-arithmetic. Thus in this case $\mathrm{Comm}^+(\Gamma)$ is the unique maximal element of the commensurability class of Γ. In the arithmetic case Borel showed that there are infinitely many maximal elements in the commensurability class (**[Bo]**).

As several people have observed, $\mathrm{Comm}(\Gamma)$ can be defined purely group-theoretically. Define a "virtual automorphism" of Γ to be an isomorphism $\phi{:}\,\Gamma_1 \to \Gamma_2$ between subgroups of finite index in Γ and define two virtual automorphisms to be "equivalent" if they agree on some subgroup of Γ of finite index. Then Mostow rigidity easily implies that $\mathrm{Comm}(\Gamma)$ is the group of equivalence classes of virtual automorphisms under composition. (We like to call this version of $\mathrm{Comm}(\Gamma)$ the "abstract commensurator" of Γ, since in geometric situations without rigidity, for instance Fuchsian groups, it is much larger than the geometric commensurator. It has also been invented by group theorists, since it is interesting for other groups too; a simple example is $\mathrm{Comm}(\mathbb{Z}^n) = \mathrm{GL}_n\mathbb{Q}$. Bass and Kulkarni (**[BK]**) have investigated it for lattices on trees.)

From a geometric point of view, a virtual automorphism $\phi{:}\,\Gamma_1 \to \Gamma_2$ for Γ represents an isometry (or just an isotopy class of homeomorphisms) between the two finite covers $\mathbb{H}^3/\Gamma_1$ and $\mathbb{H}^3/\Gamma_2$ of $M = \mathbb{H}^3/\Gamma$. We call such an isometry a "virtual symmetry" of M. We call it a "hidden symmetry" if it does not lift from a symmetry of M, i.e., ϕ lies in $\mathrm{Comm}(\Gamma) - N(\Gamma)$, where $N(\Gamma)$ is the normalizer of Γ in $\mathrm{Isom}(\mathbb{H}^3)$. Define virtual symmetries to be equivalent if they have a common lift to some mutual finite cover of M (caution: the set of equivalence classes does not form a group). Thus Margulis' theorem implies that arithmeticity of M is equivalent to the existence of infinitely many non-equivalent hidden symmetries, while for non-arithmetic M there is a finite cover of M whose symmetries give all virtual symmetries of M.

As an example, if $\mathcal{O}_d$ is the ring of integers of $\mathbb{Q}(\sqrt{-d})$, then $\mathrm{Comm}\,\mathrm{PSL}_2\mathcal{O}_d$ is $\mathrm{PGL}_2\mathbb{Q}(\sqrt{-d})$ extended by an orientation reversing involution (complex conjugation). The existence of hidden symmetries for the orbifold $M = \mathbb{H}^3/\mathrm{PSL}_2\mathcal{O}_1$ is nicely illustrated by examples occurring in the literature: the complement B of the Borromean rings in S^3 covers M in at least two inequivalent ways—an embedding of $\pi_1 B$ as a normal subgroup of $\mathrm{PSL}_2\mathcal{O}_1$ is given in **[FN]** and a non-normal embedding is given in **[Ri]**; it is not hard to check that an element of $\mathrm{Comm}\,\mathrm{PSL}_2\mathcal{O}_1$ which conjugates one of these embeddings to the other cannot generate a discrete group with $\mathrm{PSL}_2\mathcal{O}_1$.

Note that the question about whether hidden symmetries exist for $M = \mathbb{H}^3/\Gamma$ is just the question about whether $\mathrm{Comm}(\Gamma)$ (hidden symmetries) equals the normalizer of Γ (symmetries), i.e., whether Γ is normal in $\mathrm{Comm}(\Gamma)$.

In Sections 5–8 of this paper we consider the manifold $C(p, s)$ which is the complement of a p link chain in S^3 with s left half twists (or $-s$ right half twists) pictured in Fig. 1 (we also make sense of this manifold—up to commensurability—for $p = 0$). We show it has a hyperbolic structure if and only if $\{|p+s|, |s|\} \not\subseteq \{0, 1, 2\}$ and determine its commensurator (i.e., the commensurator of its fundamental group) and some arithmetic invariants in this case. These computations use that $C(p, s)$ is commensurable with the result of $(p, s/2)$ Dehn surgery on one component of the Whitehead link (one can make sense of (p, q) surgery for q a half-integer; see Sect. 5). These Dehn surgeries on the Whitehead link are also discussed in **[HMW]**.

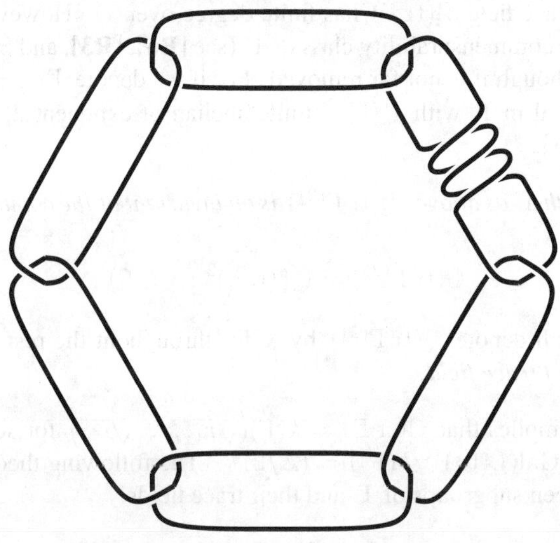

Figure 1. $C(p, s)$ for $p = 6$, $s = -3$

We give the finite list of (p, s) for which $C(p, s)$ is arithmetic—it is then commensurable with $\mathrm{PSL}_2\mathcal{O}_d$ with $d \in \{1, 2, 3, 7, 15\}$ so the commensurator is $\mathrm{PGL}_2\mathbb{Q}(\sqrt{-d})$ extended by complex conjugation. In all other cases we show that $\pi_1 C(p, s)$ is normal in its orientation preserving commensurator, with quotient the "obvious" group of orientation preserving symmetries of $C(p, s)$, and an orientation reversing commensuration exists if and only if $p + s = \pm s$ or $p + s = 0$ or $s = 0$.

We say Γ (or $\mathbb{H}^3/\Gamma$) has ***integral traces*** if $\{\mathrm{tr}\,\gamma : \gamma \in \Gamma\}$ consists of algebraic integers; it then follows that Γ can be conjugated to a subgroup of $\mathrm{PSL}_2\mathbb{A}$, where $\mathbb{A}$ is the ring of algebraic integers. In **[Ba2]** Bass shows that if a hyperbolic manifold $M = \mathbb{H}^3/\Gamma$ has no closed incompressible surface of genus > 1 then Γ has integral traces. Note that the property of having integral traces is a commensurability invariant (since an element $\gamma \in \mathrm{PSL}_2\mathbb{C}$ has algebraic integer trace if and only if it has algebraic integer eigenvalues, which is true if it is true for some power of γ). In Theorem 6.3 we determine which $C(p, s)$ have non-integral traces, namely those with $p + s = \pm s$ and

s equal to ± 2 times an odd prime power. These appear to be the first examples in the literature of hyperbolic manifolds with non-integer traces.

In Sections 9 and 10 we collect some questions and comments.

Acknowledgments. The first author acknowledges support of the NSF for this research. Both authors acknowledge useful conversations and correspondence with C. Adams and R. Riley. They are also most grateful to T. Chinburg for several pages of useful comments to the first version of this paper.

2. Trace-fields

2.1. Let Γ be a Kleinian group of finite covolume. It is well known, and follows from rigidity, that the trace field $\mathbb{Q}(\operatorname{tr}\Gamma)$ has finite degree over $\mathbb{Q}$. However, $\mathbb{Q}(\operatorname{tr}\Gamma)$ is not an invariant of the commensurability class of Γ (see **[Bo]**, **[R3]**, and Sect. 6 of this paper for examples), although it is not far removed. For if we denote $\Gamma^{(2)} = gp\{\gamma^2 : \gamma \in \Gamma\}$, then $\Gamma^{(2)}$ is normal in Γ with $\Gamma/\Gamma^{(2)}$ finite abelian of exponent 2, and the following was proved in **[R3]**.

Theorem 2.1. *With Γ as above, $\mathbb{Q}(\operatorname{tr}\Gamma^{(2)})$ is an invariant of the commensurability class of Γ. Moreover*

$$\mathbb{Q}(\operatorname{tr}\Gamma^{(2)}) = \mathbb{Q}\left((\operatorname{tr}\gamma)^2 : \gamma \in \Gamma\right). \qquad \square$$

Notation. We shall denote $\mathbb{Q}(\operatorname{tr}\Gamma^{(2)})$ by $k(\Gamma)$ throughout the rest of this paper. We call it ***the invariant trace field***.

Theorem 2.1 implies that $\mathbb{Q}(\operatorname{tr}\Gamma) = k(\Gamma)(\sqrt{r_1}, \ldots, \sqrt{r_m})$ for some $r_1, \ldots, r_m \in k(\Gamma)$, and hence $\operatorname{Gal}(\mathbb{Q}(\operatorname{tr}\Gamma)/k(\Gamma)) = (\mathbb{Z}/2)^m$. The following theorem describes the relationship between subgroups of Γ and their trace fields.

Theorem 2.2. *Assume Γ has no 2-torsion. Let K be a field satisfying $k(\Gamma) \subseteq K \subseteq \mathbb{Q}(\operatorname{tr}\Gamma)$.*

(1) *$\Gamma_K := \{\gamma \in \Gamma : \operatorname{tr}\gamma \in K\}$ is a normal subgroup of Γ with Γ/Γ_K finite abelian of exponent 2.*

(2) *There is a non-singular bilinear pairing*

$$\begin{array}{ccc} \Gamma/\Gamma_K \otimes \operatorname{Gal}(\mathbb{Q}(\operatorname{tr}\Gamma)/K) & \to & \mathbb{Z}/2 = \{\pm 1\}; \\ \gamma \otimes g & \mapsto & \frac{(\operatorname{tr}\gamma)^g}{\operatorname{tr}\gamma} \end{array}$$

in particular, $\Gamma/\Gamma_{k(\Gamma)} = (\mathbb{Z}/2)^m$.

(3) *The above results hold in the presence of 2-torsion if, when $\operatorname{tr}\gamma = 0$ (i.e., $\gamma^2 = 1$), one replaces $\operatorname{tr}\gamma$ in the above definitions by $\operatorname{tr}\gamma'$ for an element $\gamma' \in \gamma\Gamma^{(2)}$ with non-zero trace.*

Recall that any cusped hyperbolic 3-manifold $M = \mathbb{H}^3/\Gamma$ is topologically the interior of a compact manifold-with-boundary $\overline{M}$, whose boundary consists of tori. The following result was proved in **[R3]** for knot complements.

Corollary 2.3. *Let $M = \mathbb{H}^3/\Gamma$ be a manifold (that is, Γ has no torsion). If* $\operatorname{Cok}(H_1(\partial\overline{M};\mathbb{Z}) \to H_1(\overline{M};\mathbb{Z}))$ *is finite of odd order then $k(\Gamma) = \mathbb{Q}(\operatorname{tr}\Gamma)$. In particular, this holds if M is the complement of a link in a $\mathbb{Z}/2$-homology sphere.*

Proof of Corollary. Note that if P is the subgroup of Γ generated by parabolic elements, then $\Gamma/P = \operatorname{Cok}(H_1(\partial\overline{M};\mathbb{Z}) \to H_1(\overline{M};\mathbb{Z}))$, so $\Gamma/\Gamma^{(2)}P$ is the largest quotient of $\operatorname{Cok}(H_1(\partial\overline{M};\mathbb{Z}) \to H_1(\overline{M};\mathbb{Z}))$ of exponent 2. Thus the condition of the corollary is equivalent to: $\Gamma = \Gamma^{(2)}P$. But certainly $P \subseteq \Gamma_{k(\Gamma)}$, so $\Gamma = \Gamma_{k(\Gamma)}$, so $k(\Gamma) = \mathbb{Q}(\operatorname{tr}\Gamma)$. □

Proof of Theorem 2.2. (1) For any $B \in \mathrm{SL}_2\mathbb{C}$ one has the matrix equation $B^2 - \operatorname{tr}(B)B + I = 0$. Left-multiplying this by A and taking trace gives

$$\operatorname{tr}(B)\operatorname{tr}(AB) = \operatorname{tr}(AB^2) + \operatorname{tr}A.$$

If $A \in \Gamma^{(2)}$ then this equation implies $\operatorname{tr}(B)\operatorname{tr}(AB) \in k(\Gamma)$, so if $B \in \Gamma_K$ then $AB \in \Gamma_K$ (no element of Γ has trace 0 by our exclusion of 2-torsion). Thus $\Gamma^{(2)}\Gamma_K \subseteq \Gamma_K$. Now suppose both A and B are in Γ_K. Then $B^2A \in \Gamma^{(2)}\Gamma_K \subseteq \Gamma_K$, so $AB^2 \in \Gamma_K$ (since B^2A and AB^2 have the same trace). The above equation thus implies $\operatorname{tr}(AB) \in K$, so $AB \in \Gamma_K$. Since trivially $A \in \Gamma_K$ implies $A^{-1} \in \Gamma_K$, Γ_K is a subgroup. It is normal with Γ/Γ_K finite abelian of exponent 2 because this is so for any supergroup of $\Gamma^{(2)}$ in Γ.

(2) It is trivial that the pairing of (2) is linear in g if one fixes γ. For the linearity in γ, let α and β be elements of Γ represented by matrices A and B. By Theorem 2.1, $\operatorname{tr}A = \sqrt{a}$ and $\operatorname{tr}B = \sqrt{b}$ for some $a, b \in k(\Gamma)$. Moreover, $AB^2 \in \Gamma_{\mathbb{Q}(\sqrt{a})}$, so $\operatorname{tr}AB^2 = c\sqrt{a}$ for some $c \in k(\Gamma)$. The above trace equation gives $\operatorname{tr}AB = ((c+1)/b)\sqrt{ab}$, from which the desired linearity follows. The pairing of (2) is nonsingular by definition of Γ_K, and part (3) of the theorem is also clear. □

2.2. In this subsection we give a geometric description of the invariant trace field $k(\Gamma)$ in case $M = \mathbb{H}^3/\Gamma$ is a cusped hyperbolic manifold.

By **[EP]**, M has at least one triangulation by ideal tetrahedra:

$$M = S_1 \cup S_2 \cup \ldots \cup S_n,$$

where each S_j is an ideal tetrahedron in $\mathbb{H}^3$. As discussed in **[T**, chapt. 4**]**, the tetrahedron S_j is described up to isometry by a single complex number z_j with positive imaginary part (the ***tetrahedral parameter*** of S_j) such that the Euclidean triangle cut off any vertex of S_j by a horosphere section is similar to the triangle in $\mathbb{C}$ with vertices 0, 1, z_j. Alternatively, z_j is the cross-ratio of the vertices of S_j (considered as points of $\mathbb{P}^1(\mathbb{C}) = \mathbb{C} \cup \{\infty\}$). This tetrahedral parameter depends on a choice (an edge of S_j or an oriented ordering of its vertices); changing the choice replaces z_j by $1/(1 - z_j)$ or $1 - 1/z_j$. Denote the field $\mathbb{Q}(z_j : j = 1, \ldots, n)$ by $k_\Delta M$ or $k_\Delta\Gamma$. *A priori* $k_\Delta\Gamma$ might depend on the choice of triangulation, but we have:

Theorem 2.4. $k_\Delta\Gamma = k(\Gamma)$.

Proof. Denote $k_\Delta\Gamma$ by k_Δ for short. If we lift the triangulation of M to $\mathbb{H}^3$ we get a tesselation of $\mathbb{H}^3$ by ideal tetrahedra. Let V be the set of vertices of these tetrahedra in the sphere at infinity. Let k_1 be the field generated by all cross ratios of 4-tuples of points of V. Position V by an isometry of $\mathbb{H}^3$ (upper half-space model) so that three of its points are at $0,\ 1,$ and ∞, and let k_2 be the field generated by the remaining points of V. This k_2 does not depend on which three points we put at $0,\ 1,\ \infty$; in fact:

Lemma 2.5. $k_1 = k_2 = k_\Delta$.

Proof. $k_1 \subseteq k_2$ since k_1 is generated by cross-ratios of elements of k_2 while $k_2 \subseteq k_1$ because the cross-ratio of $0,\ 1,\ \infty$, and z is just z. $k_\Delta \subseteq k_1$ is trivial. Finally, put three vertices of one tetrahedron of our tesselation at $0,\ 1,$ and ∞, and then $k_2 \subseteq k_\Delta$ is a simple induction on noting that, for any field l, if three vertices and the tetrahedral parameter of an ideal tetrahedron $S \subseteq \mathbb{H}^3$ are in $l \cup \{\infty\}$, then so is the fourth vertex. □

Now suppose we have positioned V as above. Any element $\gamma \in \Gamma$ maps $0,\ 1,$ and ∞ to points $w_1,\ w_2,$ and w_3 of $V \subseteq k_\Delta \cup \{\infty\}$. Thus γ is given by a matrix $\left(\begin{smallmatrix} a & b \\ c & d \end{smallmatrix}\right)$ whose entries satisfy

$$\begin{aligned} b \qquad\qquad - dw_1 &= 0, \\ a + b - cw_2 - dw_2 &= 0, \\ a \qquad - cw_3 \qquad\quad &= 0. \end{aligned}$$

We can solve this for $a,\ b,\ c,\ d$ in k_Δ and then γ^2 is represented by the element

$$\frac{1}{ad-bc}\begin{pmatrix} a & b \\ c & d \end{pmatrix}^2 \in \mathrm{PSL}_2 k_\Delta .$$

Thus $k(\Gamma) = \mathbb{Q}(\operatorname{tr}\Gamma^{(2)}) \subseteq k_\Delta$.

For the reverse inclusion we shall use the following lemma which was observed in **[R4]**. Indeed, given the lemma, which says that Γ can be conjugated to lie in $\mathrm{PSL}_2\mathbb{Q}(\operatorname{tr}\Gamma)$, the points of V, which are the fixed points of parabolic elements of Γ, lie in $\mathbb{Q}(\operatorname{tr}\Gamma)$, since the fixed point of a parabolic element $\left(\begin{smallmatrix} a & b \\ c & d \end{smallmatrix}\right)$ is $b/(1-a) = (1-d)/c$. Thus, by Lemma 2.5, $k_\Delta \subseteq \mathbb{Q}(\operatorname{tr}\Gamma)$. On the other hand, k_Δ is clearly an invariant of the commensurability class of Γ, so we can apply this to $\Gamma^{(2)}$ to see $k_\Delta \subseteq \mathbb{Q}(\operatorname{tr}\Gamma^{(2)}) = k(\Gamma)$. □

Lemma 2.6. *A non-cocompact finite volume Kleinian group* Γ *has a faithful discrete representation in* $\mathrm{PSL}_2\mathbb{Q}(\operatorname{tr}\Gamma)$.

Proof. By putting a lift of a cusp at ∞, another lift of a cusp at 0, and the image of 0 under a parabolic element T_1 which fixes ∞ at 1, we arrange that Γ contains the elements

$$T_1 = \begin{pmatrix} 1 & 1 \\ 0 & 1 \end{pmatrix} \quad \text{and} \quad T_2 = \begin{pmatrix} 1 & 0 \\ \alpha & 1 \end{pmatrix}$$

(T_2 is any parabolic which fixes 0). Since T_1T_2 has trace $2+\alpha$ we see that $\alpha \in \mathbb{Q}(\operatorname{tr}\Gamma)$. The lemma now follows from **[Mac**, Prop. 3.1**]** which says that the coefficient field of a non-abelian subgroup Γ of $\mathrm{PSL}_2\mathbb{C}$ is generated by $\mathbb{Q}(\operatorname{tr}\Gamma)$ and the coefficients of any two non-commuting elements of Γ. □

2.3. A horospherical section of a cusp of a hyperbolic manifold M is a flat torus. This torus is isomorphic to $\mathbb{C}/\Lambda$, for some lattice $\Lambda \subset \mathbb{C}$, and the ratio of two generators of Λ is the ***conformal parameter*** of the flat torus, which we call the ***cusp parameter*** of the cusp of M. It depends on the choice of generators of Λ, but a different choice changes it by an integral Möbius transformation, so the field it generates is independent of choices. The field generated by the cusp parameters of all the cusps of $M = \mathbb{H}^3/\Gamma$ is called the ***cusp field*** of M (or of Γ). The cusp field is clearly a commensurability invariant, so we define the cusp field of an orbifold to be the cusp field of some manifold cover.

Since an ideal triangulation of M induces triangulations of the cusp tori, from which the cusp parameters may be computed, Theorem 2.4 has the corollary:

Proposition 2.7. *The cusp field of a hyperbolic orbifold is contained in the invariant trace field.* □

As we point out in Sect. 10, the cusp field can be smaller than the invariant trace field, even for hyperbolic knot complements.

2.4. The following theorem is from **[NR1]**. Our formulation here is stronger, but the proof in **[NR1]** applies verbatim.

Theorem 2.8. *If the Kleinian group Γ is generated by subgroups Γ_1 and Γ_2 and $\Gamma_1 \cap \Gamma_2$ is non-elementary then $k(\Gamma) = k(\Gamma_1)k(\Gamma_2)$ (join of fields).* □

3. Quaternion algebras

3.1. In this section we associate to a Kleinian group Γ a quaternion algebra over the trace field $\mathbb{Q}(\operatorname{tr}\Gamma)$. Namely, let $\bar{\Gamma}$ be the inverse image of $\Gamma \subset \mathrm{PSL}_2\mathbb{C}$ in $\mathrm{SL}_2\mathbb{C}$ and denote by $\mathbb{Q}\Gamma$ the algebra of all $\mathbb{Q}$-linear combinations of elements of $\bar{\Gamma}$. It follows from the characteristic equation

$$A^2 - (\operatorname{tr} A)A + I = 0$$

that $\mathbb{Q}\Gamma$ contains τI for any τ in the trace field $\mathbb{Q}(\operatorname{tr}\Gamma)$. Hence $\mathbb{Q}\Gamma$ is an algebra over $\mathbb{Q}(\operatorname{tr}\Gamma)$ (and equals $k\Gamma$ for any subfield k of $\mathbb{Q}(\operatorname{tr}\Gamma)$). It is in fact a quaternion algebra over this field (cf. **[Ba1**, Prop. 2.2] or **[Ba2]**).

Note that the result of Macbeath quoted in the proof of Lemma 2.6 follows from the observation that a quaternion algebra is generated over its base field by any two non-commuting elements. For the same reason, although $\mathbb{Q}\Gamma$ is not a commensurability invariant, $\mathbb{Q}\Gamma^{(2)}$ is. Together with the trace field $\mathbb{Q}(\operatorname{tr}\Gamma)$, this commensurability invariant determines $\mathbb{Q}\Gamma$ by the equation

$$\mathbb{Q}\Gamma = \mathbb{Q}\Gamma^{(2)} \otimes_{k(\Gamma)} \mathbb{Q}(\operatorname{tr}\Gamma).$$

Notation. We shall denote $\mathbb{Q}\Gamma^{(2)}$ by $A(\Gamma)$ and call it the ***invariant quaternion algebra*** of Γ.

Lemma 2.6 tells us that the invariant quaternion algebra $A(\Gamma)$ gives no more information than the invariant trace field $k(\Gamma)$ when Γ has cusps—$A(\Gamma)$ then equals the matrix algebra $M_2(k(\Gamma))$. However, in general $A(\Gamma)$ is more interesting, for instance $A(\Gamma)$ is a complete commensurability invariant for arithmetic Γ and equals $M_2(k(\Gamma))$ in this case if and only if Γ has cusps, as we discuss in the next section.

Recall (**[Vig]**) that a quaternion algebra A over a number field k is determined up to isomorphism by the set $S(A)$ of primes of k (finite or infinite) at which A is ramified; $S(A)$ has a finite even number of elements and any finite even set of primes of k is the ramification set of some quaternion algebra over k.

Little seems to be known about what the invariant $A(\Gamma)$ of Γ says about the geometry of $\mathbb{H}^3/\Gamma$. If Γ is arithmetic then Clozel **[Cl]** has shown

Theorem 3.1. *If Γ is arithmetic and $S(A(\Gamma))$ contains no finite primes of $k = k(\Gamma)$ (or, more generally, if for each finite $\nu \in S(A(\Gamma))$, the local field k_ν contains no quadratic extension of $\mathbb{Q}_p$, where p is the rational prime that ν divides), then Γ has a subgroup of finite index with infinite abelianization—in particular, $\mathbb{H}^3/\Gamma$ is almost sufficiently large.* □

On the other hand, using an observation of **[KLS]**, the existence of finite ramification for $A(\Gamma)$ also implies strong results about abelian quotients of finite index subgroups of Γ, as we now describe. We formulate the result in terms of $\mathbb{Q}\Gamma$, so let Γ be a Kleinian group and denote $\mathbb{Q}(\operatorname{tr}\Gamma) = k$ and $\mathbb{Q}\Gamma = A$.

Theorem 3.2. *Suppose A is ramified at the finite prime ν and let p be the rational prime that ν divides. Then Γ has a normal subgroup Δ with finite cyclic quotient (of order dividing $p^t - 1$ for some $t \le 2[k{:}Q]$) which is residually finite p, that is, the intersection of all normal subgroups of Δ of p-power index is trivial.*

Before we prove this theorem, we point out that being residually finite p is a strong property. Since finite p-groups are nilpotent, it implies that the lower central p-series of Δ intersects in the trivial group (this is the series of subgroups

$$\Delta = \Delta^0(p) \supset \Delta^1(p) \supset \Delta^2(p) \supset \dots,$$

where $\Delta^i(p)/\Delta^{i+1}(p)$ is defined inductively to be the largest exponent p quotient of $\Delta^i(p)/[\Delta, \Delta^i(p)]$). Moreover

Proposition 3.3. *Every non-abelian subgroup of a residually finite p group has a $\mathbb{Z}/p \times \mathbb{Z}/p$ quotient.*

Proof. The subgroup is residually finite p, so it has a non-abelian p-group quotient. But any non-abelian p-group G has a $\mathbb{Z}/p \times \mathbb{Z}/p$ quotient—if $G/[G,G]$ were cyclic, then $G/[G,[G,G]]$ would be a central extension of a cyclic group, hence abelian, so $[G,G] = [G,[G,G]]$, which contradicts the fact that a finite p-group is nilpotent. □

Proof of Theorem 3.2. (Following **[KLS]**, see also **[Vig]**.) Γ is represented into A^1, the group of elements of A of reduced norm 1, and hence into A^1_ν. The discrete valuation on the local field k_ν extends to one on A_ν. Let $\mathcal{O}_\nu$ be the valuation ring (elements of non-negative valuation in A_ν) and $\mathcal{P}_\nu$ its maximal ideal. Then A^1_ν is contained in the multiplicative group $\mathcal{O}^\times_\nu$ of $\mathcal{O}_\nu$ and this multiplicative group has a Hausdorff filtration

$$\mathcal{O}^\times_\nu \supset (1 + \mathcal{P}_\nu) \supset (1 + \mathcal{P}^2_\nu) \supset \dots$$

whose quotients after the first are $\mathbb{Z}/p$-vector spaces and whose first quotient is the multiplicative group of the residue class field $\bar{l_\nu}$ of a certain quadratic extension l_ν of k_ν (the unique unramified extension of k_ν). $\bar{l_\nu}$ is an extension of degree $t \leq 2[k:\mathbb{Q}]$ of the field of p elements, so its multiplicative group is cyclic of order $p^t - 1$. □

A result of G. Mess **[Me]** is interesting in reference to Theorem 3.2 and Proposition 3.3—he shows that a 3-manifold group with a $\mathbb{Z}/p \times \mathbb{Z}/p$ quotient must either have a finite index subgroup with infinite abelianization or the sizes of the successive quotients of the derived series must grow rapidly.

3.2. The trace field, invariant trace field, and invariant quaternion algebra are quite strong invariants. The following proposition is trivial (and also applies to the cusp field).

Proposition 3.4. *Let* $M = \mathbb{H}^3/\Gamma$ *be a hyperbolic orbifold. If* M *is amphicheiral (has an orientation reversing symmetry) then* $\mathbb{Q}(\operatorname{tr}\Gamma) = \overline{\mathbb{Q}(\operatorname{tr}\Gamma)}$, *and if* M *is commensurably amphicheiral (has an orientation reversing commensurability) then* $k(\Gamma) = \overline{k(\Gamma)}$ *and* $A(\Gamma) = \overline{A(\Gamma)}$, *where the bar is complex conjugation.* □

For arithmetic Γ, commensurable amphicheirality is equivalent to conjugation symmetry of the invariant quaternion algebra (and is also equivalent to the existence of Fuchsian subgroups—see **[R1]**), but this is probably false in general. Nevertheless, in practice the proposition gives an effective test for amphicheirality and commensurable amphicheirality.

4. Short geodesics in arithmetic orbifolds

The main results of this section are Theorem 4.6 and its Corollary, which set strong limits on the possible lengths of short geodesics in arithmetic cusped hyperbolic orbifolds, thus giving a necessary condition for arithmeticity. We also describe a conjectural statement of the same type for closed arithmetic orbifolds, dependent on the "Lehmer Conjecture" of number theory. We first recall some basic relevant facts about arithmeticity of Kleinian groups, including the algebraic characterization of arithmeticity given in **[MR]** and **[R1**, Chap. 2].

4.1. Arithmetic Kleinian groups are obtained as follows (cf., **[Bo]** and **[Vig**, Chap. 4], also **[Vig]** for relevant details on quaternion algebras).

Let k be a number field with one complex place and A a quaternion algebra over k ramified at all real archimedean places of k (this means $A \otimes_k \mathbb{R}$ is the algebra of hamiltonian quaternions for each real embedding $k \hookrightarrow \mathbb{R}$). Let $\mathcal{O}$ be an order of A (a finitely generated subring of the set of A-integers, which contains the ring R_k of k-integers and generates A as a k-vectorspace) and $\mathcal{O}^1$ the group of elements of norm 1 in $\mathcal{O}$. The complex place $k \hookrightarrow \mathbb{C}$ induces an embedding $\rho\colon A \hookrightarrow M_2(\mathbb{C})$ which restricts to $\rho\colon \mathcal{O}^1 \hookrightarrow \mathrm{SL}_2\mathbb{C}$. Then $\mathrm{P}\rho(\mathcal{O}^1)$ is a Kleinian group of finite covolume. An ***arithmetic Kleinian group*** is one commensurable with a group of the type $\mathrm{P}\rho(\mathcal{O}^1)$. We say it is ***derived from a quaternion algebra*** if it is actually a subgroup of some $\mathrm{P}\rho(\mathcal{O}^1)$. We call

$M = \mathbb{H}^3/\Gamma$ arithmetic or derived from a quaternion algebra if Γ is arithmetic or derived from a quaternion algebra.

The field k and quaternion algebra A associated as above with an arithmetic Kleinian group Γ are the same as the invariant trace field $k(\Gamma)$ and invariant quaternion algebra $A(\Gamma)$ of the previous section (see Theorem 4.3 below and subsequent remarks). The arithmetic Kleinian group Γ has cusps if and only if A is $\mathrm{M}_2(k)$, which holds if and only if A is unramified at all places (see e.g., **[Vig**, Chap. 3**]**). We thus obtain (see also **[H]**):

Proposition 4.1. *A non-cocompact Kleinian group Γ of finite covolume is arithmetic if and only if Γ is conjugate in $\mathrm{PSL}_2\mathbb{C}$ to a group commensurable with some "Bianchi group" $\mathrm{PSL}_2\mathcal{O}_d$, where $\mathcal{O}_d$ is the ring of integers in $\mathbb{Q}(\sqrt{-d})$.* □

4.2. We shall make use of the characterization theorems for arithmetic Kleinian groups given in **[MR]** and **[R1]** in the special case of cusped orbifolds. We first recall these results in the general case.

Theorem 4.2. *Let Γ be a Kleinian group of finite covolume. Then Γ is arithmetic if and only if $\Gamma^{(2)}$ is derived from a quaternion algebra.* □

Theorem 4.3. *With Γ as in Theorem 4.2, Γ is derived from a quaternion algebra if and only if the following conditions hold:*
(i) $k = \mathbb{Q}(\mathrm{tr}\,\Gamma)$ *has just one complex place;*
(ii) *the set* $\mathrm{tr}\,\Gamma$ *consists of algebraic integers;*
(iii) *for every real place* $\sigma\colon k \hookrightarrow \mathbb{R}$ *the set* $|\sigma(\mathrm{tr}\,\Gamma)|$ *is bounded.* □

Remarks. 1. The arguments of **[R1]** and **[Tak]** show that $|\sigma(\mathrm{tr}\,\Gamma)|$ bounded can be replaced by $|\sigma(\mathrm{tr}\,\Gamma)| \leq 2$.
2. It follows that when Γ is derived from a quaternion algebra, the associated quaternion algebra is defined over $\mathbb{Q}(\mathrm{tr}\,\Gamma)$.

When Γ contains parabolic elements we may deduce from these theorems (see **[H]** and **[R1]** for details):

Proposition 4.4. *Let Γ be a non-cocompact Kleinian group of finite covolume. Then:*
(a) Γ *is arithmetic if and only if* $k(\Gamma) = \mathbb{Q}(\sqrt{-d})$ *for some square-free* $d \in \mathbf{N}$ *and* $\mathrm{tr}\,\Gamma$ *consists of algebraic integers.*
(b) Γ *is derived from a quaternion algebra if and only if* $\mathrm{tr}\,\Gamma \subset \mathcal{O}_d$ *for some* d. □

Corollary 4.5. *If Γ is arithmetic and satisfies the condition of Corollary 2.3 (or, more generally, $\Gamma = \Gamma^{(2)}P$), then Γ is derived from a quaternion algebra. In particular, this holds if $M = \mathbb{H}^3/\Gamma$ is a link complement in a $\mathbb{Z}/2$-homology sphere.* □

4.3. The main result of this section is the following theorem and its corollary.

Theorem 4.6. *Let $M^3 = \mathbb{H}^3/\Gamma$ be a cusped arithmetic hyperbolic orbifold. If M contains a geodesic of length less than $\operatorname{arccosh}\frac{3}{2} = 0.9624236501\ldots$ then Γ is commensurable with $\mathrm{PSL}_2\mathcal{O}_d$ with $d \in \{1, 2, 3, 7, 11, 15, 19\}$ and the geodesic has length one-half a value listed in Table 1.*

Moreover, if Γ *is derived from a quaternion algebra, then the above can be improved by a factor of* 2*: a geodesic of length less than* $2\operatorname{arccosh}\frac{3}{2} = 1.9248473002\ldots$ *has length listed in Table 1 and* Γ *is commensurable with a* $\mathrm{PSL}_2\mathcal{O}_d$ *as above.*

Table 1

length	d
0.8625546276620610	3
0.9624236501192069	1
1.0612750619050357	1
1.0870701449957391	3
1.2659486384018949	7
1.3169578969248167	2
1.4183161349689732	2
1.4657153519472905	1
1.4860221248769271	7
1.5343944365026389	11
1.5667992369724111	3
1.6628858910586211	3
1.7251092553241221	3
1.7365960799226493	11
1.7400216453048509	15
1.7627471740390861	1 or 2
1.8522660627003648	1
1.9079255392337773	19

Corollary 4.7. *If the cusped hyperbolic 3-orbifold* $M = \mathbb{H}^3/\Gamma$ *contains a geodesic of length less than* 0.431277313 (0.862554627 *if* $\Gamma = \Gamma^{(2)}P$) *then* M *is non-arithmetic.* □

Remarks. 1. Let $\mathcal{O}$ be any maximal order of $\mathrm{M}_2(\mathbb{Q}(\sqrt{-d}))$. Using the standard description of types of maximal orders in $\mathrm{M}_2(\mathbb{Q}(\sqrt{-d}))$ (see **[Vig**, p. 100, ex. 4.7]), it follows that $\mathbb{H}^3/\mathbb{P}\mathcal{O}^1$ always contains a geodesic of length $2\operatorname{arccosh}\frac{3}{2}$, so Theorem 4.6 cannot be improved.

2. The proof of Theorem 4.6 gives the same results for ***closed*** arithmetic 3-orbifolds $M = \mathbb{H}^3/\Gamma$ with invariant trace field $k(\Gamma) = \mathbb{Q}(\sqrt{-d})$.

Proof of Theorem 4.6. Since for any arithmetic Γ, $\Gamma^{(2)}$ is derived from a quaternion algebra, we need only consider the case that Γ is derived from a quaternion algebra.

We first introduce some notation. Let $M^3 = \mathbb{H}^3/\Gamma$ have finite volume and let $\gamma \in \Gamma$ be a loxodromic or hyperbolic element with $\operatorname{tr}\gamma = \lambda + \lambda^{-1}$ with $|\lambda| > 1$.

Notation. The length of the geodesic in M determined by γ is $2\log|\lambda|$ and will be denoted by $\ell_0(\gamma)$. It is the real part of the "complex length" $\ell(\gamma) = 2\log\lambda$ whose imaginary part is the holonomy of the geodesic determined by γ.

Note that, with γ as above, $\operatorname{tr}\gamma = \pm 2\cosh\frac{\ell(\gamma)}{2}$, so

$$|\operatorname{tr}\gamma| = \left|e^{\frac{1}{2}\ell(\gamma)} + e^{-\frac{1}{2}\ell(\gamma)}\right| \leq \left|e^{\frac{1}{2}\ell(\gamma)}\right| + \left|e^{-\frac{1}{2}\ell(\gamma)}\right|$$
$$= e^{\frac{1}{2}\ell_0(\gamma)} + e^{-\frac{1}{2}\ell_0(\gamma)}$$
$$= 2\cosh\frac{\ell_0(\gamma)}{2}.$$

Hence, if $|\operatorname{tr}\gamma| \geq 2$,

$$\ell_0(\gamma) \geq 2\operatorname{arccosh}\frac{|\operatorname{tr}\gamma|}{2},$$

so that for γ with $|\operatorname{tr}\gamma| \geq 3$ we have $\ell_0(\gamma) \geq 2\operatorname{arccosh}\frac{3}{2}$.

Now suppose Γ contains an element γ with $\ell_0(\gamma) < 2\operatorname{arccosh}\frac{3}{2}$. Then $|\operatorname{tr}\gamma| < 3$. But by Proposition 4.4 (a), $\operatorname{tr}\gamma \in \mathcal{O}_d$. Thus if $\operatorname{tr}\gamma$ where real, it would be in $\mathbb{Z}$, and γ would be parabolic or elliptic, contrary to assumption. The existence of a non-real $\tau \in \mathcal{O}_d$ with $|\tau| < 3$ imposes strong conditions on d and τ, as we now describe.

If $d \not\equiv -1 \pmod 4$ then $\tau = a + b\sqrt{-d}$ with $a, b \in \mathbb{Z}$ and $b \neq 0$. Then $|\tau|^2 = a^2 + db^2$, whence $d < 9$. If $d \equiv -1 \pmod 4$ then $\tau = (a + b\sqrt{-d})/2$ with $a \equiv b \pmod 2$ and $b \neq 0$. In this case $|\tau|^2 = (a^2 + db^2)/4$, and it follows that $d < 35$. We thus get a finite list of possible d and for each such d there is a short finite list of $\tau \in \mathcal{O}_d$ with $|\tau| < 3$. It is thus a simple matter to list the d and τ with $\operatorname{Re}(2\operatorname{arccosh}\frac{\tau}{2}) < 2\operatorname{arccosh}\frac{3}{2}$, giving Table 1. □

4.4. For closed hyperbolic 3-orbifolds we conjecture a similar result to Corollary 4.7:

Conjecture 4.8. *There is a universal lower bound for the lengths of geodesics in closed arithmetic hyperbolic orbifolds. (The current guess is approximately* 0.09174218, *or twice this,* 0.18348436, *if the orbifold is derived from a quaternion algebra, see Sect. 4.5 below.)*

This conjecture would be a consequence of a classical conjecture of number theory due to D.H. Lehmer **[Le]** (see also **[Boy]** for a discussion). Suppose $P(x)$ is a monic integral polynomial of degree n with roots $\theta_1, \ldots, \theta_n$. The ***Mahler measure*** of P is

$$M(P) = \prod_{i=1}^{n} \max(1, |\theta_i|).$$

Lehmer's Conjecture. *There exists $m > 1$ such that $M(P) \geq m$ for all non-cyclotomic P. The currently conjectured value of m is approximately* 1.176780821 *(see Sect. 4.5 below).*

That Lehmer's conjecture implies Conjecture 4.8 follows from the following lemma, part of which is implicit in **[R1]** and **[Tak]**.

Lemma 4.9. *Let Γ be a Kleinian (or Fuchsian) group derived from a quaternion algebra. Let $\gamma \in \Gamma$ be a loxodromic or hyperbolic element, and write $\operatorname{tr}\gamma = u + u^{-1}$ with $|u| > 1$. Then u is an algebraic integer and u^{-1} is a conjugate of u. Moreover,*

(a) *if γ is loxodromic, then u is not real, and exactly four conjugates of u lie off the unit circle;*
(b) *if γ is hyperbolic, then u is real, and exactly two conjugates of u lie off the unit circle;*
(c) $\ell_0(\gamma) = \log M(P)$ *or* $2\log M(P)$ *according as γ is loxodromic or hyperbolic, where $P = P(x)$ is the monic integral polynomial of u.*

Proof. We prove the Kleinian case; the Fuchsian case is similar. Note that $u + u^{-1} = \operatorname{tr}\gamma \in k(\Gamma) = k$. On the other hand, u is not in k: $u \in k$ implies that the quaternion algebra is a matrix algebra, so $k = \mathbb{Q}(\sqrt{-d})$ and u is a unit in this field, which contradicts the assumption $|u| > 1$. Thus u has degree 2 over k and u^{-1} is the other root of the minimal polynomial $u^2 - u\operatorname{tr}\gamma + 1 = 0$.

For any real embedding $\sigma: k \hookrightarrow \mathbb{R}$, extend σ to $\psi: k(u) \to \mathbb{C}$. We claim that $|\psi(u)| = 1$. Indeed, if $|\psi(u)| \neq 1$ then for sufficiently large positive integers t we have $|\sigma(\operatorname{tr}\gamma^t)| = |\psi(u^t + u^{-t})| > 2$, which contradicts Theorem 4.3 and its following Remark 1. Parts (a) and (b) of the Lemma now follow. Moreover, $M(P) = |u|^2$ or $|u|$ according as u is complex or real, so part (c) also follows. □

4.5. Lemma 4.9 has a converse, as pointed out to us by Ted Chinburg:

Lemma 4.10. *Suppose u is an algebraic integer such that $|u| > 1$, u^{-1} is a conjugate of u, and u satisfies one of conditions* (a) *or* (b) *of Lemma 4.9. Then, in case* (a) (*resp. case* (b)) *there is a Kleinian* (*resp. Fuchsian*) *group Γ derived from a quaternion algebra and a loxodromic* (*resp. hyperbolic*) *element $\gamma \in \Gamma$ with $\operatorname{tr}\gamma = u + u^{-1}$. Moreover, in case* (b) *we can require Γ to be a subgroup of a Kleinian group Γ_0 derived from a quaternion algebra.*

Before we prove this converse, we describe its implications for short geodesics. The conjectured smallest Mahler measure of approximately 1.176780821 is attained by an example $P_1(x) = x^{10} + x^9 - x^7 - x^6 - x^5 - x^4 - x^3 + x + 1$ due to Lehmer **[Le]** given in 1933 (see **[Boy]** for a discussion). It has just one real root outside the unit circle. The smallest known Mahler measure for a polynomial with a complex root outside the unit circle is approximately 1.2013961862, attained by an example $P_2(x) = x^{18} + x^{17} + x^{16} - x^{13} - x^{11} - x^9 - x^7 - x^5 + x^2 + x + 1$ of D. W. Boyd **[Boy]**. Lemma 4.10 implies the existence of a hyperbolic 3-orbifold M_1 containing a hyperbolic 2-suborbifold N_1 with a hyperbolic geodesic of length $2\log M(P_1) \approx 0.324715228$, and also a hyperbolic 3-orbifold M_2 with a loxodromic geodesic of length $\log M(P_2) \approx 0.18348436$; all these orbifolds being derived from quaternion algebras. By Lemma 4.9, these are the conjectural shortest lengths for hyperbolic resp. loxodromic geodesics in hyperbolic 2- or 3-orbifolds derived from quaternion algebras. The conjectural shortest lengths in arithmetic orbifolds would be half these (approx. 0.162357614 resp. 0.09174218), and can also be realized in examples by the following theorem of Chinburg (private communication), which we prove below.

Theorem 4.11. *Suppose Γ is a Kleinian or Fuchsian group which is derived from a quaternion algebra and $\gamma \in \Gamma$ is a loxodromic or hyperbolic element. Then there is a group Γ_0 commensurable with Γ containing an element γ_0 with $\gamma_0^2 = \gamma$ in* $\mathrm{PSL}_2\mathbb{C}$.

Proof of Lemma 4.10. As in the proof of Lemma 4.9 we deal with the Kleinian group case; the Fuchsian case is analogous. Thus assume that u is not real and has exactly four conjugates off the unit circle. Since u^{-1} is assumed to be a conjugate of u, the conjugates of u which are not on the unit circle are exactly u, u^{-1} and their complex conjugates. Hence the only non-real conjugates of $\theta = u + u^{-1}$ are itself and $\bar{u} + \bar{u}^{-1}$. In particular the field $l_0 = \mathbb{Q}(\theta)$ has exactly one complex place whilst the field $l = \mathbb{Q}(u)$ is totally imaginary as u has no real conjugates. Also note that l has degree 2 over l_0. We now construct a quaternion algebra over l_0 in which l embeds.

Let S be the set of real places of l_0 and ν a finite place of l_0 which is inert in l. Now define the set S_1 to be S when $|S|$ is even and $S \cup \{\nu\}$ otherwise. Since $|S_1|$ is even, the classification theorem for quaternion algebras over number fields (see for example [**Vig**, Chapter 3]) guarantees the existence of a quaternion algebra A over l_0 with S_1 as ramification set. Moreover, by construction, l embeds in A—since the real places of l_0 are ramified in l and ν is inert in l (see [**Vig**, Theorems 1.2.8 and 3.3.8]).

Thinking of l as a subfield of A, u is an algebraic integer in l. Thus, to complete the proof we need to find an order of A containing u. By assumption u^{-1} is the distinct conjugate of u over l_0. Hence u has l/l_0-norm 1. Now A is a 2-dimensional l-vector space, so let τ_1, τ_2 be a basis. Then $\Lambda = R_l\tau_1 \oplus R_l\tau_2$ is an ideal of A (see [**Vig**, Chap. 1]), where R_l is the ring of integers of l, so $\mathcal{O}_\Lambda = \{x \in A : x\Lambda \subset \Lambda\}$ (the ***left order*** of Λ in the terminology of [**Vig**, Chap. 1]) is an order of A which clearly contains R_l and hence u. The final sentence of Lemma 4.10 follows from [**MR**] or [**R1**]. □

Proof of Theorem 4.11 (Chinburg). We may assume $\Gamma = \mathrm{P}\rho(\mathcal{O}^1)$, where $\mathcal{O}$ is an order in the quaternion algebra $A = A(\Gamma)$ (cf. Sects. 4.1 and 3.1). Let $\gamma = \mathrm{P}\rho(u)$ for some $u \in \mathcal{O}^1$. Since γ is loxodromic or hyperbolic, $\operatorname{tr}\gamma = u + u^{-1}$, and we can assume $|u| > 1$. As in the proof of Lemma 4.9, u has degree 2 over the center k of A. Since $u \in \mathcal{O}^1$, the $k(u)/k$-norm of u is 1. By Hilbert's Theorem 90, there is a $u_0 \in k(u)$ with $u_0/\sigma(u_0) = u$, where σ is the non-trivial element of $\mathrm{Gal}(k(u)/k)$. By multiplying u_0 by a suitable element of R_k (the ring of integers of k), we may assume $u_0 \in R_{k(u)}$. Now $\sigma(u_0^2/u) = (\sigma(u_0))^2/\sigma(u) = (u_0/u)^2/u^{-1} = u_0^2/u$, so $u_0^2/u \in k^*$. It will now suffice to find an arithmetic group Γ_0 commensurable with Γ and containing $\gamma_0 = \mathrm{P}\rho(u_0)$, since $\gamma_0^2 = \gamma$ in A^*/k^*.

As in the proof of Lemma 4.10, we can find a maximal order $\mathcal{O}$ in A containing u_0. Let R_f be the (finite) set of finite places of k over which A ramifies. Let S be a finite set of places of k such that u_0 is an S-unit. For $\nu \notin R_f \cup S$ and ν finite, u_0 is in the unit group of $\mathcal{O}_\nu = \mathcal{O} \otimes_{R_k} R_\nu$, where R_ν is the ring of integers of the completion k_ν. Hence u_0 fixes the vertex P_ν of the Bruhat-Tits building at ν which is fixed by $\mathcal{O}_\nu^*$. Suppose now that $\nu \in S - R_f$. Because u has reduced norm 1, u fixes the vertex P_ν in the building at ν which is fixed by $\mathcal{O}_\nu^*$, though u_0 need not fix P_ν. However, since $u_0^2 = u$ in A^*/k^*, the action of u_0 on the building at ν exchanges the vertices P_ν and u_0P_ν. Since the building is a tree, u_0 fixes the mid-point T_ν of the shortest path between these vertices. (Note that T_ν is either a vertex or a midpoint of an edge.) From Borel's description ([**Bo**, Prop. 4.4]) of the maximal elements in the commensurability class of Γ, there is a group Γ_0 in this commensurability class consisting of precisely those elements of A^*/k^* which locally fix the same points as described above for u_0 on Bruhat-Tits buildings, i.e., they fix P_ν (resp. T_ν) if ν is a finite place not in $R_f \cup S$ (resp. in $S - R_f$). □

5. Commensurators of chain links

Let $C(p,s)$ denote the complement of the link in S^3 consisting of a p-link chain with s left half-twists (if s is negative we mean $-s$ right half-twists, cf. Fig. 1 for $(p,s) = (6,-3)$ or in Fig. 2 for $(p,s) = (7,0)$). $C(p,s)$ has several symmetries: α the symmetry that rotates the chain clockwise, taking each link into the next; β the rotation by $180°$ about the circular axis of Fig. 2; and γ the $180°$ rotation about the horizontal axis. These elements generate a group of symmetries of order $4p$:

$$\begin{aligned} G(p,s) &= \langle \alpha, \beta, \gamma : \alpha^p = \beta^s, \beta^2 = 1, \gamma^2 = 1, \alpha\beta = \beta\alpha, \gamma\beta = \beta\gamma, \gamma^{-1}\alpha\gamma = \alpha^{-1} \rangle \\ &\cong D_{2p} \times C_2 \quad \text{if } s \text{ is even} \\ &\cong D_{4p} \quad \text{if } s \text{ is odd,} \end{aligned}$$

where D_{2n} and C_2 denote the dihedral group of order $2n$ and the cyclic group of order 2.

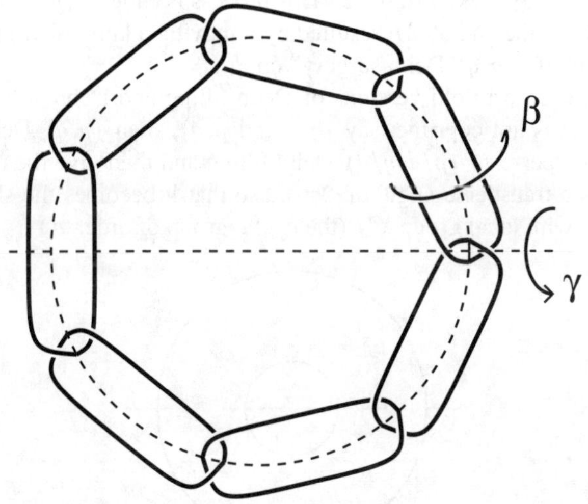

Figure 2

Theorem 5.1.

(i) *$C(p',s')$ is commensurable with $C(p,s)$ if $(p'+s',s') = \pm(p+s,s)$ or $\pm(-s,p+s)$ and is commensurable with $C(p,s)$ with reversed orientation if $(p'+s',s') = \pm(p+s,-s)$ or $\pm(s,p+s)$.*

(ii) *$C(p,s)$ has a hyperbolic structure (complete of finite volume) if and only if $\{|p+s|,|s|\} \not\subset \{0,1,2\}$.*

(iii) $C(p,s)$ *is arithmetic (and hence has infinitely many hidden symmetries) if and only if* $(|p+s|,|s|)$ *or* $(|s|,|p+s|)$ *is in* $(3,0)$, $(3,1)$, $(3,2)$, $(3,3)$, $(4,0)$, $(4,2)$, $(4,4)$, $(6,0)$. *The field of definition* $k(C(p,s))$ *is then* $\mathbb{Q}(\sqrt{-d})$ *with* d *respectively* 7, 1, 3, 1, 3, 7, 2, *and* 15.

(iv) *If* $C(p,s)$ *is hyperbolic and non-arithmetic then it has no orientation preserving hidden symmetries and its full orientation preserving symmetry group is* $G(p,s)$ *(so* $\pi_1 C(p,s)$ *is normal in its orientation preserving commensurator with quotient* $G(p,s)$ *).*

(v) *If* $C(p,s)$ *is hyperbolic and non-arithmetic then it has an orientation reversing commensurability if and only if* $p+s=\pm s$ *or* $p+s=0$ *or* $s=0$.

In the remainder of this section we give the topological part of the proof—part (i), the "only if" of (ii), and the "if" of (v). In the next section we describe the hyperbolic structures on chain link complements by means of Dehn surgery on a component of the Whitehead link and compute some of their arithmetic invariants; (ii) of the theorem is a side-product. In the subsequent two sections we prove parts (iii) and (iv) and (v).

We shall be using Dehn surgery a lot, so we briefly recall the basic terminology. Let M be a 3-manifold with an end homeomorphic to $T^2\times[0,\infty)$, where T^2 is the 2-torus. We can cut off the end to get a 3-manifold $M_0=M-T^2\times(1,\infty)$ with a boundary component $T^2=T^2\times\{1\}$. Let $\mathbf{m},\mathbf{l}$ be some chosen basis of $H_1(T^2;\mathbb{Z})$ and let (p,q) be a coprime integer pair. We can paste a solid torus onto M_0 at the boundary component T^2 to kill the homology class $p\mathbf{m}+q\mathbf{l}$. This process is called "(p,q) Dehn filling" the chosen end of M. If the end of M resulted by removing a knot K from a manifold N, then we also speak of "(p,q) Dehn surgery" on K.

We shall also need orbifold versions of Dehn filling and Dehn surgery. If (p,q) is an integer pair that is not coprime, say $d=\gcd(p,q)$, then (p,q) Dehn filling means the following: first perform $(p/d,q/d)$ Dehn filling and then give the core circle of the added solid torus a transverse angle of $2\pi/d$, so that it becomes the singular set of an orbifold structure with local group C_d (the cyclic group of order d).

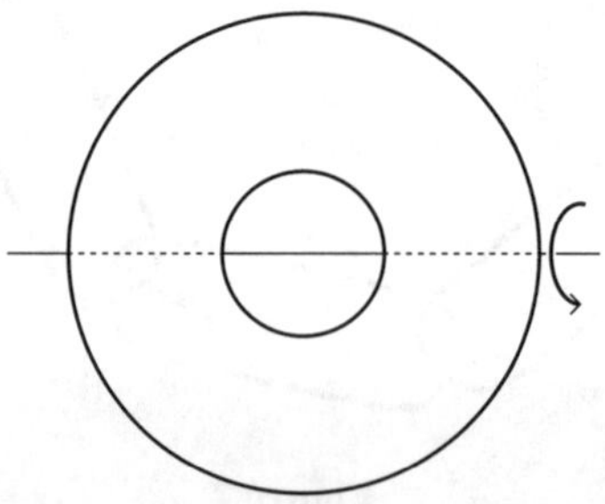

Figure 3

Let τ be the involution on T^2 which acts by multiplication by -1 on $H_1(T^2)$ (Fig. 3). The orbit space $P=T^2/\tau$ is a 2-sphere with four order 2 orbifold points (points with cone angle π). The involution τ extends to any solid torus $D^2\times S^1$ that T^2 bounds and $(D^2\times S^1)/\tau$ is the orbifold Q of Fig. 4 with $\partial Q=P$. Thus, given a 3-orbifold M with an end homeomorphic to $P\times[0,\infty)$, and a chosen homology basis

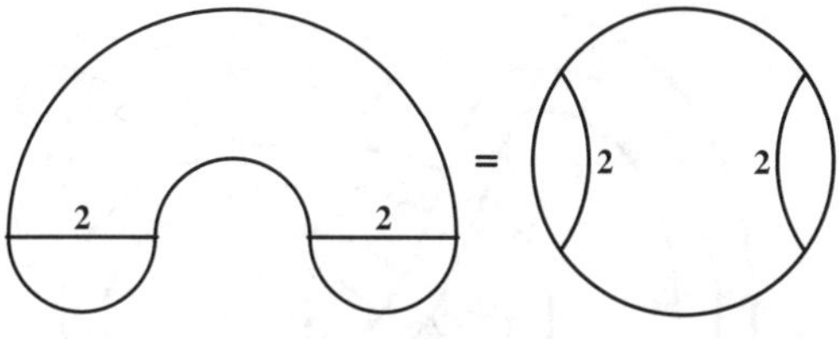

Figure 4

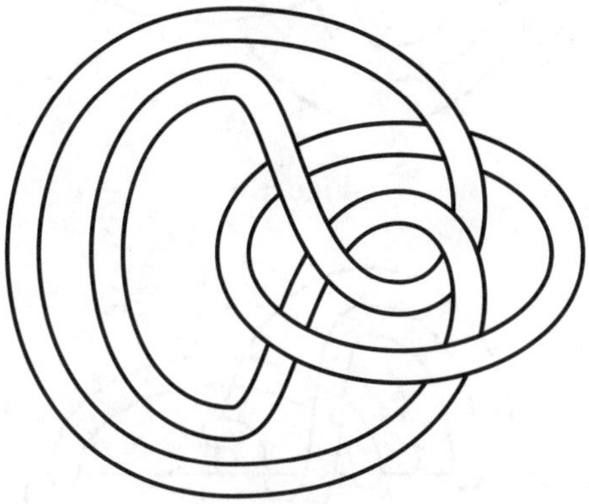

Figure 5

for the double cover T^2 of P, we can define (p,q) Dehn filling the end of M to mean the process of cutting off the end and replacing it by Q in a way that induces (p,q) Dehn filling by a solid torus in the double cover of the end.

Proof of (i) and "if" of (v) of Theorem 5.1. Let W denote the complement of the Whitehead link pictured in Fig. 5. We can obtain $C(p,2q)$ by performing (p,q) Dehn filling of W at one component of the Whitehead link (with respect to the standard meridian/longitude pair) and then taking the p-fold cover of the resulting manifold or orbifold (it is an orbifold if $\gcd(p,q) \neq 1$). To obtain $C(p,s)$ this way for s odd, one can give an interpretation of (p,q) Dehn filling for q half-integral, but instead we shall use a suitable quotient of W.

W double covers the orbifold W' pictured in Fig. 6, with one toral end and one end whose cross section is a sphere with four order 2 orbifold points. (p,s) Dehn filling the toral end of W' lifts to $(p,s/2)$ Dehn filling of one end of W. In fact, the result $W'(p,s)$ of this (p,s) Dehn filling is simply the quotient of $C(p,s)$ by the subgroup $G_0 = \langle \alpha, \beta \rangle$ of $G(p,s)$. If we quotient W' by the symmetry γ we obtain the orbifold W'' pictured in Fig. 7. Its underlying space is S^3 with two balls removed, i.e., $S^2 \times I$. We have re-drawn W'' in Fig. 8 in a more symmetric fashion; it has an order 2 rotational symmetry about its vertical axis and reflection symmetries across appropriate vertical planes.

Let $W''(p,s)$ be the corresponding quotient of $W'(p,s)$ (so $W''(p,s)$ results by (p,s) Dehn filling one end of W'—the end is filled by the orbifold of Fig. 4). By drawing

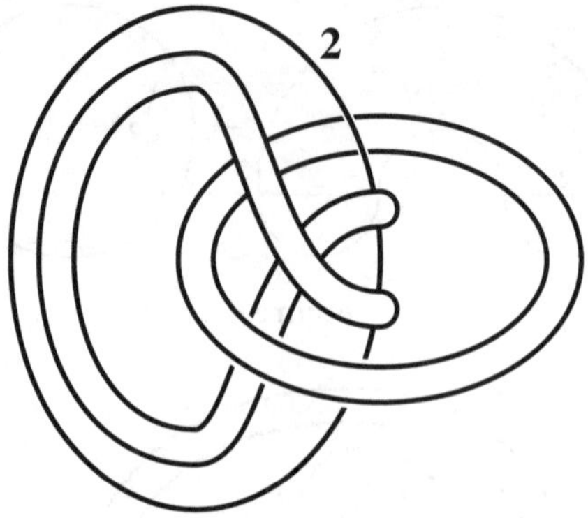

Figure 6

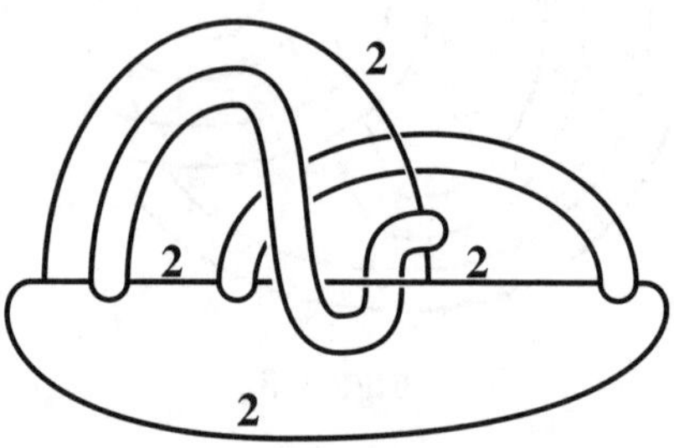

Figure 7

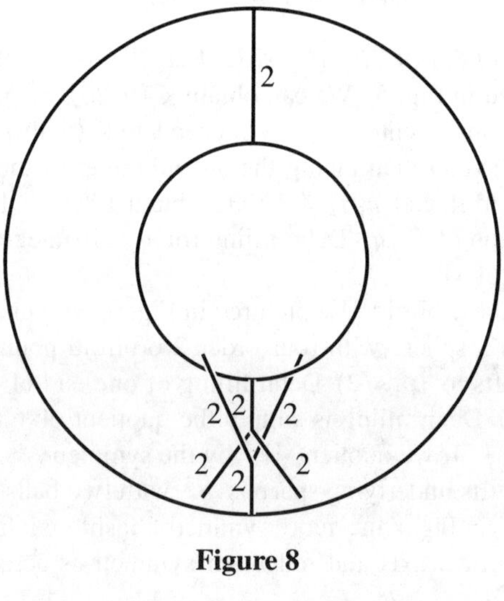

Figure 8

the curves killed by the Dehn filling (or by a homological calculation) it is not hard to see that the rotational symmetry of W'' takes (p, s) Dehn filling to $(p + 2s, -p - s)$ Dehn filling, so $W''(p,s) \cong W''(p+2s,-p-s)$. Note that (p,s) and $(-p,-s)$ Dehn filling mean the same thing, and $(p',s') = (p+2s,-p-s)$ implies $(p'+s',s') = (s,-p-s)$, so the orientation preserving commensurability of the theorem is proved. The reflection symmetries of W'' take (p,s) Dehn filling to $(p,-p-s)$ and $(p+2s,-s)$ Dehn filling, so $W''(p,s) \cong W''(p,-p-s)$ and $W''(p+2s,-s)$. This gives the orientation reversing commensurabilities of the theorem. □

If $W(p_1,q_1;\, p_2,q_2)$ denotes the result of Dehn filling at both cusps of W (that is, (p_1,q_1) and (p_2,q_2) Dehn surgery on the two components of the Whitehead link) and similarly for W'' then the same argument shows:

Proposition 5.2. $W''(p_1,s_1;\, p_2,s_2) \cong -W''(p_1,-p_1-s_1;\, p_2+2s_2,-s_2)$, *where the minus sign signifies reversed orientation, so* $W(p_1,q_1;\, p_2,q_2)$ *is commensurable with* $-W(p_1,-p_1/2-q_1;\, p_2+4q_2,-q_2)$. □

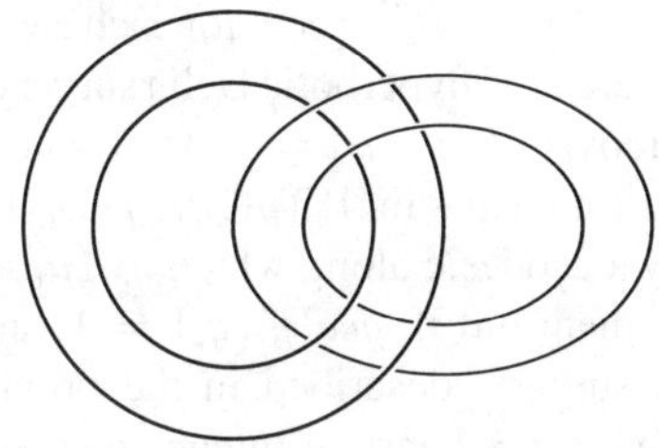

Figure 9

Proof of "only if" of part (ii) of Theorem 5.1.. We must show non-hyperbolicity for the cases claimed. For this it suffices, by the commensurability statement of part (i), to consider the five cases $(p,s) = (1,0)$, $(2,0)$, $(2,-1)$, $(3,-1)$, and $(4,-2)$. The (p,s) chain link is then respectively the unknot, the $(4,2)$ cable link, the two component unlink, the $(3,3)$ cable link, and the link pictured in Fig. 9, which has an essential torus separating the pairs of parallel components. None of these admits a complete finite volume hyperbolic structure. □

6. Surgery on one component of the Whitehead link complement

6.1. In **[T]** Thurston describes how to obtain the complement W of the Whitehead link (Fig. 5) by identifying faces of an ideal octahedron in pairs. The identification matches face A with A', B with B', etc., in Fig. 10, so as to respect the labeling of the edges.

If the octahedron is taken to be a regular ideal octahedron in hyperbolic space, then one obtains the complete finite volume hyperbolic structure on W. As described in **[T]** (see also **[NZ]**), by deforming the octahedron to differently shaped ideal hyperbolic octahedra one obtains incomplete hyperbolic structures on W, whose metric completions are hyperbolic Dehn surgeries and generalized Dehn surgeries on the Whitehead link.

We refer the reader to **[T]** and **[NZ]** for details on generalized hyperbolic Dehn surgery. The essential point is that a generalized Dehn surgery on W is parameterized by

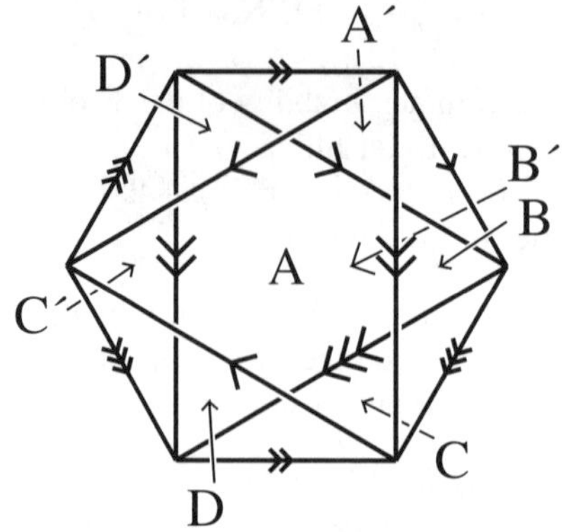

Figure 10

a "Dehn surgery parameter" $(p_i, q_i) \in R^2 \cup \{\infty\}$ for each cusp of W. The result, denoted $W(p_1, q_1; p_2, q_2)$, of this generalized hyperbolic Dehn surgery is defined if the (p_i, q_i) are sufficiently close to ∞. Moreover, if $(p_i, q_i) = \infty$ for $i = 1$ or 2 then the corresponding cusp of W is complete—still a cusp—in $W(p_1, q_1; p_2, q_2)$, and if $(p_i, q_i) \in \mathbb{Z}^2$ then the cusp has been filled in by a geodesic along which $W(p_1, q_1; p_2, q_2)$ has the structure of a hyperbolic orbifold, or manifold if $gcd(p_i, q_i) = 1$, and the underlying topology is that of topological Dehn surgery, described in the previous section. (Precisely, if $(p_i, q_i) = (rp'_i, rq'_i)$ with (p'_i, q'_i) a coprime integer pair and $r > 0$, then the cusp is filled in by a geodesic with a transverse cone angle of $2\pi/r$, while if $p_i/q_i \notin \mathbb{Q}$ then the cusp is filled by a single point at which $W(p_1, q_1; p_2, q_2)$ is not topologically a manifold.)

As in the above references, it is convenient to use an ideal triangulation of W to discuss the deformations, since the shape of an ideal hyperbolic tetrahedron is determined by a single complex parameter in the upper half plane. By subdividing the octahedron of Fig. 10 as in Fig. 11, we obtain an ideal triangulation of W with four simplices.

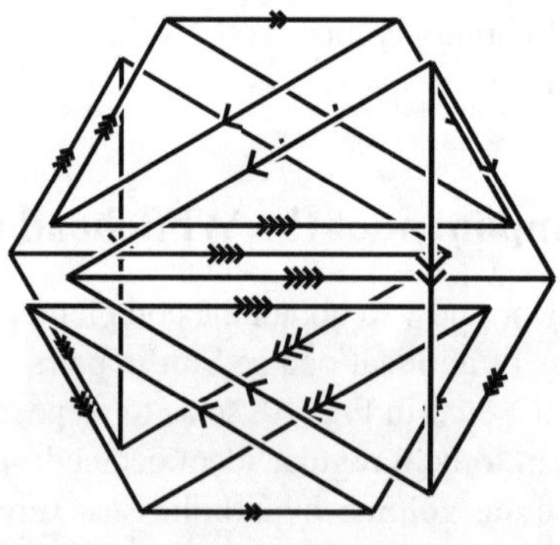

Figure 11

By cutting off the ends of W one obtains a compact manifold-with-boundary $\overline{W}$ which can be obtained by identifying truncated tetrahedra as in Fig. 12. The two boundary tori of $\overline{W}$ are triangulated as in Fig. 13, where the vertices are labeled according to the

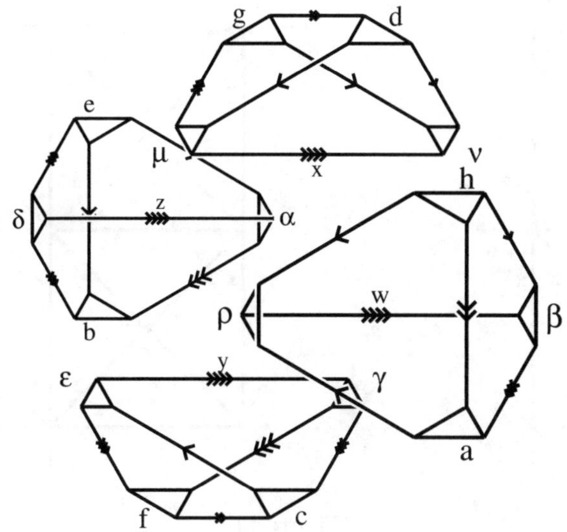

Figure 12

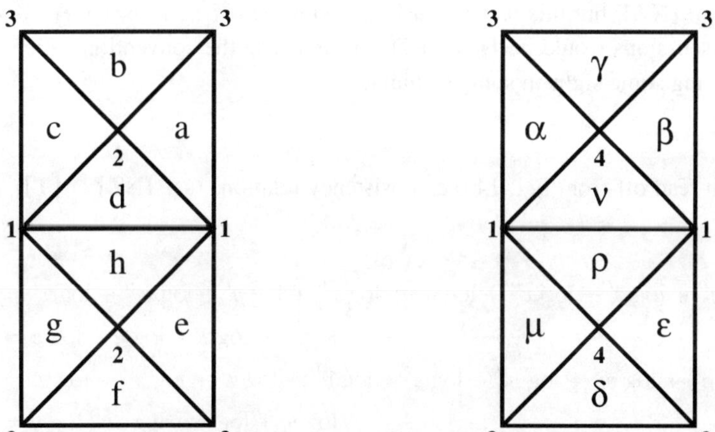

Figure 13

edges of the triangulation of W. Careful inspection shows that the standard topological meridian and longitude of each component of the Whitehead link are as indicated in Fig. 14, where we have also included labels for the complex parameters of the four tetrahedra.

A note on orientations. In the discussion of Dehn surgery on the figure eight knot complement in **[T]**, and also throughout **[NZ]**, a non-standard orientation convention was used for the longitude/meridian pair at a cusp, due to the cusp torus being viewed from inside the manifold rather than outside (this pair is drawn with standard orientation

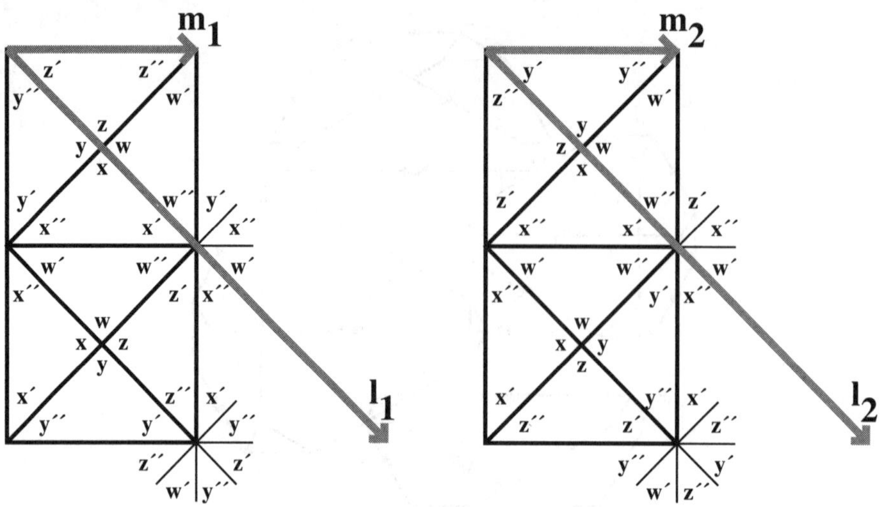

Figure 14

in Fig. 18 of **[NZ]**, but this was inconsistent with the text). Thus (p, q) Dehn surgery in those discussions would be $(p, -q)$ Dehn surgery in the convention which we follow here, affecting some signs in some formulae.

We can read off from Fig. 14 the consistency relations (see **[NZ]** or **[T]**) at the four edges:

$$\log w'' + \log z' + \log x'' + \log w' + \log x'' + \log y' + \log w'' + \log x' = 2\pi i$$
$$\log w + \log x + \log y + \log z = 2\pi i$$
$$\log z'' + \log w' + \log y'' + \log z' + \log y'' + \log x' + \log z'' + \log y' = 2\pi i$$
$$\log w + \log x + \log z + \log y = 2\pi i.$$

Here $\log$ denotes the standard branch of natural log on the complex plane split along $(-\infty, 0]$. Since $x' = \frac{x-1}{x}$ and $x'' = \frac{1}{1-x}$ and similarly for $w,\ y,\ z$, these simplify to the two relations:

$$\log w + \log x + \log y + \log z = 2\pi i \qquad (6.1a)$$
$$\log(1-w) + \log(1-x) - \log(1-y) - \log(1-z) = 0 \qquad (6.1b)$$

Similarly, we can read off the parameters $u_1,\ v_1,\ u_2,\ v_2$, which describe the holonomy of the meridians and longitudes $\mathbf{m}_1,\ \mathbf{l}_1,\ \mathbf{m}_2,\ \mathbf{l}_2$, at the two cusps. They are

$$\begin{aligned}
u_1 &= \log y'' + \log z' + \log z'' + \log w' - \pi i \\
v_1 &= \log y'' + \log y + \log x + \log x' + \log w'' + \log z' + \log x'' \\
&\quad + \log x + \log y + \log y' + \log z'' + \log w' - 4\pi i \\
u_2 &= \log z'' + \log y' + \log y'' + \log w' - \pi i \\
v_2 &= \log z'' + \log z + \log x + \log x' + \log w'' + \log y' + \log x'' \\
&\quad + \log x + \log z + \log z' + \log y'' + \log w' - 4\pi i
\end{aligned}$$

which, written in terms of x, y, z, w and simplified by (6.1), becomes

$$u_1 = \log(w-1) + \log x + \log y - \log(y-1) - \pi i \tag{6.2a}$$
$$v_1 = 2\log x + 2\log y - 2\pi i \tag{6.2b}$$
$$u_2 = \log(w-1) + \log x + \log z - \log(z-1) - \pi i \tag{6.2c}$$
$$v_2 = 2\log x + 2\log z - 2\pi i. \tag{6.2d}$$

(u_1, u_2) can be taken as the "analytic Dehn surgery parameter," as discussed in **[T]** and **[NZ]**, in which case $w,\ x,\ y,\ z$, constrained by equations (6.1), become complex analytic functions of this parameter.

The real Dehn surgery parameters (p_1, q_1) and (p_2, q_2) are determined by the equations

$$\begin{aligned}
p_1 u_1 + q_1 v_1 &= 2\pi i \\
p_2 u_2 + q_2 v_2 &= 2\pi i.
\end{aligned} \tag{6.3}$$

For the complete structure on W we have $u_1 = v_1 = u_2 = v_2 = 0$, which, with (6.1) and (6.2), easily implies $x = y = z = w = i$ (this is also clear from the regularity of the octahedron for the complete structure). We shall thus deform from these values of x, y, z, w, maintaining the consistency relations (6.1).

6.2. We are only interested in generalized Dehn surgeries on one cusp of W, so we always have $(p_1, q_1) = \infty$. The requirement that the first cusp is complete means $u_1 = v_1 = 0$. From $v_1 = 0$ and equation (6.2b) we deduce $x = -y^{-1}$. Equation (6.1a) then gives $z = -w^{-1}$ and (6.1b) then gives $y = w$, so

$$(x, y, z, w) = (x, -x^{-1}, x, -x^{-1}).$$

Conversely, $(x, y, z, w) = (x, -x^{-1}, x, -x^{-1})$ satisfies the consistency relations (6.1) and the cusp relations $u_1 = v_1 = 0$ and thus gives a structure on W which is complete at the first cusp. We thus just have the one complex parameter x, in the upper half plane, describing our deformations.

We abbreviate $(u_2, v_2) = (u, v)$ and $(p_2, q_2) = (p, q)$, and we shall write $W(p, q)$ for $W(\infty; p, q)$. The formulae (6.2) simplify to

$$\begin{aligned}
u &= \log x + \log(x+1) - \log(x-1) \\
v &= 4\log x - 2\pi i
\end{aligned} \tag{6.4}$$

and the real Dehn surgery parameter (p,q) is determined by the equation

$$pu + qv = 2\pi i. \tag{6.5}$$

Thurston's complex analytic parameter for hyperbolic Dehn surgery on one component of the Whitehead link is u. We shall use the parameter x instead. Since x describes the shape of an ideal simplex, which degenerates for x real, the natural domain for the parameter x is the complex upper half plane U. The maps $x \mapsto u$ and $x \mapsto v$ take U biholomorphically to the domains $\mathcal{U}$ and $\mathcal{V}$, where $\mathcal{U}$ is the complex plane split along the rays $(-\infty, 2\log(\sqrt{2}-1)]$ and $[2\log(\sqrt{2}+1),\infty)$ and $\mathcal{V} = \{v \in \mathbb{C} : -2\pi i < \operatorname{Im} v < 2\pi i\}$. Thus $x \in U$, $u \in \mathcal{U}$, and $v \in \mathcal{V}$ are all equally good complex analytic parameters for generalized hyperbolic Dehn surgery on one component of the Whitehead link.

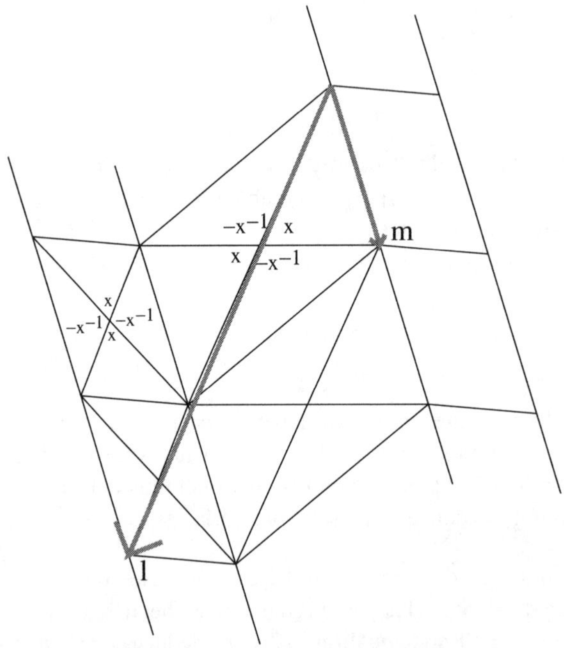

Figure 15

The triangulation of the cusp torus of $W(p,q)$ is shown in Fig. 15 with the meridian $\mathbf{m}$ and longitude $\mathbf{l}$ drawn in. Its isomorphism type as a complex torus $\mathbb{C}^2/\Lambda$ is described by the complex parameter $\tau = \lambda_2/\lambda_1$ with positive imaginary part, after choosing an oriented base λ_1, λ_2 for the lattice Λ. We can consider $\mathbf{m}$ and $\mathbf{l}$ to represent generating translations in the universal cover of the cusp torus, and we define:

Definition. The ***cusp parameter*** of $W(p,q)$ is the parameter τ of the cusp torus with respect to the oriented basis $\mathbf{m}, -\mathbf{l}$. We denote the cusp parameter for $W(p,q)$ by $\tau(W(p,q))$. (We make this choice of basis for consistency with **[T]** and **[NZ]**, as well as because of Theorem 6.3 below, which says this τ is often an algebraic integer; the seemingly "more natural" choice $\mathbf{l}, \mathbf{m}$ replaces τ by $-\tau^{-1}$.)

Our next theorem says that $\tau(W(p,q))$ determines $W(p,q)$. To state it we need some notation. Let $\mathcal{N}$ denote the closed parallelogram in $\mathbb{R}^2$ with vertices $\pm(-4,1), \pm(0,1)$ (Fig. 16).

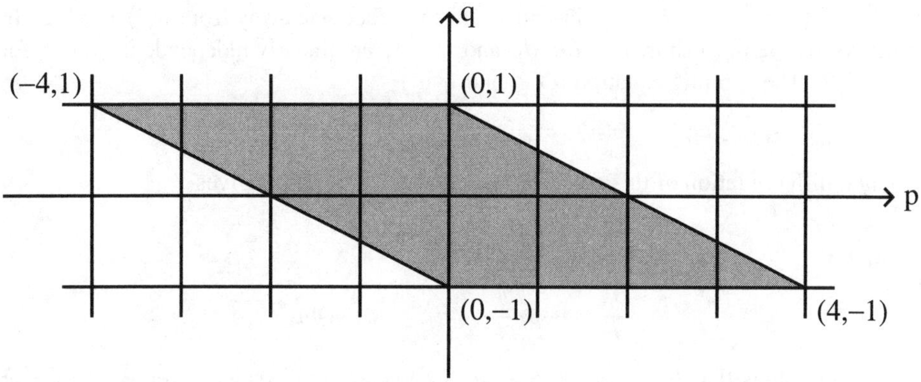

Figure 16

It is known (and we shall show) that this is the "non-hyperbolic" domain in the space of real Dehn surgery parameters. Since (p,q) Dehn surgery and $(-p,-q)$ Dehn surgery give the same result, the space of real surgery parameters for hyperbolic Dehn surgeries is

$$\mathcal{H} = (\mathbb{R}^2 \cup \{\infty\} - \mathcal{N})/\pm 1.$$

Theorem 6.1. *The map $(p,q) \mapsto \tau(W(p,q))$ is a homeomorphism $\mathcal{H} \to U$, where U is the complex upper half plane.*

Proof. By elementary calculation from Fig. 15 one obtains τ in terms of x as

$$\tau = \frac{4x}{1-x^2} - 2. \tag{6.6}$$

By solving equation (6.6) for x in terms of τ one verifies that this map $x \mapsto \tau$ is a 2-fold branched cover $U \to U$, branched at $x = i$, and that x and $-x^{-1}$ both map to the same τ. Note that replacing x by $-x^{-1}$ replaces u by $-u$, v by $-v$, and hence (p,q) by $(-p,-q)$. To prove the theorem we must thus show that the map $x \mapsto (p,q)$ is a homeomorphism of U to $\mathbb{R}^2 \cup \{\infty\} - \mathcal{N}$. We shall denote the map $x \mapsto (p,q)$ by $\phi: U \to \mathbb{R}^2 \cup \{\infty\}$.

Thurston's hyperbolic Dehn surgery theorem (cf., **[T]** or **[NZ]**) asserts that ϕ is a homeomorphism from a neighborhood of the parameter value $x = i$ corresponding to the complete structure on W to a neighborhood of ∞. Elementary computation shows that ϕ takes the boundary $\mathbb{R} \cup \{\infty\}$ of $\bar{U}$ homeomorphically to the boundary of $\mathcal{N}$ (specifically, the intervals $[-\infty,-1]$, $[-1,0]$, $[0,1]$, $[1,\infty]$ go to the segments $[(4,-1),(0,1)]$, $[(0,1),(-4,1)]$, $[(-4,1),(0,-1)]$, $[(0,-1),(4,-1)]$ respectively). It therefore suffices to show that ϕ has non-vanishing Jacobian at all $x \in U - \{i\}$.

Denote

$$a(x) = \log x, \qquad b(x) = \log \frac{x-1}{x+1}.$$

Since a is a biholomorphic map $a\colon U \to \mathcal{A} = \{z \in \mathbb{C} : 0 < \operatorname{Im} z < \pi\}$, we may show instead that $\phi \circ a^{-1}\colon a \mapsto (p,q)$ has non-vanishing Jacobian away from $a(i) = \pi i/2$. In other words, we must show that $\partial a/\partial p$ and $\partial a/\partial q$ are linearly independent over $\mathbb{R}$ for $a \neq \pi i/2$. The defining equation for p and q is

$$p(a-b) + q(4a - 2\pi i) = 2\pi i.$$

Implicit differentiation of this equation with respect to p and q gives

$$\begin{aligned} \frac{\partial a}{\partial p}\left(p - p\frac{\partial b}{\partial a} + 4q\right) &= b - a \\ \frac{\partial a}{\partial q}\left(p - p\frac{\partial b}{\partial a} + 4q\right) &= 2\pi i - 4a, \end{aligned}$$

so we must show that $b - a$ and $2\pi i - 4a$ are linearly independent over $\mathbb{R}$. However the map $b \circ a^{-1}\colon a(x) \mapsto b(x)$ is conjugate by $\log$ to the order 4 Möbius transformation $x \mapsto (x-1)/(x+1)$ and it is thus easy to see that if one divides $\mathcal{A}$ into four "quadrants" by the vertical and horizontal lines through $\pi i/2$, then $a \mapsto b$ rotates each quadrant clockwise into the next. It follows that the line from b to a never points in the same direction as the line from the center $\pi i/2$ of rotation to a (unless $a = \pi i/2$). That is, $b - a$ and $\pi i/2 - a$ are independent over $\mathbb{R}$, as was to be shown. □

We now restrict to the case that $(p,q) \notin \mathcal{N}$ and $p \in \mathbb{Z}$ and $q \in \frac{1}{2}\mathbb{Z}$. Recall that W covers the orbifolds W' and W'' of Figs. 6 and 7 with covering groups C_2 and $C_2 \times C_2$. $W(p,q)$ thus branched covers $W'(p,2q)$ and $W''(p,2q)$ in the notation of Sect. 5, and these are hyperbolic orbifolds under our assumptions on p and q.

Note that $(p,q) \mapsto (p+2q, 2q)$ transforms the domain $\mathcal{N}$ into the square with vertices $(\pm 2, \pm 2)$, so part (ii) of Theorem 5.1 is proven. We next compute some arithmetic invariants of our orbifolds.

Theorem 6.2. *Denote* $z = x - x^{-1}$ *(so* $\tau = -4/z - 2$ *and* $\mathbb{Q}(z) = \mathbb{Q}(\tau)$*). The invariant trace fields and trace fields for* $W(p,q)$, $W'(p,2q)$, *and* $W''(p,2q)$ *are as follows:*

$$\left.\begin{aligned} k(\pi_1 W(p,q)) &= \mathbb{Q}(z) \\ \mathbb{Q}(\operatorname{tr}(\pi_1 W(p,q))) &= \mathbb{Q}(\sqrt{z}) \end{aligned}\right\} q \in \mathbb{Z},$$

$$\left.\begin{aligned} k(\pi_1 W'(p,2q)) = k(\pi_1 W''(p,2q)) &= \mathbb{Q}(z) \\ \mathbb{Q}(\operatorname{tr}(\pi_1 W'(p,2q))) = \mathbb{Q}(\operatorname{tr}(\pi_1 W''(p,2q))) &= \mathbb{Q}(\sqrt{z}, i) \end{aligned}\right\} q \in \frac{1}{2}\mathbb{Z}.$$

In particular, the cusp field equals the invariant trace field for all these examples.

Remark. By Corollary 2.3 it follows that $\mathbb{Q}(z) = \mathbb{Q}(\sqrt{z})$ if $q \in \mathbb{Z}$ and p is odd. In the examples we have checked, $\mathbb{Q}(z) \neq \mathbb{Q}(\sqrt{z})$ otherwise.

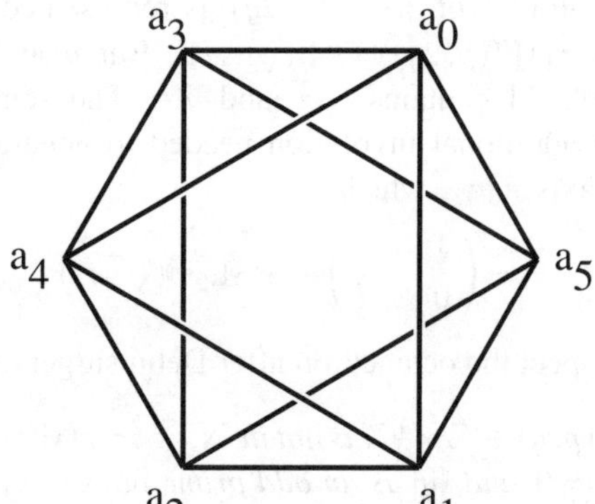

Figure 17

Proof. Denote the vertices of our original octahedron O by $a_0, \ldots, a_5$ as in Fig. 17. If we position O in $\mathbb{H}^3$ with these vertices at $0,\ 1,\ \infty,\ -1,\ x,\ x^{-1}$, respectively, then it is easy to compute that the four constituent tetrahedra (Fig. 12) have parameters $x,\ -x^{-1},\ x,\ -x^{-1}$, so O has the correct shape for the value x of our deformation parameter.

Denote by a the element of $\mathrm{PGL}_2\mathbb{C} = \mathrm{Isom}(\mathbb{H}^3)$ which translates O across its face A, taking face A' to A. Thus a takes $0,\ -1,\ x^{-1}$ to $1,\ 0,\ x$ respectively. Solving for an element of $\mathrm{PGL}_2\mathbb{C}$ shows that a can be represented by the matrix

$$a = \begin{pmatrix} 1 & 1 \\ 1-z & 1 \end{pmatrix} \in \mathrm{PGL}_2\mathbb{C}.$$

Similarly, if $b,\ c,\ d$ are the translations across faces $B,\ C,\ D$ respectively, one computes that

$$b = \begin{pmatrix} 1 & 1 \\ 1 & 1+z \end{pmatrix}, \quad c = \begin{pmatrix} 1 & 1-z \\ 1 & 1 \end{pmatrix}, \quad d = \begin{pmatrix} 1+z & 1 \\ 1 & 1 \end{pmatrix}.$$

If $q \in \mathbb{Z}$ then $\pi_1 W(p,q)$ is generated by $a,\ b,\ c$, and d. $\pi_1 W'(p,2q)$ is generated by the above elements and the additional element

$$\beta = \begin{pmatrix} 0 & 1 \\ 1 & 0 \end{pmatrix}$$

which rotates the octahedron about the axis through a_1 and a_3.

The vertices $a_0, \ldots, a_3$ lie at the cusp of $W'(p,2q)$ and are all in $\mathbb{Q}(z) \cup \{\infty\}$ and the generators $a,\ b,\ c,\ d,\ \beta$, of $\pi_1 W'(p,2q)$ have entries in $\mathbb{Q}(z)$. As in the proof of Lemma 2.5, it follows that all lifts of the cusp are in $\mathbb{Q}(z)$, so $k(\pi_1 W'(p,2q)) \subset \mathbb{Q}(z)$. But $k(\pi_1 W'(p,2q))$ must contain τ, which generates $\mathbb{Q}(z)$, so it equals this field.

Suppose $q \in \mathbb{Z}$. To compute $\mathbb{Q}(\mathrm{tr}\ \pi_1 W(p,q))$ we must put the generators $a,\ b,\ c,\ d$ in $\mathrm{SL}_2\mathbb{C}$ by dividing each by the square root of its determinant, namely $\sqrt{z}$. Their traces then evidently generate $\mathbb{Q}(\sqrt{z})$, so $\mathbb{Q}(\mathrm{tr}\ \pi_1 W(p,q))$ contains $\mathbb{Q}(\sqrt{z})$. On the other hand our representation of $\pi_1 W(p,q)$ then has image in $\mathrm{PSL}_2\mathbb{Q}(\sqrt{z})$, so $\mathbb{Q}(\mathrm{tr}\ \pi_1 W(p,q))$ is at most $\mathbb{Q}(\sqrt{z})$.

The additional generator β of $\pi_1 W'(p,2q)$ is represented by the matrix $\left(\begin{smallmatrix} 0 & i \\ i & 0 \end{smallmatrix}\right) \in \mathrm{PSL}_2\mathbb{Q}(\sqrt{z}, i)$, so $\mathbb{Q}(\operatorname{tr} \pi_1 W'(p,2q)) \subseteq \mathbb{Q}(\sqrt{z}, i)$. But a and βb have traces $1/\sqrt{z}$ and $i\sqrt{z}$, so the trace field contains $\sqrt{z}$ and i. The same argument applies to $\pi_1 W''(p,2q)$, since the additional involution needed to generate $\pi_1 W''(p,2q)$ is the $180°$ rotation about the axis $a_0 a_2$, which is

$$\gamma = \begin{pmatrix} i & 0 \\ 0 & -i \end{pmatrix} \in \mathrm{PSL}_2\mathbb{Q}(\sqrt{z}, i).$$

(Remark: γ does not respect the octahedron after Dehn surgery.) □

Theorem 6.3. *Suppose $(p,q) \in \mathbb{Z} \times \frac{1}{2}\mathbb{Z}$ is not in $\mathcal{N}$. $\tau = \tau(W(p,q))$ is not an algebraic integer if and only if $p = 0$ and $|q|$ is an odd prime power. $\pi_1 W'(p,2q)$ fails to have integral traces if and only if $p = 0$ or $-4q$ and $|q|$ is an odd prime power.*

Proof. Since $\tau = -4/z - 2$, the integrality of τ is equivalent to that of $4/z$. Moreover, to check if $\operatorname{tr} \pi_1 W'(p,2q)$ consists of algebraic integers, it suffices to check a set of generators and their pairwise products (cf. **[Mag**, p. 148**]**). Using the generators in the above proof one obtains traces $2/\sqrt{z}$, $\sqrt{z} + 2/\sqrt{z}$, multiples of these by i, and rational integral linear combinations of 1, z, and $4/z$. So $\operatorname{tr} \pi_1 W(p,q)$ consists of algebraic integers if and only if z and $4/z$ (and hence also their square roots) are algebraic integers.

By (6.4) and (6.5), x satisfies the equation

$$\left(\frac{x(x+1)}{x-1}\right)^p x^{4q} = 1, \tag{6.7}$$

which multiplies out to a monic polynomial with rational integer coefficients and constant term ± 1 unless $p + 4q = 0$. Thus x is a unit, whence $z = x - x^{-1}$ is an algebraic integer, if $p + 4q \neq 0$. On the other hand, if x satisfies equations (6.4) and (6.5) for (p,q) then one checks that $(1-\bar{x})/(1+\bar{x})$ satisfies (6.4) and (6.5) for $(p+4q, -q)$. But replacing x by $(1-\bar{x})/(1+\bar{x})$ replaces $z = x - x^{-1}$ by $4/\bar{z}$. Thus $4/\bar{z}$—and hence also $4/z$—is an algebraic integer if $p \neq 0$.

To complete the proof it suffices to show that when $p = 0$, z is an algebraic integer divisor of 4 if and only if $|q|$ is not an odd prime power. (Recall, that since $q \in \frac{1}{2}\mathbb{Z}$, $2q$ is an arbitrary integer.) Assume $p = 0$. Then (6.4) and (6.5) imply that $x/i = \exp(2\pi i/4q)$, so $z/i = \exp(2\pi i/4q) + \exp(-2\pi i/4q)$. Now z will be a divisor of 4 if and only if $\zeta := (z/i)\exp(2\pi i/4q)$ is. But ζ divides 4 if and only if its norm divides 4. The norm of $\zeta = \exp(2\pi i/2q) + 1$ is $\Phi(1)$, where Φ is the minimal polynomial for $-\exp(2\pi i/2q)$. Now, $-\exp(2\pi i/2q)$ is a primitive r-th root of unity with $r = |4q|$, $|q|$, or $|2q|$, according as $2q$ is an odd integer, twice an odd integer, or divisible by 4. Thus Φ is the r-th cyclotomic polynomial, and it is known that $\Phi(1) = 1$ unless r is the power of a prime, in which case $\Phi(1)$ equals this prime (cf. e.g., **[W]**). Thus $\Phi(1)$ fails to divide 4 if and only if r is an odd prime power, in which case $q = \pm r$. □

An interesting example is $W(0,3)$ (or $W(12,-3)$, which is commensurable with it). The invariant trace field $k(\pi_1 W(0,3))$ is $\mathbb{Q}(\sqrt{-3})$, but $\pi_1 W(0,3)$ has non-integral traces, so it cannot be commensurable with $\mathrm{PSL}_2\mathcal{O}_3$. However its volume turns out to be 20 times the volume of $\mathbb{H}^3/\mathrm{PSL}_2\mathcal{O}_3$—this is $10/3$ times the volume of a regular ideal tetrahedron.

7. Arithmeticity of chain links

Proof of part (iii) of Theorem 5.1. Recall that $C(p,s)$ is a cover of $W'(p,s)$ and if $W'(p,s)$ is hyperbolic, then it is double branched covered by $W(p,q)$ with $s=2q$. We claim that the geodesic $\gamma(p,s)$ added by hyperbolic Dehn filling W' to form $W'(p,s)$ has length

$$\ell_0(\gamma(p,s)) = -\frac{d}{s}\operatorname{Re}(u) = \frac{d}{2p}\operatorname{Re}(v), \quad d = \gcd(p,s), \tag{7.1}$$

where u and v are as in equations (6.4) and (6.5). Indeed, if $\ell(\gamma(p,s))$ is the complex length of the added geodesic then, by equation (6.5) and by **[Th]** or **[NZ]**, we have equations

$$pu + s\frac{v}{2} = 2\pi i$$
$$p'u + s'\frac{v}{2} = \ell(\gamma(p,s)),$$

where p' and s' are integers satisfying $ps' - sp' = d$. Multiplying the first of these two equations by s' and the second by s and subtracting gives $du = 2\pi i s' - s\ell(\gamma(p,s))$, from which the first equality of (7.1) follows by taking real part; the second equality follows similarly.

The computations that we now describe were computer-aided, so we omit some details. We wish to find the (p,s) for which $\ell_0(\gamma(p,s)) \geq 0.431277313$ (cf. Corollary 4.7). We use equations (7.1). First, by plotting level curves in the $(p+s,s)$-plane for $\operatorname{Re}(u)$ and $\operatorname{Re}(v)$ one finds that the conditions $|\operatorname{Re}(u)| \geq 0.43$ and $|\operatorname{Re}(v)/2| \geq 0.43$ force $(p+s,s)$ to lie in a bounded domain (which lies entirely within the circle of radius 15). It is then quick to verify that the (p,s) for which $\ell_0(\gamma(p,s)) \geq 0.431277313$ are those with $(|p+s|,|s|)$ or $(|s|,|p+s|)$ in $\{(n,0) : 3 \leq n \leq 14\} \cup \{(n,n) : 3 \leq n \leq 7\} \cup \{(3,1),(3,2),(4,2)\}$. Thus, only in these cases might $W'(p,s)$ be arithmetic.

The table of Theorem 4.6 of possible short geodesic lengths in cusped arithmetic orbifolds now eliminates all cases but those of part (iii) of Theorem 5.1, namely $(|p+s|,|s|)$ or $(|s|,|p+s|)$ equal to one of $(3,0)$, $(3,1)$, $(3,2)$, $(3,3)$, $(4,0)$, $(4,2)$, $(4,4)$, $(6,0)$. (We could also have eliminated all but these and $(6,6)$ by the observation that a cusped arithmetic orbifold can have torsion only of order 2, 3, 4, or 6, since the trace of the square of the torsion element must be in some $\mathcal{O}_d$.)

To show that the above cases do give arithmetic orbifolds we use the characterization of arithmeticity in Proposition 4.4. We must thus verify that in these cases the invariant trace field $k(\pi_1 W'(p,s))$ is $\mathbb{Q}(\sqrt{-d})$ with d respectively 7, 1, 3, 1, 3, 7, 2, 15, and that $W'(p,s)$ has integral traces. By Theorem 6.2 and the proof of Theorem 6.3 we must check that z and $4/z$ lie in the appropriate $\mathcal{O}_d$. This is evident from Table 2, which includes also the non-arithmetic case $(6,6)$. □

Table 2. z **and** $4/z$

$\{\lvert p+s\rvert, \lvert s\rvert\}$	z	$4/z$
$\{3,0\}$	$\frac{-3+\sqrt{-7}}{2}$	$\frac{-3-\sqrt{-7}}{2}$
$\{3,1\}$	$-1+\sqrt{-1}$	$2-2\sqrt{-1}$
$\{3,2\}$	$\frac{-1+\sqrt{-3}}{2}$	$-2-2\sqrt{-3}$
$\{3,3\}$	$+\sqrt{-1}$	$-4\sqrt{-1}$
$\{4,0\}$	$-1+\sqrt{-3}$	$-1-\sqrt{-3}$
$\{4,2\}$	$\frac{-1+\sqrt{-7}}{2}$	$-1-\sqrt{-7}$
$\{4,4\}$	$+\sqrt{-2}$	$-2\sqrt{-2}$
$\{6,0\}$	$\frac{-1+\sqrt{-15}}{2}$	$\frac{-1-\sqrt{-15}}{2}$
$\{6,6\}$	$+\sqrt{-3}$	$\frac{-4\sqrt{-3}}{3}$

It is remarkable that the arithmetic $W'(p,s)$ are picked out exactly by Theorem 4.6. We had conjectured the following alternate characterization of the arithmetic cases, and T. Chinburg showed us a proof, which we sketch.

Proposition 7.1. *$k(\pi_1 W'(p,s))$ is a quadratic imaginary field if and only if (p,s) is one of the arithmetic cases above or $(|p+s|,|s|)=(6,6)$ (for which $W'(p,s)$ does not have integral traces).*

Sketch Proof. Suppose $\mathbb{Q}(z)$ is imaginary quadratic. Since $z=x-x^{-1}$, x has degree at most 4 over $\mathbb{Q}$. If $p=0$ or $p+4q=0$ then equation (6.7) shows that x, respectively $(x+1)/(x-1)$, is a root of unity of degree at most 4 over $\mathbb{Q}$. This gives finitely many cases to check. If neither p nor $p+4q$ is zero, then the proof of Theorem 6.3 shows that z is an algebraic integer divisor of 4 in an imaginary quadratic number field. This again gives finitely many cases to check. □

We have computed the minimal polynomial for the generating element z of $k(\pi_1 W'(p,s))$ whenever $|p+s|$ and $|s|$ are at most 13. The degree is always at most $\max\{|p+s|,|s|\}-1$, as is not hard to prove, but is often smaller, as Table 3 shows.

8. Commensurators of chain links: proofs

In this section we prove parts (iv) and (v) of Theorem 5.1.

Lemma 8.1. *Let (p,s) be such that $C(p,s)$ is non-arithmetic. The cusp parameter τ is not in $\mathbb{Q}(\sqrt{-1})$ or $\mathbb{Q}(\sqrt{-3})$, unless $(|p+s|,|s|)=(6,6)$.*

Proof. Since $\tau+2=-4/z$ we must just show that z is not in $\mathbb{Q}(\sqrt{-1})$ or $\mathbb{Q}\sqrt{-3})$. If $p=0$ then equations (6.4) and (6.5) give $x=\exp((s+2)\pi i/2s)$, and since $z=x-x^{-1}$, it is easy to see that z is in $\mathbb{Q}(\sqrt{-1})$ or $\mathbb{Q}(\sqrt{-3})$ if and only if $|s|=3$, 4, or 6. The

Table 3. $[k(W'(p,s)):\mathbb{Q}]$

$\|p+s\|$ \ $\|s\|$	0	1	2	3	4	5	6	7	8	9	10	11	12	13
0				2	2	4	2	6	4	6	4	10	4	12
1				2	3	4	5	6	7	8	9	10	11	12
2				2	2	4	3	6	4	8	5	10	6	12
3	2	2	2	2	3	4	4	6	7	6	9	10	8	12
4	2	3	2	3	2	4	3	6	4	8	5	10	6	12
5	4	4	4	4	4	4	5	6	7	8	8	10	11	12
6	2	5	3	4	3	5	2	6	4	6	5	10	4	12
7	6	6	6	6	6	6	6	6	7	8	9	10	11	12
8	4	7	4	7	4	7	4	7	4	8	5	10	6	12
9	6	8	8	6	8	8	6	8	8	6	9	10	8	12
10	4	9	5	9	5	8	5	9	5	9	4	10	6	12
11	10	10	10	10	10	10	10	10	10	10	10	10	11	12
12	4	11	6	8	6	11	4	11	6	8	6	11	4	12
13	12	12	12	12	12	12	12	12	12	12	12	12	12	12

cases $|s| = 3$ or 4 are arithmetic, so the only case is $|s| = 6$. If $p + 2s = 0$ then $C(p,s)$ is commensurable with $C(0,s)$ by part (i) of Theorem 5.1, so again, the only non-arithmetic case with z in $\mathbb{Q}(\sqrt{-1})$ or $\mathbb{Q}(\sqrt{-3})$ is $|s| = 6$. Finally, if $p \neq 0$ and $p + 2s \neq 0$ then the proof of Theorem 6.3 shows that z is an algebraic integer divisor of 4. The only algebraic integer divisors of 4 with positive imaginary part in $\mathbb{Q}(\sqrt{-1})$ or $\mathbb{Q}(\sqrt{-3})$ are $z = i,\ 2i,\ 4i,\ 1+i,\ -1+i,\ 2+2i,\ -2+2i,\ \omega,\ 2\omega,\ 4\omega,\ -\bar{\omega},\ -2\bar{\omega}$, and $-4\bar{\omega}$, where $\omega = \exp(2\pi i/3)$. One computes directly that these correspond to $(p+s,s) = (3,3),\ (\infty,\infty),\ (3,-3),\ (1,3),\ (3,1),\ (1,-3),\ (3,-1),\ (2,3),\ (0,4),\ (2,-3),\ (3,2),\ (4,0),\ (3,-2)$, respectively, which are arithmetic cases, and thus excluded in the lemma. □

Proof of Theorem 5.1 (iv). An orbifold cusp is called ***rigid*** if its fundamental group is the non-abelian split extension of $\mathbb{Z}/n$ by $(\mathbb{Z} \oplus \mathbb{Z})$ with $n = 3$, 4, or 6. In this case the parameter τ of a flat torus covering a horosphere section of the cusp must be in $\mathbb{Q}(\sqrt{-1})$ or $\mathbb{Q}(\sqrt{-3})$.

Suppose $C(p,s)$ is nonarithmetic and $(|p+s|,|s|) \neq (6,6)$. By Lemma 8.1, $C(p,s)$ cannot cover an orbifold with a rigid cusp. On the other hand, by the Thurston/Jørgenson theory (cf., **[T**, Chapter 5]) the quotient $W''(p,s)$ of $C(p,s)$ has volume less than $\mathrm{vol}(W'') \approx 0.918$. If $W''(p,s)$ non-trivially covered the orientable commensurator quotient of $C(p,s)$, then this commensurator quotient would have volume at most half this value, namely $<\approx 0.459$. But this volume bound puts the commensurator quotient on Adams' list of smallest orientable orbifolds with non-rigid cusps (**[A1]**), and these orbifolds are all arithmetic. Thus $W''(p,s)$ is the orientable commensurator quotient.

Finally, suppose $|(p+s|,|s|) = (6,6)$. By Table 2, the cusp parameter of $W''(p,s)$ lies in $\mathbb{Q}(\sqrt{-3})$. Now, $W''(p,s)$ cannot cover an orbifold with a non-rigid cusp by

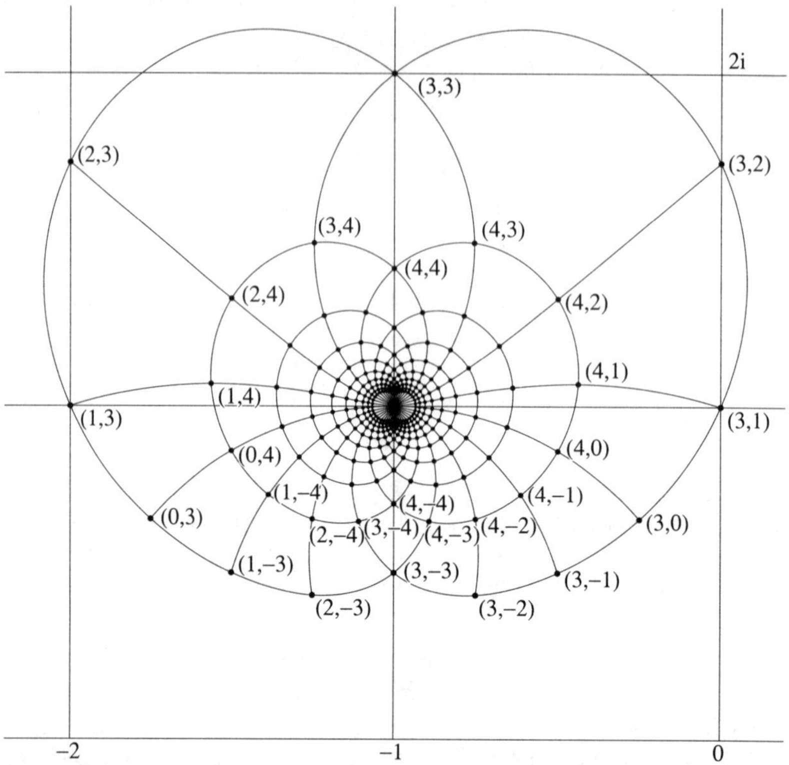

Figure 18

the argument of the previous paragraph, so if it non-trivially covers some orbifold, that orbifold must have a $(2,3,6)$ or $(3,3,3)$ cusp. But $W''(p,s)$ has volume $5v_0/6$, where v_0 is the volume of the regular ideal tetrahedron, and Adams, in these Proceedings (**[A2]**), shows that the four orbifolds with rigid cusps of type $(2,3,6)$ or $(3,3,3)$ and with volume less than $v_0/4$ have volumes $v_0/12$, $v_0/6$, $v_0/6$, or $5v_0/24$. $W''(p,s)$ cannot cover any of these because they are arithmetic (cf. **[NR2]**). Thus the orientable commensurator quotient of $W''(p,s)$ has volume at least $v_0/4$, so $\pi_1 W''(p,s)$ has index 1, 2, or 3 in its orientation preserving commensurator. The cusp of $W''(p,s)$ cannot 2-fold cover a $(2,3,6)$ or $(3,3,3)$ cusp, because of the 3-torsion. If the commensurator had index 3 then either $W''(p,s)$ or some 2-fold cover of $W''(p,s)$ would have a 3-fold symmetry. By considering the set of 3-orbifold points, it is not hard to eliminate the existence of such a symmetry. □

Proof of Theorem 5.1 (v). Recall that we have already shown that the $C(p,s)$ of part (v) of Theorem 5.1 have orientation reversing commensurabilities, so we must just show that the others do not. If $C(p,s)$ has an orientation reversing commensurability, then its orientable commensurator quotient $W''(p,s)$ admits an orientation reversing symmetry, so the cusp of $W''(p,s)$ also does. The result thus follows from the following lemma. □

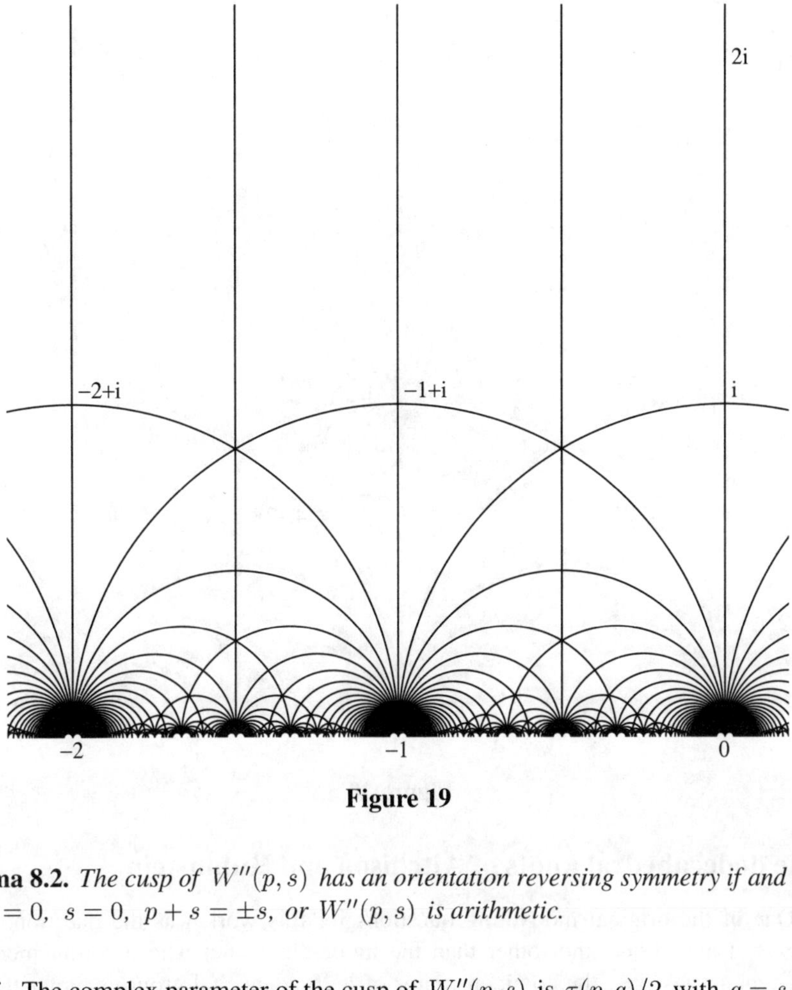

Figure 19

Lemma 8.2. *The cusp of $W''(p,s)$ has an orientation reversing symmetry if and only if $p+s=0,\ s=0,\ p+s=\pm s$, or $W''(p,s)$ is arithmetic.*

Proof. The complex parameter of the cusp of $W''(p,s)$ is $\tau(p,q)/2$ with $q=s/2$ (by the parameter for a flat $(2,2,2,2)$-orbifold we mean the parameter of its unique 2-fold torus covering space). In Fig. 18 we show the image in $(\tau/2)$-space of the integer lattice in $(p+s,s)$-space (excluding the non-hyperbolic region $\mathcal{N}$). The moduli space of flat $(2,2,2,2)$-orbifolds is the moduli space of flat tori, namely $U/\mathrm{PSL}_2\mathbb{Z}$, where U is complex upper half-space. Orientation reversal induces an involution on $U/\mathrm{PSL}_2\mathbb{Z}$ (induced by $\tau\mapsto-\overline{\tau}$) whose fixed point set gives the set of moduli of orbifolds which admit an orientation reversing isometry. The lift of this set to U is shown in Fig. 19. Fig. 20 superimposes the previous two figures—it is clear that the only $(p+s,s)$ lattice points for which the cusp of $W''(p,s)$ admits an orientation reversing isometry are those with $p+s=0,\ s=0$, or $p+s=\pm s$, or $\{|p+s|,|s|\}\in\{\{3,1\},\{3,2\},\{4,2\}\}$. By Theorem 5.1 part (iii), this proves the lemma. □

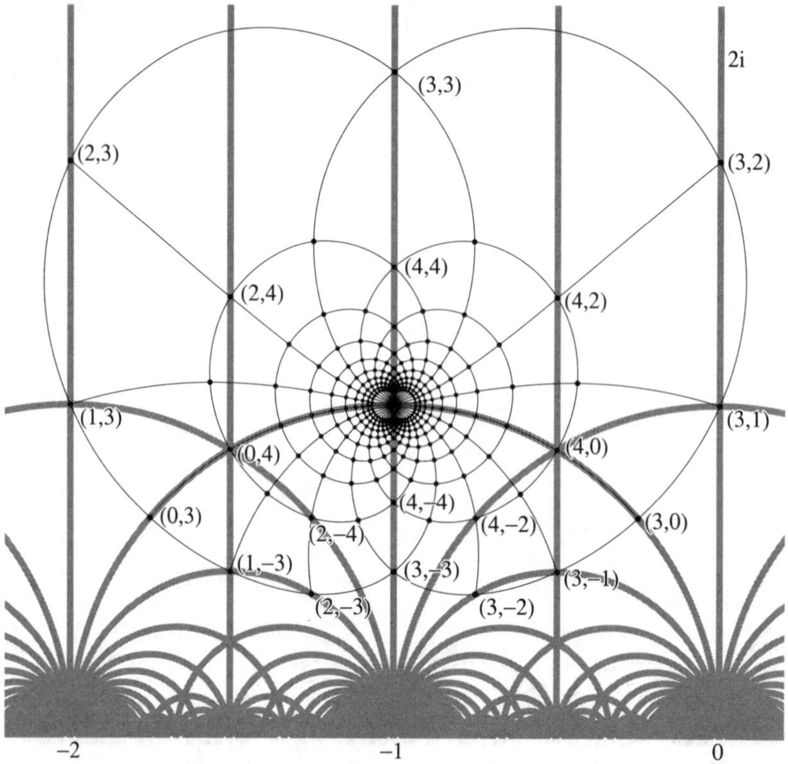

Figure 20

9. The dodecahedral knots of Aitchison and Rubinstein

9.1. One of the original motivating questions for this work was the question: does there exist a hyperbolic knot other than the figure-eight knot whose complement has hidden symmetries (i.e., the commensurator of its fundamental group is larger than its normalizer)?

Of course an arithmetic knot complement would have hidden symmetries, but by **[R2]**, the figure-eight knot complement is the only such.

Proposition 9.1. *The following are equivalent for a hyperbolic knot complement other than the figure-eight knot complement:*
(1) *it has hidden symmetries;*
(2) *its orientable commensurator quotient has a rigid cusp;*
(3) *the knot complement non-normally covers some orbifold.*

Proof. Clearly (1) is equivalent to (3). A hyperbolic knot complement (even in a rational homology sphere) has a homologically determined longitude, which must be preserved up to sign by any isometry. Since the orientable isometry group must act effectively on the cusp torus, it can only be cyclic or $\mathbb{Z}/2$ extended by a cyclic group, so the quotient has a non-rigid cusp. Thus (2) implies (1). If a knot complement covers an orbifold with a non-rigid cusp then the covering must be normal by Lemma 4 of **[R2]**, so (1) implies (2). □

We now describe a pair of knot complements, constructed by Aitchison and Rubinstein in their paper **[AR]** in these Proceedings, which are mutually commensurable and which do have hidden symmetries.

Let D denote the regular ideal dodecahedron in hyperbolic 3-space. All its dihedral angles are $\pi/3$. As described in **[AR]** there exist two distinct hyperbolic knot complements D_f and D_s obtained by identifying faces of two copies of D. Let Γ be the group generated by the order 120 group of symmetries of the ideal dodecahedron D and all reflections in faces of D. This Γ is clearly the tetrahedral group determined by the tetrahedron $T[5,2,2;2,3,6]$, which is a fundamental domain for the symmetry group of D. Let Γ^+ be the orientation preserving subgroup of Γ. Clearly, the hyperbolic 3-orbifold $\mathbb{H}^3/\Gamma^+$ has a rigid cusp of type $(2,3,6)$ and is covered by D_f and D_s. It is in fact the orientable commensurator quotient of D_f and D_s (and $\mathbb{H}^3/\Gamma$ is therefore the non-orientable commensurator quotient), for the following reason: by **[Mey]**, $\mathbb{H}^3/\Gamma^+$ has volume approximately 0.343003, and if there were a smaller quotient, it would have to have volume at most half this, which would contradict the results of Adams **[A2]**.

The question at the beginning of this section may now be updated:

Question 1. *Does there exist a hyperbolic knot other than the figure-eight knot and the two dodecahedral knots with hidden symmetries*?

As pointed out in Sect. 8, if a hyperbolic manifold covers an orbifold with a rigid cusp then the cusp parameter is in $\mathbb{Q}(\sqrt{-1})$ or $\mathbb{Q}(\sqrt{-3})$. The only knot complements with this property that we know are the complements of the above three knots.

Question 1′. *Is there a knot other than one of the above three whose complement is hyperbolic with cusp parameter in* $\mathbb{Q}(\sqrt{-1})$ *or* $\mathbb{Q}(\sqrt{-3})$?

As described above, the dodecahedral knot groups are subgroups of a polyhedral group, so they also answer the question studied by Y. W. Lee in **[L1]** and **[L2]**: can a polyhedral group contain a knot group of finite index other than the figure-eight knot group (which is in the polyhedral group $\mathrm{PGL}_2\mathcal{O}_3$)? One can ask if these three knots are the only examples of this phenomenon.

10. Further comments

10.1. We first discuss hyperbolic knot complements.

Question 2. *When is the trace field equal to the cusp field for hyperbolic knot complements*?

The only known examples where this fails are the two dodecahedral knots. They have cusp field $\mathbb{Q}(\sqrt{-3})$ and trace field $\mathbb{Q}(\sqrt{-3},\sqrt{5})$. Indeed, the cusp field is obvious. The trace field certainly contains $\sqrt{5}$ because of the element of order 5 in $(\Gamma^+)^{(2)}$ and is no bigger than $\mathbb{Q}(\sqrt{-3},\sqrt{5})$ because $(\Gamma^+)^{(2)}$ can be generated by the three matrices

$$\begin{pmatrix} 0 & -1 \\ 1 & 0 \end{pmatrix}, \quad \begin{pmatrix} 0 & -\omega^{-1} \\ \omega & (1+\sqrt{5})/2 \end{pmatrix}, \quad \begin{pmatrix} 0 & -\omega \\ \omega^{-1} & 0 \end{pmatrix},$$

where $\omega = e^{\pi i/3}$.

Question 3. *Do hyperbolic knot complements always have integral traces?*

We know no counterexamples. R. Riley has informed us that none of the examples he has computed gives a counterexample. In many cases Bass' theorem **[Ba2]**, quoted in the introduction, forces integral traces. Bass' theorem does not apply to the knot 8_{17} (this knot complement has a closed incompressible surface of genus 2, cf. **[Lo]**), but Riley informs us that it nevertheless has integral traces. Note that, by Theorem 6.3, the *link* complements $C(p, 2q)$ with $p = -4q$ and q an odd prime power have non-integral traces. These chain links belong to a general class of links, called "star links" in **[O]**, which also includes knots. Thus the class of star knots might be a place to look for counterexamples, but we doubt this.

If the answer to both questions 2 and 3 is positive for a particular knot complement other than the figure-eight complement, then this knot complement has no hidden symmetries, for if it did, then Proposition 4.4(a) would imply arithmeticity by the comments preceding Question $1'$.

In contrast to the situation for knot complements, it is easy to construct many 1-cusp orbifolds with cusp field smaller than the invariant trace field. For example, take an orbifold with two non-rigid cusps which are separated by a 2-sphere with three orbifold points and perform Dehn filling of one of the cusps. This will not affect the cusp parameter of the other one, so all the Dehn fillings will have the same cusp field, but their invariant trace fields will differ. See **[NR1]** for more details.

10.2. We do not yet know an example of a cocompact Kleinian group Γ with invariant quaternion algebra $A(\Gamma)$ equal to the matrix algebra $\mathrm{M}_2(k(\Gamma))$, but they probably exist.

We have already seen an example of non-commensurable groups with the same $A(\Gamma)$, namely, $\pi_1 W''(0,6)$ is non-arithmetic but has $A(\Gamma) = \mathrm{M}_2(\mathbb{Q}(\sqrt{-3}))$, the same as $\mathrm{PGL}_2\mathcal{O}_3$. However, knot complement examples of this exist also. Namely, the knot 5_2 and the $(-2,3,7)$-pretzel knot both have invariant trace field equal to $\mathbb{Q}(\theta)$ where $\theta^3-\theta^2+2\theta-1 = 0$, but they are non-commensurable because they have the same volume, ≈ 2.82812208, but different cusp volumes (cf. **[HMW]**; commensurable non-arithmetic one-cusp orbifolds of the same volume must have the same cusp volume because they have a mutual orbifold quotient—the commensurator). J. Weeks has pointed out that they are cut-and-paste equivalent by cutting and pasting along an immersed geodesic thrice punctured sphere.

References

[A1] Adams, C., Limit volumes of hyperbolic 3-orbifold. Preprint.

[A2] Adams, C., Non-compact hyperbolic 3-orbifolds of small volume. These Proceedings.

[AR] Aitchison, I.R. and Rubinstein, J.H., Combinatorial cubings, cusps and the dodecahedral knots. These Proceedings.

[B1] Bass, H., Groups of integral representation type. Pacific J. Math. 86 (1980), 15–51.

[B2] Bass, H., Finitely generated subgroups of GL_2. In: The Smith Conjecture, editors, J.W. Morgan and H. Bass, 127–136. Academic Press, 1984.

[BK] Bass, H. and Kulkarni, R., Uniform tree lattices. (Preprint)

[Bo] Borel, A., Commensurability classes and volumes of hyperbolic 3-manifolds. Ann. Scuola Norm. Sup. Pisa 8 (1981), 1–33.

[Boy] Boyd, D.W., Reciprocal polynomials having small measure. Math. Comp. 35 (1980), 1361–1377, ibid (1989).

[Cl] Clozel, L., On the cuspidal cohomology of arithmetic subgroups of SL(2n) and the first Betti number of arithmetic 3-manifolds. Duke Math. J. 55 (1987), 475–486.

[EP] Epstein, D.B.A. and Penner, R.C., Euclidean decompositions of non-compact hyperbolic manifolds. J. Diff. Geom. 27 (1988), 67–80.

[FN] Fine, B and Newman, M., Normal subgroup structure of the Picard group. Trans. A.M.S., 302 (1987),769–786.

[H] Helling, H., Bestimmung der Kommensurabilitätsklasse der Hilbertschen Modulgruppe. Math. Z. 92 (1966) 269–280.

[HMW] Hodgson, C.D., Meyerhoff, G.R. and Weeks, J.R., Surgeries on the Whitehead link yield geometrically similar manifolds. These Proceedings.

[KLS] Kronheimer, P.B., Larsen, M.J. and Scherk, J., Casson's invariant and quadratic reciprocity. Topology (to appear).

[L1] Lee, Y.W., Abelian rank of normal torsion-free finite index subgroups of polyhedral groups. Trans. A.M.S. 290 (1985), 734–745.

[L2] Lee, Y.W., On normal torsion-free finite index subgroups of polyhedral groups. Topology Appl. 32 (1989), 25–47.

[Le] Lehmer, D.H., Factorizations of certain cyclotomic functions. Ann. of Math. 34 (1933), 461–479.

[Lo] Lozano, M-T., Arcbodies. Math. Proc. Camb. Phil. Soc. 94 (1983), 253–260.

[Mac] Macbeath, A.M., Commensurability of cocompact three-dimensional hyperbolic groups. Duke Math. J. 50 (1983), 1245–1253.

[MR] Maclachlan, C. and Reid, A.W., Commensurability classes of arithmetic Kleinian groups and their Fuchsian subgroups. Math. Proc. Camb. Phil. Soc. 102 (1987), 251–258.

[Mag] Magnus, W., Non-euclidean tesselations and their groups. Academic Press, New York, 1974.

[Me] Mess, G., Finite covers of 3-manifolds, and a theorem of Lubotzky. (Preprint).

[Mey] Meyerhoff, R., A lower bound for the volume of hyperbolic 3-orbifolds. Duke Math. J. 57 (1988), 185–203.

[NR1] Neumann, W.D. and Reid, A.W., Amalgamations and the invariant trace field. Math. Proc. Camb. Philos. Soc. 109 (1991), 509–515

[NR2] Neumann, W.D. and Reid, A.W., Notes on Adams' small volume orbifolds. These Proceedings.

[NZ] Neumann, W.D. and Zagier, D., Volumes of hyperbolic 3-manifolds. Topology 24 (1985), 307–332.

[O] U. Oertel, Closed incompressible surfaces in complements of star links. Pacific J. Math. 111 (1984), 209–230.

[R1] Reid, A.W., Arithmetic Kleinian groups and their Fuchsian subgroups. Ph.d thesis, Univ. of Aberdeen, 1987.

[R2] Reid, A.W., Arithmeticity of knot complements. To appear in J. London Math. Soc.

[R3] Reid, A.W., A note on trace-fields of Kleinian groups. Bull. London Math. Soc. 22 (1990), 349–352.

[R4] Reid, A.W., Totally geodesic surfaces in hyperbolic 3-manifolds. To appear in Proc. Edinburgh Math. Soc.

[Ri] Riley, R., An elliptical path from parabolic representations to hyperbolic structures. In: Topology of Low-dimensional manifolds, ed. R. Fenn, L.N.M. Vol. 722, Springer-Verlag 1979.

[Tak] Takeuchi, K., A characterization of arithmetic Fuchsian groups. J. Math. Soc. Japan 27 (1975), 600–612.

[T] Thurston, W.P., The geometry and topology of 3-manifolds. Mimeographed lecture notes, Princeton Univ., 1977.

[Vig] Vignéras, M-F., Arithmétique des algèbres de Quaternions. L.N.M. 800, Springer-Verlag, 1980.

[W] Weber, H., Lehrbuch der Algebra, 2. Auflage, Bd. I, (Vieweg, Braunschweig 1898).

[Z] Zimmer, R., Ergodic Theory and Semi-simple Lie groups. Birkhauser, Boston, 1984.

Department of Mathematics, The Ohio State University, Columbus, OH 43210
Email: neumann@mps.ohio-state.edu

Department of Mathematics, The Ohio State University, Columbus, OH 43210
Mathematical Sciences Research Institute, Berkelely, CA 94720
Email: reid@mps.ohio-state.edu

Notes on Adams' Small Volume Orbifolds

Walter D. Neumann and Alan W. Reid

In **[A2]** Colin Adams shows there are just six non-compact orientable hyperbolic 3-orbifolds of volume less than $v_0/4$, where $v_0 = 1.01494146\ldots$ is the volume of a regular ideal tetrahedron in $\mathbb{H}^3$. We give various descriptions of them and their fundamental groups. In particular, we will see that they are all arithmetic, and we give explicit commensurabilities between some of them. Their fundamental groups can be described as symmetry groups of tesselations, so we will introduce terminology for this. Many of them are also "tetrahedral groups," so we introduce notation for these too.

These orbifolds show that "cusp density" is not a commensurability invariant of one-cusped orbifolds, answering a question that has been asked by several people. As we discuss in Sect. 3, cusp density *is* a commensurability invariant for non-arithmetic one-cusped orbifolds.

1. Preliminaries: tesselations and tetrahedral groups

We will only need tesselations of a rather special type. Let P be an ideal polyhedron in $\mathbb{H}^3$ whose dihedral angles are all integer submultiples of 2π and such that the two faces meeting at an edge are always congruent by a rotation about that edge. For each edge e of P let g_e be a rotation about e by the dihedral angle of P at e. Then the elements g_e generate a discrete group, and the images of P by this group tesselates $\mathbb{H}^3$. We call this tesselation ***the tesselation generated by*** P and denote it $\mathcal{T}(P)$. We denote the group of orientation-preserving symmetries of this tesselation by $\mathrm{Isom}^+(\mathcal{T}(P))$.

We denote by $T(\alpha_1, \alpha_2, \alpha_3; \beta_1, \beta_2, \beta_3)$ the hyperbolic tetrahedron (if it exists) with dihedral angles π/α_i, $i = 1, 2, 3$, at the three edges of one face and dihedral angle π/β_i at the edge opposite the π/α_i edge for each i. The ***tetrahedral group*** $\Gamma(\alpha_1, \alpha_2, \alpha_3; \beta_1, \beta_2, \beta_3)$ is the orientation-preserving subgroup of the group generated by reflections in the faces of $T(\alpha_1, \alpha_2, \alpha_3; \beta_1, \beta_2, \beta_3)$.

2. Adams' orbifolds

The following geometric descriptions of Adams' orbifolds are all easily verified by comparing with Adams' descriptions, or alternatively, by just checking that volumes are correct, which is how the arithmetic descriptions can be checked. We omit the details.

$Q_0 = \mathbb{H}^3/\Gamma_0$, of volume $v_0/12$, was proved to be the smallest orientable non-compact hyperbolic 3-orbifold by Meyerhoff **[M]**. Γ_0 has an arithmetic description as $\mathrm{PGL}(2, \mathcal{O}_3)$. It has geometric descriptions as $\Gamma(3, 2, 2; 6, 2, 3)$ and $\mathrm{Isom}^+(\mathcal{T}_0)$, where $\mathcal{T}_0$ is the tesselation $\mathcal{T}(T(3, 3, 3; 3, 3, 3))$ of $\mathbb{H}^3$ by regular ideal tetrahedra.

$Q_1 = \mathbb{H}^3/\Gamma_1$ has volume $v_1/6$, where $v_1 = 0.91596544\ldots$ is the volume of an ideal tetrahedron with dihedral angles $\pi/2,\ \pi/2,\ \pi/4$. This Γ_1 is $\mathrm{PGL}(2, \mathcal{O}_1)$. It also equals $\Gamma(3,2,2;4,2,4)$ and $\mathrm{Isom}^+(\mathcal{T}_1)$, where $\mathcal{T}_1$ is the tesselation $\mathcal{T}(\mathrm{Oct})$ of $\mathbb{H}^3$ by regular ideal octahedra.

$Q_2 = \mathbb{H}^3/\Gamma_2$ has volume $v_0/6$ and a rigid cusp of type $(3,3,3)$. It is a 2-fold cover of Q_0. In fact, $\Gamma_2 = \mathrm{PSL}(2, \mathcal{O}_3)$. It is the tetrahedral group $\Gamma(3,2,2;3,3,3)$. If one colors the tesselation $\mathcal{T}_0$ above in checker-board fashion, then Γ_2 is the orientation-preserving symmetry group of this colored tesselation. Alternatively, stellate each of the "white" tetrahedra in this colored tesselation by adding a quarter of each of the four adjacent "black" tetrahedra, to obtain a tesselation $\mathcal{T}_2$ by polyhedra with twelve triangular faces and all dihedral angles $2\pi/3$. Then $\Gamma_2 = \mathrm{Isom}^+(\mathcal{T}_2)$.

$Q_3 = \mathbb{H}^3/\Gamma_3$ also has volume $v_0/6$ and has a rigid cusp of type $(2,3,6)$. As an arithmetic group, in Borel's notation in **[B]**, it is the maximal group $\Gamma_{\{\sqrt{-3}\},\emptyset}$ commensurable with $\mathrm{PGL}(2, \mathcal{O}_3)$. It is not tetrahedral, but it contains $\Gamma(2,2,3;2,6,3)$ with index 2 and is generated by this tetrahedral group together with the obvious rotational symmetry of order 2 of the tetrahedron $T(2,2,3;2,6,3)$. It is also $\mathrm{Isom}^+(\mathcal{T}_3)$, where $\mathcal{T}_3$ is the tesselation generated by an ideal tetrahedron with angles $\pi/6,\ \pi/6,\ 2\pi/3$. The union A of two regular ideal tetrahedra pasted along a face can be cut into three ideal tetrahedra, each with angles $\pi/6,\ \pi/6,\ 2\pi/3$, so the tesselations $\mathcal{T}_0$ and $\mathcal{T}_3$ are both refinements of $\mathcal{T}(A)$. The intersection of Γ_0 and Γ_3 is $\mathrm{Isom}^+(\mathcal{T}(A)) = \Gamma(2,2,3;2,6,3)$.

$Q_4 = \mathbb{H}^3/\Gamma_4$ has volume $5v_0/24$. As an arithmetic group it is the maximal group $\Gamma_{\{2\},\emptyset}$ commensurable with $\mathrm{PGL}(2, \mathcal{O}_3)$, in Borel's notation. It is tetrahedral: $\Gamma_4 = \Gamma(4,2,2;6,2,3)$. It is also $\mathrm{Isom}^+(\mathcal{T}_4)$, where $\mathcal{T}_4 = \mathcal{T}(C)$ is the tesselation of $\mathbb{H}^3$ by regular ideal cubes. Note that the regular ideal cube C is the result of pasting four regular ideal tetrahedra to the four faces of a fifth regular ideal tetrahedron. Thus the tesselation $\mathcal{T}_0$ of $\mathbb{H}^3$ by ideal tetrahedra refines the tesselation $\mathcal{T}_4$. There are 5 ways of positioning $\mathcal{T}_4$ over $\mathcal{T}_0$ while there are 2 ways of positioning $\mathcal{T}_0$ as a refinement of $\mathcal{T}_4$. It follows that $\mathrm{Isom}^+(\mathcal{T}_4) \cap \mathrm{Isom}^+(\mathcal{T}_0)$ has index 5 in $\Gamma_0 = \mathrm{Isom}^+(\mathcal{T}_0)$ and index 2 in $\Gamma_4 = \mathrm{Isom}^+(\mathcal{T}_4)$ (proof: Γ_0 acts transitively on the five positions of $\mathcal{T}_4$ over $\mathcal{T}_0$ and $\Gamma_0 \cap \Gamma_5$ is the subgroup which fixes one of them).

$Q_5 = \mathbb{H}^3/\Gamma_5$ has volume $v_1/4$. As an arithmetic group it is the maximal group $\Gamma_{\{1+i\},\emptyset}$ commensurable with $\mathrm{PGL}(2, \mathcal{O}_1)$, in Borel's notation. It is not tetrahedral but it contains $\Gamma(4,2,2;4,2,4)$ with index 2 and is generated by this tetrahedral group and the order 2 rotational symmetry of the tetrahedron $T(4,2,2;4,2,4)$. Γ_5 is also $\mathrm{Isom}^+(\mathcal{T}_5)$, where $\mathcal{T}_5$ is the tesselation generated by an ideal tetrahedron with dihedral angles $\pi/2$, $\pi/4$, and $\pi/4$. A regular ideal octahedron can be cut into four such tetrahedra, so the tesselation $\mathcal{T}_5$ refines the octahedral tesselation $\mathcal{T}_1$. There are two ways of positioning $\mathcal{T}_1$ over $\mathcal{T}_5$, while there are three ways of positioning $\mathcal{T}_5$ as a refinement of $\mathcal{T}_1$. It follows that $\mathrm{Isom}^+(\mathcal{T}_1) \cap \mathrm{Isom}^+(\mathcal{T}_5)$ has index 2 in $\mathrm{Isom}^+(\mathcal{T}_5) = \Gamma_5$ and index 3 in $\mathrm{Isom}^+(\mathcal{T}_1) = \Gamma_1$.

Remark. In **[A1]** Adams computes the three smallest volume orientable hyperbolic 3-orbifolds with non-rigid cusps. These also all turn out to be arithmetic. Indeed, the smallest such orbifold double covers Q_1.

3. Commensurability and cusp density

An ***open horoball neighborhood*** of a cusp in a hyperbolic orbifold is a neighborhood of the cusp whose inverse image in $\mathbb{H}^3$ is a disjoint union of open horoballs. The maximum total volume of a family of disjoint open horoball neighborhoods of the cusps of a hyperbolic orbifold M is called the ***maximal cusp volume*** of M. With the added condition that the horoball neighborhoods have pairwise equal volume, we call the maximal total volume the ***uniform cusp volume*** of M. The ratio of maximal resp. uniform cusp volume to total volume of M is the ***maximal*** resp. ***uniform cusp density***. Maximal and uniform cusp density are equal if M has at most one cusp; in this case we can just speak of "cusp density." Adams' orbifolds each have just one cusp.

A one-cusped orbifold clearly has the same uniform cusp density as any of its one-cusped covers and any of its regular covers. (But note that a regular cover of a one-cusped orbifold may have larger *maximal* cusp density.)

In the non-arithmetic case, any two commensurable orbifolds cover a common orbifold (cf. [**NR**, Introduction]) and therefore have a common regular cover (orbifolds have a common regular cover if and only if they cover some common orbifold). Thus commensurable non-arithmetic one-cusped orbifolds have the same cusp density.

In [**A2**] Adams computes the cusp volumes of his orbifolds. In the cases of Q_3 and Q_4, this yields cusp densities of $3/4v_0$ and $2\sqrt{3}/5v_0$ respectively. These orbifolds are commensurable with Q_0 and Q_2 which both have a cusp density of $\sqrt{3}/2v_0$. Furthermore, Q_1 and Q_5 have cusp densities of $4/3v_1$ and $1/\sqrt{2}v_1$. It follows that the only pair of orbifolds from $Q_0, \ldots, Q_5$ that have a common regular cover or that cover a common orbifold are Q_0 and Q_2.

Thus, summarizing the above, we have:

Proposition 1. *Commensurable non arithmetic one cusped hyperbolic 3 orbifolds have the same cusp density. However this may fail if the orbifolds are arithmetic. This failure is an obstruction to the orbifolds having a common regular cover or covering a common orbifold.* □

We expect that the failure of cusp density to be a commensurability invariant in the case of one-cusped arithmetic orbifolds to be the rule rather than the exception.

Even when G and H are commensurable Kleinian groups, that is, H can be conjugated so that $G \cap H$ has finite index in each of G and H, there is no canonical way of doing this. In fact

Proposition 2. *Suppose G and H are commensurable Kleinian groups. If they are arithmetic, one can find conjugates H^g of H with $[G : G \cap H^g]$ finite but arbitrarily large. If they are non-arithmetic, there are only finitely many conjugates of H which intersect G in a finite index subgroup.*

Proof. In the non-arithmetic case this is a restatement of the fact that H has finite index in its commensurator (cf. e.g., the introduction of [**NR**]). It suffices to prove the statement in the arithmetic case when $G = H$ (since replacing H by a commensurable group changes $[G : G \cap H]$ by a bounded factor). Consider the intersection K of all conjugates G^g of G for which $[G : G \cap G^g]$ is finite. If this index is always bounded then K has finite index in G, so its normalizer in $\mathrm{PSL}(2, \mathbb{C})$ is discrete. However, K is normal in the commensurator of G, so G has discrete commensurator and is therefore non-arithmetic. □

We positioned the Γ_i in the previous section so that the intersection $\Gamma_i \cap \Gamma_j$ always has as small index as possible in Γ_i and Γ_j. (The only non-obvious part of this is that Γ_2 and Γ_3 cannot be positioned so their intersection has index 3 in each—index 2 is prohibited by the different cusp densities.) The other intersections among Γ_0, Γ_2, Γ_3, and Γ_4, positioned as above, are also easily computed. The full list, identifying all that are tetrahedral, is:

$$\begin{aligned}
\Gamma_0 \cap \Gamma_3 &= \Gamma(2,2,3;2,6,3) && \text{of index 4 in } \Gamma_0,\\
\Gamma_2 \cap \Gamma_3 & && \text{of index 8 in } \Gamma_0,\\
\Gamma_0 \cap \Gamma_4 &= \Gamma(2,2,3;6,3,2) && \text{of index 5 in } \Gamma_0,\\
\Gamma_2 \cap \Gamma_4 &= \Gamma(2,3,3;3,3,3) && \text{of index 10 in } \Gamma_0,\\
\Gamma_3 \cap \Gamma_4 & && \text{of index 5 in } \Gamma_3,\\
\Gamma_0 \cap \Gamma_3 \cap \Gamma_4 &= \Gamma(6,2,3;6,2,3) && \text{of index 10 in } \Gamma_3,\\
\Gamma_2 \cap \Gamma_3 \cap \Gamma_4 &= \Gamma(6,2,3;6,2,3) && \text{of index 20 in } \Gamma_3.
\end{aligned}$$

Acknowledgments. The first author acknowledges the support of the NSF for this research and the second author the SERC.

References

[A1] Adams, C., Limit volumes of hyperbolic 3-orbifolds. To appear in J. Diff. Geom.

[A2] Adams, C., Non-compact hyperbolic 3-orbifolds of small volume. These Proceedings.

[B] Borel, A., Commensurability classes and volumes of hyperbolic 3-manifolds. Ann. Scuola Norm. Sup. Pisa 8 (1981), 1-33.

[M] Meyerhoff, R., The cusped hyperbolic 3-orbifold of minimum volume. Bull. A. M. S. 13 (1985), 154-156.

[NR] Neumann, W. D. and Reid, A. W., Arithmetic of hyperbolic manifolds. These Proceedings.

Department of Mathematics, The Ohio State University, Columbus, OH 43210
Email: neumann@mps.ohio-state.edu

Department of Mathematics, The Ohio State University, Columbus, OH 43210
Mathematical Sciences Research Institute, Berkelely, CA 94720
Email: reid@mps.ohio-state.edu

Skein module of links in a handlebody

*Józef H. Przytycki**

Abstract. We prove that the skein module of a handlebody is a free module, solving conjectures by V. Tureav and the author.

0. Introduction

J. W. Alexander found in 1928 the Laurent polynomial invariant of oriented links in S^3, and proved that if three links are identical except the parts shown on Fig. 0.1, then their polynomials are linearly related.

Figure 0.1

J. Conway in 1969 rediscovered the Alexander formula and normalized the Alexander polynomial so that it satisfies

$$\Delta_{L_+}(t) - \Delta_{L_-}(t) = \left(\sqrt{t} - \frac{1}{\sqrt{t}}\right) \Delta_{L_0}(t) . \qquad (0.1)$$

The Alexander polynomial, as normalized by Conway, is an element of the ring $Z[t^{\mp 1/2}]$.

In the spring of 1984 V. Jones discovered an invariant of links, $V_L(t) \in Z[t^{\mp 1/2}]$, and it was proved by Jones and Lickorish and Millett in July 1984 that the Jones polynomial satisfies the relation

$$t^{-1} V_{L_+}(t) - t V_{L_-}(t) = \left(\sqrt{t} - \frac{1}{\sqrt{t}}\right) V_{L_0}(t) . \qquad (0.2)$$

In summer and fall of 1984 the Alexander and Jones polynomials were generalized to the skein (also called Flypmoth, generalized Jones, Homfly, Jones-Conway, oriented, oriented skein, Thomflyp, twisted Alexander, two-variable Jones) polynomial $P_L(v, z) \in$

* The paper (in various stages) was presented at IAS member seminar (Feb. 26, 1990), OSU research seminar (April 9, 1990) and Columbia University topology seminar (May 2, 1990). The author gratefully acknowledges the support of the Institute for Advanced Study during the period when the paper was prepared.

$Z[v^{\mp 1}, z^{\mp 1}]$, which satisfies the skein relation

$$v^{-1}P_{L_+} - vP_{L_-} = zP_{L_0}. \tag{0.3}$$

Equations described above, together with the initial condition $P_{T_1} = V_{T_1} = \Delta_{T_1} = 1$ for an unknot T_1, suffice to compute polynomials P, V and Δ.

The skein module $\mathcal{S}_3(M)$ is a straightforward generalization of the skein polynomial to any oriented 3-manifold. J. Conway, in his famous paper **[C]** considered the abstract idea of the $Z[t^{\mp 1/2}]$-module generated by isotopy classes of oriented links modulo the Conway skein relation. In that paper, Conway does not develop the idea and plays only with S^3 and D^3. However, a decade later he was advocating that idea in the broader context of any oriented 3-manifold. Conway's ideas have been developed by Giller **[G]**, Lickorish and Millett (linear skein) **[L-M]**, and the author (skein modules) **[P-1]** (see also **[L]** and **[T-1]**).

Let $\mathcal{L}$ denote (ambient) isotopy classes of oriented links in an oriented 3-manifold M. Let $R_z = Z[v^{\mp 1}, z^{\mp 1}]$, $R_z\mathcal{L}$ be a free R_z-module with basis $\mathcal{L}$. Finally, let S_3 be the submodule of $R_z\mathcal{L}$ generated by skein expressions $v^{-1}L_+ - vL_- - zL_0$. Then we define the skein module $\mathcal{S}_3(M)$ as $R_z\mathcal{L}/S_3$. The main purpose of this paper is to prove that for F a compact planar surface, $\mathcal{S}_3(F \times I)$ is an infinitely generated free R_z-module (with naturally described basis). It generalizes the results for $D^2 \times I$ (see **[FYHLMO]**, **[P-T-1]**) and $S^1 \times D^2$ (see **[H-K]**, **[T-1]**) and solves Conjecture 7 of **[P-1]**.

Let $\pi = \pi_1(F)$, $\pi^\circ = \pi - \{1\}$, $\widehat{\pi}$ denote conjugacy classes of π and $\widehat{\pi}^\circ = \widehat{\pi} - \{1\}$. Further, let $R_z\widehat{\pi}^\circ$ be the free R_z-module over $\widehat{\pi}^\circ$ and $S(R_z\widehat{\pi}^\circ)$ the symmetric tensor algebra over $R_z\widehat{\pi}^\circ$. Then:

Theorem 0.4. *There is a (module) isomorphism* $i : S(R_z\widehat{\pi}^\circ) \to \mathcal{S}_3(F \times I)$ *such that for* $w \in \widehat{\pi}^\circ$, $i(w)$ *is represented by a knot* $K \subset F \times I$ *such that the class of* K *in* $\widehat{\pi}^\circ$ *is equal to* w.

The proof of Theorem 0.4 consists of a delicate, multistep induction which incorporates the ideas of **[B-M]**, **[FYHLMO]**, **[P-T-1]**, **[H-K]**, **[H-P-2]** and **[T-2]**, and the method used in the classical proof of the Poincaré-Birkhoff-Witt theorem on universal enveloping algebras of Lie algebras. In the first step of the proof (Chapter 1) we construct an invariant of descending diagrams of links, then in Chapter 2, we induct on the power of z and then on the number of crossings (using a suitably constructed Lie bracket).

In the third chapter we analyze structures of skein modules related to $\mathcal{S}_3(F \times I)$, solving the conjectures by V. Turaev about structures of his skein modules $A_\circ(F \times I)$ and $\mathcal{A}(F \times I)$, **[T-3]**.

In the fourth chapter we generalize the notion of the Conway algebra, introduced by P. Traczyk and the author in **[P-T-1]**, from S^3 to $F \times I$ and show that the skein modules can be thought as linear models of the algebra.

In the fifth chapter we describe the skein module $\mathcal{S}$ as a quantization of the symmetric tensor algebra $S(Z[v^{\mp 1}]\widehat{\pi}^\circ)$ (isomorphic to the anti-homotopy skein module **[H-P-2]**) and describe its structure as a Hopf algebra (following **[T-3]**).

In the sixth chapter we sketch the method (using Hass–Scott's **[H-S]** generalization of a result of Grayson **[Gr]**) of extending our results to closed surfaces.

I would like to thank Jim Hoste for fruitful cooperation which resulted in the proof of two important simplifications of Theorem 0.4: structure of the homotopy skein module (**[H-P-1]**) and the anti-homotopy skein modules (**[H-P-2]**).

1. Algebra of descending diagrams

In this chapter $M = F \times I$ where F is a surface and I is the unit interval. Let $\mathcal{D}$ denote the set of oriented link diagrams in F (up to ambient isotopy of F). $\mathcal{D}$ has a natural multiplication, where $D_1 \cdot D_2$ denotes the diagram obtained by placing D_1 above D_2. The empty link $\emptyset$ is the neutral element.

Definition 1.1. Let D be an oriented knot diagram and $b \in D$ a basepoint (outside crossings). We say that D is descending with respect to b if each crossing which is met for the first time is crossed by an overcrossing, when one travels along D (according to the orientation of D) starting from b. D is said to be descending if it is descending with respect to some base point b. A link diagram D is said to be descending if it is composed of a layered family of descending knot diagrams (i.e., $D = D_1 \cdot \ldots \cdot D_k$ for D_i descending for any i).

Let $\mathcal{D}_{des}$ be a set of descending diagrams (up to ambient isotopy of F). Let $\mathcal{L}_{des}$ denote the set of descending links, i.e., descending diagrams up to Reidemeister moves which preserve "descending". Let $\bar{R} = Z[v^{\mp 1}]$ and $\bar{R}\mathcal{L}_{des}$ be a free $\bar{R}$-module with basis $\mathcal{L}_{des}$. The multiplication extends from $\mathcal{L}_{des}$ to $\bar{R}\mathcal{L}_{des}$ (which becomes a ring). Consider the ideal $\bar{S}$ of $\bar{R}\mathcal{L}_{des}$ generated by the following skein expressions on descending diagrams of knots (1.2). Let D be a descending knot diagram with respect to b and c the first crossing on D (after b). Let b' be a point on D just after c. Furthermore, we will use the notation $D^c_{-sgn\, c}$ (resp. $D^c_{\circ}$) for the diagram obtained from D by changing overcorssing to undercrossing (resp. smoothing) at c. If for example $sgn\, c = +$ then $D = D^c_+$, $D^c_{-sgn\, c} = D^c_-$ (compare Fig. 1.1).

The skein expression has the form:

$$v^{-1} D^c_+ - v D^c_- - D^c_{\circ}. \tag{1.2}$$

Notice that in 1.2 all diagrams are descending: D with respect to b, $D^c_{-sgn\, c}$ with respect to b' and $D^c_{\circ}$ with respect to b' and b. The quotient algebra $\mathcal{S}_{des} = \bar{R}\mathcal{L}_{des}/\bar{S}$ is called the skein algebra of descending diagrams.

To analyze $\mathcal{S}_{des}$ closer we have to translate it into algebraic language.

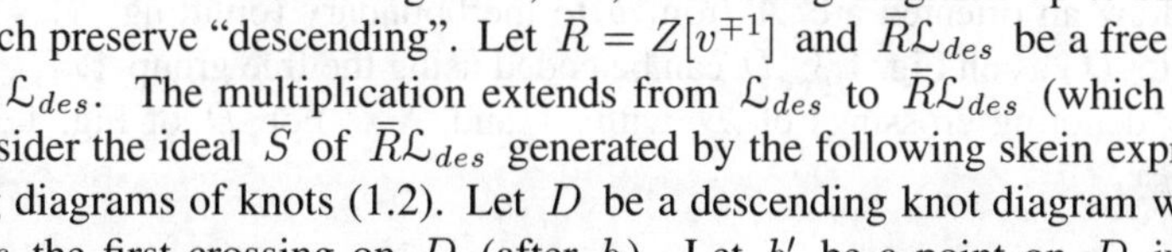

Figure 1.1

In this paper we will deal exclusively with the case of compact planar surface F (except Chapter 6). So F is a disc with k holes and $\pi_1(F) = \{x_1, \ldots, x_k: \}$. Consider the system of oriented cut curves, $X_1, \ldots, X_k$ on F as shown on Fig. 1.2.

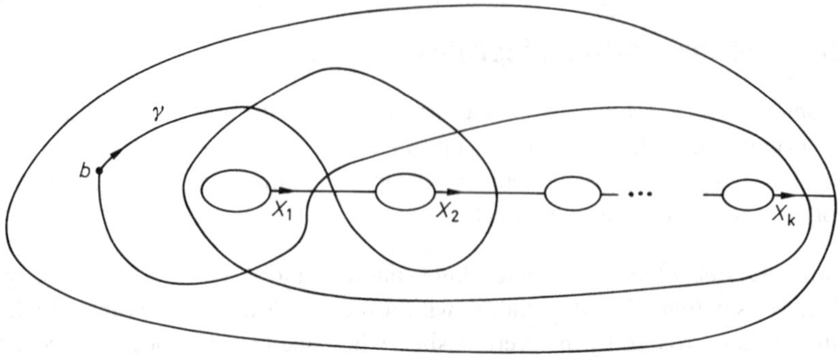

Figure 1.2

For any curve γ transversal to X_i's with base point b disjoint from X_i's, the cut system allows to express $[\gamma] \in \pi_1(F, b)$ uniquely in the alphabet $\{x_i^{\mp 1}\}$. For γ of Fig. 1.2 one gets: $[\gamma] = x_1 x_2^{-1} x_k^{-1} x_1$. Descending diagrams carry more information then elements of $\pi_1(F)$; to code this consider a descending diagram D (with respect to b) and draw an oriented arc A from b to the boundary (omitting X_i's and b and transversal to D) as on Fig. 1.3. D can be coded using the free group $\{x_1, \ldots, x_k, a: \}$ by properly denoting crossings of D with A and X_i. For D of Fig. 1.3 one gets: $D \to x_1 a x_k x_1^{-1} a^2$.

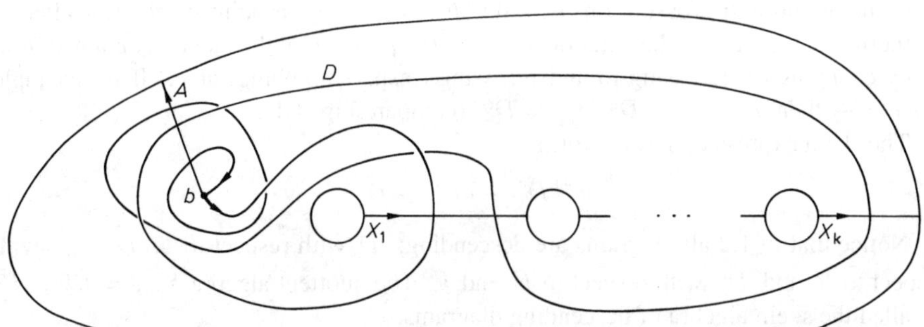

Figure 1.3

One can now make an important observation:

1.3. Let $b \in F$. Descending knot diagrams with base point b, up to ambient isotopy of F fixing b and Reidemeister moves preserving "descending" and performed outside b, are in bijection with elements of the free group $\pi(X_a) = \{x_1, \ldots, x_k, a: \}$ up to equivalence $a^{\mp 1} w \sim w \sim w a^{\mp 1}$ (the bijection depends on the choice of cuts X_i but is independent of A). In particular, we have the surjection, t, from $\pi(X_a)$ to descending diagrams described above.

Having in mind the above correspondence, we can start purely algebraic considerations.

Let $X = \{x_1, x_1^{-1}, \ldots, x_k, x_k^{-1}\}$ and $W(X)$ denote the semigroup of words over the alphabet X with neutral element (empty word) 1. Let $\pi = \pi(X)$ be a free group generated by X (i.e., $\pi = \{x_1, \ldots x_k : \}$), and $\pi^\circ = \pi^\circ(X)$ to be $\pi - \{1\}$. Similarly $W^\circ(X) = W(X) - \{1\}$. For a commutative ring R and a set Y, let RY denote the free R-module over Y. $R\pi$ is additionally a "group ring". Let $T(RY) = \bigcup_{i=0}^{\infty} T^i(RY)$ denote the tensor algebra over RY. Here $T^\circ(RY) = R$, $T^1(RY) = RY$ and $T^{i+1}(RY) = T^i(RY) \otimes RY$. The symmetric tensor algebra $S(RY)$ is defined by $T(RY)/(a \otimes b - b \otimes a)$.

From now on the ring R is fixed to be $Z[v^{\mp 1}]$ and denoted by $\bar{R}$.

Let $X_a = X \cup \{a, a^{-1}\}$ and $T(X_a)$ denote the tensor algebra $T(\bar{R}\pi(X_a))$. To formulate our first lemma we have to define an ideal J in the algebra $T(X_a)$ generated by two types of expressions:

$$\begin{array}{lll} \text{(i)} & \text{initial expressions} & w - wa^{\mp 1}, \quad w - a^{\mp} w \\ \text{(ii)} & \text{skein expressions} & \begin{cases} v^{-1}(w_1 a w_2) - v(w_1 w_2) - w_1 \otimes w_2 \\ v^{-1}(w_1 w_2) - v(w_1 a^{-1} w_2) - w_1 \otimes w_2 \end{cases} \end{array} \tag{1.4}$$

and additionally

$$v^{-1}(\emptyset) - v(\emptyset) = 1$$

where $\emptyset$ is a multiplicative unit of $T(X_a)$.

Lemma 1.5. *Consider an $\bar{R}$-algebra homomorphism*

$$\varphi : T(\bar{R}\pi^\circ(X)) \to T(X_a)/J$$

given by $\varphi(w) = w$ for $w \in \pi^\circ(X)$, then φ is an isomorphism.

Proof. (1) φ is onto.

Skein relations alone suffice to show that $\pi(X)$ generates the algebra $T(X_a)/J$. Now, the skein relation for aw gives in $T(X_a)/J$

$$v^{-1}(aw) - v(w) = 1 \otimes w,$$

so by (i)

$$1 \otimes w = (v^{-1} - v)w.$$

Similarly

$$w \otimes 1 = (v^{-1} - v)w.$$

This together with the relation

$$1 = (v^{-1} - v)\emptyset$$

shows that $\pi^\circ(X)$ generates the algebra $T(X_a)/J$, so φ is onto.

(2) φ is a monomorphism.

We will show an equivalent fact that the multiplicative subset generated by $\pi^\circ(X)$ (i.e., monomials in $\pi^\circ(X)$) is a (linear) basis of $T(X_a)/J$.

We use the standard procedure of building a binary resolving tree for $w \in T(X_a)$ eliminating a letter $a^{\mp 1}$ at each branch and getting at leaves elements of $T(\bar{R}\pi(X))$ (which are reduced to elements of $T(\bar{R}\pi^\circ(X))$ by the identity $1 \otimes w = w \otimes 1 = (v^{-1} - v)w)$; compare Fig. 1.4. We easily prove by induction on the number of letters $a^{\mp 1}$ involved in w that the result (the formula for w) does not depend on the choice of resolving tree. So the multiplicative semigroup generated by $\pi^\circ(X)$ (i.e., monomials in $\pi^\circ(X)$) is a (linear) basis of $T(X_a)/J$. □

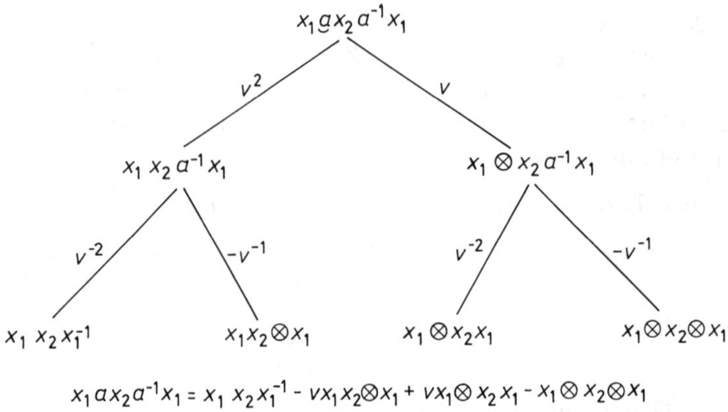

Figure 1.4

Consider a family of algebra homomorphisms $\varphi_{x_i^{\mp 1}}\colon T(X_a) \to T(X_a)$ given by

$$\varphi_{x_i^{\mp 1}}(w) = f_{x_i^{\mp 1}}(w)$$

where

$$f_{x_i^{\mp 1}}\colon \pi(X_a) \to \pi(X_a)$$

is a group homomorphism given on the free generators by formulae:

$$f_{x_i^{\mp 1}}(a) = a$$

$$f_{x_i}(x_k) = \begin{cases} x_i^{-1} x_k x_i & \text{for } i \geq k \\ a^{-1} x_i^{-1} x_k a x_i & \text{for } i \leq k \end{cases}$$

$$f_{x_i^{-1}}(x_k) = \begin{cases} x_i x_k x_i^{-1} & \text{for } i \geq k \\ x_i a x_k x_i^{-1} a^{-1} & \text{for } i \leq k\,. \end{cases}$$

Geometrically f_{x_i} is realized by pushing the base point b on the second side of X_i, without changing the isotopy type of the underlying descending diagram; see Fig. 1.5 for $f_{x_1}(x_2)$.

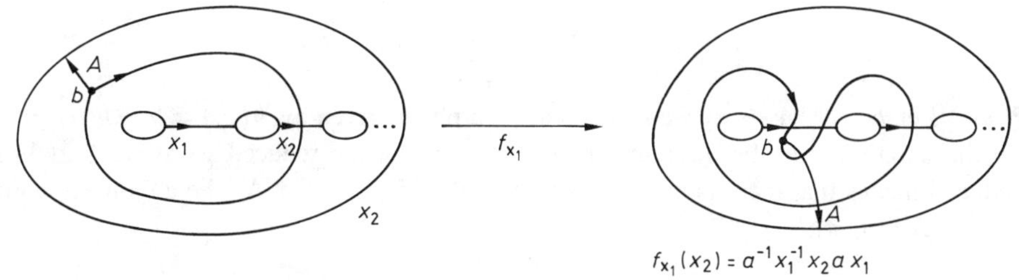

Figure 1.5

Lemma 1.6.
(a) f_{x_i} *and* $f_{x_i^{-1}}$ *are inverses;*
(b) $\varphi_{x_i^{\pm}}$ *are isomorphisms and* $\varphi_{x_i}^{-1} = \varphi_{x_i^{-1}}$*;*
(c) $\varphi_{x_i}(J) = J$ *and therefore* $\varphi_{x_i}^{\mp 1}$ *yields isomorphisms* $\widehat{\varphi}_{x_i^{\mp}}: T(X_a)/J \to T(X_a)/J$.

Proof. (a) is an easy check, for example for $k \geq i$,

$$f_{x_i^{-1}}(f_{x_i}(x_k)) = f_{x_i^{-1}}(a^{-1}x_i^{-1}x_k a x_i) = a^{-1}(x_i^{-1})(x_i a x_k x_i^{-1} a^{-1}) a x_i = x_k .$$

(b) follows immediately from (a). (c) holds because $f_{x_i}(a) = a = f_{x_i^{-1}}(a)$. □

Consider the ideal $J_c \subset T(X_a)$ generated by "conjugation" expressions

$$w - f_{x_i}(w) \text{ for any } i \text{ and } w \in X_a .$$

It will be an important task to analyze the module $T(X_a)/(J \cup J_c)$ (or equivalently $T(\bar{R}\pi^\circ(X))/\varphi^{-1}(J_c)$). The first step in doing this will be the following lemma for which we require some notation:

Let $S(X_a)$ (resp. $S(X), \ldots$) be the symmetric tensor algebra, i.e., $S(X_a) = T(X_a)/(w_1 \otimes w_2 - w_2 \otimes w_1)$.

Consider a group homomorphism

$$f: \pi(X_a) \to Aut(T(X_a))$$

given by $f_{x_i} = \varphi_{x_i}$, $f_a = Id$. We use the notation f_w for $f(w)$.

Assume that $x_1 < x_1^{-1} < x_2 < x_2^{-1} < \cdots < x_k < x_k^{-1}$ and extend this ordering to a lexicographical ordering of $\pi = \pi_1(F)$ (we compare reduced words). Let $\widehat{\pi}$ denote the conjugacy classes of π and $\psi: \widehat{\pi} \to \pi$ be an embedding such that $\psi(w_c)$ for $w_c \in \widehat{\pi}$ is the smallest in the lexicographical order, cyclically reduced word in π representing w_c, finally $\widehat{\pi}^\circ = \widehat{\pi} - \{1\}$.

Lemma 1.7.
(a) $\psi(\widehat{\pi}^\circ)$ *generates (algebraically)* $T(X_a)/(J \cup J_c)$*; i.e., the algebra homomorphism* $g: T(\bar{R}\widehat{\pi}^\circ) \to T(X_a)/(J \cup J_c)$ *given by* $g(w_c) = \psi(w_c)$ *for* $w_c \in \widehat{\pi}^\circ$*, is onto.*

(b) *When we reduce a word $w \in \pi(X)$ to its smallest representative in a conjugacy class, the only ambiguity can arise if we make a full circle of conjugations, and then we deal with the relation $w - f_w(w)$.*

(c) *g descends to an algebra isomorphism of symmetric algebras:*

$$g^S\colon S(\bar{R}\widehat{\pi}^\circ) \to S(X_a)/(J \cup J_c)\,.$$

Proof. Let $h\colon \pi(X_a) \to \pi(X)$ be the homomorphism given by $h(x_i) = x_i,\ h(a) = 1$. Let the length $\ell_x(w)$ be the number of letters $x_i^{\mp}$ in the reduced word $w \in \pi(X_a)$ (reduced means that there is no subword yy^{-1}). Let $yuy' \in \pi(X)$ be a reduced word $(y, y' \in X\ u \in \pi(X))$.

1.8. Then there is a word $w \in h^{-1}(yuy')$, unique up to the relations $wa^{\mp 1} \sim w \sim a^{\mp 1}w$, such that $\ell_x(w) = \ell_x(yuy')$ and $f_y(w)$ satisfies $\ell_x(f_y(w)) = \ell_x(h(f_y(w))) = \ell_x(uy'y)$. We think of uyy' as a reduced word so if $y' = y^{-1}$ then $\ell_x(uy'y) = \ell_x(u)$.

For example if $y = x_i$ and a subword of yuy' is $x_j x_k$ $(i < j, k)$ then we have to put "a" between x_j and x_k in w (then $f_{x_i}(x_j a x_k) = a^{-1}x_i^{-1}x_j a x_i a a^{-1} x_i^{-1} x_k a x_i = a^{-1}x_i^{-1}x_j a x_k a x_i$).

The above observation (1.8) allows an induction on $\ell_x(w)$ to prove Lemma 1.7(a). First we extend ℓ_x to monomials in $T(X_a)$ by putting $\ell_x(w_1 \otimes \cdots \otimes w_k) = \max \ell_x(w_i)$.

1.9. To prove Lemma 1.7(a), it suffices to show that monomials from $\psi(\widehat{\pi}(X))$ with $\ell_x \le k$ generate $T_k(X_a)/(J \cup J_c)_k$ for any k, where $T_k(X_a)$ is a subalgebra of $T(X_a)$ generated by words w with $\ell_x(w) \le k$ and the ideal $(J \cup J_c)_k$ is generated by skein relations for words with $\ell_x(w) \le k$ and "conjugation" relations $w - f_{x_i}(w)$ for $\ell_x(w)$, $\ell_x(f_{x_i}(w)) \le k$ (note that $J \cup J_c = \bigcup_{k=1}^{\infty} (J \cup J_c)_k$).

For $k = 1$, 1.9 is obvious. Assume it holds for $i < k$ and let $\ell_x(w) = k$. By 1.4 and the inductive assumption we can eliminate $a^{\mp 1}$ from w. So we can assume $\ell_x(h(w)) = \ell_x w$ (of course we can consider only w with $h(w) = w$ but there is no need to do it). By 1.8, w can be expressed as a linear combination of monomials with smaller ℓ_x and any cyclic conjugation of w. Then either $h(w)$ can be conjugated into ℓ_x shorter word, in which case $w \in T_{k-1}(X_a)$, or $h(w)$ is cyclically shortest, in which case w is a linear combination of elements of $T_{k-1}(X_a)$ and w' where w' is cyclically conjugated to $h(w)$ and the smallest (in the lexicographical order) in the conjugacy class, i.e. $w' \in \psi(\widehat{\pi}(X))$. This completes the inductive step and therefore the proof of Lemma 1.7(a).

(b) follows by careful analysis of the proof of (a). Introducing more terminology we can give a more precise version of (b). Let $\psi^T\colon T(\bar{R}\widehat{\pi}^\circ) \to T(\bar{R}\pi^\circ)$ be the natural extension of ψ, and let $T_k(\bar{R}\widehat{\pi}^\circ)$ be a subalgebra of $T(\bar{R}\widehat{\pi}^\circ)$ generated by monomials w of length not exceeding k (i.e., $\ell(\psi(w)) \le k$) and $g_k\colon T_k(\bar{R}\widehat{\pi}^\circ) \to T_k(X_a)/(J \cup J_c)_k$ the $\bar{R}$-algebra homomorphism induced by g. Let $(J)_k$ be the ideal in $T_k(X_a)$ generated by skein relations on $T_k(X_a)$ and $\varphi_k\psi_k^T\colon T_k(\bar{R}\widehat{\pi}^\circ) \to T_k(X_a)/(J)_k$ be the $\bar{R}$-algebra homomorphism induced by $\varphi\psi^T$. Then:

(b′) $\ker g_{k+1} = (\varphi_{k+1}\psi^T_{k+1})^{-1}((J \cup J_c)_{k+1})$ *is generated by* $\ker g_k$, $\ker(\varphi_{k+1}\psi^T_{k+1})$ *and "conjugation expressions"* $(\varphi_{k+1}\psi^T_{k+1})^{-1}(w - f_w(w))$ *for* $w \in \psi(\widehat{\pi}^\circ)$ *and* $\ell(\psi(w)) = k+1$.

(b′) also follows from an analysis of the proof of (a). In particular if $w \in T_{k+1}(X_a)$, then $(w - f_w(w)) \in T_{k+1}(X_a)/(J)_{k+1}$ can be represented by an element of $T_k(X_a)$.

For (c) one has to show that the expressions $(\varphi\psi^T)^{-1}(w - f_w(w))$ are linear combinations of commutators $a \otimes b - b \otimes a$ in $T(\bar{R}\widehat{\pi}^\circ)$.

The simple proof which we will present involves the geometric interpretation of $w \in \pi(X_a)$ and of $w - f_w(w)$. If γ_w is a descending diagram represented by w (see 1.3) then $w - f_w(w)$ represents the linear combination of all smoothings of γ_w performed twice as one travels along γ_w. That is

$$w - f_w(w) = \sum_{p_i} \mp v^{\alpha_i}(w_{i_1} \otimes w_{i_2} - w_{i_2} \otimes w_{i_1}) \tag{1.10}$$

where w_{i_1} and w_{i_2} are curves (words) obtained from γ_w by smoothing γ_w at p_i. α depends on the position of p_i. We can also assume $w_{i_j} \neq 1$ because $1 \otimes w - w \otimes 1 = 0$, so $\ell_x(w_{i_j} \otimes w_{i_s}) < \ell_x(w)$.

This allows an induction on $\ell_x(w)$ because the words w_{i_j} are well defined up to skein relations and "conjugation" relations $u - f_{x_i}(u)$ for $\ell_x(u)$, $\ell_x(f_x(u)) < \ell_x(w)$. □

One should mention that the formula of type of 1.10 appears for the first time in **[T-2]** (see also **[T-3]**) and is used to define a cobracket in the Wolpert-Goldman Lie algebra **[Go]**.

We have not determined fully the structure of the algebra of descending diagrams, so we finish this chapter by the following conjecture:

Conjecture 1.11. $\mathcal{S}_{des} = T(X_a)/(J \cup J_c)$ *is a free* $\bar{R}$*-module.*

2. Construction of the invariant $P(L) \in S(R\widehat{\pi}^\circ)$

Let $\mathcal{L}$ denote the set of (ambient) isotopy classes of oriented links in an oriented 3-manifold M. Let $R = Z[v^{\mp 1}, h]$, $R\mathcal{L}$ be a free R-module with basis $\mathcal{L}$. Finally, let S be the submodule of $R\mathcal{L}$ generated by skein expressions 2.1. Then we define a skein module $\mathcal{S}(M) = R\mathcal{L}/S$.

$$v^{-1}L_{\times} - vL_{\times} = \begin{cases} hL_{\asymp} & \text{for } \times \text{ a mixed crossing} \\ L_{\asymp} & \text{for } \times \text{ a self-crossing} \end{cases} \tag{2.1}$$

The main purpose of this paper is to prove that for F a compact planar surface, $\mathcal{S}(F \times I)$ is an infinitely generated free R-module (with naturally described basis). This generalizes the results for $D^2 \times I$ (**[FYHLMO]**, **[P-T-1]**) and $S^1 \times D^2$ (**[H-K]**, **[T-1]**). Let $\pi = \pi_1(F)$, $\pi^\circ = \pi - \{1\}$, $\widehat{\pi}$ denote conjugacy classes of π and $\widehat{\pi}^\circ = \widehat{\pi} - \{1\}$. Further let $R\widehat{\pi}^\circ$ be a free R-module over $\widehat{\pi}^\circ$ and $S(R\widehat{\pi}^\circ)$ the symmetric tensor algebra over $R\widehat{\pi}^\circ$. Then

Theorem 2.2. *There is an (R-module) isomorphism $i\colon S(R\widehat{\pi}^\circ) \to \mathcal{S}(F \times I)$ such that for $w \in \widehat{\pi}^\circ$, $i(w)$ is represented by a knot $K \subset F \times I$ so that the class of K in $\widehat{\pi}^\circ$ is equal to w.*

Let $\pi = \pi_1(F) = \{x_1, \dots, x_k\colon\}$ and let $\psi = \widehat{\pi} \to \pi$ be an embedding defined as in Chapter 1: consider lexicographic order among elements of π ($w \in \pi$ is assumed to be a reduced word), among letters one has $x_1 < x_1^{-1} < x_2 < x_2^{-1}, \dots$. For $w_c \subset \widehat{\pi}$ let $\psi(w_c)$ be the element of π, say w, which is the smallest, according to lexicographical order among elements of shortest length (i.e., cyclically reduced) representing w_c.

Let $\mathcal{D}$ be a family of oriented link diagrams on F.

2.3. We will construct now the family of $Z[v^{\mp 1}]$-module homomorphisms $P^i\colon R\mathcal{D} \to S(R\widehat{\pi}^\circ)$, $i \geq -1$ which have the following properties:

(i) $P^{-1}(R\mathcal{D}) = 0 \in S(R\widehat{\pi}^\circ)$;

(ii) $P^i(D) = P^i(D')$ if D and D' represent ambient isotopic links;

(iii) $P^i(t\psi(w_c^1) \cdot \ldots \cdot t\psi(w_c^j)) = w_c^1 \otimes \cdots \otimes w_c^j$ if $\psi(w_c^1) \leq \psi(w_c^2) \leq \cdots \leq \psi(w_c^j)$, where t is the map from 1.3;

(iv) $v^{-1}P^i(D_+) - vP^i(D_-) = \begin{cases} hP^{i-1}(D_0) & \text{for a mixed crossing} \\ P^i(D_0) & \text{for a self-crossing} \end{cases}$;

(v) $P^i(hC) = hP^{i-1}(C)$ for $i \geq 0$.

We proceed by induction on i. For $i = -1$ there is nothing to do, so assume that P^i is constructed for $i < m$ ($m \geq 0$). Before constructing P^m we will show the following fact (Proposition 2.4).

Consider the following bilinear operation $[\ \ ]'\colon R\mathcal{D} \times R\mathcal{D} \to R\mathcal{D}$ defined on pairs of diagrams in general position as follows:

$$[D_1, D_2]' = \sum_{\emptyset \neq \{c_{i_1}, \dots, c_{i_j}\} = C \in D_1 \cap D_2} h^{\alpha(C)} v^{pos(C)} (-v)^{-neg(C)} (D_1 \cdot D_2)_{0,\dots,0}^{c_{i_1}, \dots, c_{i_j}}$$

with $\alpha(C) = \alpha(c_{i_1}, \dots, c_{i_j}) = \frac{1}{2}(j - 2 + com(D_1 \cdot D_2) - com(D_1 \cdot D_2)_{0,\dots,0}^{c_{i_1}, \dots, c_{i_j}})$ where $com(D)$ denotes the number of components of D and $pos(C)$ (resp. $neg(C)$) denotes the number of positive (resp. negative) crossings in the family $\{c_{i_1}, \dots, c_{i_j}\} = C$. $D_1 \cdot D_2$ refers to the product of diagrams and the sub- and super-scripts refer to smoothing the crossings at the indicated points.

Let $[\ ,\ ]_m\colon R\mathcal{D} \times R\mathcal{D} \to P^{m-1}(R\mathcal{D}) \subset S(R\widehat{\pi}^\circ)$ be given by

$$\begin{aligned} [D_1, D_2]_m &= hP^{m-1}([D_1, D_2]') \\ &= h \sum_{\substack{\emptyset \neq \{c_{i_1}, \dots, c_{i_j}\} \\ = C \in D_1 \cap D_2}} h^{\alpha(c_{i_1}, \dots, c_{i_j})} v^{pos(C)} (-v)^{-neg(C)} P^{m-1-\alpha(C)} (D_1 \cdot D_2)_{0,\dots,0}^{c_{i_j}, \dots, c_{i_j}} . \end{aligned}$$

Proposition 2.4.

(i) $[D_1, D_2]_m$ *is unchanged by Reidemeister moves, so it defines the* $Z[v^{\mp 1}]$ *bilinear form* $[\,,\,]_m\colon R\mathcal{L} \times R\mathcal{L} \to S(R\widehat{\pi}^\circ)$,

(ii) *If* X, X^1, $Y \in R\mathcal{L}$ *and* $X = X^1$ *in* $R\mathcal{L}/S = \mathcal{S}(F \times I)$ *then* $[X, Y]_m = [X^1, Y]_m$ *and* $[Y, X]_m = [Y, X^1]_m$ *that is* $[\,,\,]_m$ *yields the* $Z[v^{\mp 1}]$ *bilinear form* $[\,,\,]_m\colon \mathcal{S}(F \times I) \times \mathcal{S}(F \times I) \to S(R\widehat{\pi}^\circ)$,

(iii) $[\ ,\ ]_m$ *defines an analogue of a Lie bracket on* $R\mathcal{L}$ *with Leibniz rule. That is*
(a) $[L_1, L_2]_m = -[L_2, L_1]_m$*, in particular* $[L, L]_v = 0$*;*
(b) $[L_1, [L_2, L_3]_m] + [L_2, [L_3, L_1]_m] + [L_3, [L_1, L_2]_m] = 0$*;*
where $[L_1, [L_2, L_3]_m] = h^2 P^{m-2}[D_1, [D_2, D_3]']'$ *for any diagrams* $D_1,\ D_2,\ D_3$ *(in general position) representing* $L_1,\ L_2$ *and* L_3*, or equivalently*

$$[L_1, [L_2, L_3]_m] = hP^{m-1}(D_1 \cdot [D_2, D_3]' - [D_2, D_3]' D_1)$$

(c) $[L_1 \cdot L_2, L_3]_m = hP^{m-1}(L_1[L_2, L_3]' + [L_1, L_3]' \cdot L_2)$.

Proof. We have to first rewrite the formula for $[\ ,\ ]_m$ in a more useful way(depending on the order of crossings between D_1 and D_2):

Lemma 2.5. *Let* $c_1, c_2, \ldots, c_j$ *be a fixed ordering of all crossings between* D_1 *and* D_2*, then*

$$[D_1, D_2]_m = hP^{m-1}\left(\sum_{i=1}^{j} sgn\, c_i v^{sgn\, c_i} v^{2\sum_{t=1}^{i-1} sgn\, c_t} (D_1 \cdot D_2)^{c_1,\ldots,c_{i-1},c_i}_{-sgn\, c_1,\ldots,-sgn\, c_{i-1},0}\right).$$

Proof. Consider

$$P^{m-1}((D_1 \cdot D_2)^{c_1,\ldots,c_{i-1},c_i}_{-sgn\, c_1,\ldots,-sgn\, c_{i-1},0})$$

and build the binary resolving tree for it, using skein relation for all crossings $c_1, \ldots, c_{i-1}$, and getting a combination of 2^{i-1} diagrams. From this tree, the original formula for $[D_1, D_2]_m$ can be easily obtained. □

Proof of Proposition 2.4 continued. (i). For Reidemeister moves, in which only D_1 or D_2 is involved (but not both), the claim is obvious because P^{m-1} is an isotopy invariant (and $D_1 \cap D_2$ is unchanged). So consider Ω_2 and Ω_3 in which D_1 and D_2 are involved.

(Ω_2) $[D_1, D_2]_m$ is unchanged by Ω_2 because $sgn\, c_1 v^{sgn\, c_1} \Omega_2(D_1 \cdot D_2)^{c_1}_0 + sgn\, c_2\, v^{sgn\, c_2} v^{2 sgn\, c_1} \Omega_2(D_1 \cdot D_2)^{c_1,c_2}_{-sgn\, c_1,0} = 0$ (to see this, notice that $sgn\, c_1 = -sgn\, c_2$ and $\Omega_2(D_1 \cdot D_2)^{c_1}_0 = \Omega_2(D_1 \cdot D_2)^{c_1,c_2}_{-sgn\, c_1,0}$ up to Ω_1 as shown in Fig. 2.1).

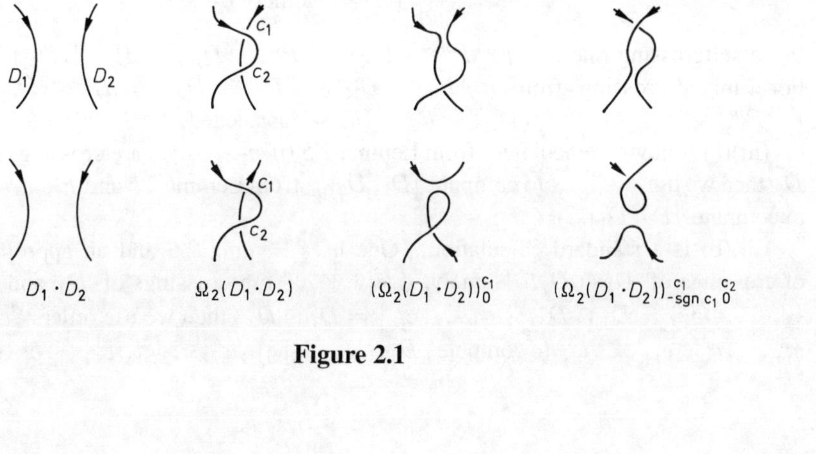

Figure 2.1

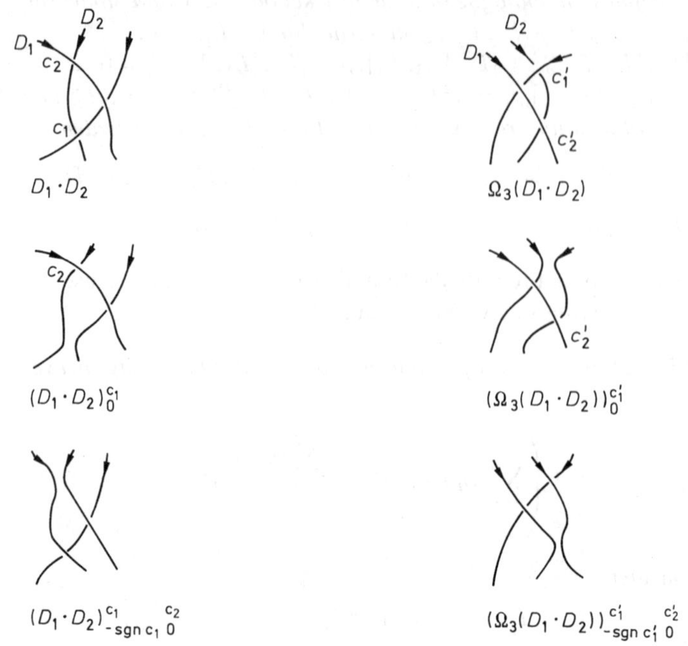

Figure 2.2

(Ω_3) It suffices to consider the case as in Fig. 2.2 and again one easily sees that

$$(D_1 \cdot D_2)_0^{c_1} = (\Omega_3(D_1 \cdot D_2))_0^{c_1'}$$
$$(D_1 \cdot D_2)_{-sgn\, c_1,0}^{c_1,c_2} = (\Omega_3(D_1 \cdot D_2)_{-sgn\, c_1',0}^{c_1',c_2'} .$$

(ii). Consider a skein relation

$$v^{-1}D_{+} - vD_{-} = \begin{cases} hD_{0} & \text{for a mixed crossing} \\ D_{0} & \text{otherwise.} \end{cases}$$

For a selfcrossing one has $v^{-1}P^{m-1}(D_{+}) - vP^{m-1}(D_{-}) = P^{m-1}(D_{0})$ as needed. For a mixed crossing, from $v^{-1}P^{m-1}(D_{+}) - vP^{m-1}(D_{-}) = hP^{m-2}(D_{0})$ follows $v^{-1}P^{m-1}(D_{+}) - vP^{m-1}(D_{-}) = P^{m-1}(hD_{0})$ as needed.

(iii)(a) follows immediately from Lemma 2.5 (if $c_1, \ldots, c_j$ are crossings of D_1 and D_2 then we use $c_1 \ldots c_j$ to compute $[D_1, D_2]_m$ from Lemma 2.5 and $c_j,\ c_j - 1, \ldots, c_1$ to compute $[D_2, D_1]_m$).

(iii)(b) is a standard calculation. One uses Lemma 2.5 and an appropriate order of crossings of $D_1 \cap D_j$. Namely, if $c_1^1, \ldots c_{i_1}^1$ are crossings of D_2 and D_3 (resp. $\{c_1^2, \ldots, c_{i_2}^2\} = D_1 \cap D_3$, $\{c_1^3, \ldots, c_{i_3}^3\} = D_1 \cap D_2$) then we use orders: $c_1^1, \ldots, c_{i_1}^1$, $c_1^2, \ldots, c_{i_2}^2$, $c_1^3, \ldots, c_{i_3}^3$ to compute $[D_1, [D_2, D_3]_m]$, $c_{i_2}^2, \ldots c_1^2$, $c_1^1, \ldots c_{i_1}^1$, $c_{i_3}^3, \ldots c_1^3$

to compute $[D_2,[D_3,D_1]_m]$ and $c_3^1,\dots,c_{i_3}^3,\ c_{i_2}^2,\dots,c_1^2,\ c_{i_1}^1,\dots c_1^1$ to compute $[D_3,[D_1,D_2]_m]$.

(iii)(c) follows immediately from Lemma 2.5. □

Similarly for $B_1,\ D_1,\ D_2,\ B_2 \in R\mathcal{D}$ we use notation $B_1\cdot[D_1,D_2]_m\cdot B_2$ for $hP^{m-1}(B_1\cdot[D_1,D_2]'\cdot B_2)$.

We can extend Proposition 2.4 and Lemma 2.5 to this situation. In particular, the Jabobi identity will have the form:

$$B_1\cdot([L_1,[L_2,L_3]_m]+[L_2,[L_3,L_1]_m]+[L_3,[L_1,L_2]_m])\cdot B_2=0 \tag{2.6}$$

where $B_1\cdot[L_1,[L_2,L_3]_m]\cdot B_2$ is by defnition $hP^{m-1}(B_1\cdot L_1\cdot[L_2,L_3]'\cdot B_2 - B_1\cdot[L_2,L_3]'L_1\cdot B_2)$.

We will first define P^m on descending diagrams; to do so we need some preliminary work. The free group $\pi(X_a)=\{x_1,\dots,x_k,a\colon\}$ has a natural projection onto the set of based descending diagrams of knots, up to Reidemeister moves preserving "descending" and not involving base points (it becomes a bijection if we divide $\pi(X_a)$ by relations $a^{\mp1}w\sim w\sim wa^{\mp1}$, see 1.3). So we have an R-algebra homomorphism $t\colon T(X_a)\to R\mathcal{L}$. Relations in $J\cup J_c$ are all special cases of isotopies and skein relations, so t yields an R-algebra homomorphism $\widetilde{t}\colon T(X_A)/(J\cup J_c)\to \mathcal{S}(F\times I)$.

By Lemma 1.7 we have an R-algebra epimorphism

$$g\colon T(R\widehat{\pi}^\circ)=T(R\widehat{\pi}^\circ(X))\to T(X_a)/(J\cup J_c)\,.$$

We have constructed $Z[v^{\mp1}]$-module homomorphisms (for $i<m$) $P^i\colon R\mathcal{L}\to S(R\widehat{\pi}^\circ)$.

We can uniquely define $Z[v^{\mp1}]$-module homomorphisms $\widetilde{P}^m\colon T(R\widehat{\pi}^\circ)\to S(R\widehat{\pi}^\circ)$ by using the following rules:

(i) for $\alpha_i\in\widehat{\pi}^\circ$, $\alpha_1\le\alpha_2\le\dots\le\alpha_s$ (we define $\alpha\le\beta$ if $\psi(\alpha)\le\psi(\beta)$):

$$\widetilde{P}^m(\alpha_1\otimes\dots\otimes\alpha_s)=\alpha_1\otimes\dots\otimes\alpha_s;$$

(ii) $\widetilde{P}^m(hC)=hP^{m-1}(\widetilde{t}g(C))$ for $C\in T(R\widehat{\pi}^\circ)$, notice that $P^{m-1}\colon R\mathcal{D}\to S(R\widehat{\pi}^\circ)$ factors through $\mathcal{S}(F\times I)$;

(iii) for $w_i\in\widehat{\pi}^\circ$

$$\begin{aligned}\widetilde{P}^m(w_1\otimes\dots\otimes w_i\otimes w_{i+1}\dots\otimes w_s)&=\widetilde{P}^m(w_1\otimes\dots\otimes w_{i+1}\otimes w_i\otimes\dots\otimes w_s)\\&\quad+hP^{m-1}(\widetilde{t}g(w_1)\cdot\ \dots\cdot[\widetilde{t}g(w_i),\widetilde{t}g(w_{i+1})]'\cdot\ \dots\cdot\widetilde{t}g(w_s))\,.\end{aligned}$$

Uniqueness of $\widetilde{P}^m$ is obvious. The proof that $\widetilde{P}^m$ is well defined is essentially a repetition of the classical proof of the Poincaré-Birkhoff-Witt theorem on universal enveloping algebras. The crucial step is to go from $w_1\otimes\dots\otimes w_i\otimes w_{i+1}\otimes w_{i+2}\otimes\dots\otimes w_s$ to $w_1\otimes\dots\otimes w_{i+2}\otimes w_{i+1}\otimes w_i\otimes\dots\otimes w_s$ using two different ways and to get the same results (due to Jacobi identity). I will perform this computation below.

(i)
$$\begin{aligned}&\widetilde{P}^m(w_1\otimes\dots\otimes w_i\otimes w_{i+1}\otimes w_{i+2}\otimes\dots\otimes w_s)\\&=\widetilde{P}^m(w_1\otimes\dots\otimes w_i\otimes w_{i+2}\otimes w_{i+1}\otimes\dots\otimes w_s)\\&\quad+hP^{m-1}(\widetilde{t}g(w_1)\cdot\ \dots\cdot\widetilde{t}g(w_i)\cdot[\widetilde{t}g(w_{i+1}),\widetilde{t}g(w_{i+2})]'\cdot\ \dots\cdot\widetilde{t}g(w_s))\\&=\widetilde{P}^m(w_1\otimes\dots\otimes w_{i+2}\otimes w_i\otimes w_{i+1}\otimes\dots\otimes w_s)\end{aligned}$$

$$
\begin{aligned}
&+hP^{m-1}(\widetilde{tg}(w_1)\cdot\ldots\cdot\widetilde{tg}(w_i)\cdot[\widetilde{tg}(w_{i+1}),\widetilde{tg}(w_{i+2})]'\cdot\ldots\cdot\widetilde{tg}(w_s))\\
&+hP^{m-1}(\widetilde{tg}(w_1)\cdot\ldots\cdot[\widetilde{tg}(w_i),\widetilde{tg}(w_{i+2})]'\cdot\widetilde{tg}(w_{i+1})\cdot\ldots\cdot\widetilde{tg}(w_s))\\
&=\widetilde{P}^m(w_1\otimes\cdots\otimes w_{i+2}\otimes w_{i+1}\otimes w_i\otimes\cdots\otimes w_s)\\
&+hP^{m-1}(\widetilde{tg}(w_1)\cdot\ldots\cdot(\widetilde{tg}(w_i)))[\widetilde{tg}(w_{i+1}),\widetilde{tg}(w_{i+2})]'\\
&+[\widetilde{tg}(w_i),\widetilde{tg}(w_{i+2})]'\cdot\widetilde{tg}(w_{i+1})\\
&+\widetilde{tg}(w_{i+2})\cdot[\widetilde{tg}(w_i),\widetilde{tg}(w_{i+1})]'\cdot\ldots\cdot\widetilde{tg}(w_s)).
\end{aligned}
$$

(ii) Similarly

$$
\begin{aligned}
&\widetilde{P}^m(w_1\otimes\cdots\otimes w_i\otimes w_{i+1}\otimes w_{i+2}\otimes\cdots\otimes w_2)\\
&\quad=\widetilde{P}^m(w_1\otimes\cdots\otimes w_{i+1}\otimes w_i\otimes w_{i+2}\otimes\cdots\otimes w_s)\\
&\quad+hP^{m-1}(\widetilde{tg}(w_1)\cdot\ldots\cdot[\widetilde{tg}(w_i),\widetilde{tg}(w)_{i+1}]'\cdot\widetilde{tg}(w_{i+2})\cdot\ldots\cdot\widetilde{tg}(w_s))\\
&\quad=\widetilde{P}^m(w_1\otimes\cdots\otimes w_{i+1}\otimes w_{i+2}\otimes w_i\otimes\cdots\otimes w_s)+hP^{m-1}(()+())\\
&\quad=\widetilde{P}^m(w_1\otimes\cdots\otimes w_{i+2}\otimes w_{i+1}\otimes w_i\otimes\cdots\otimes w_s)\\
&\quad+hP^{m-1}(\widetilde{tg}(w_1)\cdot\ldots\cdot([\widetilde{tg}(w_i),\widetilde{tg}(w_{i+1})]'\cdot\widetilde{tg}(w_{i+2})\\
&\quad+\widetilde{tg}(w_{i+1})\cdot[\widetilde{tg}(w_i),\widetilde{tg}(w_{i+2})]'\\
&\quad+[\widetilde{tg}(w_{i+1}),\widetilde{tg}(w_{i+2})]'\widetilde{tg}(w_i))\cdot\ldots\cdot\widetilde{tg}(w_s)).
\end{aligned}
$$

The difference (i)–(ii) should give 0, so let's compute the difference between the right sides of (i) and (ii). We get

$$
\begin{aligned}
&hP^{m-1}(\widetilde{tg}(w_1)\cdot\ldots\cdot(\widetilde{tg}(w_i)[\widetilde{tg}(w_{i+1}),\widetilde{tg}(w_{i+2})]'-[\widetilde{tg}(w_{i+1}),\widetilde{tg}(w_{i+2})]'\cdot\widetilde{tg}(w_i)\\
&\quad+[\widetilde{tg}(w_i),\widetilde{tg}(w_{i+2})]'\cdot\widetilde{tg}(w_{i+1})-\widetilde{tg}(w_{i+1})\cdot[\widetilde{tg}(w_i),\widetilde{tg}(w_{i+2})]'\\
&\quad+\widetilde{tg}(w_{i+2})[\widetilde{tg}(w_i),\widetilde{tg}(w_{i+1})]'-[\widetilde{tg}(w_i),\widetilde{tg}(w_{i+1})]'\widetilde{tg}(w_{i+2}))\cdot\ldots\cdot\widetilde{tg}(w_s))\\
&\quad=\widetilde{tg}(w_1)\cdot\ldots\cdot([\widetilde{tg}(w_i)[\widetilde{tg}(w_{i+1})),\widetilde{tg}(w_{i+2}]_m]-[\widetilde{tg}(w_{i+1})[\widetilde{tg}(w_i),\widetilde{tg}(w_{i+2})]_m]\\
&\quad+[\widetilde{tg}(w_{i+2})[\widetilde{tg}(w_i),\widetilde{tg}(w_{i+1}]_m])\cdot\ldots\cdot\widetilde{tg}(w_s)=0
\end{aligned}
$$

by the Jacobi identity (2.6).

The $Z[v^{\mp 1}]$-homomorphism $\widetilde{P}^m$ factors through $T(R\widehat{\pi}^\circ)/\ker g$. To see this we use Lemma 1.7 and the interpretation of $w-f_w(w)$ given in the proof of Lemma 1.7(c); see the formula 1.10. Consider a descending diagram γ_w, represented by $w\in\pi(X_a)$ and let $c_1,c_2,\ldots,c_{2s}$ be crossings of γ_w counted starting from the base point $b\in\gamma_w$ (of course each crossing is listed twice). Now let's build a binary resolving tree for γ_w using all crossings in a given order. We will get 3^s leaves and they will all cancel one with another, except the leaves with no smoothings involved, to equal to γ_w. Therefore, $\widetilde{P}^m$ of an element of $T(R\widehat{\pi}^\circ)$ representing $w-f_w(w)$ is equal to 0. The cancellation of leaves is easy to see from the resolving tree, or to get it more easily, one can use an often used fact that leaves of a resolving tree (up to permutation) do not depend on the order of applied crossings. So, we can use the ordering with $c_1=c_2,c_3=c_4,\ldots,c_{2s-1}=c_{2s}$ and then the cancellation of leaves is obvious; see Fig. 2.3 and 2.4 for an example.

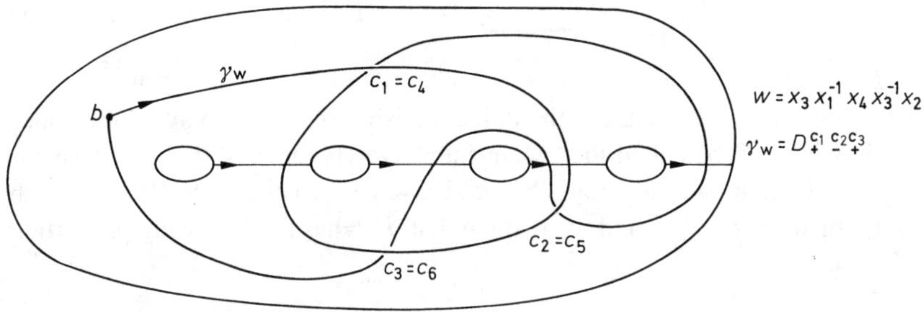

Figure 2.3

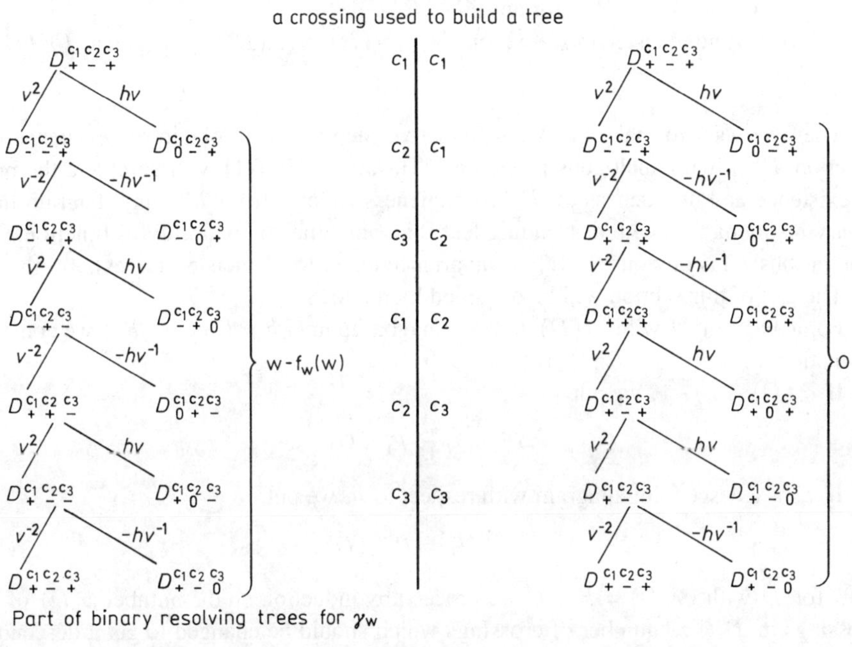

Figure 2.4

$\widetilde{P}^m$ yields a $Z[v^{\mp 1}]$-homomorphism $P^m_{des}\colon \bar{R}\mathcal{L}_{des} \to S(R\widehat{\pi}^\circ)$ which factors through $\mathcal{S}_{des} = \bar{R}\mathcal{L}_{des}/S_{des}$.

Now we construct P^m by induction on the number $cr(D)$ of the crossing points of D. For each $k \geq 0$ we define a function P^m_k which assigns an element of $S(R\widehat{\pi}^\circ)$ to each diagram of an oriented link with no more than k crossings. Then P^m will be defined for every diagram by $P^m(D) = P^m_k(D)$ for $k \geq cr\, D$. Of course the functions P^m_k must satisfy certain coherence conditions. We begin from the definition of $P^m_\circ$. For a diagram D of n components with $cr(D) = 0$ we can write $D =$

$t(\psi(w_c^1) \otimes \cdots \otimes \psi(w_c^{n-s}) \otimes \underbrace{1 \otimes \otimes 1}_{s \text{ times}})$ with $\psi(w_c^1) \leq \cdots \leq \psi(w_c^{n-s})$ and we put $P_\circ^m(D) = (v^{-1} - v)^s w_c^1 \otimes w_c^2 \otimes \cdots \otimes w_c^{n-s}$.

To define P_{k+1}^m we will use the induction several times. To avoid misunderstanding, the following will be called the "Main Inductive Hypothesis" MIH: we assume that we have already defined a function P_k^m attaching an element of $S(R\widehat{\pi}^\circ)$ to each diagram D for which $cr(D) \leq k$ and we assume that P_k^m has the following properties:

2.7.

(a) For a descending diagram D, $P_k^m(D) = P_{des}^m(D)$.

(b) $v^{-1}P_k^m(D_{+}) = vP_k^m(D_{-}) = \begin{cases} hP^{m-1}(D_{0}) & \text{for a mixed crossing} \\ P_k^m(D_{0}) & \text{otherwise} \end{cases}$.

(c) $P_k^m(h\,C) = hP^{m-1}(C)$ for any $C \in R\mathcal{D}$.

(d) If for a Reidemeister move Ω on D, $(cr(D), cr(\Omega(D)) \leq k$, then $P_k^m(D) = P_k^m(\Omega(D))$.

Then we want to make the Main Inductive Step (MIS) to obtain the existence of a function P_{k+1}^m with analogous properties. This, as in **[P-T-1]**, will complete the proof of existence and uniqueness of P^m (uniqueness follows from 2.7: any diagram has a binary resolving tree with descending leaf diagrams; the compatibility of functions P_k^m (for various k) is obvious as well as invariance under Reidemeister moves).

The rest of this section will be occupied by the MIS.

For a diagram D with $cr(D) \leq k+1$ and basepoints $b = (b_1, \ldots, b_n)$ we construct a function $P_b^m(D) \in S(R\widehat{\pi}^\circ)$ as follows.

If $cr(D) < k+1$, we put

$$P_b^m(D) = P_k^m(D)\,. \tag{2.8}$$

If D is a descending diagram with respect to b, we put

$$P_b^m(D) = P_{des}^m(D)\,. \tag{2.9}$$

Now for D with $cr(D) = k+1$, we proceed by induction on the number $b(D)$ of bad crossings in D (i.e., number of crossings which should be changed to get a descending diagram). For a different choice of basepoints $b' = (b_1', \ldots, b_n')$ we will write $b'(D)$. Assume that P_b^m is defined for all diagrams $D \in \mathcal{D}$ with basepoints such that $b(D) < t$. Then for D with $b(D) = t$, let p be the first bad crossing of D (starting from b_1 and walking along the diagram). Depending on p being positive or negative we have $L = L_+^p$ or L_-^p. We define $P_b^m(L)$ using the equation

$$v^{-1}P_b^m(L_+^p) - vP_b^m(L_-^p) = \begin{cases} hP^{m-1}(L_\circ^p) & \text{for mixed crossing} \\ P_b^m(L_\circ^p) & \text{otherwise} \end{cases}.$$

$P_b^m(L)$ is well defined because $cr(L_\circ^p) \leq k$ and $b(L_{-sgn\,p}^p) < t$.

We will show that P_b^m is independent of the choice of basepoints b, and that it has the properties required for P_{k+1}^m.

Skein relations for P_b^m. We have to prove 2.7(b) for P_b^m. Let p be the considered crossing of D. We proceed by induction on $b(D)$ exactly as in **[P-T-1]**, so we omit here the details.

Changing Base Points. We will show that P_b^m does not depend on the choice of b, provided the order of components is not changed. This amounts to the verification that we may replace b_i by b_i' taken from the same component in such a way that b_i' lies after b_i and there is exactly one crossing point, say p, between b_i and b_i' (Fig. 2.5).

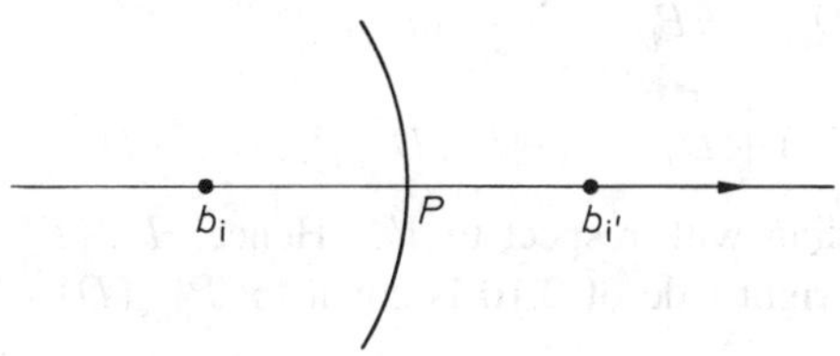

Figure 2.5

Let $b' = (b_1, \dots, b_i', \dots, b_n)$. We want to show that $P_b^m(D) = P_{b'}^m(D)$ for any diagram D with $cr(D) = k+1$.

We use induction on $B(L) = \max(b(L), b'(L))$. The inductive step (in the case $b(L)$ and $b'(L)$ have a common bad crossing) is exactly the same as in **[P-T-1]**, so we omit it. There are two kinds of initial steps:

(1) Assume $B(D) = 0$. Then D is descending with respect to both choices of base points and p is the crossing between different components. Therefore, $P_b^m(D) = P_{des}^m(D) = P_{b'}^m(D)$.

(2) Assume that $B(D) = 1$ and $b(D) \neq b'(D)$. This is possible only when p is a self-crossing of the i'th component of D. There are two cases.

(a) $b(D) = 1$ and $b'(D) = 0$. Therefore,

$$P_{b'}^m(D) = P_{des}^m(D).$$

We can use the crossing p to compute $P_b^m(D)$ by applying the rule

$$v^{-1}P_b^m(D_+^p) - vP_b^m(D_-^p) = P_k^m(D_\circ^p).$$

Furthermore, $D_{-sgn\,p}^p$ is descending (with respect to b), and $D_\circ^p$ is descending as well. Therefore, $P_b^m(D_{-sgn\,p}^p) = P_{des}^m(D_{-sgn\,p}^p)$ and

$$P_k^m(D_\circ^p) = P_{des}^m(D_\circ^p).$$

Finally, the descending diagrams D_+^p, D_-^p, $D_\circ^p$ are related by

$$v^{-1}P_{des}^m(D_+^p) - vP_{des}^m(D_-^p) = hP_{des}^m(D_0^p),$$

so $P_b^m(D) = P_{des}^m(D) = P_{b'}^m(D)$ as required.

(b) $b(D) = 0$ and $b'(D) = 1$. This case can be dealt with like 2(a).

Independence of order of components. It is enough to verify that for a given diagram D ($cr(D) = k + 1$) and fixed basepoints $b = (b_1, \ldots, b_i, b_{i+1}, \ldots, b_n)$ we have $P_b^m(L) = P_{b'}^m(D)$ where $b' = (b_1, \ldots, b_{i+1}, b_i, \ldots, b_n)$. This is easily reduced by the usual induction on $b(L)$ to the case of a descending diagram $D_1 \cdot \ldots \cdot D_i \cdot D_{i+1} \ldots D_n$. Then $P_b^m(D) = P_{des}^m(D)$. If $c_1, \ldots, c_s$ are all crossings between D_i and D_{i+1}, then one can easily find that for

$$D' = D_1, \ldots, D_{i+1} \cdot D_i \ldots D_n$$

$$\begin{aligned} P_{b'}^m(D) &= P_{b'}^m(D') + hP_{b'}^{m-1} \sum_{i=1}^{s} sgn\, c_i v^{sgn\, c_i} v^{2\sum_{j=1}^{i-1} sgn\, c_j} (D)^{c_1 \ldots c_i}_{-sgn\, c_1 \ldots 0} \qquad (2.10) \\ &= P_{b'}^m(D') + D_1 \cdot \ldots \cdot [D_i, D_{i+1}]_m \cdot \ldots \cdot D_n, \end{aligned}$$

because D' is descending with respect to b'. Hence, $P_{b'}^m(D') = P_{des}^m(D')$ and by definition of P_{des}^m the right side of 2.10 is equal to $P_{des}^m(D)$. Therefore $P_b^m(D) = P_{b'}^m(D)$.

Let's define P_{k+1}^m to be equal to $P_b^m(D)$.

Independence of P_{k+1}^m of Reidemeister moves. Let $cr(D) = k + 1$. Let Ω be a Reidemeister move which do not increase the number of crossings. We will show that $P_{k+1}^m(D) = P_{k+1}^m(\Omega(D))$.

Reidemeister moves which we have to consider are shown in Fig. 2.6.

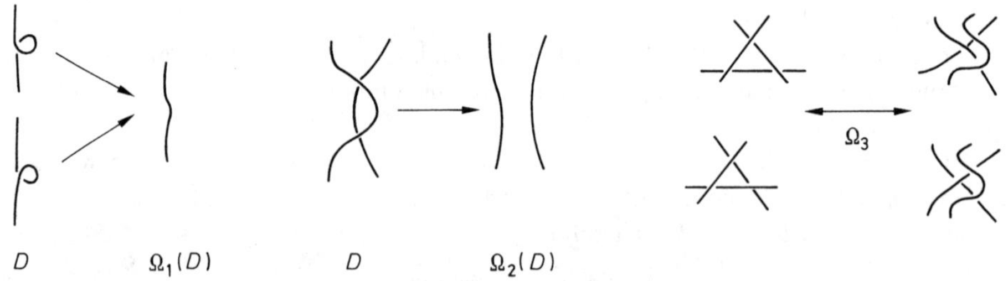

Figure 2.6

It is convenient to work with based diagrams (D, b) and $P_b^m(D)$. We will always assume that base points lie outside the part of the diagram involved in the considered Reidemeister move Ω. We will also assume (and there are such choices of b) that no bad crossing lies on the part of D involved in the move except, possibly, the crossing between middle and bottom arcs in Ω_3.

We use induction on $b(D)$.

If $b(D) = 0$, then it is easily seen that also $b(\Omega(D)) = 0$ and for these descending diagrams $P_{des}^m(D) = P_{des}^m(\Omega(D))$, so $P_b^m(D) = P_b^m(\Omega(D))$. We assume by induction that $P_b^m(D) = P_b^m(\Omega(D))$ for $b(D) < t$ and furthermore that $b(D) = t$. There are two cases to consider:

(i) there is a bad crossing p in (D, b) which is different from all crossings involved in the considered Reidemeister move;

(ii) the only bad crossings are involved in the Reidemeister move.

In case (i), we proceed exactly as in **[P-T-1]**, so we omit the details.

In case (ii), the only bad crossing, say p, is the crossing between middle and lower arcs involved in Ω_3. We use the skein relation for p and $b(D^p_{-sgn\,p}) = 0$ and $cr(D^p_\circ) = k$, so we can apply Ω_3 to $D^p_{-sgn\,p}$ and any Reidemeister move to $D^p_\circ$ (MIH) and show as in **[P-T-1]** that $P^m_b(D) = P^m_b(\Omega_3(D))$.

This completes the construction of P^m_{k+1}, hence of P^m and the family of $Z[v^{\mp 1}]$-homomorphisms $P^i \colon R\mathcal{D} \to S(R\widehat{\pi}^\circ)$ satisfying 2.3(i)–(v). □

Theorem 2.11. *There is an R-module homomorphism $P \colon R\mathcal{D} \to S(R\widehat{\pi}^\circ)$ which satisfies $P(C) \equiv P^i(C) \bmod h^{i+1}$. Furthermore*

(i) *P is independent of isotopy class of a diagram D in $F \times I$*

(ii) *P is independent of skein class of $C \in R\mathcal{D}$ so*

$$v^{-1}P(D_{+}) - vP(D_{-}) = \begin{cases} hP(D_{0}) & \text{for a mixed crossing} \\ P(D_{0}) & \text{for a self-crossing} \end{cases}$$

and P yields an R-isomorphism of modules $\widetilde{P} \colon \mathcal{S}(F \times I) \to S(R\widehat{\pi}^\circ)$.

We start the proof with the following lemma.

Lemma 2.12.

(a) *For any $C \in R\mathcal{D}$, $P^i(C) \equiv P^{i+1}(C) \bmod h^{i+1}$ or equivalently $P^i(C) = P^{i+1}(C)$ in $S(R\widehat{\pi}^\circ)/h^{i+1}S(R\widehat{\pi}^\circ)$.*

(b) *Coefficients of any element of $P^i(R\mathcal{D})$ are polynomials of h-degree at most i.*

Proof. (a),(b): We proceed by induction on i. For $i = -1$ it holds, so assume it holds for $i < m$ and consider $i = m$. Because $P^m(hC) = hP^{m-1}(C)$ it is sufficient to check for a diagram D. Now we use the fact that (a) and (b) hold for D^p_+ iff they hold for D^p_- assuming they hold for D^p_0. This completes the proof of Lemma 2.12. □

Proof of Theorem 2.11. P is a well-defined R-isomorphism by Lemma 2.12 and the simple observation that for any $C \in R\widehat{\pi}^\circ$ there is N_c such that for $i \geq N_c$, $P^i(C) = P^{i+1}(C)$ (and so $P(C) = P^i(C)$).

$\widetilde{P}$ is well defined by corresponding properties of P^i. $\widetilde{P}$ is an isomorphism because $\mathcal{S}(F \times I)$ is generated (as an algebra) by $\widetilde{t\psi}(\widehat{\pi}^\circ)$ (and $\widetilde{P}\widetilde{t\psi} = id$) and monomials in $\widehat{\pi}^\circ$ freely generate the R-module $S(R\widehat{\pi}^\circ)$. This completes the proof of Theorem 2.11. □

3. Corollaries

As a corollary to Theorem 2.11, we have the following result which is equivalent to Theorem 0.4.

Theorem 3.1. *Let $R' = Z[v^{\mp 1}, h^{\mp 1/2}]$ and S' be a subalgebra of $R'\mathcal{L}$ generated by skein expressions $v^{-1}D_{+} - vD_{-} = h^{1/2}D_{0}$ and let $\mathcal{S}'(M) = R'\mathcal{L}/S'$. Then $\mathcal{S}'(F \times I)$ is an infinitely generated free R'-module for F a compact planar surface. The images under $\widetilde{t\psi}$ of monomials in $\widehat{\pi}^\circ$ form a basis of $\mathcal{S}'(F \times I)$.*

Proof. Consider an R-algebra homomorphism $H: S(R\widehat{\pi}^\circ) \to S(R'\widehat{\pi}^\circ)$ given by $H(w) = h^{1/2}w$ for $w \in \widehat{\pi}^\circ$. We define an R'-module homomorphism $P': R'\mathcal{D} \to S(R'\widehat{\pi}^\circ)$ by putting $P'(D) = h^{-c(D)/2}H(P(D))$ where $c(D)$ is the number of components of the link defined by D. By definition P' is an isotopy invariant and we will check in a moment that it satisfies $v^{-1}P'(D_+) - vP'(D_-) = h^{1/2}P'(D_0)$. So P' yields the R'-module homomorphism $\widetilde{P}': \mathcal{S}'(F \times I) \to S(R'\widehat{\pi}^\circ)$.

$$
\begin{aligned}
&v^{-1}P'(D_+) - vP'(D_-) - h^{1/2}P'(D_0) \qquad (3.2)\\
&= H(h^{-c(D_+)/2}v^{-1}P(D_+) - h^{-c(D_-)/2}vP(D_-) - h^{1/2}h^{-c(D_0)/2}P(D_0))\\
&= \begin{cases} h^{-c(D_+)/2}H(v^{-1}P(D_+) - vP(D_-) - hP(D_0)) & \text{for a mixed crossing}\\ h^{-c(D_+)/2}H(v^{-1}P(D_+) - vP(D_-) - P(D_0)) & \text{otherwise}\end{cases}\\
&= 0\,.
\end{aligned}
$$

$\widetilde{P}'$ is an isomorphism because monomials $w_{i_1} \otimes \cdots \otimes w_{i_s}$ ($w_i \in \widehat{\pi}^\circ$) form an R'-basis of $S(R'\widehat{\pi}^\circ)$; its images in $\mathcal{S}'(F \times I)$ ($\widetilde{t\psi}(w_{i_1} \otimes \cdots \otimes w_{i_s})$ generates $\mathcal{S}'(F \times I)$ and $\widetilde{P}'\widetilde{t\psi} = id_{S(R'\widehat{\pi}^\circ)}$. This completes the proof of Theorem 2.13 (as well as 0.4, put $z = h^{1/2}$). □

The same method as used in the proofs of Theorems 2.2 and 2.11 can be used to analyze another version of the skein module of $F \times I$, the one analyzed by V. Turaev in **[T-3]**. We use his notation in the following.

Let K be a commutative ring with 1. The skein module $\mathcal{A}(M)$ is a module over the commutative polynomial ring $\Lambda = K[v^{\mp 1}, h, \hbar]$ defined as the quotient of the free module $\Lambda\mathcal{L}$ by the submodule generated by the skein expressions

$$
\begin{array}{ll}
v^{-1}D_+ - vD_- - hD_0 & \text{if } \times \text{ is a mixed crossing}\\
v^{-1}D_+ - vD_- - \hbar D_0 & \text{if } \times \text{ is a self-crossing}\\
v^{-1}\emptyset - v\emptyset - \hbar 1 & \text{where } \emptyset \text{ (resp. 1) represents}\\
& \text{an empty (resp. trivial) knot.}
\end{array}
\qquad (3.3)
$$

Theorem 3.4. *There is an invariant of links in $F \times I$; $\widehat{P}: \Lambda\mathcal{D} \to S(\Lambda\widehat{\pi})/J_\hbar$ which descends to an isomorphism $\widetilde{P}: \mathcal{A}(F \times I) \to S(\Lambda\widehat{\pi})/J_\hbar$ where $J_\hbar$ is the ideal generated by expressions $(v^{-1} - v)C - \hbar C \otimes 1$ for $C \in \widehat{\pi}$ and 1 representing the empty word in $\widehat{\pi}$.*

This answers a question raised by V. Tureav in **[T-3]**.

Proof. We proceed exactly as in the proof of Theorem 2.11, i.e., we construct inductively a family of $\overline{\Lambda} = K[v^{\mp 1}, \hbar]$-module homomorphisms $\widehat{P}^i: \Lambda\mathcal{D} \to S(\Lambda\widehat{\pi})/J_\hbar$, $i \geq -1$ as follows:

1. We construct a $\overline{\Lambda}$-homomorphism for descending diagrams $\widehat{P}_{des}: \overline{\Lambda}\mathcal{D}_{des} \to T(\overline{\Lambda}\widehat{\pi})/J_1 \cup J_2$ where J_1 is the ideal generated by $(v^{-1} - v)C - hC \otimes 1$ and $(v^{-1} - v)C - \hbar 1 \otimes C$ and J_2 is generated by "conjugacy expression" (as ker g of Lemma 1.7).
2. Assuming $\widehat{P}^i$ is constructed for $i < m$ we construct $\widehat{P}^m$ inducting on the number of crossings of the diagram. The initial step is to define $\widehat{P}^m$ on descending diagrams.

For this we define a "Lie bracket" $[\ ,\ ]_m: \Lambda\mathfrak{D} \times \Lambda\mathfrak{D} \to \widehat{P}^{m-1}(\Lambda\mathfrak{D}) \subset S(\Lambda\widehat{\pi})/J_\hbar$ and use it, together with $\widehat{P}_{des}$, to define $\widehat{P}^m$ on descending diagrams. The details are the same as in the proof of Theorem 2.11, so we omit them. □

Corollary 3.5. *The skein module $\mathcal{A}(F \times I)$ of Theorem 3.4 gives rise to the following interesting free modules.*

1. $\hbar = 1$*: the skein module $\mathcal{S}(F \times I)$ of Theorems 2.2 and 2.11.*
2. $v = 1,\ \hbar = 0$*: the homotopy skein module* **[H-P-1]**, **[T-3]**.
3. $h = 0,\ \hbar = 1$ *(or $\hbar$ invertible): the anti-homotopy skein module* **[H-P-2]**, *compare Chapter 5.*
4. $v = 1,\ w \otimes 1 = 0$*: skein module (bialgebra) $A_\circ$ of* **[T-3]**. *It was conjectured to be free in* **[T-3; 10.3]**.
5. *If we exclude $\emptyset$ from $\mathcal{L}$ and the skein relation involving $\emptyset$, and put $v = 1,\ h = \hbar$, $w \otimes 1 = 0$, then we get the free $Z[h]$-module corresponding to the Alexander polynomial of S^3.* □

Remark 3.6. The skein module $\mathcal{A}(F \times I)$ is essentially no more general than $\mathcal{S}(F \times I)$ and $\mathcal{S}_3(F \times I)$ (similarly as in the case of links in S^3), in particular the link invariant $\widehat{P}$ does not distinguish more links than P, namely: let $H_1: S(Z[v^{\mp 1}, h]\widehat{\pi}^\circ) \to S(Z[v^{\mp 1}, \hbar]\widehat{\pi}^\circ)$ be a $Z[v^{\mp 1}]$-algebra homomorphism given by

$$H_1(h^i w_1 \otimes \cdots \otimes w_k) = h^i \hbar^{i+k} w_1 \otimes \cdots \otimes w_k \quad (w_j \in \widehat{\pi}^\circ).$$

Let $H_2: S(Z[v^{\mp 1}, h, \hbar, \frac{v^{-1}-v}{\hbar}]\widehat{\pi}^\circ) \to S(Z[v^{\mp 1}, h, \hbar]\widehat{\pi})/J_\hbar$ be a $Z[v^{\mp 1}, h, \hbar]$-algebra homomorphism given by

$$H_2\left(\left(\frac{v^{-1}-v}{\hbar}\right)^i \alpha\right) = a \otimes \underbrace{1 \otimes \cdots \otimes 1}_{i \text{ times}}, \qquad i \geq 0$$

where $\alpha \in S(Z[v^{\mp 1}, h, \hbar]\widehat{\pi}^\circ)$. Then

$$\widehat{P}(D) = H_2(\hbar^{-com\, D} \cdot H_1(P(D)))\,. \tag{3.7}$$

Proof. 3.7 holds for a generating set of $\mathcal{A}(F \times I)$, i.e., for links corresponding to $w_1 \otimes \cdots \otimes w_k \otimes 1 \otimes \cdots \otimes 1$ where $w_i \in \widehat{\pi}^\circ$, $\psi(w_i) \leq \psi(w_i) \leq \psi(w_{i+1})$. Then if 3.7 holds for D_0 and $D_\pm$ then it holds for $D_\mp$ so it holds always. □

4. The Conway algebra of $F \times I$

We can formulate our results in a slightly more general form using the concept of the Conway algebra (**[P-T-1]**, **[P-T-2]**, **[P-1]**) **[H-W]** instead of the more restrictive skein modules.

Definition 4.1. The skein relation $\sim_s$ is the smallest equivalence relation on isotopy classes of oriented links in an oriented 3-manifold which satisfies the following condition: Let p_1 and p_2 be two crossings of the same sign of links L_1 and L_2, respectively, and $(L_1)^{p_1}_{-sgn\, p_2} \sim_s (L_2)^{p_2}_{-sgn-p_2}$, $(L_1)^{p_1}_0 \sim_s (L_2)^{p_1}_0$, then $L_1 \sim_s L_2$.

Definition 4.2. Let $\pi = \pi_1(F)$, $\widehat{\pi}$ denote conjugacy classes of π, $\widehat{\pi}^\circ = \widehat{\pi} - \{1\}$, and $M_s(\widehat{\pi})$ denote the set of commutative monomials in $\widehat{\pi}$ including the empty one $\emptyset$ (i.e., $M_s(\widehat{\pi})$ forms a basis of $S(R\widehat{\pi})$ for any ring R). We will write monomials as $w_1 \otimes w_2 \otimes \cdots \otimes w_k$.

The Conway algebra $\mathcal{A}(\pi)$ is an abstract algebra $(A; a_\alpha, |, *)$ where A is the universe, a_α are constants (0-argument operations) with $\alpha \in M_s(\widehat{\pi})$ and $|, *$ are 2-argument operations $A \times A \to A$ and the following conditions hold:

$$a_w \mid a_{w\otimes 1} = a_w \qquad (C1)$$
$$a_w * a_{w\otimes 1} = a_w \qquad (C2)$$
$$(a \mid b) \mid (c \mid d) = (a \mid c) \mid (b \mid d) \qquad (C3)$$
$$(a \mid b) * (c \mid d) = (a * c) \mid (b * d) \qquad (C4)$$
$$(a * b) * (c * d) = (a * c) * (b * d) \qquad (C5)$$
$$(a \mid b) * b = a \qquad (C6)$$
$$(a * b) \mid b = a \qquad (C7)$$

Theorem 4.3. *For a given Conway algebra $\mathcal{A}(\pi)$ there exists a uniquely determined invariant Q which attaches an element $Q(L)$ from A to every isotopy class L of oriented links in $F \times I$ and satisfies the conditions (we use the notation of Chapters 1 and 2):*

$$\left.\begin{array}{l} Q(t\psi(v_1 \otimes \cdots \otimes v_k)) = a_{v_1 \otimes \cdots \otimes v_k} \\ \text{for } \psi(v_1) \le \psi(v_2) \le \ldots \le \psi(v_k) \end{array}\right\} \begin{array}{l}\text{initial}\\ \text{conditions}\end{array} \qquad (1)$$

$$\left.\begin{array}{l} Q(D_{+}) = Q(D_{-}) \mid Q(D_{0}) \qquad (2) \\ Q(D_{-}) = Q(D_{+}) * Q(D_{0}) \qquad (3) \end{array}\right\} \begin{array}{l}\text{skein}\\ \text{relations}\end{array}$$

Notice that Q is an invariant of skein equivalence classes of links.

Proof. To prove Theorem 4.3, it suffices to consider only the universal Conway algebra, i.e., the algebra in which the universe A consists of terms (that is words over the alphabet $\{a_\alpha\}$, $|$, $*$, (,) which are sensible) with obvious operations $|$ and $*$, divided by axioms C1–C7. We will follow the proof of Theorem 2.11 but we have to use a different device instead of induction on the power of h.

Let $com(w)$ for $w \in A$ denote the number of factors in the index of the first letter (not parenthesis) of w (e.g., $com((a_{v_1\otimes v_2 \otimes v_3} \mid a_{v_1 \otimes v_2}) \mid (a_{v_4} * a_{v_1 \otimes v_5})) = 3$). We will distinguish operations $a \mid b$ and $a * b$ such that $com(a) = com(b) + 1$. We call such operations mixed and denote them by $\bar{|}$ and $\bar{*}$.

Now instead of using h we perform an induction on the "nesting" of mixed operations. The number $mon_w a$, the mixed "nesting" of a in $w \in A$, is defined by:

(1) $mon_w a = 0$ if w does not use $\bar{|}$ and $\bar{*}$.

(2) If $a \in w_1$ then $mon_{w_1 | w_2} a = mon_{w_1 * w_2} a = mon_{w_1} a$.

(3) If $a \in w_2$ then $mon_{w_1|w_2}a = \begin{cases} mon_{w_2}a & \text{if } | \text{ is not a mixed operation} \\ mon_{w_2}a + 1 & \text{if } |=\bar{|} \end{cases}$ and

similarly $mon_{w_1 * w_2}a = \begin{cases} mon_{w_2}a & \text{if } * \text{ is not a mixed operation} \\ mon_{w_2}a + 1 & \text{if } * = \bar{*} . \end{cases}$

In the inductive process we construct the sequence of invariants $Q^i\colon \mathcal{D} \to A$ where $Q^i(\mathcal{D})$ has no word w with letter a and $mon_w a > i$. We leave the details to the reader. □

Example 4.4. Set $A = S(Z[v^{\mp 1}, z^{\mp 1}]\widehat{\pi}^\circ)$,

$$a_{v_1 \otimes \cdots \otimes v_s \otimes \underbrace{1 \otimes \cdots \otimes 1}_{k \text{ times}}} = \left(\frac{v^{-1} - v}{2}\right)^k v_1 \otimes \cdots \otimes v_s \quad (v_i \in \widehat{\pi}^\circ),$$

and $|$ and $*$ are defined as follows:

$$b \mid c = v^2 b + v\, z\, c$$
$$a * c = v^{-2} a - v^{-1} z\, c$$

i.e., if $v^{-1}a - vb = zc$ then $b \mid c = a,\ a * c = b$.

The invariant links yielded by this Conway algebra correspond to that of Theorem 0.4, or, more precisely, $Q(D) = P'(D)$ for $z = h^{1/2}$ where P' is the invariant of Theorem 3.1.

In order to get a link invariant as in Theorem 4.3, it is not necessary to have the operations $|$ and $*$ defined on the whole product $A \times A$, and relations C3–C5 need not to be satisfied by all elements of $A \times A \times A \times A$. We deal then with a partial Conway algebra. A useful example (which produces a link invariant) is the algebra of terms over alphabet $\{a_\alpha\}$, $|$, $*$, $(,)$, where a product $a \mid b$ (and $a * b$) is performed iff $com(a) = com(b) \mp 1$. For a detailed discussion of partial Conway algebras, we refer to **[P-T-1]**, **[P-T-2]** and **[P-1]**.

Here we only mention that the skein equivalence classes of links in $F \times I$ lead to the universal partial Conway algebra (this algebra was described by Conway in **[C]** in the case of S^3).

Remark 4.5. The method described in the paper works also for F being a torus and Theorems 0.4, 2.11 and 4.3 are still true (the proofs are even shorter because $\pi_1(F)$ is abelian). The case of higher genus closed surfaces F is much more complicated, the conjugacy problem in $\pi_1(F)$ should be dealt with delicately. Still Theorems 0.4, 2.11 and 4.3 are valid but we postpone the discussion of the combinatorial proofs to a future paper. In Chapter 6 we sketch the proofs based on **[H-S]**.

5. Quantization and the Hopf algebra structure of $\mathcal{S}(F \times I)$

The quantum nature of skein modules of $F \times I$ has been analyzed in detail by V. Turaev **[T-2]**, **[T-3]**, see also **[H-P-1]**. Our previous results allow us to use the more restrictive version of Turaev's definition of quantization which is closer to the original definition by V. Drinfeld **[D]**.

Definition 5.1. Let P be a Poisson algebra over a commutative ring $\bar{R}$ and let A be an algebra over $\bar{R}[h]$ which is free as an $\bar{R}[h]$-module. An $\bar{R}$-module epimorphism $q\colon A \to P$ is called a quantization of P if
(i) $q(p(h)a) = p(0)q(a)$ for all $a \in A,\ p(h) \in \bar{R}[h]$,
(ii) $ab - ba \in hq^{-1}([q(a), q(b)])$ for all $a, b \in A$.

Recall that a Poisson algebra is a commutative $\bar{R}$-algebra equipped with a Lie bracket which satisfies the following Leibniz rule:

$$[a \cdot b, c] = a[b, c] + [a, c]b\,.$$

Theorem 5.2. *The skein module $\mathcal{S}(F \times I)$ of Theorem 2.2 is a quantization of the anti-homotopy skein module of* **[H-P-2]**. *More exactly, let $\bar{R} = Z[v^{\mp 1}],\ R = \bar{R}[h] = Z[v^{\mp 1}, h]$. Let anti-homotopy skein module $\mathcal{H}^a\mathcal{S}(M)$ be defined as a quotient of $\bar{R}\mathcal{L}/S_a$ where S_a is a submodule generated by the "anti-homotopy skein expressions"*

$$v^{-1}L_{+} - vL_{-} = \begin{cases} 0 & \textit{for a mixed crossing} \\ L_{0} & \textit{for a self-crossing} \end{cases}.$$

$q\colon \mathcal{S}(M) \to \mathcal{H}^a\mathcal{S}(M)$ is defined to be the identity of links. The Lie bracket on $\mathcal{H}^a\mathcal{S}(F \times I)$ is defined to be $[D_1, D_2]'$ for $h = 0$; see the formula before Proposition 2.4 (one can also use a formula similar to that of Lemma 2.5). Checking that $\mathcal{H}^a\mathcal{S}(F \times I)$ with $[\ ,\ ]'$ is a Poisson algebra is similar to the proof of Proposition 2.4 (in fact, Proposition 2.4 is more delicate than Theorem 5.2). □

V. Turaev has shown **[T-3; §9]** that skein modules of $F \times I$ have a structure of a bialgebra and conjectured that together with the antipodal map related to the mirror image operation, skein modules are Hopf algebras. We prove this conjecture in **[P-4]**. An independent proof has been also obtained by P. Cotta-Ramusino and M. Rinaldi. The construction of the co-multiplication is due to Turaev and was partially inspired by the composition product of polynomial invariants of links in S^3 introduced by F. Jaeger **[Ja]**.

We follow the construction of **[T-3]**. We treat $\mathcal{S}(F \times I)$ as a $Z[h]$-module. Assume that F is a planar surface with fixed embedding into $\mathbb{R}^2$. Then the $Z[h]$-module homomorphism $\nabla\colon \mathcal{S}(F \times I) \to \mathcal{S}(F \times I) \otimes_{Z[h]} \mathcal{S}(F \times I)$ is specified by the following 3 properties.
(i) $\nabla(\emptyset) = \emptyset \otimes \emptyset$;
(ii) $\nabla(va) = (v \otimes v)\nabla a$ for any $a \in \mathcal{S}(F \times I)$ (as usual $(a \otimes b)\cdot(c \otimes d) = (a\,c) \otimes (b\,d)$);
(iii) for any oriented link diagram D,

$$\nabla(D) = \sum_{f \in \ell b\ell(D)} \langle D \mid f\rangle (v^{-\tau_1(f)} D_{f,1}) \otimes v^{-\tau_2(f)} D_{f,2})\,. \tag{5.3}$$

Now we will explain formula 5.3. Let $\mathrm{Tait}(D)$ (or $\mathrm{writhe}(D)$) be the sum over all crossings of D of the sign of crossings ($\mathrm{Tait}\,D = \sum_{p_i} sgn\,p_i$). Let $\tau(D)$ be the rotational angle (or Whitney degree) of D, i.e., $\tau(D)$ is the sum of the signs of the Seifert circles of D where the sign of such a circle is 1 if it is oriented counterclockwise and -1 otherwise (i.e., $\tau(\circlearrowleft) = 1$, $\tau(\circlearrowright) = -1$). $Com(D)$ denotes the number of components of the link represented by D. D is a 4-valent graph with signed vertices (the sign corresponds to the sign of the crossing) (see Fig. 5.1).

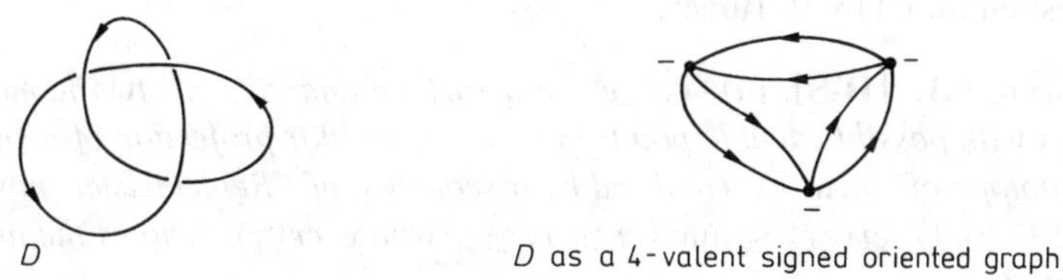

Figure 5.1

Let $Edg(D)$ denote the set of edges of the graph D. By a 2-labelling of D we mean a function $f\colon Edg(D) \to \{1,2\}$ such that around vertex v the labellings of Fig. 5.2 are allowed.

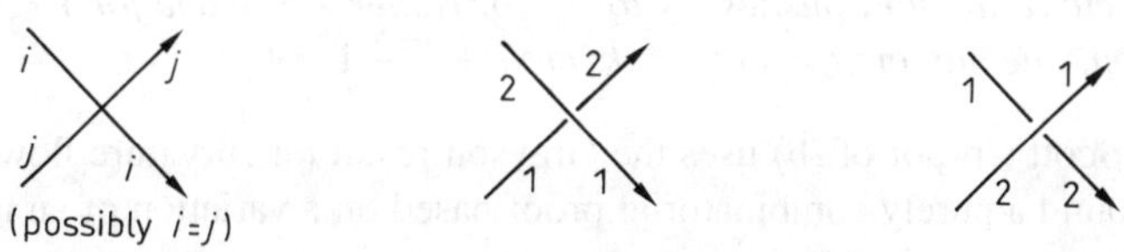

Figure 5.2

The set of 2-labellings of D is denoted by $lbl(D)$. The edges of D with label i form an oriented link diagram denoted by $D_{f,i}$.

Put $\|f\| = com\,D - \sum_{i=1}^{2} com\,D_{f,i}$. The vertices of D which are neither in $D_{f,1}$ nor $D_{f,2}$ are called f-smoothing vertices of D. Let $|f|_-$ (resp. $|f|_+$) denote the number of negative (resp. positive) f-smoothing vertices of D. Let $|f| = |f|_- + |f|_+$. Define $\langle D \mid f\rangle = (-1)^{|f|_-} h^{(|f|+\|f\|)/2}$. Finally let

$$\tau_1(F) = \mathrm{Tait}\,(D_{f,1}) - \mathrm{Tait}\,(D) + \tau(D_{f,2})$$
$$\tau_2(f) = \mathrm{Tait}\,(D_{f,2}) - \mathrm{Tait}\,(D) - \tau(D_{f,1}).$$

This completes the definition of the co-multiplication ∇. It is checked in **[T-3]** that $\nabla(D_1 \cdot D_2) = \nabla(D_1)\cdot\nabla(D_2)$ and that $\mathcal{S}(F\times I), \nabla$ is a bialgebra with co-unit $\varepsilon\colon \mathcal{S}(F\times I) \to Z[h]$ given by $\varepsilon(D) = 0$ for any nonempty diagram D, $\varepsilon(v) = 1$ and $\varepsilon(\emptyset) = 1$. It is checked in **[P-4]** that the bialgebra $\mathcal{S}(F\times I)$ together with the antipode $S\colon \mathcal{S}(F\times I) \to \mathcal{S}(F\times I)$ given by $S(v) = v^{-1}$, $S(D) = (-1)^{com\,D}\,\overline{D}$, where $\overline{D}$ dentoes the mirror image of D, is a Hopf algebra.

6. The case of a closed surface F

The recent result of J. Hass and P. Scott **[H-S]** allows a quick proof of the results concerning the structure of the algebra of descending diagrams on $F \times I$ for any (including closed) oriented surface, in particular to prove Lemma 1.7 for any surface (allowing hyperbolic structure). This in turn allows to extend Theorems 2.11, 0.4 and 4.3 to any surface (as noted before, for F a torus the proof of Theorems 2.11, 0.4 and 4.3 is even shorter than for surfaces with boundary). The Hass-Scott theorem answers affirmatively a question raised by V. Turaev.

Theorem 6.1. [H-S]. *(a) Let F be a surface and γ_0, γ_1 two homotopic immersed curves with possibly double points (so γ_i is a regular projection of a knot) then there is a homotopy of γ_0 to γ_1 (realized by a sequence of "Reidemeister moves") so that for any $t \in \langle 0,1 \rangle$ $cr(\gamma_t) \leq \max(cr\gamma_0, cr\gamma_1)$, where $cr(\gamma)$ denotes the number of double points of γ.*

(b) Let F be a surface with hyperbolic structure, γ_1 a geodesic in F or an infinitesimally small circle bounding a disc. Then for any curve γ_0 homotopic to γ_1 with $cr\gamma_0 = cr\gamma_1$ the curvature flow pushes γ_0 to γ_1. This homotopy can be thought of as a finite sequence of "Reidemeister moves" and for any γ_t (outside singularities), $cr(\gamma_t) \leq cr(\gamma_0)$. If γ_1 is a geodesic but γ_0 is homotopic to $k \cdot \gamma_1$ (i.e., γ_0 is not a primitive element of $\pi_1(F)$) and γ_0 has the minimal number of crossings in its homotopy class, then the curvature flow pushes γ_0 to γ_1 (or rather $k\gamma_1$), and for $t < 1$ and γ_t in regular position, one has $cr\gamma_t = cr\gamma_0 = k^2 cr\gamma_1 + k - 1$. □

Hass' and Scott's proof of (b) uses the Grayson result on curvature flows **[Gr]**. For (a) they also found a purely combinatorial proof based on a variation of an idea of G. D. Birkhoff **[B]**.

Theorem 0.4 has now the form:

Theorem 6.2. *Let F be any compact surface and $\leq$ a linear ordering of elements of $\widehat{\pi}^\circ$ as in Chapter 1 for a nonclosed surface. Then there is an R_z-module isomorphism $i: S(R_z\widehat{\pi}^\circ) \to \mathcal{S}_3(F \times I)$ such that for $w_1 \leq w_2 \leq \cdots \leq w_s$, $w_i \in \widehat{\pi}^\circ$ $i(w_1 \otimes \cdots \otimes w_s) = v^{-alg(D_i \cdot \ldots \cdot D_s)} D_1 \cdot \ldots \cdot D_s$ for some diagrams $D_1, \ldots, D_s$ where D_i represents $\psi(w_i)$ in $\widehat{\pi}^\circ$ and $alg(D_1 \cdot \ldots \cdot D_s) = \sum_{i<j} alg(D_i, D_j)$ and $alg(D_i, D_j)$ is an algebraic crossing number between D_i and D_j.*

The assumption that F is compact is not essential because for any F, Reidemeister moves suffices to realize an ambient isotopy of links in $F \times I$ (compare **[Hu**, Corollary 6.3**]**) and the method for a compact F with boundary extends to any nonclosed surface.

The computation of $\mathcal{S}(F \times I)$ is the first step of computing it for more general 3-manifolds. The next step would be the computation for lens spaces using their genus 1 Heegaard decomposition (compare **[H-P-3]**). A different possible generalization of the $F \times I$ case is to project a 3-manifold into a nice spine and to consider diagrams on the spine. I have been informed by D. Rolfsen that a similar idea has been considered by O. Ya Viro. One can also consider a simplification of $\mathcal{S}_3(M)$ which uses only diagrams ✕

and  (the second skein module $\mathcal{S}_2(M)$; **[P-1]**). The computation is then fairly easy but the consequences are still interesting (see **[P-3]**).

Let us finish with the following conjecture.

Conjecture 6.3. $\mathcal{S}_3(M)$ is a torsion free $Z[v^{\mp 1}, z^{\mp 1}]$-module if and only if M does not contain a nonseparating embedded S^2.

References

[Ab] E. Abe, Hopf Algebras, Cambridge University Press, 1977.

[A] J. W. Alexander, Topological invariants of knots and links, Trans. Amer. Math. Soc. 30 (1928), 275–306.

[B-M] R. Ball, R. L. Metha, Sequence of invariants for knots and links, J. Physique 42 (1981), 1193–1199.

[B] G. D. Birkhoff, Dynamical systems with two degrees of freedom, Trans. Amer. Math. Soc. 18 (1917), 199–300.

[Bi] G. Birkhoff, Representability of Lie algebras and Lie groups by matrices, Ann. of Math (2) 38, (1937), 526–532.

[C] J. H. Conway, An enumeration of knots and links, Computational Problems in Abstract Algebra (ed. J. Leech, Pergamon Press, 1969), 329–358.

[C-R] P. Cotta-Ramusino, M. Rinaldi, On the algebraic structure of link-diagrams on a 2-dimensional surface, preprint, 1990.

[D] V. G. Drinfield, Quantum groups, Proc. Internat. Congress Math. Berkeley 1986 (Amer. Math. Soc., Providence, RI, 1987), 798–820.

[FYHLMO] P. Freyd, D. Yetter, J. Hoste, W. B. R. Lickorish, K. Millett, A. Ocneanu, A new polynomial invariant of links, Bull. Amer. Math. Soc. 12 (2), (1985), 239–246.

[G] C. A. Giller, A family of links and the Conway calculus, Trans. Amer. Math. Soc. 270 (1982), 75–109.

[Go] W. Goldman, Invariant functions of Lie groups and Hamiltonian flows of surface group representations, Invent. Math. 85 (1986), 263–302.

[Gr] M. Grayson, to appear in Annals of Math.

[H-S] J. Hass, P. Scott, Flows and intersections of curves on surfaces, Lecture by J. Hass at Research Seminar in Low Dimensional Topology at Ohio State Univ., April 10, 1990.

[H-W] J. Hass, B. Wajnryb, Invariants of oriented knots and links in S^3, preprint 1986.

[H] J. Hoste, A polynomial invariant of knots and links, Pacific J. Math. 124 (1986), 295–320.

[H-K] J. Hoste, M. Kidwell, Dichromatic link invariants, Trans. Amer. Math. Soc., 321, (1990), 197–229.

[H-P-1] J. Hoste, J. H. Przytycki, Homotopy skein modules of orientable 3-manifolds, Math. Proc. Camb. Phil. Soc. 108 (1990), 475–488.

[H-P-2] J. Hoste, J. H. Przytycki, Anti-homotopy skein modules of 3-manfolds, in preparation (compare Abstracts of AMS 10 (2) 1989, p. 165, 848–57–47).

[H-P-3] J. Hoste, J. H. Przytycki, The $(2, \infty)$-skein module of lens spaces: a generalization of the Jones polynomial, preprint (see Abstracts of AMS 11 (3) 1990, 291).

[Hu] J. P. Hudson, Piecewise Linear Topology, Benjamin, Inc., N.Y. 1969.

[J] V. F. R. Jones, A polynomial invariant for knots via von Neumann algebras, Bull. Amer. Math. Soc. 12 (1985), 103–111.

[Ja] F. Jaeger, Composition products and models for the Homfly polynomial, preprint, 1988.

[L] W. B. R. Lickorish, Linear skein theory and link polynomials, Topology and its Applications 27 (1987), 265–274.

[L-M] W. B. R. Lickorish and K. C. Millett, A polynomial invariant of oriented links, Topology 26 (1987), 107–141.

[Po] H. Poincaré, Sur les groupes continus, Cambridge Philos. Trans. 18 (1899), 220–255.

[P-1] J. H. Przytycki, Skein modules of 3-manifolds, Bull. Polon. Acad. Sci.: Math. 39 (1991).

[P-2] J. H. Przytycki, Survey on recent invariants in classical knot theory, Warsaw University preprints, 6, 8, 9 (1986).

[P-3] J. H. Przytycki, The second skein module of 3-manifolds and Lickorish-Millett formula on Kauffman polynomial, preprint 1990.

[P-4] J. H. Przytycki, Quantum group of links in a handlebody, Contemporary Math., to appear.

[P-T-1] J. H. Przytycki, P. Traczyk, Invariants of links of Conway type, Kobe J. Math. 4 (1987), 115–139.

[P-T-2] J. H. Przytycki, P. Traczyk, Conway algebras and skein equivalence of links, Proc. Amer. Math. Soc. 100 (4) (1987), 744–748.

[T-1] V. G. Turaev, The Conway and Kauffman modules of the solid torus, Zap. Nauchn. Sem. LOMI 167 (1988), 79–89. English translation: J. Soviet Math.

[T-2] V. G. Turaev, Algebras of loops on surfaces, algebra of knots, and quantization, In "Braid Group, Knot Theory and Statistical Mechanics" Advanced Series in Mathematical Physics, Vol. 9 (1989), 59–95.

[T-3] V. G. Turaev, Skein quantization of Poisson algebras of loops on surfaces, preprint, 1989.

[T-4] V. G. Tureav, Problem 10 in: Problem list from a workshop on low dimensional topology, Luming 1989.

[W] E. Witt, Treue Darstellung Liescher Ringe, J. reine angew. Math. 177 (1937), 152–160.

Department of Mathematics, University of California at Riverside, Riverside, CA 92521

Totally Tangential Links of Intersection of Complex Plane Curves with Round Spheres

Lee Rudolph[1)]

§0. Introduction

How can a piece of plane curve intersect a round sphere?

More precisely, a ***piece of plane curve*** $\Gamma \subset \mathbb{C}^2$ is a 2-manifold-with-boundary, smoothly embedded as a closed set, such that $\operatorname{Int}\Gamma$ is $\mathbb{C}$-analytic. If $\partial\Gamma$ is disjoint from the round sphere $S^3 := S^3((z_0, w_0); r)$, then $L := S^3 \cap \Gamma$ is a compact $\mathbb{R}$-analytic variety of dimension ≤ 1. The question is: What is the topology of $\mathcal{L} := (S^3, L)$?

Consideration of the "generic" case leads, depending on interpretation, to two readily proved statements. Let $\mathcal{S}(L)$ be the $\mathbb{R}$-analytic variety $\{(z,w) \in L : T_{(z,w)}(\Gamma) \subset T_{(z,w)}(S^3)\}$. Then: (A) for all Γ, for all (z_0, w_0), for almost all r with $S^3 \cap \partial\Gamma = \emptyset$, $\mathcal{S}(L) = \emptyset$; (B) for all Γ, for almost all (z_0, w_0), for all r with $S^3 \cap \partial\Gamma = \emptyset$, $\mathcal{S}(L)$ is finite.

In case (A), $\mathcal{L}$ is a smooth link, naturally oriented by the complex orientation of Γ; call $\mathcal{L}$ a ***transverse*** $\mathbb{C}$***-link***. The work of classifying transverse $\mathbb{C}$-links is not complete: known examples include all links of singularities, all links-at-infinity, and – including both these classes – all quasipositive links, where (cf. **[10]**–**[18]**) a link is ***quasipositive*** if it can be realized as the closure $\widehat{\beta}$ of a braid β which is a product of conjugates of the standard generators σ_i of B_n; it is reasonable to conjecture (cf. **[12]**) that the transverse $\mathbb{C}$-links are precisely the quasipositive links.

In case (B), if $\mathcal{S}(L) \neq \emptyset$, then L is not a 1-manifold (it is a 1-complex, but has isolated vertices, vertices of even degree ≥ 4, or both); its 1-dimensional strata are naturally oriented. I am not aware of any work on non-manifold L.

But cases (A) and (B) are not exhaustive: they leave open a possibility, that (C): for some Γ, for some (z_0, w_0), for some r with $S^3 \cap \partial\Gamma = \emptyset$, $\dim \mathcal{S}(L) = 1$.

In fact, (C) can happen, and in a nice way. Call $\mathcal{L}$ a ***totally tangential*** $\mathbb{C}$***-link*** if: (1) $L = \mathcal{S}(L)$, (2) L is a 1-manifold, (3) the tangency of Γ to S^3 at each point of L is "simple"; equivalently, if L is a 1-dimensional non-degenerate critical manifold of $\rho_\Gamma := \rho \mid \Gamma$, $\rho(z,w) := |z - z_0|^2 + |w - w_0|^2$. (Note that L has no natural orientation.)

1) Research partially supported by NSF grant DMS-8801959.

Example. Let m and n be relatively prime positive integers. Let Γ be the affine algebraic curve $\{(z,w) : z^m w^n = 1\}$. Let $(z_0, w_0) = (0,0)$. Then, from

$$\frac{\partial}{\partial\tau}\rho_\Gamma(\tau^m, \tau^{-n}) = 0 \Leftrightarrow |\tau|^{2(m+n)} = \frac{n}{m} \Leftrightarrow \rho_\Gamma(\tau^m, \tau^{-n})$$
$$= \left(\frac{n}{m}\right)^{\frac{m}{m+n}} + \left(\frac{m}{n}\right)^{\frac{n}{m+n}} =: r^2,$$

it is immediate that L satisfies (1) and (2), nor is (3) difficult to verify. In this case, the totally tangential $\mathbb{C}$-link $\mathcal{S}$ is a negative torus knot $\mathcal{O}\{m, -n\}$.

Since torus knots are rather special, on the strength of this example alone it might be thought that totally tangential $\mathbb{C}$-links are rare. The contrary is true.

Theorem. *Every unoriented link is isotopic to a totally tangential $\mathbb{C}$-link.*

The proof (given in §1) is to show that a totally tangential $\mathbb{C}$-link is the same as an $\mathbb{R}$-analytic Legendrian link (in S^3 with its standard contact structure) and then to appeal to practitioners of "symplectic topology" **[0]** who have proved that every unoriented link is isotopic to an $\mathbb{R}$-analytic Legendrian link **[3]**, **[5]**.

A Legendrian link has a canonical framing. In §2, I define three invariants of an unoriented link $\mathcal{L}$ which are (componentwise) suprema of sets of framings of $\mathcal{L}$, and discuss their interrelationships and their relationships with various hypotheses and conjectures in the knot theory of complex plane curves.

§1. Proof of the theorem

By definition, a smooth link $\mathcal{L} = (L, S^3)$ is ***Legendrian*** (with respect to the standard contact structure on $S^3 = S^3((z_0, w_0); r))$ if and only if, for every $(z,w) \in L$, the real tangent line to L at (z,w) is contained in the unique complex line tangent to S^3 at (z,w), namely $(z,w) + \mathbb{C}\cdot(\overline{w} - \overline{w}_0, -\overline{z} + \overline{z}_0)$.

A totally tangential $\mathbb{C}$-link is certainly $\mathbb{R}$-analytic and Legendrian. To show that, conversely, every $\mathbb{R}$-analytic Legendrian link is a totally tangential $\mathbb{C}$-link, it will suffice to show that if $\mathcal{K} = (S^3, K)$ is an $\mathbb{R}$-analytic Legendrian knot (i.e., if $K \neq \emptyset$ is connected) in $S^3 = S^3((0,0); 1)$, then $\mathcal{K}$ is a totally tangential $\mathbb{C}$-link. We show first that $K \subset \operatorname{Int}\Gamma$ for some piece of plane curve Γ, then that Γ is simply tangent to S^3 along K.

If $f(s)$ is a $\mathbb{C}$-valued $\mathbb{R}$-analytic function defined on a neighborhood of 0 in $\mathbb{R}$ by a convergent power series $\sum a_n s^n$, then the series $\sum a_n(s+it)^n$ converges on a neighborhood of 0 in $\mathbb{C}$ to a $\mathbb{C}$-analytic extension of f; two $\mathbb{C}$-analytic extensions of f agree on their common domain. Let $(p,q) : S^1 \to S^3 \subset \mathbb{C}^2$ be an $\mathbb{R}$-analytic map on the unit circle $S^1 \subset \mathbb{C}$. The preceding remarks on existence and (germwise) uniqueness of $\mathbb{C}$-analytic extensions, applied in local $\mathbb{C}$-analytic coordinates near enough points of S^1, show that (p,q) has a $\mathbb{C}$-analytic extension (which we also denote by (p,q)) on a neighborhood of S^1 in $\mathbb{C}$. Let $\mathcal{K}$ be an $\mathbb{R}$-analytic knot, $(p,q) \mid S^1$ an embedding onto K and $A \subset \mathbb{C}$ a sufficiently small closed annulus with $S^1 \subset \operatorname{Int} A$. Then $(p,q) \mid A$ is an embedding, and $\Gamma := (p,q)(A)$ is a piece of plane curve with $K \subset \operatorname{Int}\Gamma$. If $\mathcal{K}$ is also Legendrian, then Γ is tangent to S^3 at each point of K.

It remains to show that K is a non-degenerate critical manifold of ρ_Γ. This is a local condition at each point of K, and can be verified by a calculation in local coordinates, or as follows. Apply the Laplacian Δ to $\rho_\Gamma \circ (p,q) = |p|^2 + |q|^2 \colon A \to \mathbb{R}$. Using the $\mathbb{C}$-analyticity of (p,q) on A, we calculate $\Delta(|p|^2 + |q|^2) = 4\|\nabla(p,q)\|^2 > 0$; since along $K \subset S^3$ the second partial derivative of ρ_Γ in the tangential direction is strictly negative, the second partial derivative in the normal direction (on Γ) must be strictly positive, and this is precisely the condition that K be a non-degenerate critical manifold of index 0 for ρ_Γ, so $\mathcal{K}$ is indeed a totally tangential $\mathbb{C}$-link.

To complete the proof of the Theorem we must show that any unoriented link $\mathcal{L}$ is isotopic to an $\mathbb{R}$-analytic Legendrian link $\mathcal{L}'$. But this is well-known to symplectic topologists, **[0]**, **[3]**, **[5]**. One may construct a smooth Legendrian link $\mathcal{L}''$ isotopic to $\mathcal{L}$ "by hand". There exist $\mathbb{R}$-analytic links $\mathcal{L}'''$ which approximate $\mathcal{L}''$ as closely as desired. A particular such $\mathcal{L}'''$ may not be Legendrian with respect to the standard contact structure on S^3, but it is Legendrian with respect to a deformation of that structure, and the required deformation can be made as small as desired by making the approximation of $\mathcal{L}''$ by $\mathcal{L}'''$ sufficiently close; then there is an $\mathbb{R}$-analytic autodiffeomorphism of the sphere that carries the deformed structure onto the standard structure and $\mathcal{L}'''$ onto the required link $\mathcal{L}'$. (I thank Y. Eliashberg for this proof.)

Question. Call a transverse or totally tangential $\mathbb{C}$-link ***algebraic*** (resp. ***entire***) if it is cut out of a round sphere, not simply by a piece of plane curve, but by a whole curve $f^{-1}(0)$, where f is a polynomial (resp., an entire function). It is easy to show (using Taylor series and the rudiments of Stein theory) that every transverse $\mathbb{C}$-link is isotopic to an algebraic, and *a fortiori* entire, transverse $\mathbb{C}$-link. The example in §0 shows that any negative torus knot is isotopic to an algebraic totally tangential $\mathbb{C}$-link. Are other knot types realizable as algebraic, or as entire, totally tangential $\mathbb{C}$-links?

§2. Three invariants of unoriented links defined in terms of framings

A ***framing*** of a link $\mathcal{L} = (S^3, L)$ is a function f from the set of components of L to $\mathbb{Z}$; we identify a framing of a knot with the integer which is its value. Write $f \preceq g$ if and only if $f(K) \leq g(K)$ for every component K of L.

A field ν of nonzero normal vectors of L determines a framing f_ν: let the component K flow along ν onto a disjoint curve K^+, and set $f_\nu(K) := \ell k(K, K^+)$. A framed link $(\mathcal{L}, f)$ determines (up to isotopy) a Seifert surface $A(\mathcal{L}, f) \subset S^3$; each component of $A(\mathcal{L}, f)$ is an annulus which contains one component K of L, and the linking number of the two boundary circles of that annulus is $-f(K)$.

The ***canonical framing*** $f_\mathcal{L}$ of a Legendrian link $\mathcal{L}$ is $f_{\nu(\mathcal{L})}$, where $\frac{1}{i}\nu(\mathcal{L})$ is the unit tangent vector. If $\mathcal{L}$ is $\mathbb{R}$-analytic, and Γ is a piece of plane curve simply tangent to S^3 along $L = S^3 \cap \Gamma$, then the transverse $\mathbb{C}$-link obtained by intersecting Γ and a slightly larger round sphere concentric with S^3 is of type $(S^3, \partial A(\mathcal{L}, f_\mathcal{L}))$.

Example (continued). For the Legendrian torus knot $\mathcal{O}\{m,-n\}$ constructed in the introduction, all critical points of ρ_Γ lie on L, so $(S^3, \partial A(\mathcal{O}\{m,-n\},\ f_{\mathcal{O}\{m,-n\}}))$ is the link-at-infinity of Γ; by **[8]** or an *ad hoc* argument; $f_{\mathcal{O}\{m,-n\}} = -mn$.

Let $M(\mathcal{L})$ be the $\preceq$-supremum of $\{f\colon (\mathcal{L}, f) \simeq (\mathcal{L}', f_{\mathcal{L}'}),\ \mathcal{L}'$ *Legendrian* $\}$.

Proposition 1. $M(\mathcal{L})$ *is a framing.*

Proof. If $\mathcal{K} = (S^3, K)$ is a knot, then $-\infty = \sup\emptyset < M(\mathcal{K})$ by the references quoted in §1, and $M(\mathcal{K}) < \infty$ by **[2**, Theorem 12**]**, so $M(\mathcal{K})$ is a framing. Since a link $\mathcal{L}$ is Legendrian if and only if its component knots are, and the restriction of the canonical framing is canonical, the general case follows. □

A framed link $(\mathcal{L}, f)$ is ***embedded*** on a Seifert surface $F \subset S^3$ if $A(\mathcal{L}, f) \subset F$; the embedding is ***full*** if no component of $F - A(\mathcal{L}, f)$ is contractible (if no split summand of $\mathcal{L}$ is an unknot $\mathcal{O}$, then any embedding is full). Let $q(\mathcal{L})$ be the $\preceq$-supremum of $\{f$: *there is a full embedding of* $(\mathcal{L}, f)$ *on some fiber surface of* $\mathcal{O}\{n,n\}\}$. (The fibered link $\mathcal{O}\{n,n\}$ has many descriptions: a torus link of type (n,n); n consistently oriented fibers of a Hopf fibration $S^3 \to S^2$; the link of the singularity at $(0,0)$ of the plane curve $\{(z,w)\colon z^n - w^n = 0\}$; the link-at-infinity of the same curve. The latter two descriptions show that one fiber surface is $\{(z,w) \in S^3\colon z^n - w^n \geq 0\}$.)

Proposition 2. $q(\mathcal{L})$ *is a framing.*

Proof. According to **[16]**, $q(\mathcal{L})$ is also the $\preceq$-supremum of $\{f\colon A(\mathcal{L}, f)$ *is a quasi-positive Seifert surface* $\}$. A construction **[13]** shows this set of framings is not empty; a calculation with link polynomials **[15]** shows the supremum is finite. □

Example (continued). By **[1]**, $M(\mathcal{O}\{m,-n\}) \leq mn - m - n$; the canonical framing of the Legendrian $\mathcal{O}\{m,n\}$ found earlier is $-mn$, so $-mn \leq M(\mathcal{O}\{m,-n\})$. Unless $m = n = 1$, there is a lot of daylight between these bounds; in that case, we find $M(\mathcal{O}\{1,-1\}) = M(\mathcal{O}) = -1$.

The constructions in **[17]** show that $-mn \leq q(\mathcal{O}\{m,-n\})$. (For instance, to see that $A(q(\mathcal{O}\{4,-5\},-20)$ is quasipositive, let b be the quasipositive embedded band representation $(\sigma_{1,5}, \sigma_{2,6}, \sigma_{3,7}, \sigma_{4,8}, \sigma_{5,9}, \sigma_{1,6}, \sigma_{2,7}, \sigma_{3,8}, \sigma_{4,9})$ in B_9. The quasipositive braided Seifert surface $S(b)$ is certainly some annulus $A(q(\mathcal{O}\{m,-n\}, f)$, for $S(b)$ has a spine which is a knot with the "fence diagram" illustrated in Figure 1; to convert a fence diagram to a link diagram of the usual sort, impose the convention that horizontal segments always lie over vertical ones. The framing f is the writhe of the knot diagram minus the number of "northeast corners", here, $-15 - 5 = -20$.)

Let $P_\mathcal{K}(\ell, m) \in \mathbb{Z}[\ell^{\pm 1}, m^{\pm 1}]$ (resp., $F_\mathcal{K}(a,x) \in \mathbb{Z}[a^{\pm 1}, x^{\pm 1}]$) be the ***oriented*** **[7]**, or FLYPMOTH (resp., ***semi-oriented*** **[7]**, or Kauffman) polynomial of the knot $\mathcal{K}$, **[4]**, **[6]**, **[9]**; then $q(\mathcal{K}) \leq -1 - \deg_a(F_\mathcal{K} \bmod 2)$ and $q(\mathcal{K}) \leq \frac{1}{2}\deg_\ell(-1 + m(\ell + \ell^{-1})P_{\partial A(\mathcal{K},0)})$, according to **[15]**. From these we can deduce, e.g., that $q(\mathcal{O}) \leq -1$ (and so $q(\mathcal{O}) = -1$), and similarly that $q(\mathcal{O}\{2,-n\}) \leq -2n$ (and so $q(\mathcal{O}\{2,-n\}) = -2n$) for $n = 2k+1 > 1$. However, the polynomial invariants, and the corresponding upper bounds for q, remain elusive for other negative torus knots.

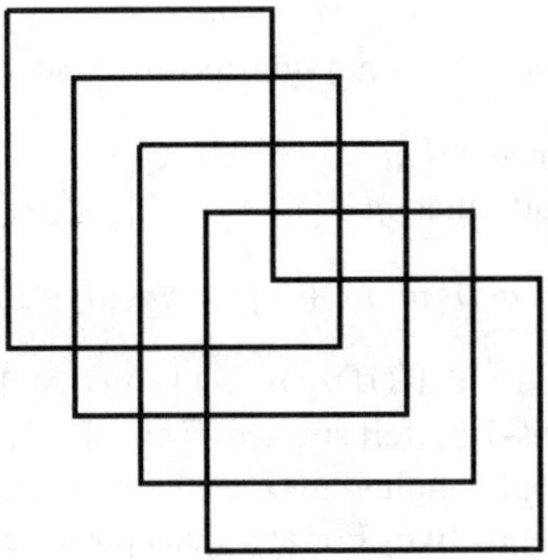

Figure 1

Let $q^*(\mathcal{L})$ be the $\preceq$-supremum of $\{f\colon (S^3, \partial A(\mathcal{L}, f))$ *is isotopic to a transverse* $\mathbb{C}$-*link*$\}$. Note that q^* might conceivably not be a framing (it might take the value ∞). We have $M(\mathcal{L}) \preceq q^*(\mathcal{L})$ (by the Theorem) and $q(\mathcal{L}) \preceq q^*(\mathcal{L})$ (by **[10]** and **[11]**).

Various related hypotheses in the knot theory of complex plane curves have an impact on the interrelationships of the invariants $M(\mathcal{L})$, $q(\mathcal{L})$, and $q^*(\mathcal{L})$.

TC. *If S is an oriented, closed, smooth 2-submanifold of $\mathbb{CP}^2$ representing $d \geq 1$ times $[\mathbb{CP}^1] \in H_2(\mathbb{CP}^2; \mathbb{Z})$, then the genus of S is at least $\frac{1}{2}(d-1)(d-2)$ (which is the genus of any smooth projective complex algebraic curve homologous to S).*

MQ. *The unknotting number of a torus knot $\mathcal{O}\{m,n\}$ is $\frac{1}{2}(|m|-1)(|n|-1)$.*

QP1. *Every transverse $\mathbb{C}$-link is quasipositive.*

QP2. *Every amphicheiral quasipositive knot is a slice knot.*

QP3. *If $\mathcal{L}$ is a quasipositive link and its mirror image* $\operatorname{Mir}\mathcal{L}$ *is also quasipositive, then $\mathcal{L} = \mathcal{O}\{0,n\}$ is an unlink.*

The conjecture "TC is true" is often attributed to Thom. The question "Is MQ true?" is a special case of "the question of Milnor on the unknotting number of algebraic knots" **[2]** (and an affirmative answer in this special case would imply the widely conjectured affirmative answer in the general case). The questions "Is QP1 true?" and "Is QP3 true?" are essentially posed in **[12]** and **[19]**, respectively; I conjecture that both answers are "yes". By a proof-reading oversight, **[19]** asserts that QP2 has been shown to be true; in fact, I conjecture that QP2 is true, but I do not know how to prove it. It is known (cf. **[1]**, **[2]**, **[12]**) that TC implies both MQ and QP2. Of course QP3 implies QP2. By a remark of H. Morton **[19]**, if no knot but the unknot has oriented polynomial equal to 1, then QP3 is true.

Clearly, QP1 implies that $q^*(\mathcal{L}) = q(\mathcal{L})$ for all $\mathcal{L}$. In particular, QP1 implies that $q^*(\mathcal{L})$ is a framing, and that $M(\mathcal{L}) \preceq q(\mathcal{L})$, for all $\mathcal{L}$. I would conjecture both these statements, and even that $M(\mathcal{L}) = q(\mathcal{L}) = q^*(\mathcal{L})$ for all $\mathcal{L}$.

Proposition 4. *Assume QP3. Then for any knot $\mathcal{K}$, $q(\mathcal{K}) + q(\operatorname{Mir}\mathcal{K}) < 0$.*

Proof. The link $\mathcal{L} := (S^3, \partial A(\mathcal{K}, q(\mathcal{K})))$ is quasipositive, so if we assume QP3, then $\operatorname{Mir}\mathcal{L} = (S^3, \partial A(\operatorname{Mir}\mathcal{K}, -q(\mathcal{K})))$ is not quasipositive, so $q(\operatorname{Mir}\mathcal{K}) < -q(\mathcal{K})$. □

Example (continued). Since $\mathcal{O}\{m,n\} = \operatorname{Mir}\mathcal{O}\{m,-n\}$, if we assume QP3, then $q(\mathcal{O}\{m,n\}) < mn$ and in particular $q(\mathcal{O}\{m,m+1\}) < m(m+1)$.

Scholium. *Assume TC. Then* $q(\mathcal{O}\{m,m+1\}) < m(m+1)$.

Sketch Proof. Suppose that $A := A(\mathcal{O}\{m,m+1\}, m(m+1))$ is a quasipositive Seifert surface; then by **[18]** the genus-1 Seifert surface $B := A * A(\mathcal{O},-1)$ constructed from Q by plumbing on a positive Hopf annulus $A(\mathcal{O},-1)$ (along an arc from one component of ∂A to the other) is also quasipositive. For any quasipositive Seifert surface $F \subset S^3$, if $F' \subset D^4$ is a Seifert ribbon with $\partial F' = \partial F$ obtained by "pushing $\operatorname{Int} F$ into $\operatorname{Int} D^4$," then there exists a piece of plane curve Γ transverse to S^3 such that (D^4, F') is isotopic to $(D^4, D^4 \cap \Gamma)$, cf. **[10]** and **[11]**. Applying this with $F = B$, we see that the knot $(S^3, \partial B)$ can be realized as a transverse $\mathbb{C}$-link bounding a plane curve of genus 1, and in fact (cf. the Question at the end of §1) we can realize ∂B as a transverse intersection of $S^3 \subset \mathbb{C}^2 \subset \mathbb{CP}^2$ with a smooth projective complex algebraic curve Γ. But ∂B is the $m(m+1)$-twisted double of $\mathcal{O}\{m,m+1\}$ with a negative clasp; by an old, clever observation of Casson, Gordon, and Litherland, ∂B is a ribbon knot. Replace $D^4 \cap \Gamma$ by a ribbon disk in D^4 with boundary ∂B; the resulting smooth surface is homologous to Γ in $\mathbb{CP}^2$ and of genus $\frac{1}{2}(d-1)(d-2)-1$. □

Acknowledgements. Much of this material was presented in a seminar talk at The Ohio State University Research Semester in Topology; I thank its organizers for their invitation. Professor Y. Eliashberg kindly provided references to **[3]** and **[5]**.

References

[0] V. I. Arnol'd, First steps in symplectic topology, Uspekhi Mat. Nauk, 41:6 (1986), 3–18 (= Russian Math. Surveys, 41:6 (1986), 1–21).

[1] D. Bennequin, Entrelacements et équations de Pfaff, Astérisque, 107–8 (1983), 87-161.

[2] M. Boileau et C. Weber, Le problème de J. Milnor sur le nombre gordien des noeuds algébriques, in: "Noeuds, Tresses, et Singularités", Monographie No. 31 de L'Enseignement Mathématique (1983), 49–98.

[3] Y. Eliashberg, Topological characterization of Stein manifolds, International J. Math. (1990).

[4] P. Freyd, D. Yetter, J. Hoste, W. B. R. Lickorish, K. Millett, and A. Ocneanu, A new polynomial invariant of knots and links, Bull. Am. Math. Soc., 12 (1985), 239–246.

[5] M. Gromov, Partial Differential Relations, Springer Verlag, 1986.

[6] L. Kauffman, "On Knots", Princeton University Press, 1987.

[7] W. B. R. Lickorish, A relationship between link polynomials, Math. Proc. Camb. Phil. Soc., 100 (1986), 109–112.

[8] W. D. Neumann, Complex plane curves via their links at infinity, Invent. Math. 98 (1989), 445–489.

[9] J. Przytycki and P. Traczyk, Invariants of links of Conway type, Kobe J. Math. 4 (1987), 115–139.

[10] Lee Rudolph, Algebraic functions and closed braids, Topology 22 (1983), 191–202.

[11] Lee Rudolph, Braided surfaces and Seifert ribbons for closed braids, Comm. Math. Helv. 58 (1983), 1–37.

[12] Lee Rudolph, Constructions of quasipositive knots and links, I, in Noeuds. Tresses, et Singularités (Geneva, 1983), 233–246.

[13] Lee Rudolph, Some knot theory of complex plane curves, L'Ens. Math. 29 (1983), 185–208.

[14] Lee Rudolph, Constructions of quasipositive knots and links, II, Contemp. Math. 35 (1984), 485–491.

[15] Lee Rudolph, A congruence for link polynomials, Math. Proc. Camb. Phil. Soc. 107 (1990), 319–327.

[16] Lee Rudolph, A characterization of quasipositive Seifert surfaces, to appear in Topology (1991).

[17] Lee Rudolph, Quasipositive annuli, preprint (1990).

[18] Lee Rudolph, Quasipositive plumbing, preprint (1990).

[19] Lee Rudolph, Problem 9, Contemp. Math. 78 (1988), 562.

Clark University, Department of Mathematics and Computer Science, Worcester, Massachusetts 01610
Email: lrudolph@clarku.bitnet

Some Pictorial Remarks on Suzuki's Brunnian Graph

Martin Scharlemann[1)]

A graph in S^3 is *minimally knotted* or *Brunnian* if every proper subgraph can be isotoped into the plane, but the graph itself cannot be. In **[Su]** Suzuki introduces a family $\{\theta_n\}$ of Brunnian graphs (Figure 1).

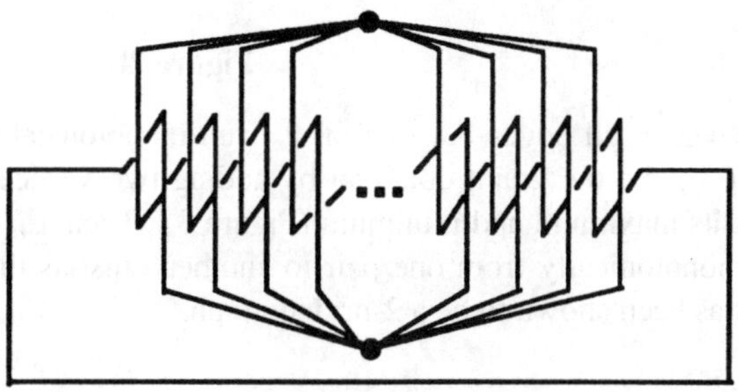

Figure 1

Suzuki proves that θ_n cannot be isotoped into the plane by showing that the $(n-1)$ st elementary ideal of $\pi_1(S^3 - \theta_n)$ differs from that of the free group F_{n-1}. The complement of a graph in the plane has free fundamental group, so θ_n cannot be isotoped into the plane. The purpose of this note is to give an alternate, perhaps more geometric proof that θ_n is not planar. I would like to thank Darren Long and Abigail Thompson for helpful conversations.

θ_n can be viewed (Figure 2a) as the lift of a certain arc in S^3 to the n-fold cyclic branched cover of the unknot (the vertical axis in Figure 2a is the branch set). As such, it can be thought of as a generaliztion of the trefoil knot θ_2. If we fully exploit the dihedral symmetry of θ_n, we may even view θ_2 as the lift of an arc in S^3 to the $2n$-fold dihedral cover of a standard θ-curve in S^3 (the T-shaped Figure in Figure 2b).

Besides dihedral symmetry, there are two other properties which θ_n shares with θ_2. First, it has unknotting number one. That is, after the change of a single crossing θ_n becomes planar. (Pick any crossing in Figure 1.) Our interest will focus on a second common property. Recall

Definition. **[Sc**, 2.1]: A *2-bridge graph* is a graph in $S^2 \times I \subset S^3$ in which there are four vertices, two in $S^2 \times \{0\}$ and two in $S^2 \times \{1\}$, and on each edge the height function $S^2 \times I \to I$ is monotonic.

1) Research supported in part by a grant from the National Science Foundation.

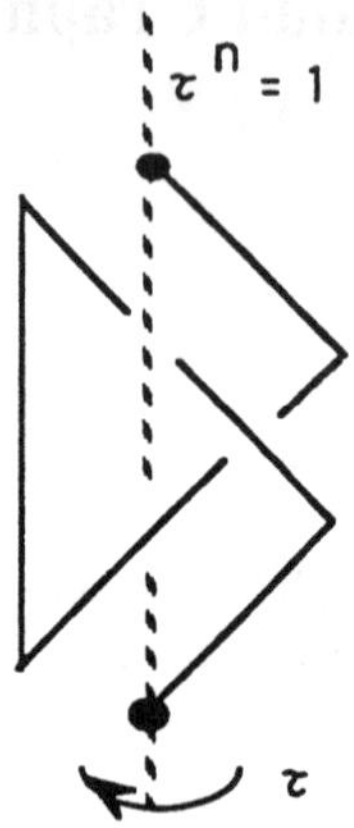

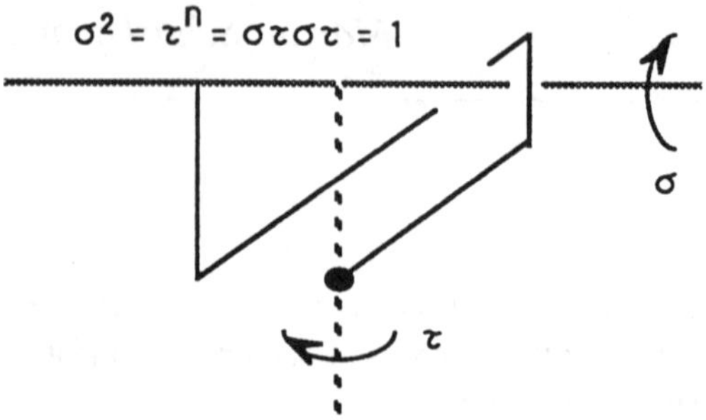

Figure 2a **Figure 2b**

After a minor change, all edges but one of θ_n run monotonically from upper left to lower right. Let Γ_n be the graph obtained by adding two vertices to the one non-monotonic edge, at its maximum and minimum (Figure 3). Then Γ_n has four vertices, and all edges run monotonically from one pair to another. Just as the trefoil knot is a 2-bridge knot, θ_n has been shown to be a 2-bridge graph.

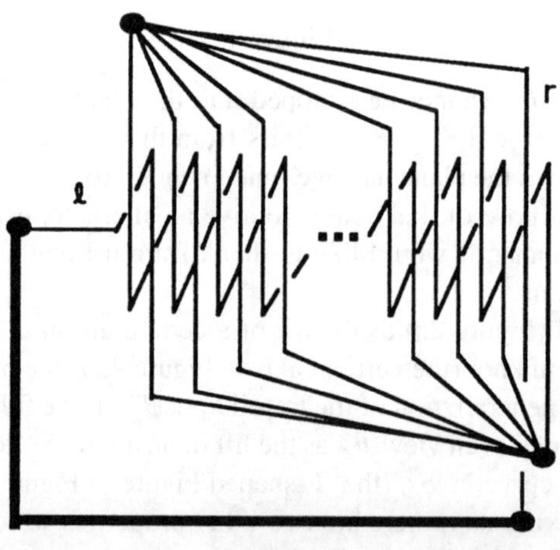

Figure 3

For 2-bridge graphs, there is a simple criterion for planarity:

Theorem [Sc, 2.4]. *A 2-bridge graph is planar if and only if there is a level-preserving isotopy to a graph lying in* $S^1 \times I \subset S^2 \times I$.

In particular Γ_n is planar only if it can be deformed preserving the braid structure (i.e., the monotonicity of the height function on the edges) into the standard planar imbedding of Figure 4. That is, only if Γ can be made planar by "combing" the braid into neighborhoods of the vertices, where it is absorbed. It seems visibly apparent that this cannot be done; we'll present an easy proof which uses the Burau representation of the braid group.

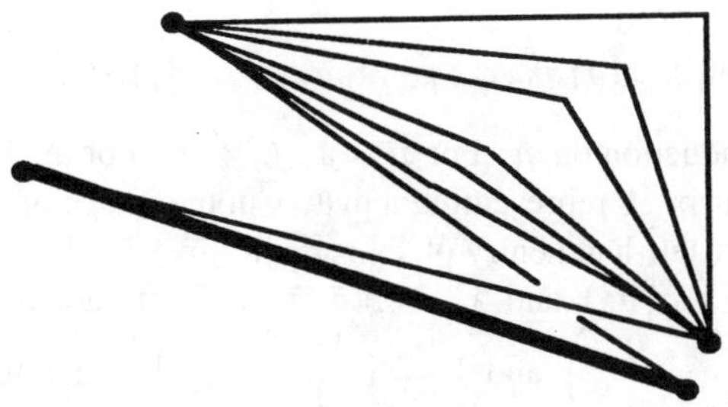

Figure 4

Γ_n has $n+2$ edges: $n-1$ edges run between the old vertices, one edge (darkened in Figure 3) runs between the new vertices, two edges (denoted ℓ and r in Figure 3) run between a new and an old vertex. To apply **[Sc**, 2.4**]** we should regard the braid as lying in $S^2 \times I$ but, if we remove the single edge connecting the two new vertices (darkened in Figure 3), we can regard the edges of Γ_n as a braid in $D^2 \times I$, i.e., as an element of the braid group B_{n+1}. As an element of B_{n+1} the braid is easy to read off from Figure 3:

$$\sigma_1^2\sigma_2^2 \ldots \sigma_{n-1}^2\sigma_n\overline{\sigma}_{n-1} \ldots \overline{\sigma}_1 .$$

Put formally (cf. **[Sc**, 2.4**]**), to show that Γ_n is not planar we need to show that the braid (in $S^2 \times I$) given by the strands of Γ_n cannot be obtained from the braid $\overline{\sigma}_1$ (Figure 4) by pre- and post-multiplication by elements of B_n acting on the last n strands and B_2 acting on the first two. Viewed as a braid in $S^2 \times I$, it is easy to trade post-multiplication by B_2 (i.e., twists in the first two strands) for pre-multiplication. Viewed as a braid in $D^2 \times I$ by removing the darkest arcs, pre-multiplication by a generator of B_2 changes the algebraic crossing number of the edges ℓ and r by 2. For both the planar graph (Figure 4) and Γ the crossing number is -1, so no multiplication by elements of B_2 need be considered.

We must, then, solve the following problem: Show that, for any elements ρ, τ in $B_n \subset B_{n+1}$ (acting on the last n strands), $\rho\overline{\sigma}_1\tau \neq \sigma_1^2\sigma_2^2 \ldots \sigma_{n-1}^2\sigma_n\overline{\sigma}_{n-1} \ldots \overline{\sigma}_1$.

Recall the unreduced Burau representation $\varphi\colon B_{n+1} \to GL_{n+1}(Z[t,t^{-1}])$ (cf. **[Bi**, p. 118**]**). For each generator σ_i of B_{n+1}, $1 \leq i \leq n$,

$$\varphi(\sigma_i) = \begin{pmatrix} I_{i-1} & 0 & 0 \\ 0 & T & 0 \\ 0 & 0 & I_{n-i} \end{pmatrix},$$

where $T = \begin{pmatrix} 1-t & t \\ 1 & 0 \end{pmatrix}$ with inverse $T^{-1} = \begin{pmatrix} 0 & 1 \\ t^{-1} & 1-t^{-1} \end{pmatrix}$.

Claim 1. For any ρ, τ in $B_n \subset B_{n+1}$, let $A = \varphi(\rho\overline{\sigma}_1\tau)$. Then $A_{11} = 0$.

Proof. We are regarding $B_n \subset B_{n+1}$ as acting on the last n strands, so the elements ρ, τ are represented by matrices of the form $\begin{pmatrix} 1 & 0 \\ 0 & * \end{pmatrix}$. Multiplication on the left by $\varphi(\rho)$ thus leaves the first column of $\varphi(\overline{\sigma}_1)$ unchanged and multiplication on the right by $\varphi(\tau)$ leaves the first row unchanged, so the combination leaves the upper left entry unchanged at 0. □

Claim 2. If C is the matrix $\varphi(\sigma_1^2\sigma_2^2 \dots \sigma_{n-1}^2\sigma_n\overline{\sigma}_{n-1} \dots \overline{\sigma}_1)$, $C_{11} = (1-t)^{2n-1}$.

Proof. The proof is by induction on n. For $n = 1$, $C = T$. For the inductive step, we will show that for *any* matrix A representing a matrix in the image of B_{n-1} (generated by $\sigma_3, \dots, \sigma_n$), the upper left-hand entry of $\varphi(\sigma_1^2\sigma_2^2 A\overline{\sigma}_2\overline{\sigma}_1)$ is $(1-t)^2 A_{33}$.

Let $\alpha = A_{33}$, $U = \varphi(\sigma_1^2\sigma_2^2)$ and $V = \varphi(\overline{\sigma}_2\overline{\sigma}_1)$. There are 3×3 matrices U' and V' so that $U = \begin{pmatrix} U' & 0 \\ 0 & I_{n-2} \end{pmatrix}$ and $V = \begin{pmatrix} V' & 0 \\ 0 & I_{n-2} \end{pmatrix}$. Therefore, the upper left 3×3 submatrix of UAV depends only on the upper left 3×3 submatrix of A, i.e., $\begin{pmatrix} I_2 & 0 \\ 0 & \alpha \end{pmatrix}$. So we may disregard A entirely, after multiplying the third row of V' by α. Furthermore, since we're only interested in $(UAV)_{11}$, we need only consider the first row of U' and the first column of V'. Since $V' = \begin{pmatrix} 1 & 0 \\ 0 & T^{-1} \end{pmatrix} \cdot \begin{pmatrix} T^{-1} & 0 \\ 0 & 1 \end{pmatrix}$, and $T^{-1} = \begin{pmatrix} 0 & 1 \\ t^{-1} & 1-t^{-1} \end{pmatrix}$, the only nonzero entry in the first column of V' is $(V')_{31} = t^{-2}$. Since $U' = \begin{pmatrix} T^2 & 0 \\ 0 & 1 \end{pmatrix} \cdot \begin{pmatrix} 1 & 0 \\ 0 & T^2 \end{pmatrix}$, $(U')_{13} = [(T^2)_{12}]^2 = [t(1-t)]^2$. Hence $(UAV)_{11} = [t(1-t)]^2 = (1-t)^2\alpha$. □

The two claims imply that $\rho\overline{\sigma}_1\tau \neq \sigma_1^2\sigma_2^2 \dots \sigma_{n-1}\sigma_n\overline{\sigma}_{n-1} \dots \overline{\sigma}_1$ so Γ_n is not planar. □

References

[Bi] J. Birman, Braids, links, and mapping class groups, Ann. Math. Studies 82, Princeton Univ. Press 1974.

[Sc] M. Scharlemann, Planar graphs, family trees, and braids, preprint, Bochum 1989.

[Su] S. Suzuki, Almost unknotted θ_n-curves in the 3-sphere, Kobe J. Math. 1 (1984), 19–22.

Department of Mathematics, University of California at Santa Barbara, Santa Barbara, CA 93106

Automatic Structure and Graphs of Groups

*Michael Shapiro**

Automatic groups were first defined by Cannon, Epstein, Holt, Paterson and Thurston (**[CEHPT]**†). As we shall see shortly, an automatic group is a group with an automatic structure, and an automatic structure is a choice of a particular kind of normal form for the elements of the group. This normal form is required to be a regular language, and thus the definition of an automatic group rests on formal language theory. However, from the very first, the discussion of automatic groups has been strongly geometric in flavor. To this end, we define the Cayley graph of a group.

We start with a finitely generated group G. We will take a free monoid $\mathcal{A}^\star$ generated by $\mathcal{A} = \{g_1, \ldots, g_k\}$. Thus $\mathcal{A}^\star = \cup_{m=0}^{\infty}\{a_1 \ldots a_m \mid a_i \in \mathcal{A}\}$. We refer to the elements of $\mathcal{A}^\star$ as ***words*** or ***strings*** on $\mathcal{A}$. Multiplication in $\mathcal{A}^\star$ is given by concatenation of these words, and the identity element of $\mathcal{A}^\star$ is the empty word which we denote by ι. The ***length*** of a word $w = a_1 \ldots a_m$ is m. This is denoted by $\ell(w)$.

We will assume that we have a monoid map of $\mathcal{A}^\star$ onto G which we denote by $w \mapsto \overline{w}$. Thus $\{\overline{g_1}, \ldots, \overline{g_k}\}$ generates G as a monoid. We call $\mathcal{A}$ together with the map into G, a ***monoid generating set*** for G.

Assume now that we have an involution on $\mathcal{A}$, denoted by $g_i \mapsto g_i^{-1}$ so that $\overline{g_i^{-1}} = \overline{g_i}^{-1}$. The ***Cayley graph*** of G with respect to $\mathcal{A}$ is a labelled directed graph whose vertices are the elements of G and whose edges are $\{(h, h', g) \mid h, h' \in G,\ g \in \mathcal{A},\ h = h'\overline{g}\}$. The edge (h, h', g) is labelled by g. We say that h and h' are ***adjacent***. We denote the Cayley graph by $\Gamma_{\mathcal{A}}(G)$, or $\Gamma(G)$ when there is no fear of confusion.

Our definition of the Cayley graph differs from the usual one by taking the labels in $\mathcal{A}$ rather than in G. The only practical effect of this comes from the fact that the map $\mathcal{A} \mapsto G$ may not be an injection. Thus, there may be multiple edges between adjacent vertices. We are always free to choose $\mathcal{A}$ so that $\mathcal{A} \mapsto G$ is an injection.

Since $\overline{\mathcal{A}}$ generates G, $\Gamma_{\mathcal{A}}(G)$ is connected. We consider each edge of $\Gamma_{\mathcal{A}}(G)$ to be isometric to the unit interval, and endow $\Gamma_{\mathcal{A}}(G)$ with the path metric. We denote this metric (and its restriction to G) by $d_{\mathcal{A}}(\cdot,\cdot)$. The restriction of $d_{\mathcal{A}}(\cdot,\cdot)$ to G is often referred to as the ***word metric***. This is because $d_{\mathcal{A}}(h, h') = \min\{\ell(w) \mid w \in \mathcal{A}^\star,\ h = h'\overline{w}\}$. We define a length function on G, $\ell_{\mathcal{A}}(\cdot)$, given by $\ell_{\mathcal{A}}(h) = d_{\mathcal{A}}(1, h) = \min\{w \mid w \in \mathcal{A}^\star, \overline{w} = h\}$. A word w is called ***geodesic*** if $\ell(w) = \ell_{\mathcal{A}}(\overline{w})$ Notice that each word w labels a unique edge path starting at 1 and ending at $\overline{w}$. Thus we may consider w a map of the interval $[0, \ell(w)]$ into $\Gamma_{\mathcal{A}}(G)$. This map is an isometry if w is geodesic. In any case, we extend this map to the non-negative reals by setting $w(t) = \overline{w}$ for $t > \ell(w)$.

---

* Partially supported by NSF grants DMS #850-5550 and DMS #900-7959

† This has gone through numerous revisions and is currently **[ECHLPT]**.

Given $\mathcal{A}$, a ***language*** over $\mathcal{A}$ is a subset of $\mathcal{A}^\star$. Given two languages $L, M \subset \mathcal{A}^\star$, we may perform the standard Boolean operations to find languages $L \cup M$, $L \cap M$ and $\mathcal{A}^\star \setminus L$. In addition, the multiplication operator in $\mathcal{A}^\star$, namely, concatenation allows us to define $LM = \{uv \mid u \in L,\ v \in M\}$, and for $n \geq 1$, $L^n = \{u_1 \dots u_n \mid u_i \in L\}$. We take $L^0 = \{\iota\}$, and set $L^\star = \cup_{n=0}^{\infty} L^n$. This latter agrees with our usage that $\mathcal{A}^\star$ is the free monoid on $\mathcal{A}$.

There are several ways to define the class of regular languages over $\mathcal{A}$. Perhaps the simplest is to say that it is the smallest class which contains all the finite subsets of $\mathcal{A}^\star$ and is closed under all the operations of the previous paragraph. Another way is to say that a language is regular if it is the language of a finite state automaton, which we now define.

A ***finite state automaton*** is a tuple $(\mathcal{A}, S, s_0, \tau, Y)$ where $\mathcal{A}$ is a finite set called an ***alphabet***, its elements are called ***letters***, S is a finite set of ***states***, $s_0 \in S$ is called the ***start state***, τ is a function from $\mathcal{A} \times S$ to S, called the ***transition function***, and $Y \subset S$ is the set of ***accept states***. Such a finite state automaton is sometimes referred to as a deterministic finite state automaton. It is the existence of the function τ which makes this automaton deterministic, as we shall see below.

Each finite state automaton with alphabet $\mathcal{A}$ defines a language over $\mathcal{A}$ in the following way. Let $w = a_1 \dots a_n \in \mathcal{A}^\star$ with each $a_i \in \mathcal{A}$. We let $t_0 = s_0$, and for $1 \leq j \leq n$ define t_j inductively by $t_j = \tau(a_j, t_{j-1})$. We accept w if and only if $t_n \in Y$.

If we like, we may see a finite state automaton as a finite labelled directed graph with a base point and a set of preferred vertices. The vertices of this graph are the states of the machine, the base point corresponds to s_0. There is an edge from s_i to s_j labelled by $g \in \mathcal{A}$ exactly when $\tau(g, s_i) = s_j$. The preferred vertices are the accept states. In this model, the language accepted by a finite state automaton consists of those words which label edge paths starting at the base point and ending at preferred vertices.

The existence of the function τ determines that each vertex of the graph of a finite state automaton has exactly one edge coming out of it labelled by each letter. An arbitrary finite directed labelled graph need not have this property. Such a graph, together with a base point and a set of preferred vertices is a ***non-deterministic finite state automaton***. Such a graph also determines a language, namely the set of words occurring on edge paths from the base point to preferred vertices. The surprising result here is that any language accepted by a non-deterministic finite state automaton is also accepted by a deterministic finite state automaton. (See **[HU]**.)

Before defining an automatic structure, we will need the notion of a product language. We will wish to study pairs of words (w, w') where $w, w' \in \mathcal{A}^\star$. To do this, we suppose that $\$ \notin \mathcal{A}$ and consider $\mathcal{A}_\$ = (\mathcal{A} \cup \{\$\}) \times (\mathcal{A} \cup \{\$\}) \setminus \{(\$, \$)\}$. We include $\mathcal{A}^\star \times \mathcal{A}^\star$ into $\mathcal{A}_\$^\star$ as follows. Given $w = a_1 \dots a_m$, $w' = b_1 \dots b_n$, we let $p = \max\{m, n\}$. If $m < p$, for $m < j \leq p$ we let $a_j = \$$. If $n < p$, for $n < j \leq p$ we let $b_j = \$$. We now take $(w, w') \in \mathcal{A}^\star \times \mathcal{A}^\star$ to $(a_1, b_1) \dots (a_p, b_p) \in \mathcal{A}_\$^\star$. Note that this inclusion is not a monoid map as its image need not be closed under concatenation. We will suppress reference to $\mathcal{A}_\$^\star$. However, whenever we state that a collection of pairs of words is regular, we will mean that its image is regular in $\mathcal{A}_\$^\star$.

It is now easy to state the definition of an automatic structure.

Definition ([CEHPT]). Given a finite set $\mathcal{A}$ together with a monoid map $\mathcal{A}^\star \mapsto G$ and a language L over $\mathcal{A}$, we will say that $(\mathcal{A}, L)$ is an ***automatic structure*** for G if

(1) L is regular;

(2) $\overline{L} = G$;

(3) $\{(w, w') \mid w, w' \in L, Overlinew = \overline{w'}\}$ is regular;

(4) For each $g \in \mathcal{G}$, $\{(w, w') \mid w, w' \in L, \overline{w} = \overline{w'g}\}$ is regular.

If G has an automatic structure, we say G is ***automatic***.

One may strengthen the notion of an automatic structure to that of a ***bi-automatic structure*** by demanding that in addition to (4), we have

(4′) For each $g \in \mathcal{G}$, $\{(w, w') \mid w, w' \in L, \overline{w} = \overline{gw'}\}$ is regular.

A group is said to be ***bi-automatic*** if it has a bi-automatic structure.

A major class of automatic (indeed bi-automatic) groups is the ***hyperbolic*** or ***negatively curved*** groups of Gromov (**[Gr]**). For discussions of these, see for example **[Ca2]**, **[CDP]**, **[Gh]**, **[GH]** or **[Sh]**. Automatic structures for these arise in a very natural way. If G is a negatively curved group, and $\mathcal{G}$ is any monoid generating set, and $L \subset \mathcal{G}^\star$ is the set of geodesics, then $(\mathcal{G}, L)$ is an automatic structure for G.

The notion of an asynchronous automatic structure (**[CEHPT]**) generalizes that of an automatic structure. Rather than demanding that the languages in (3) and (4) be regular, we demand only that they be accepted by asynchronous or two-tape automata. We will not give a formal definition of these automata and their languages here. For this the reader may see **[RS]** or **[BGSS]**. A group is said to be asynchronously automatic if it has an asynchronous automatic structure.

It is a theorem of **[CEHPT]** that if G is automatic then for any monoid generating set $\mathcal{A}$, we can find an automatic structure $(\mathcal{A}, L)$ for G. Hence for any automatic group G, we may always choose an automatic structure $(\mathcal{A}, L)$ so that there is an appropriate involution on $\mathcal{A}$ and hence $\Gamma_{\mathcal{A}}(G)$ is defined.

It is also a theorem of **[CEHPT]** that if $(\mathcal{A}, L)$ is an automatic structure for G, then we may pass to an automatic structure $(\mathcal{A}, L')$ where L' bijects to G under the monoid map. We call such an automatic structure an ***automatic structure with uniqueness***.

Asynchronous versions of both these theorems also hold.

If $\mathcal{A}$ is a monoid generating set for G and K is a constant, we say that $u, v \in \mathcal{A}^\star$ are ***asynchronous K-fellow travellers*** if there are monotone functions ϕ and θ so that for all t, $d_{\mathcal{A}}(u(\phi(t)), v(\theta(t))) \leq K$. We say that u, v are ***K-fellow travellers*** if for all t, $d_{\mathcal{A}}(u(t), v(t)) \leq K$.

Perhaps the strongest link between automatic structure (asynchronous automatic structure) and the geometry of the Cayley graph is given by the following.

Theorem ([CEHPT]). *Let $\mathcal{G}$ be a monoid generating set for G, and let $L \subset \mathcal{G}^\star$ be a regular language which surjects onto G. Suppose that for each $h \in G$ there are only finitely many $w \in L$ such that $\overline{w} = h$. Then L is the language of an (asynchronous) automatic structure for G if and only if there is a constant K such that whenever $u, v \in L$, $g \in \mathcal{G} \cup \{\iota\}$ with $\overline{u} = \overline{vg}$, it follows that u and v are (asynchronous) K-fellow travellers.*

Our statement of the asynchronous version of this Theorem more closely parallels that of **[BGSS]** than that of **[CEHPT]**. Since we have not given a definition of an asynchronous automatic structure, we will use the previous theorem as our means of proving that a group is asynchronously automatic.

Another important link between automatic structure and geometry occurs in **[GS]** where Gersten and Short, partly inspired by Gilman **[Gi]** study the rational subgroups of hyperbolic, automatic, bi-automatic and asynchronously automatic groups.

Definition ([GS]). Suppose that $\mathcal{G}$ is a monoid generating set for G and that $L \subset \mathcal{G}^\star$ is a regular language which surjects onto G. Then $H \subset G$ is said to be ***L-rational*** if $\{w \mid w \in L,\ \overline{w} \in H\}$ is regular. H is said to be ***L-quasiconvex*** if there is a constant K such that if $w \in L$ and $\overline{w} \in H$, then for all t, $d(w(t), H) \leq K$.

Theorem ([GS]). *Let L, G, H be as above. Then H is L-rational if and only if H is L-quasiconvex.*

It is worth noting that if G is negatively curved then for any monoid generating set $\mathcal{G}$, the set of geodesics in $\mathcal{G}^\star$ is regular. If L is the set of geodesics in $\mathcal{G}^\star$ and M is the set of geodesics with respect to some other monoid generating set, then H is L-rational if and only if H is M-rational. If G is negatively curved, we will say that $H \subset G$ is rational if it is rational with respect to the set of geodesics for some (hence every) generating set.

Our interest is in the case where H is a subgroup.

In **[BGSS]**, Baumslag, Gersten, Shapiro and Short study amalgams of automatic groups. They give the following tool for finding automatic and asynchronous automatic structures for these amalgams.

Theorem([BGSS]). *Let A and B be automatic groups, each containing C as a subgroup. Let*

$$G = A \star_C B\,.$$

Let $L(A) \subset \mathcal{A}_A^\star$ and $L(B) \subset \mathcal{A}_B^\star$ be the languages of automatic structures for A and B respectively, and suppose further that the following hold.

(1) *C is rational with respect to $L(A)$. (Hence we may pass to a regular language $L(C) \subset L(A)$ which bijects to C.)*
(2) *There are regular sublanguages $L(A/C)$ and $L(B/C)$ of $L(A)$ and $L(B)$ which biject to the set of cosets $\{gC\}$ of C in A and B.*
(3) *There is a constant K so that if $u \in L(C) \subset L(A)$, and $v \in L(B)$ with $\overline{u} = \overline{v}$, then u and v are (asynchronous) K-fellow travellers in $\Gamma(G)$.*
(4) *There is a constant K so that if $u \in L(A/C)$ (respectively $L(B/C)$), $v \in L(C)$, and $w \in L(A)$, (respectively $L(B)$), with $\overline{uv} = \overline{w}$, then uv and w are (asynchronous) K-fellow travellers.*

Then G is (asynchronously) automatic.

They then show that these conditions are synchronously satisfied when A and B are abelian, or when they are negatively curved and C is infinite cyclic. The conditions are asynchronously satisfied when A and B are negatively curved and C is rational in each.

In this paper, we first extend this result to finite trees of groups (Theorem 1) and then to finite graphs of groups (Theorem 3). (For a definition of these, see **[Se]**.) As an application of Theorem 1, we show that the fundamental groups of closed $PSL_2(\mathbb{R})$

manifolds are automatic. The amalgamation here corresponds to the decomposition of the manifold into two Seifert fibered manifolds, each with torus boundary. Theorem 4 and its corollaries deal with finite graphs of groups in which each vertex group is negatively curved and includes the image of each edge group as a rational subgroup. In general, we can only give an asynchronous automatic structure for these. However, if the elements of each edge group have approximately the same length in each vertex group, we can give an automatic structure. Theorem 5 tells us that if the vertex groups are negatively curved, the edge groups are infinite cyclic, and the graph is a tree, then the resulting group is automatic.

We then study finite graphs and trees of abelian groups. As we have mentioned, amalgams of finitely generated abelian groups are automatic. Now, HNN extensions of $\mathbb{Z} \times \mathbb{Z}$ include fundamental groups of closed Nil and Solv geometry 3-manifolds. These Solv geometry groups are not automatic (**[Ge1]**, or the most recent version of **[CEHPT]**) and these Nil groups are not even asynchronously automatic (**[CEHPT]**). We show in Theorem 6 that if the edge groups "intersect well" inside the vertex groups, a finite graph of abelian groups is asynchronously automatic. If the graph is a tree, the resulting group is automatic.

Automatic structures for geometrically finite hyperbolic groups are given in **[CEHPT]**, and **[Ta]** gives automatic structures in the special case of finite volume hyperbolic groups. These structures allow us to choose at will the automatic structures on each of the parabolic subgroups. We study finite graphs of groups in which each vertex group is a finite volume hyperbolic group and each edge group is abelian and injects as a parabolic subgroup of its vertex groups. Theorem 7 shows that if the edge groups "intersect well" the resulting group is asynchronously automatic. Under a stronger assumption the group is automatic. This will include the case where the graph of groups corresponds the fundamental group of a manifold formed by gluing together finite volume hyperbolic pieces along peripheral tori.

We have discussed finite graphs of groups in which the vertex groups are the fundamental groups of Seifert fibered manifolds and finite volume hyperbolic groups, and the edge groups correspond to boundary tori. This suggests the toral decomposition of an irreducible Haken 3-manifold, and suggests that other than the obvious counter-examples — closed Nil and Solv manifolds — these might have automatic fundamental groups. This is in fact the case as has been shown by **[CEHPT]** and independently by myself. While my proof is very much in the spirit of Theorem 3, I do not at the present time know how to make this an application of this theorem. Consequently, the present paper ends, like the I Ching, at a point before completion.

I wish to thank Boris Apanasov, Gilbert Baumslag and Martin Bridson for helpful conversations. Walter Neumann's support and encouragement were especially helpful while I was working on the applications of these methods to Haken 3-manifolds.

Theorem 1. *Let T be a finite tree of groups, with vertex set $\{v_1, \ldots, v_n\}$, and oriented edges $\{e_{ij}\}$. Let the vertex group of v_i be G_i, and the edge group of e_{ij} be C_{ij}. Let $G_1, \ldots, G_n$ be generated by $\mathcal{G}_1, \ldots, \mathcal{G}_n$. For each i, $0 \leq i \leq n$, let $L_i \subset \mathcal{G}_i^\star$ be the regular language of an automatic structure with uniqueness for G_i. Suppose further that the following hold.*

(1) *For each edge* e_{ij}, $\partial_0(C_{ij})$ *is rational with respect to* L_i. *(Hence we may pass to a regular language* $L_i(\partial_0 C_{ij}) \subset L_i$ *which bijects to* $\partial_0 C_{ij}$*.)*
(2) *For each edge* e_{ij} *there is a regular sublanguage* L_{ij} *of* L_i *which bijects to the set of cosets* $\{g\partial_0(C_{ij}) \mid g \in G_i\}$ *of* $\partial_0(C_{ij})$ *in* G_i, *the bijection being given by* $w \mapsto \overline{w}\partial_0(C_{ij})$.
(3) *There is a constant* K *so that if* $u \in L_i(\partial_0 C_{ij}) \subset L_i$, *and* $v \in L_j$ *with* $\overline{u} = \overline{v}$, *then* u *and* v *are (asynchronous)* K*-fellow travellers in* $\Gamma(G)$.
(4) *There is a constant* K *so that if* $u \in L_{ij}$, $v \in L_i(\partial_0 C_{ij})$, *and* $w \in L_i$, *with* $\overline{uv} = \overline{w}$, *then* uv *and* w *are (asynchronous)* K*-fellow travellers.*

Then $G = G(T)$ *is (asynchronously) automatic.*

Proof. Note that we may assume that in each L_{ij}, ι is the representative of the trivial coset $\partial_0(C_{ij})$. For if this coset is represented by $x_{ij} \neq \iota$, we replace L_{ij} with $L_{ij} \cup \{\iota\} \setminus \{x_{ij}\}$, and this new language is also regular. This may force us to replace K in condition (4) with $K + \max\{\ell(x_{ij})\}$. We assume that this has been done.

Our proof proceeds by induction on the number of vertices in the tree. For $n = 0$, there is nothing to prove, so we assume $n \geq 1$. Given a tree, T, we take as our generating set

$$\mathcal{G} = \coprod_{i=0}^{n} \mathcal{G}_i .$$

For each tree, T, and each choice of vertex, $\star$, we define the language

$$\begin{aligned}\hat{L}_\star(T) = \{w = w_1 \dots w_m \mid \\ &(1)\ \ w_k \in L_{i(k)j(k)},\\ &(2)\ \ j(k) = i(k+1) \text{ for } k = 1, \dots, m-1 \text{ and } i(1) = j(m) = \star\\ &(3)\ \ \text{if } i(k) = j(k+1), \text{ then } w_{k+1} \neq \iota\}.\end{aligned}$$

We define

$$L_\star(T) = \hat{L}_\star(T) L_\star .$$

Condition (2) says that each $w \in \hat{L}_\star(T)$ corresponds to a loop in T based at $\star$, while condition (3) insures that this loop is uniquely determined by w.

Lemma 1.1. *For each* T *and choice of* $\star$, $\hat{L}_\star(T)$ *is regular. (Hence* $L_\star(T)$ *is regular.)*

Proof. We give two proofs. In the first proof, we construct a generalized finite state automaton, M, which accepts $\hat{L}(T)$. Such a machine is a finite labelled directed graph. The edges of this graph are labelled with regular languages. A vertex is designated a start state, and one or more vertices are designated accept states. Such a machine accepts a word $u = u_1 \dots u_k$ if each u_i lies in the language labelling edge $e(i)$ and $e(1) \dots e(k)$ form a path from the start state to an accept state. It is a theorem that the language accepted by such a generalized finite state automaton is regular. (See for example, **[CEHPT]**.)

For each vertex v_i of T and each vertex v_j adjacent to v_i, M has a vertex, s_{ij}. M also has vertices s_0 and s_f. For each edge $(\star, i)$, M has an edge from s_0 to $s_{\star i}$ labelled by $L_{\star i}$. For each edge (j, k) and each state s_{ij} with $k \neq i$, M has an edge from s_{ij} to s_{jk} labelled by L_{jk}, and an edge from s_{ij} to s_{ji} labelled by L^-_{ji}. (Given a language L, we use L^- to denote $L \setminus \{\iota\}$.) For each edge $(j,\star)$ and each state s_{ij} with $i \neq \star$, M has an edge from s_{ij} to s_f labelled by $L_{j\star}$. For each edge $(j,\star)$ and each state $s_{\star j}$, M has an edge from $s_{\star j}$ to s_f labelled by $L^-_{j\star}$. The start state is s_0, while s_f is the sole accept state. We have constructed M so that a word is accepted by M if and only if that word labels a path in T as in the definition of $\hat{L}_\star(T)$.

The second proof is by induction on n. For $n=0$, $\star = v_0$, and $\hat{L}_{v_0}(T) = \{\iota\}$.

For $n \geq 1$, we renumber the vertices so that $\star = v_0$ and v_1 is connected to v_0 by an edge. Then $T \setminus \{e_{01}, e_{10}\}$ has components $T_0 \ni v_0$ and $T_1 \ni v_1$. Each of these has fewer vertices than T, so we may assume by induction that $\hat{L}_{v_0}(T_0)$ and $\hat{L}_{v_1}(T_1)$ are regular.

Then $\hat{L}_\star(T)$ consists of exactly those alternating words from $\hat{L}_{v_0}(T_0)$ and $L_{01}\hat{L}_{v_1}(T_1)L_{10}$ which obey (2) and (3). This is because each word in $\hat{L}_\star(T)$ corresponds to a path in T and any such path may be decomposed into paths in T_0 and T_1 together with traverses of the edges e_{01} and e_{10}. Thus, the reader may see that

$$\hat{L}_\star(T) = (\hat{L}_0(T_0)L_{01}((\hat{L}_1(T_1)L_{10})^-(\hat{L}_0(T_0)L_{01})^-)^\star \hat{L}_1(T_1)L_{10}) \setminus \mathcal{G}^\star L^-_{01}\,.$$

Since the class of regular languages is closed under union, intersection, complementation, concatenation and $\star$, this shows that $\hat{L}_\star(T)$ is regular. □

Lemma 1.2. *For each T and choice of $\star$, $\hat{L}_\star(T)$ bijects to $G(T)/G_\star$. Hence $L_\star(T)$ bijects to $G(T)$.*

Proof. The proof is by induction on n. For $n=0$, $\hat{L}_\star(T) = \{\iota\}$ and $L_\star(T) = L_\star$.

For $n \geq 1$, we decompose T into T_0 and T_1 as in the previous lemma. Notice that

$$G(T) = G(T_0) \star_{C_{01}} G(T_1)\,.$$

By induction, we may assume for $k = 0,1$ that $\hat{L}_k(T_k)$ bijects to $G(T_k)/G_k$ and $L_k(T_k)$ bijects to $G(T_k)$. Thus, $\hat{L}_0(T_0)L_{01}$ and $\hat{L}_1(T_1)L_{10}$ biject to $G(T_0)/\partial_0 C_{01}$ and $G(T_1)/\partial_1 C_{01}$ respectively. Thus, by the normal form for free products with amalgamation,

$$(\hat{L}_0(T_0)L_{01}((\hat{L}_1(T_1)L_{10})^-(\hat{L}_0(T_0)L_{01})^-)^\star \hat{L}_1(T_1)L_{10})$$

bijects to $G(T)/C_{01}$. Since

$$\hat{L}_\star(T) = (\hat{L}_0(T_0)L_{01}((\hat{L}_1(T_1)L_{10})^-(\hat{L}_0(T_0)L_{01})^-)^\star \hat{L}_1(T_1)L_{10}) \setminus \mathcal{G}^\star L^-_{01}\,,$$

we see that $\hat{L}_0(T) = \hat{L}_\star(T)$ bijects to $G(T)/G_0 = G(T)/G_\star$, as required. □

Lemma 1.3. *For any choice of vertex, $\star$, $L_\star(T)$ is the language of an (asynchronous) automatic structure for $G(T)$.*

Proof. The proof is by induction on n. In the case $n=0$, we have $L_\star(T) = L_0$ and this is the language of an automatic structure on $G(T) = G_0$.

We pass to the case $n > 0$. Since $L_\star(T)$ has unique representation, we need only prove the (asynchronous) fellow traveller property for multiplication by a generator $g \in \mathcal{G}$. That is to say, we must show that there is a constant M so that if $w, w' \in L_\star(T)$, $g \in \mathcal{G}$ and $\overline{w} = \overline{w'g}$, then w and w' are (asynchronous) M-fellow travellers.

The proof of this will employ the following two observations.

(1) Suppose that for $0 \le j < m$, x_j and x_{j+1} are (asynchronous) K-fellow travellers. Then x_0 and x_m are (asynchronous) mK-fellow travellers.
(2) Suppose that x_0 and x_1 are (asynchronous) K-fellow travellers. Then for any u, ux_0 and ux_1 are (asynchronous) K-fellow travellers.

As usual, we assume the vertices of T are numbered so that $\star = v_0$, and v_0 and v_1 are joined by an edge, e_{01}. T_0 and T_1 are the components of $T \setminus \{e_{01}, e_{10}\}$.

Case 1 w' has the form $w' = xy'z'$ with $y' \in \hat{L}_0(T_0)$, $z' \in L_0$ and $\overline{g} \in G(T_0)$. (Note: we are only assuming that the image of g in $G(T)$ lies in the image of $G(T_0)$ in $G(T)$. We are not assuming that $g \in \mathcal{G}_i$ with $v_i \in T_0$.)

In this case $y'z' \in L_0(T_0)$, and there is a unique $y \in L_0(T_0)$ with $\overline{y} = \overline{y'z'g}$. By induction, $L_0(T_0)$ is the regular language of an (asynchronous) automatic structure on $G(T_0)$ with fellow traveller constant (say) M_0. Let

$$N = \max\{\ell_{G(T_0)}(\overline{g}) \mid g \in \mathcal{G} \text{ and } \overline{g} \in G(T_0)\}.$$

Thus we may write $g = \overline{u}$ where $\ell(u) = m \le N$ and u is written in generators for the groups of T_0. For $0 \le j \le m$, we let y_j be the accepted word for $\overline{y'z'u(j)}$. Now for $0 \le j < m$, y_j and y_{j+1} are M_0-fellow travellers. Since $y_0 = y'z'$ and $y = y_m$, y and $y'z'$ are NM_0-fellow travellers. But $xy \in L_0(T)$ and $\overline{xy} = \overline{xy'z'g}$. Thus, by uniqueness, $w = xy$ and hence w and w' are (asynchronous) NM_0-fellow travellers.

Case 2 $w' = xy'z'$ with $y' \in \hat{L}_0(T_0)$, $z' \in L_0$, and $\overline{g} \notin G(T_0) < G(T)$.

Let v_j be the vertex nearest to v_0 such that $\overline{g} \in G_j < G(T_1) < G(T)$. There is a unique geodesic running from v_0 to v_j, and by renumbering if necessary, we may assume the vertices on this path are numbered $v_0, v_1, \ldots, v_j$.

Let $z \in L_{01}$ be the unique representative for $\overline{z'}\partial_0 C_{01}$. We then have a unique $u \in L_0(\partial_0 C_{01})$ so that $\overline{z'} = \overline{zu}$ with $z \in L_0$ and $\overline{u} \in C_{01}$. By assumption, z' (asynchronously) K-fellow travels with zu. Now consider the word $u_0 \ldots u_j$ where $u_i \in L_{ii+1}$ for $0 \le i < j$, $u_j \in L_j$, and $\overline{u} = \overline{u_0 \ldots u_j}$. This uniquely defines $u_0 \ldots u_j$.

We claim that u and $u_0 \ldots u_j$ are (asynchronous) $(2jK)$-fellow travellers. To see this, write $\overline{u} = \overline{u_0 r_0}$ with $u_0 \in L_{01}$, $r_0 \in L_0$. Then by assumption, u and $u_0 r_0$ are (asynchronous) K-fellow travellers. We then write $\overline{r_0} = \overline{w_1}$ with $w_1 \in L_1$. By assumption, r_0 and w_1 are (asynchronous) K-fellow travellers. Thus, $u_0 r_0$ and $u_0 w_1$ are (asynchronous) K-fellow travellers. This makes u and $u_0 w_1$ (asynchronous) $2K$-fellow travellers. We write $\overline{w_1} = \overline{u_1 r_1}$ with $u_1 \in L_{12}$, $r_1 \in L_1$. Continuing in this way, we use $2j$ applications of assumptions (3) and (4) to see that u and $u_0 \ldots u_j$ are (asynchronous) $(2jK)$-fellow travellers; see Fig. 1.

Now there is a unique v in L_j so that $\overline{v} = \overline{u_j g}$. Now, v and $u_j g$ are $\ell_{\mathcal{G}_j}(\overline{g})M_j$-fellow travellers, where M_j is the fellow traveller constant of the automatic structure L_j on G_j. We may now find $t_j \ldots t_0$ so that for $1 \le i \le j$, $t_i \in L_{ii-1}$, $t_0 \in L_0$ and $\overline{v} = \overline{t_j \ldots t_0}$. As with u and $u_0 \ldots u_j$, v and $t_j \ldots t_0$ are (asynchronous) $((2j+1)K)$-fellow travellers. By our choice of j, t_j is non-trivial. Thus $xy'z'u_0 \ldots u_{j-1}t_j \ldots t_0 \in L_{t_0}(T)$. But by construction $\overline{xy'z'u_0 \ldots u_{j-1}t_j \ldots t_0} = \overline{w'g}$ and thus by uniqueness, $xy'z'u_0 \ldots u_{j-1}t_j \ldots t_0 = w$. We thus see that w and w' are (asynchronous) $((4j +$

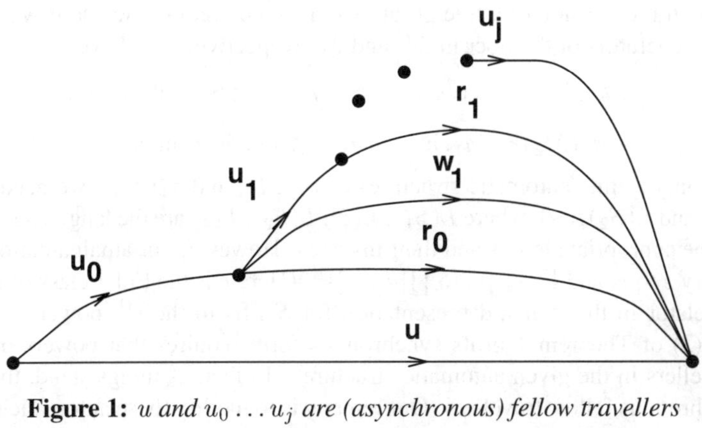

Figure 1: *u and $u_0 \dots u_j$ are (asynchronous) fellow travellers*

$3)K) + \max_j\{\ell_{\mathcal{G}_j}(\overline{g})M_j\})$- fellow travellers. Since $j \leq n$ and there are only finitely many $g \in \mathcal{G}$, this gives a global bound.

The remaining cases follow by exchanging the roles of T_0 and T_1. Taking the maximum over all cases gives a fellow traveller constant for all $g \in \mathcal{G}$ This concludes the proof of Theorem 1. □

As an application of this Theorem, we give a proof that the fundamental group of a closed $PSL_2(\mathbb{R})$ manifold is automatic. This has been known for some time by a remark of Thurston, and appears in the latest version of **[CEHPT]**.

Theorem 2. *Suppose M is a closed manifold with $PSL_2(\mathbb{R})$ geometry. Then $\pi_1(M)$ is automatic.*

Proof. Suppose M is a closed $PSL_2(\mathbb{R})$ manifold. Then M projects to a closed hyperbolic surface orbifold S. We then have

$$1 \longrightarrow \mathbb{Z} \longrightarrow \pi_1(M) \longrightarrow \pi_1^{\text{orb}}(S) \longrightarrow 1.$$

The $\mathbb{Z}$ is central. (See **[Sc]**.)

For convenience we pass to a finite index torsion-free subgroup of $\pi_1^{\text{orb}}(S)$ and thus to a finite sheeted cover of M which is a closed $PSL_2(\mathbb{R})$ manifold which fibers over a closed hyperbolic surface. We now call these M and S.

S may be decomposed over an embedded separating curve, and the pre-image decomposes M into two pieces M_1 and M_2 which fiber over surfaces with boundary S_1 and S_2, each of genus at least 1. This gives $\pi_1(M)$ the structure of a free product with amalgamation. We will apply Theorem 1.

Each S_i has free fundamental group, the number of generators being twice the genus. We thus have, say,

$$\pi_1(S_1) = \langle x_1, y_1, \dots x_m, y_m \rangle,$$

$$\pi_1(S_2) = \langle u_1, v_1, \dots u_n, v_n \rangle.$$

Now any central extension of a free group is in fact a direct product, so if we take z and w to be the generators of the fiber in M_1 and M_2 respectively, we have

$$\pi_1(M_1) = \langle x_1, y_1, \ldots x_m, y_m, z \mid z \text{ is central} \rangle,$$

$$\pi_1(M_2) = \langle u_1, v_1, \ldots u_n, v_n, s \mid s \text{ is central} \rangle.$$

If we seek only to find automatic structures of $\pi_1(M_1)$ and $\pi_1(M_2)$, we need only take $L(S_1)L(z)$ and $L(S_2)L(s)$, where $L(S_1)$, $L(z)$, $L(S_2)$, $L(s)$ are the languages of reduced words on the appropriate letters and their inverses. However, our amalgamation requires that we carry z to s, and $\prod[x_i, y_i]$ to $\prod[u_i, v_i]s^e$. (Here e is the Euler class of the bundle M. The relator in the standard presentation for S lifts to the eth power of the fiber.) Condition (3) of Theorem 1 in its synchronous form requires that powers of these be fellow travellers in the given automatic structures. In fact, as things stand, they are not even asynchronous fellow travellers. Our remedy is to modify these automatic structures. We use the following tool.

Let $\mathcal{A}$ be an alphabet. Let $p \in \mathcal{A}^\star$. Let P be a symbol not in $\mathcal{A}$. We define a function $\varphi_p : \mathcal{A}^\star \longrightarrow (\mathcal{A} \cup \{P\})^\star$ as follows. Given a word $w \in \mathcal{A}^\star$, start at the beginning of w, and working forwards, replace every occurrence of p by the symbol P. The stipulation that these replacements be made by proceeding from left to right ensures that φ_p is well defined. We call φ_p a ***replacement function***, and each instance of p replaced by P a ***replacement***.

Lemma 2.1. *φ_p is an injection. If $L \subset \mathcal{A}^\star$ is regular, then so is $\varphi_p(L)$. If $u = xy$, $u' = xy'$ with $\max\{\ell(y), \ell(y')\} \leq K$, then $\varphi(u)$ and $\varphi(u')$ agree except on suffixes of length at most $K + \ell(p)$.*

Proof. The first statement is obvious, for the inverse of φ_p (on its image) is given by replacing each P with the word p.

To see that $\varphi_p(L)$ is regular, we start with a finite state automaton for L. We think of the states of this machine as vertices of a labelled directed graph. The edges encode the transition function of the machine and are labelled by the appropriate letters of $\mathcal{A}$. We modify the machine in the following way. From each vertex of the machine there is at most one path bearing the label p. Append an edge labelled P from the beginning to the end of this path. Our new machine accepts all of L together with words in L where some (or all) occurrences of p are replaced by P. Now delete from this language all words containing p and all words containing $p'p$ where $p'p'' = p$ and p'' is an initial segment of p. These last two sets are regular, hence $\varphi_p(L)$ is regular.

The last part of the lemma is obvious.

Notice that these properties persist under composition. That is, if $\varphi = \varphi_{p_1} \circ \ldots \circ \varphi_{p_k}$ is the composition of such functions, then $\varphi(L)$ bijects to L and $\varphi(L)$ is regular if L is regular. If u and u' agree on all but the last K letters, $\varphi(u)$ agrees with $\varphi(u')$ except possibly on the final $K + \sum \ell(p_i)$ letters. □

We now let $p = \prod[x_i, y_i]$, $p^{-1} = \prod[y_i, x_i]$, $q = \prod[u_i, v_i]$, $q^{-1} = \prod[v_i, u_i]$. Thus $\overline{p}^{-1} = \overline{p^{-1}}$, and $\overline{q}^{-1} = \overline{q^{-1}}$. We set $\varphi_p^{\pm} = \varphi_{p^{-1}} \circ \varphi_p$ and $\varphi_q^{\pm} = \varphi_{q^{-1}} \circ \varphi_q$. We take as our languages,

$$L(M_1) = \varphi_p^{\pm}(L(S_1))L(z)$$

$$L(M_2) = \varphi_q^{\pm}(L(S_2))L(s).$$

We set

$$\overline{P} = \overline{p}$$

but

$$\overline{Q} = \overline{qs^e},$$

and similarly for their inverses, P^{-1} and Q^{-1}. We must show that these give automatic structures for $\pi_1(M_1)$ and $\pi_1(M_2)$ and that these satisfy the conditions of Theorem 1.

To see that these give automatic structures, we must see that they surject to their respective groups and enjoy the fellow traveller property. For each u, $\overline{u} = \overline{\varphi_p^{\pm}(u)}$, so $L(M_1)$ bijects to $\pi_1(M_1)$. We check the fellow traveller property in $L(M_1)$. To this end, let $w, w' \in L(M_1)$ with $\overline{w} = \overline{w'g}$ and g one of our chosen generators for $\pi_1(M)$. In fact, we need not consider the case where g is either P or P^{-1}, for each of these may be expressed as a product of the remaining generators. Now w and w' each have the form $\varphi_p^{\pm}(u)v$ and $\varphi_p^{\pm}(u')v'$, where $v, v' \in L(z)$ Since $L(S_1)$ bijects to $\pi_1(S_1)$ and $\overline{\varphi_p^{\pm}(u)} = \overline{u}$ for each $u \in L(S_1)$, we see that $\overline{w} = \overline{w'g}$ if and only if $\overline{uv} = \overline{u'v'}$. Thus, if $g = z^{\pm 1}$, we have $u = u'$ and v and v' differ only by a final letter. If g is one of the surface generators, then $v = v'$ and u and u' are fellow travellers in $\pi_1(S_1)$. But since $\pi_1(S_1)$ is free, this means that u and u' agree on all but a final letter. This, together with Lemma 2.1, implies that $\varphi_p^{\pm}(u)$ and $\varphi_p^{\pm}(u')$ agree on all but the final $8m + 1$ letters. This shows that w and w' are fellow travellers. Hence $L(M_1)$ is an automatic structure for $\pi_1(M_1)$.

The proof for $L(M_2)$ involves only a minor complication. Notice that while $\varphi_q^{\pm}$ may very well change the value of u, it does not change the value of its projection in $\pi_1(S_2)$. Thus $\varphi_q^{\pm}(L(S_2))$ bijects to the $\pi_1(M_2)/\langle s\rangle$. Consequently $L(M_2)$ bijects to $\pi_1(M_2)$. Suppose we are given $w = \varphi_q^{\pm}(u)v \in L(M_2)$ and $w' = \varphi_q^{\pm}(u')v'$ with $\overline{w} = \overline{w'g}$ and g a generator. We may assume that $g \neq Q^{\pm 1}$ since each of these may be written as a product of the remaining generators. Since w and w' end at distance one from each other in $\pi_1(M_2)$, u and u' still end within an edge of each other and hence agree on all but a final letter. Thus $\varphi_q^{\pm}(u)$ and $\varphi_q^{\pm}(u')$ agree as before on all but the last $8n + 1$ letters. But then these differ at most by one in the number of replacements carried out by $\varphi_q^{\pm}$, and thus differ in $\pi_1(M_2)$ by at most s^{2e}. Thus, if $g \neq s^{\pm 1}$, it is possible that v and v' still differ, but at most by s^{2e}. This shows that $L(M_2)$ is an automatic structure.

The point here is that while $\varphi_p^{\pm}$ is acting only to control length in the automatic structure of the boundary of S_1, $\varphi_q^{\pm}$ is acting to control the length of the boundary, and the lift of $\pi_1(S_2)$ to $\pi_1(M_2)$. Nearby elements in $\pi_1(S_2)$ are taken to lifts whose difference in the fiber is bounded.

We now proceed to verify the conditions of Theorem 1.

Clearly the amalgamating subgroup is rational on each side, because it is given by $\{P^{\pm i}z^{\pm j}\}$ and $\{Q^{\pm i}s^{\pm j}\}$. This verifies condition (1).

To verify condition (2), notice that $L(S_1)$ and $L(S_2)$ contain regular sublanguages which biject to the set of cosets of their respective boundary subgroups. This follows from the fact that the groups involved are free, and therefore negatively curved, and the boundary subgroups are cyclic, and therefore rational. See **[BGSS]**. More explicitly, one may take this language $L(S_1/\partial)$ to be the set of words in $L(S_1)$ which do not end in more than one half of the word p and which end in strictly less than one half of the word p^{-1}, and similarly for $L(S_2/\partial)$. Then $\varphi_p^{\pm}(L(S_1/\partial))$ and $\varphi_q^{\pm}(L(S_2/\partial))$ are regular sublanguages of $L(M_1)$ and $L(M_2)$ bijecting to the set of cosets of the boundary subgroups in $\pi_1(M_1)$ and $\pi_1(M_2)$ respectively. The bijections are given by $w \mapsto \overline{w}\langle\overline{p},\overline{z}\rangle$ and $w \mapsto \overline{w}\langle\overline{q},\overline{s}\rangle = \overline{w}\langle\overline{Q},\overline{s}\rangle$

Condition (3) follows immediately from the fact that the languages involved are $\{P^{\pm i}z^{\pm j}\}$ and $\{Q^{\pm i}s^{\pm j}\}$ and we are amalgamating $\overline{P}$ to $\overline{Q}$ and $\overline{z}$ to $\overline{s}$.

Finally, we wish to verify condition (4). We first establish that it holds in $L(M_1)$. Suppose $w = \varphi_p^{\pm}(w')z^k \in L(M_1)$, $u = \varphi_p^{\pm}(u')$ with $u' \in L(S_1/\partial)$, and that $v = P^i z^j = \varphi_p^{\pm}(p^i)z^j$ and that $\overline{w} = \overline{uv}$. We must show that there is a constant K independent of w, u, and v so that w and uv are K-fellow travellers. Now $\overline{w} = \overline{u'p^iz^j}$. If $u'p^i$ is reduced as written, then $w' = u'p^i$ so that in fact, $w = uv$. Now it may be that there is cancellation between u' and p^i. Thus (in the case where $i > 0$) we may have $u' = u''r^{-1}$ where $p = rt$. Then $w' = u''tp^{i-1}$. Hence $w = \varphi_p^{\pm}(u''t)P^{i-1}z^j$. Since $u''t$ and u' agree on all but the final $6m$ letters, $\varphi_p^{\pm}(u''t)$ and $\varphi_p^{\pm}(u)$ agree on all but the last $10m$ letters. Now $\overline{\varphi_p^{\pm}(u''t)} = \overline{\varphi_p^{\pm}(u)P}$, so w and uv are $(10m+1)$-fellow travellers. The case where $i < 0$ is similar.

The argument in M_2 poses only one additional complication. Here we have $w = \varphi_q^{\pm}(w')s^k \in L(M_2)$, $u = \varphi_q^{\pm}(u')$ with $u' \in L(S_2/\partial)$, $v = Q^i s^j = \varphi_q^{\pm}(q^i)s^j$ and $\overline{w} = \overline{uv}$. Again, if there is no cancellation between u' and q^i, we have $w = uv$ as before. On the other hand, if there is cancellation, we have $w = \varphi_q^{\pm}(u''t)Q^{i-1}z^{j+e}$. Thus w and uv are $(10n + e)$-fellow travellers.

Now invoking Theorem 1 completes our proof that the fundamental group of a closed $PSL_2(\mathbb{R})$ manifold is automatic. □

We now extend Theorem 1 to a finite graph of groups.

Theorem 3. *Let Y be a finite graph of groups, with vertex set $\{v_1, \ldots, v_n\}$, and oriented edge set E. Let the vertex group of v_i be G_i, and the edge group of e be C_e. Let $G_1, \ldots, G_n$ be generated by $\mathcal{G}_1, \ldots, \mathcal{G}_n$. For each vertex v_i, let $L_i = L_{v_i} \subset \mathcal{G}_i^\star$ be the regular language of an automatic structure with uniqueness for G_i. Fix a maximal tree T of Y, and let $Y \setminus T = \{s_j\}$. Suppose further that the following hold.*

(1) *For each edge e and $k = 0, 1$, $\partial_k C_e$ is rational with respect to $L_{\partial_k e}$. (Hence we may pass to a regular language $L_{\partial_k e}(\partial_k C_e) \subset L_{\partial_k e}$ which bijects to $\partial_k C_e$.)*

(2) *For each edge e, and $k = 0,1$ there is a regular sublanguage $L(G_{\partial_k e}/\partial_k C_e)$ of $L_{\partial_k e}$ which bijects to the set of cosets $\{g\partial_k(C_e) \mid g \in G_{\partial_k e}\}$ of $\partial_k(C_e)$ in $G_{\partial_k e}$, the bijection being given by $w \mapsto \overline{w}\partial_k(C_e)$.*

(3) *There is a constant K so that if $\alpha \in C_e$, $u \in L_{\partial_0 e}$ and $v \in L_{\partial_1 e}$ then if $e \in T$ and $\overline{u} = \partial_0\alpha$, $\overline{v} = \partial_1\alpha$, then u and v are (asynchronous) K-fellow travellers in $\Gamma(G)$. If $e = s_j$ and $\overline{u} = \partial_0\alpha$, $\overline{v} = \partial_1\alpha$, then u and $s_j v s_j^{-1}$ are (asynchronous) K-fellow travellers in $\Gamma(G)$. (We use s_j to denote the stable letter corresponding to the edge s_j.)*

(4) *There is a constant K so that if $u \in L(G_{\partial_k e}/\partial_k C_e)$, $v \in L_{\partial_k e}$ with $\overline{v} \in \partial_k(C_e)$, and $w \in L_{\partial_k e}$, with $\overline{w} = \overline{uv}$, then w and uv are (asynchronous) K-fellow travellers.*

Then $G = G(Y)$ is (asynchronously) automatic.

Proof. As before, we may assume that in each instance, the trivial coset is represented by the empty word.

We have fixed T, a maximal subtree of Y, and set $\{s_j\} = Y \setminus T$. Hence $G(Y)$ is an HNN extension of $G(T)$ with stable letters $\{s_j\}$. We start by jacking up our results about T. For this purpose, we continue with the notation of Theorem 1.

For each pair of vertices a, c, of T, we define the language

$\hat{L}_{ac}(T) = \{w = w_1 \dots w_m \mid$
(1) $w_k \in L_{i(k)j(k)}$,
(2) $j(k) = i(k+1)$ for $k = 1, \dots, m-1$ and $i(1) = a$, $j(m) = c$,
(3) if $i(k) = j(k+1)$, (1) then $w_{k+1} \neq \iota\}$.

We define

$$L_{ac}(T) = \hat{L}_{ac}(T) L_c .$$

Lemma 3.1. *For each tree T and choice of vertices a, c, $\hat{L}_{ac}(T)$ is regular. (Hence $L_{ac}(T)$ is regular.)*

Proof. As in Lemma 1.1, we give two proofs. In the first, we construct a generalized finite state automaton, M. For each vertex v_i of T and each vertex v_j adjacent to v_i, M has a vertex, t_{ij}. M also has vertices t_0 and t_f. For each edge e_{ai}, M has an edge from t_0 to t_{ai} labelled by L_{ai}. For each edge e_{jk} and each state t_{ij} with $k \neq i$, M has an edge from t_{ij} to t_{jk} labelled by L_{jk}, and an edge from t_{ij} to t_{ji} labelled by L_{ji}^-. For each edge e_{jc} and each state t_{ij} with $i \neq c$, M has an edge from t_{ij} to t_f labelled by L_{jc}. For each edge e_{jc} and each state t_{cj}, M has an edge from t_{cj} to t_f labelled by L_{jc}^-. The start state is t_0, while t_f is the sole accept state. We have constructed M so that a word is accepted by M if and only if that word labels a path in T as in the definition of $\hat{L}_\star(T)$.

The second proof is by induction on the size of T. We already know the result in the case where $a = c$. Thus, we may assume $a \neq c$. There is a unique geodesic connecting a to c in T. Let b be the vertex at distance one from a along this geodesic. $T \setminus \{e_{ab}, e_{ba}\}$ has two components, $T_a \ni a$ and $T_b \ni b$. $\hat{L}_{ac}(T)$ is then given by

$$\hat{L}_{ac} = \hat{L}_a(T_a) L_{ab} ((\hat{L}_b(T_b) L_{ba})^- (\hat{L}_a(T_a) L_{ab})^-)^\star \hat{L}_{bc}(T_b). \qquad \square$$

Lemma 3.2. *$\hat{L}_{ac}(T)$ bijects to $G(T)/G_c$. (Hence $L_{ac}(T)$ bijects to $G(T)$.)*

Proof. Again, we know this when $a = c$, so we may assume they are different, and decompose T into T_a and T_b as above. We know that $\hat{L}_{aa}(T_a)$ and $\hat{L}_{bb}(T_b)$ biject to $G(T_a)$ and $G(T_b)$, and we may assume by induction that $\hat{L}_{bc}(T_b)$ bijects to $G(T_b)/G_c$. Hence $\hat{L}_{aa}(T_a) L_{ab}$ and $\hat{L}_{bb}(T_b) L_{ba}$ biject to $G(T_a)/C_{ab}$ and $G(T_b)/C_{ab}$. Now $G(T) = G(T_a) \star_{C_{ab}} G(T_b)$, and thus $G(T)/G(T_b)$ bijects to $\hat{L}_a(T_a) L_{ab} ((\hat{L}_b(T_b) L_{ba})^- (\hat{L}_a(T_a) L_{ab})^-)^\star$. Thus, $G(T)/G_c$ bijects to the language $\hat{L}_{ac}(T)$, as required. $\square$

We are now prepared to define a language for $G(Y)$. We take as our generating set

$$\mathcal{G} = \{s_j^{\pm 1}\} \cup \coprod_{i=0}^{n} \mathcal{G}_i.$$

For each edge $s_j \notin T$, let C_j be the edge group carried by s_j. Then for each s_j and $k = 0, 1$ the language $L(G_{\partial_k s_j}/\partial_k C_j)$ is regular by assumption. Notice that

$$L(G(T)/\partial_k C_j) = \hat{L}_{a\partial_k s_j}(T) L(G_{\partial_k s_j}/\partial_k C_j)$$

bijects to $G(T)/\partial_k C_j$.

Let $\delta_0 = 1$, $\delta_1 = -1$. We may now write

$$\begin{aligned} L(G(Y)) = &\Big\{ \Big(\bigcup_{j,k} L(G(T)/\partial_k C_j)(\{s_j^{\delta_k}\}^\star \setminus \{\iota\}) \Big)^\star L_{aa}(T) \Big\} \\ &\setminus (\mathcal{G}^\star (\bigcup_j \{s_j s_j^{-1}, s_j^{-1} s_j\}) \mathcal{G}^\star). \end{aligned}$$

Since $G(Y)$ is an HNN extension of $G(T)$ with stable letters $\{s_j\}$, by Britton's lemma, $L(G(Y))$ bijects to $G(Y)$ (see, for example, **[LS]**). By construction, $L(G(Y))$ is regular. It remains to show that it is the language of an (asynchronous) automatic structure for $G(Y)$.

In order to prove this, we will require the following lemma.

Lemma 3.3. *If $y \in L_{aa}(T)$ and $y' \in L_{ac}(T)$ with $\overline{y} = \overline{y'}$ then y and y' are (asynchronous) (pK)-fellow travellers, where $p \leq n$ is the distance from a to c in T.*

Proof. We will proceed by induction on the length of the geodesic in T from a to c. There is nothing to prove in the case $a = c$. Hence we may assume that $a \neq c$. We let b be the vertex between a and c at distance 1 from c. Consider $y'' \in L_{ab}(T)$ with $\overline{y} = \overline{y''}$. By induction, we may assume that y and y'' are (asynchronous) $(p-1)K$-fellow travellers. y'' has the form $w_1 \dots w_m$ where $w_m \in L_b$, and $w_{m-1} \in L_{db}$. (d is a vertex adjacent to b.) We distinguish two cases:

Case 1. $d \neq c$. In this case, there is $w'_m \in L_{bc}$ and $w_{m+1} \in L_c$ so that $\overline{w_m} = \overline{w'_m w_{m+1}}$. By the assumptions of the theorem, w_m and $w'_m w_{m+1}$ are (asynchronous) K-fellow travellers. Further, $w_1 \dots w_{m-1} w'_m w_{m+1} \in L_{ac}(T)$, so $y' = w_1 \dots w_{m-1} w'_m w_{m+1}$. Thus y' and y'' are (asynchronous) K-fellow travellers. Hence, y' and y are (asynchronous) pK-fellow travellers.

Case 2. $d = c$. If $\overline{w_m}$ lies in $\partial_0 C_{bc}$, then let $w'_{m-1} \in L_c$ with $\overline{w'_{m-1}} = \overline{w_{m-1} w_m}$. As usual, these (asynchronously) K-fellow travel, and $y' = w_1 \dots w_{m-2} w'_{m-1}$.

If $\overline{w_m} \notin \partial_0 C_{bc}$, rewrite $\overline{w_m}$ as $\overline{w'_m w_{m+1}}$ with $w'_m \in L_{bc}$, $w_{m+1} \in L_c$. Again, these (asynchronously) K-fellow travel, and $y' = w_0 \dots w'_m w_{m+1}$. □

Lemma 3.4. *$L(G(Y))$ is the language of an (asynchronous) automatic structure.*

Proof. Let $w, w' \in L(G(Y))$, with $g \in \mathcal{G}$, and $\overline{w} = \overline{w' g}$. Then w' has the form xy' where x is either empty, or ends in some $s_j^{\pm 1}$, and $y' \in L_{aa}(T)$. We distinguish two cases.

Case 1. $g \notin \{s_j^{\pm 1}\}$. There is a unique $y \in L_{aa}(T)$ so that $\overline{y} = \overline{y'g}$. See Fig. 2. By Theorem 1, y and y' are (asynchronous) fellow travellers. Clearly $w = xy$, so that w and w' are (asynchronous) fellow travellers.

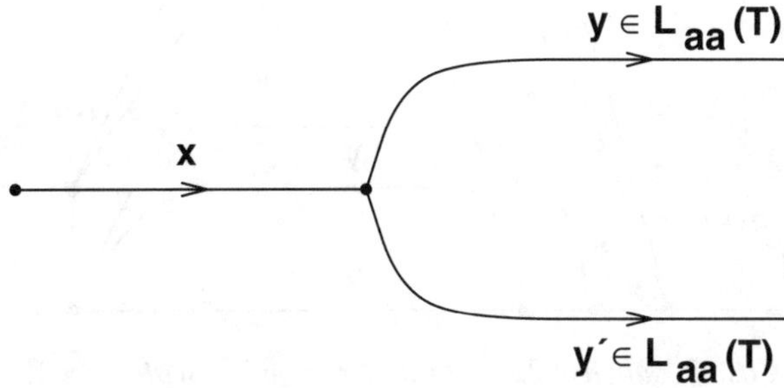

Figure 2: *Case 1.* $g \notin \{s_j^{\pm 1}\}$.

Case 2. $g = s_j$. (The case $g = s_j^{-1}$ follows similarly.) Let $b = \partial_0 s_j$. There is a unique $y \in L_{ab}(T)$ such that $\overline{y} = \overline{y'}$ and by Lemma 3.3, y and y' (asynchronously) fellow travel. We may write $\overline{y} = \overline{uv}$ where $u \in \hat{L}_{ab}(T)L(G_b/\partial_0 C_j)$, $\overline{v} \in \partial_0 C_j$, and $v \in L_b$. This determines u and v, and by the assumptions of the theorem, y and uv are (asynchronous) fellow travellers. Say $\overline{v} = \partial_0 \alpha$. Let $t \in L_{\partial_1 s_j}$ be the word such that $\overline{t} = \partial_1 \alpha$. By assumption, v and $s_j t s_j^{-1}$ are (asynchronous) K-fellow travellers, and hence, so are v and $s_j t$. Let $z \in L_{aa}(T)$ with $\overline{t} = \overline{z}$. By Lemma 3.3, these are (asynchronous) nK-fellow travellers.

Subcase 1. $u \neq \iota$ or x does not end in s_j^{-1}. See Fig. 3. Then $xus_jz \in L(G(Y))$ and since $\overline{w's_j} = \overline{xus_jz}$, this latter is w.

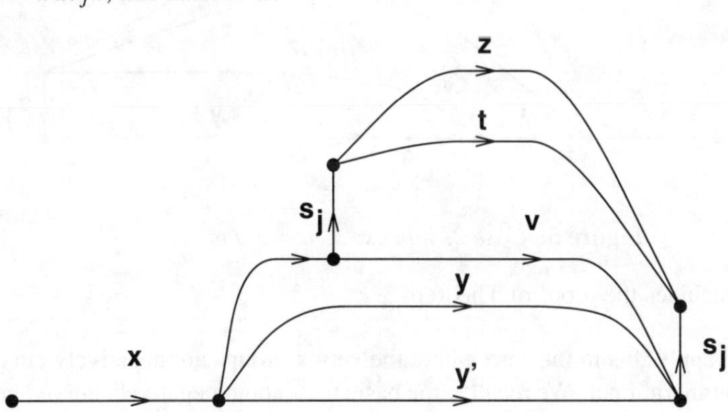

Figure 3: *Case 2, subcase 1.* $u \neq \iota$*, or* x *does not end in* s_j^{-1}.

Subcase 2. $u = \iota$ and $x = ps_j^{-m}$ with $m > 1$. See Fig. 4. Then $w = ps_j^{-m+1}z$.

Subcase 3. $u = \iota$ and $x = ps_j^{-1}$. See Fig. 5. We rewrite $p = qr$ where q is either empty, or ends in some $s_j^{\pm 1}$, and $r \in \hat{L}_{ab}(T)L(G_b/\partial_1 C_j)$. Notice that

$$\overline{w} = \overline{qrz}\,.$$

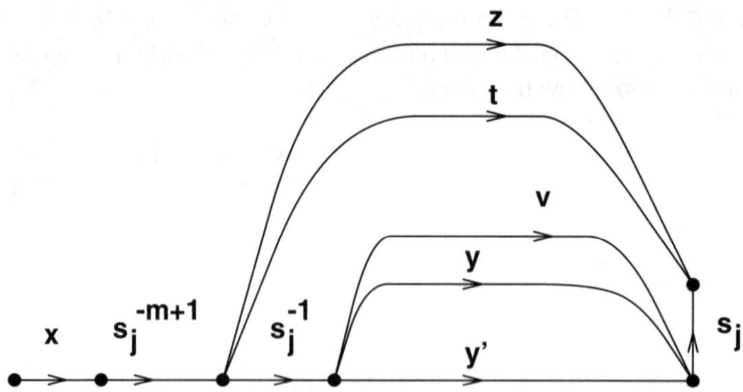

Figure 4: *Case 2, subcase 2.* $u = \iota$ *and* $x = ps_j^{-m}$ *with* $m > 1$.

(Again, $\overline{z} = \overline{s_j^{-1} y' s_j}$ and these are (asynchronous) fellow travellers.) There is a unique word $\beta \in L_{ab}(T)$ so that $\overline{\beta} = \overline{rz}$, and by the assumptions of the theorem, these are (asynchronous) fellow travellers. Finally, if $\gamma \in L_{aa}(T)$ with $\overline{\beta} = \overline{\gamma}$, then by Lemma 3.3, β and γ are fellow travellers.

Now $q\gamma \in L(G(Y))$, and $\overline{q\gamma} = \overline{w' s_j}$. Hence $w = q\gamma$, and w and w' are (asynchronous) fellow travellers as required.

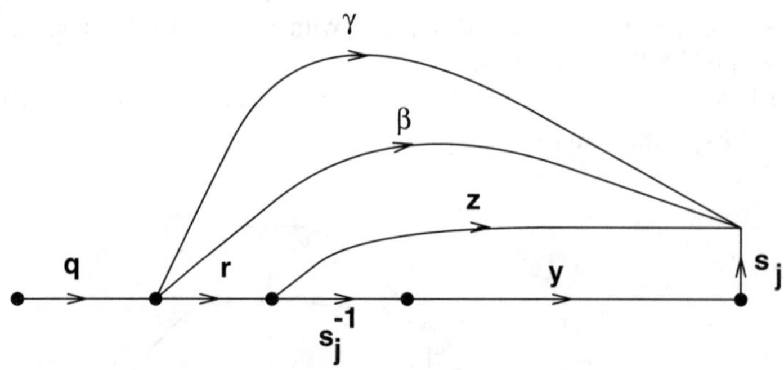

Figure 5: *Case 2, subcase 3.* $u = \iota$ *and* $x = ps_j^{-1}$.

This concludes the proof of Theorem 3. □

We will apply this in the case where the vertex groups are negatively curved and the edge groups are rational. We recall some basic facts about negatively curved groups. If G is a negatively curved group, and $\mathcal{G}$ is a generating set for G, then the set of all geodesics in G with respect to this generating set is regular. (See, for example, **[Gr]**.) If C is a subgroup of G, then C is rational with respect to the set of all geodesics in G if and only if the embedding of C into G is a quasiisometry for some (hence any) generating set. In particular, this implies that C is rational with respect to the set of geodesics for one generating set for G if and only if C is rational with respect to the set of geodesics for any generating set for G. (See **[BGSS]** or **[GS]**.)

Theorem 4. *Let Y be a finite graph of groups with vertex set $\{v_1, \ldots, v_n\}$ such that the vertex groups $G_1, \ldots, G_n$ are negatively curved and for each edge group C_e and for $k = 0, 1$, $\partial_k(C_e)$ is a rational subgroup of $G_{\partial_k e}$. Then $G(Y)$ is asynchronously automatic. Further, if there exist generating sets $\mathcal{G}_1, \ldots, \mathcal{G}_n$ and a constant, ϵ, such that for each edge e with (say) $\partial_0(e) = i$, $\partial_1(e) = j$ and for each $\alpha \in C_e$,*

$$| \ell_{\mathcal{G}_i}(\alpha) - \ell_{\mathcal{G}_j}(\alpha) | \leq \epsilon,$$

then $G(Y)$ is automatic.

Proof. We check that Y satisfies the assumptions of the preceding theorem. We fix generating sets $\mathcal{G}_1, \ldots, \mathcal{G}_n$ for the vertex groups $G_1, \ldots, G_n$. If possible, we choose these so that for each edge e with (say) $\partial_0(e) = i$, $\partial_1(e) = j$ and for each $\alpha \in C_e$,

$$| \ell_{\mathcal{G}_i}(\alpha) - \ell_{\mathcal{G}_j}(\alpha) | \leq \epsilon.$$

We take a lexicographic ordering for the elements of each $\mathcal{G}_i$. For each i, we take L_i to be the language of lexicographically least geodesics for each element of G_i. It is a standard result that this gives an automatic structure for G_i.

Condition (1) is satisfied by assumption.

If $v_i = \partial_k e$, we take $L(G_i/\partial_k C_e)$ to consist of the shortest and lexicographically least element of each coset. It is shown in **[BGSS]** that this language is regular, thus satisfying condition (2) and that it satisfies condition (4).

We must check condition (3). For this, we will use results on quasigeodesics in negatively curved groups. We recall some definitions and basic results. Given a pair of metric spaces (A, d_A) and (B, d_B), we say a map f from A to B is a (λ, ϵ)-***quasiisometry*** if there are positive constants $\lambda \leq 1$ and ϵ so that for each $x, y \in A$, $\lambda d_B(f(x), f(y)) - \epsilon \leq d_A(x, y) \leq \lambda^{-1} d_B(f(x), f(y)) + \epsilon$. A (λ, ϵ) -quasiisometry of the interval $[0, t]$ (into some metric space, (X, d_X)) is called a (λ, ϵ)-quasigeodesic. A standard result says that if (X, d_X) is negatively curved (for our purposes, the Cayley graph of a negatively curved group), there is an ϵ' depending only on (λ, ϵ) and X such that if σ is a geodesic in X and γ is a (λ, ϵ)-quasigeodesic in X so that $\sigma(0) = \gamma(0)$ and $\sigma(\ell(\sigma)) = \gamma(\ell(\gamma))$, then for all t, $d_X(\sigma(t), \gamma) \leq \epsilon'$ and $d_X(\sigma, \gamma(t)) \leq \epsilon'$. Thus each of σ and γ lie entirely in an ϵ' neighborhood of each other. Now this suggests that there should be K depending only on X, λ and ϵ so that σ and γ should be asynchronous K-fellow travellers. In fact, this is so, and can be shown using Cannon's **[Ca]** notion of "progression in geodesic corridors". (See **[BGSS]**.)

To verify condition (3), we pick a maximal tree, T, and first check when $e \in T$. We let $\partial_0 e = i$, $\partial_1 e = j$. We take an arbitrary $\alpha \in C_e$ and let $u \in L_i$, $v \in L_j$ with $\overline{u} = \partial_0 \alpha$, $\overline{v} = \partial_1 \alpha$. We must show that u and v are asynchronous fellow travellers. See Fig. 6

Since C_e is quasiconvex in G_j, there is ϵ_0, independent of v, so that for all T, $d_{\mathcal{G}_j}(v(t), \partial_0 C_e) \leq \epsilon_0$. Thus, for each integer t we may pick $c(t) \in \partial_1 C_e = \partial_0 C_e$ so that $d_{\mathcal{G}}(c(t), v(t)) \leq \epsilon_0$. We may pick $c(0) = 1$ and $c(\ell(v)) = \overline{v}$. We form a path γ in $\Gamma_{\mathcal{G}_i}(G_i)$ by connecting successive points $c(t)$ by geodesics in $\Gamma_{\mathcal{G}_i}(G_i)$. Clearly v and γ are asynchronous fellow travellers in $\Gamma_{\mathcal{G}}(G)$. We will show that u and γ are asynchronous fellow travellers in $\Gamma_{\mathcal{G}_i}(G_i)$. To do this, it suffices to show that there are λ' and ϵ' independent of α so that γ is a (λ', ϵ')-quasigeodesic.

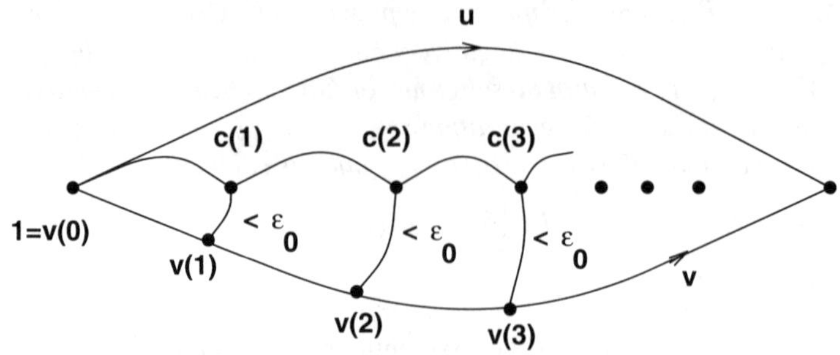

Figure 6: *u and v are asynchronous fellow travellers.*

Now, since C_e is quasiconvex in G_i and G_j, C_e is quasiisometrically embedded in each of these. (See **[GS]** or **[BGSS]**.) That is to say, given some (hence any) generating set $\mathcal{C}$ for C_e, there are constants λ_C and ϵ_C such that the inclusions of $(C_e, d_{\mathcal{C}})$ into $(G_i, d_{\mathcal{G}_i})$ and $(G_j, d_{\mathcal{G}_j})$ are (λ_C, ϵ_C)-quasiisometries. (These inclusions are defined only on vertices and we take the metrics as restricted to these.) It follows that there are constants λ_1 and ϵ_1 so that (taking the metrics restricted to C_e), the identity map between $(C_e, d_{\mathcal{G}_i})$ and $(C_e, d_{\mathcal{G}_j})$ is a (λ_1, ϵ_1)-quasiisometry.

Now let a and b be points on γ so that (say) a lies between $c(t_0)$ and $c(t_0+1)$ and b lies between $c(t_0+k+1)$ and $c(t_0+k+2)$. For each t and r, $r - 2\epsilon_0 \leq d_{\mathcal{G}_j}(c(t), c(t+r)) \leq r + 2\epsilon_0$. Thus $\lambda_1 r - \epsilon_2 \leq d_{\mathcal{G}_i}(c(t), c(r)) \leq \lambda_1^{-1} r + \epsilon_2$, where $\epsilon_2 = \lambda_1^{-1}\epsilon_0 + \epsilon_1$. In particular, $d_{\mathcal{G}_i}(c(t), c(t+1)) \leq \lambda^{-1} + \epsilon_2$, and thus $\lambda k - 2(\lambda^{-1} + \epsilon_2) \leq d_{\mathcal{G}_i}(a, b) \leq \lambda^{-1} k + 2(\lambda^{-1} + \epsilon_2)$. On the other hand, the length in $\Gamma_{\mathcal{G}_i}(G_i)$ of the segment of γ between a and b is at least $d_{\mathcal{G}_i}(a, b)$, and at most $(k+2)(\lambda_1^{-1} + \epsilon_2)$. Thus γ is a (λ', ϵ')-quasigeodesic, where $\lambda' = (\lambda_1^{-1} + \epsilon_2)/\lambda_1$ and $\epsilon' = 2 + 2(\lambda_1^{-1} + \epsilon_2)$. In particular, u and γ are asynchronous fellow travellers, and thus so are u and v.

We must show that under the additional hypothesis on lengths, u and v are synchronous fellow travellers. Now for each t, $v(t)$ is within ϵ_0 of $c(t)$ in $\Gamma_{\mathcal{G}_j}(G_j)$. Since γ and u asynchronously fellow travel, there is some s so that $u(s)$ is within (say) ϵ_0' of $c(t)$. But this means $\mid \ell_{\mathcal{G}_i}(c(t)) - s \mid \leq \epsilon_0'$. Similarly $\mid \ell_{\mathcal{G}_j}(c(t)) - t \mid \leq \epsilon_0$. But since $\mid \ell_{\mathcal{G}_i}(c(t)) - \ell_{\mathcal{G}_j}(c(t)) \mid \leq \epsilon$, it follows that $\mid t - s \mid \leq \epsilon + \epsilon_0 + \epsilon_0'$ Thus $d_{\mathcal{G}}(v(t), u(t)) \leq \epsilon + 2(\epsilon_0 + \epsilon_0')$. Thus u and v synchronously fellow travel.

We must now consider the case where $e = s_m \notin T$. We must show that u and $s_m v s_m^{-1}$ are asynchronous fellow travellers, and that they are synchronous fellow travellers under the additional hypothesis on lengths. Here we may take vertices $c(t)$ to "shadow"$s_m v$ in $\partial_1 C_e$ as before. We take $c'(t) = c(t)s_m^{-1}$, and let γ be the path formed by connecting successive $c'(t)$ by geodesics in $\Gamma_{\mathcal{G}_i}(G_i)$. As before, u and γ are asynchronous fellow travellers. Since $d_{\mathcal{G}}(c(t), c'(t)) = 1$, γ stays close to $\{c(t)\}$ and hence also asynchronously fellow travels with $s_m v$. Thus u asynchronously fellow travels with $s_m v$ and hence with $s_m v s_m^{-1}$. We argue as before under the additional hypothesis on lengths. □

It is worth pointing out that Theorem 3 and its special case Theorem 4 do not always give the best possible result. For example, suppose $G_0 = \pi_1(S)$ is a closed hyperbolic surface group, and G is the HNN extension of this group arising from a pseudo Anosov map $f : S \longrightarrow S$. Then by a well known result of Thurston, G is the fundamental group of a closed hyperbolic 3-manifold, M. In particular, G is negatively curved and hence automatic. However, as g ranges over the elements of G_0 the words for g and $tf_\star(g)t^{-1}$ are only asynchronous fellow travellers. (Here t is the stable letter.) Thus Theorem 4 only tells us that G is asynchronously automatic. It is interesting to compare the asynchronous automatic structure $L(G(Y))$ with an automatic structure for G consisting of geodesics. Since G is an HNN extension of G_0 with all of G_0 as the base, the automatic structure $L(G(Y))$ has the form $(\{t\}^\star \cup \{t^{-1}\}^\star)L_0$. L_0, the sublanguage for $\pi_1(S)$ in $L(G(Y))$ is regular, hence $\pi_1(S)$ is $L(G(Y))$-rational. (In fact, in any application of Theorem 3, each vertex group (and hence, each edge group) will be $L(G(Y))$ rational. We leave the details to the reader.)

But while $\pi_1(S)$ is an $L(G(Y))$ rational subgroup, it is not quasiisometrically embedded in $\pi_1(M)$ and hence not $L(M)$ rational, where $L(M)$ is the language of geodesics with respect to an arbitrary generating set $\mathcal{M}$ for $\pi_1(M)$. To see this, we need a result of **[GS]** which states that a subgroup G_0 of a negatively curved group G is rational with respect to a language of geodesics in G if and only if the embedding of G_0 into G is a quasiisometry. Now G_0 and G act on $\mathbb{H}^2$ and $\mathbb{H}^3$ respectively by isometries. Each element g of G and G_0 corresponds to a unique shortest closed geodesic in S or M respectively, and we take the ***translation lengths*** $t_S(g)$ and $t_M(g)$ to be the lengths of these geodesics. (We overburden the wonderful letter, ℓ, by denoting the length of a curve σ in $\mathbb{H}^2$ or $\mathbb{H}^3$ by $\ell(\sigma)$.) These translation lengths are comparable to word lengths in the respective groups. That is to say, given generating sets $\mathcal{M}$ and $\mathcal{S}$ for $\pi_1(M)$ and $\pi_1(S)$, there are λ_S, ϵ_S, λ_M and ϵ_M so that for each $g \in G_0$, $\lambda_S \ell_{\mathcal{S}}(g) - \epsilon_S \leq t_S(g) \leq \lambda_S^{-1}\ell_{\mathcal{S}}(g) + \epsilon_S$, and for each $g \in G$, $\lambda_M \ell_{\mathcal{M}}(g) - \epsilon_M \leq t_M(g) \leq \lambda_M^{-1}\ell_{\mathcal{M}}(g) + \epsilon_M$.

A pseudo-Anosov map f gives rise to a pair of measured foliations $\mathcal{F}$ and $\mathcal{F}'$ by geodesics, and a $\mu > 1$ so that f stretches each leaf of $\mathcal{F}$ by a factor of μ and contracts each leaf of $\mathcal{F}'$ by a factor of μ. (See **[FLP]**.) We may find an arbitrarily long segment σ' of $\mathcal{F}'$ whose endpoints are separated by an arbitrarily short segment σ of $\mathcal{F}$. These two together form an embedded curve in S representing (say) $g \in G_0$. Now $\ell(\sigma') - \ell(\sigma) \leq t_S(g) \leq \ell(\sigma') + \ell(\sigma)$. For each n, we have $\overline{g} = t^n f_\star^n(g)t^{-n}$ and $t_S(f_\star^n(g)) \leq \ell(f^n(\sigma'\sigma)) \leq \mu^{-n}\ell(\sigma') + \mu^n\ell(\sigma)$. Thus, while $\ell_S(g) \geq \lambda_S(\ell(\sigma') - \ell(\sigma)) - \epsilon_S$, $\ell_M(g) \leq 2n + \lambda_S^{-1}(\mu^{-n}\ell(\sigma') + \mu^n\ell(\sigma)) + \epsilon_S$. If we fix n, we may choose σ and σ' so that $\lambda_S\mu^{-n}\ell(\sigma') \geq 2n + \lambda_S\mu^n\ell(\sigma)$. We then have $\ell_M(g) \leq 2\lambda_S^{-2}\mu^{-n}\ell_S(g)$. Letting n increase without bound shows the embedding of G_0 into G is not a quasiisometry.

The situation for free products with amalgamation is much better behaved. Suppose $G = A \star_C B$, and suppose the inclusions of C into A and B are quasiisometries. Then A is quasiisometrically embedded in G. (And hence, C is quasiisometrically embedded in G.) To see this, fix generating sets $\mathcal{A}$ and $\mathcal{B}$ for A and B, and suppose w is a geodesic in $\Gamma_{\mathcal{A}}(A)$. Let v be a geodesic in $\Gamma_{\mathcal{A}\cup\mathcal{B}}(G)$ with $\overline{w} = \overline{v}$. Since $\overline{v} \in A$, v alternates substrings which lie in $\Gamma_{\mathcal{A}}(A)$ with substrings representing elements of C. That is to say, $v = a_1c_1 \ldots a_nc_n$ where each $a_i \in \mathcal{A}^\star$, each $c_i \in \mathcal{B}^\star$ with $\overline{c_i} \in C$. Then we can rewrite each c_i as $c_i' \in \mathcal{A}^\star$, and since C is quasiisometrically embedded in A and B, we have $\ell(c_i') \leq \lambda\ell(c_i) + \epsilon$. Consequently $\ell(w) \leq \lambda\ell(v) + n\epsilon \leq (\lambda + \epsilon)\ell(v)$.

Theorem 4 has the following corollaries.

Corollary 4.1. *Let Y be a finite graph of groups in which each edge group and each vertex group is a finitely generated free group. Then $G(Y)$ is asynchronously automatic.*

Proof. This follows by noticing that a finitely generated subgroup of a finitely generated free group is quasiconvex and hence rational. □

Corollary 4.2. *Let H be a rational subgroup of a negatively curved group K. Let Y be a finite graph of groups such that for each vertex v there is a fixed isomorphism $f_v : G_v \to K$. Suppose that for each edge e, $f_{\partial_0 e}$ is an isomorphism from $\partial_0 C_e$ to H, and that for each $\alpha \in C_e$, $f_{\partial_0}(\partial_0 \alpha) = f_{\partial_1}(\partial_1 \alpha)$. Then $G(Y)$ is automatic.*

Proof. Since the vertex groups are all isomorphic, we take disjoint copies of "the same" generating set for each group. Now each element of an edge group injects as "the same" element of each vertex group and thus has the same length in each vertex group. Thus by Theorem 4 (with $\epsilon = 0$) $G(Y)$ is automatic. □

Theorem 5. *Let T be a finite tree of groups in which each vertex group is negatively curved and each edge group is infinite cyclic. Then $G(T)$ is automatic.*

Proof. By a result of **[Gr]**, an infinite cyclic subgroup of a negatively curved group is quasigeodesic, and hence, rational. Consequently, T satisfies the asynchronous conditions of Theorem 4. We take the vertices of T to be $\{v_1, \ldots, v_n\}$, the vertex groups to be $G_1, \ldots, G_n$ and the edge groups to be $\{C_{ij}\}$ with $C_{ij} = \langle c_{ij} \rangle$. We need only prove the following lemma.

Lemma 5.1. *We may find generating sets $\mathcal{G}_1, \ldots, \mathcal{G}_n$ for $G_1, \ldots, G_n$ and ϵ such that if $\alpha \in C_{ij}$, then*

$$| \ell_{\mathcal{G}_i}(\partial_0 \alpha) - \ell_{\mathcal{G}_j}(\partial_1 \alpha) | \leq \epsilon .$$

Proof. There is nothing to prove if $n = 0$. Hence, we may assume $n \geq 1$ and by renumbering if necessary we may assume that v_0 and v_1 are connected by an edge. Deleting this edge disconnects T into two components, say $T_0 \ni v_0$ and $T_1 \ni v_1$. Each of T_0 and T_1 has fewer vertices than T. Hence, by induction, we may assume that generating sets $\mathcal{H}_1, \ldots, \mathcal{H}_n$ and ϵ' have been chosen and that these satisfy the lemma for T_0 and T_1. Suppose $\alpha = c_{01}^m \in C_{01}$. By a lemma of **[BGSS]**, there are $\lambda_0, \lambda_1 \in \mathbb{Q}$ and ϵ'' so that for all m, and $k = 0, 1$,

$$| \ell_{\mathcal{H}_k}(\partial_k c_{01}^m) - \frac{m}{\lambda_k} | \leq \epsilon'' .$$

We may find integers $\rho_0, \rho_1 \geq 1$ so that $\lambda_0 \rho_0 = \lambda_1 \rho_1$. We set

$$\mathcal{G}_i = \mathcal{H}_i \cup \ldots \cup (\mathcal{H}_i)^{\rho_k}$$

where $v_i \in T_k$.

If $x \in G_i$ we then have

$$| \ell_{\mathcal{G}_i}(x) - \frac{\ell_{\mathcal{H}_i}(x)}{\rho_k} | \leq 1 .$$

(Again $v_i \in T_k$.) Thus, for $\alpha \in C_{01}$,

$$| \ell_{\mathcal{G}_0}(\partial_0 \alpha) - \ell_{\mathcal{G}_1}(\partial_1 \alpha) | \leq \epsilon'' + 2 ,$$

while for $\alpha \in C_{ij} \neq C_{01}$,

$$| \ell_{\mathcal{G}_i}(\partial_0 \alpha) - \ell_{\mathcal{G}_j}(\partial_1 \alpha) | \leq \epsilon' + 2.$$

Thus $\mathcal{G}_1, \ldots, \mathcal{G}_n$ satisfy the lemma with $\epsilon = \max\{\epsilon' + 2, \epsilon'' + 2\}$. This completes the proof of Lemma 5.1 and Theorem 5. □

Suppose Y is a finite graph of groups in which each edge group and each vertex group is infinite cyclic. By Theorem 4, $G(Y)$ is asynchronously automatic. Let ℓ be a loop based at vertex v_1 of Y. We may find an element $c_1^p \in C_1 = \langle c_1 \rangle$ such that carrying c_1^p around ℓ takes c_1^p to c_1^q. Thus $G(Y)$ contains a Baumslag-Solitar group $\langle c, t \mid tc^p t^{-1} = c^q \rangle$. If, for every simple loop of Y (and hence for every loop of Y), $p = \pm q$, then applying Lemma 5.1 to a maximal sub-tree of Y shows $G(Y)$ is automatic. Gersten **[Ge2]** has shown that this result is best possible. That is $G(Y)$ is automatic only if there is no loop in Y for which $p \neq \pm q$.

With proper conditions, Theorem 3 may be applied to a graph product of finitely generated abelian groups.

Definition. Let $G(Y)$ be a graph of groups in which each vertex (and hence each edge) group is finitely generated abelian. We may write each vertex group G_i and each edge group C_e as $G_i = F_i \times H_i$, and $C_e = F_e \times H_e$ where each F_i and F_e are free abelian, and each H_i and H_e are finite. Suppose we may find ordered bases $F_i = \langle x_{i1}, \ldots, x_{im(i)} \rangle$, $F_e = \langle y_{e1}, \ldots, y_{em(e)} \rangle$ and that for each e, j and $k = 0, 1$, we have $\partial_k y_{ej} \in \langle x_{\partial_k e j} \rangle \times H_{\partial e}$ We then say that the edge groups ***intersect well***.

Theorem 6. *Let Y be a finite graph of groups in which each vertex group is finitely generated abelian. If the edge groups intersect well, then $G(Y)$ is asynchronously automatic. If Y is a tree, then $G(Y)$ is automatic.*

Proof. For each vertex group G_i we take generating set $\mathcal{G}_i = \{x_{i1}^{\pm 1}, \ldots, x_{im(i)}^{\pm 1}\} \cup H_i$. Let $X_{ip} = \{x_{ip}\}^{\star} \cup \{x_{ip}^{-1}\}^{\star}$. Then $L_i = X_{im(i)} \ldots X_{i1} H_i$ is the language of an automatic structure for G_i. It is easy to check that these languages satisfy the requirements of Theorem 3. In the case that Y is a tree, one modifies these languages, proceeding as in Lemma 5.1. □

Having discussed graph products of negatively curved groups, and graph products of finitely generated abelian groups, one might ask whether the methods here can be combined for graph products of finite volume hyperbolic groups or geometrically finite hyperbolic groups. Indeed they can.

Automatic structures for geometrically finite hyperbolic groups appear in **[CEHPT]**, and **[Ta]** gives similar methods in the case of finite volume automatic groups. His methods are quite similar to those of **[CEHPT]** and generalize to the larger case. We will give our treatment in terms of finite volume hyperbolic groups.

Before describing these automatic structures, we recall briefly the geometric situation which gives rise to them. Let G be a finite volume hyperbolic group. Then G acts via isometries on $\mathbb{H}^n$ with quotient M. The cusps of M are covered by horoballs in $\mathbb{H}^n$. By taking a thick/thin decomposition of M, one obtains a G equivariant system of these horoballs. Let X be the space obtained from $\mathbb{H}^n$ by deleting the interiors of these horoballs. We measure lengths of paths in X using the Riemannian metric of $\mathbb{H}^n$, and endow X with the path metric. Then G acts co-compactly on X by isometries. Consequently the Cayley graph of G (with respect to any finite generating set) is quasiisometric to X. In fact, one may take this quasiisometry to be an embedding of the vertices of $\Gamma(G)$ into X.

Now the geodesics of X consist of what Tatsuoka terms "glancing geodesics". These are smooth paths which are made up piecewise of hyperbolic geodesics in the interior of X alternating with Euclidean geodesics lying on the horospherical boundaries of X. Given such a glancing geodesic, γ, there is a retraction r_γ of X onto γ which is strongly distance decreasing away from the horospheres. One simply retracts X orthogonally onto γ where possible and takes the remainder of X to one or the other of the end points of γ. It follows that for any quasigeodesic w in the Cayley graph of G, that w follows close to the hyperbolic segments of the corresponding geodesic γ, and that w travels close to the appropriate horospheres. One then finds an automatic structure for G by choosing a bi-automatic structure for each parabolic subgroup and looking at those words which follow the chosen structure in the parabolic subgroup whenever they are close to the appropriate horosphere, and taking the shortest representative for each element from among these words.

We summarize these results in the following paraphrase.

Theorem ([CEHPT], [Ta]). *Let G be a finite volume hyperbolic group. Let $P_1, \ldots, P_k$ be parabolic groups for the cusps of $\mathbb{H}^n/G$, and let $\mathcal{P}_1, \ldots, \mathcal{P}_k$ be generating sets for these groups. For each j, $1 \leq j \leq k$, suppose that $(\mathcal{P}_j, L_j)$ is a bi-automatic structure with uniqueness for P_j. Suppose $\mathcal{G} \supset \cup_{j=1}^k \mathcal{P}_j$ is a generating set for G. Then there is an automatic structure with uniqueness $(\mathcal{G}, L)$ for G. There is a constant K so that L consists of the shortest and lexicographically least words among those words w in $\mathcal{G}^\star$ with the property that if $w = xyz$, $\ell(y) \geq K$, and y is a maximal length subword of xyz so that $\overline{y} \in P_j$ then $y \in L_j$.*

If we wish to find automatic structures for amalgams of these groups, we need to take care that the edge groups intersect well inside the parabolic subgroups of the vertex groups.

Definition. Let Y be a finite graph of groups in which each vertex group G_i is finite volume hyperbolic, each edge group C_e is abelian. Suppose further that for $\partial_0 e = v_i$ that $\partial_0 C_e < P_j < G_i$ where P_j is a maximal parabolic subgroup of G_i. Suppose Z is a graph of groups whose underlying graph is a connected subgraph of the underlying graph of Y. Suppose each edge group of Z is the corresponding edge group of Y and that each vertex group of Z is a maximal parabolic subgroup of the corresponding vertex group of Y. We call such a graph an ***abelian subgraph*** of Y. We say the edge groups ***intersect well*** in Y if the edge groups intersect well in each abelian subgraph of Y.

Theorem 7. *Let Y be a graph of groups in which each vertex group G_i is finite volume hyperbolic, and each edge group C_e is free abelian and for $k = 0, 1$, $\partial_k(C_e)$ consists of parabolics in $G_{\partial_k e}$. Suppose that the edge groups intersect well. Then $G(Y)$ is asynchronously automatic. If each abelian subgraph of Y is a tree, $G(Y)$ is automatic.*

Proof. We start by picking automatic structures for the parabolic subgroups of the vertex groups. For each maximal parabolic subgroup, P_j, examine the maximal abelian subgraph Z of Y which contains it. Since the edge groups intersect well here, we may find automatic structures for each vertex group of Z which are asynchronously compatible as in Theorem 6. If each abelian subgraph is a tree, we may choose these structures to be synchronously compatible as in Theorem 6. In this way, we choose automatic structures for each maximal parabolic subgroup P_j of G_i such that for each edge group C_e such that each $\partial_0 C_e < P_j$, C_e is rational with respect to this structure and admits the usual language of coset representatives. Let the structure on P_j be $(\mathcal{P}_j, L(P_j))$. Now for each G_i, let $(\mathcal{G}_i, L_i)$ be an automatic structure with uniqueness for G_i determined by a generating set $\mathcal{G}_i$ containing the appropriate $\mathcal{P}_j$ together with $\{L(P_j) \mid P_j < G_i\}$ as described above. We start by showing that each P_j is L_i rational.

Lemma 7.1. *Let L_i be as above. Then P_j is L_i rational.*

Proof. Let G_i act on X_i as above. We check that P_j is L_i quasiconvex. This follows easily. For if $w \in L_i$ and $\overline{w} \in P_j$ then w stays within a bounded distance of the appropriate horosphere. But P_j acts co-compactly on this horosphere. Hence for each t, $w(t)$ is within a bounded X_i distance, and hence a bounded Cayley graph distance, of a point of P_j. Hence P_j is L_i quasiconvex. By **[GS]**, P_j is L_i-rational. □

Lemma 7.2. *Let $L(G_i/P_j)$ consist of those words of $w \in L_i$ that are shortest and lexicographically least in their respective cosets $\overline{w}P_j$. Then $L(G_i/P_j)$ is regular.*

Proof. The proof follows along the lines of Lemmas 1–3 of the proof of Theorem C of **[BGSS]**. We start by proving that there exists a K such that if $w \in L_i$ is not shortest in $\overline{w}P_j$ then there is some g with $\ell_{\mathcal{G}_i}(g) \le K$, $\overline{g} \in P_j$, and the representative w' in L_i for $\overline{wg}$ is shorter than w.

Suppose this is not so. Then for every K there is a $w \in L_i$ such that for any $w' \in L_i$ with $\ell_{\mathcal{G}_i}(w') < \ell_{\mathcal{G}_i}(w)$, and $\overline{w}P_j = \overline{w'}P_j$, we must have $d_{\mathcal{G}_i}(\overline{w}, \overline{w'}) \ge K$. In particular, this distance is large when we take w' to be the shortest representative of this coset.

Now, $\overline{w}$ and $\overline{w'}$ differ by an element of P_j. Hence w and w' both end near a common horosphere. Now only a very short portion of w' lies near this horosphere, for otherwise, we could find a shorter representative for the coset. Similarly, only a short portion of w lies near the horosphere, for if not, we could find a representative of the coset which is shorter than w and differs from w by a small amount, which contradicts our supposition.

We will now show that $\overline{w}$ and $\overline{w'}$ differ by a bounded amount. To see this, let α and α' be the glancing geodesics from 1 to $\overline{w}$ and $\overline{w'}$ respectively. Let β and β' be the initial segments of α and α' reaching from 1 to the horosphere. Then by the same reasoning as above, β and β' contain all but a bounded amount of α and α'. On the other hand, Lemma 3.3 of **[Ta]** tells us the following. We paraphrase.

Lemma [Ta]. *There is a constant K' such that if β and β' are glancing geodesics from a common point to a common horosphere, and each of β and β' contacts that horosphere only at their endpoints, then β and β' end within distance K' of each other.*

Hence there is a bound on the distance from the end of β to the end of β'. Thus the distance from $\overline{w}$ to $\overline{w'}$ is bounded in the X_i metric, and hence is bounded in the Cayley graph metric. This contradicts the assumption that w and w' end arbitrarily far apart.

A similar argument shows that there is a constant K such that if $w, w' \in L_i$ are both shortest representatives of $\overline{w}P_j = \overline{w'}P_j$, then $d(\overline{w}, \overline{w'}) \leq K$. Now the language

$$L_{2K} = \{(w, w') \in L_i \times L_i \mid \overline{w} = \overline{w'g}\ (1) \text{ with } \overline{g} \in P_j \text{ and } \ell(g) \leq 2K\}$$

is regular. Hence, so is the language consisting of those $w \in L_i$ such that if $(w, w') \in L_{2K}$ then either $w = w'$, $\ell(w) < \ell(w')$, or w is lexicographically prior to w'. But this latter is exactly $L(G_i/P_j)$. □

Lemma 7.3. *There is a constant K such that if $\overline{w} = \overline{uv}$ with $w, v \in L_i$, $u \in L(G_i/P_j)$ and $\overline{v} \in P_j$ then w and uv are K-fellow travellers.*

Proof. We know that only a short portion of u lies near the horosphere. On the other hand, the language $L(G_i)$ is constructed so that any portion of w lying near the horosphere must travel via the automatic structure for P_j. Hence, we may write $w = xy$ with y an element of the automatic structure for P_j, and only a short portion of x lies near the horosphere. Now x is a representative of the same coset as u. Let α_u and α_x be the glancing geodesics for $\overline{u}$ and $\overline{x}$ respectively. Let β_u and β_x be the initial segments of α_u and α_x running from 1 to the horosphere. Then by the lemma of **[Ta]**, these end a bounded distance apart. Consequently u and x end a bounded distance apart, and hence are fellow travellers. Now v and y start a bounded distance apart and end at the same place. Since the automatic structure for P_j is bi-automatic, this makes v and y fellow travellers. Consequently, w and uv are fellow travellers. □

Now to complete the proof of the Theorem 7, we must see that we have fulfilled the conditions of Theorem 3. Suppose $\partial_0 C_e < P_j < G_i$. Then since $\partial_0 C_e$ is $L(P_j)$ rational and P_j is L_i rational, it follows that $\partial_0 C_e$ is L_i rational. Thus condition (1) is fulfilled.

We consider conditions (2) and (3) together. Since the edge groups intersect well, $L(P_j)$ contains regular sublanguages $L(P_j/\partial_0 C_e)$ and $L(\partial_0 C_e)$. Further, these have the property that if $w \in L(P_j)$, $u \in L(P_j/\partial_0 C_e)$, and $v \in L(\partial_0 C_e)$ with $\overline{w} = \overline{uv}$ then w and uv are fellow travellers. Now the regular language $L(G_i/P_j)L(P_j/\partial_0 C_e)$ bijects to $G_i/\partial_0 C_e$, though it may not lie inside of L_i. However, if $w \in L_i$ and $w' \in L(G_i/P_j)L(P_j/\partial_0 C_e)$, with $\overline{w} = \overline{w'}$ then w and w' fellow travel. Consequently,

$$\{(w, w') \mid w \in L_i,\ w' \in L(G_i/P_j)L(P_j/\partial_0 C_e), \text{ and } \overline{w} = \overline{w'}\}$$

is regular. Its projection on the first factor gives us the regular language $L(G_i/\partial_0 C_e) \subset L_i$ which we seek. This fulfills condition (2). By construction it also fulfills condition (3). Finally, condition (4) is guaranteed by the supposition that the edge groups intersect well. □

Notice that in the case where the graph of groups Y arises from gluing n-dimensional finite volume hyperbolic manifolds along peripheral $(n-1)$-tori, that each maximal abelian subgraph is either an edge or a single vertex, so that the synchronous version applies.

Like the I Ching, the present memoir ends at a point before completion. We have discussed automatic structures for amalgams of Seifert fibered spaces and amalgams of finite volume hyperbolic groups. This raises the question as to whether one can use the torus decomposition of a Haken 3-manifold into Seifert fibered and hyperbolic pieces to find automatic structures for the fundamental groups of these manifolds. The answer is yes, the sole exceptions being closed Nil geometry manifolds (**[CEHPT]**) and closed Solv geometry manifolds (**[Ge1]**, **[CEHPT]**). This is shown in the most recent version of **[CEHPT]** and was done at around the same time by ourselves. One of the main ingredients of the proof is to see how to pick compatible structures on the hyperbolic and Seifert fibered pieces. This entails picking structures on the pieces so that the structures fellow travel when they are glued along the boundary tori. There is no problem with the hyperbolic pieces because, as we have seen, here one may pick structures for the boundary tori at will. However the gluings of Seifert fibered pieces are somewhat more delicate since they interchange the fiber and "horizontal" directions. This seems to require a choice of structures which do not exactly fulfill the requirements of Theorem 3. However, these structures may be glued together using a slightly more delicate application of the same methods or the somewhat different methods of **[CEHPT]**. Time does not allow an account of this work to appear in the present volume.

Finally, I would like to mention that Apanasov **[Ap]** has suggested that one can easily find automatic structures for fundamental groups of manifolds with conformal structures. This provides yet another approach in the case where the gluing together of the pieces can be carried out by conformal maps.

References

[Ap] B. Apanasov, (private communication, 1990).

[BGSS] G. Baumslag, S.M. Gersten, M. Shapiro and H. Short, Automatic groups and amalgams, Journal of Pure and Applied Algebra 76 (1991), 229–316.

[Ca] J.W. Cannon, The combinatorial structure of co-compact discrete hyperbolic groups, Geometrae Dedicata 16, 1984, 197–210.

[Ca2] J.W. Cannon, Lectures at Trieste, in "Proceedings of the I.C.T.P. meeting 1989, Trieste," World Scientific.

[CEHPT] J.W. Cannon, D.B.A. Epstein, D.F. Holt, M.S. Paterson, and W.P. Thurston, Word Processing and group theory, (preprint, 1990).

[CDP] M. Coonaert, T. Delzant and A. Papadopoulos, "Notes sur les groupes hyperboliques de Gromov," Lecture Notes in Mathematics #1441, Springer-Verlag, 1990.

[ECLHPT] D.B.A. Epstein, J.W. Cannon, D.F. Holt, S.V.F. Levy M.S. Paterson, and W.P. Thurston, "Word Processing in Groups" to appear: Jones and Bartlett Publishers, Boston, 1992 (current version of **[CEHPT]**).

[FLP] A. Fathi, F. Laudenbach and V. Poenaru, Travaux de Thurston, Astérisque 66–67, Société Mathématique de France, 1979.

[Ge] S.M. Gersten, Dehn functions and lp norms of finite presentations, to appear in Proceedings of a Workshop on Algorithmic Problems, (C.F. Miller and G. Baumslag, eds.) Springer-Verlag, MSRI series.

[GS] S.M. Gersten and H. Short, Rational subgroups of bi-automatic groups, Annals of Math. **134** (1991), 125–158.

[Gh] E. Ghys, Les groupes hyperboliques, Seminaire Bourbaki, 722, Astérisque 189-190, Société Mathématique de France, 1990.

[GH] E. Ghys and P. de la Harpe (editors), "Sur les groupes hyperboliques d'apres Mikhael Gromov," Birkhäuser, Progress in Mathematics series, vol. 83, 1990.

[Gi] R. Gilman, Groups with rational cross-section, in: "Combinatorial group theory and topology", Annals of Math. Study 111 edited by S. Gersten and J. Stallings, 1987, Princeton University Press, 175–183.

[Gr] M. Gromov, Hyperbolic groups, 1987, 75–263, Essays in Group Theory, Springer Verlag, M.S.R.I. series Vol 8, edited by S.M.Gersten.

[LS] R.C. Lyndon and P.E. Schupp, Combinatorial group theory, 1977, Springer Verlag.

[RS] M.O. Rabin and D. Scott, I.B.M. Journal of Research and Development (1959), "Sequential Machines: selected papers," edited by E.F. Moore, Addison-Wesley.

[Sc] P. Scott, The Geometries of 3-manifolds, Bull. London Math. Soc. 15, 1983, 401–487.

[Se] J.-P. Serre, Trees, Springer Verlag, 1980.

[Sh] H. Short (editor), Notes on word hyperbolic groups, in Group theory from a geometric viewpoint, conference held in I.C.T.P. Trieste, March 1990, edited by E. Ghys, A. Haefliger and A. Verjovsky, World Scientific, 1991.

[Ta] K. Tatsuoka, Finite volume hyperbolic groups are automatic, (preprint, 1990).

The Ohio State University, Department of Mathematics, Columbus, Ohio 43210

Quantum Representations of Modular Groups and Maslov Indices

V. G. Turaev

1. Introduction

Let Mod_g be the Teichmüller modular group of genus $g \geq 1$. Recall that Mod_g is the isotopy group of orientation preserving self-diffeomorphisms of the oriented smooth closed surface of genus g. The group Mod_g is equipped with the discrete topology.

With each modular Hopf algebra A over a field K one associates a projective linear representation e of Mod_g acting in a certain finite dimensional K-linear space $\overline{\Psi}_g = \overline{\Psi}_g(A)$ (see **[11**, §4**]**). Similar representations were introduced and studied in the frameworks of quantum field theory, conformal field theory and algebraic geometry (see **[1]**, **[15]** and references therein).

Projectivity of e means that for any $\alpha, \beta \in \mathrm{Mod}_g$ there exists a non-zero element $k = k(\alpha, \beta)$ of the field K such that

$$e(\alpha \circ \beta) = k\, e(\alpha) \circ e(\beta)\,.$$

The aim of this paper is to give an explicit formula for $k(\alpha, \beta)$. This formula involves the Maslov indices of Lagrangian planes. Using this formula we compute the cohomology class $[e] \in H^2(\mathrm{Mod}_g, K^*)$ of the 2-dimensional cocycle

$$(\alpha, \beta) \mapsto k(\alpha, \beta) \in K^*$$

where K^* denotes the multiplicative group $K \backslash \{0\}$. (This class is the sole obstruction for e to be a genuine linear representation up to multiplication of the values of e by non-zero scalars.)

We remark that the Maslov indices come up in a similar way in the theory of the Shale-Weil representation of the symplectic group. This suggests a certain parallelism between these two theories.

This paper was completed while the author was participating in the Research Semester in Low Dimensional Topology at The Ohio State University. The author thanks the staff of the Mathematical Department of this University for hospitality.

2. Organization of the paper

In Section 3 we recall the definition of the Maslov indices. In Section 4 the main results of the paper (Theorems I and II) are formulated. In Section 5 we introduce notation related to surgeries on links in S^3. In Section 6 we recall the technique (introduced in **[11]**) of presenting elements of Mod_g by framed links in S^3. In Sections 7 and 8 we prove respectively Theorems I and II.

3. Maslov indices

Let H be a finite-dimensional real vector space equipped with a non-degenerate skew symmetric form

$$B : H \times H \to \mathbb{R} .$$

Note that $\dim H$ is necessarily even. A linear subspace L of H is a Lagrangian subspace if

$$B(L, L) = 0 \quad \text{and} \quad \dim L = (\dim H)/2 .$$

For any three Lagrangian subspaces L_1, L_2 and L_3 of H one defines the Maslov index

$$\tau(L_1, L_2, L_3) \in \mathbb{Z}$$

to be the signature of the symmetric bilinear form A on $(L_1 + L_2) \cap L_3$ defined by the following formula: if $a, b \in (L_1 + L_2) \cap L_3$ and $a = x_1 + x_2$ with $x_1 \in L_1$, $x_2 \in L_2$, then

$$A(a, b) = B(x_2, b) .$$

That A is well-defined, symmetric and bilinear follows from the fact that if $b = y_1 + y_2$ with $y_1 \in L_1$, $y_2 \in L_2$, then

$$B(x_2, b) - B(y_2, a) = B(x_2 - a, b - y_2) = B(x_1, y_1) = 0 .$$

Recall that the signature of a (possibly degenerate) symmetric bilinear form A over $\mathbb{R}$ is defined to be the number of positive entries minus the number of negative entries in a diagonal form of A.

When L_1 and L_2 are transverse, the definition of the Maslov index given above coincides with the standard one (see **[4]**, **[6]**). In general it is equivalent to the definition of Kashiwara (cf. **[7]**).

Here are fundamental properties of the Maslov index:

$$\tau(L_1, L_2, L_3) = -\tau(L_2, L_1, L_3) = -\tau(L_1, L_3, L_2);$$

$$\tau(f(L_1), f(L_2), f(L_3)) = \tau(L_1, L_2, L_3)$$

where f is an arbitrary linear automorphism of (H, B);

$$\tau(L_1, L_2, L_3) \equiv (\dim H)/2 + \dim(L_1 \cap L_2) + \dim(L_1 \cap L_3) + \dim(L_2 \cap L_3) \pmod 2 ; \tag{3.1}$$

$$\tau(L_1, L_2, L_3) - \tau(L_1, L_2, L_4) + \tau(L_1, L_3, L_4) - \tau(L_2, L_3, L_4) = 0 \tag{3.2}$$

for any 4 Lagrangian subspaces L_1, L_2, L_3, L_4 of H. For proofs and further properties of the Maslov index see **[4]**, **[6]**, [**7**, §1.5, 1.9].

4. Main results

Let ∇ be the rectangle with g attached bands in $\mathbb{R}^2$, as in Fig. 1. Let U be a small closed neighborhood of ∇ in $\mathbb{R}^3 \supset \mathbb{R}^2$. In what follows U will be considered as a model handlebody of genus g. The group Mod_g will be identified with the isotopy group of a degree 1 diffeomorphisms $\partial U \to \partial U$. Put

$$H = H_1(\partial U; \mathbb{R})$$

and denote by L the kernel of the inclusion homomorphism $H \to H_1(U; \mathbb{R})$. Provide H with the intersection form corresponding to the orientation of ∂U induced by the right-hand orientation of U.

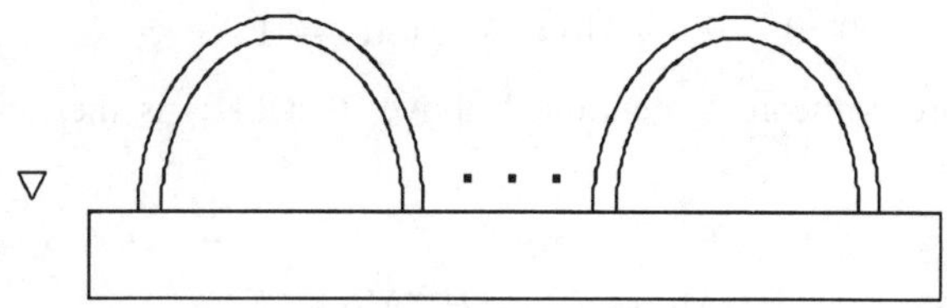

Figure 1

For any linear endomorphism f on H set $|f| = \dim(L \cap f(L)) - g$. For $\alpha \in \mathrm{Mod}_g$ denote by α_* the induced automorphism of H. For $\alpha, \beta \in \mathrm{Mod}_g$ set

$$n(\alpha, \beta) = \frac{|\alpha_*| + |\beta_*| - |(\alpha\beta)_*| - \tau(L, \alpha_*(L), (\alpha\beta)_*(L))}{2}.$$

It follows easily from the formula (3.1) that $n(\alpha, \beta)$ is an integer.

Theorem I. *Let A be a modular Hopf algebra over a field K. Let $c = c_A$ be the element of K^* associated with A (see* **[11**, §3]). *Let e be the projective linear representation of Mod_g associated with A. Then for any $\alpha, \beta \in \mathrm{Mod}_g$*

$$e(\alpha\beta) = c^{n(\alpha,\beta)} e(\alpha) e(\beta).$$

The proof of Theorem I, given in Section 7, is based on the results of **[11]** and the non-additivity of signature due to Wall **[14]**.

Theorem I enables us to compute the cohomology class $[e] \in H^2(\mathrm{Mod}_g; K^*)$ mentioned in the Introduction. It is well known that the first Chern class

$$c_1 \in H^2(BU(g); \mathbb{Z}) \tag{4.1}$$

may be transferred to Mod_g to give rise to a certain class

$$u \in H^2(\mathrm{Mod}_g; \mathbb{Z})$$

(see **[2]**, **[9]**, **[10]**, **[12]**, **[13]**). Namely, consider the group homomorphisms

$$\mathrm{Mod}_g \overset{\alpha \mapsto \alpha_*}{\longrightarrow} Sp(2g; \mathbb{R}) \hookleftarrow U(g)$$

where we have identified the group of intersection preserving automorphisms of H with $Sp(2g; \mathbb{R})$ and where the unitary group $U(g)$ lies in the symplectic group as the maximal compact subgroup. We have the induced mappings of the classifiying spaces:

$$B\,\mathrm{Mod}_g = K(\mathrm{Mod}_g, 1) \xrightarrow{F} B\,Sp(2g;\mathbb{R}) \longleftarrow BU(g)\,.$$

The latter mapping is known to be a homotopy equivalence. Therefore the first Chern class (4.1) uniquely lifts to a class

$$\widetilde{c}_1 \in H^2(B\,Sp(2g;\mathbb{R});\mathbb{Z})\,.$$

Set

$$u = F^*(\widetilde{c}_1) \in H^2(\mathrm{Mod}_g;\mathbb{Z})\,.$$

Note that according to Harer **[5]** the group $H^2(\mathrm{Mod}_g;\mathbb{Z})$ is the infinite cyclic group generated by u. The class u determines a central exentsion

$$0 \to \mathbb{Z} \to \overline{\mathrm{Mod}_g} \to \mathrm{Mod}_g \to 1\,. \tag{4.2}$$

It follows from Harer's theorem mentioned above, that (4.2) is the universal central extension of Mod_g.

Theorem II. *Let*

$$\varphi\colon H^2(\mathrm{Mod}_g;\mathbb{Z}) \to H^2(\mathrm{Mod}_g;K^*)$$

be the linear homomorphism induced by the coefficient homomorphism $n \mapsto c^n$. *Then*

$$[e] = \varphi(-2u)\,.$$

Recall that two projective linear representations e_1, e_2, of a group G in a K-linear space are considered to be equivalent if for any $\alpha \in G$ there exists $k_\alpha \in K^*$ such that $e_1(\alpha) = k_\alpha e_2(\alpha)$.

Corollary I. *Let* r *be the projection* $\overline{\mathrm{Mod}_g} \to \mathrm{Mod}_g$. *The composition*

$$e \circ r : \overline{\mathrm{Mod}_g} \to \mathrm{Aut}\,\overline{\Psi}_g$$

is equivalent to a (genuine) linear representation. □

Of course this corollary directly follows from the universality of the extension (4.2). However, we may apply Theorem II avoiding Harer's results. Indeed, $r^*(u) = 0$ by the very definition of the extension (4.1). Theorem II implies that $r^*([e]) = 0$, and we are done.

Corollary II. *If* $c^{2r} = 1$ *with integer* $r \geq 1$, *then* e *lifts (up to equivalence) to a linear representation of* $\overline{\mathrm{Mod}_g}/r\mathbb{Z}$.

Proof. The class u lifts to $\overline{\mathrm{Mod}_g}/r\mathbb{Z}$ as a coboundary mod r. Therefore $[e] = \varphi(-2u)$ lifts to $\overline{\mathrm{Mod}_g}/r\mathbb{Z}$ as a coboundary. This implies our claim. □

Example. If the modular Hopf algebra A is the algebra U_t considered in **[11**, §8**]** with $t = \exp(\pi\sqrt{-1}/2r)$ then

$$c = c_A = \exp\left(\frac{3\pi\sqrt{-1}(r-2)}{2r}\right)\,.$$

Clearly $c^{4r} = 1$ for odd r and $c^{2r} = 1$ for even r.

Remark. If $d \in K^*$ is a square root of $c = c_A$, then the representation e of Mod_g is equivalent to the projective representation e' defined by the formula $e'(\alpha) = d^{|\alpha_*|}e(\alpha)$ for $\alpha \in \mathrm{Mod}_g$. Theorem I implies that for any $\alpha, \beta \in \mathrm{Mod}_g$

$$e'(\alpha\beta) = d^{-\tau(L,\alpha_*(L),(\alpha\beta)_*(L))}e'(\alpha)e'(\beta).$$

This formula recalls the corresponding formula for the Shale-Weil representation R_L of $Sp(2g;\mathbb{R})$ (see **[7]**):

$$R_L(f_1f_2) = k_L(f_1,f_2)R_L(f_1)R_L(f_2)$$

where

$$k_L(f_1,f_2) = \exp\left(-\frac{\sqrt{-1}\pi}{4}\tau(L,f_1(L),(f_1f_2)(L))\right).$$

There are some other similarities between the Shale-Weil representation and the representation e. For example, the symplectic group acts as a group of automorphisms of the Heisenberg group whereas Mod_g is the group of outer automorphisms of $\pi_1(\partial U)$.

5. Notation

Let ℓ be a framed link in the 3-sphere S^3. (The sphere is provided with the right-hand orientation.) Denote by $M(\ell)$ the oriented 3-manifold obtained by surgery on S^3 along ℓ. Denote by $\mathcal{D}(\ell)$ the oriented 4-manifold obtained by gluing 2-handles to the 4-ball B along the components of $\ell \subset S^3 = \partial B$. Clearly, $\partial\mathcal{D}(\ell) = M(\ell)$. The bilinear intersection form

$$H_2(\mathcal{D}(\ell);\mathbb{R}) \times H_2(\mathcal{D}(\ell);\mathbb{R}) \to \mathbb{R} \tag{5.1}$$

may be diagonalized with respect to some basis in $H_2(\mathcal{D}(\ell);\mathbb{R})$. Denote by $\sigma_-(\ell)$ the number of non-positive entries on the diagonal. It is easy to see that

$$\sigma_-(\ell) = \frac{\dim(H_1(M(\ell);\mathbb{R}) + \#\ell - \mathrm{ind}\,\mathcal{D}(\ell)}{2} \tag{5.2}$$

where $\#\ell$ is the number of components of ℓ and $\mathcal{D}(\ell)$ is the signature of the form (5.1). (The equality (5.2) directly follows from the equality $\dim H_2(\mathcal{D}(\ell);\mathbb{R}) = \#\ell$ and the fact that the annihilator of (5.1) is isomorphic to $H_2(M(\ell);\mathbb{R}) \approx H_1(M(\ell);\mathbb{R})$.)

6. Presentation of surface homeomorphisms via links

It is observed in **[11]** that self-diffeomorphisms of oriented closed surfaces may be described in terms of links and tangles in $\mathbb{R}^3 = S^3\backslash\{\infty\}$. We recall here this description.

Let ∇ and U be the same objects as in Section 4. Let ∇' and U' be the mirror images of ∇ and U with respect to a horizontal plane (see Fig. 2). Let ℓ be a framed link in $\mathbb{R}^3\backslash(U \cup U')$ located above the rectangle of ∇ and below the rectangle of ∇' (in the encircled and shaded region in Fig. 2). The components of ℓ are permitted to link with each other and with the bands of ∇, ∇'. Since ℓ is disjoint from $U \cup U'$ we may assume that U and U' survive the surgery along $\ell \subset S^3$ and lie in $M(\ell)$. Suppose that the complement of $\mathrm{Int}(U \cup U')$ in $M(\ell)$ is a cylinder with the basis ∂U and $\partial U'$. Fix a diffeomorphism

$$r : \partial U \times [0,1] \to M(\ell) \backslash \mathrm{Int}(U \cup U')$$

which is the identity on $\partial U \times 0 = \partial U$. Let $\alpha : \partial U \to \partial U$ be the composition

$$\partial U \overset{a \mapsto a \times 1}{\longrightarrow} \partial U \times 1 \overset{r}{\longrightarrow} \partial U' \to \partial U\,,$$

where the latter mapping is the mirror reflection in the horizontal plane. It is easy to see that α is an orientation preserving diffeomorphism, determined by the triple (∇, ∇', ℓ) uniquely up to isotopy. We will say that this triple presents α.

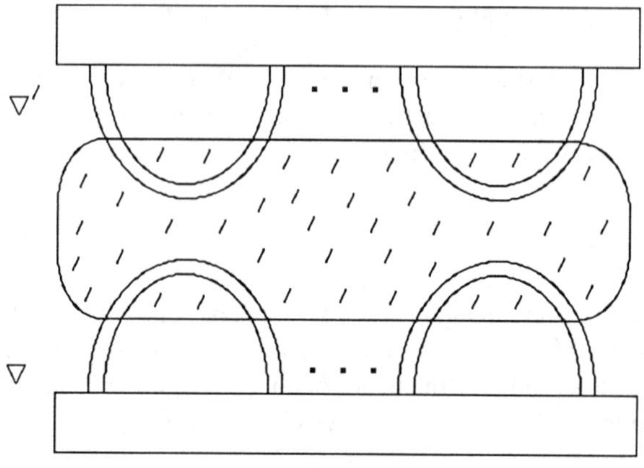

Figure 2

Each element α of Mod_g may be presented by such a triple. Indeed, let M be the closed 3-manifold obtained by gluing 2 copies U_1, U_2 of U along $\alpha : \partial U_1 \to \partial U_2$. Present M as the result of surgery on S^3 along a framed link $\ell \subset S^3$. Push the copies ∇_1, ∇_2 of ∇ lying respectively in U_1, U_2 into $S^3 \backslash \ell$. Applying Kirby-Fenn-Rourke moves (see **[3]**) to $\nabla_1 \cup \nabla_2 \cup \ell \subset S^3$ to make bands of ∇_1, ∇_2 unlinked and unknotted (at the price of adding new components to ℓ) we get $\nabla_1 = \nabla$, $\nabla_2 = \nabla'$. Then the triple (∇, ∇', ℓ) presents α.

Note that if a triple (∇, ∇', ℓ) presents $\alpha \in \mathrm{Mod}_g$, then

$$H_1(M(\ell); \mathbb{R}) = H/(L + \alpha_*^{-1}(L))$$

where H and L are the linear spaces defined in Section 4. Therefore

$$\dim H_1(M(\ell); \mathbb{R}) = \dim(L \cap \alpha_*(L)) = |\alpha_*| + g\,. \tag{6.1}$$

The language of triples is sufficiently convenient to provide a description of the composition law in Mod_g. Namely, let the triples $(\nabla, \nabla', \ell_1)$ and $(\nabla, \nabla', \ell_2)$ present respectively $\alpha, \beta \in \mathrm{Mod}_g$. (Here ℓ_1, ℓ_2 are framed links in $\mathbb{R}^3 \backslash (U \cup U')$ located in the encircled region in Fig. 2.) Put the whole picture of $(\nabla, \nabla', \ell_1)$ on the top of the picture of $(\nabla, \nabla', \ell_2)$ (see Fig. 3). Replace the rectangle in the center of the picture by $2g$ vertical untwisted unlinked bands. This results in a new framed link, say, ℓ which consists of $\ell_1 \coprod \ell_2$ and g unlinked components, and which lie again between ∇ and ∇', being possibly linked with ∇ and ∇'. Then the triple (∇, ∇', ℓ) presents $\alpha \circ \beta$. This follows from Lemma 4.4 of **[11]** (or may be easily verified directly).

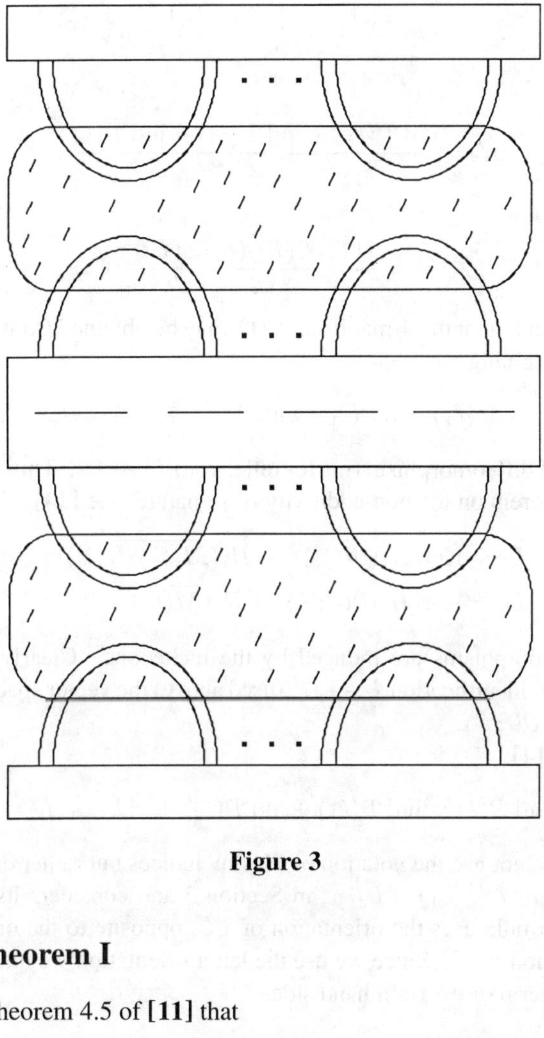

Figure 3

7. Proof of Theorem I

It follows from Theorem 4.5 of **[11]** that

$$e(\alpha\beta) = c^n e(\alpha)e(\beta)$$

with $n \in \mathbb{Z}$. The integer n is actually computed in **[11**, §4.8**]**. Namely, if the triples $(\nabla, \nabla', \ell_1)$, $(\nabla, \nabla', \ell_2)$ represent α, β and if the triple (∇, ∇', ℓ) is formed as in Section 6 to present $\alpha \circ \beta$, then

$$n = \sigma_-(\ell_1) + \sigma_-(\ell_2) - \sigma_-(\ell)\,. \tag{7.1}$$

We shall prove that $n = n(\alpha, \beta)$ where $n(\alpha, \beta)$ is the integer defined in Section 4.

Note that

$$\#\ell = \#\ell_1 + \#\ell_2 + g\,.$$

This formula together with (5.2), (6.1), and (7.1) imply that

$$n = \frac{|\alpha_*| + |\beta_*| - |(\alpha\beta)_*|}{2} - n'$$

where

$$n' = \frac{\operatorname{ind}\mathcal{D}(\ell_1) + \operatorname{ind}\mathcal{D}(\ell_2) - \operatorname{ind}\mathcal{D}(\ell)}{2}\,.$$

It remains to show that

$$n' = \frac{\tau(L, \alpha_*(L), (\alpha\beta)_*(L))}{2}\,.$$

It is easy to observe that the 4-manifold $\mathcal{D}(\ell)$ may be obtained from the disjoint union $\mathcal{D}(\ell_1) \coprod \mathcal{D}(\ell_2)$ by gluing

$$U \subset M(\ell_1) = \partial\mathcal{D}(\ell_1) \text{ with } U' \subset M(\ell_2) = \partial\mathcal{D}(\ell_2)$$

along the canonical diffeomorphism (mirror reflection) $U \to U'$. Thus we are in position to apply Wall's theorem on the non-additivity of signature (see **[14]**). Put

$$L_1 = \operatorname{Ker}(H_1(\partial U; \mathbb{R}) \to H_1(\overline{M(\ell_1)\backslash U}; \mathbb{R}))\,,$$
$$L_2 = \operatorname{Ker}(H_1(\partial U'; \mathbb{R}) \to H_1(\overline{M(\ell_2)\backslash U'}; \mathbb{R}))\,,$$

where both homomorphisms are induced by the inclusions. Clearly, $L_1 = \alpha_*^{-1}(L)$. Under the canonical identification $U \to U'$ (used above) the vector space L_2 is identified with $\beta_*(L) \subset H_1(\partial U; \mathbb{R})$.

According to **[14]**

$$\operatorname{ind}\mathcal{D}(\ell) = \operatorname{ind}\mathcal{D}(\ell_1) + \operatorname{ind}\mathcal{D}(\ell_2) + \tau(L, L_1, L_2)\,.$$

Note that Wall does not use the notation of Maslov indices but rather defines directly the symmetric form on $(L + L_1) \cap L_2$ as in Section 3 and considers its signature. Note also that Wall's formula uses the orientation of ∂U opposite to the one induced by the right-hand orientation in U. Since we use the latter orientation we have replaced minus by plus in the last term of the right hand side.

Finally, we have

$$n' = -\frac{\tau(L, L_1, L_2)}{2} = -\frac{\tau(L, \alpha_*^{-1}(L), \beta_*(L))}{2}$$
$$= -\frac{\tau(\alpha_*(L), L, (\alpha\beta)_*(L))}{2} = \frac{\tau(L, \alpha_*(L), (\alpha\beta)_*(L))}{2}\,.$$

This finishes the proof of Theorem I. □

8. Proof of Theorem II

Denote the symplectic group $Sp(2g;\mathbb{R})$ by G and denote the same group with the discrete topology by G^δ. The commutative diagram of group homomorphisms

$$\begin{array}{ccc} \mathrm{Mod}_g & \overset{\alpha\mapsto\alpha_*}{\longrightarrow} & G \\ P\downarrow & \nearrow id & \\ G^\delta & & \end{array}$$

(where $P(\alpha)=\alpha_*\in G^\delta$) induces a commutative diagram of linear homomorphisms

$$\begin{array}{ccc} H^2(\mathrm{Mod}_g;\mathbb{Z}) & \overset{F^*}{\longleftarrow} & H^2(BG;\mathbb{Z}) \\ P^*\uparrow & \swarrow (id)^* & \\ H^2(G^\delta;\mathbb{Z}) & & \end{array}$$

where F^* is the homomorphism used in Section 4 to define the class u. Thus

$$u=(P^*\circ(id)^*)(\widetilde{c}_1)\,.$$

Consider the group of Borelian cohomologies $H^2_b(G;\mathbb{Z})$. This group is known to be isomorphic to $\mathrm{Hom}(\pi_1(G),\mathbb{Z})=\mathbb{Z}$ (see **[8]**). Moreover, it follows from the results of **[8]** that the natural homomorphism

$$I:H^2_b(G;\mathbb{Z})\to H^2(G^\delta;\mathbb{Z})$$

maps a certain generator, say, μ of $H^2_b(G;\mathbb{Z})=\mathbb{Z}$ into $(id)^*(\widetilde{c}_1)$:

$$I(\mu)=id^*(\widetilde{c}_1)\in H^2(G^\delta;\mathbb{Z})\,.$$

It was shown by Lion (see **[7]**) that the Borelian 2-cocycle of G

$$(f_1,f_1)\mapsto\tau(L,f_1(L),(f_1f_2)(L)) \tag{8.1}$$

presents 4μ. (That this cochain is a cocycle directly follows from the formula (3.2).) The formula (3.1) implies that the Borelian 2-cocycle N of G defined by the rule

$$N(f_1,f_2)=\frac{|f_1|+|f_2|-|f_1\circ f_2|-\tau(L,f_1(L),(f_1\circ f_2)(L))}{2}$$

attains only integer values. Multiplying this cocycle by -2 we get a cocycle cohomological to (8.1). Therefore, $-2N$ presents 4μ in $H^2_b(G;\mathbb{Z})=\mathbb{Z}\mu$. Hence N presents -2μ.

Lifting the cocycle N to Mod_g via the homomorphism $P\colon\mathrm{Mod}_g\to G^\delta$ we get the 2-cocycle $(\alpha,\beta)\mapsto n(\alpha,\beta)$ of Mod_g introduced in Section 4. This cocycle presents the cohomology class

$$P^*(I(-2\mu))=-2(P^*\circ(id)^*)(\widetilde{c}_1)=-2u\,.$$

This implies the claim of Theorem II. □

References

[1] Atiyah, M., Topological quantum field theories. Publ. Math. IHES 68 (1989), 175–186.

[2] Dupont, J. L., Curvature and characteristic classes. Lect. Notes Math. 640 (1978).

[3] Fenn, R., Rourke, C., On Kirby's calculus of links. Topology 18 (1979), 1–15.

[4] Guillemin, V., Sternberg, S., Geometric asymptotics. Math. Surveys 14, A.M.S., Providence, RI (1977).

[5] Harer, J., The second homology group of the mapping class group of an orientable surface. Invent. Math. 72 (1983), 221–239.

[6] Leray, J., Analyse lagrangienne et mécanique quantique. Publ. Math. I.R.M.A. 25, Strasbourg (1978).

[7] Lion, G., Vergne, M., The Weil representation, Maslov indices and theta series. Progress in Math. 6 Birkhäuser, Boston (1980).

[8] Mackey, G. W., Les ensembles boréliens et les extensions des groupes. J. Math. Pures Appl. 36 (1957), 171–189.

[9] Meyer, W., Die Signatur von Flächenbündeln. Math. Ann. 201 (1973), 239–264.

[10] Morita, S., Characteristic classes of surface bundles. Invent. Math. 90 (1987), 551–577.

[11] Reshetikhin, N. Yu., Turaev, V. G., Invariants of 3-manifolds via link polynomials and quantum groups. Invent. Math. 103 (1991), 547–598.

[12] Turaev, V. G., Cocycle for the first symplectic Chern class and the Maslov indices, Funkts. Anal. Prilozen. 18 (1984), 43–48.

[13] Turaev, V. G., First symplectic Chern class and Maslov indices. J. Soviet Math. 37 (1987), 1115–1127.

[14] Wall, C. T. C., Non-additivity of the signature. Invent. Math. 7 (1969), 269–274.

[15] Witten, E., Quantum field theory and Jones polynomial. Comm. Math. Phys. 121 (1989), 351–399.

Institute of Mathematics, University of Strasbourg, 7 Rue Rene Descartes, 67084 Strasbourg, France

The p-torsion of the Farrell–Tate Cohomology of the Mapping Class Group $\Gamma_{(p-1)/2}$

Yining Xia

Let Γ_g denote the mapping class group of genus g, i.e., $\pi_0\mathrm{Diff}^+(S_g)$, where S_g is the closed orientable surface of genus g and $\mathrm{Diff}^+(.)$ is the group of orientation preserving diffeomorphisms. It is well-known that Γ_g is of finite virtual cohomological dimension (vcd) and that the $vcd(\Gamma_g) = 4g - 5$, if $g > 1$ **[H2]**.

Since the 1980's, computations of cohomology (or homology) of Γ_g have been worked on by many people. For example, Miller **[Mi]**, Morita **[Mo]**, Harer **[H1]**, Charney and Cohen **[CC]**, Charney and Lee **[CL]**, Glover and Mislin **[GM]**, Benson and Cohen **[BC]**, **[C]**.

Let Γ be a group of finite vcd. Recall that the Farrell-Tate cohomology groups $\widehat{H}^i(\Gamma; M)$ are torsion groups for any $i \in \mathbb{Z}$. They agree with the ordinary cohomology groups $H^i(\Gamma; M)$ for dimensions i greater than $vcd(\Gamma)$ **[Br]**.

Definition. The group Γ is called p-periodic for a prime p if there exists a positive integer m such that the functors $\widehat{H}^i(\Gamma; -)_{(p)}$ and $\widehat{H}^{i+m}(\Gamma; -)_{(p)}$ are naturally equivalent for all $i \in \mathbb{Z}$. Here $\widehat{H}^i(\Gamma; -)_{(p)}$ denotes the p-primary component of $\widehat{H}^i(\Gamma; -)$. The least positive such integer m is called "the p-period" of Γ.

In this paper we give a complete calculation of the p-torsion of the Farrell-Tate cohomology of the mapping class group $\Gamma_{(p-1)/2}$. Therefore, the p-torsion of the ordinary cohomology of the mapping class group $\Gamma_{(p-1)/2}$ is also determined for dimensions greater than $2p - 7$.

Theorem 1. *(The main result)*

a) $\widehat{H}^i(\Gamma_{(p-1)/2}; \mathbb{Z})_{(p)} = \prod_{1\le n\le k} \widehat{H}^i(\mathbb{Z}_p; \mathbb{Z})_{(p)}$, *if* $p = 6k - 1$.

b) $\widehat{H}^i(\Gamma_{(p-1)/2}; \mathbb{Z})_{(p)} = \widehat{H}^i(\mathbb{Z}_p \rtimes \mathbb{Z}_3; \mathbb{Z})_{(p)} \times \prod_{1\le n\le k} \widehat{H}^i(\mathbb{Z}_p; \mathbb{Z})$ *if* $p = 6k + 1$.

Corollary 2. *(Ordinary cohomology version)*

a) $H^{2i}(\Gamma_{(p-1)/2}; \mathbb{Z})_{(p)} = \prod_{1\le n\le k} H^{2i}(\mathbb{Z}_p; \mathbb{Z})_{(p)} = \prod_{1\le n\le k} \mathbb{Z}_p$, *if* $p = 6k - 1$ *and* $2i > 2p - 7$.

b) $H^{6i}(\Gamma_{(p-1)/2}; \mathbb{Z})_{(p)} = H^{6i}(\mathbb{Z}_p \rtimes \mathbb{Z}_3; \mathbb{Z})_{(p)} \times \prod_{1\le n\le k} H^{6i}(\mathbb{Z}_p; \mathbb{Z}) = \prod_{1\le n\le k+1} \mathbb{Z}_p$, *if* $p = 6k + 1$ *and* $6i > 2p - 7$.

c) $H^{2i}(\Gamma_{(p-1)/2}; \mathbb{Z})_{(p)} = H^{2i}(\mathbb{Z}_p \rtimes \mathbb{Z}_3; \mathbb{Z})_{(p)} \times \prod_{1\le n\le k} H^{2i}(\mathbb{Z}_p; \mathbb{Z}) = \prod_{1\le n\le k} \mathbb{Z}_p$, *if* $p = 6k + 1$, $i \not\equiv 0 \pmod 3$ *and* $2i > 2p - 7$.

d) $H^{2i-1}(\Gamma_{(p-1)/2};\mathbb{Z})_{(p)} = 0$ *if* $2i-1 > 2p-7$.

1. The number of conjugacy classes of order p elements in the mapping class group $\Gamma_{(p-1)/2}$

For p an odd prime, the surface $S_{(p-1)/2}$ may be viewed as a p-fold branched cover of the 2-sphere with 3 branch points. Hence the cyclic group $\mathbb{Z}_p$ acts on the closed surface $S_{(p-1)/2}$ with 3 fixed points, whereas $\mathbb{Z}_p \times \mathbb{Z}_p$ can not act on $S_{(p-1)/2}$ since the Riemann-Hurwitz formula fails. It follows from **[Br**, p. 290**]** that the mapping class group $\Gamma_{(p-1)/2}$ is p-periodic.

Since $\Gamma_{(p-1)/2}$ has p-periodicity, we can use the Brown theorem **[Br**, p. 293**]**: $\widehat{H}^i(\Gamma_{(p-1)/2};\mathbb{Z})_{(p)} = \prod_{P_j \in S} \widehat{H}^i(N(P_j);\mathbb{Z})_{(p)}$ where S is the set of conjugacy classes of $\mathbb{Z}_p$ in $\Gamma_{(p-1)/2}$, and $N(P_j)$ is the normalizer of P_j in $\Gamma_{(p-1)/2}$.

Lemma 1.1. *The number of conjugacy classes of elements of order* p *in* $\Gamma_{(p-1)/2}$ *equals* $(p-1)(p+1)/6$.

Proof. The set of conjugacy classes of elements of order p in $\Gamma_{(p-1)/2}$ is in one to one correspondence with all possible fixed point data $(\beta_1, \beta_2, \beta_3)$, $\beta_i \in \mathbb{Z}_p - \{0\}$, $\beta_1 + \beta_2 + \beta_3 = 0 \pmod p$ **[S]**. Recall that for x an orientation–preserving periodic diffeomorphism of a closed orientable surface S_g of prime period p, the fixed point datum of x is an (unordered) set $\sigma(x) = (\beta_1, \beta_2, \ldots, \beta_q)$, where q is the number of fixed points of x and β_i is the integer $(\mathrm{mod}\ p)$ such that $x^{\beta} i$ acts as multiplication by $e^{2\pi/p}$ in the local invariant complex structure at the i-th fixed point.

Arrange all fixed point data for order p elements of $\Gamma_{(p-1)/2}$ as in Table 1.

Notice.

1) Given two numbers β_1 and β_2, the third number β_3 is uniquely determined.
2) In the i–th column the first number $\beta_1 = i$, $i = 1, 2, \ldots, p-1$.
3) The inequality $\beta_1 \leq \beta_2$ is always true.

We observe the following facts:

(a) The i–th column has an "empty box" when $i \leq (p-1)/2$, the i–th column has no "empty box" when $i > (p-1)/2$.
(b) If we count the number of fixed point data in Table 1 ignoring order, $(\beta_1, \beta_2, \beta_3)$ occurs twice if $\beta_i = \beta_j$ for some $1 \leq i < j \leq 3$. Otherwise it occurs three times. For example, $(1, 1, 5)$ occurs twice and $(1, 2, 4)$ occurs three times.
(c) Each fixed point datum must appear in some column.
(d) The number of fixed point data in the i–th column $= p - 1 - i$ for $i \leq (p-1)/2$.
(e) The number of fixed point data in the i–th column $= p - i$ for $i > (p-1)/2$.

It follows that the number of conjugacy classes of elements of order p equals $1/3[(p-2)(p-1)/2 + (p-1)/2 + p - 1] = (p+1)(p-1)/6$.

Remark, $(p+1)(p-1)/6$ is always an integer. □

Table 1

[1]	[2]	[3]	[4]	$\cdots$
$(1,1,p-2)$				
$(1,2,p-3)$	$(2,2,p-4)$			
$(1,3,p-4)$	$(2,3,p-5)$	$(3,3,p-6)$		
$(1,4,p-5)$	$(2,4,p-6)$	$(3,4,p-7)$	$(4,4,p-8)$	
$\vdots$	$\vdots$	$\vdots$	$\vdots$	
$(1,p-2,1)$	EMPTY	$(3,p-2,p-1)$	$(4,p-2,p-2)$	$\cdots$
EMPTY	$(2,p-1,p-1)$	$(3,p-1,p-2)$	$(4,p-1,p-3)$	$\cdots$

$[(p-3)/2]$	$[(p-1)/2]$	$[(p+1)/2]$
$((p-3)/2,(p-3)/2,3)$		
$((p-3)/2,(p-1)/2,2)$	$((p-1)/2,(p-1)/2,1)$	
$((p-3)/2,(p+1)/2,1)$	EMPTY	$((p+1)/2,(p+1)/2,1)$
EMPTY	$((p-1)/2,(p+3)/2,p-1)$	$(p+1)/2,(p+3)/2,p-2)$
$\vdots$	$\vdots$	$\vdots$
$((p-3)/2,p-1,(p+5)/2)$	$((p-1)/2,p-1,(p+3/2)$	$((p+1)/2,p-1,(p+1)/2)$

$[(p+3)/2]$	$\cdots$	$[p-2]$	$[p-1]$
$((p+3)/2,(p+3)/2,p-3)$			
$\vdots$			
$((p+3)/2,p-2,(p+1)/2)$	$\cdots$	$(p-2,p-2,4)$	
$((p+3)/2,p-1,(p-1)/2)$	$\cdots$	$(p-2,p-1,3)$	$(p-1,p-1,2)$

2. The normalizer of $\mathbb{Z}_p$ in $\Gamma_{(p-1)/2}$

Lemma 2.1. *Let $\mathbb{Z}_p$ be a subgroup of $\Gamma_{(p-1)/2}$. Then there exists an exact sequence $1 \to \mathbb{Z}_p \to N(\mathbb{Z}_p) \to \Gamma^3$, where Γ^3 denotes $\pi_0 \mathrm{Diff}^+(S^2;3)$, the group of path components of orientation preserving diffeomorphisms of S^2 which permute three distinguished points. (In fact, $\Gamma^3 = \Sigma_3$, the symmetry group of 3 letters* **[Bi]**.*)*

Proof. We will construct an injective homomorphism $I\colon N(\mathbb{Z}_p)/\mathbb{Z}_p \to \Gamma^3$.
a) Let $\mathbb{Z}_p = \langle x \rangle$, $h \in N(\langle x \rangle)$, $hxh^{-1} = x^k$ in $\Gamma_{(p-1)/2}$. By the result of Birman and Hilden (1972), **[BH2]**, we can represent x and h by elements of $\mathrm{Diff}^+(S_{(p-1)/2})$ (which we also denote x and h) such that $hxh^{-1} = x^k$.

Consider the following diagram:

$$\begin{array}{ccccc} S_{(p-1)/2} & \overset{h}{\longrightarrow} & S_{(p-1)/2} & \overset{h^{-1}}{\longrightarrow} & S_{(p-1)/2} \\ \downarrow p & & \downarrow p & & \downarrow p \\ S^2 = S_{(p-1)/2}/\mathbb{Z}_p & \underset{f}{\longrightarrow} & S^2 = S_{(p-1)/2}/\mathbb{Z}_p & \underset{f^{-1}}{\longrightarrow} & S^2 = S_{(p-1)/2}/\mathbb{Z}_p \end{array}$$

The map h takes fiber to fiber, so h induces a map f from S^2 to S^2. Obviously f is an orientation preserving homeomorphism since h is.

For the fixed points b_1, b_2, b_3, of the diffeomorphism x, since $x^k(h(b_i)) = hx(b_i) = h(x(b_i)) = h(b_i)$, $h(b_i)$ is also a fixed point of x, i.e., $h(b_i) = b_j$, $1 \le i, j \le 3$. Therefore, f permutes the three points $p(b_1)$, $p(b_2)$ and $p(b_3)$ in S^2. This implies $f \in \Gamma^3$. Define $I(h) = f$.

b) $I: N(\langle x \rangle)/\mathbb{Z}_p \to \Gamma^3$ is well-defined.

If $h_1 = h_2$ in $N(\langle x \rangle)$, i.e., h_1 is isotopic to h_2 in $\mathrm{Diff}^+(-)$ and $h_1 x h_1^{-1} = x^k$, $h_2 x h_2^{-1} = x^k$, then by Birman and Hilden's result **[BH2]**, there exists an isotopy $H: S_{(p-1)/2} \times I \to S_{(p-1)/2}$ such that $H_0 = h_1$, $H_1 = h_2$, $H_s x H_s^{-1} = x^k$. So, the diffeomorphisms f_1 and f_2 of S^2 induced by h_1 and h_2 are isotopic in $\mathrm{Diff}^+(S^2, 3)$. This implies $f_1 = f_2$ in Γ^3. If $h = x^k$, then f induced by h must be the identity map from S^2 to S^2.

c) I is a homomorphism by the definition of I.

d) I is a one to one homomorphism.

Suppose f, induced by h, is isotopic to the identity map in $\mathrm{Diff}^+(S^2, 3)$. The homotopy lifting theorem shows that h is isotopic to the identity map up to the choice of initial point, i.e., I is a one to one. □

3. The Action of $N(\mathbb{Z}_p)$ on $\mathbb{Z}_p$

Lemma 3.1. *For each $\mathbb{Z}_p$ of $\Gamma_{(p-1)/2}$, consider the subgroup $\mathrm{Im}(I) \subseteq \Gamma^3$ of Lemma 2.1.*

a) *If $p = 6k - 1$, $\mathrm{Im}(I)$ acts trivially on $\mathbb{Z}_p$.*

b) *If $p = 6k + 1$, there exists exactly one conjugacy class of $\mathbb{Z}_p$ in $\Gamma_{(p-1)/2}$ such that the order 3 elements in $\mathrm{Im}(I)$ act nontrivially on that $\mathbb{Z}_p$. For other $\mathbb{Z}_p$'s in $\Gamma_{(p-1)/2}$, $\mathrm{Im}(I)$ acts trivially on $\mathbb{Z}_p$.*

A basic number theoretic sublemma is needed before we complete the proof of Lemma 3.1.

Sublemma. *There is no integer m such that $m^2 + m + 1 = 0 \pmod{p}$ for $p = 6k - 1$, $1 < m < p$. There are two integers m_1 and m_2 such that $m_i^2 + m_i + 1 = 0 \pmod{p}$ for $p = 6k + 1$ and $1 < m_i < p$, and these satisfy $m_2 = m_1^2$.*

Proof. If m is a solution of $m^2 + m + 1 = 0 \pmod p$, then $m \neq 1 \pmod p$. Otherwise, $m = 1 \pmod p$ and $m^2 + m + 1 = 0 \pmod p$ implies $3 = 0 \pmod p$, a contradiction.

The fact that $m \neq 1 \pmod p$ and $m^2 + m + 1 = 0 \pmod p$ also implies $m^3 = 1 \pmod p$, i.e., 3 divides $p - 1$. So, for $p = 6k - 1$ there is no solution; for $p = 6k + 1$, there are exactly two solutions of the quadratic equation. In fact, if m is a solution, m^2 is also a solution. □

The Proof of Lemma 3.1. Let us rearrange all the fixed point data $(\beta_1, \beta_2, \beta_3)$ for $\Gamma_{(p-1)/2}$ as follows:

Table 2

(1)	(2)	(3)	$\cdots$	$(p-1)$
$(1,1,p-2)$	$(1,2,p-3)$	$(1,3,p-4)$	$\cdots$	$(1,p-2,1)$
$(2,2,p-4)$	$(2,4,p-6)$	$(2,6,p-8)$	$\cdots$	$(2,2p-4,2)$
$(3,3,p-6)$	$(3,6,p-9)$	$(3,9,p-12)$	$\cdots$	$(3,3p-6,3)$
$\vdots$	$\vdots$	$\vdots$		$\vdots$
$(p-1,p-1,2)$	$(p-1,2p-2,3)$	$(p-1,3p-3,4)$	$\cdots$	$(p-1,(p-2)(p-1),p-1)$

Note that the i–th row of the table is i times the first row $\pmod p$.

We observe two facts:

(1) Any fixed point datum is contained in Table 2.

(2) If $(\alpha_1, \beta_1, \gamma_1) = (\alpha_2, \beta_2, \gamma_2)$ as unordered triples, where $(\alpha_1, \beta_1, \gamma_1)$ and $(\alpha_2, \beta_2, \gamma_2)$ are in different columns, then these two columns must be the same up to permutation.

To see (1), note that any fixed point data (α, β, γ) can be multiplied by m such that $m\alpha = 1 \pmod p$, so (α, β, γ) must be in the $m\beta$–th column.

For (2), as unordered triples, $(\alpha_1, \beta_1, \gamma_1) = (\alpha_2, \beta_2, \gamma_2)$ implies $m(\alpha_1, \beta_1, \gamma_1) = m(\alpha_2, \beta_2, \gamma_2)$, i.e., $(m\alpha_1, m\beta_1, m\gamma_1) = (m\alpha_2, m\beta_2, m\gamma_2)$ for any m, $1 \leq m \leq p-1$. Therefore, these two columns are the same.

Claim. a) Assume $p > 3$. Then any two fixed point data in the m–th column are different if $m^2 + m + 1 \neq 0 \pmod p$.

b) Three sets of fixed point data are the same in the m-th column in Table 2, if $m^2 + m + 1 = 0 \pmod p$.

In fact, suppose $(1, m, n) = (h, hm, hn)$, $1 \neq h \pmod p$.

Case 1. $1 = hm$, $m = hn$, $n = h \pmod p$. In this case $1 = n^3$, $m = n^2 \pmod p$, and $m^3 = 1 \bmod (p)$, so $m^2 + m + 1 = 0 \pmod p$.

Case 2. $1 = hn$, $m = h$, $n = hm \pmod p$. In this case $1 = m^3 \pmod p$ so $m^2 + m + 1 = 0 \pmod p$.

Note that in either case, $n = m^2 \pmod p$.

Conversely, if there is an m such that $m^2+m+1=0 \pmod p$, $1<m\leq p-1$, then $(1,m,m^2)$ is a fixed point datum in the m–th column, and $(1,m,m^2)=m(1,m,m^2)=m^2(1,m,m^2)$.

Therefore, three sets in the m–th column are the same. This proves the claim.

Let x be an element of order p in $\Gamma_{(p-1)/2}$. If $\sigma(x)$, the fixed point datum of x, lies in column i of Table 2, then all the sets in the i–th column of table 2 represent the fixed point data $\sigma(x^k)$, $1\leq k\leq p-1$. Therefore, by using the claim above, if $\sigma(x)$ is in the m-column, where $m^2+m+1=0 \pmod p$, then $\sigma(x)=\sigma(x^m)=\sigma(x^{m^2})$. Otherwise, $\sigma(x^i)\neq\sigma(x^j)$ for $i\neq j$.

It is well-known that $|N(\langle x\rangle)/C(\langle x\rangle)|=$ the number of i such that x^i is conjugate with x, where $1\leq i\leq p-1$. Here $N(\langle x\rangle)$ (resp. $C(\langle x\rangle)$) denotes the normalizer (resp. centralizer) of the cyclic subgroup generated by x.

The sublemma and argument above imply that:

a) If $p=6k-1$, $|N(\mathbb{Z}_p)/C(\mathbb{Z}_p)|=1$ for all $\mathbb{Z}_p$ in $\Gamma_{(p-1)/2}$.

b) If $p=6k+1$, there exists exactly one conjugacy class of $\mathbb{Z}_p$ in $\Gamma_{(p-1)/2}$ such that $|N(\mathbb{Z}_p)/C(\mathbb{Z}_p)|=3$. For other $\mathbb{Z}_p$ in $\Gamma_{(p-1)/2}$, $|N(\mathbb{Z}_p)/C(\mathbb{Z}_p)|=1$.

The Lemma 2.1 together with a) and b) above implies Lemma 3.1. □

Corollary 3.2. *There are k different conjugacy classes of $\mathbb{Z}_p$ in $\Gamma_{(p-1)/2}$ for $p=6k-1$. There are $k+1$ different conjugacy classes of $\mathbb{Z}_p$ in $\Gamma_{(p-1)/2}$ for $p=6k+1$.*

Proof. Lemmas 1.1 and 3.1 imply this corollary. □

4. The proof of Theorem 1

There is a short exact sequence for an odd prime $p>3$ by Lemma 2.1.

$$1\to\mathbb{Z}_p\to N(\mathbb{Z}_p)\to\operatorname{Im}(I)\to 1\,,$$

where $\operatorname{Im}(I)$ is a subgroup of Σ_3.

Case 1. $\operatorname{Im}(I)$ acts trivially on $\mathbb{Z}_p$. By **[Br**, p. 84], for $i>0$, $\widehat{H}^i(N(\mathbb{Z}_p);\mathbb{Z})_{(p)}=H^i(N(\mathbb{Z}_p);\mathbb{Z})_{(p)}=H^i(\mathbb{Z}_p;\mathbb{Z})^{\operatorname{Im}(I)}=H^i(\mathbb{Z}_p;\mathbb{Z})$. Since the p-period of $N(\mathbb{Z}_p)=2|N(\mathbb{Z}_p)/C(\mathbb{Z}_p)|=2$ **[Sw]**, $\widehat{H}^i(N(\mathbb{Z}_p);\mathbb{Z})_{(p)}=\widehat{H}^i(\mathbb{Z}_p;\mathbb{Z})$ for all integer i.

Case 2. $\operatorname{Im}(I)$ acts nontrivially on $\mathbb{Z}_p$. By **[Br**, p. 84] and Lemma 3.2, for $i>0$, $\widehat{H}^i(N(\mathbb{Z}_p);\mathbb{Z})_{(p)}=H^i(N(\mathbb{Z}_p);\mathbb{Z})_{(p)}=H^i(\mathbb{Z}_p;\mathbb{Z})^{\operatorname{Im}(I)}=H^i(\mathbb{Z}_p;\mathbb{Z})^{\mathbb{Z}_3}=H^i(\mathbb{Z}_p\rtimes\mathbb{Z}_3;\mathbb{Z})_{(p)}$. Again by **[Sw]**, the p-period of $N(\mathbb{Z}_p)=2|N(\mathbb{Z}_p)/C(\mathbb{Z}_p)|=6=$ the p-period of $\mathbb{Z}_p\rtimes\mathbb{Z}_3$, $\widehat{H}^i(N(\mathbb{Z}_p);\mathbb{Z})_{(p)}=\widehat{H}^i(\mathbb{Z}_p\rtimes\mathbb{Z}_3;\mathbb{Z})_{(p)}$ for all integers i.

Finally, Brown's decomposition theorem, Lemma 3.1 and Corollary 3.2 imply Theorem 1. □

Appendix

Lemma 2.1 is a special case of the statement by Birman and Hilden that $N(\mathbb{Z}_p)/\mathbb{Z}_p$ is isomorphic to $\Gamma^n = \pi_0\mathrm{Diff}^+(S^2;n)$ if the orbit space $S_g/\mathbb{Z}_p = S^2$, where n is the number of fixed points of $\mathbb{Z}_p$ acting on S_g.

However, it seems that there exists a gap on the proof of Lemma 5.1 **[BH2]**. In particular, the homomorphism I may not be a surjection of $N(\mathbb{Z}_p)/\mathbb{Z}_p$ into Γ^n.

The claim in Lemma 5.1 **[BH2]** that a closed curve lifts to a closed curve if and only if it encircles a multiple of p branch points is false. Actually,

Proposition. *For a prime p, let $p: S_g \to S^2$ be a p-sheeted branched covering map with ramification points $b_1, b_2, \ldots, b_n \in S_g$, $\mathbb{Z}_p = \langle T\rangle$, T the deck transformation of order p with n fixed points in S_g. Denote the fixed point datum of T as $\sigma(T) = (\beta_1, \beta_2, \beta_3, \ldots, \beta_n)$. Let γ be a closed curve in $S^2 - \{p(b_1), p(b_2), \ldots, p(b_n)\}$, then γ lifts to a closed curve in $S_g - \{b_1, b_2, \ldots, b_n\}$ if and only if the homotopy class $[\gamma] \in \pi_1(S^2 - \{p(b_1), p(b_2), \ldots, p(b_n)\}) = \langle x_1, x_2, \ldots, x_n \mid x_1x_2\ldots x_n = 1\rangle$ can be written in the form $[\gamma] = \Pi x_i^n i \Pi x_i^m i \ldots \Pi x_i^k i$ such that $\Sigma(n_i\beta_i + m_i\beta_i + \cdots + k_i\beta_i) = 0 \pmod p$, where x_i is represented by a simple closed curve c_i such that $p(b_i)$ is in one complementary component of c_i and $p(b_j)$, $j \neq i$, are in the other.*

Proof. In fact, the covering map $p : S_g - \{b_1, b_2, \ldots, b_n\} \to S^2 - \{p(b_1), p(b_2), \ldots, p(b_n)\}$ is classified by the homomorphism $f: \pi_1(S^2 - \{p(b_1), p(b_2), \ldots, p(b_n)\}) = \langle x_1, x_2, \ldots, x_n \mid x_1x_2\ldots x_n = 1\rangle$ into $\mathbb{Z}_p = \langle T\rangle$, where $f(x_i) = T_i^\beta$, $1 \leq i \leq n$ **[E]**.

Recall that a closed curve γ lifts to a closed curve if and only if $[\gamma] \in p^*[\pi_1(S_g - \{b_1, b_2, \ldots, b_n\})]$, i.e., $f([\gamma]) = 0$ in $\mathbb{Z}_p$. Suppose $[\gamma] = \Pi x_i^{n_i} \Pi x_i^{m_i} \cdots \Pi x_i^{k_i}$. Then $f([\gamma]) = T^{\Sigma(n_i\beta_i + m_i\beta_i + \cdots + k_i\beta_i)} = 0$ if and only if $\Sigma(n_i\beta_i + m_i\beta_i + \cdots + k_i\beta_i) = 0 \pmod p$.

Example 1. Let $p: S_2 \to S^2$ be a 5-sheeted branched covering map with 3 branch points b_1, b_2 and b_3, T the deck transformation of order 5 with the fixed points b_1, b_2, and b_3 in S_2, and the fixed point datum $\sigma(T) = (1, 3, 1)$. If γ, an orientable closed curve in $S^2 - \{p(b_1), p(b_2), p(b_3)\}$ goes around $p(b_1)$ twice, around $p(b_2)$ once and not around $p(b_3)$, then γ lifts to $s_2 - \{b_1, b_2, b_3\}$, although γ encircles the two branched points (or three if counted with multiplicity).

Consider the diffeomorphism w of S^2, which switches $p(b_1)$ and $p(b_2)$ in a disk D containing $p(b_1)$ and $p(b_2)$, and $w = Id$ in $S^2 - D$. Then $w(\gamma)$ can not lift to $S_2 - \{b_1, b_2, b_3\}$ since $w(\gamma)$ goes around $p(b_1)$ once, around $p(b_2)$ twice and not around $p(b_3)$, this implies $n_1\beta_1 + n_2\beta_2 + n_3\beta_3 = 1 + 6 = 7 \neq 0 \pmod 5$. It follows that w cannot lift to a diffeomorphism of S_2. So w, as an element of Γ^3, can not be in the image $\mathrm{Im}(I)$ of Lemma 2.1.

Example 2. The special case when $p = 2$ **[BH1]** is valid. Consider the 2-sheeted covering $p: S_g \to S^2$ with n branch points. Now $\beta_1 = \beta_2 = \cdots = \beta_n = 1$ and a closed curve γ in $S^2 - \{p(b_1), p(b_2), \ldots, p(b_n)\}$ lifts if and only if $[\gamma] = \Pi x_i^n i \Pi x_i^m i \cdots \Pi x_i^k i$, $\Sigma(n_i + m_1 + \cdots + k_i) = 0 \pmod 2$. For any diffeomorphism w of S^2 permuting the set $\{p(b_1), p(b_2), \ldots, p(b_n)\}$, the closed curve $w(\gamma)$ lifts if and only if γ lifts. Therefore w lifts to a diffeomorphism of S_g permuting the set $\{b_1, b_2, \ldots, b_n\}$. This shows that $N(\mathbb{Z}_2)/\mathbb{Z}_2 = C(\mathbb{Z}_2)/\mathbb{Z}_2 = \Gamma^n$ by an argument similar to Lemma 2.1.

Acknowledgments. I would like to thank Professor H. Glover and Professor G. Mislin for their direction of this work which is part of my Ohio State University Ph.D. thesis. I would also like to thank the referee for helpful suggestions. I also wish to thank the International Mathematical Research Institute at Ohio State for their support.

References

[BC] D. Benson and F.R. Cohen, Mapping class groups of low genus and their cohomology, Memoirs of the A.M.S., 443 (1991)

[Bi] J. Birman, Braids, links and mapping class groups, Ann. of Math. Studies 82, Princeton Univ. Press, 1975.

[BH1] J. Birman and H. Hilden, Mapping class groups of closed surfaces as covering spaces, Advances in the theory of Riemann Surface. Ann. of Math. 66 (1971), 81–115.

[BH2] J. Birman and H. Hilden, Isotopies of homeomorphisms of Riemann surfaces, Ann. of Math. 97 (1973), 424–439.

[Br] K. S. Brown, Cohomology of Groups, Graduate Texts in Math. 87, Springer-Verlag, 1982.

[CC] R. Charney and F. Cohen, A stable splitting for the mapping class group, Michigan Math. J. 35 (1988), 269–284.

[CL] R. Charney and R. Lee, An application of homotopy theory to mapping class groups, J. Pure and Applied Algebra 44 (1987) 125–137.

[C] F. R. Cohen, Homology of mapping class groups for surfaces of low genus, Contemp. Math. 58 (1987), 21–30.

[E] J. Ewing, Automorphisms of surfaces and class numbers: An illustration of the G-index theorem, London Math. Soc. Lecutre Note Series 86 (1983), 120–127.

[GM] H. Glover and G. Mislin, Torsion in the mapping class group and its cohomology, J. Pure and Applied Algebra 44 (1987) 177–189.

[H1] J. Harer, Stability of the homology of the mapping class groups of orientable surfaces, Ann. of Math. 121 (1985), 215–249.

[H2] J. Harer, The virtual cohomological dimension of the mapping class group of an orientable surface, Invent. Math. 84 (1986), 157–176.

[Mi] E. Miller, The homology of the mapping class group, J. Differential Geom. 24 (1986), 1–14.

[Mo] S. Morita, Characteristic classes of surface bundles, Bulletin of AMS 11 (1984), 386–388.

[Sw] R. G. Swan, The p-period of a finite group. Illinois J. Math. 4 (1960), 341–346.

[S] P. Symonds, The cohomology representation of an action of C_p on a surface, Trans. Amer. Math. Soc. 306 (1988), 389–400.

[X] Y. Xia, Farrell–Tate cohomology of the mapping class group, The Ohio State University, Ph.D. Thesis, 1990.

The Ohio State University, Department of Mathematics, Columbus, OH 43210

Tangles in Prisms, Tangles in Cobordisms

*David N. Yetter**

Abstract. The algebraic structures underlying the coupled topological field theories used by Witten **[18]** (cf. also Atiyah et al. **[1]**) to give invariants of links in 3-manifolds are examined in terms of the structure of categories of tangles in prisms over compact oriented surfaces and tangles in cobordisms of oriented surfaces.

The results for tangles in prisms yield as corollaries presentations of the braid groups of oriented surfaces (in the case of the sphere, yielding the usual presentations due to Fadell and VanBuskirk **[4]**, in the case of the nth braid group of the torus, a presentation with three generators and $2n - 2$ relations).

0. Introduction

Recent work of Freyd, Joyal, Shum, Street and Yetter **[5]**, **[6]**, **[9a-b]**, **[10]**, **[17]**, **[19]**, **[20]** has shown that there is an intimate relationship between certain non-symmetric generalizations of compact closed categories (cf. Kelly and Laplaza **[13]**) and classical knot theory (or more generally the theory of isotopy classes of embeddings of certain types of spaces in $\mathbb{R}^3$ or $\mathbb{S}^3$). These results may be viewed on the one hand as "geometrizing" coherence theorems in monodial category theory, or on the other as "algebraicizing" knot theory: a generalization of the braid group approach (cf. Artin **[2]**, Birman **[3]**). It is the purpose of this paper to address the problem of "algebraicizing" the theory of knots in more general 3-manifolds. A complete treatment of knot theory in oriented 3-manifolds from this point of view would require a complete description of the algebraic structure of the category of oriented cobordisms of 2-manifolds.

Witten's **[18]** approach to the Jones polynomial **[8]** and its generalizations **[7]**, **[11]**, **[12]** in terms of quantum field theory, and the subsequent exposition of his ideas by Atiyah's seminar **[1]** make it clear that the problems addressed herein should properly be viewed as contributions to our understanding of topological (and possibly conformal) field theories.

We begin by establishing a general framework for our constructions and reviewing "categorical knot theory" as previously developed. Then we deal with the structure of categories of tangles in prisms on compact surfaces. Finally, we describe the algebraic structure of the categories of tangles in (oriented) cobordisms of oriented surfaces in terms of the structure of the categories of tangles on prisms over oriented surfaces and Morse functions on the underlying cobordisms.

* Supported by a grant from the Ohio Board of Regents, administered through The Ohio State University, Office of Research and Graduate Study's Seed Grant Program, and by a Program Grant from the Australian Research Council.

Throughout, compositions of maps are written in diagrammatic order (thus fg denotes f followed by g), and the statement of a diagram means the commutativity of the diagram unless it is explicitly stated to the contrary.

For topological background, the reader is referred particularly to **[1]**, **[3]**, **[19]**, **[20]**. For categorical background, the reader is referred particularly to **[5]**, **[9b]**, **[13a]**, **[15a]**, **[17]**, **[19]**.

1. A general framework and categorical knot theory in prisms on prisms

Given any space, M, we can consider the isotopy classes of subspaces (possibly equipped with some auxiliary structure) of some particular type in $M \times [0,1]$. Provided the collection of permitted subspaces satisfies some closure conditions, this set of subspaces becomes a category, with composition given by pasting copies of $M \times [0,1]$ (equipped with their subspaces) end-to-end. Indeed, when this is formalized we will in this way obtain a functor from the category with spaces as objects and embeddings as maps to ***Cat***.

Throughout most of this work we will assume that the category of spaces is some full subcategory of the category of p.l. spaces with embeddings as maps. Usually we will restrict our attention to the subcategory of 2-manifolds. We will for the sake of generality denote this category by ***Emb***. Most (and in dimensions less than three all) of the results will also hold for smooth (i.e., C^∞-) spaces.

To formalize which types of subspaces give rise to categories as sketched above, we make

Definition 1.1. A ***formal cobordism family*** (for brevity an FCF) is a collection, $\mathbb{F}$, of pairs of spaces (X, X') (X' should be thought of as the boundary of X, but need not be the geometrical boundary of X), possibly equipped with compatible orientations or other structure, and satisfying

FCF 0: the pair of empty spaces is in $\mathbb{F}$;

FCF 1: X' has a neighborhood of the form $X' \times [0,1)$ in X for all (X, X') in $\mathbb{F}$;

FCF 2: if (X, X') and (Y, Y') are in $\mathbb{F}$, so is $(X+Y, X'+Y')$;

FCF 3: if (X, X') and (Y, Y') are in $\mathbb{F}$, and if C is a union of connected components of X', D a union of connected components of Y', and $\phi : C \xrightarrow{\sim} D$ is a p.l. homeomorphism (reversing orientation in the case where the pairs are oriented), then $(X +_\phi Y, (X' - C) + (Y' - D))$ is in $\mathbb{F}$;

FCF 4: if (X, X') is in $\mathbb{F}$, and C is a union of components of X', then $(C \times [0,1], C \times \{0,1\})$ is in $\mathbb{F}$;

FCF 5: if (X, X') is in $\mathbb{F}$, and $S \subset X$ is a union of connected components, then $(S, S \cap X')$ is in $\mathbb{F}$;

and if the spaces are oriented

FCF 6: if (X, X') is in $\mathbb{F}$, then so is (X^*, X'^*), where $(\ )^*$ denotes reversal of orientation.

To any formal cobordism family, $\mathbb{F}$, we can associate a functor from ***Emb*** to ***Cat***:

Definition 1.2. For any space M, and an FCF, $\mathbb{F}$, the $\mathbb{F}$-***isotopy category of*** M, denoted $\{\mathbb{F}, M\}$, is the category whose objects are all isomorphism classes of embeddings $A: C \hookrightarrow M$, where C is a union of components of some X' for some (X, X') in $\mathbb{F}$. A map from $A: C \hookrightarrow M$ to $B: D \hookrightarrow M$ is named by an embedding $f: X \hookrightarrow M \times [0,1]$ so that $X' = C \cup D$ (with the orientation on D reversed in the oriented case) and $f|_{C\cup D} = A \times 0 \cup B \times 1$. Two embeddings $f: X \hookrightarrow M \times [0,1]$ and $g: Y \hookrightarrow M \times [0,1]$ name the same map if they are isotopic in the sense that there is a map $\phi: X \times [0,1] \to M \times [0,1]$ and an isomorphism (preserving orientation and any auxiliary structure) $\psi: Y \xrightarrow{\sim} X$ so that $\phi(-,0) = f$, $\phi(\psi(-),1) = g$, $\phi(-,t)$ is an embedding for all t, and $\phi(\xi, t)$ is independent of t if $\xi \in X'$. In cases where the pairs of $\mathbb{F}$ have some structure, we require ϕ and ψ to preserve it in some obvious sense. If we are in the smooth setting, we require moreover that the image of f meet $M \times \{0,1\}$ properly.

The identity of $A: C \hookrightarrow M$ is then named by the map $A \times Id: C \times [0,1] \hookrightarrow M \times [0,1]$. If $f: X \hookrightarrow M \times [0,1]$ (resp. $g: Y \hookrightarrow M \times [0,1]$) names a map from $A: C \hookrightarrow M$ to $B: D \hookrightarrow M$ (resp. from $B: D \hookrightarrow M$ to $P: E \hookrightarrow M$), then the composite of f and g is the map from $A: C \hookrightarrow M$ to $P: E \hookrightarrow M$ named by the embedding

$$f(Id \times (-)\cdot\frac{1}{2}) +_D g(Id \times (\frac{1}{2} + (-)\cdot\frac{1}{2}) : (X +_D Y, (X' - D) + (Y' - D)) \hookrightarrow M \times [0,1].$$

For any embedding $\imath: M \hookrightarrow N$, there is a functor $\{\mathbb{F}, \imath\}: \{\mathbb{F}, M\} \to \{\mathbb{F}, N\}$ induced by composing the embeddings naming objects with $\imath$, and those naming maps with $\imath \times Id$.

Indeed one can do more. For any embeddings $\imath, \nu: M \hookrightarrow N$, and any isotopy $\omega: M \times [0,1] \to N$ between them, there is a natural isomorphism $\{\mathbb{F}, \omega\}: \{\mathbb{F}, \imath\} \to \{\mathbb{F}, \nu\}$, given, at any object $A: C \hookrightarrow M$, by the map named by the embedding $(A \times Id)(\omega, p_2): C \times [0,1] \to N \times [0,1]$.

Of course, some work is required to prove that the construction of Definition 1.2 does define categories, functors, and natural isomorphisms. Before outlining the proof of the results implicit in Definition 1.2, let us restate them explicitly and succinctly. To do so, note that ***Emb*** can be made into a 2-category (actually a category enriched over the category of groupoids) by taking isotopies as 2-cells (cf. Kelly and Street **[13a]**, Kelly **[12a]**). Letting $Cat_{\#}$ denote the 2-category of categories, functors and natural isomorphisms, we have

Theorem 1.3. *For any formal cobordism family,* $\mathbb{F}$, $\{\mathbb{F}, -\}$ *as defined in 1.2 above is a 2-functor from* ***Emb*** *to* $Cat_{\#}$.

Proof. It is evident that the construction given assigns to each space, M, a directed graph whose vertices are all embeddings of C's into M (for C a union of components of some X', and (X, X') in $\mathbb{F}$), and whose edges are all isotopy classes of embeddings of X's into $M \times [0,1]$. The source (resp. target) of an edge is given by the inverse image of $M \times \{0\}$ (resp. $M \times \{1\}$, with orientation reversed in the oriented case).

A composition is induced on this directed graph by pasting representative embeddings end-to-end using FCF 3. This can be seen to be associative by applying an ambient isotopy of $M \times [0,1]$, which simply rescales $[0,1]$-coordinates.

The identity map on an object, $A: C \hookrightarrow M$, is named by $A \times Id: C \times [0,1] \hookrightarrow M \times [0,1]$ (which names a map by FCF 4). That this names an identity map can be verified by using an ambient isotopy which rescales the $[0,1]$ coordinate of $M \times [0,1]$, and whose restriction to the embedded subspace merely stretches the collar neighborhood of C which exists by FCF 1.

That an embedding $\imath: M \hookrightarrow N$ induces a functor $\{\mathbb{F}, \imath\}$ is immediate, since composing with $\imath$ or $\imath \times Id_{[0,1]}$ carries names of objects, names of maps, and isotopies (that is, equations between names of maps) in $\{\mathbb{F}, M\}$ to the same in $\{\mathbb{F}, N\}$, plainly preserving the sources and targets and composition.

To see that the construction given carries an isotopy, $\omega: M \times [0,1] \to N$, from $\imath: M \hookrightarrow N$ to $\nu: M \hookrightarrow N$ to a natural isomorphism, we must verify the naturality condition, since the components are plainly invertible by applying the same construction to the reversed isotopy gives the inverse.

Now given a map from $A: C \hookrightarrow M$ to $B: D \hookrightarrow M$ named by $F: X \hookrightarrow M \times [0,1]$ (that is, $F\,|_{X'} = C + D = A \times 0 + B \times 1$, $F^{-1}(M \times \{0\}) = C$, and $F^{-1}(M \times \{1\}) = D$), we must show that

$$\{\mathbb{F}, \imath\}(F)\{\mathbb{F}, \omega\}_D = \{\mathbb{F}, \omega\}_C \{\mathbb{F}, \nu\}(F).$$

In particular, we must give an isotopy between the embeddings which name the two composites. To do this, let X_o be a subspace on X obtained by deleting a collar neighborhood of X' in such a way that X_o is p.l. homeomorphic to X, and is a strong deformation retract on X_o. On X_o, the required isotopy will be the restriction of $(F \times Id_{[0,1]})(\omega \times Id_{[0,1]})(Id_M \times \psi)$, where $\psi: [0,1]^2 \to [0,1]^2$ is given by $\psi(s,t) = (s, (s+t)/2)$. Off of X_o, the isotopy stretches (resp. compresses) the collar neighborhood of the source (resp. target). Schematically, the isotopy is illustrated in Figure 1.3.1.

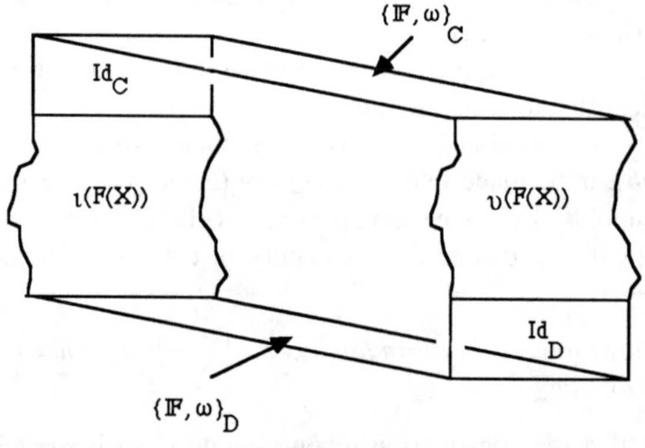

Figure 1.3.1

The preservation of the 2-dimensional composition of 2-cells follows from the trivial observation that the construction of the components of $\{\mathbb{F}, \omega\}$ respects pasting isotopies end-to-end. The 1-dimensional composition of 2-cells is given in ***Emb*** by composing the one isotopy with the restriction of the other, as in Figure 1.3.2. It is immediate from the construction of $\{\mathbb{F}, \omega\}_C$ that this composition is preserved. □

Before proceeding to examine more properties of this construction, we should point out several examples of formal cobordism families.

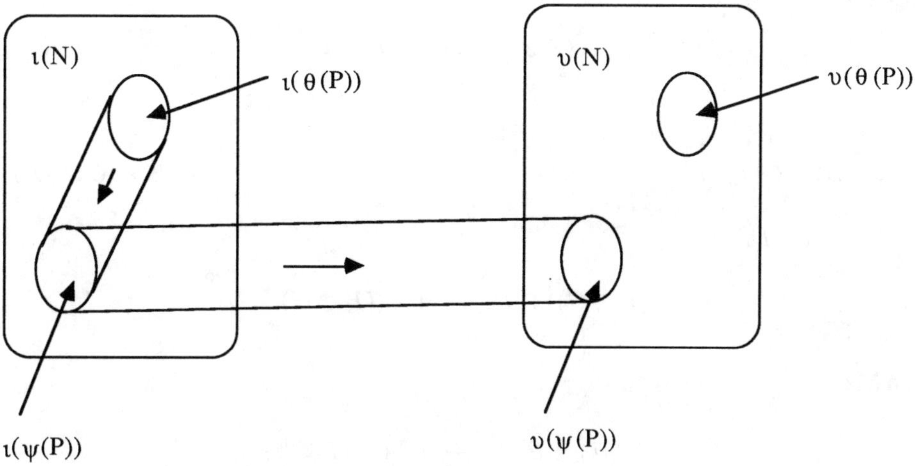

Figure 1.3.2

Example 1.4. $\mathbb{T}$ (for tangles) is the formal cobordism family consisting of all pairs $(X, \partial X)$ where X is a compact 1-manifold.

Example 1.5. $\mathbb{OT}$ (for oriented triangles) is the formal cobordism family consisting of all pairs $(X, \partial X)$, where X is a compact oriented 1-manifold, and ∂X is regarded as a signed set. (Unions defining compositions must thus identify points with opposite signs.)

Example 1.6. $\mathbb{GR}$ (for graphs) consisting of all pairs (X, X'), where X is a compact 1-complex, $X' \subset \partial X$, and X is equipped with a finite subset of $X - X'$, which includes all points of $\partial X - X'$ and all singular points (called the set of vertices).

Example 1.7. $\mathbb{RT}$ (for ribbon tangles) consists of all (X, X'), where X is of the form $X_0 \times [0, 1]$ for X_0 some compact oriented 1-manifold, and X' is $\partial X_0 \times [0, 1]$.

Example 1.8. n-**cobord** consists of all $(X, \partial X)$, where X is an oriented compact $(n + 1)$-dimensional manifold (regarded as giving cobordisms between any partition of ∂X into a pair of sets of components).

Most of the results in this paper will hold for Examples 1.4–1.7. The difficulties in extending them to Example 1.8 are analogous to the difficulties in extending techniques which work well for the fundamental group to higher homotopy groups.

To proceed much further, we will require some category theoretic preliminaries (cf. Mac Lane **[15a]**).

Definition 1.9. A ***monoidal category*** $\mathbf{V} = (\mathcal{V}, \otimes, I, \alpha, \rho, \lambda)$ consists of a category, $\mathcal{V}$, a functor $\otimes: \mathcal{V} \times \mathcal{V} \to \mathcal{V}$ (written in infix notation), and natural isomorphisms $\alpha_{A,B,C}: (A \otimes B) \otimes C \to A \otimes (B \otimes C)$, $\rho_A: A \otimes I \to A$, and $\lambda_A: I \otimes A \to A$ such that **M1**:

$$\begin{array}{ccccc}
 & & (A \otimes B) \otimes (C \otimes D) & & \\
 & \alpha \nearrow & & \searrow \alpha & \\
((A \otimes B) \otimes C) \otimes D & & & & A \otimes (B \otimes (C \otimes D)) \\
 & \alpha \otimes D \searrow & & \nearrow A \otimes \alpha & \\
 & (A \otimes (B \otimes C)) \otimes D & \underset{\alpha}{\longrightarrow} & A \otimes ((B \otimes C) \otimes D) &
\end{array}$$

and **M2**:

$$\begin{array}{ccc}
(A \otimes I) \otimes B & \overset{\alpha}{\longrightarrow} & A \otimes (I \otimes B) \\
\rho \otimes B \searrow & & \swarrow A \otimes \lambda \\
 & A \otimes B &
\end{array}$$

V is ***strict*** if all components of α, ρ, and λ are identity maps.

A ***(strong) monoidal functor*** between monoidal categories **V** and **W** is a triple $(F, F^{\sim}, F^0)$, where F is a functor from $\mathcal{V}$ to $\mathcal{W}$, $F^{\sim}$ is a natural isomorphism between $\otimes_{\mathbf{V}} F$ and $(F, F)\otimes_{\mathbf{W}}$ (as functors from $\mathcal{V} \times \mathcal{V}$ to $\mathcal{W}$), and F^0 is an isomorphism (in $\mathcal{W}$) between $F(I)$ and I, such that **MF1**:

$$\begin{array}{ccc}
F((A \otimes B) \otimes C) & \overset{F(\alpha)}{\longrightarrow} & F(A \otimes (B \otimes C)) \\
F^{\sim}\downarrow & & \downarrow F^{\sim} \\
F(A \otimes B) \otimes F(C) & & F(A) \otimes F(B \otimes C) \\
F^{\sim} \otimes F(C)\downarrow & & \downarrow F(A) \otimes F^{\sim} \\
(F(A) \otimes F(B)) \otimes F(C) & \underset{\alpha}{\longrightarrow} & F(A) \otimes F(B) \otimes F(C))
\end{array}$$

and **MF2**:

$$\begin{array}{ccccccc}
F(A \otimes I) & \overset{F(\rho)}{\longrightarrow} & F(A) & \quad & F(I \otimes A) & \overset{F(\lambda)}{\longrightarrow} & F(A) \\
F^{\sim}\downarrow & & \uparrow \rho & & F^{\sim}\downarrow & & \uparrow \lambda \\
F(A) \otimes F(I) & \underset{F(A) \otimes F^0}{\longrightarrow} & F(A) \otimes I & & F(I) \otimes F(A) & \underset{F^0 \otimes F(A)}{\longrightarrow} & I \otimes F(A)
\end{array}$$

Abusing notation, we use F to denote either the functor or the whole triple $(F, F^{\sim}, F^0)$. A monoidal functor is ***strict*** if all components of $F^{\sim}$ and F^0 are identity maps.

A ***monodial natural transformation*** between two monoidal functors $F,\ G: \mathbf{V} \to \mathbf{W}$ is a natural transformation $\psi: F \to G$ such that **MN1**:

$$\begin{array}{ccc} F(A \otimes B) & \xrightarrow{F^{\sim}} & F(A) \otimes F(B) \\ \psi \downarrow & & \downarrow \psi \otimes \psi \\ G(A \otimes B) & \xrightarrow[G^{\sim}]{} & G(A) \otimes G(B) \end{array}$$

and **MN2**:

$$\begin{array}{ccc} F(I) & & \\ & \searrow^{F^0} & \\ \psi \downarrow & & I \\ & \nearrow_{G^0} & \\ G(I) & & \end{array}$$

In general, monodial categories and functors arising "in nature" are not strict, while strict monoidal categories and functors are easier to work with. However, the following theorem of Mac Lane **[15]** obviates the effects of this dichotomy:

Theorem 1.10. *For any (small) monoidal category,* $\mathbf{V}$*, there is a strict monoidal category,* $\mathbf{V}_{st}$*, which is monoidally equivalent to* $\mathbf{V}$*. (That is there are monodial functors in both directions, which are adjoint with monoidal natural isomorphisms as unit and counit.)* □

The following generalization of the more classical notion of symmetry, introduced by Joyal and Street **[9a-b]**, will apply to the monoidal structure of the isotopy categories of the plane.

Definition 1.11. A ***braiding*** in a monoidal category is a natural isomorphism

$$\sigma_{A,B}: A \otimes B \to B \otimes A$$

satisfying **B1**:

$$\begin{array}{ccccc} & & (B \otimes A) \otimes C \xrightarrow{\alpha} B \otimes (A \otimes C) & & \\ & \sigma \otimes C \nearrow & & \searrow B \otimes \sigma & \\ (A \otimes B) \otimes C & & & & B \otimes (C \otimes A) \\ & \alpha \searrow & & \nearrow \alpha & \\ & & A \otimes (B \otimes C) \xrightarrow[\sigma]{} (B \otimes C) \otimes A & & \end{array}$$

and **B2**:

$$\begin{array}{ccccc} & & A\otimes(C\otimes B) \xrightarrow{\alpha^{-1}} (A\otimes C)\otimes B & & \\ & A\otimes\sigma \nearrow & & \searrow \sigma\otimes B & \\ A\otimes(B\otimes C) & & & & (C\otimes A)\otimes B \\ & \alpha^{-1}\searrow & & \nearrow \alpha^{-1} & \\ & & (A\otimes B)\otimes C \underset{\sigma}{\longrightarrow} C\otimes(A\otimes B) & & \end{array}$$

A monoidal category equipped with a choice of braiding is called a ***braided monoidal category***. A braided monoidal category equipped with a natural automorphism of the identity functor, $\theta_A \colon A \to A$, satisfying **BAL**:

$$\begin{array}{ccc} A\otimes B & \xrightarrow{\sigma} & B\otimes A \\ \theta\downarrow & & \downarrow\theta\otimes\theta \\ A\otimes B & \underset{\sigma^{-1}}{\longrightarrow} & A\otimes A \end{array}$$

is said to be ***balanced***.

As noted in Kelly and Laplaza **[13]**, a monoidal category, $\mathbf{V}$, can be regarded in an obvious way as bicategory with one 0-cell, so that it makes sense to speak of a ***left*** (resp. ***right***) ***adjoint*** for an object, A: to wit an object ${}^*A$ (resp. A^*) equipped with unit and counit maps $h_A \colon I \to A\otimes{}^*A$ and $e_A \colon {}^*A\otimes A \to I$ (resp. $\eta_A \colon I \to A^*\otimes A$ and $\varepsilon_A \colon A\otimes A^* \to I$) satisfying the usual "triangular equations":

$$A \xrightarrow{\lambda^{-1}} I\otimes A \xrightarrow{h\otimes 1} (A\otimes{}^*A)\otimes A \xrightarrow{\alpha} A\otimes({}^*A\otimes A) \xrightarrow{1\otimes e} A\otimes I \xrightarrow{\rho} A,$$

$${}^*A \xrightarrow{\rho^{-1}} {}^*A\otimes I \xrightarrow{1\otimes h} {}^*A\otimes(A\otimes{}^*A) \xrightarrow{\alpha^{-1}} ({}^*A\otimes A)\otimes{}^*A \xrightarrow{e\otimes 1} I\otimes{}^*A \xrightarrow{\lambda} {}^*A,$$

$$A \xrightarrow{\rho^{-1}} A\otimes I \xrightarrow{1\otimes\eta} A\otimes(A^*\otimes A) \xrightarrow{\alpha^{-1}} (A\otimes A^*)\otimes A \xrightarrow{\varepsilon\otimes 1} I\otimes A \xrightarrow{\lambda} A,$$

$$A^* \xrightarrow{\lambda^{-1}} I\otimes A^* \xrightarrow{\eta\otimes 1} (A^*\otimes A)\otimes A^* \xrightarrow{\alpha} A^*\otimes(A\otimes A^*) \xrightarrow{1\otimes\varepsilon} A^*\otimes I \xrightarrow{\rho} A^*,$$

are all identities.

Using terminology suggested by Joyal and Street (see also Freyd and Yetter **[6]**) we make the following definition.

Definition 1.12. An ***autonomous category*** is a monoidal category equipped with a choice for each object A of a left adjunction $({}^*A, h_a, e_A)$ and a right adjunction $(A^*, \eta_A, \varepsilon_A)$.

As was observed in Freyd and Yetter [6], each of $^*(\ )$ and $(\ )^*$ extends to a strong contravariant monoidal endofunctor with canonical isomorphisms:

$$c:^*(A \otimes B) \xrightarrow{\sim} {}^*B \otimes {}^*A,$$

$$n:^* I \xrightarrow{\sim} I,$$

$$\gamma: (A \otimes B)^* \xrightarrow{\sim} B^* \otimes A^*,$$

and

$$\nu: I^* \xrightarrow{\sim} I.$$

There are, moreover, natural isomorphisms of monoidal functors

$$k:^*(A^*) \xrightarrow{\sim} A$$

and

$$\kappa: ({}^*A)^* \xrightarrow{\sim} A,$$

each defined in the obvious way in terms of the monoidal structure and the units and counits of the adjuctions. The reader can easily verify that $({}^*(\ ), (\ )^*, k, \kappa)$ gives an adjoint equivalence between the autonomous category and its opposite category. (Recall that if $\mathbf{V}$ is a monoidal category, $\mathcal{V}^{op}$ becomes a monoidal category when equipped with the same $\otimes$ and I, and with α^{-1}, ρ^{-1} and λ^{-1} from $\mathbf{V}$ as its α, ρ and λ.)

We have particular call to consider those categories in which the adjoints to an object coincide, and can be regarded in a reasonable way as a dual to the object. Following Freyd and Yetter [6] we make the following definition.

Definition 1.13. A ***sovereign category*** is an autonomous category equipped with a natural isomorphism of (contravariant) monoidal functors $\phi: (\ )^* \rightarrow^* (\ )$, satisfying moreover,

$$\begin{array}{ccc} A^{**} & \xrightarrow{\phi_{A^*}} & {}^*(A^*) \\ (\phi_A^{-1})^* \downarrow & & \downarrow k \\ ({}^*A)^* & \xrightarrow[k]{} & A \end{array}$$

S

Diagram 1.13.1

A ***pivotal category*** is a sovereign category in which all components of ϕ are identity maps (and thus necessarily ${}^*X = X^*$ for all objects X).

The following is then equivalent to a definition in Shum [17].

Definition 1.14. A ***tortile category*** is a balanced braided sovereign category in which $\phi = \theta$, and in which θ satisfies

$$\theta_{A^*} = (\theta_A)^*.$$

Finally, we make a relaxed version of a definition from Yetter [20].

Definition 1.15. A ***system of vertices*** in a braided pivotal category is a family of maps, $V_{A,B}: A \to B$, indexed by pairs of objects, and satisfying
V1.1: $fV_{B,C} = V_{A,C}$, whenever $f: A \to B$ is nameable in terms of identity maps, components of σ, α, ρ, λ and their inverses, $\otimes$, and composition;
V2.1: $(A \otimes \eta_B)(V_{A\otimes B^*, C\otimes B}) = V_{A,C\otimes B}$;
V2.2: $(\eta_A \otimes B)(A^* \otimes V_{A\otimes B,C}) = V_{B,A^*\otimes C})$.
A ***graphical category*** is a braided pivotal category together with a system of vertices.

The following propositions explain why monoidal categories should arise as isotopy categories for formal cobordism families.

Proposition 1.16. *For any formal cobordism family,* $\mathbb{F}$, *there is an isomorphism of categories between* $\{\mathbb{F}, M + N\}$ *and* $\{\mathbb{F}, M\} \times \{\mathbb{F}, N\}$.

Proof. Consider a map in $\{\mathbb{F}, M + N\}$ named by an embedding $F: X \hookrightarrow (M + N) \times [0, 1]$. Now $X_M = F^{-1}(M \times [0,1])$ (resp. $X_N = F^{-1}(N \times [0,1])$) is a union of components of X, and thus $(X_M, X' \cap X_M)$ (resp. $(X_N, X' \cap X_N)$) is in $\mathbb{F}$ by FCF 5. The restriction of F to X_M (resp. X_N) thus names a map in $\{\mathbb{F}, M\}$ (resp. $\{\mathbb{F}, N\}$).

In the other direction, given a pair of maps named by embeddings $A: X \hookrightarrow M \times [0, 1]$ and $B: Y \hookrightarrow N \times [0, 1]$, we have that $(X + Y, X' + Y')$ is in $\mathbb{F}$ by FCF 2, and thus $A + B: X + Y \hookrightarrow (M + N) \times [0, 1]$ names a map in $\{\mathbb{F}, M + N\}$. It is clear that these constructions provide a bijection between the set of names of maps in $\{\mathbb{F}, M + N\}$ and the set of pairs of names of maps the first from $\{\mathbb{F}, M\}$, the second from $\{\mathbb{F}, N\}$.

A similar argument establishes a bijection between sets of isotopies, and pairs of isotopies, thus showing that there is a bijection between maps in $\{\mathbb{F}, M\}$ and $\{\mathbb{F}, N\}$.

It is easy to show that this bijection is functorial. (Essentially, composition, source, target and identity maps can all be computed separately in each summand of the space.) □

Proposition 1.17. *Let* S *be any space, then* $\{\mathbb{F}, S \times (0, 1)\}$ *is a pivotal category as follows: Identify* $\{\mathbb{F}, (S\times(0,1))+(S\times(0,1))\}$ *with* $\{\mathbb{F}, (S\times(0,1))\}\times\{\mathbb{F}, (S\times(0,1))\}$ *by the isomorphism of the previous proposition; then* $\otimes$ *is induced by the inclusion* $\mu: (S \times (0, 1)) + (S \times (0, 1)) \hookrightarrow (S \times (0, 1))$ *given by the identity on the* S*-coordinates, multiplication by* $\frac{1}{2}$ *on the first summand's* $(0, 1)$*-coordinate, and multiplying by* $\frac{1}{2}$ *and adding* $\frac{1}{2}$ *to the second summand's* $(0, 1)$*-coordinate.* I *is named by the unique inclusion of the empty space (which is in* $\mathbb{F}$ *by FCF 0) into* $S \times (0, 1)$. α *(resp.* ρ, λ*) is the natural isomorphism induced by the ambient isotopy* $Id_S \times a$ *(resp. the isotopy* $Id_S \times r$, *the isotopy* $Id_S \times \ell$*), where* $a: (0, 1) \times (0, 1) \to (0, 1)$ *is given by*

$$a(s,t) = \begin{cases} s(1+t) & \text{if } s \leq \frac{1}{4} \\ s + t/4 & \text{if } \frac{1}{4} \leq s \leq \frac{1}{2} \\ (s-1)/(1+t) + 1 & \text{if } \frac{1}{2} \leq s \end{cases}$$

(resp. $r(s, t) = s(1 + t)$, $\ell(s, t) = (s - 1)/(t + 1) + 1$*) (or rather by p.l. or smooth approximations of these piecewise smooth isotopies).*

The duality operation is induced on names of objects by composition with the map $Id_S \times n$, *where* $n(s) = 1 - s$, *and on names of maps by* $Id_S \times n \times n$. *The units and counits of the adjunctions at any object* $A: C \hookrightarrow S \times (0, 1)$ *are given by composing* *the map naming Id_A with the product of the identity map on S with the map given by mapping simplexes of the subdivision of $[0,1] \times [0,1]$ in the first drawing of Figure 1.17.1 linearly to the simplexes given in the other drawings as labelled.*

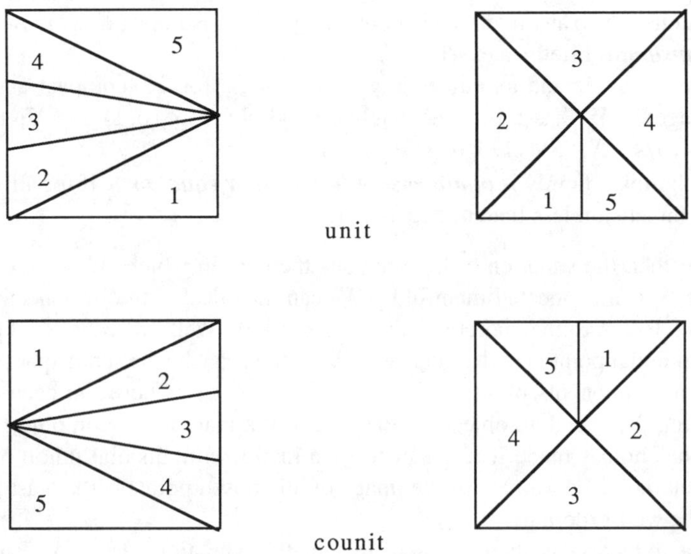

Figure 1.17.1

Proof. This is essentially the proof of one direction of the coherence theorem for pivotal categories given by Freyd and Yetter **[5]**, **[6]**. The computationally-minded can amuse themselves by writing explicit isotopies to verify the coherence and adjunction conditions. □

Previous work on categories of tangles has relied on a reduction to smaller equivalent categories with strict monoidal structures. This reduction is analogous to the reduction from knots modulo isotopy to knot diagrams modulo the Reidemeister moves. This reduction is set forth most explicitly for a variety of cases in Joyal and Street **[10]**. We now sketch an adaptation of their technique to the situation above.

To do this we make some definitions (one of which will be needed only later).

Definition 1.18. An FCF is ***compact*** if every pair in it is a pair of compact spaces. An FCF is ***object essentially 0-dimensional*** (briefly o.e. -0) if it is compact and for every component C of any X', there is a homotopy $h: C \times [0,1] \to C$, so that

$$h(c,0) = c,$$
$$h(c,1) = *,$$

and the restriction of h to any proper closed subinterval not containing 1 is an isotopy. An FCF, $\mathbb{F}$, is ***regular*** if

R1: $\mathbb{F}$ is o.e. -0;

R2: for every (X, X') in $\mathbb{F}$, there exist (X_1, X_1') and (X_2, X_2') in $\mathbb{F}$ and unions of connected components C_i of X_i' and an isomorphism (of p.l. or smooth spaces as needed) $\phi: C_1 \to C_2$ so that $(X, X') = (X_1 +_\phi X_2, (X_1' - C_1) + (X_1' - C_1))$, each (X_1, X_1') admits a contraction, h, onto a pair (S_i, S_i'), where S_i is a contractible 1-complex, and S_i' is a finite set of points. Moreover, h may be chosen so that the restriction of h to any interval not containing 1 is an isotopy ($\mathbb{F}$ is ***arrow essentially*** 1-***dimensional***, briefly, a.e. -1);

R3: for any (X, X'), and an embedding $\psi: Y' \hookrightarrow X$ for Y' compact and such that the image of Y' has a bicollar neighborhood $\Psi: Y' \times (0,1) \hookrightarrow X$, then $(X - \Psi((1/4, 3/4), X' \cup \Psi\{1/4, 3/4\}))$ is in $\mathbb{F}$.

A formal cobordism family is ***of dimension less than or equal to*** n if for all (X, X') in $\mathbb{F}$, X is of dimension less than or equal to n.

Consider then the situation of the previous theorem in which $M = S \times [0,1]$, and assume that S is a connected manifold. (We can then deal with disconnected S using Proposition 1.16.) Assume, moreover, that $\mathbb{F}$ is o.e. -0. Well-order the set of points of S so that the minimal point is in the interior of S. Let $\rho r(\mathbb{F})$ be the set of spaces occurring as connected components of some X' for $(X, X') \in \mathbb{F}$. Choose in each element of $\rho r(\mathbb{F})$ a point. Now call an object or component of a map naming an object ***good*** with respect to the choices made if it is named by a map from a disjoint union of spaces in $\rho r(\mathbb{F})$ so that the S-coordinate of the image of all chosen points is the least point (with respect to the well-ordering).

The first reduction is then the note that in this situation, $\{\mathbb{F}, M\}$ is monoidally equivalent to its full subcategory consisting of good objects. To see this, consider a map, A, from a disjoint union of spaces, D, in $\rho r(\mathbb{F})$ to M. Now, order the components of D by the images of their chosen points using the lexicographic order on M induced by the ordering on $[0,1]$ and the well-ordering of S, with the ordering on $[0,1]$ dominant. For each component, C, of D which is not already good choose (in order) a sufficiently fine triangulation of M so that a star neighborhood of the chosen point of the component is isotopic in $M - A(D - C)$ to a star neighborhood of some (chosen) point with minimal S-coordinate. Note that this can be done so as to preserve the induced order on the chosen points of the components. Now, construct an isotopy which fixes all of the components except C and takes the chosen point to the new chosen point with minimal S-coordinate by first applying the restriction of the isotopy of the star neighborhood, and extending to the trivial isotopy on the remaining components of D. Composing the maps induced by these isotopies gives an isomorphism, Φ_A, between the original object, A, and a good object, $G(A)$. We can extend G to a monoidal functor by taking G on maps to be given by $G(F: A \to B) = \Phi_A^{-1} F \Phi_B$. It is then easy to see that Φ and Φ^{-1} become the unit and counit of an adjuncton between G and the inclusion of the full subcategory of good objects (by construction $G(A) = A$ whenever A is already good). That this is monoidal requires a little more work: F^o is trivial, but $F^\sim$ requires us to construct some new isotopies using similar techniques to those above.

We can now perform a further reduction to obtain a strict monoidal subcategory whose objects are just words on $\rho r(\mathbb{F})$. Choose a strictly increasing sequence

$$0 < x_1 < x_2 < \cdots < x_n < \cdots < 1$$

and choose for each i a sufficiently fine triangulation so that a closed star neighborhood of (s, x_i) (for s the minimal S-coordinate) can be chosen disjoint from all previously chosen star neighborhoods and all remaining x_j's. Now given any good object, choose a small star neighborhood of each chosen point and construct an isomorphism to a good object with name factoring through these neighborhoods by applying an isotopic contraction of each component of the name of the object. Applying a further isotopy to slide and dilate or contract the star neighborhoods of the chosen points to star neighborhoods in the chosen sequence ***in order*** then gives an isomorphism with an object with chosen point mapped to the chosen sequence. Finally, choose a "standard" embedding of each space in $pr(\mathbb{F})$ into a standard star neighborhood of a point in $int(S) \times (0,1)$, and apply an isotopy of the given embedding to this standard embedding (this of course requires our contractibility assumption). Call such objects ***standard***. An argument similar to that for "good" objects then shows that the full subcategory of standard objects is equivalent to the full subcategory of good objects and hence to $\{\mathbb{F}, M\}$. This equivalence can be used to ***induce*** a monoidal structure on the standard objects. (The reader doubtless sees that concatenation of words on $pr(\mathbb{F})$, identified with their standard embeddings will be $\otimes$ on standard objects, but it is easier to induce the monoidal structure, thereby getting the monoidal equivalence for free, then check that the induced structure is the obvious one.)

Most of the following theorem then follows from the reduction outlined above and results of Freyd and Yetter **[5]**, Yetter **[20]**, and Shum **[17]**, effectively summarizing these results in our current language. The cases with target spaces of dimensions greater than two are essentially results occuring in unpublished work of Joyal and Street, or an adaptation of those results to graphical categories.

Theorem 1.19. *$\{\mathbb{OT}, (0,1)\}$ is equivalent to the (strict) pivotal category freely generated by a single object.*

$\{\mathbb{T}, (0,1) \times (0,1)\}$ is equivalent to the category of unoriented tangles of **[19]**.

$\{\mathbb{OT}, (0,1) \times (0,1)\}$ is equivalent to the category of oriented tangles, that is to the braided pivotal category freely generated by a single object, X, modulo the relations

$$(A \otimes \eta_{A^*})(\sigma_{A,A} \otimes A^*)(A \otimes \varepsilon_A) = 1_A = (A \otimes \eta_{A^*})(\sigma^{-1}_{A,A} \otimes A^*)(A \otimes \varepsilon_A).$$

$\{\mathbb{RT}, (0,1) \times (0,1)\}$ is equivalent to the category of ribbon tangles, that is to the tortile category freely generated by a single object.

$\{\mathbb{GR}, (0,1) \times (0,1)\}$ is equivalent to the category of branched tangles, that is the graphical category freely generated by a single object, X, modulo the relations

$$(A \otimes \eta_{A^*})(\sigma_{A,A} \otimes A^*)(A \otimes \varepsilon_A) = 1_A = (A \otimes \eta_{A^*})(\sigma^{-1}_{A,A} \otimes A^*)(A \otimes \varepsilon_A).$$

$\{\mathbb{OT}, (0,1)^n\}$, for any $n \geq 3$, is equivalent to the compact closed (i.e., symmetric pivotal) category freely generated by a single object.

$\{\mathbb{GR}, (0,1)^n\}$, for any $n \geq 3$, is equivalent to the symmetric graphical category freely generated by a single object. □

One last result at this level of generality:

Theorem 1.20. *If M is a manifold with boundary, then $\{\mathbb{F}, M\}$ is equivalent to $\{\mathbb{F}, M - \partial M\}$. If moreover, $\{\mathbb{F}, M - \partial M\}$ has a monoidal structure then this equivalence induces a monoidal structure on $\{\mathbb{F}, M\}$.*

Proof. There is an obvious inclusion of categories, i, from $\{\mathbb{F}, M - \partial M\}$ into $\{\mathbb{F}, M\}$. We construct a functor in the reverse direction first on objects by taking the identity on objects whose names factor through $M - \partial M$, and for the other objects taking their images under an ambient isotopy which contracts a collar neighborhood of ∂M. For each object A we then have an isomorphism ϕ_A between it and an object $r(A)$ of $\{\mathbb{F}, M - \partial M\}$. Defining the image of maps so that this isomorphism and its inverse become the unit and counit of an adjunction between i and r, then gives the desired result.

To define the monoidal structure on $\{\mathbb{F}, M\}$, we let $A \otimes B$ be $i(r(A) \otimes r(B))$ (where the first tensor is in $\{\mathbb{F}, M\}$ and the second is in $\{\mathbb{F}, M - \partial M\}$). In either category the object I is named by the map from the empty space. Once it is observed that for objects of $\{\mathbb{F}, M - \partial M\}$, r and i act as the identity, and ϕ at such objects is an identity map, it follows that $i(\alpha)$, $i(\rho)\phi^{-1}$, and $i(\lambda)\phi^{-1}$ are the structural natural transformations for $\{\mathbb{F}, M\}$. □

To give some idea where all this is leading, an analogy is in order: if we regard monoidal categories as being like rings, all of this so far has been analogous to considering rings of polynomials. To accomplish our goal of understanding the structure of $\{\mathbb{F}, M\}$ for M a general surface, we will require the analogs of other rings, modules, and tensor-products of modules.

2. The annulus

Before going on to consider more general surfaces, we examine the structure of the isotopy categories associated to the annulus, $S^1 \times (0, 1)$.

First observe that by Proposition 1.16, for any FCF, $\mathbb{F}$, $\{\mathbb{F}, S^1 \times (0, 1)\}$ is a pivotal category. Next, observe that there is a non-trivial isotopy from the identity map on $S^1 \times (0, 1)$ to itself, given by (a p.l. approximaton of) $b \colon S^1 \times (0, 1) \times [0, 1] \to S^1 \times (0, 1)$ by $b(z, s, t) = (ze^{2\pi i t}, s)$. Thus by the 2-functoriality of $\{\mathbb{F}, -\}$ there is a natural automorphism of the identity functor on $\{\mathbb{F}, S^1 \times (0, 1)\}$, β, induced by b.

Proposition 2.1. *For any FCF, $\mathbb{F}$, the natural automorphism $\beta \colon Id_{\{\mathbb{F}, S^1 \times (0,1)\}} \to Id_{\{\mathbb{F}, S^1 \times (0,1)\}}$ is a monoidal natural transformation.*

Proof. The condition MN2 on the component at the object I is immediate, since this is named by the map from the empty space. The condition MN1 reduces in this case to $\beta_{A \otimes B} = \beta_A \otimes \beta_B$. This follows from the observation that each is named by the embedding of $(C + D) \times [0, 1]$ into $S^1 \times (0, 1) \times [0, 1]$ by the map $\beta_!$, where $\beta_!(\gamma, t) = ((AB)_1(\gamma)e^{2\pi i t}, (AB)_2(\gamma), t)$, where $(AB) = ((AB)_1, (AB)_2) \colon (C + D) \to S^1 \times (0, 1)$ is the map which names $A \otimes B$ (or in the p.l. case by p.l. approximations of this map). In the p.l. case, these names may be different approximations, but they are certainly isotopic and thus name the same map. □

Our main result in this section is that in the case where $\mathbb{F}$ is regular the preceding essentially determines $\{\mathbb{F}, S^1 \times (0, 1)\}$. To be precise:

Theorem 2.2. *If* $\mathbb{F}$ *is a regular FCF, then* $\{\mathbb{F}, S^1 \times (0,1)\}$ *is monoidally equivalent to the category obtained from* $\{\mathbb{F}, (0,1) \times (0,1)\}$ *by freely adjoining a monoidal natural automorphism of the identity functor.*

Proof. Call the monoidal category obtained from $\{\mathbb{F}, (0,1) \times (0,1)\}$ by freely adjoining a monoidal natural automorphism of the identity functor, $\mathbf{V}$. Now there is an evident monoidal functor, Φ, from $\mathbf{V}$ to $\{\mathbb{F}, S^1 \otimes (0,1)\}$ induced by mapping the objects and maps of $\{\mathbb{F}, (0,1) \times (0,1)\}$ by the monoidal functor induced by the map $\phi: (0,1) \times (0,1) \hookrightarrow S^1 \times (0,1)$ given by $\phi(r,s) = (e^{\pi i r}, s)$ (or a p.l. approximation of this if we are in the p.l. case), and mapping components of the natural automorphism to corresponding components of β.

We must now show two things: First that $\{\mathbb{F}, S^1 \times (0,1)\}$ is equivalent to the image of this functor, and second that the functor is fully faithful.

Now, we begin by showing that Φ is full: a fact we will need both for its fully-faithfulness, and for the equivalence. Notice that the image of the functor consists of precisely those maps whose source and target lie in the image of $\{\mathbb{F}, \phi\}$, and which admit decompositions under composition and $\otimes$ into maps in the image of $\{\mathbb{F}, \phi\}$ and components of β and β^{-1} at such objects.

Now, we claim that if $\mathbb{F}$ is regular, then any map in $\{\mathbb{F}, S^1 \times (0,1)\}$ between objects in the image of $\{\mathbb{F}, \phi\}$ admits such a decomposition. To verify the claim, suppose that $F: X \hookrightarrow S^1 \times (0,1) \times [0,1]$ names some map from one object in the image of $\{\mathbb{F}, \phi\}$ to another. Now by the regularity of $\mathbb{F}$, X admits a decomposition as $X_1 +_C X_2$, where each X_i has components which admit contractions to a point which are isotopies on any proper initial segment. Choosing a sufficiently fine triangulation on $S^1 \times (0,1) \times [0,1]$, we can replace F with another name for the same map by applying the contraction to all of each component of each X_i except for small collar neighborhoods of the source and target of F and a small bicollar neighborhood of C and stretching the collars and bicollar, until all of X_i is mapped into the interior of a star neighborhood of the vertex of the triangulation, except for the (stretched) collars and bicollar.

Now choose a section of the projection onto $(0,1) \times [0,1]$ lying in the interior of the complement of the image of ϕ, and intersecting the new name of the map only in points of the stretched (bi)collar(s). Finally, apply an ambient isotopy which fixes $S^1 \times (0,1) \times \{0,1\}$, stretches a neighborhood of the chosen section, and moves all of the image of the new name into the image of ϕ. This new name is then an embedding $F_+: X \hookrightarrow S^1 \times (0,1) \times [0,1]$, which factors through ϕ, except for some (bi)collars on components of X''s.

Applying another ambient isotopy, which is non-trivial only outside slightly larger neighborhoods of the non-trivial parts of F than the star neighborhoods used above, we can assume that the (bi)collar(s) leave these neighborhoods on the top or bottom faces (i.e., extreme in the ordering on the $[0,1]$ coordinate).

In this circumstance, consider (a p.l. approximation of) the map from $q: S^1 \times (0,1) \times [0,1] \to S^1$ given by

$$q(z,s,t) = \begin{cases} 1 & \text{if } z = e^{\pi i r},\ r \in (0,1) \\ z^2 & \text{otherwise} \end{cases}.$$

Then for any of the (bi)collar(s), $\theta\colon C \times [0,1] \to S^1 \times (0,1) \times [0,1]$, θq is homotopic to a map which factors through the second projection, and thus determines an element, ε, of $\pi_1(S^1, 1) = \mathbb{Z}$. It is then easy to see that the (bi)collar names the component of β^ε at the object named by the embedding of C, followed by an isomorphism in the image of $\{\mathbb{F}, \phi\}$.

From this last name for our map, and this observation about the maps named by the (bi)collar(s), we can provide an expression for the map in terms of components of β and β^{-1}, maps in the image of $\{\mathbb{F}, \phi\}$, composition, and $\otimes$, thus demonstrating that Φ maps onto the full subcategory on objects in the image of $\{\mathbb{F}, \phi\}$.

For the monoidal equivalence of the entire category to this full subcategory, choose a triangulation fine enough that the star neighborhood of any vertex has S^1-coordinates differing in argument by less than $\pi/6$. Now for any object, and each component of the name of the object which does not factor through ϕ, choose an isotopy shrinking the component into the interior of a star neighborhood of one of its points. If the shrunken version still does not factor through ϕ, choose a section of the projection onto $(0,1)$ lying outside the image of ϕ and not intersecting the name of the "shrunken" object (if necessary shrink the name further). Finally, apply an isotopy which contracts all of the complement of this section into the image of ϕ via (a p.l. approximation of) a contraction along fibres over $(0,1)$.

Now the isomorphisms induced by the composition of these isotopies (the identity if the object already lies in the image of $\{\mathbb{F}, \phi\}$) and their inverses can be seen to be the components of an adjoint equivalence between the entire category and the full subcategory on objects in the image of $\{\mathbb{F}, \phi\}$, that is between the entire category and the image of Φ.

It now remains to show that Φ is faithful. To do this it suffices to show that given two expressions for maps in terms of components of β and β^{-1}, maps in the image of $\{\mathbb{F}, \phi\}$, composition and $\otimes$, and an isotopy between the names constructed in the obvious way from the expressions, that there are equations in $\{\mathbb{F}, (0,1) \otimes (0,1)\}$ which together with the monoidal naturality and invertibility of β verify the equality of the two maps.

Now, so long as the isotopy only moves things within the image of ϕ, any relation induced is an equation in $\{\mathbb{F}, (0,1) \times (0,1)\}$. Now, consider any occurence of β or β^{-1} in either expression. This is named by an embedding of $C \times [0,1]$ as given in the proof of Proposition 2.1. Now, so long as the 0 and 1 sections of this embedding are not removed from the image of ϕ by the isotopy, the isotopy replaces a component of β (or β^{-1}) with composites of $\{\mathbb{F}, (0,1) \times (0,1)\}$-isomorphisms and other components of β (or β^{-1}). Any such relation can be verified by commuting the isomorphisms with β (or β^{-1}) by the naturality of β.

If on the other hand, one of the ends (i.e., 0 or 1 sections) of the embedding naming the component of β (or β^{-1}) leaves the image of ϕ during the isotopy, we must consider the image of its motion under q as an element of $\pi_1(S^1)$. If it is trivial, we can replace this isotopy by one in which the end does not leave the image of ϕ. Otherwise consider that map in the expression for the whole map whose target or source has a name one or more components of which are the source or target for the instance of β (or β^{-1}) under consideration. Now, as a result of the isotopy, some factor of this map (provided by condition R3 in the definition of regularity) must be moved out of the image of ϕ, and back into the image of ϕ, along a path whose image under q determines the same element of π_1 as the motion of the end of the name of the instance of β (or β^{-1}) under consideration. At the other end of this map, the isotopy must leave a copy of β to the power determined by the element of π_1. Now, any such motion can be accounted for by the naturality and invertibility of the powers of β (in particular, the commutativity of the square associated to the factor map whose name is being moved around the annulus by the isotopy). □

Essentially this monoidal structure has been used by V. Turaev to prove the freeness of the skein module (see J. Przytycki's contribution to this volume) of the solid torus.

In our examples, where we have an explicit description of $\{\mathbb{F}, (0,1) \times (0,1)\}$ (up to monoidal equivalence), this provides an explicit description of $\{\mathbb{F}, S^1 \times (0,1)\}$ up to monoidal equivalence. For more general surfaces, we no longer have a monoidal structure. To describe the isotopy category associated to more general surfaces, we require more category theory.

3. Actions of monoidal categories

In terms of the analogy suggested at the end of Section 1, we now turn to the analog in the theory of monoidal categories of modules over a monoidal category.

Despite the obvious analogy to ring theory, it appears (on no less authority than that of Mac Lane) that the idea of actions of monoidal categories on other categories is new*. It is perhaps not surprising that new categorical ideas should be required now — the previously unguessed at role of categories in topology and physics being the driving force in this development. This section is highly categorical, but the analogy with modules over rings should provide a compass for the non-categorical reader.

Definition 3.1. If **V** and **W** are monoidal categories, a ***left (strong)* V*-module***, M, is a category, $\mathcal{M}$, equipped with a functor $\#\colon \gamma \times \mathcal{M} \to \mathcal{M}$ (written in infix notation), and natural isomorphisms, $\underline{a}_{A,B,X}\colon (A \otimes B)\#X \to A\#(B\#X)$ and $\underline{\ell}_X\colon I\#X \to X$, satisfying **LM1**:

$$
\begin{array}{ccccc}
 & & (A \otimes B)\#(C\#X) & & \\
 & \underline{a} \nearrow & & \searrow \underline{a} & \\
((A \otimes B) \otimes C)\#X & & & & A\#(B\#(C\#X)) \\
 & \alpha\#X \searrow & & \nearrow A\#\underline{a} & \\
 & (A \otimes (B \otimes C))\#X & \underset{\underline{a}}{\longrightarrow} & A\#((B \otimes C)\#X) &
\end{array}
$$

* Note added in proof: Kapranov and Voevodsky (unpublished work) considered module categories over **Vect** in connection with the Zamolodnikov tetrahedral equation.

and **LM2**:

$$\begin{array}{ccc} (A \otimes I)\#X & \overset{\underline{a}}{\longrightarrow} & A\#(I\#X) \\ \rho\#X \searrow & & \swarrow A\#\underline{\ell} \\ & A\#X & \end{array}$$

A ***right (strong)* W*-module***, **N**, is a category $\mathcal{N}$, equipped with a functor $\&: \mathcal{N} \times \mathcal{W} \to \mathcal{N}$, (written in infix notation) and natural isomorphisms, $a_{Y,C,D}: (Y\&C)\&D) \to Y\&(C \otimes D)$ and $r_Y: Y\&I \to Y$ satisfying **RM1**:

$$\begin{array}{ccccc} & & (Y\&C)\&(D \otimes E) & & \\ & a \nearrow & & \searrow a & \\ ((Y\&C)\&D)\&E & & & & Y\&(C \otimes (D \otimes E)) \\ & a\&E \searrow & & \nearrow Y\&\alpha & \\ & & (Y\&(C \otimes D))\&E \underset{a}{\longrightarrow} Y\&((C \otimes D) \otimes E) & & \end{array}$$

and **RM2**:

$$\begin{array}{ccc} (Y\&I)\&E & \overset{a}{\longrightarrow} & Y\&(I \otimes E) \\ r\&E \searrow & & \swarrow Y\&\lambda \\ & Y\&E & \end{array}$$

and a ***(strong)* V, W*-bimodule***, **P**, is a category $\mathcal{P}$ which is both a left **V**-module, and a right **W**-module, and which is moreover equipped with a natural isomorphism $\mathbf{a}_{A,Z,C}: (A\#Z)\&C \to A\#(Z\&C)$ satisfying **BM1**:

$$\begin{array}{ccccc} & & (A \otimes B)\#(X\&C) & & \\ & \mathbf{a} \nearrow & & \searrow \underline{a} & \\ ((A \otimes B)\#X)\&C & & & & A\#(B\#(X\&C)) \\ & \underline{a}\&C \searrow & & \nearrow A\#\mathbf{a} & \\ & & (A\#(B\#X))\&C \underset{\mathbf{a}}{\longrightarrow} A\#((B\#X)\&C & & \end{array}$$

and **BM2**:

$$\begin{array}{ccccc}
 & & (A\#X)\&(C\otimes D) & & \\
 & a \nearrow & & \searrow \mathbf{a} & \\
((A\#X)\&C)\&D & & & & A\#(X\&(C\otimes D)) \\
 & \mathbf{a}\&D \searrow & & \nearrow A\#a & \\
 & & (A\#(X\&C))\&D \xrightarrow[\mathbf{a}]{} A\#((X\&C)\&D) & &
\end{array}$$

A left or right $\mathbf{V}$-module, or a $\mathbf{V},\mathbf{W}$-bimodule is ***strict*** if all components of the structural natural transformations are identity maps.

It is easy to find examples of modules and bimodules over monoidal categories. Given any monoidal category, $\mathbf{U}$, and monoidal subcategories $\mathbf{V}$ and $\mathbf{W}$, $\mathcal{U}$ becomes a $\mathbf{V},\mathbf{W}$-bimodule when equipped with restrictions of the monoidal structure. Similarly given two monoidal categories $\mathbf{V}$ and $\mathbf{W}$, the product category, $\mathcal{V}\times\mathcal{W}$ becomes a $\mathbf{V},\mathbf{W}$-bimodule when equipped with the left action of $\mathbf{V}$ on the first factor by $\otimes$, and the right action of $\mathbf{V}$ on the second factor by $\otimes$, module structures induced by the monoidal structures on the factors, and $\mathbf{a}$ in the bimodule structure taken to be the identity natural transformation. One can also, of course, construct "syntatic" examples by performing free constructions.

This last raises the question of coherence for modules and bimodules over monoidal categories. Fortunately, no real new work is required to deal with this coherence problem: the proof of Mac Lane's coherence theorem for monoidal categories **[15]** adapts directly to give the following coherence theorems for modules over monoidal categories.

Theorem 3.2. *Any formal diagram constructed from object variables of two sorts,* $\mathbf{V}$ *(to be instantiated by objects in a monoidal category) and* $\mathbf{M}$ *(to be instantiated by objects in a left module over the monoidal category),* $\otimes$, $\#$, *composition, identity maps, and instances of* α, ρ, λ *and their inverses and of* $\underline{a}$, $\underline{l}$ *and their inverses names a commutative diagram for any instantiations of the object variables by objects in a monoidal category and a left module over the monoidal category.* □

A similar statement holds for right modules over monoidal categories.

For bimodules over a pair of monoidal categories we have the following.

Theorem 3.3. *Any formal diagram constructed from object variables of three sorts,* $\mathbf{V}$ *and* $\mathbf{W}$ *(to be instantiated by objects in two monoidal categories) and* $\mathbf{M}$ *(to be instantiated by objects in a bimodule over the monoidal categories),* $\otimes$, $\#$, $\&$, *composition, identity maps, and instances of* α, ρ, λ *and their inverses and of* $\underline{a}$, $\underline{l}$, a, r *and* $\mathbf{a}$ *and their inverses and constants* $I_{\mathbf{V}}$ *and* $I_{\mathbf{W}}$ *of types* $\mathbf{V}$ *and* $\mathbf{W}$ *respectively names a commutative diagram in the bimodule for any instantiation of the types* $\mathbf{V}$ *and* $\mathbf{W}$ *by monoidal categories, any bimodule over these monoidal categories, and any instantiation of the object variables by objects in the appropriate monoidal category and the bimodule.* □

The following definition then gives the analog of linear maps.

Definition 3.4. A ***(strong) left* $\mathbf{V}$*-module functor*** from a left $\mathbf{V}$-module M to a left $\mathbf{V}$-module N is a functor $F\colon \mathcal{M} \to \mathcal{N}$, together with isomorphisms $F^{\sim}_{A,X}\colon F(A\#X) \to A\#F(X)$ natural in both variables, and satisfying **LMF1**:

$$\begin{array}{ccc}
F((A\otimes B)\#X) & \stackrel{F(\underline{a})}{\longrightarrow} & F(A\#(B\#X)) \\
F^{\sim}\downarrow & & \downarrow F^{\sim} \\
(A\otimes B)\#F(X) & & A\#F(B\#X) \\
\underline{a}\searrow & & \swarrow A\#F^{\sim} \\
 & A\#(B\#F(X)) &
\end{array}$$

and **LMF2**:

$$\begin{array}{ccc}
F(I\#X) & & \\
 & \searrow F(\underline{\ell}) & \\
F^{\sim}\downarrow & & F(X) \\
 & \nearrow \underline{\ell} & \\
I\#F(X) & &
\end{array}$$

(Strong) right* $\mathbf{V}$*-module functors are defined similarly, with corresponding conditions RMF1 and RMF2.

If $\mathbf{V}$ and $\mathbf{W}$ are monoidal categories, a ***(strong)* $\mathbf{V},\mathbf{W}$*-bimodule functor*** from a $\mathbf{V},\mathbf{W}$-bimodule M to a $\mathbf{V},\mathbf{W}$-bimodule N is a functor $F\colon \mathcal{M} \to \mathcal{N}$, together with natural transformations ${}^{\sim}F$ and $F^{\sim}$, so that $(F, {}^{\sim}F)$ is a left $\mathbf{V}$-module map, $(F, F^{\sim})$ is a right $\mathbf{W}$-module map, and satisfying **BMF**:

$$\begin{array}{ccc}
F((A\#X)\&C) & \stackrel{F(\mathbf{a})}{\longrightarrow} & F(A\#(X\&C) \\
F^{\sim}\Big\downarrow & & \Big\downarrow {}^{\sim}F \\
F(A\#X)\&C & & A\#F(X\&C) \\
{}^{\sim}F\&C\Big\downarrow & & \Big\downarrow A\#F^{\sim} \\
(A\#F(X))\&C & \underset{\mathbf{a}}{\longrightarrow} & A\#(F(X)\&C)
\end{array}$$

As with monoidal functors, we will call left or right module functors, or bimodule functors ***strict*** if the components of $F^{\sim}$ and ${}^{\sim}F$ (as appropriate) are all identity maps.

Definition 3.5. The ***left* V*-modular transformation*** from a left **V**-module functor, F, to another, G, is a natural transformation Ψ from F to G so that **LMT**:

$$\begin{array}{ccc} F(A\#X) & \overset{\sim F}{\longrightarrow} & A\#F(X) \\ \psi\big\downarrow & & \big\downarrow A\#\psi \\ G(A\#X) & \underset{\sim G}{\longrightarrow} & A\#G(A) \end{array}$$

A ***right* W*-modular transformation*** between right **W**-module functors is defined similarly (the diagrammatic condition being denoted RMT). A **V**, **W*-bimodular transformation*** is a natural transformation between bimodule functors which is both a left **V**-modular transformation and a right **W**-modular transformation.

We can, of course, organize left **V**-modules, left **V**-module functors, and left **V**-modular transformations into a 2-category, **V**-***Mod***. Likewise we can form 2-categories, ***Mod***-**W** and **V**-***Mod***-**W**. As in the theory of modules over rings, we have change of base functors.

Proposition 3.6. *If $F{:}\,\mathbf{V} \to \mathbf{W}$ is a monoidal functor, then there is a 2-functor $F^*{:}\,\mathbf{W}$-**Mod**$\to$ $\mathbf{V}$-**Mod**, given on objects by mapping $\mathbf{M} = (\mathcal{M}, \#)$ to $F^*(\mathbf{M}) = (\mathcal{M}, F(-)\#(-))$; on 1-cells by mapping $(\Phi, \Phi^\sim)$ to (Φ, Φ'), where $\Phi'_{A,X} = \Phi^\sim_{F(A),X}$; and on 2-cells by taking a natural transformation to itself. Similarly we can define base change functors F^*:**Mod**-$\mathbf{W}$ $\to$**Mod**-$\mathbf{V}$given a monoidal functor $F{:}\,\mathbf{V} \to \mathbf{W}$, and $(F,G)^*{:}\,\mathbf{W}$-**Mod**-$\mathbf{Y}$ $\to$ $\mathbf{V}$-**Mod**-$\mathbf{X}$ given monoidal functors $F{:}\,\mathbf{V} \to \mathbf{W}$ and $G{:}\,\mathbf{X} \to \mathbf{Y}$.*

Moreover if $\phi{:}\,F \to G$ is a monoidal natural transformation for $F,G{:}\,\mathbf{V} \to \mathbf{W}$, then there is a 2-natural transformation $\phi^*{:}\,F^* \to G^*$. The component $\phi^*_{\mathbf{M}}$ is the $\mathbf{V}$-module functor $(Id_{\mathcal{M}}, \phi^\sim)$, where $\phi^\sim_{A,X} = \phi_A\#Id_X$. Similar statements hold for right modules and bimodules.

*In particular, the change of base functors and change of base transformations give us the 1-cell and 2-cell parts of 2-functors $(-)$-**Mod** and **Mod**-$(-)$ from **MonCat**^{op} to 2-**Cat** and $(-)$-**Mod**-$(-)$ from **MonCat**^{op} $\times$ **MonCat**^{op} to 2-**Cat**.*

The proof is standard and left to the reader. □

(This immediately raises the question of an analog to Morita equivalence. As we will not require it in what follows, we will not address the general question. The following theorem is the obvious case: monoidal equivalence implies equivalent categories of modules.) It immediately follows from the fact that 2-functors preserve adjunctions that:

Theorem 3.7. *If $\mathbf{V}_i$ and $\mathbf{W}_i$ are monoidally equivalent via (F_i, G_i, ϕ_i) for $(i = 1, 2)$, then*

$$(F_i^*, G_i^*, \phi_i^*)$$

*are equivalences of categories between $\mathbf{V}_i$-**Mod** and $\mathbf{W}_i$-**Mod**, and*

$$((F_1, F_2)^*, (G_1, G_2)^*, (\phi_1, \phi_2)^*)$$

is an equivalence of categories between $\mathbf{V}_1$-***Mod***-$\mathbf{V}_2$ *and* $\mathbf{W}_1$-***Mod***-$\mathbf{W}_2$. □

The chief point of this result is that so long as one is only concerned with 2-categorical data, one may as well (by the coherence theorem of Mac Lane) work with modules and bimodules over strict monoidal categories. (This will not apply if one is so perverse as to remember that 2-***Cat*** is a 3-category and inquire about the 3-cells.)

We will apply these notions to isotopy categories in two ways: one using the monoidal structure on $\{\mathbb{F}, (0,1) \times (0,1)\}$ (or equivalently on $\{\mathbb{F}, [0,1] \times [0,1]\}$), the other using the monoidal structure on $\{\mathbb{F}, S^1 \times (0,1)\}^n$ (or equivalently on $\{\mathbb{F}S^1 \times [0,1]\}^n$). We now state a general result concerning module structures over $\{\mathbb{F}, N \times [0,1]\}$ for any manifold, N.

Proposition 3.8. *If* M *is any* n*-manifold with boundary, and* $\xi\colon N \hookrightarrow \partial M$ *is any embedding of an* $(n-1)$*-manifold (possibly with boundary) into the boundary of* M*, then any pair,* $\Xi\colon N \times [0, 1+\varepsilon) \hookrightarrow M$*,* $\imath\colon M \times [0,1] \to M$ *(resp.* $\Xi\colon N \times (0-\varepsilon, 1] \hookrightarrow M$*,* $\imath\colon M \times [0,1] \to M$*) induces a left (resp. right)* $\{\mathbb{F}, N \times [0,1]\}$*-module structure on* $\{\mathbb{F}, M\}$*, where* Ξ *is an embedding whose restriction to* $N \times \{0\}$ *is* ξ *(resp. whose restriction to* $N \times \{1\}$ *is* ξ*) and whose image is an open neighborhood of the image of* ξ*, and* $\imath$ *is an isotopy, trivial on* $M - Im(\Xi)$*, from the identity on* M *to an inclusion of* M *into itself which maps* $\xi(N)$ *to* $\Xi(N \times \{1\})$ *(resp. to* $\Xi(N \times \{0\})$*).*

Proof. The natural isomorphism $\underline{l}$ (resp. r) is induced by the reverse of $\imath$. The natural isomorphism $\underline{a}$ (resp. a) is induced by an isotopy represented schematically as in Fig. 3.8.1. □

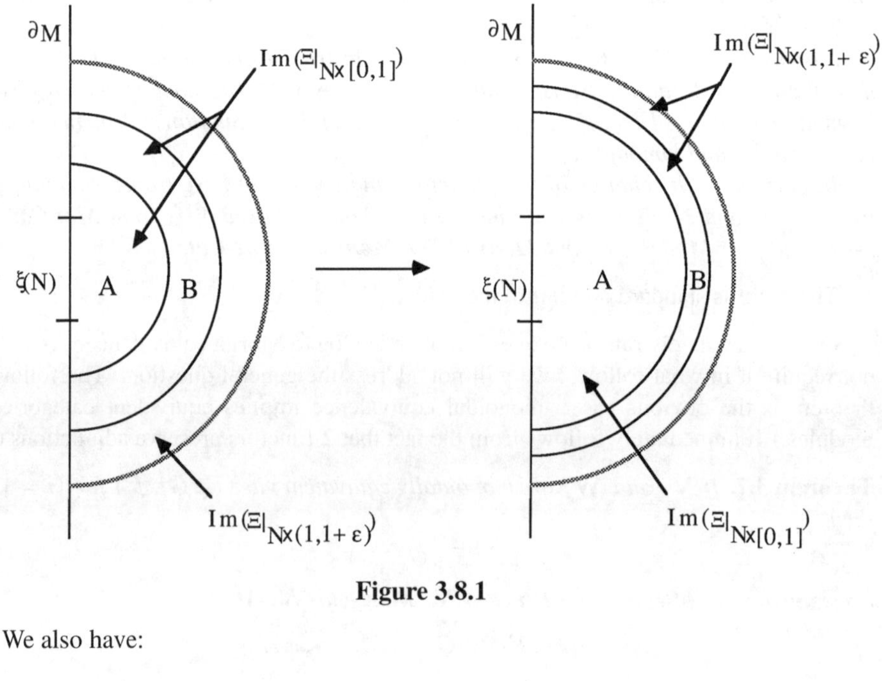

Figure 3.8.1

We also have:

Proposition 3.9. *If* ξ_l Ξ_l *and* $\imath_l$ *induce a left* $\{\mathbb{F}, N \times [0,1]\}$*-module structure on* $\{\mathbb{F}, M\}$ *as in the previous proposition, and likewise* ξ_r, Ξ_r *and* $\imath_r$ *induce a right* $\{\mathbb{F}, P \times [0,1]\}$*-module structure on* $\{\mathbb{F}, M\}$*, and if, moreover,* $Im(\Xi_l)$ *and* $Im(\Xi_r)$ *are disjoint, then* $\{\mathbb{F}, M\}$ *is a* $\{\mathbb{F}, N \times [0,1]\}, \{\mathbb{F}, P \times [0,1]\}$*-bimodule with* $\mathbf{a}$ *given by identity maps.*

Proof. Immediate from the previous proposition and the fact that $\imath_l$ is trivial on $Im(\Xi_r)$ and $\imath_r$ is trivial on $Im(\Xi_l)$. □

If we are concerned with the categorical structure only up to equivalence of categories, we need only know the maps inducing the module structure up to ambient isotopy in M. To be precise:

Proposition 3.10. *If* M *is any* n*-manifold with boundary, and* $(\xi_{l,i}, \Xi_{l,i}, \imath_{l,i})$ *(resp.* $(\xi_{r,i}, \Xi_{r,i}, \imath_{r,i})$*, resp. both together) induce a left* $\{\mathbb{F}, N_{l,i} \times [0,1]\}$*-module (resp. a right* $\{\mathbb{F}, N_{r,i} \times [0,1]\}$*-module, resp.* $\{\mathbb{F}, N_{l,i} \times [0,1]\}$*,* $\{\mathbb{F}, N_{r,i} \times [0,1]\}$*-bimodule) structure on* $\{\mathbb{F}, M\}$ *for* $i = 0, 1$*, and* $H: M \times [0,1] \to M$ *is an ambient isotopy of* M *with* $H(m,0) = m$ *and which induces an isotopy of* $(\xi_{l,0}, \Xi_{l,0}, \imath_{l,0})$ *with* $(\xi_{l,1}, \Xi_{l,1}, \imath_{l,1})$ *(resp.* $(\xi_{r,0}, \Xi_{r,0}, \imath_{r,0})$ *with* $(\xi_{r,1}, \Xi_{r,1}, \imath_{r,1})$*, resp. both), then there is an isomorphism of categories* $\Phi: \{\mathbb{F}, M\} \to \{\mathbb{F}, M\}$*, where* Φ *is a left-module (resp. right module, resp. bimodule) functor from* $\{\mathbb{F}, M\}$ *with the structure induced by the triples with subscript* 0 *to* $\{\mathbb{F}, M\}$ *with the structure induced by the triples with subscript* 1*.*

Proof. The functor Φ is induced by the (p.l.) homeomorphism $m \mapsto H(m,1)$, and its inverse, Ψ, is induced by the inverse homeomorphism. The structural transformations for these to be module or bimodule functors are induced by the isotopy, H. The coherence conditions on the structural transformations follow immediately from the fact that H induces an isotopy of the data used to define the module and bimodule structures. □

In the case of regular FCF's, we can use these actions to determine isotopy categories of surfaces obtained by pasting surfaces with boundary together along portions of their boundaries. To do this, however, we need the analog of bilinear maps and tensor products of bimodules.

Definition 3.11. If $\mathbf{V}, \mathbf{W}$, and $\mathbf{X}$ are monoidal categories, and $\mathbf{M}, \mathbf{N}$, and $\mathbf{P}$ are a $\mathbf{V}, \mathbf{W}$-bimodule, a $\mathbf{W}, \mathbf{X}$-bimodule, and a $\mathbf{V}, \mathbf{X}$-bimodule respectively, then a $\mathbf{V}, \mathbf{W}, \mathbf{X}$-***bilinear functor***, $B: \mathbf{M} \times \mathbf{N} \to \mathbf{P}$, is a functor $B: \mathcal{M} \times \mathcal{N} \to \mathcal{P}$, which is a $\mathbf{V}, \mathbf{X}$-bimodule functor with left structural structural transformation $:B$, and right structural transformation $B:$, when $\mathcal{M} \times \mathcal{N}$ is given the bimodule structure induced by those on its factors (with $\mathbf{a}$ given by identities), together with a natural isomorphism, $B^{\sim}_{S,\Omega,T}: B(S \& \Omega, T) \to B(S, \Omega \# T)$, where $\&$ denotes the right $\mathbf{W}$-module structure on $\mathbf{W}$ and $\#$ denotes the left $\mathbf{W}$-module structure on $\mathbf{N}$, and satisfying **BLF1**:

$$\begin{array}{ccccc}
 & & B^{\sim} & & \\
 & B((\Phi\#S)\&\Omega,T) & \longrightarrow & B(\Phi\#S,\Omega\#T) & \\
B(\mathbf{a},T)\nearrow & & & & \searrow :B \\
B(\Phi\#(S\&\Omega),T) & & & & \Phi\#B(S,\Omega\#T) \\
:B\searrow & & & & \nearrow \Phi\#B^{\sim} \\
 & & \Phi\#B(S\&\Omega,T) & &
\end{array}$$

BLF2:

$$\begin{array}{ccccc}
 & & B(S\&\Omega,T\&\Psi) & & \\
B{:}\nearrow & & & & \searrow B^{\sim} \\
B(S\&\Omega,T)\&\Psi & & & & B(S,\Omega\#(T\&\Psi)) \\
B^{\sim}\&\Psi\searrow & & & & \nearrow B(S,\mathbf{a}^{-1} \\
 & B(S,\Omega\#T)\&\Psi & \longrightarrow & B(S,(\Omega\#T)\&\Psi) & \\
 & & B{:} & &
\end{array}$$

BLF3:

$$\begin{array}{ccccc}
 & & B^{\sim} & & \\
 & B(S\&(\Omega\otimes\Xi),T) & \longrightarrow & B(S,(\Omega\otimes\Xi)\#T) & \\
B(a,T)\nearrow & & & & \searrow B(S,\underline{a}) \\
B((S\&\Omega)\&\Xi,T) & & & & B(S,\Omega\#(\Xi\#T)) \\
B^{\sim}\searrow & & & & \nearrow B^{\sim} \\
 & & B(S\&\Omega,\Xi\#T) & &
\end{array}$$

and **BLF4**:

$$\begin{array}{ccc}
 & B^{\sim} & \\
B(S\&I,T) & \longrightarrow & B(S,I\#T) \\
B(r,T)\searrow & & \swarrow B(S,\underline{l}) \\
 & B(S,T) &
\end{array}$$

A ***strict*** $\mathbf{V},\mathbf{W},\mathbf{X}$-***bilinear*** functor is a bilinear functor which is strict as a left $\mathbf{V}$-module functor and as a right $\mathbf{X}$-module functor. A $\mathbf{V},\mathbf{W},\mathbf{X}$-***strictly bilinear functor*** is a bilinear functor in which all components of $B^{\sim}$ are identity maps. A ***strict* V, W, X*-strictly bilinear functor*** is a bilinear functor which is both.

Just as in linear algebra, there is a universal bilinear map.

Definition 3.12. If $\mathbf{V}$, $\mathbf{W}$, and $\mathbf{X}$ are monoidal categories and $\mathbf{M}$ and $\mathbf{N}$ are a $\mathbf{V},\mathbf{W}$-bimodule and a $\mathbf{W}$, $\mathbf{X}$-bimodule, respectively, then a $\mathbf{V},\mathbf{X}$-bimodule, $\mathbf{M}\underline{\otimes}\mathbf{N}$, equipped with a $\mathbf{V},\mathbf{W},\mathbf{X}$-bilinear functor, $\underline{\otimes}$, from $\mathbf{M}\times\mathbf{N}$ is ***a tensor product of* M *and* N *over* W** if given any bilinear functor $B\colon \mathbf{M}\times\mathbf{N}\to\mathbf{P}$ for any $\mathbf{V},\mathbf{X}$-bimodule, $\mathbf{P}$, there is a unique $\mathbf{V},\mathbf{X}$-bimodule functor "B" : $\mathbf{M}\underline{\otimes}\mathbf{N}\to\mathbf{P}$ so that $\underline{\otimes}$ "B" $= B$ and $\underline{\otimes}^{\sim}$ "B" $= B^{\sim}$.

Note that this is only one of several universal properties depending on the strictness of the bilinear maps involved and the strictness required of the unique map, "B". It is, however, the most natural and useful for what follows. It turns out that the universal bilinear map is strict, but not strictly-bilinear.

Proposition 3.13. *If* $\mathbf{V}$, $\mathbf{W}$, *and* $\mathbf{X}$ *are monoidal categories, and* $\mathbf{M}$ *and* $\mathbf{N}$ *are a* $\mathbf{V},\mathbf{W}$*-bimodule and a* $\mathbf{W},\mathbf{X}$*-bimodule, respectively, then there exists a tensor product over* $\mathbf{W}$, $\mathbf{M}\underline{\otimes}\mathbf{N}$, *which is moreover unique up to* $\mathbf{V},\mathbf{X}$*-module isomorphism.*

Proof. Since the tensor product was defined by a universal mapping property, uniqueness follows easily by standard techniques once existence has been established.

The existence proof is essentially the same as in multilinear algebra. The one slightly different point is that the relationship with the various module structures is imposed by adjoining maps and quotienting to obtain the naturality and coherence conditions, rather than by merely quotienting.

Let $\mathcal{C}$ be the category constructed from $\mathcal{M}\times\mathcal{N}$ by freely adjoining, for all triples (S,Ω,T) of objects, S in $\mathcal{M}$, Ω in $\mathbf{W}$, and T in $\mathcal{N}$, maps $\phi_{S,\Omega,T}\colon (S\&\Omega,T)\to(S,\Omega\# T)$ and $\psi_{S,\Omega,T}\colon (S,\Omega\# T)\to(S\&\Omega,T)$, and imposing the conditions that ϕ and ψ be inverses, and natural in all three variables, as well as the commutativity of all diagrams of the forms BLF1-4 with components of ϕ taking the place of $B^{\sim}$.

First note that $\mathcal{C}$ can be given a $\mathbf{V}$, $\mathbf{X}$-bimodule structure. On objects and maps which come from maps in $\mathcal{M}\times\mathcal{N}$, we can let $\#$ and $\&$ be given by the functors of the same name on $\mathcal{M}\times\mathcal{N}$. Likewise we can take the structural natural transformations to have the same components as the transformations of the same name in $\mathcal{M}\times\mathcal{N}$. Those instances of functoriality and naturality which do not involve components of ϕ or ψ and all instances of the coherence conditions then follow from instances of the corresponding condition in $\mathcal{M}\times\mathcal{N}$. For the actions on components of ϕ and ψ, let $\Phi\#\phi_{S,\Omega,T}\colon (\Phi\#(S\&\Omega),T)\to(\Phi\# S,\Omega\# T)$ be $(\mathbf{a}_{\Phi,S,\Omega},T)\phi_{\Phi\# S,\Omega,T}$, and $\Phi\#\psi_{S,\Omega,T}$ be $\psi_{\Phi\# S,\Omega,T}(\mathbf{a}^{-1}_{\Phi,S,\Omega},T)$. Similarly let $\phi_{S,\Omega,T}\&\Psi$ be given by $\phi_{S,\Omega,T\&\Psi}(S,\mathbf{a}_{\Omega,T,\Psi})$ and $\psi_{S,\Omega,T}$ by $(S,a^{-1}_{\Omega,T,\psi})$.

We must now check the instances of functoriality and naturality involving these maps (notice, coherence is not an issue here, since we are only establishing the bimodule structure, and the commutativity of the necessary diagrams follows from that in $\mathbf{M}\times\mathbf{N}$). To see that compositions with ϕ's and ψ's are preserved, first notice that compositions of ϕ's and ψ's only must be preserved by invertibility if they are compatible, and by the freeness condition on them if there is a coincidence of source and target between non-inverses. Functoriality in Φ or Ψ (the variable of type $\mathbf{V}$ or $\mathbf{X}$) follows from the naturality of ϕ, ψ and $\mathbf{a}$. The preservation of compositions of ϕ's and ψ's with maps from $\mathcal{M} \times \mathcal{N}$ follows from the freeness and naturality condition on ϕ and ψ.

It is then immediate by construction that the functor $\imath: \mathbf{M} \times \mathbf{N} \to \mathbf{C}$ is strict $\mathbf{V}, \mathbf{W}, \mathbf{X}$-bilinear, and where $\mathbf{C}$ is $\mathcal{C}$ equipped with the $\mathbf{V}, \mathbf{X}$-bimodule structure given above, and the structural natural transformation, $\imath^{\sim}$, is given by ϕ.

Finally, we claim that $\mathbf{C}$ is a tensor product of $\mathbf{M}$ and $\mathbf{N}$ over $\mathbf{W}$.

Now suppose $\mathbf{P}$ is a $\mathbf{V}, \mathbf{X}$-bimodule, and $B: \mathbf{M} \times \mathbf{N} \to \mathbf{P}$ is a $\mathbf{V}, \mathbf{W}, \mathbf{X}$-bilinear functor. Let "$B$"$(\phi_{S,\Omega,T}) = B^{\sim}_{S,\Omega,T}$, and "$B$"$(\psi_{S,\Omega,T}) = B^{\sim -1}_{S,\Omega,T}$. It follows by the coherence conditions on both $\mathbf{C}$ and $\mathbf{P}$ for the bilinear functors $\imath$ and B that this assignment is a $\mathbf{V}, \mathbf{X}$-bimodule functor. (As in the construction of the bimodule structure on $\mathcal{C}$, the only thing to check are those cases of the conditions involving ϕ or ψ.)

Finally, as to the uniqueness of "B", on objects and maps on $\mathbf{M} \times \mathbf{N}$, any fill in must agree with "B" in order that the diagram commute. On components of ϕ and ψ, any such functor must agree in order that the structure map be preserved. □

4. Gluing manifolds and tensoring categories

We now turn to the principal result which enables us to carry out our program of describing the structure of isotopy categories of surfaces up to equivalence.

This result should be seen as a kind of analog of the VanKampen Theorem or Meyer-Vietoris Theorem, as it gives us the ability to compute the isotopy category of certain pushouts of spaces from the isotopy categories of the spaces in the pushout diagram. The reader will notice that the formulation is in terms of tensor products. This is hardly surprising in as much as the theory of isotopy categories is far more closely related to topological quantum field theories than to homotopy or homology theory.

The proof of this theorem is given in the p.l. setting. For the isotopy categories of spaces of dimension 3 or less, it will also hold in the smooth setting by triangulating the smooth structure, applying the given proof, then smoothing all of the p.l. maps in the proof. We will be concerned in what follows only with the isotopy categories of surfaces, so we shall use the result in the smooth setting whenever necessary.

Theorem 4.1. *If X_i, $i = 0, 1$, are n-manifolds with boundary and Z is an $(n-1)$-manifold (possibly with boundary), then given embeddings $\xi_i: Z \times [0,1] \to X_i$ with $\xi_i|_{Z\times\{i\}}$ factoring through ∂X_i and isotopies of the sort required in the hypothesis of Proposition 3.8, then for any regular FCF, $\mathbb{F}$, of dimension less than or equal to n, there is an equivalence of categories between $\{\mathbb{F}, X_0\} \underline{\otimes}_{\{\mathbb{F}, Z\times[0,1]\}} \{\mathbb{F}, X_1\}$ and $\{\mathbb{F}, X_0 +_Z X_1\}$, where the pushout is the pushout of $\xi_0|_{Z\times\{0\}}$ along $\xi_1|_{Z\times\{1\}}$. Here the tensor is the target of the universal $\mathbf{1}, \{\mathbb{F}, Z \times [0,1]\}, \mathbf{1}$-bilinear functor, where the category $\mathbf{1}$ is given the obvious trivial monoidal structure.*

If moreover, ξ_i has an image disjoint from that of some embedding $\theta_i: Y_i \times [0,1] \to X_i$ which induces a $\{\mathbb{F}, Y_i \times [0,1]\}$-module structure on the other side from the $\{\mathbb{F}, Z \times [0,1]\}$-module structure on $\{\mathbb{F}, X_i\}$ by Proposition 3.8, then the tensor product can be *taken as the target of the universal* $\{\mathbb{F}, Y_0 \times [0,1]\}$, $\{\mathbb{F}, Z \times [0,1]\}, \{\mathbb{F}, Y_1 \times [0,1]\}$*-bilinear map.*

Proof. We begin by showing that there is an equivalence of categories between $\{\mathbb{F}, X_0 +_Z X_1\}$ and its full subcategory, $\mathcal{C}$, whose objects are named by embeddings each component of which factors through one of $X_0 - \partial X_0$, $X_1 - \partial X_1$, or $Im(\xi_0) \cup Im(\xi_1)$. As in the proof of Theorems 1.20 and 2.2, we construct for each object an isotopy from the embedding naming the object to an embedding naming an object of $\mathcal{C}$: if the object is in $\mathcal{C}$, take the trivial isotopy; otherwise, choose a triangulation of $X_0 +_Z X_1$ subdividing the coarsest one for which the embedding naming the object is given by linear maps of simplexes, and fine enough that the star neighborhood of any vertex is contained in one of the three subspaces. Choosing a vertex in each component of the name of the map, apply an isotopy to contract the name of the object into a star neighborhood of the chosen vertex (the existence of such an isotopy is guaranteed by the regularity of $\mathbb{F}$). The object named by this "shrunken" embedding is then the image of the object in the full subcategory, and the isotopy names an isomorphism in each direction between the original object and the "shrunken" object. A standard argument then lets us extend this function on objects to a functor in such a way that the isomorphisms become the unit and counit of an adjoint equivalence (whose other functor is the full inclusion).

Now in the case where we are considering the "moreover" statement in the theorem, note that the actions of $\{\mathbb{F}, Y_0 \times [0,1]\}$ and $\{\mathbb{F}, Y_1 \times [0,1]\}$ restrict to actions on $\mathcal{C}$ since by the separation of the images and the triviality of the isotopies used to define the action outside of a neighborhood of $Y_i \times [0,1]$, whenever C is an object of $\mathcal{C}$, so are $A_0 \# C$ and $C \& A_1$ and A_i any object of $\{\mathbb{F}, Y_i \times [0,1]\}$ for $i = 0, 1$.

What has been shown thus far depends only on the formal cobordism family being o.e.-0. The rest of the proof uses the full force of regularity. We claim now that $\mathcal{C}$ is equivalent (as a $\{\mathbb{F}, Y_0 \times [0,1]\}, \{\mathbb{F}, Y_1 \times [0,1]\}$-bimodule) to $\{\mathbb{F}, X_0\} \underline{\otimes}_{\{\mathbb{F}, Z \times [0,1]\}} \{\mathbb{F}, X_1\}$.

To verify the claim, first observe that the functor $\{\mathbb{F}, f\}$ from $\{\mathbb{F}, X_0 + X_1 - \partial(X_0 + X_1)\}$ to $\{\mathbb{F}, X_0 +_Z X_1\}$, induced by the canonical map, f, from the disjoint union to the pushout, factors through $\mathcal{C}$, and that this factorization is $\{\mathbb{F}, Y_0 \times [0,1]\}, \{\mathbb{F}, Y_1 \times [0,1]\}$-bilinear when we use the equivalence of Theorem 1.20 to induce bimodule structures on $\{\mathbb{F}, X_0 - \partial X_0\}$ and $\{\mathbb{F}, X_1 - \partial X_1\}$ and identity $\{\mathbb{F}, X_0 - \partial X_0\} \times \{\mathbb{F}, X_1 - \partial X_1\}$ with $\{\mathbb{F}, X_0 + X_1 - \partial(X_0 + X_1)\}$ by the isomorphism of Proposition 1.16.

Here $\{\mathbb{F}, f\}^\sim$ is the natural isomorphism induced by applying the isotopy which induces ℓ^{-1} on $\{\mathbb{F}, X_1\}$, while fixing $\{\mathbb{F}, X_0\}$, followed by an ambient isotopy which fixes all of $X_0 +_Z X_1$ outside of the neighborhoods of $Im(\xi_0)$ and $Im(\xi_1)$ required in the construction of Proposition 3.8, and moves $Im(\xi_0)$ into $Im(\xi_1)$, in such a way that the Z-coordinates in the preimage in $Z \times [0,1]$ (under ξ_0 or ξ_1) remain constant throughout the motion and finally by applying the isotopy which induces r on $\{\mathbb{F}, X_0\}$, while fixing X_1.

The coherence conditions BLF1 and BLF2 follow from the triviality of the isotopy outside the neighborhood of $Im(\xi_0) \cup Im(\xi_1)$, while BLF4 follows from the construction of $\{\mathbb{F}, f\}^\sim$ and the fact that the middle isotopy in the construction does nothing, since the name of the I is the empty embedding. The proof of BLF3 is indicated schematically by Fig. 4.1.1.

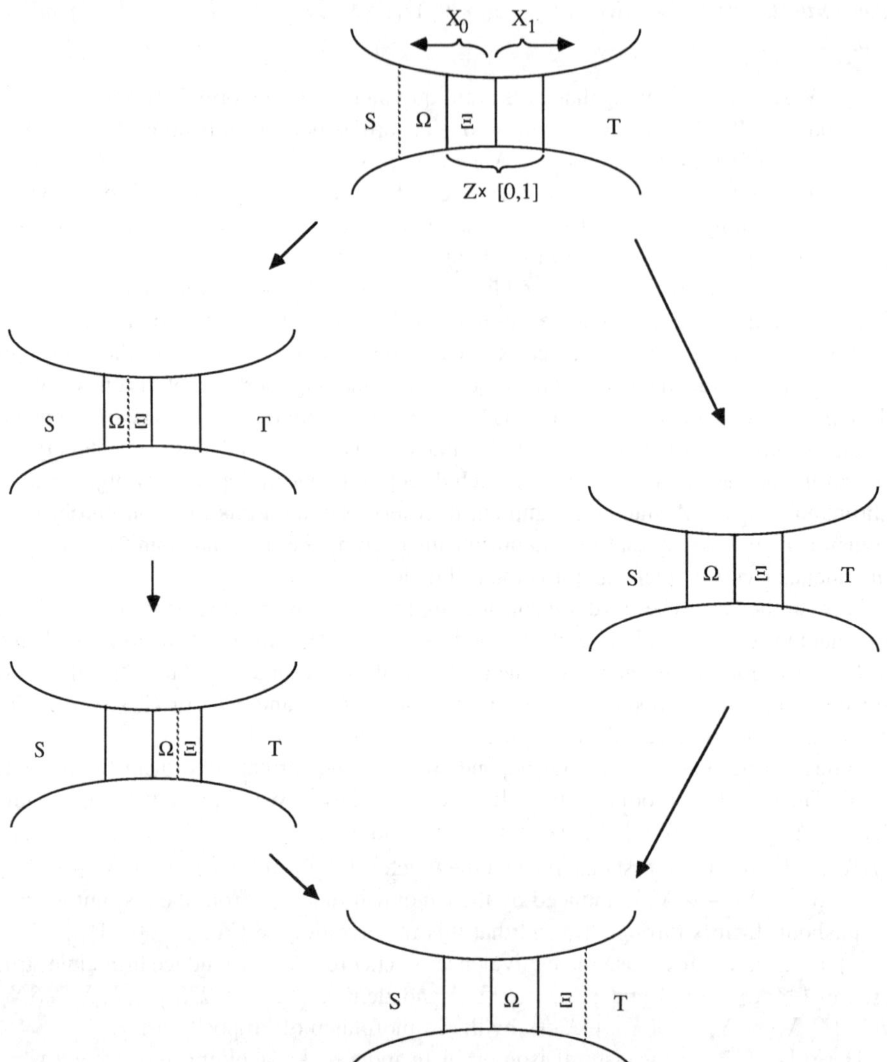

Figure 4.1.1

Thus there is a functor from the tensor product bimodule to $\mathcal{C}$ induced by the universal property. We must now construct a functor in the reverse direction and show that this pair of functors induces an equivalence.

To do this we will use the regularity of $\mathbb{F}$ to provide "small" names for all maps. Given a map $F: A \to B$ in $\mathcal{C}$, consider an embedding $[F]: (X, X') \hookrightarrow X_0 +_Z X_1$ which names F. Now choose a triangulation of the target space sufficiently fine that $[F]$ is given by linear maps to simplexes of the triangulation and the star neighborhood of any vertex is contained in one of $X_0 - Z$, $X_1 - Z$, or $Im(\xi_0) \cup Im(\xi_1)$, and then pass to the barycentric subdivision of this triangulation. The subdivision will throughout the rest of the proof be referred to as "the triangulation".

Now choose an open collar neighborhood, N, in (X, X') of X'; then by R3 $(X - N, X'')$ is in $\mathbb{F}$ (where X'' is $\partial N \cap (X - N)$). Futher choose a decomposition of $(X - N, X'')$ as provided by R2, and a contraction, h, of each piece to a 1-complex as provided by R2 and let t be the earliest time for which the image of the contraction is contained in the union of the star neighborhoods in the triangulation of the vertices of the 1-complex. Finally, choose an ambient isotopy whose restriction to the image of the pieces of X is a rescaling of h restricted to times between 0 and t inclusive, and which is trivial outside some neighborhood of the image of $X - N$ whose components separate the components of the image of $X - N$, and which separates the image of $X - N$ from the image of X'.

Applying this ambient isotopy to $[F]$ gives us another embedding, $\{F\}$, which also names F, and which is contained in a star neighborhood of a 1-complex. Finally apply an ambient isotopy to move any local maxima, local minima (these taken with respect to the $[0,1]$-coordinate) and singularities of the 1-complex off of $\partial(Im(\xi_0) \cup Im(\xi_1))$, and to arrange that all levels intersecting $\partial(Im(\xi_0) \cup Im(\xi_1))$ have bicollar neighborhoods in X.

The resulting embedding will be denoted $|F|$, and is a name for F which is contained in a star neighborhood of a 1-complex which enters or leaves $Im(\xi_0) \cup Im(\xi_1)$ at points with a manifold neighborhood (in the 1-complex) whose projection onto $[0,1]$ is invertible, whose level in X has a bicollar neighborhood in X.

Now, decompose X at all levels, t, where its image intersects $\partial(Im(\xi_0) \cup Im(\xi_1))$ but for which the image is disjoint from $\partial(Im(\xi_0) \cup Im(\xi_1))$ at levels in $(t - \varepsilon, t)$ or $(t, t + \varepsilon)$ but not both, for some $\varepsilon > 0$.

We now have expressed F by a composition of maps each of which is named by an embedding each component of which is contained in one of $X_0 - \partial X_0$, $X_1 - \partial X_1$, or $Im(\xi_0) \cup Im(\xi_1)$. We use this composite to give a preimage for the given map in the tensor product. We distinguish those maps in the expression one or more of whose components are of the form $(C \times [0,1], C \times \{0,1\})$ and intersect $\partial Im(\xi_1 \cap X_1)$, briefly "with components entering (resp. exiting) X_1", if those levels for $(t, t+\varepsilon)$ (resp. $(t - \varepsilon, t)$) for some $\varepsilon > 0$ also intersect the boundary. For any other of the maps in the decomposition, we can choose as a preimage in the tensor product the image of $(A\&\Omega, B)$ in the tensor product, where A (resp. Ω, B) is the map in $\{\mathbb{F}, X_0\}$ (resp. $\{\mathbb{F}, Z \times [0,1]\}$, $\{\mathbb{F}, X_0\}$) named by the embeddings of those components whose image lies entirely in $X_0 - Im(\xi_0)$ (resp. $\{\mathbb{F}, Z \times [0,1]\}$, $\{\mathbb{F}, X_0\}$), where we have identified $Z \times [0,1]$ with $Im(\xi_0) \cup Im(\xi_1)$ by rescaling the $[0,1]$-coordinates in the standard way for piecing intervals together. For the maps in the composite which have names with components entering (resp. exiting) X_1, we choose as a preimage $\phi_{A,\Omega,B}$ followed by the map which is the image of $(G\#H, K)$ in the tensor product, where G is the map in $\{\mathbb{F}, X_0\}$ named by the inclusion of those components factoring through $X_0 - Im(\xi_0)$, and H is the map in $\{\mathbb{F}, Z \times [0,1]\}$ named by components other than those entering X_1, and K is the map in $\{\mathbb{F}, X_1\}$ named by the other components (resp. such a map followed by $\psi_{A,\Omega,B}$).

Once we know that this construction is independent of the name of F to which it is applied and of the choices made in the construction, it will follow almost immediately that it gives rise to a functor, and that in the "moreover" statement the functor is $\{\mathbb{F}, Y_0\}, \{\mathbb{F}, Y_1\}$-bilinear. To see that this construction is independent of the name of F and of the choices made, it suffices to show that given any two names for F satisfying the conditions on the one used to construct the preimage, and an isotopy between them, that the preimages constructed are equal in the tensor product. An argument similar to that given above allows us to replace this isotopy by one which at every time t maps X into a star neighborhood of the images of the 1-complexes of condition R2. Such an isotopy can then be decomposed as a composite of isotopies in each of $X_0 - \partial X_0$, $X_1 - \partial X_1$, or $Im(\xi_0) \cup Im(\xi_1)$. The relation it induces must then hold in the tensor product by the relations in each of the categories $\{\mathbb{F}, X_0\}$, $\{\mathbb{F}, X_1)$, and $\{\mathbb{F}, Z \times [0,1]\}$ and the naturality of ϕ and ψ.

Let this functor be denoted by L and the canonical functor from the tensor product to $\mathbf{C}$ by C. It follows from the universal property of the tensor product that CL is the identity functor on the tensor product. However, the construction of L was given in terms of constructing a preimage under C for maps in $\mathbf{C}$, and thus LC is the identity functor on $\mathbf{C}$. □

5. Disks and n-holed annuli

We will apply the results of the preceding section to describe the isotopy categories of all orientable surfaces in terms of those of the disk and the annulus. To do this, we must first relate the isotopy category of the disk to that of the n-holed annulus, and describe various possible $\{\mathbb{F}, S^1 \times [0,1]\}$-module structures on these categories.

To do this, we require a definition which is of interest in its own right, but which we use only to derive some technical results.

Definition 5.1. If $\mathbb{F}$ and $\mathbb{G}$ are formal cobordism families, the ***separated joint*** of $\mathbb{F}$ and $\mathbb{G}$, $\mathbb{F} \oplus \mathbb{G}$ is the formal cobordism family consisting of all disjoint unions of members of $\mathbb{F}$ and members of $\mathbb{G}$ with components colored by the name of the family from which they came.

The reader can easily verify

Lemma 5.2. *If $\mathbb{F}$ and $\mathbb{G}$ are compact (resp. o.e.-0, regular), then so is $\mathbb{F} \oplus \mathbb{G}$.* □

Now let $\mathbb{D}$ denote the regular formal cobordism family $\{(D \times [0,1] \times [n], D \times \{0,1\} \times [n]) \mid n \in \mathbb{N}\}$, where D is the disk. (The reader can easily verify the axioms.)

Similarly, it is easy to prove

Lemma 5.3. *If $\mathbb{F}$ and $\mathbb{G}$ are formal cobordism families, then there is a functor $U_F(X)$: $\{\mathbb{F} \oplus \mathbb{G}, X\} \to \{\mathbb{G}, X\}$ for any space X, given both on objects and maps by taking the restriction of an embedding naming the map to those components in $\mathbb{G}$.* □

Now consider $\mathbb{F} \oplus \mathbb{D}$ for any formal cobordism family $\mathbb{F}$. Then in particular we have a functor $U_F(D)$: $\{\mathbb{F} \oplus \mathbb{D}, D\} \to \{\mathbb{D}, D\}$. A moment's consideration is enough to see that an object, C, of $\{\mathbb{D}, D\}$ is a choice of n holes in the disk for some n, and that thus $U_F(D)^{-1}(C)$ is equivalent as a category to $\{\mathbb{F}, A[n]\}$, where $A[n]$ is an n-holed annulus.

For brevity let us denote $\{\mathbb{F}, S^1 \times [0,1]\}$ by $\mathbf{A}\mathbb{F}$. We can then sharpen the observation above to state

Theorem 5.4. *For any formal cobordism family, $\mathbb{F}$, the category $\{\mathbb{F}, A[n]\}$ is equivalent to $U_F(D)^{-1}(C)$, where C is an object of $\{\mathbb{D}, D\}$ named by an embedding of n disks into the disk, D, and $U_F(D)\colon \{\mathbb{F} \oplus \mathbb{D}, D\} \to \{\mathbb{D}, D\}$ is the functor of Lemma 5.3.*

If moreover, $A[n]$ is endowed with an orientation whose restiction to the boundary orients the boundary components of the n disks counterclockwise (resp. clockwise) and the other boundary component clockwise (resp. counterclockwise), then $\{\mathbb{F}, A[n]\}$ has n right (resp. left) $\mathbf{A}\mathbb{F}$-module structures, corresponding to the n disks, and a left (resp. right) $\mathbf{A}\mathbb{F}$-module structure corresponding to the remaining boundary component. Moreover for any choice of m $(\leq n)$ of the right (resp. left) module structures, there is an $\mathbf{A}\mathbb{F}$, $\mathbf{A}\mathbb{F}^m$-bimodule (resp. $\mathbf{A}\mathbb{F}^m$, $\mathbf{A}\mathbb{F}$-bimodule) structure.

Proof. This follows immediately from the observation above the Propositions 3.8 and 3.9. □

Now the preceding gives us even more actions when it is observed that a left (resp. right) $\mathbf{A}\mathbb{F}$-module structure is the same as a right (resp. left) $\mathbf{A}\mathbb{F}_{\text{rev}}$-module structure, where the subscript "rev" indicates reversing the order of $\otimes$, and the roles of α and α^{-1} and of p and λ, while leaving composition in the same order.

The reader who has carefully followed the discussion up to this point will immediately observe that the preceding observation together with Theorems 2.2, 4.1, and 5.4 allow us to give a description of the isotopy category, $\{\mathbb{F}, \Sigma\}$ for regular $\mathbb{F}$ and Σ any compact orientable surface with or without boundary in terms of isotopy categories for the disk. (For the sphere, we do not even need Theorem 5.4, we can piece disks together along their boundary to see that $\{\mathbb{F}, S^2\}$ is equivalent to $\{\mathbb{F}, D\} \underline{\otimes}_{\mathbf{A}\mathbb{F}} \{\mathbb{F}, D\}$.)

As it stands, this description is not very explicit. To make it more explicit, we now describe the various $\mathbf{A}\mathbb{F}$-module structures on the isotopy categories of the disk and of $A[n]$ in terms of the structure of the isotopy categories of the disk, which for Examples 1.4-1.7 has been described at length in other works (cf. Freyd and Yetter **[5]**, **[6]**, Joyal and Street **[9a-b]**, **[10]**, Shum **[17]** and Yetter **[20]**).

We now consider those parts of the structures given in those works which exist on the isotopy category of the disk for general $\mathbb{F}$.

Theorem 5.5. *If $\mathbb{F}$ is any formal cobordism family, then $\{\mathbb{F}, D\}$ (equivalently $\{\mathbb{F}, (0,1) \times (0,1)\}$) admits the structure of a braided pivotal category and of a tortile category. The structures agree except for the unit and counit of the left adjunction for each object. The pivotal structure is given by that of Proposition 1.17. the braiding is induced by the isotopy illustrated schematically in Fig. 5.5.1. The twist map θ is induced by the isotopy (of the disk) given by (a p.l. approximation of) $\Theta\colon D \times [0,1] \to D$ by $\Theta(z,t) = ze^{\pi it}$.*

Proof. Essentially as in Shum **[17]**. □

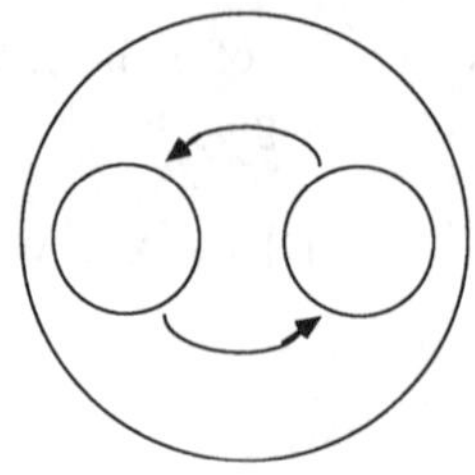

Figure 5.5.1

A little more work is needed to be able to state in a unified way the explicit constructions.

Definition 5.6. A ***word model*** for an isotopy category $\{\mathbb{F}, X \times (0,1)\}$ is a full subcategory, $\mathcal{W}$, so that for each word of components of X''s in $\mathbb{F}$ there is a unique object in $\mathcal{W}$ named by an embedding of those components ordered by their $(0,1)$-coordinates in the order given by the word, and there is a pivotal structure on $\mathcal{W}$ given by concatenation on words naming objects, and a monoidal equivalence of $\mathcal{W}$ with $\{\mathbb{F}, X \times (0,1)\}$.

The discussion immediately before Theorem 1.9 then gives a proof of

Theorem 5.7. *If $\mathbb{F}$ is o.e.-0, and X is a path-connected space, then $\{\mathbb{F}, X \times (0,1)\}$ admits a word model $\mathcal{W}_{\mathbb{F}, X\times(0,1)}$.* □

Lemma 5.8. *The word model for $\{\mathbb{F}, (0,1) \times (0,1)\}$ admits a braiding and tortile and pivotal structures compatible with the inclusion.*

Sketch of Proof. Let E denote the inclusion of the word model, and W denote the (co)reflection. Once it is noted that $W(E(A)^*)$ is a dual in either the sense required for a tortile or pivotal structure, the rest of the conditions follow from elementary properties of categorical equivalences. □

Now, note that for regular $\mathbb{F}$, identifying $\{\mathbb{F}, (0,1) \times (0,1)\}$ with a subcategory of $\mathbf{A}\mathbb{F}$ as in the proof of Theorem 2.2 allows us to identify the word model $\mathcal{W}_{\mathbb{F},(0,1)\times(0,1)}$ with a submonoidal category of a word model, $\mathcal{A}\mathbb{F}$, for $\mathbf{A}\mathbb{F}$. In terms of this identification we then have

Proposition 5.9. *If D is oriented so that the boundary of D is oriented counterclockwise (resp. clockwise) then the left (resp. right) action of $\mathcal{A}\mathbb{F}$ on $\mathcal{W}_{\mathbb{F},D}$ induced by the action of the full isotopy category on the full isotopy category, which is in turn induced by gluing the annulus to the boundary of the disk and shrinking back into the original disk, is given on objects and maps contained in the image of $\mathcal{C}_{\mathbb{F},D}$ in $\mathcal{A}\mathbb{F}$ by*

$$A\#B = A \otimes B \qquad (\textit{resp. } B\&A = B \otimes A)\,,$$

and on components of β by

$$\beta_A\#B = \sigma_{A,B}\sigma_{B,A}(\theta_A \otimes B) \quad (\textit{resp. } B\&\beta_A = \sigma^{-1}_{B,A}\sigma^{-1}_{A,B}(B \otimes \theta^{-1}_A))\,.$$

Likewise, if we let $U_{\mathbb{F}}: \mathcal{W}_{\mathbb{F}\oplus\mathbb{D},D} \to \mathcal{W}_{\mathbb{D},D}$ be the functor induced by restricting embeddings to components in $\mathbb{D}$, we have

Proposition 5.10. *Let A be an object of $\mathcal{W}_{\mathbb{D},D}$ induced by an embedding, $[A]$, of $[n] \times D$ into D, and let $\{i\} \times D$ be the image of one of the n disks. Suppose, moreover, that $D - \mathrm{int}(Im([A]))$ is oriented in such a way that ∂C is oriented clockwise (resp. counterclockwise). Then there is a right (resp. left) $\mathcal{A}\mathbb{F}$-module structure induced on $U_{\mathbb{F}^{-1}}(A)$ by the action of the full isotopy category for the annulus induced by gluing a suitably oriented annulus to $\partial(\{i\} \times D)$ and rescaling. Explicitly, this action is given on objects and maps contained in the image of $\mathcal{W}_{\mathbb{F},D}$ in $\mathcal{A}\mathbb{F}$ by*

$$\begin{aligned}(X_0 \otimes P \otimes X_1 \otimes P \otimes \ldots \otimes P \otimes X_n)\&B \\ = (X_0 \otimes P \otimes X_1 \otimes P \otimes \ldots \otimes X_{i-1} \otimes B \otimes P \otimes \ldots \otimes P \otimes X_n)\end{aligned}$$

(resp. $B\#(X_0 \otimes P \otimes X_1 \otimes P \otimes \ldots \otimes P \otimes X_n) = (X_0 \otimes P \otimes X_1 \otimes P \otimes \ldots \otimes P \otimes B \otimes X_i \otimes \ldots \otimes P \otimes X_n)$), and on components of β by

$$\begin{aligned}(X_0 \otimes P \otimes X_1 \otimes P \otimes \ldots \otimes X_n)\&\beta_B \\ = (X_0 \otimes P \otimes X_1 \otimes P \otimes \ldots \otimes X_{i-1} \otimes (\sigma_{B,P}\sigma_{P,B}(\theta_B \otimes P)) \otimes \ldots \otimes P \otimes X_n)\end{aligned}$$

(resp. $(X_0 \otimes P \otimes X_1 \otimes P \otimes \ldots \otimes P \otimes X_n)\&\beta_B = (X_0 \otimes P \otimes X_1 \otimes P \otimes \ldots \otimes X_{i-1} \otimes (\sigma^{-1}_{B,P}\sigma^{-1}_{P,B}(\theta^{-1}_B \otimes P)) \otimes \ldots \otimes P \otimes X_N)$) where P is the unique object of the word model named by an embedding of D, and the $\otimes$-products are unparenthesized since the word model is strict monoidal. □

This is now an explicit enough description to allow us to give succinct descriptions of isotopy categories of specific surfaces up to equivalence.

Example 5.11. S^2: We regard the sphere as being obtained by gluing two disks together along their boundary. Thus we have an equivalence of categories between $\{\mathbb{F}, D\}\underline{\otimes}_{\mathcal{A}\mathbb{F}}\{\mathbb{F}, D\}$ and $\{\mathbb{F}, S^2\}$. We then have that $\{\mathbb{F}, S^2\}$ is equivalent to $\mathcal{W}_{\mathbb{F},D}\underline{\otimes}_{\mathcal{A},\mathbb{F}}\mathcal{W}_{\mathbb{F},D}$. Finally, an examination of the construction of the tensor product of modules over monoidal categories, shows that $\{\mathbb{F}, S^2\}$ is, in fact, equivalent to the quotient of $\mathcal{W}_{\mathbb{F},D}$ by the congruence of categories (***not*** of monoidal categories) generated by the relations

$$A \otimes (\sigma_{B,C}\sigma_{C,B}(\theta_B \otimes C)) = \sigma^{-1}_{A,B}\sigma^{-1}_{B,A}(A \otimes \theta^{-1}_B)) \otimes C$$

for all objects A, B, and C of $\mathcal{W}_{\mathbb{F},D}$. (There is a trick here: the tensor product is at the level of the full isotopy category, identifiable with the subcategory of objects of the forms $(A \otimes I) \otimes (B \otimes C)$ or $(A \otimes B) \otimes (I \otimes C)$, but when one cuts down to word models, one strictifies and the parenthesization and I's disappear, and ϕ and ψ in the construction of the tensor product become identities.)

In our next example, this trick can be used only along one of the components of the cut along which pieces are being glued, so one must actually adjoint maps as well as quotienting.

Example 5.12. $S^1 \times S^1$: We regard $S^1 \times S^1$ as having been obtained by gluing two copies of $A[1]$ along their boundaries. Thus we have that $\{\mathbb{F}, S^1 \times S^1\}$ is equivalent to $\mathbf{A}\mathbb{F}\underline{\otimes}_{\mathbf{A}\mathbb{F}\times\mathbf{A}\mathbb{F}_{\text{rev}}}\mathbf{A}\mathbb{F}$, and hence to $\mathcal{A}\mathbb{F}\underline{\otimes}_{\mathcal{A}\mathbb{F}\times\mathcal{A}\mathbb{F}_{\text{rev}}}\mathcal{A}\mathbb{F}$. One can use the trick of the previous example to deal with the $\mathcal{A}\mathbb{F}$-actions giving a covering of the category not by $\mathcal{W}_{\mathbb{F},D}$, but by $\mathcal{A}\mathbb{F}$ so that the left (resp. right) $\mathcal{A}\mathbb{F}_{\text{rev}}$ module structure can be identified with tensoring on the right (resp. left) in $\mathcal{A}\mathbb{F}$. It then follows from the construction in the proof of Proposition 3.13 that $\{\mathbb{F}, S^1 \times S^1\}$ is equivalent to the category obtained from $\mathcal{A}\mathbb{F}$ by adjoining isomorphisms natural in both variables, $\phi\colon A \otimes B \to B \otimes A$, and quotienting by the conditions that all diagrams of the forms **T1**:

$$\begin{array}{ccccc}
 & S\otimes(T\otimes(C\otimes D)) & \xrightarrow[\phi]{} & ((C\otimes D)\otimes S)\otimes T & \\
S\otimes\alpha \nearrow & & & & \\
S\otimes((T\otimes C)\otimes D) & & & \Big\downarrow \alpha\otimes T & \\
\phi \searrow & & & & \\
 & (D\otimes S)\otimes(T\otimes C) & \xrightarrow[\phi]{} & (C\otimes(D\otimes S))\otimes T &
\end{array}$$

which corresponds to condition BLF3 in the definition of bilinear functors. Condition BLF4 imposes no condition since the word models are strict. Conditions BLF1 and BLF2 become merely the definition of how $\#$ and $\&$ act on the maps ϕ. (Note: this last step, adjoining ϕ and quotienting is done as a category, ***not*** as a monoidal category.)

Example 5.13. $A[n]$ (again): We have already observed that $\{\mathbb{F}, A[n]\}$ is equivalent to the inverse image of the object, $P^{\otimes n}$, of $\mathcal{W}_{\mathbb{D},D}$ under the functor $U_{\mathbb{F}}\colon \mathcal{W}_{\mathbb{F}\otimes\mathbb{D},D} \to \mathcal{W}_{\mathbb{D},D}$ obtained by composing the relevant functors in the equivalences of word models with the full isotopy categories with the functor induced by forgetting component of maps names coming from $\mathbb{F}$.

Example 5.14. The general case: Let Σ be a compact orientable surface of genus g with n circles as boundary components. Then Σ can be regarded as having been constructed by gluing copies of $A[g]$ and $A[g+n]$ along their outer edges and along the boundaries of g of their holes. We can use the same trick as in Examples 5.11 and 5.12 to glue the outer edges of the multi-holed annuli together, while gluing the g holes of $A[g]$ to the first g holes of $A[g+n]$ is done as was done for the two holes in Example 5.12. Thus, we obtain a category equivalent to $\{\mathbb{F}, \Sigma\}$, whose objects can be identified with the objects of $\mathcal{W}_{\mathbb{F}\oplus\mathbb{D},D}$ of the form $X_0 \otimes P \otimes \ldots \otimes P \otimes X_{2g+n}$, and whose maps are obtained from the subcategory of those maps in $\mathcal{W}_{\mathbb{F}\otimes\mathbb{D},D}$ with names whose components in $\mathbb{D}$ are of the form $\phi \times \mathrm{Id}_{[0,1]}\colon D \times [0,1] \to D \times [0,1]$, by freely adjoining for $i = 1, \ldots, g$ isomorphisms $\beta^{(i)}$ from

$$X_0 \otimes P \otimes \ldots \otimes P \otimes (A \otimes X_i') \otimes P \otimes \ldots \otimes P \otimes X_{2g-i} \otimes P \otimes \ldots \otimes P \otimes X_{2g+n}$$

to

$$X_0 \otimes P \otimes \ldots \otimes P \otimes X_i' \otimes P \otimes \ldots \otimes P \otimes (X_{2g-i} \otimes A) \otimes P \otimes \ldots \otimes P \otimes X_{2g+n}$$

natural in all variables, and each satisfying the condition given by the diagram in Example 5.12, and finally quotienting by

$$A \otimes (\sigma_{B,C}\sigma_{C,B}(\theta_B \otimes C)) = (\sigma_{A,B}^{-1}\sigma_{B,A}^{-1}(A \otimes \theta_B^{-1})) \otimes C$$

where B is a factor of any of the X_i 's under $\otimes$, and $A \otimes B \otimes C = X_0 \otimes P \otimes \ldots \otimes P \otimes X_{2g+n}$.

6. Braid groups revisited

The original motivation (Yetter **[19]**) for considering the algebraic structure of the category of tangles was to provide a setting in which the Kauffman polynomial (cf. **[11]**, **[12]**) could be understood as arising from representations of an algebraic object in the same way that the homfly polynomial **[7]** arose from representations of the Artin braid groups.

We now turn to consider how the methods contained herein can be applied to the braid groups of surfaces. We may regard braids as special maps in $\{\mathbb{T}, \Sigma\}$, namely those which admit names of the form $(\phi, p_{[0,1]})\colon [n] \times [0,1] \to \Sigma \times [0,1]$ for some map $\phi\colon [n] \times [0,1] \to \Sigma$.

In the case of $\{\mathbb{T}, D\}$, braids are precisely the automorphisms, an observation which follows from the prime decomposition theorem for knots (by which components of invertible tangles must be unknotted), the proof in **[19]** that "uniform tangles" are invertible if and only if they are braids, and a simple inductive argument.

We conjecture that braids are precisely the automorphism in $\{\mathbb{T}, \Sigma\}$ for any orientable surface Σ but do not need this for what follows.

Instead, beginning with the monoidal equivalence between the category of ***Braids*** as described by Joyal and Street **[9a]**, **[9b]** and the subcategory of $\{\mathbb{T}, D\}$ described above, the reader can copy the proofs of Theorem 2.2, Proposition 3.8, and Theorems 4.1 and 5.4 to show that the category of braids on the annulus is the monoidal category obtained by freely adjoining a monoidal natural transformation to the free braided monoidal category, that each circle boundary component of a surface with boundary gives rise to an action of the category of braids on the annulus on the category of braids of the surface, that gluing surfaces along circles carries over to tensoring braid categories, and that the braid category for an n-holed annulus can be obtained as the inverse image of an identity map under a "forgetful functor".

We provide examples of how this method can be used to produce presentations of braid groups.

Example 6.2. S^2: Fix n and choose an embedding of $[n]$ into the disk considered as a hemisphere in S^2. Now conceived of as an object in $\{\mathbb{T}, D\}$, we have $\mathrm{Aut}_{\{\mathbb{T},D\}} = B_n$, which admits the standard presentation

$$\langle \sigma_1, \ldots, \sigma_{n-1} \,|\, \sigma_i\sigma_j = \sigma_j\sigma_i \quad i = 1, \ldots, n-3, \;\; j = i+2, \ldots, n-1; \\ \sigma_i\sigma_{i+1}\sigma_i = \sigma_{i+1}\sigma_i\sigma_{i+1} \quad i = 1, \ldots, n-2\rangle,$$

where, in a word, model, $[n] = [1] \otimes \ldots \otimes [1]$ (n-copies) and σ_i is $[1] \otimes \ldots \otimes \sigma_{[1],[1]} \otimes \ldots \otimes [1]$ (with the σ involving the ith and $i+1^{\text{st}}$ factors). Any instance of the extra relation imposed by passing to the sphere (see Example 5.11) is then of the form

$$[p] \otimes (\sigma_{[q],[r]}\sigma_{[r],[q]}(\theta_{[q]} \otimes [r])) = \sigma^{-1}_{[p],[q]}\sigma^{-1}_{[q],[p]}([p] \otimes \theta^{-1}_{[q]})) \otimes [r]$$

for $[a] = [1] \otimes \ldots \otimes [1]$ (a factors), and $p+q+r=n$. A little fiddling with the coherence conditions of σ and θ and the fact that in $\mathbf{Tang} = \{\mathbb{T}, D\}$, $\theta_{[1]} = \mathrm{Id}_{[1]}$, shows that these relations all follow from those where $q=1$, which simplify as

$$[p] \otimes (\sigma_{[1],[r]}\sigma_{[r],[1]}) = (\sigma^{-1}_{[p],[1]}\sigma^{-1}_{[1],[p]}) \otimes [r] .$$

Using the coherence conditions on σ to render these in terms of $\sigma_{[1],[1]}$, its inverse and $\otimes$-ing with identity maps, then rewriting in the usual group generator notation, this becomes

$$\sigma_{p+1}\ldots\sigma_{n-1}\sigma_{n-1}\ldots\sigma_{p+1} = \sigma_p^{-1}\ldots\sigma_1^{-1}\sigma_1^{-1}\ldots\sigma_p^{-1} .$$

All of these relations reduce by conjugation to

$$\sigma_1\ldots\sigma_n^2\ldots\sigma_1 = 1 ,$$

and thus the nth braid group of the sphere admits the presentation

$$\langle \sigma_1\ldots\sigma_{n-1} \,|\, \sigma_i\sigma_j = \sigma_j\sigma_i \quad i=1,\ldots,n-3,\ j=i+2,\ldots,n-1;$$
$$\sigma_i\sigma_{i+1}\sigma_i = \sigma_{i+1}\sigma_i\sigma_{i+1} \quad i=1,\ldots,n-2;\quad \sigma_1\ldots\sigma_n^2\ldots\sigma_1 = 1\rangle .$$

This is the course the usual presentation of these groups first discovered by Fadell and VanBuskirk **[4]**.

For other surfaces, one enumerates the components of the various natural isomorphisms generating the structure of the category (and their tensor products with identities if they are monoidal natural transformations) which are applicable to the object which is the image of $[n]$ under the embedding of a disk and takes these as generators, then one sets down all of the commutative diagrams among these which define the functoriality and naturality conditions and the various coherence conditions. One can then use the coherence conditions to cut down the number of generators and relations. (It is helpful to remain in categorical notation until this stage.)

We carry out this program only in the simplest case of the torus.

Example 6.2. $S^1 \times S^1$: Using the method outlined above, one can easily use the coherence conditions (including monoidal naturality of β) on the natural isomorphisms σ, β and ϕ in the structure of $\{\mathbb{T}, S^1 \times S^1\}$ to reduce the description of the nth braid group of the torus to examining the functoriality, naturality and coherence conditions applicable to the maps σ_i, $i=1,\ldots,n-1$, β_i, $i=1,\ldots,n$, and ψ, where in categorical notation, these are $[1]\otimes\ldots\otimes\sigma_{[1],[1]}\otimes\ldots\otimes[1]$, $[1]\otimes\ldots\otimes\beta_{[1]}\otimes\ldots\otimes[1]$, and $\phi^{-1}_{[n-1],[1]}$ respectively. (It is purely a convenience to what follows to take a component of ϕ^{-1} as a generator rather than a component of ϕ.)

The naturality and coherence conditions, together with the functoriality of $\otimes$ as applied to instances of σ are equivalent to the usual Artin relations on the σ_i's. Functoriality of $\otimes$ as applied to instances of β are equivalent to the commutativity of the β_i's. The instances of naturality involving σ's and β's are equivalent (in the presence of functoriality) to $\beta_i\sigma_i = \sigma_i\beta_{i+1}$ and $\beta_{i+1}\sigma_i = \sigma_i\beta_i$ while functoriality of $\otimes$ applied to σ's and β's is equivalent to $\beta_j\sigma_i = \sigma_i\beta_j$ for $i = 1, \ldots, n-1$, and $j = 1, \ldots, i-1$ or $j = i+1, \ldots, n$. It is convenient to define σ_n to be $\psi\sigma_1\psi^{-1}$, since the naturality of ψ then translates easily as $\beta_i\psi = \psi\beta_{i+1}$ and $\sigma_i\psi = \psi\sigma_{i+1}$, where subscripts are taken $\bmod n$. We have now given translations of all of the naturality, functoriality and coherence conditions satisfied by the generating maps. (Note: coherence conditions, except those used on σ, are used only to reduce to "one string" instances, and thus at the level of group generators contribute no relations.)

To summarize, we thus obtain the following presentation for the nth braid group of the torus:

$$\begin{aligned}\langle \sigma_1, \ldots, \sigma_n, \beta_1, \ldots, \beta_n, \psi \mid \sigma_i\sigma_j = \sigma_j\sigma_i \quad & i = 1, \ldots, n-3, \;\; j = i+2, \ldots, n-1;\\ & \sigma_i\sigma_{i+1}\sigma_i = \sigma_{i+1}\sigma_i\sigma_{i+1} \quad i = 1, \ldots, n-2;\\ & \beta_i\sigma_i = \sigma_i\beta_{i+1}\,; \quad \beta_{i+1}\sigma_i = \sigma_i\beta_i \quad i = 1, \ldots, n-1;\\ & \beta_j\sigma_i = \sigma_i\beta_j \quad i = 1, \ldots, n-1, \text{ and } j = 1, \ldots, i-1\\ & \text{or } j = i+1, \ldots, n; \quad \beta_i\psi = \psi\beta_{i+1} \quad i = 1, \ldots, n;\\ & \sigma_i\psi = \psi\sigma_{i+1} \quad i = 1, \ldots, n\rangle\,.\end{aligned}$$

We can do even better, by using the last two relations repeatedly, we get that $\psi^{-i+1}\sigma_1\psi^{i-1} = \sigma_i$ and $\psi^{-i+1}\beta_1\psi^{i-1} = \beta_i$. Letting $\sigma = \sigma_1$ and $\beta = \beta_1$, we obtain a presentation with three generators:

$$\begin{aligned}\langle \sigma, \beta, \psi \mid, [\psi\sigma\psi^{-1}\sigma\psi, \sigma] = 1; \quad & [\sigma^{\psi^i}, \sigma] = 1\,, \quad i = 2, \ldots, n-2;\\ [\sigma, \beta^{\psi^i}] = 1\,, \quad i = 2, \ldots, n-1; \quad & [\sigma\psi^{-1}, \beta] = 1; \quad [\beta, \psi\sigma] = 1\rangle\,.\end{aligned}$$

7. Out of prisms

The construction of Witten **[18]** for invariants of (framed) links in arbitrary (compact oriented) 3-manifolds depends upon considering tangles in non-trivial cobordisms between oriented surfaces. We are now concerned primarily with the smooth case, and many of our proofs will involve Morse functions.

Tangles in cobordisms more general than prisms can be viewed in terms of some notions which arise naturally from our notion of formal cobordism families.

Definition 7.1. If $\mathbb{F}$ is a formal cobordism family, the ***category of*** $\mathbb{F}$, $\mathfrak{C}\mathbb{F}$, is the category whose objects are all spaces which are union of connected components of X''s as X' ranges over all $(X, X') \in \mathbb{F}$, and in which a map $F: C \to D$ is named by a pair $(X, X') \in \mathbb{F}$, and an isomorphism $\phi: C \coprod D \to X$ (reversing orientation on D and preserving it on C if the spaces are oriented, and preserving any other structure in the definition of $\mathbb{F}$). Two such pairs $((X, X'), \phi)$ and $((Y, Y'), \psi)$ name the same map in $\mathfrak{C}\mathbb{F}$ if there is an isomorphism (preserving any auxiliary structure) $\xi: X \to Y$ such that $\phi\xi = \psi$. Identity maps are named by pairs $((C \times [0,1], C \times \{0,1\}), (\mathrm{Id}, 0) \coprod (\mathrm{Id}, 1): C \coprod C \to C \times \{0,1\})$, where $C \times \{0\}$ is interpreted as the source and $C \times \{1\}$ is interpreted as the target. The composite of maps named by $((X, X'),\ \phi: C \coprod D \to X')$ and $((Y, Y'),$

$\psi\colon D\coprod E \to Y')$ is named by

$$((X +_D Y, (X' - D) \cup (Y' - D)), (\phi\mid_C \coprod \psi\mid_E)\colon C\coprod E \to (X' - D)\cup(Y' - D)).$$

Proposition 7.2. *For any formal cobordism family* $\mathbb{F}$, *the category* $\mathfrak{C}\mathbb{F}$ *admits a symmetric pivotal category structure (that is a compact closed category structure in the terminology of Kelly and Laplaza* **[13]**), *with* $\otimes$ *given by disjoint union of names, the symmetry having components named by*

$$\begin{aligned}(((C\coprod D)\times[0,1],(C\coprod D)\times\{0,1\}),(\mathrm{Id},0)+(s,1))\colon&\\ (C\coprod D)\coprod(D\coprod C)\to(C\coprod D)&\times\{0,1\}),\end{aligned}$$

where $s\mid_D$ *is* i_D *and* $s\mid_C$ *is* i_C, *the canonical inclusions. Similarly, the units and counits of the adjunctions are named by the same pairs as the identities, but with the source or target regarded as empty, and the other taken as* $C\coprod C$.

Proof. A direct proof is easy, computational, and left to the reader. Alternatively, one can adapt arguments from Joyal and Street **[10]**). □

Notice that in the case of n-**cobord**, $\mathfrak{C}n$-**cobord** is the domain category for the functorial definition of a topological quantum field theory given by Atiyah, et al. **[1]**, and that an $(n+1)$-dimensional topological field theory is nothing more than a tensor preserving functor from $\mathfrak{C}n$-**cobord** to $\mathbb{C}-vs$. To extend this observation to Witten's coupled field theories, we use

Definition 7.3. The ***coupled isotopy category*** of two formal cobordism families $\mathbb{F}$ and $\mathbb{G}$, denoted $\{\mathbb{F},\mathbb{G}\}$ is the category whose objects are embeddings $A\colon X'\to Y'$, where X' (resp. Y') is the second member of a pair $(X,X')\in\mathbb{F}$ (resp. $(Y,Y')\in\mathbb{G}$), and with a map from $A\colon C\to\Gamma$ to $B\colon D\to\Delta$ named by a 5-tuple, $((X,X'), z\colon C\coprod D\to X', (Y,Y'), \xi\colon\Gamma\coprod\Delta\to Y', f\colon(X,X')\to(Y,Y'))$, where $((X,X'), z\colon C\coprod D\to X')$ names a map in $\mathfrak{C}\mathbb{F}$, $((Y,Y'),\xi\colon\Gamma\coprod\Delta\to Y')$ names a map in $\mathfrak{C}\mathbb{G}$, and f is an embedding of pairs such that

$$\begin{array}{ccccc} C & \xrightarrow{Z\mid_C} & X' & \xleftarrow{Z\mid_D} & D\\ \Big\downarrow{\scriptstyle\exists} & & \Big\downarrow{\scriptstyle f} & & \Big\downarrow{\scriptstyle\exists}\\ \Gamma & \xrightarrow[\zeta\mid_\Gamma]{} & Y' & \xleftarrow[\zeta\mid_\Delta]{} & \Delta \end{array}$$

(We usually write $f\colon X\to Y$ instead of explicitly writing it as a map of pairs.) Two such 5-tuples,

$$((X,X'), z\colon C\coprod D\to X', (Y,Y'), \xi\colon\Gamma\coprod\Delta\to Y', f\colon X\to Y)$$

and

$$((U,U'), w\colon C\coprod D\to U', (V,V'), \omega\colon\Gamma\coprod\Delta\to V', g\colon U\to V)$$

name the same map if there is an isomorphism $\Phi\colon (X,X') \to (U,U')$, and an isomorphism $\Psi\colon (Y,Y') \to (V,V')$ such that $z\Phi = w$, $\xi\Psi = \omega$, $z\Phi g = zf\Psi$, and Φg is isotopic to $f\Psi$ by an isotopy which fixes the image of $z\Phi g = zf\Psi$.

$$((C\times[0,1], C\times\{0,1\}),\ (\mathrm{Id},0)\coprod(\mathrm{Id},1)\colon C\coprod C \to C\times\{0,1\},\ (\Gamma\times[0,1],\Gamma\times\{0,1\}),$$

$$(\mathrm{Id},0)\coprod(\mathrm{Id},1)\colon \Gamma\coprod\Gamma \to \Gamma\times\{0,1\},\ A\times\mathrm{Id}\colon C\times[0,1]\to\Gamma\times[0,1])$$

names the identity map for $A\colon C\to\Gamma$, and the composite of

$$((X,X'), z\colon C\coprod D \to X', (Y,Y'), \xi\colon \Gamma\coprod\Delta \to Y', f\colon X\to Y)$$

and

$$((U,U'), w\colon D\coprod E \to U', (V,V'), \omega\colon \Delta\coprod\Upsilon \to V', g\colon U\to V)$$

is named by

$$((X+_D U, (X'-D)\cup(U'-D)),\quad z\,|_C + w\,|_E\colon C\coprod E \to (X'-D)\cup(Y'-D),$$

$$(Y+_\Delta V, (Y'-\Delta)\cup(V'-\Delta)),\quad \zeta\,|_\Gamma + w\,|_\Upsilon\colon \Gamma\coprod\Upsilon \to (Y'-\Delta)\cup(V'-\Delta),$$

$$f +_D g\colon X +_D U \to (Y +_\Delta V).$$

It is not difficult to prove

Proposition 7.4. *There is a functor $U_{\mathbb{F},\mathbb{G}}\colon \{\mathbb{F},\mathbb{G}\} \to \mathfrak{C}\mathbb{G}$, which takes the object named by $A\colon C\to\Gamma$ to the object of $\mathfrak{C}\mathbb{G}$ named by Γ, and carries the map named by*

$$(X,X'), z\colon C\coprod D \to X', (Y,Y;), \xi\colon \Gamma\coprod\Delta \to Y', f\colon X\to Y)$$

to the map named by $((Y,Y';), \xi\colon \Gamma\coprod\Delta\to Y')$. □

The main use of this proposition is to reduce the question of describing $\{\mathbb{F},\mathbb{G}\}$ to the questions of describing the fibres of $U_{\mathbb{F},\mathbb{G}}$, describing $\mathfrak{C}\mathbb{G}$, and describing the composite maps in $\{\mathbb{F},\mathbb{G}\}$ lying over a composite in $\mathfrak{C}\mathbb{G}$. (The categorically minded reader should be warned that the word fibre is used in a loose suggestive manner, since the functor $U_{\mathbb{F},\mathbb{G}}$ is neither a fibration nor an op-fibration.)

We will now examine the question of describing $\{\mathbb{F}, 2\text{-}\mathbf{cobord}\}$ in the case where $\mathbb{F}$ is a regular formal cobordism family. This case is of particular interest as Witten's **[18]** invariants of (framed, oriented) links in 3-manifolds arise from a functor from $\{\mathbb{RT}, 2\text{-}\mathbf{cobord}\}$ to the category of complex vector spaces by specializing to endomorphisms of $(\emptyset,\emptyset,\mathrm{Id}\colon\emptyset\to\emptyset)$.

The second part of the program, describing $2\text{-}\mathbf{cobord}$, is essentially equivalent to describing all orientable 3-manifolds with boundary, and is thus only partly within the realm of current knowledge. We can, however, readily describe those fibres over maps of $2\text{-}\mathbf{cobord}$ which are named by cobordisms which admit a Morse function with a single critical point having the source (resp. target) as the level set of 0 (resp. 1). The reader should remember that the number of components is not fixed.

We consider critical points classified by their degrees. Degrees 0 and 3 are formally dual as are degrees 1 and 2, thus it suffices to describe the maps of $\{\mathbb{F}, 2\text{-}\mathbf{cobord}\}$

lying above maps of $\mathcal{C}2$-**cobord** named by cobordisms of surfaces with single critical points of the degrees 0 and 1.

For degree 0, it is easy to see that there is a one-to-one correspondence between maps named by embeddings into the 3-ball regarded as a corbordism from the empty space to S^2 with target named by some embedding $A: X' \hookrightarrow S^2$, and $\mathrm{Hom}_{\{\mathbb{F},S^2\}}(I, A)$, where I is the object named by the embedding of the empty space. (Recall we are assuming that $\mathbb{F}$ is regular, so any name or any isotopy of names can easily be replaced by one which misses the critical point.) As any cobordism with only this critical point is a disjoint union of the 3-ball with a prism on surface Σ, any map lying over it consists of a pair (ϕ, ψ) for ϕ a map in $\{\mathbb{F}, S^2\}$ with empty source, and ψ a map in $\{\mathbb{F}, \Sigma\}$.

Similarly, for degree 1, one can move any name of a map (or any isotopy of the same) out of a small ball about the critical point bounded away from the boundary of the cobordism by choosing a 1-complex to which the name of the map contracts, moving it (or its image under an isotopy) out of a larger ball about the critical point, and then applying a contraction of the sort in the definition of regularity until the name of the map (or the isotopy) no longer intersects the first ball.

Now, suppose that we have a cobordism in this case, with underlying 3-manifold, M, with Morse function $\Phi: M \to [0,1]$, $\Sigma_i = \Phi^{-1}(i)$, and critical level, c. Choose a small ball, B, about the critical point as in the last paragraph, and a larger neighborhood, N, so that $\Sigma_c \cap B = \Sigma_c \cap N$, and $M - N$ is diffeomorphic over $[0,1]$ to $(\Sigma_c - N) \times [0,1]$. We can then see, by applying an isotopy away from the critical point so that the name intersects the critical level with a bicollar neighborhood, that the maps in $\{\mathbb{F}, 2\text{-}\mathbf{cobord}\}$ lying above cobordisms of the second form have names which may be regarded as a pair of maps $f \in \{\mathbb{F}, \Sigma_0\}$, $g \in \{\mathbb{F}, \Sigma_1\}$, where the target of f and the source of g are the same object of $\{\mathbb{F}, \Sigma_c - N\}$ when $\{\mathbb{F}, \Sigma_c - N\}$ is identified with $\{\mathbb{F}, \Sigma_i - N\}$ for $i = 0, 1$ by translating in the $[0,1]$-coordinate.

An argument reminiscent of the proof of Theorem 4.1 then shows that two such pairs (f, g) and (h, k) name the same map if and only if there is a ϕ in $\{\mathbb{F}, \Sigma_c - N\}$ such that $f = h\phi$ and $k = \phi g$.

Summarizing these results in terms of individual hom-sets, we have

Proposition 7.5. *If $A: X' \to \Sigma$ and $B: Y' \to \Xi$ are names of objects in $\{\mathbb{F}, 2\text{-}\mathbf{cobord}\}$ for $\mathbb{F}$ some regular formal cobordism family, and $(M, \partial M)$ is some cobordism from Σ to Ξ with exactly one critical point of degree 1, then $\mathrm{Hom}(A, B) \cap U^{-1}_{\mathbb{F},2-\mathbf{cobord}}(M)$ is canonically isomorphic to the set*

$$[\coprod_{C \in \mathrm{Ob}(\{\mathbb{F},\Sigma_c - N\})} \mathrm{Hom}_{\{\mathbb{F},\Sigma\}}(A, i(C)) \times \mathrm{Hom}_{\{\mathbb{F},\Xi\}}(\imath(C), B)]/\equiv,$$

where i and $\imath$ are the inclusions induced by the translation, and $\equiv$ is the equivalence relation induced by $(h\phi, g) \equiv (h, \phi g)$. Composition with an endomorphism $e: A \to A$ (resp. $e: B \to B$) considered as a map in the fibre over Id_Σ (resp. Id_Ξ) is given on representatives by $e(f, g) = (ef, g)$ (resp. $(f, g)e = (f, ge)$). □

Notice that maps in the degree 0 case can also be written in this form by replacing (ϕ, ψ) with $(\psi, \langle \phi, \mathrm{Id}_{\mathrm{target}(\psi)} \rangle)$, where $\langle \phi, \mathrm{Id}_{\mathrm{target}(\psi)} \rangle$ is the map in $\{\mathbb{F}, S^2 \coprod \Sigma\} = \{\mathbb{F}, S^2\} \times \{\mathbb{F}, \Sigma\}$ named by the disjoint union of the embedding naming ϕ and the embedding naming $\mathrm{Id}_{\mathrm{target}(\phi)}$.

It follows by duality that maps over cobordisms with single critical points of degree 2 and 3 also have such representations.

We now take up the question of how maps in $\{\mathbb{F}, 2-\mathbf{cobord}\}$ compose. We have already dealt with the case where one of the maps lies over a prism (a cobordism with no critical points), and the other over a prism (here we are in one of our isotopy categories for surfaces) or over a cobordism with one critical point. The following proposition, which is trivial to prove, describes when two maps over composable cobordisms are composable.

Proposition 7.6. *If (f, g) names a map over a cobordism M from Σ to Ξ, and (h, k) names a map over a cobordism N from Ξ to Θ, then (f, g) and (h, k) are composable if and only if g and h are composable as maps in $\{\mathbb{F}, \Xi\}$.* □

We now consider cobordisms with a Morse function with several critical points, say M, with $\Phi\colon M \to [0,1]$, and $\Sigma_i = \Phi^{-1}(i)$, and one critical point at each level c_j for

$$0 < c_1 < \cdots < c_t < 1.$$

Now choose a level r_i, $c_i < r_i < c_{i+1}$ for $i = 1, \ldots, t-1$, and let $r_0 = 0$, $r_t = 1$. Now choose a ball, B_i about each critical point which is bounded away from the Σ_{r_j}, $j = i-1, i$. And choose a larger neighborhood N_i of each critical point so that $\Sigma_{c_i} \cap N_i = \Sigma_{c_i} \cap B_i$ and $\Phi^{-1}([r_{i-1}, r_i] - N_i)$ is diffeomorphic over $[r_{i-1}, r_i]$ to $(\Sigma_{c_i} - N_i) \times [r_{i-1}, r_i]$.

An argument similar to that preceding Proposition 7.5 then shows that any map has a name which avoids all of the B_i's and intersects all Σ_{c_i}'s and Σ_{r_i}'s in levels with bicollar neighborhoods.

Such a name can then be regarded as giving a string of maps $(f_1, g_1, f_2, g_2, \ldots, f_t, g_t)$, where f_i is a map in $\{\mathbb{F}, \Sigma_{r_{i-1}}\}$ with target in $\{\mathbb{F}, \Sigma_{r_{i-1}} - N\}$ and g_i is a map in $\{\mathbb{F}, \Sigma_{r_i}\}$ with source in $\{\mathbb{F}, \Sigma_{r-i} - N_{i-1}\}$, so that g_i is composable with f_{i+1} as maps in $\{\mathbb{F}, \Sigma_{r_i}\}$, and the target of f_i is equal as an object in $\{\mathbb{F}, \Sigma_{c_i} - B_i\}$ to the source of g_i. (Note here we are identifying $\Sigma_{r_{i-1}} - N_i$, $\Sigma_{c_i} - B_i$, and $\Sigma_{r_i} - N_i$ by translation in the $[0,1]$ direction.)

The question of when two such expressions name the same map breaks into two parts: the easy question of when two such representations for the same Morse function name the same map, and the harder question of relating names given in terms of different Morse functions.

The first is readily seen by the same argument which was given for Proposition 7.5 to reduce to the equivalence relation $\equiv$ induced by all instances of

$$(\ldots, f_i = f\phi, g_i, \ldots) \equiv (\ldots, f, \phi g_i, \ldots)$$
$$(\ldots, g_i = g\psi, f_{i+1}, \ldots) \equiv (\ldots, g, \psi f_{i+1}, \ldots)$$

for maps ϕ in $\{\mathbb{F}, \Sigma_{c_i} - N_i\}$ and ψ in $\{\mathbb{F}, \Sigma_{r_i}\}$.

Summarizing this in the form of a proposition:

Proposition 7.7. *Let M be an oriented 3-manifold considered as a cobordism between Σ_0 and Σ_i, then fixing a Morse function Φ on M with n critical points, each critical point on a distinct level, and denoting the critical levels and levels chosen between them* *as above, there is a one-to-one correspondence between maps in* $\{\mathbb{F}, 2\text{-}\mathbf{cobord}\}$ *lying over* $(M, \Sigma_0 \cup \Sigma_1)$*, the set of equivalence classes of* $2n$*-tuples*

$$(f_1, g_1, f_2, g_2, \ldots, f_t, g_t),$$

where f_i *is a map in* $\{\mathbb{F}, \Sigma_{r_{i-1}}\}$ *with target in* $\{\mathbb{F}, \Sigma_{r_{i-1}} - N_i\}$ *and* g_i *is a map in* $\{\mathbb{F}, \Sigma_{r_i}\}$ *with source in* $\{\mathbb{F}, \Sigma_{r_i} - N_i\}$*, so that* g_i *is composable with* f_{i+1} *as maps in* $\{\mathbb{F}, \Sigma_{r_i}\}$*, and the target of* f_i *is equal as an object in* $\{\mathbb{F}, \Sigma_{c_i} - B_i\}$ *to the source of* g_i*, under the equivalence relation* $\equiv$ *of the previous paragraph.* □

Unfortunately, settling questions of equality of maps in $\{\mathbb{F}, 2\text{-}\mathbf{cobord}\}$ will in general require comparing names of maps given in terms of different Morse functions on the same cobordism. We now give theorems which take a name of a map as a $2n$-tuple in terms of one Morse function on the cobordism, and give a name as a $2m$-tuple in terms of another Morse function (or a single map in $\{\mathbb{F}, \Sigma\}$ if the cobordism is trivial).

The first is a more or less trivial observation:

Proposition 7.8. *If* Φ *is any Morse function on a cobordism* $(M, \Sigma_0 \cup \Sigma_1)$ *and* Ψ *is another Morse function, so that there is a map* $\Theta\colon M \times [0,1] \to [0,1]$ *so that* $\Theta(-,t)$ *is a Morse function for all* t*,* $\Theta(-,0) = \Phi$*, and* $\Theta(-,1) = \Psi$*, then the* $2n$*-tuples of maps as in Proposition 7.7 name the same maps in terms of* Ψ *as in terms of* Φ*.* □

Slightly less trivial is the description when the Morse functions are related by commuting a pair of critical points with independent trajectories:

Proposition 7.9. *If* Φ *is any Morse function on a cobordism* $(M, \Sigma_0 \cup \Sigma_1)$*. Suppose, moreover, that in the notation above* N_i *and* N_{i+1} *can be chosen (by independent trajectories) to be disjoint, for some* i*. Then there are larger balls* P_i *and* P_{i+1} *which are disjoint, and such that* $\Phi^{-1}([r_{i-1}, r_{i+1}] - (P_i \cup P_{i+1}))$ *is diffeomorphic over* $[r_{i-1}, r_{i+1}]$ *to* $\Sigma_{r_i} - (P_i \cup P_{i+1}) \times [r_{i-1}, r_{i+1}]$*. Moreover, there is a Morse function* Ψ *equal to* Φ *outside of* $P_i \cup P_{i+1}$*, and such that inside* P_i *there is exactly one critical point at level* c_{i+1} *and inside* P_{i+1} *there is exactly one critical point at level* c_i*.*

In this situation, given a name for a map in $\{\mathbb{F}, 2-\mathbf{cobord}\}$ *lying over* $(M, \Sigma_0 \cup \Sigma_1)$ *and named by*

$$(\ldots, g_{i-1}, f_i, g_i, f_{i+1}, g_{i+1}, f_{i+2}, \ldots)$$

in terms of Φ*, then*

$$(\ldots, g_{i-1}\varepsilon_{i-1}, \phi_{i-1}(f_i), \phi_i(g_i), \phi_i(f_{i+1}), \phi_{i+1}(g_{i+1}), \eta_{i+1} f_{i+2}, \ldots)$$

is a name for the same map in terms of Ψ*, where* ϕ_j *is a reflection functor from* $\{\mathbb{F}, \Sigma_{r_j}\}$ *to its full subcategory with objects named by embeddings to* $\Sigma_{r_j} - (P_i \cup P_{i+1})$ *(for* $j = i-1, i, i+1$*), and* ε_j *(resp.* η_j*) is the unit (resp. counit) of the adjunction with the inclusion functor, and* ϕ_{i-1} *(resp.* ϕ_{i+1}*) and* ϕ_i *agree on* $\Sigma_{r_{i-1}} - N_i \cong \Sigma_{r_i} - N_i$ *(resp.* $\Sigma_{r_{i+1}} - N_{i+1} \cong \Sigma_{r_i} - N_{i+1}$*).*

Proof. The first paragraph is standard differential topology. For the second, note that the subcategory is a reflective subcategory, since names of maps in $\{\mathbb{F}, \Sigma_{r_j}\}$ can be chosen to avoid any finite set of points (in this case a point in each of P_i and P_{i+1}), and an isotopy of the complement of these points chosen to induce the reflections. The agreement of the reflection functors can be ensured simply by ensuring that the inducing isotopies agree on these sets. (Choose the isotopy to induce ϕ_i first, then choose the others to agree.)

A final observation then completes the proof: note that names for maps in the full subcategories can be chosen to avoid all of $P_i \cup P_{i+1}$, so that the change in the Morse function inside $P_i \cup P_{i+1}$ does not affect them. □

Finally, we consider the case in which one Morse function is obtained from the other by cancelling a pair of critical points (i.e., passing through a birth point or a death point).

Proposition 7.10. *Suppose Φ is a Morse function on a cobordism $(M, \Sigma_0 \cup \Sigma_1)$, and that the critical points c_i and c_{i+1} differ in degree by one, and that Ψ is a Morse function agreeing with Φ outside $N_i \cup N_{i+1}$, with no critical points inside $N_i \cup N_{i+1}$ (that is obtained by cancelling the ith and $i+1^{\text{st}}$ critical points). Then given a name*

$$(\ldots, g_{i-1}, f_i, g_i, f_{i+1}, g_{i+1}, f_{i+2}, \ldots)$$

in terms of Φ, then

$$(\ldots, g_{i-1}\varepsilon_{i-1}\phi_{i-1}(f_i)\phi_i(g_i), \phi_i(f_{i+1})\phi_{i+1}(g_{i+1})\eta_{i+1}f_{i+2}, \ldots)$$

is a name for the same map in terms of Ψ, where ϕ_j, ε_j, and η_j are as in the previous proposition with

$$P_i = P_{i+1} \supset N_i \cup N_{i+1}\,.$$

Proof. Essentially as in the previous proposition. □

These last three propositions taken together describe completely (albeit not effectively for computational purposes) when any two names of maps in $\{\mathbb{F}, 2\text{-}\mathbf{cobord}\}$ lying over a given cobordism name the same map, since it is always possible to get from one Morse function to another without going through worse degeneracies than birth points, death points or crossing points.

8. Concluding observations and unexamined questions

It is not hard to show:

Proposition 8.1. *For any regular formal cobordism family, $\mathbb{F}$, $\{\mathbb{F}, 2-\mathbf{cobord}\}$ is a symmetric pivotal category.* □

We then see that a coupled topological field theory (with the higher dimensional theory being $(2+1)$-dimensional) is precisely a $\otimes$-preserving functor from $\{\mathbb{F}, 2\text{-}\mathbf{cobord}\}$ to $\mathbb{C}\text{-}vs$ for suitable $\mathbb{F}$. The construction of Witten **[18]** then provides a family of such functors which have reasonably geometric interpretations. Having identified the underlying algebraic structure we are now in a position to consider several problems:

1. Classify coupled topological field theories in low dimensions.
2. Can one assign an interpretation of the sort provided in Witten's work to any such functor?

J. Przytycki has raised the question of whether results contained herein could be applied to prove the freeness of various skein modules.

Finally there is the question of relating the more geometrical approach to invariants of links in 3-manifolds given above to the combinatorial Kirby-tangle approach first suggested by Freyd and Yetter **[5]** as a construction for invariants of 3-manifolds (without links), and used by Reshitikhin and Tureav **[16]** to provide a mathematically rigorous version of Witten's constructions. (See also Kirby and Melvin **[14]**.)

References

[1] Atiyah, M. F., et al., Notes on the Oxford Seminar on Jones-Witten Theory, Michaelmas Term 1988.

[2] Artin, E., "The Theory of Braids", Ann. of Math. 47 (1947), 101–126.

[3] Birman, J. S., Braids, Links, and Mapping Class Groups, Princeton Univ. Press, 1974.

[4] Fadell, E. and VanBuskirk, J., "The Braid Groups of E^2 and S^2", Duke Math. J. 29 (1962), 243–258.

[5] Freyd, P. J. and Yetter, D. N., "Braided Compact Closed Categories with Applications to Low Dimensional Topology", Adv. in Math. 77 (2) (1989) 156–182.

[6] Freyd, P. J. and Yetter, D. N., "Coherence Theorems via Knot Theory", J. Pure and App. Alg. (to appear).

[7] Freyd, P. J.; Yetter, D. N., Lickorish, W. B. R., Millett, K. C., Horste, J., and Ocneanu, A., "A new polynomial invariant of knots and links", Bull. A.M.S. 12 (1985), 239–246.

[8] Jones, V. F. R., "A polynomial invariant of knots via von Neumann algebras", Bull. A.M.S. 12 (1985), 103–111.

[9a] Joyal, A. and Street, R., "Braided monoidal categories", (preprint).

[9b] Joyal, A. and Street, R., "Braided tensor categories", (preprint) (updated version of [9a]).

[10] Joyal, A. and Street, R., "Geometric Tensor Calculus", Adv. in Math. (to appear).

[11] Kauffman, L., "An invariant of regular isotopy", Trans. A.M.S. (to appear).

[12] Kauffman, L., "New Invariants in the Theory of Knots", Amer. Math. Monthly, 95 (3) (1988), 195–242.

[12a] Kelly, G. M., Basic Concepts of Enriched Category Theory, London Math. Soc. Lecture Note no. 64, Cambridge Univ. Press, 1982.

[13] Kelly, G. M. and Laplaza, M. L., "Coherence for compact closed categories", Jour. Pure and App. Alg. 19 (1980), 193–213.

[13a] Kelly, G. M. and Street, R. H., "Review of the elements of 2-categories", in Category Seminar: Sydney 1972/3, Lecture Notes in Mathematics no. 420, Springer-Verlag, 1974.

[14] Kirby, R. and Melvin, P., "Evaluations of the 3-manifold Invariants of Witten and Reshetikhin-Turaev for $sl(2, \mathbb{C})$", to appear in the Proceedings of the Conference on Low-Dimensional Topology, Durham, England, June 1989.

[15] MacLane, S., "Natural Associativity and Commutativity", Rice Univ. Studies 49 (1963) 28–46.

[15a] MacLane, S., Categories for the Working Mathematician, Springer-Verlag, 1971.

[16] Reshitikhin, N. and Turaev, V. G., "Invariants of 3-manifolds via Link Polynomials and Quantum Groups", Invent. Math. (to appear).

[17] Shum, M.-C., Tortile Tensor Categories, Ph.D. dissertation, Macquarie University (1989).

[18] Witten, E., "Quantum Field Theory and the Jones Polynomial", Comm. Math. Phys. 121 (1989) 351–399.

[19] Yetter, D. N., "Markov Algebras", in Braids, (J. S. Birman and A. Libgober, eds.), AMS Contemp. Math. Series, vol. 78, (1988) 705–730.

[20] Yetter, D. N., "Category Theoretic Representations of Embedded Graphs in S^3", Adv. in Math., 77 (2) (1989) 137–155.

Department of Mathematics, Kansas State University, Manhattan, KS 66506

A Computational Algorithm of Spectral Flow in Floer Homology

Tomoyoshi Yoshida

1. Introduction

Let M be an oriented closed Riemannian 3-manifold. Let A be a smooth connection on the principal $\mathrm{SU}(2)$-bundle $M \times \mathrm{SU}(2)$. Let $\Omega^i(M) \otimes su(2)$ be the space of smooth $su(2)$-valued i-forms on M ($i = 0, 1, 2, 3$). Let D_A be the operator on $(\Omega^1 \oplus \Omega^0)(M) \otimes su(2)$ defined by, for $(a, b) \in (\Omega^1 \oplus \Omega^0)(M) \otimes su(2)$,

$$D_A(a, b) = (*d_A a + d_A b, d_A^* a).$$

Here d_A denotes the covariant derivative with respect to the connection A and $*$ denotes the Hodge star operator. Then D_A can be extended to a self-adjoint Fredholm operator from the L_1^2-Sobolev norm completion of $(\Omega^1 \oplus \Omega^0)(M) \otimes su(2)$ to the L^2-completion of it.

Let A_0 and A_1 be smooth irreducible flat connections on $M \times \mathrm{SU}(2)$ such that $\mathrm{Ker}\, D_{A_0} = \mathrm{Ker}\, D_{A_1} = 0$. Let $\{A_t\}_{0 \le t \le 1}$ be a smooth path of smooth connections on $M \times \mathrm{SU}(2)$ connecting A_0 and A_1. Then the spectral flow of $\{D_{A_t}\}_{0 \le t \le 1}$, $SF(M, \{A_t\})$, is defined (**[A-P-S]**).

In the case that M is an integral homology 3-sphere, then $(\mathrm{mod}\, 8)$ reduction of $SF(M, \{A_t\})$ gives Floer's $(\mathrm{mod}\, 8)$ relative Morse index of the gauge equivalence classes $[A_0]$ and $[A_1]$ in the Chern-Simons Morse theory on the space of the gauge equivalence classes of the connections on $M \times \mathrm{SU}(2)$ (**[F]**).

In **[Y]**, we gave a formula for $SF(M, \{A_t\})$ under the situation in which $M = M_1 \cup M_2$ and $M_1 \cap M_2 = \partial M_1 = \partial M_2 = \Sigma$ is an orientable closed surface of genus $g \ge 2$ in terms of Maslov index of finite dimensional symplectic geometry. As an application of this formula, we calculated the Floer homology group of those homology 3-spheres obtained by performing Dehn surgeries along the figure eight knot in S^3 (**[Y]**).

The purpose of this paper is to show that the main method in the above calculation of spectral flow in the case of the figure eight knot in **[Y]** can be generalized to more general settings. In Section 2, we give the setting which we are concerned with and we give a method to reduce calculations of spectral flow in this setting to simple diagram chasing (Theorem 2.1). In Section 5 we give the actual numbers associated to diagrams.

This work was completed during my stay at The Ohio State University and the author would like to thank their Mathematical Research Institute for its hospitality.

2. The main theorem and the setting which we are concerned with

Let M be an oriented closed Riemannian 3-manifold. Let K be a knot in M. Let $U(K)$ be a closed tubular neighborhood of K in M. Let $N = \overline{M - U(K)}$ be the closure of the complement of $U(K)$.

Let

$$R(M) = \mathrm{Hom}(\pi_1(M), \mathrm{SU}(2))^* / \operatorname{ad} \mathrm{SU}(2)$$
$$R(N) = \mathrm{Hom}(\pi_1(N), \mathrm{SU}(2))^* / \operatorname{ad} \mathrm{SU}(2)$$

be the space of the conjugacy classes of the irreducible representations of $\pi_1(M)$ and $\pi_1(N)$ into $\mathrm{SU}(2)$ respectively. These spaces have the natural structures of real analytic varieties, hence are topological spaces.

The inclusion map $j\colon N \to M$ induces the natural real analytic map

$$j^*\colon R(M) \to R(N)$$

and the condition for the conjugacy class $[\rho]$ of an irreducible representation $\rho\colon \pi_1(N) \to \mathrm{SU}(2)$ to be in $\operatorname{Im} j^*$ is that $\rho(m) = 1$, where m denotes the homotopy class in $\pi_1(N)$ represented by the meridian curve in $\partial U(K) = \partial N$.

Let S^1 and D^2 be the unit circle and the unit disk respectively in the complex plane oriented by the complex structure. Let $S^1 \times S^1$ be the 2-torus. Let m and ℓ be the homotopy classes in $\pi_1(S^1 \times S^1)$ represented by $S^1 \times 1$ and $1 \times S^1$ respectively. Let

$$R(S^1 \times S^1) = \mathrm{Hom}(\pi_1(S^1 \times S^1), \mathrm{SU}(2)) / \operatorname{ad} \mathrm{SU}(2)$$

be the space of the conjugacy classes of the representations of $\pi_1(S^1 \times S^1)$ into $\mathrm{SU}(2)$. For $\rho \in R(S^1 \times S^1)$, we can write up to conjugation,

$$\rho(m) = \begin{pmatrix} e^{iu} & 0 \\ 0 & e^{-iu} \end{pmatrix} \qquad \rho(\ell) = \begin{pmatrix} e^{iv} & 0 \\ 0 & e^{-iv} \end{pmatrix}$$

for $-\pi \le u,\ v < \pi$. The correspondence $\rho \to (e^{iu}, e^{iv})$ induces a bijection between $R(S^1 \times S^1)$ and $S^1 \times S^1/\sim$, where $(z, w) \sim (\overline{z}, \overline{w})$ ($^-$ denotes the complex conjugate) for $z, w \in S^1$. Thus $R(S^1 \times S^1)$ is the 2-sphere with distinguished 4 points $Q = \{(\pm 1, \pm 1)\}$ each of which corresponds to a representation of $\pi_1(S^1 \times S^1)$ into the center of $\mathrm{SU}(2)$.

Let $P_0 = R(S^1 \times S^1) - Q$. P_0 is a four punctured sphere. Let $q\colon \widetilde{P}_0 \to P_0$ be the universal abelian covering of P_0. $\widetilde{P}_0$ can be identified with $\mathbb{R}^2 - \widetilde{Q}$, where $\widetilde{Q}$ denotes the set of those points whose coordinates are integer multiples of π. The variables (u, v) in the above can be considered as a coordinate of $\widetilde{P}_0$. Let Λ be the discrete subgroup of the euclidean motions of $\widetilde{P}_0$ generated by three motions, $(u, v) \mapsto (u + 2\pi, v)$, $(u, v) \mapsto (u, v + 2\pi)$ and $(u, v) \mapsto (-u, -v)$. Then $\widetilde{P}_0/\Lambda = P_0$.

Let $h\colon D^2 \times S^1 \to U(K)$ be a diffeomorphism. Let $i\colon \partial U(K) \to N$ be the inclusion. Then the composition $f = i(h \mid \partial D^2 \times S^1)\colon S^1 \times S^1 \to N$ induces a real analytic map

$$f^*\colon R(N) \to R(S^1 \times S^1).$$

In this paper, we consider the following setting.

$(*)$ A_0 and A_1 are two smooth irreducible flat connections on $M \times \mathrm{SU}(2)$. ρ_0 and ρ_1 are the holonomy representations of $\pi_1(M)$ into $\mathrm{SU}(2)$ associated to A_0 and A_1 respectively. We assume that $[\rho_0]$ and $[\rho_1]$ are both isolated points of $R(M)$. We denote $j^*([\rho_i]) \in \pi_1(N)$ also by $[\rho_i]$ $(i = 0, 1)$. We assume that

(i) there is a 1-dimensional connected component E of $R(N)$ such that $[\rho_0]$, $[\rho_1] \in E$,

(ii) E contains a smooth segment E_0 connecting $[\rho_0]$ and $[\rho_1]$ such that $f^*(E_0) \subset P_0$,

(iii) there is a Λ-fundamental segment C of $q^{-1}(f^*(E_0))$ which is a smoothly immersed curve in $\widetilde{P}_0$ and intersects $\mathbb{L}$ transversely at the endpoints $C \cap q^{-1}([\rho_0])$ and $C \cap q^{-1}([\rho_1])$, where

$$\mathbb{L} = \{(u, v) \in \widetilde{P}_0 \mid u = 2m\pi\,,\ m \in \mathbb{Z}\}.$$

Let $\pi_1^{\mathrm{reg}}(\widetilde{P}_0, \mathbb{L})$ denote the set of the smooth regular homotopy classes of immersed smooth curves in $\widetilde{P}_0$ with endpoints on $\mathbb{L}$ such that it intersects transversely with $\mathbb{L}$ at the endpoints. Obviously the group Λ acts on $\pi_1^{\mathrm{reg}}(\widetilde{P}_0, \mathbb{L})$.

The main theorem is the following.

Theorem 2.1. *Let A_0 and A_1 be smooth irreducible flat connections on $M \times \mathrm{SU}(2)$ such that* $\operatorname{Ker} D_{A_0} = \operatorname{Ker} D_{A_1} = 0$. *Let K be a knot in M. Let $U(K)$ be a closed tubular neighborhood of K in M. Set $N = \overline{M - U(K)}$. We assume the above setting $(*)$ holds for A_0, A_1 and N. Let C be the immersed smooth curve in P_0 mentioned in the above setting $(*)$. Let $\{A_t\}_{0 \le t \le 1}$ be a smooth path of irreducible smooth connections on $M \times \mathrm{SU}(2)$ connecting A_0 and A_1.*

Then the $(\mathrm{mod}\, 8)$ *spectral flow $SF(M, \{A_t\})$ is determined by the Λ-equivalence class of the relative smooth regular homotopy class of C in $\pi_1^{\mathrm{reg}}(\widetilde{P}_0, \mathbb{L})$.*

The proof of this theorem will be given in Section 4.

3. The invariant $\mathcal{Y}(\{A_t\})$ and a formula of spectral flow

To prove Theorem 2.1, we recall the formula of spectral flow in **[Y]**. For the details in this section, see **[Y]**.

Let M be an oriented closed Riemannian 3-manifold. Let $M = M_1 \cup M_2$ be a splitting of M, where M_1 and M_2 are codimension 0 submanifolds of M and $M_1 \cap M_2 = \partial M_1 = \partial M_2 = \Sigma$ is an orientable closed surface of genus g (≥ 2). The orientation of M induces the orientations of M_i $(i = 1, 2)$ and we orient Σ as the boundary of M_1. We assume that, near Σ, the Riemannian metric on M is the product metric $\Sigma \times [-1, 1]$.

Let A be an irreducible smooth connection on $M \times \mathrm{SU}(2)$ which restricts to a product connection $B \times 1$ on $(\Sigma \times [-1, 1]) \times \mathrm{SU}(2)$. Here B is an irreducible smooth flat connection on $\Sigma \times \mathrm{SU}(2)$ and 1 is the trivial connection on $[-1, 1] \times \mathrm{SU}(2)$.

On $\Sigma \times [-1, 1]$, we write $(a, b) \in (\Omega^1 \oplus \Omega^0)(\Sigma \times [-1, 1]) \otimes su(2)$ as

$$(a(s), b(s)) = (p(s) + q(s)ds, b(s)), \qquad -1 \le s \le 1$$

where $p(s) \in \Omega^1(\Sigma) \otimes su(2)$ and $q(s) \in \Omega^0(\Sigma) \otimes su(2)$ for $-1 \le s \le 1$, and we regard $(a(s), b(s)) = (p(s), q(s), b(s))$ to be an element of $(\Omega^1 \oplus \Omega^0 \oplus \Omega^0)(\Sigma) \otimes su(2)$.

In this notation, on $\Sigma \times [-1,1]$, the operator D_A can be written in the matrix form

$$D_A \begin{pmatrix} p(s) \\ q(s) \\ b(s) \end{pmatrix} = \sigma \left(\frac{\partial}{\partial s} - D_B \right) \begin{pmatrix} p(s) \\ q(s) \\ b(s) \end{pmatrix}$$

where

$$\sigma = \begin{pmatrix} * & 0 & 0 \\ 0 & 0 & 1 \\ 0 & -1 & 0 \end{pmatrix} \quad \text{and} \quad D_B = \begin{pmatrix} 0 & d_B & *d_B \\ d_B^* & 0 & 0 \\ -*d_B & 0 & 0 \end{pmatrix}.$$

Here $*$ denotes the Hodge star operator on $\Omega^j(\Sigma) \otimes su(2)$ $(j = 0,1)$ and d_B denotes the covariant derivative.

D_B can be extended to a self-adjoint Fredholm operator from the L_1^2-completion of $(\Omega^1 \oplus \Omega^0 \oplus \Omega^0)(\Sigma) \otimes su(2)$ to the L^2-completion of it. The set of the eigenvalues $\{\mu\}$ of D_B is a discrete subset of $\mathbb{R}$ and each eigenvalue has finite multiplicity. The set of all the normalized eigenforms $\{\psi_\mu\}$ of D_B forms a complete orthonormal basis of the L^2-completion of $(\Omega^1 \oplus \Omega^0 \oplus \Omega^0)(\Sigma) \otimes su(2)$.

Let $\mathcal{H}_B$ denote the space of the $su(2)$-valued 1-forms of Σ harmonic with respect to B, i.e.,

$$\mathcal{H}_B = \{\omega \in \Omega^1(\Sigma) \otimes su(2) \mid d_B\omega = d_B^*\omega = 0\}.$$

$\mathcal{H}_B$ is a $(6g-6)$-dimensional real vector space isomorphic to the de-Rham cohomology group $H_B^1(\Sigma, su(2))$ with $su(2)$-valued local coefficient system defined by the holonomy representation of B.

$\mathcal{H}_B$ has the non-degenerate symplectic structure defined by

$$\langle \omega_1, \omega_2 \rangle = -\int_\Sigma tr(\omega_1 \wedge \omega_2)$$

for $\omega_1,\ \omega_2 \in \mathcal{H}_B$. A Lagrangian L of $\mathcal{H}_B$ is a $(3g-3)$-dimensional subspace of $\mathcal{H}_B$ such that $\langle \omega_1, \omega_2 \rangle = 0$ for any $\omega_1,\ \omega_2 \in L$.

Each 0-eigenform ψ_0 of D_B has the form $\psi_0 = (\omega, 0, 0) \in (\Omega^1 \oplus \Omega^0 \oplus \Omega^0)(\Sigma) \otimes su(2)$ for some $\omega \in \mathcal{H}_B$. Thus $\operatorname{Ker} D_B$ can be identified with $\mathcal{H}_B$.

Let P_+ (resp. P_-) be the subspace of the L^2-completion of $(\Omega^1 \oplus \Omega^0 \oplus \Omega^0)(\Sigma) \otimes su(2)$ spanned by $\{\psi_\mu\}_{\mu>0}$ (resp. $\{\psi_\mu\}_{\mu<0}$).

For a subspace W of $(\Omega^1 \oplus \Omega^0 \oplus \Omega^0)(\Sigma) \otimes su(2)$, we set

$$\Omega(M_i, W) = \{\varphi \in (\Omega^1 \oplus \Omega^0)(M_i) \otimes su(2) \mid \varphi|\partial M_i \in W\}.$$

Definition 3.1. Let $\mathcal{E}_A^1$ (resp. $\mathcal{E}_A^2$) be the operator from the L_1^2-completion of $\Omega(M_1, P_- + \mathcal{H}_B)$ (resp. $\Omega(M_2, P_+ + \mathcal{H}_B)$) to the L^2-completion of $(\Omega^1 \oplus \Omega^0)(M_1) \otimes su(2)$ (resp. $(\Omega^1 \oplus \Omega^0)(M_2) \otimes su(2)$) defined to be the closure of D_A.

$\mathcal{E}_A^i$ is a Fredholm operator of index $3g-3$ $(i = 1,2)$. Let $\pi_A^i\colon \operatorname{Ker} \mathcal{E}_A^i \to \mathcal{H}_B$ be the map defined by, for $\varphi \in \operatorname{Ker} \mathcal{E}_A^i$,

$$\pi_A^i(\varphi) = \text{the harmonic part of } (\varphi \mid \partial M_i).$$

Set $L^i = \text{Image}\,(\pi_A^i)$. Then L^i is a Lagrangian of $\mathcal{H}_B$.

Let $\{A_t\}_{0 \le t \le 1}$ be a smooth path of irreducible smooth connections on $M \times \mathrm{SU}(2)$ connecting two irreducible flat connections A_0 and A_1. We assume that, for $0 \le t \le 1$, A_t restricts to a product connection $B_t \times 1$ on $(\Sigma \times [-1,1]) \times \mathrm{SU}(2)$, where B_t is an irreducible flat connection on $\Sigma \times \mathrm{SU}(2)$ and 1 is the trivial connection on $[-1,1] \times \mathrm{SU}(2)$.

By applying the above construction for A_t $(0 \le t \le 1)$, we obtain a one-parameter family of the Lagrangian pairs $\{(L_t^1, L_t^2)\}$.

Let $\mathbb{V}$ be a fixed non-degenerate $(6g-6)$-dimensional real symplectic vector space. Let

$$\Theta_t \colon \mathcal{H}_{B_t} \to \mathbb{V}, \qquad 0 \le t \le 1$$

be isomorphisms of symplectic vector spaces depending continuously on t.

Let $\mathcal{L}$ be the space of all the Lagrangians of $\mathbb{V}$. Set

$$\mathcal{L}^2 = \mathcal{L} \times \mathcal{L} \text{ and } \mathcal{L}_0^2 = \{(L^1, L^2) \in \mathcal{L}^2 \mid L^1 \cap L^2 = 0\}.$$

We denote $\Theta_t(L_t^i)$ also by L_t^i for $0 < t < 1$ and $i = 1,2$. Then as above, for $\{A_t\}_{0 \le t \le 1}$, we obtain a one-parameter family of points of $\mathcal{L}^2$, $p(t) = \{(L_t^1, L_t^2)\}_{0 \le t \le 1}$. As shown in Section 5 of **[Y]**, by perturbing $\{A_t\}_{0 \le t \le 1}$ and metric on M if necessary, we may assume that $p(t)$ is a continuous path in $\mathcal{L}^2$. Moreover, the perturbation can be chosen arbitrarily small. Then $p(t) = \{(L_t^1, L_t^2)\}_{0 \le t \le 1}$ determines a homotopy class $[p(t)]$ in $\pi_1(\mathcal{L}^2, \mathcal{L}_0^2)$ under the assumption that $\operatorname{Ker} D_{A_0} = \operatorname{Ker} D_{A_1} = 0$.

As shown in **[Y]**, $\pi_1(\mathcal{L}^2, \mathcal{L}_0^2) = \mathbb{Z}$ and the orientation of Σ determines the natural generator $\mathcal{Y}_0$ of this group. We define the integer $\mathcal{Y}(\{A_t\})$ by $[p(t)] = \mathcal{Y}(\{A_t\})\mathcal{Y}_0$.

In the above setting we obtain the formula of the spectral flow (**[Y]**)

$$SF(M, \{A_t\}) = \mathcal{Y}(\{A_t\}).$$

4. Proof of Theorem 2.1

Let M^1 and M^2 be closed oriented Riemannian 3-manifolds. Let K^1 and K^2 be knots in M^1 and M^2 respectively. Let $U(K^j)$ be a closed tubular neighborhood of K^j in M^j $(j = 1,2)$. Set $N^j = \overline{M^j - U(K^j)}$.

Let A_0^j and A_1^j be smooth irreducible flat connections on $M^j \times \mathrm{SU}(2)$ such that $\operatorname{Ker} D_{A_0^j} = \operatorname{Ker} D_{A_1^j} = 0$. Let ρ_0^j and ρ_1^j be the holonomy representations of A_0^j and A_1^j respectively. Let $\{A_t^j\}_{0 \le t \le 1}$ be a smooth path of smooth irreducible connections on $M^j \times \mathrm{SU}(2)$ connecting A_0^j and A_1^j.

We assume that, for $j = 1,2$, the data $(M^j, K^j, N^j, A_0^j, A_1^j)$ satisfy the setting $(*)$ in Section 2. Let E_0^j be the segment in $R(N^j)$ connecting $[\rho_0^j]$ and $[\rho_1^j]$ and let C^j be the immersed curve in $\widetilde{P}_0$ described in the setting $(*)$.

For $j = 1,2$, let $[C^j]$ be the regular homotopy class of C^j in $\pi_1^{\mathrm{reg}}(\widetilde{P}_0, \mathbb{L})$.

To prove Theorem 2.1, it suffices to prove $SF(M^1, \{A_t^1\}) = SF(M^2, \{A_t^2\})$ under the assumption $[C^1] = [C^2]$.

To apply the formula in Section 3, we need a technical modification. Since $\pi_1(N^j)$ is finitely generated, for a sufficiently large positive integer k, we can find a set of simple closed curves $\{\nu_1^j, \ldots, \nu_k^j\}$ in N^j such that $(\nu_1^j \cup \cdots \cup \nu_k^j) \cap \partial N^j = \nu_r^j \cap \nu_s^j =$ $\{\text{base point}\}$ $(1 \leq r < s \leq k)$, and such that $\pi_1((\nu_1^j \cup \ldots \cup \nu_k^j) \cup \partial N^j)$ carries the generator of $\pi_1(N^j)$. In particular we may assume that $k \geq 2$.

Let $Y = Y_1 \cup Y_0 \cup Y_2$ be the 1-complex pictured in Fig. 4.1, where Y_2 is a bouquet of k circles.

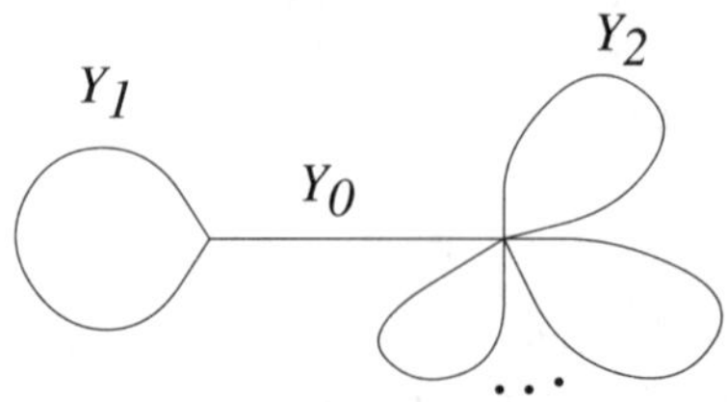

Figure 4.1

For $j = 1, 2$, there is an embedding $\Psi^j : Y \to M^j$ such that $\Psi^j(Y_1)$ is isotopic to K^j and such that $\Psi^j(Y_2)$ is isotopic to $(\nu_1^j \cup \ldots \cup \nu_k^j)$. Let $U^j(Y)$ be a regular neighborhood of $\Psi^j(Y)$ in M^j. For $j = 1, 2$, we set

$$M_1^j = U^j(Y), \quad M_2^j = \overline{M^j - U^j(Y)} \quad \text{and} \quad \Sigma^j = M_1^j \cap M_2^j .$$

Obviously we may assume $M_2^j \subset \operatorname{Int} N^j$.

By the above choice of $\nu_1^j, \ldots, \nu_k^j$, $\pi_1(\Sigma^j)$ carries the generator of $\pi_1(N^j)$. It follows that each irreducible flat connection on $N^j \times \mathrm{SU}(2)$ restricts to an irreducible flat connection on $\Sigma^j \times \mathrm{SU}(2)$. We give a Riemannian metric on M^j such that, near Σ^j, it is isometric to the product Riemannian manifold $\Sigma^j \times [-1, 1]$ for some metric on Σ^j.

The $(\mathrm{mod}\, 8)$ spectral flow $SF(M^j, \{A_t^j\})$ does not depend on the particular choice of a path $\{A_t^j\}_{0 \leq t \leq 1}$ connecting A_0^j and A_1^j (**[T]**). We choose $\{A_t^j\}_{0 \leq t \leq 1}$ in the following way. Let E_0^j be the segment in $E^j \subset R(N^j)$ connecting ρ_0^j and ρ_1^j as in the above. We write E_0^j as $E_0^j = \{\rho_t^j\}_{0 \leq t \leq 1}$, where ρ_t^j is an irreducible representation of $\pi_1(N^j)$ into $\mathrm{SU}(2)$ for $0 \leq t \leq 1$. Let $\{A_t^j\}_{0 \leq t \leq 1}$ be a smooth path of smooth connections on $M^j \times \mathrm{SU}(2)$ connecting A_0^j and A_1^j such that, for $0 \leq t \leq 1$, $A_t^j \mid N^j \times \mathrm{SU}(2)$ is the irreducible flat connection corresponding to the representation ρ_t^j and such that, near Σ^j, A_t^j restricts to $B_t^j \times 1$. Here B_t^j is the flat connection on $\Sigma^j \times \mathrm{SU}(2)$ associated to the restriction $\rho_t^j \mid \pi_1(\Sigma^j)$.

Let $U(Y)$ be the 3-dimensional thickening of the 1-complex Y in Fig. 4.1 with boundary Σ. $U(Y)$ is a handlebody of genus $k+1$ (≥ 3). For $j = 1, 2$, we extend the embedding $\Psi^j : Y \to M^j$ to a diffeomorphism $\Psi^j : U(Y) \to U^j(Y)$. We assume that $U(Y)$ is endowed with a Riemannian metric which is a product near Σ and that, for $j = 1, 2$, Ψ^j is an isometry. We consider the restriction of the connection $\{A_t^j\}$ on $U^j(Y) \times \mathrm{SU}(2)$ to be a connection on $U(Y) \times \mathrm{SU}(2)$.

Let $V(Y_1)$ be a closed tubular neighborhood of Y_1 in $\operatorname{Int} U(Y)$. We set $U_0(Y) = \overline{U(Y) - V(Y_1)}$. Then $\partial U_0(Y) = \Sigma \cup \partial V(Y_1)$ (disjoint union). We assume that $\Psi^j(V(Y_1)) = U(K^j)$.

Under the assumption of $[C^1] = [C^2]$ in $\pi_1^{\mathrm{reg}}(\widetilde{P}_0, \mathbb{L})$, we may assume that there is a smooth map $p\colon [0,1] \times [0,1] \to \widetilde{P}_0$ such that $p(t,0) = \rho_t^1$, $p(t,1) = \rho_t^2$ for $0 \le t \le 1$ and $\partial p(t,s)/\partial t = 0$ for $0 \le t,\ s \le 1$.

$U_0(Y)$ is homotopy equivalent to the one-point union $(S^1 \vee \ldots \vee S^1) \vee (S^1 \times S^1)$. For $0 \le t,\ s \le 1$, $p(t,s)$ gives a non-trivial flat connection on $(S^1 \times S^1) \times \mathrm{SU}(2)$. From this, it follows that we can choose $\{A_t^1\}$ and $\{A_t^2\}$ so that there may be smooth connections $\mathbb{A}$ on $U(Y) \times [0,1] \times [0,1] \times \mathrm{SU}(2)$ and $\mathbb{B}$ on $\Sigma \times [0,1] \times [0,1] \times \mathrm{SU}(2)$ satisfying the following conditions:

(1) $\mathbb{A} \mid U(Y) \times \{t\} \times \{0\} \times \mathrm{SU}(2) = A_t^1, \quad 0 \le t \le 1;$

(2) $\mathbb{A} \mid U(Y) \times \{t\} \times \{1\} \times \mathrm{SU}(2) = A_t^2, \quad 0 \le t \le 1;$

(3) $\mathbb{A} \mid U(Y) \times \{i\} \times \{s\} \times \mathrm{SU}(2)$ is an irreducible flat connection for $i = 0, 1$ and $0 \le s \le 1$;

(4) $\mathbb{A} \mid U_0(Y) \times \{t\} \times \{s\} \times \mathrm{SU}(2)$ is an irreducible flat connection and, on $\partial V(Y_1) \times \{t\} \times \{s\} \times \mathrm{SU}(2)$, it is the flat connection associated to $p(t,s)$, for $0 \le t, s \le 1$;

(5) $B_{t,s} = \mathbb{B} \mid \Sigma \times \{t\} \times \{s\} \times \mathrm{SU}(2)$ is an irreducible flat connection for $0 \le t, s \le 1$;

(6) $A_{t,s} = \mathbb{A} \mid U(Y) \times \{t\} \times \{s\} \times \mathrm{SU}(2)$ restricts to the product $B_{t,s} \times 1$ near $\Sigma \times \{t\} \times \{s\}$, for $0 \le t, s \le 1$.

We can construct such $\mathbb{A}$ by first constructing it on $U_0(Y) \times \mathrm{SU}(2)$ using $p(t,s)$ and its flat extension on $((S^1 \vee \ldots \vee S^1) \vee (S^1 \times S^1)) \times \mathrm{SU}(2)$ and then extending it smoothly on $V(Y_1) \times \mathrm{SU}(2)$.

We choose and fix a trivialization as in Section 3,

$$\Theta = \Theta_{t,s}\colon \mathcal{H}_{B_{t,s}} = H^1_{B_{t,s}}(\Sigma, su(2)) \to \mathbb{V}$$

where $\mathbb{V}$ is a $(6(k+1)-6)$-dimensional symplectic vector space and Θ is a symplectic isomorphism depending continuously on t, s for $0 \le t, s \le 1$.

$\mathbb{A}$ is a homotopy of smooth connections on $U(Y) \times \mathrm{SU}(2)$ connecting $\{A_t^1\}$ and $\{A_t^2\}$.

For $j = 1, 2$, let $\{(L(j)_t^1, L(j)_t^2)\}_{0 \le t \le 1}$ be the path of Lagrangian pairs associated with the path of connections $\{A_t^j\}_{0 \le t \le 1}$ on $M^j \times \mathrm{SU}(2)$ (see Section 3).

As in Step 1, we consider the restriction $A_t^j \mid U^j(Y)$ to be a connection on $U(Y) \times \mathrm{SU}(2)$ via the diffeomorphism Ψ^j for $j = 1, 2$ and $0 \le t \le 1$.

$\partial V(Y_1)$ is a 2-torus, and, for $0 \le t, s \le 1$, $H^1_{A_{t,s}}(\partial V(Y_1), su(2))$ is a 2-dimensional real vector space with non-degenerate symplectic structure defined by the cup product and taking the trace of the coefficient. Let $\mathbb{U}$ be a standard non-degenerate 2-dimensional symplectic vector space. We choose and fix a trivialization

$$\Xi = \Xi_{t,s}\colon H^1_{A_{t,s}}(\partial V(Y_1), su(2)) \to \mathbb{U}.$$

Then

$$\Theta + \Xi\colon H^1_{A_{t,s}}(\partial U_0(Y), su(2)) \to \mathbb{V} + \mathbb{U}$$

gives a trivialization of $H^1_{A_{t,s}}(\partial U_0(Y), su(2)) = \mathcal{H}_{B_{t,s}} + H^1_{A_{t,s}}(\partial V(Y_1), su(2))$.

For $0 \le t, s \le 1$, the connection $A_{t,s}$ has the property (6) and the operator $\mathcal{E}^1_{A_{t,s}}$ can be defined as in Section 3. Hence we have the Lagrangian $L^1_{t,s}$ of $\mathbb{V}$ associated to $\mathcal{E}^1_{A_{t,s}}$ as in Section 3. Since genus of $\Sigma = \partial U(Y) \ge 3$, by Lemma 5.6 in **[Y]**, by perturbing

$\{A_{t,s}\}_{0\leq t,s\leq 1}$ and metric on $U(Y)$ if necessary, we may assume that $\{L^1_{t,s}\}_{0\leq t,s\leq 1}$ is a continuous 2-parameter family of Lagrangians. Moreover, the perturbation can be chosen arbitrarily small.

This 2-parameter family of Lagrangians $\{L^1_{t,s}\}$ gives a homotopy connecting $\{L^1_{t,0}\} = \{L(1)^1_t\}$ and $\{L^1_{t,1}\} = \{L(2)^1_t\}$. For $0 \leq t, s \leq 1$, the tangent space $X_{t,s}$ of the curve $\{p(t,s)\}_{0\leq t\leq 1}$ at $p(t,s)$ is a 1-dimensional subspace of the tangent space of $\widetilde{P}_0$ at $p(t,s)$. We can consider $X_{t,s}$ to be a 1-dimensional subspace of $H^1_{A_{t,s}}(\partial V(Y_1), su(2))$. We denote $\Xi(X_{t,s}) \subset \mathbb{U}$ also by $X_{t,s}$. Then

$$\{L^1_{t,s} + X_{t,s}\}_{0\leq t,s\leq 1} \tag{4.1}$$

is a continuous 2-parameter family of Lagrangians of $\mathbb{V} + \mathbb{U}$.

Since, for $j = 1, 2$ and $0 \leq t \leq 1$, $A^j_t \mid N^j \times \mathrm{SU}(2)$ is an irreducible flat connection, by the Mayer-Vietoris theorem, we have the exact sequence

$$0 \to H^1_{A^j_t}(N^j, su(2)) \to H^1_{A^j_t}(U_0(Y), su(2)) + H^1_{A^j_t}(M^j_2, su(2)) \to \mathcal{H}_{B_t} \to 0 \tag{4.2}$$

Now, for $j = 1$ and 2,

$$\mathrm{Im}(H^1_{A^j_t}(N^j, su(2)) \to H^1_{A^j_t}(\partial N^j, su(2)) = H^1_{A^j_t}(\partial V(Y_1), su(2)))$$

is identified with $X_{t,0}$ and $X_{t,1}$ respectively for $0 \leq t \leq 1$. Let $*$ be the Hodge star operator on $H^1_{A_{t,s}}(\partial V(Y_1), su(2))$, where we consider the said space to be the space of the harmonic 1-forms on $\partial V(Y_1)$ for $0 \leq t, s \leq 1$. We denote $\Xi(*X_{t,s}) \subset \mathbb{U}$ also by $*X_{t,s}$.

By definition, for $j = 1, 2$,

$$L(1)^2_t = \mathrm{Im}(H^1_{A_{t,0}}(M^1_2, su(2)) \to \mathcal{H}_{B_{t,0}}),$$

and

$$L(2)^2_t = \mathrm{Im}(H^1_{A_{t,1}}(M^2_2, su(2)) \to \mathcal{H}_{B_{t,1}}).$$

We set, for $0 \leq t, s \leq 1$,

$$Z_{t,s} = (\Theta + \Xi)(\mathrm{Im}(H^1_{A_{t,s}}(U_0(Y), su(2)) \to H^1_{A_{t,s}}(\partial U_0(Y), su(2))),$$

where the map is the homomorphism induced by the inclusion. Then by the exact sequence (4.2) we see that both of

$$(Z_{t,0}, L(1)^2_t + *X_{t,0}) \text{ and } (Z_{t,1}, L(2)^2_t + *X_{t,1}) \tag{4.3}$$

are complementary pairs of Lagrangians of $\mathbb{V} + \mathbb{U}$ for $0 \leq t \leq 1$. Consider the 2-parameter family of Lagrangians of $\mathbb{V} + \mathbb{U}$, $\{*Z_{t,s})_{0\leq t,s\leq 1}$. Combining it with (4.1), we obtain the 2-parameter continuous family of Lagrangian pairs of $\mathbb{V} + \mathbb{U}$,

$$\{q(t,s)\}_{0\leq t,s\leq 1} = \{(L^1_{t,s} + X_{t,s}, *Z_{t,s})\}_{0\leq t,s\leq 1}.$$

Now we form the space $\mathcal{L}^2$ of all the Lagrangian pairs of $\mathbb{V}+\mathbb{U}$ and the subspace $\mathcal{L}^2_0$ as in Section 3. By the property (3) of $\mathbb{A}$, we see that $q(0,s)$, $q(1,s) \in \mathcal{L}^2_0$ for $0 \leq s \leq 1$. Hence the paths $\{q(t,0)\}_{0\leq t\leq 1}$ and $\{q(t,1)\}_{0\leq t\leq 1}$ are mutually homotopic relative $\mathcal{L}^2_0$ and they determine the same homotopy class in $\pi_1(\mathcal{L}^2, \mathcal{L}^2_0)$. From the complementary relations of (4.3), it follows that $\{q(t,0)\}_{0\le t\le 1}$ and $\{q(t,1)\}_{0\le t\le 1}$ are homotopic rel. $\mathcal{L}_0^2$ to the paths

$$\{(L(1)_t^1 + X_{t,0}, L(1)_t^2 + *X_{t,0})\}_{0\le t\le 1}$$

and

$$\{(L(2)_t^1 + X_{t,1}, L(2)_t^2 + *X_{t,1})\}_{0\le t\le 1}$$

respectively. This implies that the paths

$$\{(L(1)_t^1, L(1)_t^2)\}_{0\le t\le 1} \quad \text{and} \quad \{(L(2)_t^1, L(2)_t^2)\}_{0\le t\le 1}$$

are mutually homotopic rel. $\mathcal{L}_0^2$. Thus $\mathcal{Y}(\{A_t^1\}) = \mathcal{Y}(\{A_t^2\})$. This proves the proposition. □

5. A computational algorithm of $(\bmod\ 8)$ spectral flow

Every oriented immersed smooth curve in $\widetilde{P}_0$ with endpoints on $\mathbb{L}$ is regularly homotopic to an immersed smooth curve which is a consecutive composition of the oriented curves in $\widetilde{P}_0$ in Fig. 5.1 below (note that compositions should be made so that the tangent lines of two consecutive curves agree at each connecting point), where $\circ$ (resp. $\times$) denotes those points in the plane $\mathbb{R}^2$ whose coordinates are $(2m\pi, n\pi)$ (resp. $((2m+1)\pi, n\pi)$) for integers m, n. We associate an integer to each of the above oriented curves as indicated in parentheses in Fig. 5.1.

Example. The following two curves (Fig. 5.2) are regularly homotopic rel. boundary to each other. The number corresponding to these curves is $-2 \pmod 8$.

Our computational algorithm is given by

Theorem 5.1. *Let M be an oriented closed Riemannian 3-manifold. Let K be a knot in M. Set $N = \overline{M - U(K)}$, where $U(K)$ is a closed tubular neighborhood of K in M. Let A_0 and A_1 be smooth irreducible flat connections on $M \times \mathrm{SU}(2)$ such that $\operatorname{Ker} D_{A_0} = \operatorname{Ker} D_{A_1} = 0$. We assume that the setting $(*)$ in Section 2 holds for A_0, A_1 and N. Let C be the immersed curve in $\widetilde{P}_0$ mentioned in the setting $(*)$. Then, for any smooth path $\{A_t\}_{0\le t\le 1}$ of irreducible smooth connections on $M \times \mathrm{SU}(2)$ connecting A_0 and A_1, the $(\bmod 8)$ spectral flow $SF(M, \{A_t\})$ is given by adding the above numbers in Fig. 5.1 associated to curves which appear in the representation of the regular homotopy class $[C]$ by a consecutive composition of curves given in Fig. 5.1.*

Proof. Let $Y = Y_1 \cup Y_0 \cup Y_2$ be the 1-complex depicted in Fig. 4.1 in Section 4 for an arbitrarily fixed integer $k \ge 2$. Let $U(Y)$ be the 3-dimensional thickening of Y with $\partial U(Y) = \Sigma$. Note that genus of $\Sigma \ge 3$. Let $V(Y_1)$ be a closed tubular neighborhood of Y_1 in $\operatorname{Int} U(Y)$. We set $U_0(Y) = \overline{U(Y) - V(Y_1)}$. Then $\partial U_0(Y) = \Sigma \cup \partial V(Y_1)$ (disjoint union).

The orientation of $S^1 \times S^1$ induces the orientations of Σ, $\partial V(Y_1)$ and $U(Y)$.

Let $\{p(t)\}_{0\le t\le 1}$ be a smooth immersed curve in $\widetilde{P}_0$ such that $p(0)$, $p(1) \in \mathbb{L}$ and $p(t)$ intersects transversely with $\mathbb{L}$ at the endpoints. We regard $p(t)$ to be a smooth path

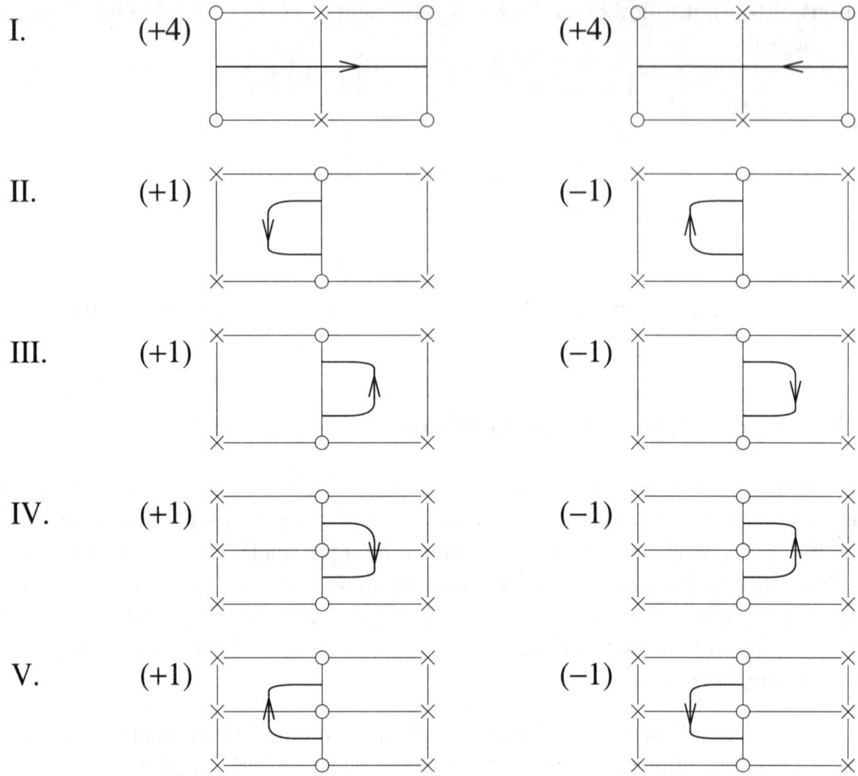

Figure 5.1

Figure 5.2

of smooth flat connections on $(S^1 \times S^1) \times \mathrm{SU}(2)$. We identify $(S^1 \times S^1) \times \mathrm{SU}(2)$ with $\partial V(Y_1) \times \mathrm{SU}(2)$. We can extend $p(t)$ to a smooth path of smooth irreducible flat connections $\{G_t\}_{0 \leq t \leq 1}$ on $U(Y) \times \mathrm{SU}(2)$ so that G_0 and G_1 are irreducible flat connections on $U(Y) \times \mathrm{SU}(2)$ as $p(0)$, $p(1) \in \mathbb{L}$. For $0 \leq t \leq 1$, let B_t be the restriction of G_t on $\Sigma \times \mathrm{SU}(2)$. B_t is an irreducible flat connection on $\Sigma \times \mathrm{SU}(2)$ for $0 \leq t \leq 1$.

For $0 \leq t \leq 1$, we can define the operator

$$\mathcal{E}_{G_t} \colon \Omega(U(Y), P_+ + \mathcal{H}_{B_t}) \to (\Omega^1 \oplus \Omega^0)(U(Y)) \otimes su(2)$$

as in Definition 3.1. Then $\mathcal{E}_{G_t}$ is a Fredholm operator of index $3k$. The map $\pi_{G_t}: \operatorname{Ker} \mathcal{E}_{G_t} \to \mathcal{H}_{B_t}$ can be defined as in Section 3 and $L_t = \operatorname{Image}(\pi_{G_t})$ is a Lagrangian of $\mathcal{H}_{B_t}$ for $0 \le t \le 1$.

Let $\mathbb{V}$ and $\mathbb{U}$ be standard non-degenerate symplectic real vector spaces of dimension $6k$ and 2 respectively. Let

$$\Theta = \Theta_t: H^1_{B_t}(\Sigma, su(2)) \to \mathbb{V}$$

and

$$\Xi = \Xi_t: H^1_{G_t}(\partial V(Y_1), su(2)) \to \mathbb{U}$$

be trivializations as in Sections 3 and 4. We denote $\Theta(L_t)$ also by L_t. Then by Section 5 in **[Y]**, by perturbing $\{G_t\}$ and metric on $U(Y)$ if necessary, we may assume that $\{L_t\}_{0 \le t \le 1}$ is a continuous one-parameter family of Lagrangians in $\mathbb{V}$. Moreover the perturbation can be chosen arbitrarily small.

For $0 \le t \le 1$, we set

$$X_t = \text{the tangent line of } p(t) \text{ at } t$$

and

$$Z_t = (\Theta_t + \Xi_t)(\operatorname{Im}(H^1_{G_t}(U_0(Y), su(2)) \to H^1_{G_t}(\partial U_0(Y), su(2)))$$

where $\Theta_t + \Xi_t: H^1_{G_t}(\partial U_0(Y), su(2)) \to \mathbb{V} + \mathbb{U}$. Then

$$\{q(t)\}_{0 \le t \le 1} = \{(L_t + X_t, *Z_t)\}_{0 \le t \le 1} \tag{5.1.1}$$

is a continuous path in the space of Lagrangian pairs in $\mathbb{V} + \mathbb{U}$, $\mathcal{L}^2$. Since G_0 and G_1 are both irreducible flat connections on $U(Y) \times \mathrm{SU}(2)$, by Mayer-Vietoris sequence of de-Rham cohomology, we can see that $q(0)$ and $q(1) \in \mathcal{L}^2_0$.

The orientation of $S^1 \times S^1$ induces the orientations of Σ and $\partial V(Y_1)$ and this determines the generator $\mathcal{Y}_0$ of $\pi_1(\mathcal{L}^2, \mathcal{L}^2_0)$ as in Section 6 of **[Y]**. We define the integer $\mathcal{Y}(\{G_t\})$ by $[q(t)] = \mathcal{Y}(\{G_t\})\mathcal{Y}_0$, where $[q(t)]$ denotes the homotopy class of $\{q(t)\}$ in $\pi_1(\mathcal{L}^2, \mathcal{L}^2_0)$. That $\mathcal{Y}(\{G_t\}) \pmod 8$ depends only on the Λ-equivalence class of the regular homotopy class of $\{p(t)\}_{0 \le t \le 1}$ in $\pi_1^{\mathrm{reg}}(\widetilde{P}_0, \mathbb{L})$ can be deduced by essentially the same argument as the proof of Theorem 2.1 in Section 4. Moreover it can be seen from the definition of $\mathcal{Y}(\{G_t\})$ that, if $\{p(t)\}_{0 \le t \le 1}$ is regularly homotopic to a consecutive composition of oriented curves in $\widetilde{P}_0$ in Fig. 5.1, then $\mathcal{Y}(\{G_t\})$ is the sum of the corresponding $\mathcal{Y}$'s of these oriented curves. Also the proof of Theorem 2.1 in Section 4 shows that, in the setting of Theorem 5.1, $SF(M, \{A_t\}) \pmod 8$ is equal to $\mathcal{Y}$ of $[C]$.

It remains to calculate $\mathcal{Y}$ for each curve in Fig. 5.1.

Let us consider the curve in the left of II in Fig. 5.1. We take a point $z \in \mathbb{L}$ as indicated in Fig. 5.3. By regular homotopy, the curve can be shrunk into an arbitrarily small half circle C_ε centered at z of radius $\varepsilon > 0$. Since $\mathcal{Y}$ is invariant under regular homotopy, it suffices to consider $\mathcal{Y}$ of C_ε for arbitrarily small $\varepsilon > 0$. For sufficiently small $\varepsilon > 0$, and for $0 \le t \le 1$, the above G_t can be chosen to be an irreducible smooth connection on $U(Y) \times \mathrm{SU}(2)$ which is very near to a constant irreducible flat connection. Hence for sufficiently small $\varepsilon > 0$, both of $\{L_t\}$ and $\{*Z_t\}$ become almost constant

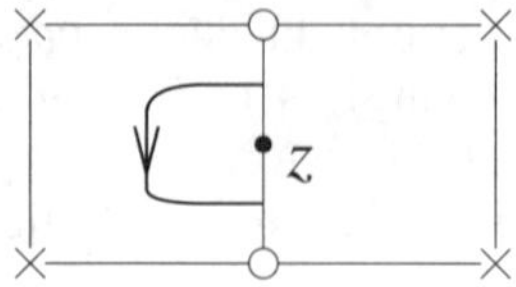

Figure 5.3

for $0 \leq t \leq 1$. On the other hand, the tangent line $\{X_t\}$ rotates by the amount π while t varies from 0 to 1. Now $\mathbb{V}$ and $\mathbb{U}$ can be considered complex vector spaces whose complex structures are compatible with the symplectic structures. Let $\overset{\text{top}}{\wedge}_{\mathbb{C}}\mathbb{V}$ and $\overset{\text{top}}{\wedge}_{\mathbb{C}}\mathbb{U}$ denote the top-dimensional exterior products (over $\mathbb{C}$) of $\mathbb{V}$ and $\mathbb{U}$ respectively. Then there is a canonical isomorphism

$$\overset{\text{top}}{\wedge}_{\mathbb{C}}\mathbb{V} \wedge_{\mathbb{C}} \overset{\text{top}}{\wedge}_{\mathbb{C}}\mathbb{U} \cong \mathbb{C}.$$

By the standard symplectic geometry, it can be seen that, under the canonical symplectic structure of $\mathbb{C}$, the $\mathcal{Y}$ of the path of Lagrangian pairs (5.1.1) in the particular case of Fig. 5.3 is equal to the corresponding $\mathcal{Y}$ of the path of the Lagrangian pairs of $\mathbb{C}$,

$$\{(\overset{\text{top}}{\wedge}_{\mathbb{R}}L_t \wedge_{\mathbb{R}} X_t,\ \overset{\text{top}}{\wedge}_{\mathbb{R}} * Z_t)\}_{0 \leq t \leq 1}. \tag{5.1.2}$$

Since $\overset{\text{top}}{\wedge}_{\mathbb{R}}L_t$ and $\overset{\text{top}}{\wedge}_{\mathbb{R}} * Z_t$ are both almost constant for $0 \leq t \leq 1$ and sufficiently small $\varepsilon > 0$, the $\mathcal{Y}$ of (5.1.2) can be read from the rotation of $\{X_t\}_{0 \leq t \leq 1}$ in $\mathbb{U}$ and it is equal to $+1$ (the sign $+$ is determined by the orientation of the curve). Thus we obtain $\mathcal{Y}(\{G_t\}) = +1$ for the curve of the left of II in Fig. 5.1.

As for the other curves in II, III of Fig. 5.1, the essentially same argument works and we get the indicated number for $\mathcal{Y}(\{G_t\})$ for these curves.

As was shown in Section 8 in **[Y]**, by Theorem 2.1 and Fintushel-Stern's result about Floer homology group of Brieskorn homology 3-sphere $\Sigma(2,3,19)$ **[F-S]**, we obtain that $\mathcal{Y}\ (\mathrm{mod}\ 8)$ is given as follows in the case of the following diagram (Fig. 5.4). The curve of the right in the above is regularly homotopic to the following curve depicted in Fig. 5.5. By Fig. 5.4, Fig. 5.5 and the above result on the curve of II, we obtain $\mathcal{Y}$ for the curve of the left of IV in Fig. 5.1. By Λ-invariance of $\mathcal{Y}$, we obtain all the values of $\mathcal{Y}$ as indicated in Fig. 5.1. This proves the theorem. □

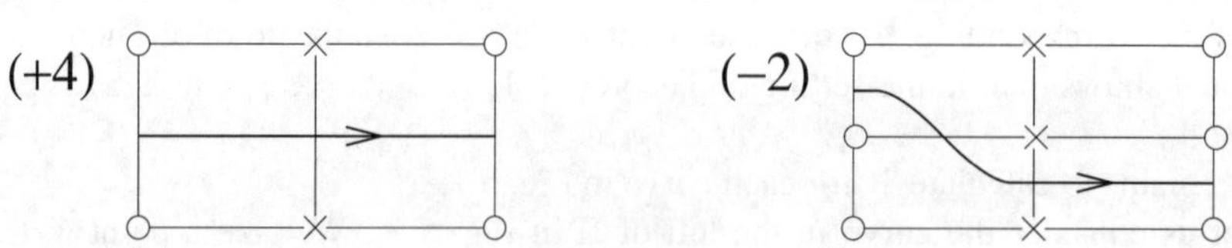

Figure 5.4

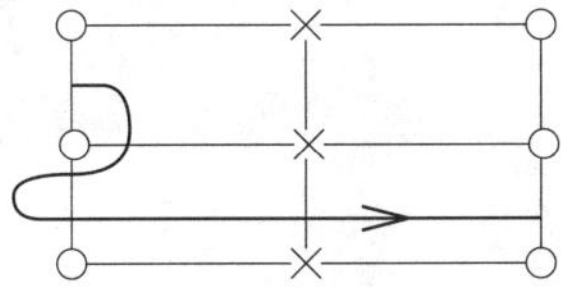

Figure 5.5

References

[A-P-S] Atiyah, M. F., Patodi, V. K., Singer, I.M.: Spectral asymmetry and Riemannian geometry III, Math. Proc. Camb. Philos. Soc. 79 (1976), 71–99.

[F] Floer, A.: An instanton-invariant for 3-manifolds, Commun. Math. Phys. 118, 21 (1988), 5–240.

[T] Taubes, C. H.: Casson's invariant and gauge theory, J. Differential Geometry 31 (1990), 547–599.

[F-S] Fintushel, R., Stern, J. R.: Instanton homology of Seifert fibered homology three spheres, to appear in Proc. London Math. Soc.

[Y] Yoshida, T.: Floer homology and splittings of manifolds, Ann. of Math. 134 (1991), 277-323. 1989.

Department of Mathematics, Tokyo Metropolitan University, Minami-Osawa 1-1, Hachioji, Tokyo 192-03, Japan